# Human Biology
## FORM, FUNCTION, AND ADAPTATION

# Human Biology
## FORM, FUNCTION, AND ADAPTATION

William DeWitt

Williams College

Scott, Foresman and Company

Glenview, Illinois    Boston    London

*For my family*

Acknowledgments for illustrations and other copyrighted materials in this book appear on the page with the copyrighted material or in the Acknowledgments section at the back of the book, which is considered an extension of the copyright page.

**Library of Congress Cataloging-in-Publication Data**
DeWitt, William.
 Human biology.

 Includes index.
 1. Human biology. I. Title.
QP34.5.D47 1989     612     88-29725
ISBN 0-673-18683-0

# PREFACE

In many years of teaching human biology to undergraduates at Williams College, I have been struck by two recurring themes: the vast majority of students are woefully ignorant of human biology, but they have a keen interest in it. For the teacher, this almost paradoxical situation is fraught with difficulty but rich in opportunity. To be successful, a beginning-level course in human biology must present the basic science in a clear and accessible form while sustaining the students' inherent interest in the subject. None of the available textbooks provides the kind of treatment that I feel accomplishes these goals. It is for that reason that I decided to write this book.

## COURSE LEVEL

This textbook is designed for use in lower-level courses in human biology. Because it presumes little or no background in biology or chemistry, it can be conveniently used for nonmajors' courses, but its level is also entirely appropriate for freshman or sophomore biology majors.

## APPROACH

I have tried to make this book fundamentally different from others by emphasizing two approaches. First, I have made a special effort to provide a broad perspective on the material, stressing the importance of comprehension over memorization. This is not to say that I have omitted the detailed factual information that is so essential to a full understanding of the complexities of human biology. Instead, I have attempted to supply a conceptual framework that will assist students in organizing and learning the essential details. Second, I have tried to spark the students' interest by including a great deal of practical information on contemporary topics relating to health and disease. I have chosen to scatter much of this information among the basic science material in the text. This format, rather than one of isolating the "interesting" material in separate sections at the end of chapters, has the advantages of enlivening the "drier" fundamentals and breaking up the pace of the discourse. Extensive discussions of material that I perceive to be of special interest to students—

AIDS, herpesviruses, obesity and anorexia, Alzheimer's disease, and the like—are written as essays, separate from the substance of the text. There are also three four-color photo essays that help highlight human embryology, circulation, and vision. Dealing with the medical aspects of human biology in these three ways provides what I hope is a valuable mix of breadth and depth.

## ORGANIZATION

The book includes six major sections. Section One ("Levels of Organization") describes the organization of the human body, from the molecular to the organismal levels. By emphasizing the interrelationship of the various levels of organization, the chapters of this section provide a comprehensive overview of body structure and function.

Section Two ("Linking Form and Function") is comprised of 12 chapters dealing with the physiological systems of the body (excluding the reproductive and immune systems). The aim of this section is not only to provide a description of the anatomy and physiology of each system, but to show how the activities of the various systems of the body are integrated with one another to form a functioning whole.

The male and female reproductive systems, pregnancy, and embryonic development are discussed in Section Three ("A New Life"). While including basic information on the reproductive systems, birth control, and reproduction, this section also provides a detailed description of the sequence of morphological events occurring during embryonic development. In addition, it considers how various environmental agents may interfere with normal embryogenesis to cause birth defects.

Section Four ("The Blueprints of Life") discusses basic mechanisms of heredity, molecular genetics, and genetic engineering. This section is designed to provide insight into the genetic revolution that has transformed our views of genetic disease and its treatment.

Section Five ("Life Threatened") includes three chapters that focus on biological threats to life. It provides a detailed account of infectious disease processes, the functioning of the body's immune system, cancer, aging, and death.

The book concludes with Section Six ("The Past, The Present, The Future"). It is composed of two chapters, the first of which, "Human Evolution," speculates on the origin of life, and traces the history of life and what we know about the course of human evolution. The second chapter in Section Six discusses a variety of topics on human ecology, focusing on trends in human population growth and the influence of humans on their environment.

## FORMAT

One of the most difficult tasks in writing a textbook is deciding on questions of organization and level. What topics should be included? In what sequence should they be presented? What should be the depth and level of coverage? I resolved these questions, as all authors presumably do, by reaching the inevitable conclusion that there are no perfect answers to them. No format will please all potential users of a textbook, precisely because there are many effective ways of teaching any subject.

In deciding on the format I finally adopted, I drew heavily from my own teaching experience in addition to consulting with human biology teachers from a wide range of undergraduate institutions. The format and organization of the final product have a rationale that I hope is compelling.

## MAJOR TOPICS COVERED

In addition to including material that falls under the general heading of basic physiology and anatomy, there were a number of additional topics that I felt were particularly relevant to a thoroughly modern course in human biology. These topics, including the immune system, cancer, aging, molecular genetics, and embryonic development, are ones in which many new and significant developments have occurred within the past few years. They are topics that students have almost certainly encountered on television or in newspapers and magazines, and about which they are particularly interested. Furthermore, research in these areas has led to medical breakthroughs that not only have begun to transform the practice of medicine, but have created ethical dilemmas with which our society must deal. For these reasons, I thought it imperative to include this subject matter. Inevitably, other material had to be omitted or condensed, but I have tried to cull evenly so that no single area is affected in a major way.

## CHAPTER SEQUENCE

I have chosen to discuss basic chemistry and metabolism (Chapter Two, "The Chemistry of Life") early in the book and in considerable detail. Although I had planned initially to omit much of this material or place it in an appendix, I soon realized that knowledge of some basic chemistry was essential for anything but a wholly superficial treatment of human biology. Whether discussing the pH and ionic composition of body fluids, the exchange of materials between cells of the body, the capture and transfer of energy, temperature regulation, nutrition, or countless other subjects, an understanding of their underlying chemical basis is crucial.

I have taken a number of steps to make the chemistry more palatable than it otherwise might be. First, I have made sure that its level is appropriate to a student with little or no chemistry background. Second, I have minimized the use of chemical notation. Third, I have included mention of only those topics that are essential for understanding other material in the text. Lastly, in discussing metabolism I have stressed concepts such as the purpose and design of metabolic systems, and I have relegated pathways and other detailed information to tables and figures.

Another major organizational question that I confronted was whether chapters on the nervous and endocrine systems should precede or follow chapters on the other physiological systems. Either way is flawed. Because of its complex levels of integration, the nervous system is by far the most difficult system to understand. Hence it is not easy for a student to encounter this material early on in the course. On the other hand, the activities of all other systems are controlled by the nervous and endocrine systems, and understanding their control is a big part of understanding how these other systems function.

Although I finally decided to place the chapters on the nervous and endocrine systems at the end of the physiology section, the solution turned out to be more of a compromise than a definitive pedagogical statement. I ended up discussing—admittedly in a superficial way—the nervous and endocrine control of each of the systems as I covered them. The later chapters on the nervous and endocrine systems describe detailed mechanisms and provide an overview of the body's control systems. In a sense, then, the material on nervous and endocrine regulation appears twice, first in outline and later in detail.

Perusal of the table of contents will reveal that I have not treated the lymphatic system as a separate chapter. For discussion purposes, I have instead chosen to separate its circulatory and immune functions. A full understanding of the circulatory system, which appears early in Section Two, requires knowledge of lymph and its circulation, particularly the anatomical connections between lymphatic vessels and blood vessels. Thus, it made sense to cover the arrangement of the lymphatic vessels and the basic anatomy of the lymphatic system in the chapters on the circulatory system. The immune function of the lymphatic system relates more appropriately to later discussions of disease, where I have included it.

Finally, I have placed chapters on human evolution and ecology at the end of the book. In this position, these subjects serve to integrate and to provide an overview of topics covered in earlier chapters.

## ACKNOWLEDGMENTS

During the years I was developing and writing this book, many experienced teachers of human biology courses furnished helpful ideas about organization, level, and style. I am particularly indebted to Patricia H. Jones, Bruce E. Grayson, Shaun McEllin, and Robert Sussman, who read the manuscript and provided thorough and insightful comments.

I am also grateful to Harriett Prentiss, Executive Editor at Scott, Foresman, who shared my concept of the book and whose hard work and wise counsel helped turn vision into reality. Jean Dal Porto, my project editor, dealt competently and efficiently with the zillions of inevitable production problems.

Last, but of course first, I thank those who bore the brunt of the many years of writing and rewriting and rewriting—my wife, Mary Lou, and Tyler and Erica.

*William DeWitt*

# CONTENTS

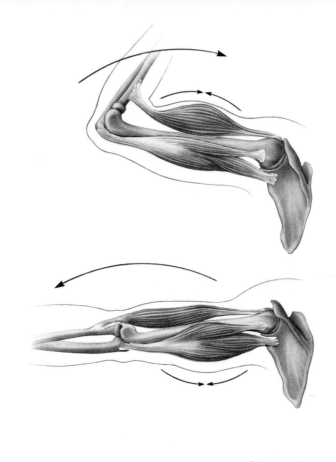

## Section Two
## LINKING FORM AND FUNCTION    81

### Chapter Five
### THE INTEGUMENTARY SYSTEM    82

### Chapter Six
### THE SKELETAL SYSTEM AND JOINTS    94

### Chapter Seven
### THE MUSCULAR SYSTEM    118

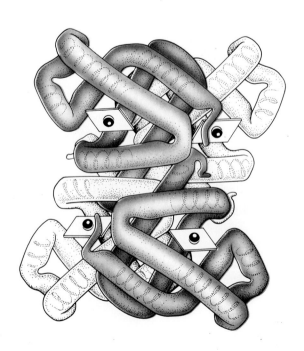

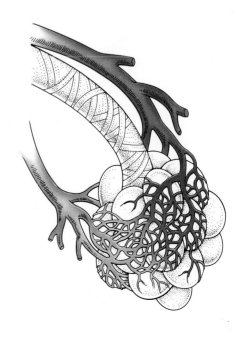

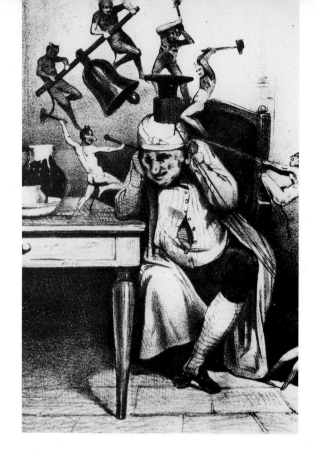

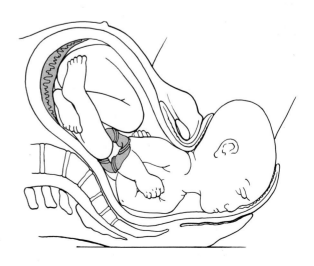

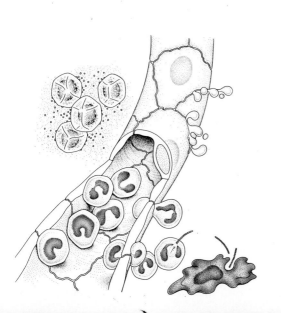

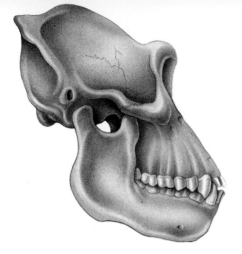
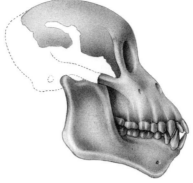
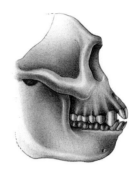

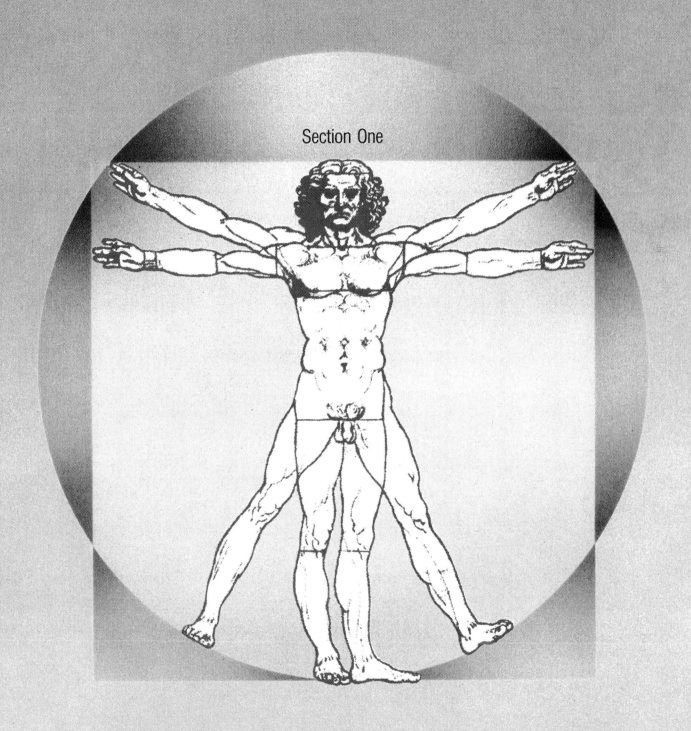

Section One

# LEVELS OF ORGANIZATION

# HUMAN BIOLOGY AND LIFE

Why study human biology? The first and most obvious reason is to satisfy our innate curiosity about ourselves. What young child has not wondered about his or her own origins, or puzzled over the biological facts of birth and death? Our fascination with things biological does not end with childhood. As we begin to question the meaning of our very existence, its biological basis becomes the touchstone of our lives. To understand how our bodies function is to begin to understand ourselves.

A second reason for learning human biology is to acquire the practical knowledge that will allow us to care for our bodies in something more than a hit-or-miss way. From understanding the need for good nutrition and adequate exercise to recognizing illness and dealing with a physician, a firm grasp of human biology is indispensible. Knowledge is our best protection against medical incompetence, quackery, and the false hopes of exaggerated claims; it frees us from the terror of superstition and provides us with a framework for making rational decisions about our health.

There is yet a third important reason for studying human biology. Recent technological developments in human biology and medicine are changing not only the kind of lives we lead and the way we view ourselves, but the very structure of our society. This "biorevolution" will come to affect virtually every aspect of our lives; its potential benefits to humankind are truly phenomenal. As we learn to manipulate biological processes that only a short time ago seemed unalterable, we will no longer be content to accept unfortunate biological circumstances as inevitable. Soon, for example, we will possess the technology that will give us extraordinary control over our own reproduction. Contraceptive methods will undoubtedly become safer and more effective. Infertility may be virtually eliminated as we invent new drugs and perfect techniques of artificial insemination and *in vitro* fertilization (by which "test tube" babies are made; see Figure 1–1).

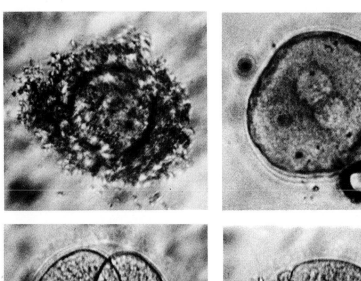
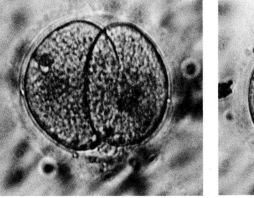
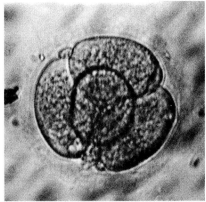

FIGURE 1–1 In producing a "test tube" baby, a mature human egg cell is removed surgically from the mother's ovary and mixed with a sample of the father's sperm in a petri dish. Following fertilization, the embryo, shown here in early stages of cell division, is implanted in the mother's uterus. The first child resulting from this method, which is now referred to as *in vitro* fertilization (*in vitro* = in a glass vessel), was born in England on July 25, 1978. Since that time, hundreds of test tube babies have been born to infertile couples.

FIGURE 1-2 The mouse on the left, nearly twice as large as its normal littermate, was created by genetic engineering. It developed from a newly fertilized egg that had been injected with multiple copies of a rat gene for growth hormone.

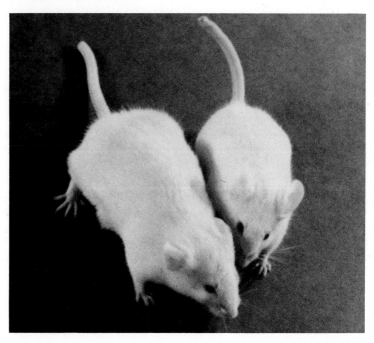

Even our heredity will not be inviolable. New diagnostic techniques have already allowed us to identify many genetic (hereditary) diseases before birth, and the list of diseases we can detect will surely continue to expand. We are now able to control or correct the symptoms of a variety of genetic diseases. Genetic engineering is on the threshold of developing ways to eliminate such diseases entirely by altering the genetic material itself (see Figure 1-2). Our life span may be extended as organ transplantation becomes increasingly routine and as artificial organs are developed and become practical to use.

But with our new-found knowledge have come social, legal, and ethical questions that we have barely recognized, much less addressed (see *Focus: Benefits and Problems from Genetic Engineering*). As we perfect methods for detecting genetic diseases and birth defects in the unborn, can we agree on what we should do about them? When we are able to alter our genetics, will we be able to define what is desirable or even what is normal? As we improve our ability to sustain life, can we learn how or when to let it end? As the ability to control the biological aspect of our lives increases, do we want to limit or prevent the use of certain technologies? Such questions — and these are only a mere sampling — must be resolved by a public that is capable of making rational decisions based upon firm knowledge of human biology.

## WHAT IS LIFE?

In 1945, the noted Austrian physicist, Erwin Schrödinger, published a little book entitled *What is Life?* In it, Schrödinger questioned the ability of the established laws of chemistry and physics to account for the events that take place within living organisms. He concluded that new chemical and physical principles — laws in addition to those already known — must be discovered before it would be possible to understand the way living systems function.

Like so many books that attempt to address questions of a profound nature, *What is Life?* poses more questions than it answers. At the time it was written, it had little impact on biologists, but it sparked intense interest among physical scientists, who were intrigued by the challenge of discovering "new principles" basic to life. As a result, many physical scientists turned their attention to biological problems. Although Schrödinger's "new principles" were never found (and we have no reason to believe they exist), these new biologists, trained in chemistry and physics, changed the focus and direction of biological research. They were, to a large extent, responsible for bringing the science of biology into a new era, one in which chemistry and physics are increasingly used to study the structure and function of living organisms.

For all the impact that Schrödinger's book had on the course of biological research, his book provided no new answers to his central question: What is life? Schrödinger was not the first to ask the question, nor was he the last; indeed, biologists are still asking the same question today.

## Life's Undefinable Nature

Biology is the study of life, of living things. This simple and seemingly straightforward definition would be neither remarkable nor worthy of further consideration if it were not for one curious fact: no one can define life. Biologists are thus placed in the unenviable position of studying something they cannot define.

The impossibility of defining life has long fascinated philosophers and biologists alike. Up until the early nineteenth century, most scientists adhered to a concept known as *vitalism,* which contended that all living things possessed a "vital force" that characterized life and that served to distinguish the living from the inanimate. A basic premise of the vitalists was that the "vital force" was necessary for the production of organic compounds, those complex carbon-containing substances of which living organisms are composed. Consequently, organic compounds were thought of as a unique characteristic of life, and life was defined as those entities capable of synthesizing organic compounds. However, when the German chemist, Friedrich Wöhler, synthesized the organic compound urea in his laboratory in 1828, vitalism had to be abandoned, and biologists were left with no clear-cut way to define life.

One of the reasons we cannot define life unequivocally is because there is not always a clear distinction between those entities we consider living and those we consider nonliving. For example, although the vast majority of scientists would not dispute the contention that a frog is living and a rock is nonliving, it is not always possible to obtain agreement when considering entities that have some characteristics we normally attribute to living systems, but also have other characteristics that we consider attributes of the nonliving world (Figure 1–3). Viruses, for instance, are borderline entities of this type.

Biologists have long disputed whether viruses should be considered living or nonliving precisely because they have some characteristics of both the animate and the inanimate. They are composed of substances—primarily proteins and nucleic acids—that are found as constituents of all living organisms, but they have no ability to function as independent entities. Instead, viruses are entirely dependent upon living organisms for their animate qualities. Their only life-like activity is reproduction, yet they are able to reproduce *only* within living organisms by using raw materials and synthetic machinery present in the cells they infect. Outside of a living organism, as when they are isolated in the virologist's test tube, viruses are essentially inanimate, and except for their chemical composition, they display **none** of the properties normally associated with life. Asking if such a borderline system is alive is much like wondering whether someone is happy—the ultimate judgment will depend on individual interpretation. In this context, it is clear that defining life becomes a philosophical argument.

How, one may ask, can we study life if we can't define what we're studying? The answer to our dilemma is to recognize that although we cannot define life in any rigorous sense, we **can** list a number of characteristics that **often** distinguish living entities from nonliving ones. It is important to realize that any one entity that we would call living will not necessarily possess all of the characteristics on our list, but it will possess most of them.

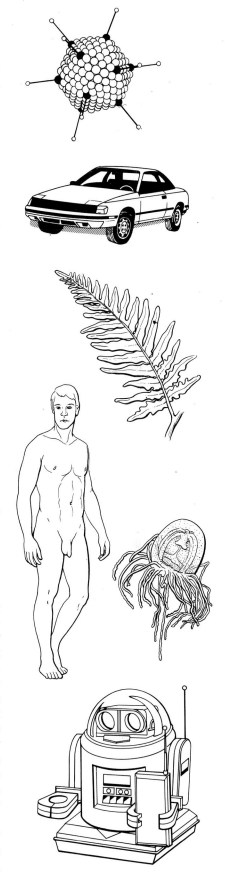

FIGURE 1–3 What characteristics distinguish the living from the nonliving? The distinction between the two is not always a clear one.

# Benefits and Problems from Genetic Engineering

Human growth hormone, once available in only limited amounts from human cadavers, is now produced commercially by techniques of genetic engineering in which human genes for growth hormone have been inserted into bacteria; the bacteria produce human growth hormone, which is extracted and purified. The ready availability of human growth hormone will greatly benefit children who do not make enough hormone of their own, but has also raised a host of ethical concerns. This issue is discussed in the following article from the Wall Street Journal of April 8, 1987.

**Synthetic Growth Hormone Raises Hopes Of Many — and Ethical Concerns Over Use**

WASHINGTON — Increasingly abundant supplies of synthetic human-growth hormone offer new hope to abnormally small children — and yet another example of biotechnology bringing fresh dilemmas.

The growing availability of the hormone means that more children can be treated, and that treatment can start earlier and last longer — slowly adding important inches to the children's ultimate height and thus enhancing their ability to cope in a large-scaled world.

But some parents have already begun to besiege doctors to prescribe the hormone for kids who ultimately will be only slightly shorter than normal or even of average height. These parents figure that even if their children don't become basketball stars or fashion models, the extra inches will contribute to career and social success. And because the hormone can't be easily detected in ordinary urinalysis tests, weight lifters and other athletes are pursuing it on the inevitable black market as a bulk-building alternative to easily spotted anabolic steroids.

Heading Into the Unknown

Many doctors, however, caution against casual use of the hormone. "Possible side effects . . . are slow to show, and that is why it is essential to be very conservative," says Raymond Hintz, a Stanford University pediatric endocrinologist. "Five to 10 years from now, we'll know more about the stuff. Use of human-growth hormone is not to be undertaken lightly."

The uses and possible abuses of biosynthetic human-growth hormone, hGH, will begin to be aired today before the health subcommittee of the House Energy and Commerce Committee. Concern over possible misuse has prompted chairman Henry Waxman (D., Calif.) to introduce a controversial bill to place synthetic hGH on Schedule II of the Controlled Substances Act, along with cocaine, morphine, some amphetamines and other medically useful substances with high potential for abuse.

Such a step would require hormone manufacturers, distributors, pharmacies and doctors to undertake extensive security, record-keeping and other measures.

Growth in children and teen-agers is stimulated by a hormone secreted in occasional spurts by the pituitary, the tiny gland at the base of the brain. Some children fail to produce this hormone or produce it only in minute quantities; some grow to an adult height of under four feet, and many others wind up well below five feet.

Since 1963, doctors have been able to obtain some additional growth in affected children — anywhere from a few inches to perhaps a foot — by injecting them with hGH from the pituitary glands of cadavers. But such supplies were extremely limited, and many affected children were under-treated or not treated at all. And in the spring of 1985, use of the chemical was halted when several young men who had received treatment died from a rare, slow-acting virus traced to contaminated natural growth hormone.

Only a few months later, though, Genentech Inc. received approval from the Food and Drug Administration to market a biosynthetic hGH called Protropin. Last month, Eli Lilly & Co. received FDA approval for its version, Humatrope; marketing began last week.

Other domestic and foreign biotechnology firms are pushing hard to develop their own synthetic hGH. Scientists say the prospect of unlimited supplies will spur research into other possible uses of the hormone: reducing weight, healing wounds, treating bone diseases, and retarding wrinkling and other symptoms of aging.

Right now, about 5,000 children are being treated for hGH deficiency; another 10,000 could easily qualify for help. Inevitably, as supplies grow, doctors will begin to relax criteria, and more and more very short children will be treated.

But doctors fear increasing parental pressure for "cosmetic" use on healthy children whose projected adult height

## Characteristics of Living Systems

### Cellular Organization

All living organisms are composed of cells and cell products (such as the hard substance of bone), and thus share a *common cellular organization.* This generalization, called the **cell theory,** has been a unifying principle of biology

may be below average but still well within the normal range. Several studies indicate that society tends to look up, figuratively as well as literally, to tall people—perceiving them as more intelligent and giving them preference in hiring and promotions.

Stanford's Dr. Hintz tells of one parent who wanted hGH for an 11-year-old son whose adult height was projected at 5 feet 10 inches, and he says "not a week goes by" that he or his colleagues aren't badgered by parents of normally growing children. He says nearly all specialists are thus far resisting such pressure.

Cost of treatment ranges from about $6,000 to $20,000 or more a year. If treatment is for true growth deficiency, medical insurance usually covers most of the expense, and Medicaid helps impoverished families in every state but Alabama.

"Two things will help us keep (cosmetic) usage down," says Martin Press, a pediatrics professor at Yale Medical School. "One is cost; if I tell parents I won't ask the insurance companies to cover it, they lose interest quite rapidly. The other is that many kids aren't too keen to have three shots a week for a good many years"—the usual treatment.

'Prospect of Two Classes'

But the cost of hGH raises an obvious concern: that only well-to-do families will be able to afford the cosmetic treatment, and thus richer kids will get yet another advantage. "The prospect of two classes—one tall and moneyed, the other short and poor—is ugly and disquieting," writes Thomas Murray, an ethicist at the Institute for the Medical Humanities at the University of Texas at Galveston.

So far, interest in Mr. Waxman's bill has come mainly from the U.S. Olympic Committee and doctors active in sports medicine. "Since the synthetic became available, there has been a fivefold increase in queries from athletes about using human-growth hormone," says William Taylor, a physician who writes widely about drug use by athletes. "And as volume goes up, you can be sure that the percentage diverted from regular distribution channels will also go up."

But the lineup of Waxman bill opponents is impressive. Genentech and Lilly both insist that their tightly controlled distribution systems confine the hormone to specialists treating children for specifically diagnosed growth problems, and that this, plus intensive education of doctors and parents, makes government controls unnecessary. An organization of families with growth-deficient children argues that the bill would raise costs, complicate distribution, and hamper further research.

Many doctors also resist the Waxman approach. "The paperwork burden is bound to result in additional costs to the patient," says Robert Blizzard, director of the University of Virginia's Children's Medical Center. "Education and tight controls by the manufacturers should be the mechanism to prevent abuse."

By Alan L. Otten. *Staff Reporter of* The Wall Street Journal.

Many doctors fear increasing parental pressure for "cosmetic" use of human growth hormone in healthy children.

since the late 1830s when it was suggested by two German biologists, Matthias Schleiden and Theodor Schwann.

Although the cell theory may seem an obvious conclusion today, it constituted a brilliant conceptual breakthrough at the time it was first conceived. For centuries, biologists had been so impressed with the vast diversity of living forms that they tended to overlook underlying similarities. Thus, it is not surprising that the cell theory was proposed long after cell structure had

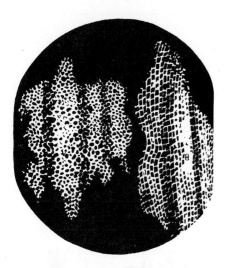

FIGURE 1–4 The structure of cork as seen by Robert Hooke. This is a reproduction of a plate from his book *Microphagia,* published in 1665.

been described by early microscopists. Indeed, Schleiden and Schwann formulated their theory nearly two hundred years after Robert Hooke first examined thin slices of cork under the microscope, observing empty spaces separated by heavy structural walls. He described the spaces as "cellulae," from the Latin word *cella* meaning small room. Although Hooke was in fact only observing the space once occupied by living material, the term *cell* has been used since to describe the fundamental unit of life (Figure 1–4).

Some organisms, such as bacteria and amoebae, are composed of a single cell and are termed **unicellular;** others, formed of many cells, are **multicellular.** The cells of all bacteria are 50 to 100 times smaller than those of other organisms and have a comparatively simple structure that is termed **prokaryotic** (*pro* = primitive; *karyon* = nucleus). All other organisms have a **eukaryotic** (*eu* = true) cell structure, which is more complex.

In terms of their organization, multicellular organisms range from the very simple to the exceedingly complex. The simplest multicellular organisms are composed of only a few cells possessing similar functional capabilities and structure, and the individual cells show little coordination in activity. In the most complex multicellular organisms such as human beings, there are trillions of cells. Among these are more than one hundred different cell types that are specialized both structurally and functionally to perform specific tasks. Moreover, the activities of the different cell types are finely coordinated with one another, forming an organism capable of performing the myriad functions required to sustain life.

It is important to stress that there is considerable structural and functional variability between cells of different organisms and between different types of cells in the same organism. Nevertheless, the fundamental similarities between all cells are far more striking than the differences.

## Chemical Composition

All living cells and the organisms they compose have a similar chemical composition. The principal constituent of all cells is water, which comprises anywhere from about 60 to 95 percent of their weight, depending upon the cell type. Water is indispensable to life, for it serves as the medium in which all cellular activities occur.

The molecules that compose cells have backbones that are made up of chains or rings of carbon atoms, to which other atoms (primarily hydrogen, oxygen, nitrogen, and phosphorus) are attached. Molecules with backbones of carbon atoms are termed **organic molecules.** There are four main classes of small organic molecules in cells: *simple sugars, amino acids, nucleotides,* and *fatty acids.* These serve as sources of energy and as building blocks for the manufacture of large organic molecules, called **macromolecules.** Cells contain primarily four types of macromolecules, including *carbohydrates, proteins, nucleic acids,* and *lipids.*

## Metabolism

The cells of all living organisms act as a kind of chemical factory that performs two interrelated functions. One of these is to convert nutrient molecules obtained from the environment, or to recycle unneeded cellular molecules, into essential cellular components. The other function is to break down nutrient molecules in order to release energy required to sustain cellular activities.

These two functions comprise a cell's **metabolism,** another common characteristic of all cell types. Metabolism includes all the chemical processes by which cellular material is produced, maintained, and broken down, and by which energy is made available to drive these processes and all other activities of the organism.

The chemical conversions that occur as a result of metabolism are not simple. They often take place in lengthy sequences of chemical reactions, called **metabolic pathways.** Metabolic pathways are similar in all cell types. The uniformity in the metabolic design of all cells, from the simplest to the most complex, is perhaps the most astonishing feature of living organisms.

## Homeostasis

Living organisms are **open systems;** that is, they regularly exchange energy and materials with their surroundings. This exchange, however, does not occur in a haphazard way. Rather, the survival of a living organism depends upon a precisely balanced flow of energy and materials into and out of the organism. Such a balanced flow of energy and materials is called a **steady state.**

Thus, living organisms are open systems maintained in a steady state. Furthermore, living organisms are such delicately tuned systems that the slightest disturbance in the steady state condition may result in the death of the organism. Life maintains this precarious position by performing a delicate balancing act. The degradation of aging or damaged cellular material, for example, must be balanced with the synthesis of new cellular material; energy storage must balance energy utilization; water loss must balance water gain; and so forth.

Living organisms must maintain their steady state within relatively narrow physiological limits, often in the face of wide fluctuations in environmental conditions. A human being living in the desert, for instance, must maintain a constant body temperature (98.6°F) that is often lower than the ambient temperature, and must prevent water loss. The maintenance of relatively constant internal conditions within living organisms requires automatic mechanisms that sense changes in the internal environment and make appropriate metabolic adjustments that minimize those changes. The noted American physiologist, W. B. Cannon, referred to such mechanisms as *homeostatic* ones. They maintain **homeostasis,** the relatively constant, but dynamic, internal environment necessary for life.

As an example of homeostatic mechanisms, Figure 1–5 describes those that control water balance in humans. Any loss or gain of water from optimum levels in body tissues triggers homeostatic mechanisms that tend to counteract

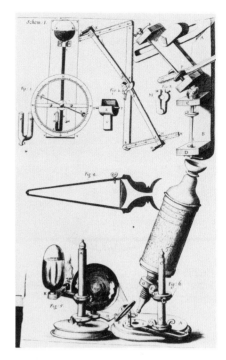

A plate from *Microphagia*, showing the instrumentation that Robert Hooke used to gather data that he observed and described.

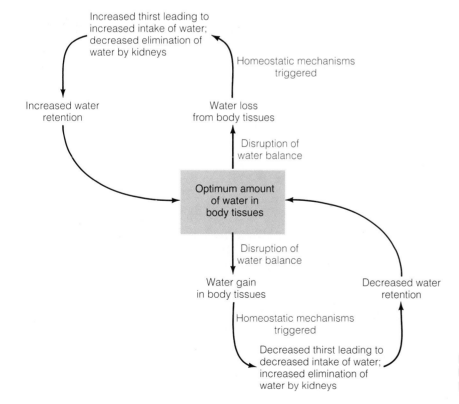

FIGURE 1–5 An example of the operation of homeostatic mechanisms that control water balance in humans.

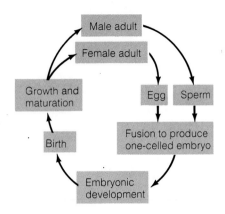

FIGURE 1–6 The life cycle of humans.

the change. Water balance is thus maintained under continuously changing (dynamic) physiological conditions.

## Irritability

An essential property of all living organisms is the ability to detect and to react to events that occur both within the organism and in its environment. This property is referred to as **irritability.** The result of irritability is that chemical, mechanical, thermal, and light stimuli, as well as others, bring about physiological responses that tend to maximize an organism's chances of survival. These responses occur on a variety of levels, ranging from the subtle metabolic responses that maintain homeostasis within living cells to the highly coordinated movements in multicellular animals that are used in obtaining food, avoiding danger, and so forth. Nonliving things, by contrast, cannot react in a *directed* way to stimuli. Thus, for example, a stone may move when it is kicked, but it cannot respond in any way to minimize the effects of being kicked.

## Life Cycle, Growth, and Reproduction

All living organisms have a **life cycle;** that is, a sequential series of events that occur during the life of the organism. For instance, the first stage in the life cycle of a human being is the fusion of a sperm and an egg cell to produce a fertilized egg (a unicellular embryo). This is followed by embryonic development, birth, attainment of sexual maturity, reproduction, and finally, death. From the beginning of embryonic development until the attainment of adult size, the organism *grows* (increases in mass). In humans as in most organisms, this growth is accompanied by a change in organization and form (Figure 1–6).

Almost every type of organism is capable of reproduction at one stage of its life cycle. Reproduction results in the production of new organisms of the same type as the parents. Although there are a few hybrid organisms such as mules and various seedless plants that are sterile and cannot reproduce, the capacity for reproduction is nevertheless considered a general and unique attribute of living organisms. In this regard, it is perhaps well to point out that reproduction is required to perpetuate the species, but it is not a requirement for the life of the organism.

There are two fundamental reproductive patterns among living organisms: **asexual** and **sexual.** Asexual reproduction is achieved by a single organism. For instance, in the division of a unicellular organism such as an amoeba, the cell grows in size and then divides into two virtually identical cells. The budding off of an organism from a parent, as in the production of new plants from strawberry runners, is another example of asexual reproduction. Sexual reproduction, on the other hand, requires two individuals and involves the

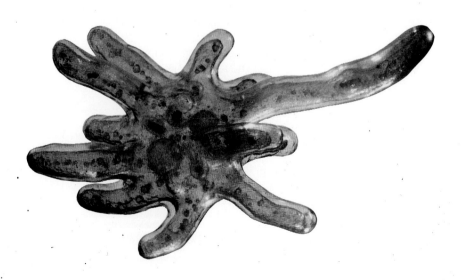

Amoebas exhibit asexual reproduction, which is achieved by a single organism and does not involve the union of genetic material from two sources.

fusion of specialized cells produced by both individuals. The fusion of sperm and egg cells in human reproduction, as noted above, illustrates sexual reproduction.

In multicellular organisms, new cells are continually produced from existing cells during growth or to replace dying cells. This type of reproduction is called *cellular division*. Asexual reproduction in unicellular organisms also occurs by cellular division but, of course, results in the production of new individuals.

## Evolution

Much of our discussion above has focused on the capacity of living organisms to make short-term adjustments in response to internal and external stimuli. These adjustments, occurring during the lifetime of the individual, result in what is called short-term *adaptation*.

All living organisms are also products of a long-term adaptation to a changing environment. This long-term adaptation is called **evolution** and is a fundamental characteristic of all living systems.

As a result of evolution, types of organisms gradually become modified in form and function through time as they interact with their environment. It is important to stress that evolution, unlike short-term adaptation, does not occur during the lifetime of the individual. Instead, the evolutionary process involves a gradual change in type over many generations. Evolutionary change is, then, a historical process, and gives rise to the many different types of organisms that inhabit the earth.

By suggesting that all living organisms are related to one another by common descent, the theory of evolution explains why all living organisms share a fundamental pattern of structural and functional organization, including such common features as cellular organization, chemical composition, metabolism, homeostasis, irritability, a life cycle, growth, and reproduction. This basic pattern of organization is the framework on which evolutionary processes have built the individual refinements that characterize each type of living organism. Thus, for example, human beings differ from cats, frogs, and jellyfish in obvious ways—the refinements of the basic pattern—but human beings continue to share with all organisms the fundamental pattern of organization that is characteristic of life.

## Life: A Complex, Dynamic System

We now know that the unique properties of living systems are *not* derived from some special characteristic of their fundamental constituents, or from some magical "vital force." Instead, life represents an exceedingly complex organization of nonliving material, operating in accordance with the fundamental physical and chemical principles of the inanimate world. A living organism is a dynamic, continuously changing entity. In order to survive—to prevent the dissolution of its organization—a living organism must be able to respond to both short-term and long-term changes in the environment in ways that serve to stabilize its entire structure. This is the essence of a living system: its stability is derived from its nature as a dynamic entity of flexible structure that can respond to a variety of environmental factors that tend to destabilize it.

## INCREASING OUR UNDERSTANDING OF HUMAN BIOLOGY

The subject of human biology encompasses a broad range of material from the disciplines of biology, chemistry, and medicine. All of this material has been accumulated through the research activity of thousands of scientists, using methods of observation, experimentation, and analysis that are the standard tools of any researcher.

TABLE 1-1 The Scientific Method

| | |
|---|---|
| Problem | What causes acquired immune deficiency syndrome (AIDS)? |
| Information Known | AIDS is prevalent among active homosexuals (primarily males), intravenous drug users, hemophiliacs receiving clotting factors manufactured from donated blood, children born to women with AIDS, and people in intimate contact with individuals with AIDS. |
| Hypothesis | AIDS is caused by a virus that is transmitted in the blood and semen of affected individuals. |
| Test of Hypothesis | Examine samples of blood and semen from (1) AIDS victims, and (2) normal healthy individuals under the electron microscope in an attempt to identify virus particles unique to AIDS victims. |
| Results | Virus particles are consistently visible in the blood and semen of AIDS victims. They are not present in the blood and semen of most healthy individuals, although they are seen in these fluids in a small minority of apparently healthy individuals. |
| Conclusions | A virus can be identified consistently in the blood and semen of individuals with AIDS. However, it is also present in these same fluids of some healthy individuals. Two possibilities suggest themselves: (1) the identified virus is not related to AIDS, *or* (2) the identified virus is the cause of AIDS, but some individuals that do not express the disease may carry the virus. |

## Scientific Research

To gain knowledge on any subject, it is necessary to make an inquiry aimed at uncovering the facts. Although the aim is obvious, the task is more easily described than done. The hard part is to distinguish fact from what may merely appear to be fact. To do this, research scientists use a systematic procedure that has come to be known as the *scientific method.* In actuality, the scientific method is used in all valid research, scientific or otherwise, and in much of our daily problem solving.

Use of the scientific method is illustrated in Table 1-1. It involves five steps: (1) identifying a problem; (2) gathering information that is known about the problem; (3) formulating a hypothesis, or tentative explanation, based upon the information that has been assembled; (4) testing the hypothesis; and (5) reaching conclusions that are consistent with the tests of the hypothesis.

After selecting a problem, information related to it is usually assembled by doing a literature review. In scientific research, this generally means reading original research journals. The formulation of a hypothesis from what is already known about the problem requires *inductive reasoning,* a thought process requiring creativity and insight, by which the accumulated information is organized in ways that suggest generalizations. In the example shown in Table 1-1, the hypothesis was formulated by looking at all groups of individuals that have contracted AIDS and attempting to discover what they all have in common. Inductive reasoning led to the hypothesis that AIDS is caused by a virus that is transmitted in the blood and semen of infected individuals.

A plausible hypothesis must be testable and permit predictions to be made from it. In order to figure out how to test the hypothesis, **deductive reasoning** is used. Deductive reasoning derives specific conclusions from generalizations, often in the form of an "if . . . then" statement. With regard to the example in Table 1-1, such a statement might be: if an AIDS virus is present in body fluids of affected individuals, then it should be visible in the electron microscope.

In science, testing a hypothesis is often done by performing experiments. These must be repeatable and designed so that all parameters that might vary (known as variables) in the experimental procedure are under the control of the experimenter. In practice, we do this by performing two parallel experiments simultaneously; the two experiments are identical except for the one factor that is being tested, and consequently, the results of one of the parallel experiments serve as a standard, or *control*, with which to compare the results of the other. In the experiment described in Table 1–1, the control experiment consists of blood and semen from normal, healthy individuals to which the same fluids from AIDS victims are compared.

All too frequently, a well-designed experiment will not yield clear-cut results that are easily summarized in definitive conclusions. Notice in Table 1–1 that the conclusions do not support the hypothesis unequivocally. Instead, they suggest further experiments that must be done to prove or disprove the hypothesis. (The hypothesis in Table 1–1 happens to be correct, as discussed in Box 21B, Chapter 21, but the experimental results have been complicated by the fact that apparently healthy, symptomless individuals may harbor the virus and transmit it to others.) It is typical for experimental research to involve a lengthy process of trial-and-error, in which the formulation and testing of an initial hypothesis is followed by successive modifications and refinements of the original hypothesis to account for ongoing experimental results.

Although all scientific research follows the scientific method, at least implicitly, it is not all strictly experimental. Research done routinely in epidemiology (the study of the incidence, distribution, and control of disease in populations) is *correlative*, meaning that it attempts to establish orderly connections between sets of data. Epidemiological studies involve the correlation of data obtained from large numbers of humans that have a specific disease. By comparing such parameters as diet, exposure to environmental agents and social contacts of healthy and diseased individuals, it is often possible to uncover patterns that will lead to an understanding of the cause of the disease. For example, early epidemiological studies of AIDS victims provided the first suggestions concerning its mode of transmission, and led directly to studies that revealed the presence of the AIDS virus.

Some of the most significant advances in scientific knowledge have derived from *descriptive research*, which involves observing and describing things and events in the natural world. In biology, descriptive research has given us most of what we know about such subjects as microscopic structure, gross anatomy, and systematics (the classification of organisms), to name only a few.

Depending on its aim, scientific research may be categorized as *basic* or *applied*. Basic research is not directed toward solving immediate practical problems, but rather seeks to uncover fundamental information about the nature of things. For example, basic research in biology might aim to discover the molecular structure of a natural antibiotic compound such as penicillin, and how its molecular structure determines biological function. The discoveries of basic research often lead to success in applied research, which attempts to solve practical needs. For instance, the basic research mentioned above might provide the knowledge necessary for conducting applied research aimed at synthesizing a new, more potent, antibiotic by slightly altering the molecular structure of the natural one.

It is important to realize that the fundamental knowledge required for applied research is derived from basic research. This is why any policy that supports applied research to the neglect of basic research—as popular as such a policy might be in times of limited resources—is ultimately a self-defeating one. A perfect example of this is AIDS research. Had we not accumulated through basic research a knowledge of the structure of viruses and their mode of reproduction, the function of the immune system, and countless other related facts, we could not possibly have come as far as we have, and as quickly, in identifying the cause of AIDS and developing methods of prevention and treatment.

## The Validity of Medical and Biological Research to Human Biology

Most of what we know about human biology has been learned from experiments using research organisms other than human beings. This is true for several reasons. In the first place, it usually is not ethically possible nor, for that

# The Thalidomide Tragedy

Thalidomide is a sedative and anti-emetic (a drug that prevents vomiting) that was sold over the counter in West Germany and was available by prescription in other European countries between late 1959 and early 1961. During that period, it was taken by thousands of pregnant women in Europe to prevent morning sickness. It was never licensed for sale in the United States because of bureaucratic delay and concern over reports that it caused polyneuritis, an inflammation of nerve cells, in a few individuals.

If thalidomide was taken by pregnant women between the end of the fourth and the sixth week of development (days 28–42 after fertilization) during the time when the limbs were forming, abnormal limb development occurred. In some cases, limbs were entirely absent, but, more commonly, they were severely deformed, often being short and flipperlike, a condition termed *phocomelia* (seal limbs). Thalidomide caused other birth defects as well, but limbs were most frequently affected. As many as ten thousand children, the majority in West Germany, were born with phocomelia or other birth defects caused by thalidomide. In fact, thalidomide was such a potent teratogen that even a single 100 mg tablet taken during the critical period could cause severe birth defects.

Although it is still not understood how thalidomide causes birth defects, we know that the thalidomide tragedy resulted because of unusual differences in the response of various experimental animals to the drug. When administered orally, thalidomide is not teratogenic in rats and mice—the laboratory animals in which it was originally tested. However, subsequent testing has shown that thalidomide is teratogenic in small oral doses in rabbits and nonhuman primates. To forestall another thalidomide tragedy, drugs are now tested for teratogenicity in several different classes of experimental animals. Even then, it is always possible that a drug may be teratogenic only in humans, and it is consequently prudent for pregnant women to avoid all drugs that are not absolutely necessary.

matter, is it desirable to perform experiments using human beings as subjects. Secondly, even in those limited situations where experimentation on humans is appropriate (for example, in the final stages of drug testing), humans are poor experimental animals. Our varied genetic backgrounds and life histories make it difficult to run controlled experiments, and our long generation time inhibits research aimed at determining genetic effects.

Some of the most fundamental and valuable knowledge we have of human biology, such as the chemical basis of heredity, the mechanism by which proteins are synthesized, the design of metabolic pathways, the role of vitamins in metabolism, and the ways that energy is stored and released, among many other things, has been derived from basic research on bacteria, yeasts, and fruit flies (*Drosophila*). These are ideal research organisms because of their ready availability, short generation time, and ease of culture.

In research with direct clinical applications, mammals are preferred as research organisms because their physiology is, in most cases, extremely similar to that of humans. Rats, mice, hamsters, and rabbits are used extensively in the testing of drugs, potential **carcinogens** (cancer-causing agents), potential **mutagens** (agents that alter the hereditary material), and potential **teratogens** (agents that cause birth defects). Cats, dogs, and monkeys are primarily used in the development of surgical techniques, such as organ transplantation, and of other medical procedures, such as the treatment of burns and fractures.

Over the years, a great deal of evidence has accumulated to support the contention that research done on experimental organisms is, in general, applicable to humans. This is not surprising, considering the fundamental similarity of all living organisms; in fact, it would be surprising were it not the case.

Carcinogen testing is a good example of the applicability to humans of research on other organisms. Epidemiological studies have identified 82 chemicals for which there is at least some evidence of human carcinogenicity.

When tested in animals, 81 of these chemicals were shown to be carcinogenic. The one exception is arsenic, which is associated in humans with skin cancer when ingested in drinking water and with lung cancer after occupational exposure. There is some evidence to suggest, however, that arsenic may act to potentiate the action of other carcinogens, but may not be carcinogenic itself. Such an effect would explain the discrepancy between epidemiological studies in humans and animal testing, since humans come in contact with a variety of carcinogens, while research organisms, kept under carefully monitored conditions, presumably do not.

Bacteria have also been used successfully in carcinogen testing, even though there is no condition in bacteria that is strictly analogous to cancer. In tests of several hundred substances, it has been shown that nearly 90 percent of those chemicals that are known carcinogens in humans are mutagens in bacteria. There is a good correlation between carcinogenicity in humans and mutagenicity in bacteria presumably because interaction of a chemical substance with the cell's hereditary material—a necessity for a mutagen—may be an initial step in carcinogenicity.

It would be a mistake to assume that the results of research on any experimental organism can always be extrapolated to humans. There have been some notable exceptions (see *Focus: The Thalidomide Tragedy*), but these do not necessarily invalidate the generalization, as long as the exceptions are infrequent and the generalization is accepted with care. In this context, it is worth remembering as you read this book that most of the fundamental mechanisms that explain how the human body functions have been derived from research on experimental organisms. Human biology as a discipline thus has its roots firmly embedded in the biological sciences.

## Study Questions

1. Describe the characteristics of living systems.
2. If mules are incapable of reproduction, why should they be considered living entities?
3. In your own words, explain the purpose of a control in experimental research.
4. Indicate the differences between experimental, correlative, and descriptive research.
5. How do the aims of basic and applied research differ?
6. Why are results of research on experimental organisms often applicable to humans?

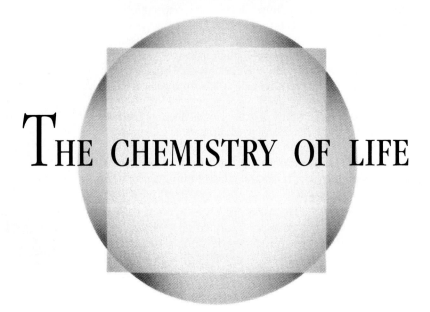

### Chapter Two

# THE CHEMISTRY OF LIFE

## THE CHEMICAL NATURE OF LIVING MATTER

Living things can seem so unique that it is easy to overlook the fact that they are composed of the same fundamental chemical particles that form inanimate objects. Consequently, in order to understand how human beings are formed and how they function, we must obtain some familiarity with the chemical structure of matter in general and with some of the chemical and physical principles that apply directly to human physiology.

### Atoms

Like all matter, living organisms are composed of small particles called **atoms.** An atom contains a **nucleus** and one or more negatively-charged particles called **electrons** that move around the nucleus in cloudlike regions, termed **atomic orbitals.** The space taken up by the rotating electrons accounts for most of the volume of an atom; most of its mass is found in the nucleus, which is composed of positively-charged particles called **protons** and uncharged particles called **neutrons** (except for the nucleus of the smallest atom, the hydrogen atom, which contains only a single proton and no neutrons). Protons and neutrons are of approximately the same mass, while electrons are vastly smaller, possessing a mass only 1/1835 that of a proton. Protons and electrons carry opposite charges of equal magnitude, arbitrarily designated as +1 and −1, respectively; hence, neutral atoms have equal numbers of protons and electrons (Figure 2–1).

### Elements

There are 92 different, naturally occurring types of neutral atoms, each of which has a characteristic number of protons and electrons and unique chemical and physical properties. Each of the different types of atoms is called an **element.**

The cells and cell products of living organisms are not composed of all 92 elements; instead, only 25 are known to be essential to at least some living organisms (Table 2–1). Of these, hydrogen (H), carbon (C), nitrogen (N), and oxygen (O) are by far the most abundant, representing more than 99 percent of the atoms present in living organisms. Together with phosphorus (P) and sulfur (S), these elements are the main constituents of the organic substances that comprise cells. Five additional elements are used in significant amounts and play central roles in cellular physiology. These are calcium (Ca), sodium (Na), potassium (K), chlorine (Cl), and magnesium (Mg); they exist primarily in the form of charged atoms, called *ions* (discussed later).

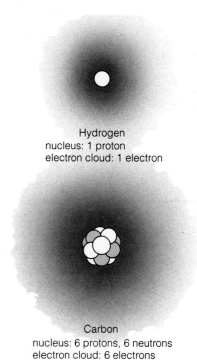

Hydrogen
nucleus: 1 proton
electron cloud: 1 electron

Carbon
nucleus: 6 protons, 6 neutrons
electron cloud: 6 electrons

FIGURE 2–1 Atomic structure of hydrogen and carbon, two elements common in living organisms. Electrons are present in cloudlike regions (atomic orbitals) surrounding the nucleus. The size of the nucleus in these drawings is exaggerated; in reality, the diameter of the electron cloud is many thousands of times the diameter of the nucleus.

TABLE 2–1 Elements Known to be Essential to Living Organisms

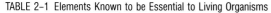

| Main Constituents of Organic Compounds that Compose Living Organisms | Other Elements Present in Substantial Amounts, Generally as Ions | Trace Elements | |
|---|---|---|---|
| Hydrogen (H) | Calcium (Ca) | Boron (B)* | Cobalt (Co) |
| Carbon (C) | Potassium (K) | Fluorine (F) | Copper (Cu) |
| Nitrogen (N) | Sodium (Na) | Silicon (Si) | Zinc (Zn) |
| Oxygen (O) | Chlorine (Cl) | Vauadium (V) | Selenium (Se) |
| Phosphous (P) | Magnesium (Mg) | Chronium (Cr) | Molybdenum (Mo) |
| Sulfur (S) | | Manganese (Mn) | Tin (Su) |
| | | Iron (Fe) | Iodine (I) |

*Boron has been shown to be essential in higher plants, but not in other organisms.

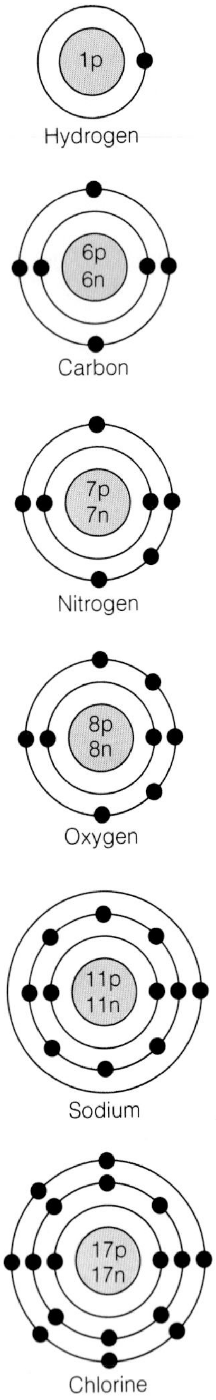

FIGURE 2–2 Diagrams showing electrons (black dots) in valence shells of six biologically important elements. The valence shell closest to the nucleus can hold two electrons, whereas the second and third valence shells can each hold eight. In atoms with more than 18 electrons, additional valence shells exist that contain electrons.

Other elements in living organisms, termed **trace elements,** are present only in minute quantities, but are nevertheless absolutely essential for life. Thirteen trace elements are known to be essential in higher animals. Boron (B) is essential to higher plants, but apparently not to other organisms. There is some evidence that aluminum (Al) and nickel (Ni) may also be required by some organisms.

Trace elements are often minor constituents of important biochemical substances. For example, thyroid hormone, essential for normal growth and development and for the regulation of the rate of metabolism, contains iodine (I); hemoglobin, the oxygen-carrying protein in red blood cells, contains iron (Fe); and vitamin $B_{12}$ contains cobalt (Co). Dietary deficiencies of trace elements are associated with a variety of diseases that we will consider in our discussion of digestion and nutrition (see Chapter 11).

## Chemical Bonding

The atoms of elements in living systems do not usually exist in a neutral, uncombined state. Rather, atoms combine with one another to form relatively stable combinations called **molecules.** Molecules containing atoms of two or more different elements are referred to as **compounds.**

The attractive force that effectively holds two atoms in a molecule together is known as a **chemical bond,** and this force is due to interactions between electrons of the two atoms. In order to understand why these interactions occur, we need to understand a bit more about atomic orbitals. Atomic orbitals constitute regions surrounding the nucleus where electrons are most likely to be located; they have been likened to flight patterns around airports. Electrons in atomic orbitals close to the nucleus contain less energy than electrons farther from the nucleus. Atomic orbitals, or groups of orbitals, that contain electrons with similar energy levels are termed **valence shells** of an atom. As shown in Figure 2–2, the first valence shell of an atom is capable of holding two electrons, the second valence shell can hold eight, the third can hold eight, and so on. Because an atom is most stable when its outermost valence shell is filled with electrons, atoms will interact with one another—form a chemical bond—by sharing or transferring electrons in order to fill their outer valence shells.

## Covalent and Ionic Bonds

Electron interactions that result in chemical bond formation are traditionally separated into two main types: **covalent bonding** and **ionic bonding.** Covalent bonding results when electrons are shared more or less equally between two atoms; that is, as the two bonding atoms approach one another, one or more electrons from each atom are attracted to the positively charged nucleus of the other atom, creating a force of attraction that holds the two atoms together. The electrons are thus, in a sense, shared between the two atoms, and this sharing of electrons is called a covalent bond. A single covalent bond is formed when two electrons, one from each atom, are shared by the two atoms (Figure 2–3). Each covalent bond, which represents a shared pair of electrons, is usually represented by a line connecting the bonded atoms, as shown in Figure 2–3. Double covalent bonds between two atoms can form by the sharing of two electron pairs, and triple covalent bonds form by the sharing of three electron pairs. Double and triple bonds are represented by two and three lines, respectively. Multiple bonds are stronger than single ones. The atoms that form the organic molecules of living organisms are held together primarily by covalent bonding.

In other types of electron interactions, electrons in one atom are so strongly attracted by another atom that one or more electrons are actually transferred to the other atom, creating a positively charged atom (by electron loss) and a negatively charged atom (by electron gain) that are then attracted to

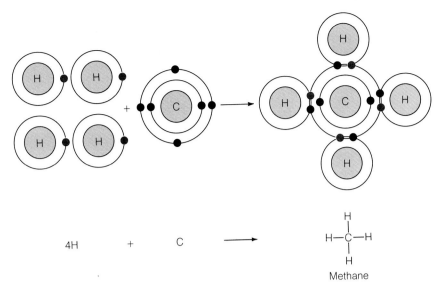

$$4H \quad + \quad C \quad \longrightarrow \quad \begin{matrix} & H & \\ H-&\overset{\displaystyle |}{\underset{\displaystyle |}{C}}&-H \\ & H & \end{matrix}$$

Methane

FIGURE 2–3 The formation of single covalent bonds between a carbon atom and four hydrogen atoms to form methane. Electrons are shared between the carbon and hydrogen atoms, completing the outer valence shells of all atoms.

one another by virtue of their opposite charges. Such an electrostatic attraction is called an **ionic bond** (Figure 2–4).

## Polar and Nonpolar Molecules

In some cases, two atoms involved in a covalent bond may not share the bonding electrons equally between them, but instead, the shared electrons are more attracted, and hence closer, to one atom than to the other. Electrons are not actually transferred from one atom to another (as in ionic bonding), but the atom that tends to attract the bonding electrons will possess a slight negative charge relative to the other atom. Depending upon the geometric shape of a molecule, such unequal sharing of electrons between two atoms may result in a molecule that possesses one end with a slight positive charge and the other end with a slight negative charge. Such a molecule is termed a **polar molecule.** Water, an important polar molecule, is shown in Figure 2–5. **Nonpolar molecules** such as methane (see Figure 2–3) do not possess negative and positive ends. Polar molecules, charged molecules, and ions tend to dissolve in water, whereas nonpolar substances do not (Figure 2–6). As we shall see, the ability of water to dissolve a wide range of compounds is crucial to its metabolic role in living systems.

## Weak Interactions Between Atoms

Covalent and ionic bonds form particularly strong interactions between atoms, and they require substantial amounts of energy to be broken. In addition to these strong types of chemical bonds, a number of relatively weak interactions between atoms can occur that are particularly important in the chemistry of

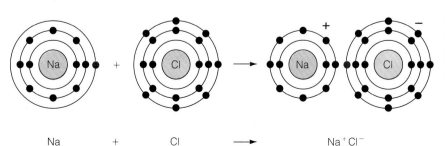

$$Na \quad + \quad Cl \quad \longrightarrow \quad Na^+Cl^-$$

FIGURE 2–4 Ionic bonds form between sodium and chlorine when the sodium actually transfers an electron to the chlorine atom. The two charged atoms, called ions, are then held together by electrostatic attraction.

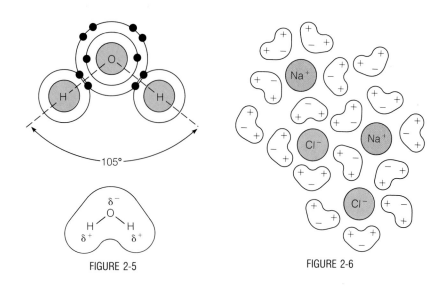
FIGURE 2-5

FIGURE 2-6

FIGURE 2-5 The water molecule has the shape of an isosceles triangle with an H-O-H angle of about 105°. In addition, the shared electron pairs between the hydrogen and oxygen atoms are located, on the average, closer to the oxygen atom than to the hydrogen atoms. This imparts a partial negative charge ($\delta^-$) to the oxygen atom and a partial positive charge ($\delta^+$) to the hydrogen atoms. The combination of the geometry of the water molecule coupled with the distribution of charge among its atoms makes water a polar molecule.

FIGURE 2-6 Sodium chloride ($Na^+Cl^-$), common table salt, dissolves in water because the attraction of the polar water molecules to the sodium and chloride ions prevents the ions from interacting electrostatically with one another. As shown here, the positive ends of water molecules cluster around negatively charged chloride ions, whereas the negative ends of water molecules are attracted to positively charged sodium ions.

biological systems. These weak interactions, which may require on the order of 10 to 100 times less energy for their disruption than the strong chemical bonds, may occur between atoms in the same molecule or in different molecules. Two important types of weak interactions are **hydrogen bonds** and **hydrophobic bonds.** Hydrogen bonds represent weak attractions between a hydrogen atom, which is covalently bonded to an oxygen or a nitrogen atom, and another oxygen or nitrogen atom in the same or a different molecule (Figure 2–7). Hydrophobic bonds are weak attractions between nonpolar molecules or nonpolar regions of the same molecule caused by the tendency of nonpolar entities to associate in an aqueous medium (e.g., cell cytoplasm) so that their contact with water is minimized.

Hydrogen bonds and hydrophobic bonds, among other weak interactions, are particularly important in stabilizing the three-dimensional structure of large organic molecules of the cell (Figure 2–8). Because the metabolic function of a molecule depends to a large extent upon its three-dimensional structure, maintenance of a particular molecular geometry is crucial to many metabolic activities. For example, a hemoglobin molecule in a red blood cell may completely lose its ability to carry oxygen simply because the geometry of the molecule has been altered. In addition to determining the shape of molecules, weak interactions influence the distribution of molecules within the cell. The formation of any cellular structure, such as the nucleus or the plasma membrane, depends upon weak interactions between its constituent molecules.

## Chemical Reactions

Under appropriate conditions, molecules may react with one another to form new molecules. Such a transformation, which occurs by the breaking of bonds in the original molecules and the formation of new bonds, is called a **chemical reaction.** For example, the equation

$$6CO_2 \quad + \quad 6H_2O \rightarrow C_6H_{12}O_6 \quad + \quad 6O_2$$
(carbon dioxide)   (water)   (glucose)   (oxygen)

represents the chemical reaction between six molecules of carbon dioxide and six molecules of water to form one molecule of glucose and six molecules of molecular oxygen. In a chemical reaction, the original molecules on the left side of the equation are called the **reactants,** and the newly formed molecules on the right side of the equation are called the **products.**

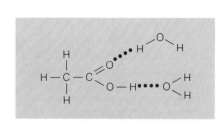

FIGURE 2-7 When acetic acid, an important ingredient of vinegar, is mixed with water, hydrogen bonds (dotted lines) form between acetic acid (left) and water molecules.

## Water

Water ($H_2O$), the main constituent of living cells, is indispensable to life as we know it on this planet. The importance of water lies in the fact that *it is the essential medium in which all chemical activities of the cell occur.* Water has a number of chemical properties that make it particularly suited for this central role in cellular metabolism.

### Solvent Properties of Water.

*Solutions.*    When a substance, such as cane sugar (sucrose), is dissolved in water, a homogeneous mixture is formed; that is, the sucrose molecules are distributed uniformly throughout the water and under normal conditions will not settle out. Such a homogeneous mixture is called a **solution.** Water, the substance in excess, is referred to as the **solvent,** and the dissolved substance (in this case, sucrose) is known as the **solute.**

Water is often referred to as the "universal solvent." Although this is an overstatement, it is certainly true that water has the ability to dissolve a large number of chemical compounds. Biological systems take advantage of the dissolving properties of water by using it to transport nutrients, waste products, and other substances into and out of cells, and into and out of the organism as a whole.

*Ions.*    Water is an important solvent in yet another respect. Many molecules, particularly those held together by ionic bonds, will dissociate (or **ionize**) when dissolved in water, forming charged atoms or groups of atoms called **ions.** For example, when solid sodium chloride (NaCl; common table salt) is dissolved in water, it dissociates to form positively charged sodium ions ($Na^+$) and negatively charged chloride ions ($Cl^-$). Sodium ions have a single positive charge because they have transferred one of their electrons to the chlorine

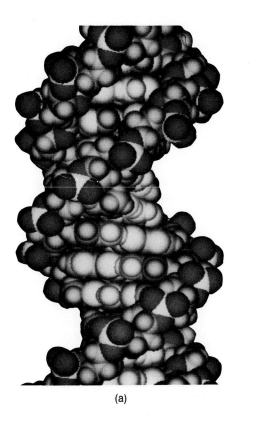

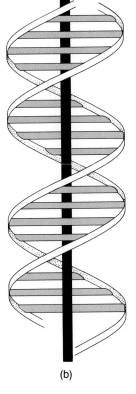

(a)                                    (b)

FIGURE 2–8 In the DNA molecule, hydrogen bonds hold two separate strands of covalently bonded atoms in a rigid helical configuration. (a) A molecular model of a portion of the DNA molecule. Each sphere represents an individual atom. (b) A schematic diagram interpreting the structure seen in (a). The ribbons correspond to strands of the molecule that are joined by hydrogen bonds.

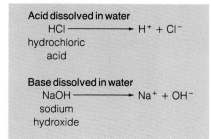

Acid dissolved in water

$$HCl \longrightarrow H^+ + Cl^-$$
hydrochloric
acid

Base dissolved in water

$$NaOH \longrightarrow Na^+ + OH^-$$
sodium
hydroxide

FIGURE 2–9 When dissolved in water, acids release hydrogen ions ($H^+$) and bases release hydroxide ions ($OH^-$).

atom, forming a negatively charged chloride ion. Molecules that ionize in water are called **electrolytes.**

Some ions of major importance in all cells are hydrogen ion ($H^+$), hydroxide ion ($OH^-$), sodium ion ($Na^+$), potassium ion ($K^+$), calcium ion ($Ca^{++}$), magnesium ion ($Mg^{++}$), chloride ion ($Cl^-$), sulfate ion ($SO_4^=$), phosphate ion ($HPO_4^=$), and bicarbonate ion ($HCO_3^-$).

An important parameter of all body fluids, including **intracellular fluids** (fluids within cells) and **extracellular fluids** (fluids outside of cells, such as blood and lymph), is their **ionic balance.** Ionic balance refers to the concentration of different ions, where concentration is defined as the number of solute particles—in this case, ions—per unit volume of solvent (water). The ionic balance of all body fluids, sometimes referred to as the *electrolyte balance,* must be maintained at extremely constant levels, and, in fact, cell damage or death may result from even minor disturbances in ionic balance. Sodium and chloride ions are the principal ions in extracellular fluids, and potassium and phosphate ions are the intracellular ions present in greatest concentration.

So-called metal ions (such as $Ca^{++}$, $Na^+$, $K^+$, $Fe^{+++}$, $Mg^{++}$) have many important functions in the organism. $Ca^{++}$ is important in muscle contraction, in blood clotting, and in the development of bones and teeth; $Na^+$, $K^+$, and $Ca^{++}$ are involved in the transmission of nerve impulses; $Fe^{+++}$ (ionic iron) is an essential constituent of the oxygen-carrying protein hemoglobin and of other important proteins; and $Mg^{++}$ is required for many metabolic reactions of the cell.

## Acids and Bases

Other ions are involved in the crucial role of determining and maintaining the proper acid-base balance in the body. There are certain substances, termed **acids,** that dissociate when placed in water, releasing hydrogen ions ($H^+$); **bases,** on the other hand, produce hydroxide ions ($OH^-$) when dissolved in water (Figure 2–9). It is the hydrogen ions that impart acidity to an aqueous solution, whereas hydroxide ions impart basicity (alkalinity). If a solution contains equal concentrations of hydrogen and hydroxide ions, the solution is called neutral; that is, it is neither acidic nor basic. A solution that contains a greater concentration of hydrogen ions than hydroxide ions is called acidic, and the degree of acidity is directly proportional to the concentration of hydrogen ions relative to hydroxide ions. Conversely, a solution that contains greater concentrations of hydroxide ions than hydrogen ions is basic, and again, the degree of basicity is directly proportional to the concentration of hydroxide ions relative to hydrogen ions.

***The pH Scale.*** In actuality, acidity and basicity are merely opposite sides of the same coin; if a solution's acidity is low, its basicity is high, and vice versa. Because of this inverse relationship between acidity and basicity, it has been possible to derive a quantitative scale, the **pH scale,** to indicate the acidity of an aqueous solution. As shown in Figure 2–10, the pH scale runs from 0 to 14, with the midpoint of the scale, a pH of 7, representing a neutral solution. Acidic solutions have pH values of less than 7, and basic solutions have pH values greater than 7. The more acidic (less basic) the solution, the lower the pH value; the more basic (less acidic) the solution, the higher the pH value.

***Regulation of pH and the Role of Buffers.*** Cellular metabolism results in the production of many acidic by-products such as sulfuric acid, phosphoric acid, and a variety of organic acids; in addition, many basic substances enter the body in our food. However, living cells cannot tolerate large fluctuations in pH, and consequently, the pH of intracellular and extracellular fluids must be maintained within a remarkably narrow range. For example, the pH of blood is maintained between 7.35 and 7.45, and fluctuations below or above these

FIGURE 2–10 The pH scale.

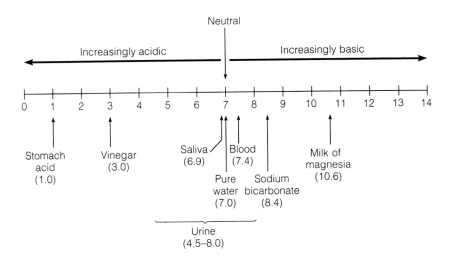

values, usually occurring only as a result of disease, can result in severe cell damage and even death.

The pH of intracellular and extracellular fluids is regulated by dissolved substances or mixtures of substances called **buffers,** which tend to resist large changes in pH when hydrogen ions or hydroxide ions are added to these fluids. Several ions, including bicarbonate ions and phosphate ions, play crucial roles as buffers.

As an example of the action of a buffer, let us consider the bicarbonate ion ($HCO_3^-$). Bicarbonate ions are maintained at reasonably high levels within cells and in the extracellular fluids that bathe the cells. When hydrogen ions are produced as a result of cellular metabolism, they are eliminated by combination with bicarbonate ions to form carbonic acid:

$$H^+ + HCO_3^- \rightarrow H_2CO_3$$

Carbonic acid, in turn, dissociates into carbon dioxide and water, as follows:

$$H_2CO_3 \rightarrow H_2O + CO_2$$

The excess carbon dioxide is picked up by the blood and travels to the lungs where it is exhaled.

### The Role of Water in Temperature Regulation

The temperature of the human body is normally maintained at about 98.6°F. Fluctuations of a few degrees may occur during illness, but wide fluctuations in body temperature, resulting in temperatures less than about 92°F or more than about 106°F, are not tolerated. This is because the chemical reactions that constitute the cell's metabolism can occur efficiently only within a narrow temperature range. Particularly at high extremes, the metabolic machinery may be permanently damaged as a consequence of the disruption of weak attractions that determine the three-dimensional structure of important metabolic molecules.

The metabolic activities of cells generate large amounts of heat, a fact that is obvious to anyone who has exercised vigorously. Most of this heat must be dissipated quickly and efficiently in order to prevent a fatal rise in temperature. Water plays a crucial role in this respect. Relative to most solvents, water can absorb greater amounts of heat without a temperature change. In addition, the evaporation of water requires substantial amounts of heat, and consequently, the evaporation of sweat on the surface of the body has a cooling effect.

## Small Organic Compounds of Living Organisms

Living organisms contain a variety of small organic compounds, most of which are dissolved in the intracellular and extracellular fluids along with inorganic ions. Although many types of small organic compounds are found in living organisms, most can be categorized into one of the following main groups.

### Simple Carbohydrates

**Simple carbohydrates,** such as *glucose, fructose, sucrose,* and other small molecules serve as nutrients to provide energy and raw materials for the manufacture of cellular constituents (Figure 2–11). Surplus glucose is stored in large molecules, called *polysaccharides* (discussed below).

### Amino Acids, Nucleotides, and Fatty Acids

Compounds such as *amino acids* (Figure 2–12), *nucleotides* (Figure 2–13), and *fatty acids* (Figure 2–14), are essential precursors for the synthesis of, respectively, proteins, nucleic acids, and fats (these complex organic com-

FIGURE 2–11 Some simple carbohydrates.

**Monosaccharides**
Monosaccharides (*mono* = one; *sakcher* = sugar) are the simplest carbohydrates. The two most important energy-yielding monosaccharides in humans are glucose and fructose.

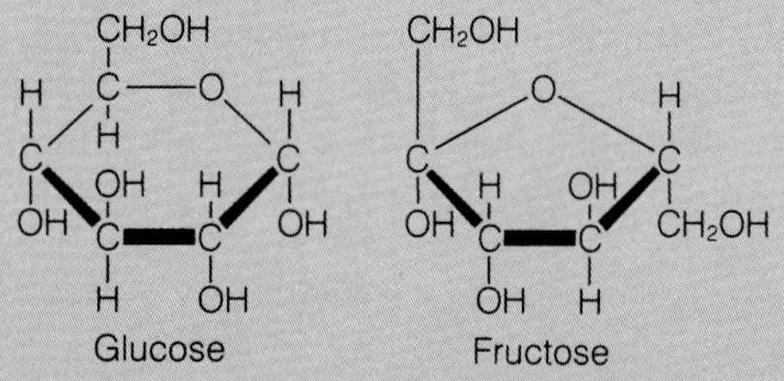

Glucose          Fructose

**Disaccharides**
Disaccharides are composed of 2 monosaccharide units, bound covalently together by the elimination of water. Sucrose, common cane, and beet sugar is composed of one molecule of glucose and one molecule of fructose.

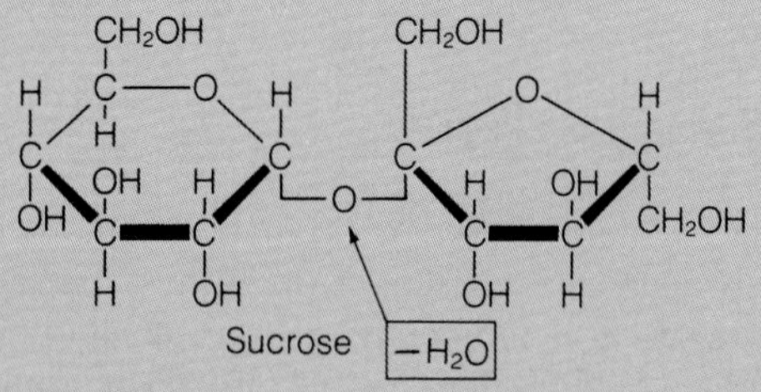

Sucrose    $-H_2O$

Another common disaccharide is lactose, the sugar in milk.

Lactose is composed of one molecule of galactose (below left) and one molecule of glucose (below right).

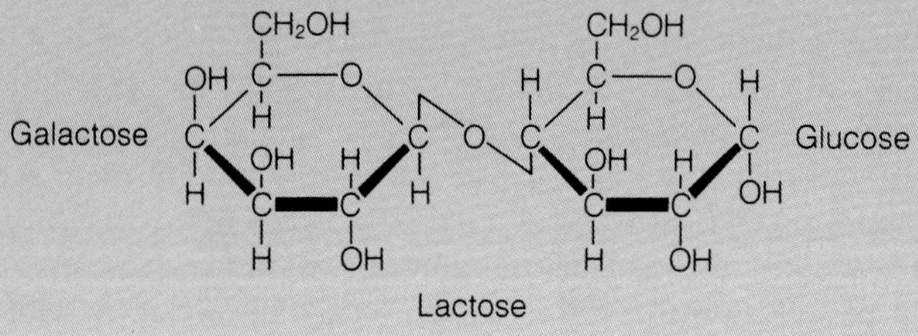

Galactose          Glucose

Lactose

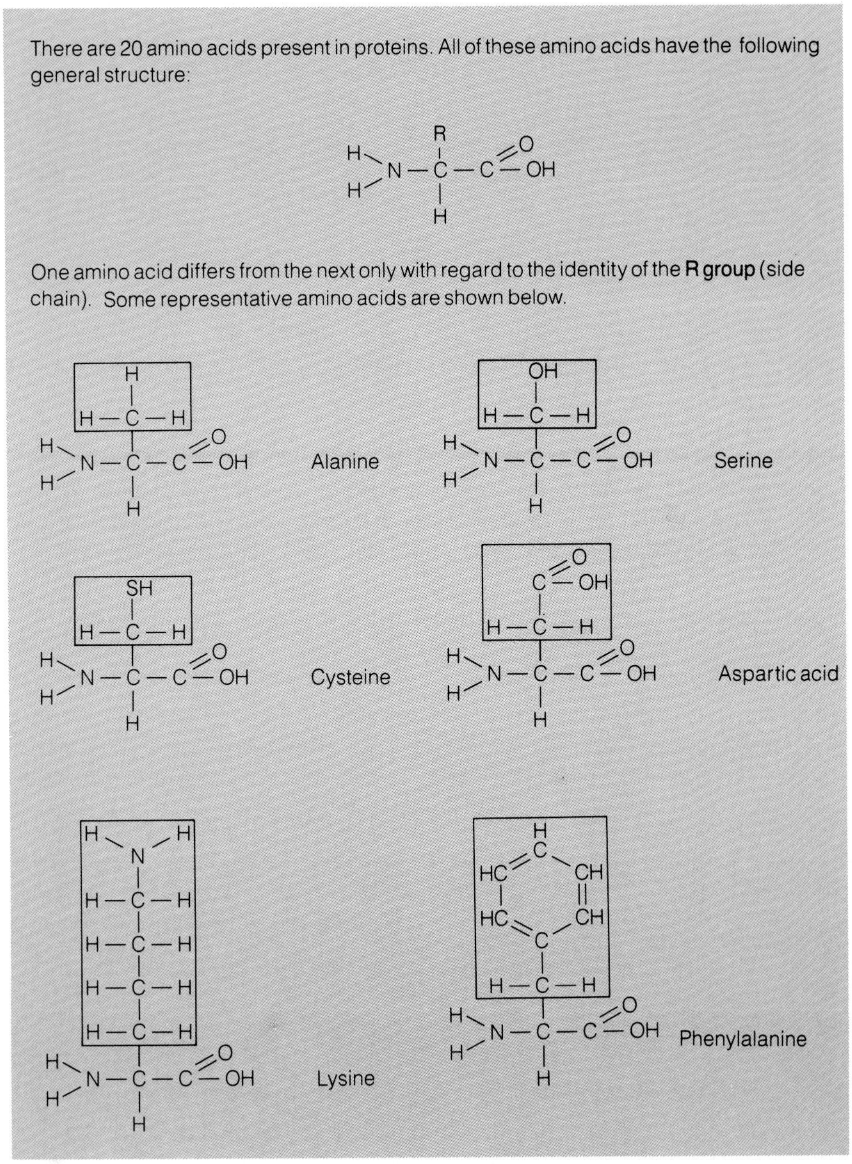
There are 20 amino acids present in proteins. All of these amino acids have the following general structure:
One amino acid differs from the next only with regard to the identity of the **R group** (side chain). Some representative amino acids are shown below.
Alanine
Serine
Cysteine
Aspartic acid
Lysine
Phenylalanine

FIGURE 2–12 The general structure of amino acids and the chemical structures of some representative types. Amino acids are joined together in chains to form proteins.

pounds are discussed below). Fatty acids and, when necessary, amino acids may also be broken down to provide energy required for a number of metabolic processes.

## Vitamins

*Vitamins* are required for many of the chemical reactions that occur within cells. In humans, vitamins must be obtained in the diet, although a few are synthesized by bacteria residing in the intestinal tract. The lack of specific vitamins can lead to *vitamin-deficiency diseases,* a topic included in our discussion of nutrition (Chapter 11).

## Energy-transferring Molecules

**Energy-transferring molecules** store energy obtained from the breakdown of nutrient molecules and then release the energy when it is subsequently needed to drive energy-requiring processes of cells. The most important energy-transferring molecule is *adenosine triphosphate* (ATP), which we will consider in our discussion of energy transfer in living cells later in this chapter. The chemical structure of ATP is shown in Figure 2–15.

FIGURE 2–13 Chemical components of
nucleotides, the subunits of which nucleic acids
(DNA and RNA) are formed.

Nucleotides, of which nucleic acids (DNA and RNA) are composed, consist of a
nitrogenous base, a pentose sugar (a sugar with 5 carbons), and a phosphate group.

**Nitrogenous bases**

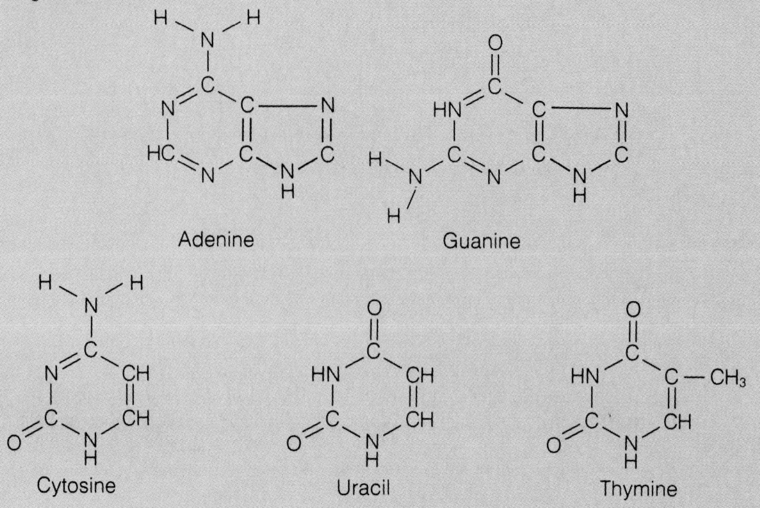

Adenine          Guanine

Cytosine          Uracil          Thymine

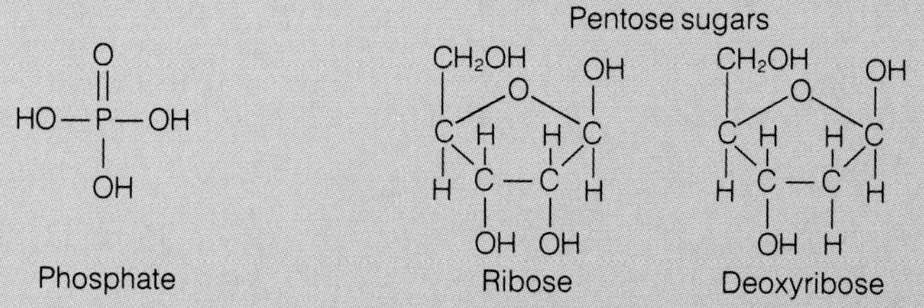

Phosphate          Ribose          Deoxyribose

Pentose sugars

Nucleotides forming DNA contain (1) either adenine, guanine, cytosine, or thymine as
the nitrogenous base, (2) deoxyribose, and (3) a phosphate group. The three components of a nucleotide are linked together covalently by the elimination of water.

**A Representative Nucleotide**

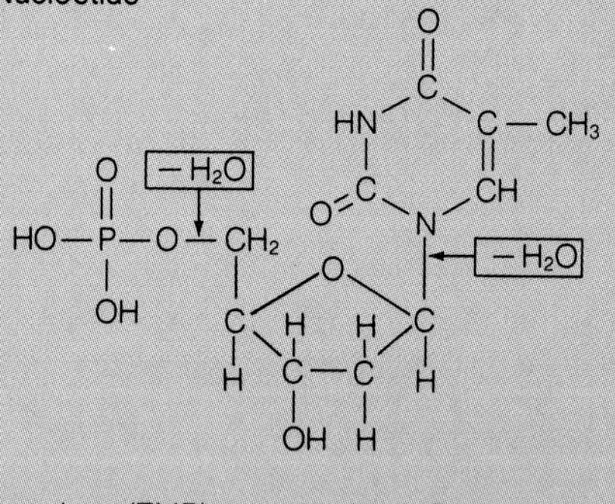

Thymidine monophosphate (TMP), a nucleotide in DNA

In nucleotides forming RNA, uracil substitutes for thymine and ribose substitutes for
deoxyribose. Other components are similar to those in DNA nucleotides.

## Metabolic Intermediate Compounds

A variety of **metabolic intermediate compounds** are formed in metabolic
pathways as a result of the multistep conversion of one substance to another.
For example, the energy-yielding breakdown of glucose to carbon dioxide and
water does not occur in one step, but in a sequence of chemical reactions
involving a large number of metabolic intermediate compounds.

Fatty acids have the general structure:

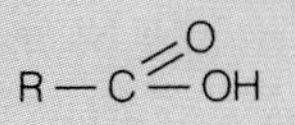

$$R - C\overset{O}{\underset{OH}{\diagdown}}$$

where R is a long chain composed of carbon and hydrogfen atoms (hydrocarbon chain). The hydrocarbon chain is nonpolar, and the carboxylic group ($- C\overset{O}{\diagdown}OH$) is polar; thus, part of the fatty acid molecule does not dissolve in water (is hydrophobic) and part of it does (is hydrophilic).

One common fatty acid in the storage fats of animals is stearic acid:

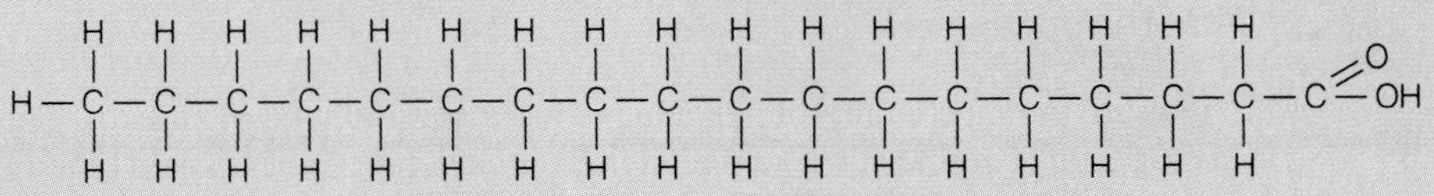

Stearic acid is an 18 carbon fatty acid. It is called a **saturated** fatty acid because the hydrocarbon chain has only single bonds (the carbon atoms are saturated with hydrogen atoms).

A common **unsaturated** fatty acid is oleic acid:

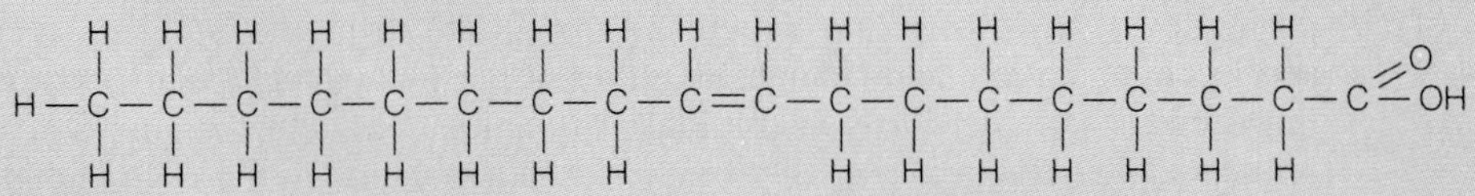

Unsaturated fatty acids with more than one double bond are termed **polyunsaturated.** One polyunsaturated fatty acid, linoleic acid, is produced in plants and is required in the diet of humans.

Saturated fatty acids, found mainly in animal fats, stimulate cholesterol synthesis by the body, whereas unsaturated fatty acids, found primarily in vegetable oils, stimulate the breakdown and excretion of cholesterol. Because a great deal of evidence suggests that high cholesterol levels in the blood are a predisposing factor to atherosclerosis, a disease characterized by the accumulation of fatty deposits on the internal walls of arteries, nutritionists recommend the substitution of unsaturated fatty acids for saturated fatty acids in the diet.

FIGURE 2-14 The fatty acids.

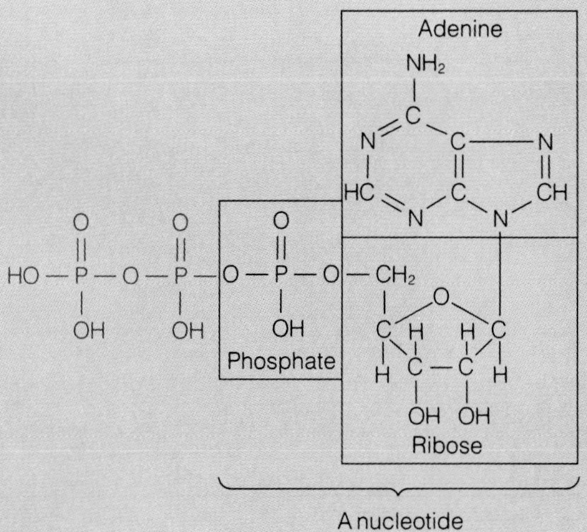

FIGURE 2-15 The chemical structure of ATP. Notice that ATP is a nucleotide on to which two additional phosphate groups have been attached. The nitrogenous base is adenine, and the pentose sugar is ribose.

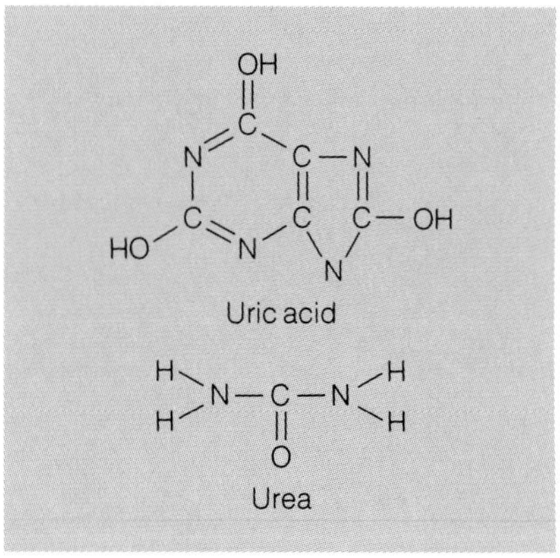

FIGURE 2-16 Urea and uric acid. In humans, urea is the primary nitrogenous waste product.

## Organic Waste Products

**Organic waste products,** particularly nitrogen-containing compounds such as urea and uric acid that cannot be used further by cells, are eliminated as waste products (Figure 2-16).

## Large Organic Molecules of Living Cells

Living organisms contain four classes of large organic compounds (also called **organic macromolecules**): *proteins, nucleic acids, polysaccharides,* and *lipids.*

## Proteins

Proteins are the most numerous organic compounds in living organisms. They are composed of amino acids, of which there are twenty types, hooked together in long chains, much like beads on a string (Figure 2-17). Such molecules, which are composed of many similar or identical subunits, are called **polymers** (*poly* = many; *meros* = parts).

Each different type of protein has a unique sequence of amino acids, which impart to the protein a specific three-dimensional shape (configuration) stabilized primarily by hydrophobic and hydrogen bonds. The chemical properties of any protein are dependent upon both its amino acid sequence and its configuration, and changes in either of these parameters may result in impaired, or complete loss of, biological function. The specific amino acid sequence of a protein may change as a result of a mutation in the DNA of cells that synthesize the protein (see "Nucleic Acids"); changes in configuration can be caused by high temperature or altered pH.

*Proteins as Enzymes.*   Many proteins function as **enzymes,** or **biological catalysts.** An enzyme speeds up a specific chemical reaction that occurs within a cell (Figure 2-18). As such, enzymes serve a vital function, for without them, most reactions within living cells would occur so slowly—in fact, at such negligible rates—as to be useless in providing the cell's metabolic requirements.

The types of enzymes present within a particular cell determine its chemical capabilities, and hence, variations in enzyme content account for functional differences between cell types. Thus, for example, the enzymes within a liver cell allow it to do things, such as produce bile, that a red blood cell or any other type of cell cannot.

FIGURE 2-17 Amino acids are hooked together in long chains to form proteins. The covalent linkage between two adjacent amino acids, formed by the elimination of a water molecule, is called a peptide bond.

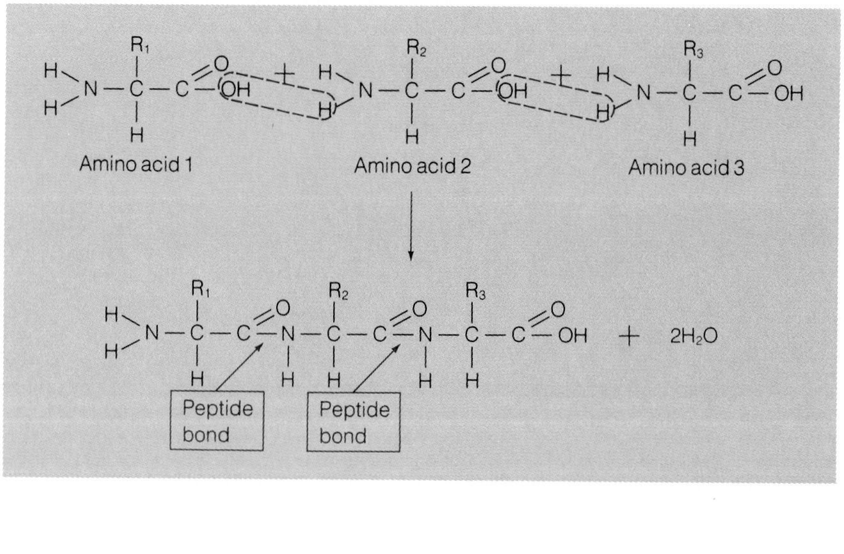

*Other Functions of Proteins.* In addition to their crucial enzymatic function, proteins play a variety of other important roles within living organisms. Proteins form much of the structural material of cells and of the connective tissue (see Chapter 4) that holds the cells of the body together. Proteins act as carriers of a variety of substances into and out of cells, and of oxygen and other substances in the blood. Many **hormones** are proteins; these serve to regulate and coordinate various physiological activities that occur in different parts of the body. **Antibodies**, substances that are important in fighting disease, are also composed of proteins.

Some proteins, called **lipoproteins,** contain lipid components. Various types of lipoproteins have been implicated as carriers of cholesterol in the blood; the importance of these lipoproteins in the occurrence of heart attacks is discussed in Chapter 9. Other proteins have carbohydrate components and are termed **glycoproteins.** Glycoproteins are involved in many cellular processes, including cellular recognition phenomena by which cells of one type recognize cells of their own and other types.

## Nucleic Acids

The two kinds of nucleic acids in living organisms are deoxyribonucleic acid (DNA) and ribonucleic acid (RNA). Like proteins, nucleic acids are long-chain polymers. They are composed of four types of nucleotides linked together in linear sequence to form tremendously long molecules, often containing millions of nucleotides (Figure 2–19).

DNA functions as the cell's hereditary material. It contains, in coded form, the information needed to synthesize all the different proteins an organism contains; these proteins, in turn, determine the structural and functional pattern of cells themselves and of the entire organism that the cells form. RNA is involved in the process by which the information in DNA is decoded for use by the cell in synthesizing proteins. The "cracking" of the DNA code—the unraveling of the molecular mechanisms by which cellular protein synthesis occurs—is one of the more fascinating and impressive achievements of modern biology. The mechanism of protein synthesis and the significance of its discovery are discussed further in Chapter 20.

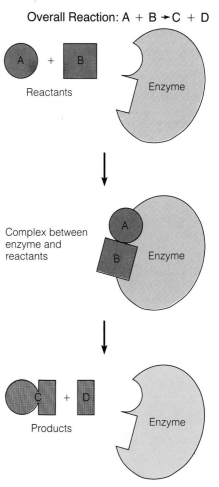

FIGURE 2–18 The lock and key mechanism of enzyme action suggests that the reactants fit specifically into indentations or grooves on the surface of the enzyme, much like a key fits a lock. By holding the reactants in close proximity and in the proper orientation to one another so that they can react chemically, the enzyme speeds up the formation of the products.

Overall Reaction: A + B → C + D

Reactants

Enzyme

Complex between enzyme and reactants

Enzyme

Products

Enzyme

FIGURE 2–19 Nucleotide chains are formed by covalent bonding between individual nucleotides, with the elimination of water. RNA is composed of a single nucleotide strand, whereas DNA contains two nucleotide strands wound together in the form of a double helix, as illustrated in Figure 2–8.

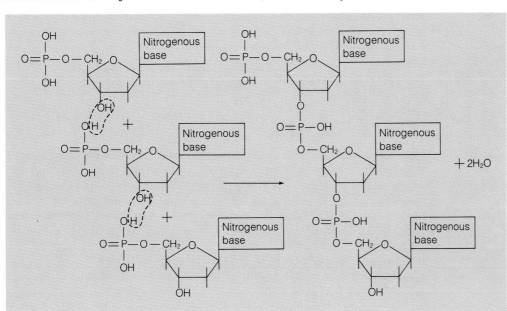

## Diet by Eating

Health fads are perennial events on the American scene, and none are more popular than those that promise weight loss without dieting. One such fad involved the use of so-called starch blockers that purported to inhibit the digestion of starch. Promotional campaigns by manufacturers suggested that an individual could eat a starchy diet, rich in bread, pasta and potatoes, and still lose weight by taking starch blockers with meals. Over a two-year span beginning in 1980, more than 200 brands of starch blockers were sold in health-food stores and pharmacies. Marketed as foods under such names as Carbo-Lite, Amyl-Lite and Amahib, starch blockers were so phenomenally popular that a million tablets a day were being consumed before they were taken off the market by the Food and Drug Administration (FDA) in July, 1982.

Starch blockers contain a protein, extracted from kidney beans, that had been shown in the test tube to be a potent inhibitor of the enzyme alpha amylase. Alpha amylase, present in human saliva and in the digestive juices that the pancreas secretes into the small intestine, normally catalyzes the digestion of starch to maltose units. Maltose is then broken down with the aid of another enzyme, maltase, to yield glucose. Starch blockers were supposed to work by inhibiting the action of alpha amylase. Problems arose, however, when many people using starch blockers experienced abdominal cramps, vomiting, and diarrhea, sometimes so severe as to require hospitalization. It was reports of these adverse reactions that prompted the FDA to ban the sale of starch blockers, pending further research. With the ban ended yet another dream of painless dieting.

There is a footnote to the story. Since the FDA ban, at least one carefully controlled research study in humans has shown that starch blockers do not inhibit the absorption of calories from starch.

---

The specific nucleotide sequence in DNA may be altered, or *mutated,* by various chemical and physical agents in the environment. Because the nucleotide sequence in DNA determines the DNA code, such mutations may result in the synthesis of altered proteins that lack normal biological function. Should a mutation occur in a **germ cell** (a sperm or egg cell) that forms a new individual, the mutation will be passed on to the next generation and may result in a *hereditary disease.* Mutations in the DNA of cells other than germ cells—so-called **somatic cells**—have been implicated in the formation of tumors.

### Polysaccharides

**Polysaccharides** serve as the main energy storage molecules of living systems. Two of these are *glycogen,* found primarily in animals, and *starch,* present mostly in plants; both are composed of long-chain glucose polymers (Figure 2–20). When glucose is needed as a nutrient source, glycogen stored in liver cells or starch obtained in the diet may be broken down to yield individual glucose molecules (see *Focus: Diet by Eating*).

### Lipids

**Lipids** comprise a diverse class of chemical substances that are characterized by their high solubility in nonpolar organic solvents (such as ether, chloroform, and benzene) and their low solubility in water. Some types of lipids, such as the so-called *fats (glycerides)* serve as energy storage compounds that can be mobilized when carbohydrate energy sources are exhausted. Fats are composed of *fatty acids* linked to the carbohydrate molecule *glycerol.* Other lipids, especially *phospholipids* and *cholesterol* are important constituents of cell membranes. Cholesterol is termed a *steroid lipid;* other steroid lipids that are manufactured from cholesterol function as *sex hormones* (estrogens, progesterones, and androgens) and *antiinflammatory hormones* (cortisone and its relatives). Some representative lipids are shown in Figure 2–21.

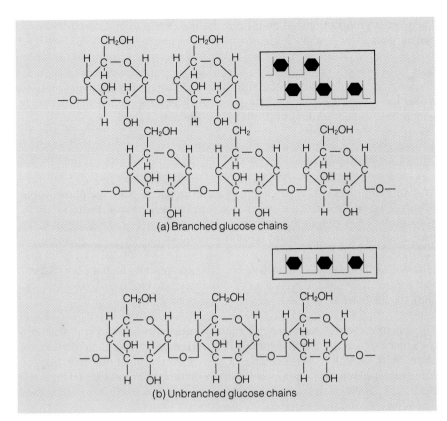

(a) Branched glucose chains

(b) Unbranched glucose chains

FIGURE 2-20 Starch consists of a mixture of (a) branched and (b) unbranched glucose chains, whereas glycogen consists of branched chains. Insets show each type of chain in abbreviated form with glucose molecules represented as solid hexagons.

There are some lipids, called **glycolipids,** that contain carbohydrate components. One class of glycolipids are the gangliosides that are abundant in the plasma membrane of nerve cells. Gangliosides are continually synthesized in normal nerve cells and excess gangliosides are degraded enzymatically.

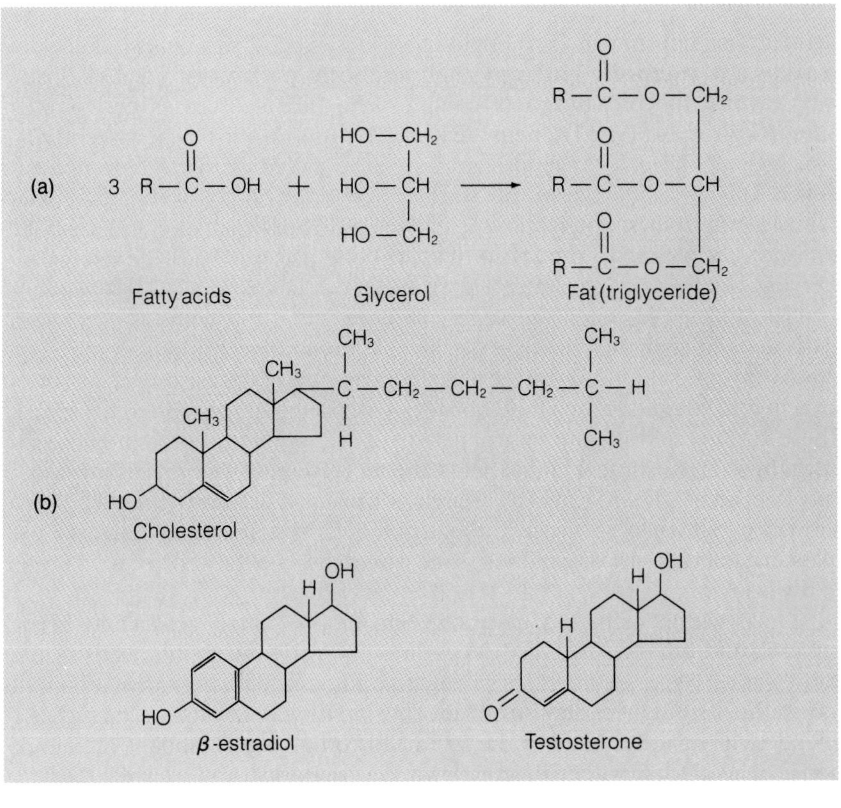

FIGURE 2-21 (a) One type of fat, the triglycerides common in animal fat, is formed by combining three fatty acid molecules with glycerol. (b) The steroids cholesterol, $\beta$-estradiol (an estrogen), and testosterone. Cholesterol is a precursor to estradiol, testosterone, and other steroid hormones. Although estradiol and testosterone have very similar structures, they show profound differences in their biological effects.

(a)

Fatty acids    Glycerol    Fat (triglyceride)

(b)    HO    Cholesterol

$\beta$-estradiol    Testosterone

There are a group of related hereditary diseases, called *lipid storage diseases*, that are caused by deficiencies in enzymes that degrade gangliosides. One lipid storage disease is *Tay Sachs disease*, in which a particular ganglioside accumulates in nerve cells of the brain, causing their eventual destruction. Tay Sachs disease, which occurs most frequently in Jews of Eastern European descent, is a particularly tragic disease, causing progressive developmental retardation in infants and usually leading to death before the age of three.

## CELLULAR METABOLISM

As mentioned previously, all chemical transformations that take place within living cells are collectively referred to as cellular metabolism. Since all physiological function ultimately derives from the activities of cellular metabolism, it is crucial to understand metabolic design.

### Characteristics of Cellular Metabolism

#### Metabolic Pathways

The chemical transformations of cellular metabolism occur in multistep sequences of reactions called **metabolic pathways.** Metabolic pathways are often quite complex. For example, a particular nutrient molecule may be converted sequentially into dozens of metabolic intermediate compounds before a final product, termed the **end-product,** is formed. Such a pathway may be represented as follows:

$$A \rightarrow B \rightarrow C \rightarrow \cdots \rightarrow R \rightarrow S \rightarrow T$$

The nutrient molecule, ingested from the environment, is represented as Compound A; B through S are metabolic intermediate compounds; and T is the end-product.

#### Catabolic and Anabolic Processes

Cellular metabolism may be thought of as consisting of two generalized types of pathways: **catabolic pathways** and **anabolic pathways.** Catabolic pathways involve the breakdown of complex chemical substances into simpler ones; they release (yield) energy that may be used to drive energy-requiring processes of the cell. Anabolic pathways, also called *biosynthetic pathways,* require energy, and involve the synthesis of more or less complex cellular components from simpler nutrient molecules. Thus, catabolic and anabolic pathways are linked in the sense that catabolic pathways provide energy for carrying out anabolic ones (see Figure 1–3).

The linkage between the supply of energy and biosynthesis—between catabolic and anabolic pathways—may best be understood by considering a simple example. Suppose you eat a meal of hamburger (consisting primarily of protein and fat) and French fried potatoes (consisting mostly of starch and fat). These are digested in your gastrointestinal tract, where the protein is broken down into its constituent amino acids, the fat (glyceride) is broken down into fatty acids and glycerol, and the starch is broken down into glucose. These substances are then absorbed through the cells that form the lining of the intestinal tract and into the bloodstream, where they are transported to the cells of the body.

Once within cells, the metabolic fate of the amino acids, fatty acids, glycerol, and glucose may vary depending upon the metabolic needs of the particular cell type. In situations where large amounts of energy are needed, all four of these substances may be broken down in catabolic reactions that release energy. On the other hand, if the cell needs only a small amount of energy, perhaps the glucose will be broken down in a catabolic pathway, while the fatty acids, glycerol, and amino acids are used as building blocks for the biosynthesis of cellular components; these biosyntheses occur in anabolic pathways that, of course, require energy that has been made available in catabolic reactions. In yet another situation in which the cell has ample energy supplies, all four substances may be converted into storage compounds, such as glycogen or fats. The biosyntheses of storage compounds are anabolic processes that require energy, but far more energy is potentially available in the storage compounds than is required for their synthesis.

As is apparent from the foregoing example, it would be naive to think of catabolic and anabolic processes as representing merely two separate metabolic subsystems, one of which provides energy for the other. In fact, the activities of catabolic and anabolic pathways are elaborately coordinated with one another, forming an integrated metabolic system. The operation of any particular metabolic pathway is governed by a variety of factors that depends ultimately upon the overall physiological requirements of the cell.

## Autotrophs and Heterotrophs

There are two general metabolic patterns that are apparent among living organisms. Those organisms that are able to synthesize all the organic compounds they need from carbon dioxide, water, and other simple inorganic compounds are called **autotrophs.** Green plants, for example, are *photosynthetic autotrophs* that can use energy in the sun's rays to synthesize all necessary organic compounds. Animals, on the other hand, are **heterotrophs** (Figure 2–22). Heterotrophs require organic substances as food sources. Through catabolic reactions, they are able to extract energy from these substances; in anabolic reactions, they synthesize from organic nutrients other organic compounds they need. In some cases, organic substances produced by autotrophs are used in the metabolic systems of heterotrophs in unaltered form. This is the case with vitamins, which are synthesized by autotrophs and must be obtained by heterotrophs in their diet. Of course, some vitamins and other kinds of essential organic compounds, including some amino acids and one type of fatty acid, may be synthesized in the laboratory for use by humans as nutritional supplements. But in nature, these substances are produced only by autotrophs and must be obtained in the diet of heterotrophs.

FIGURE 2–22 Heterotrophs can be divided into *omnivores* (which eat both autotrophs and heterotrophs), *carnivores* (which eat species that eat autotrophs), and *herbivores* (which eat autotrophs).

Since many of the organic substances essential to heterotrophs are manufactured from carbon dioxide *only* by autotrophs, humans and other heterotrophic organisms are dependent upon autotrophs—and ultimately upon solar energy—for survival. The bumper sticker that reads "Have you thanked a green plant today?" contains a biologically reasonable suggestion.

## The Efficiency of Cellular Metabolism

Cellular metabolism is performed with an efficiency and economy unmatched in processes occurring in the inanimate world. Considering their enormous complexity and vast functional capabilities, living cells contain surprisingly few chemical compounds. One way cells function so efficiently is to use one compound for a variety of purposes. For example, we have already seen that a particular nutrient molecule may be used as an energy source under one set of metabolic conditions or, under another, as a starting material for the biosynthesis of needed cellular components.

Another source of efficiency is the recycling of materials from unneeded or damaged cellular components to provide energy or raw materials for the manufacture of new cellular constituents. Indeed, such recycling occurs not only within cells, but at the level of the organism as a whole, as dead or damaged cells are broken down and their constituents are reutilized. Recycling activities reach such heights of efficiency that few substances that are potentially utilizable are discarded by the cells or by the organism.

## Energy and Metabolism

One of the main functions of cellular metabolism is to provide energy for all energy-requiring activities of the cell. In order to understand how cells capture, store, and use energy, we must first know something about energy itself.

## Energy

*Energy* is defined as the capacity to do work. Work may take a variety of forms such as lighting a house, synthesizing a protein molecule, heating a pan of water, and so forth. The types of work we have just described obviously require different forms of energy: lighting a house requires radiant, or light, energy; synthesizing a protein molecule requires chemical energy, and heating water requires heat energy.

## The Laws of Thermodynamics

Physical chemists have defined certain energy relationships known as the Laws of Thermodynamics. The **First Law of Thermodynamics** states that energy can be converted from one form into another, and in the process, energy is neither created nor destroyed. Because all forms of energy are fully interconvertible, the convention has been adopted to quantify energy on the basis of heat energy. Heat energy is measured in **calories;** one calorie is the amount of heat required to raise one gram of water one degree Centigrade (or, more precisely, from 14.5 to 15.5°C). Since one calorie turns out to be a rather small amount of energy, particularly when dealing with the energy requirements of humans, the term *Calorie,* spelled with a capital C, has been defined to represent 1000 calories. To avoid confusion between the two terms, it is preferable to refer to a Calorie as a *kilocalorie*—a term we will use throughout this book. The calories that dieters count are kilocalories.

The **Second Law of Thermodynamics** states that the universe tends spontaneously toward a state of increasing disorder or randomness. In order to quantify the energy relationships that are implicit in the Second Law, a term called **entropy** has been defined, which is a measure of randomness or lack of order, expressed in heat units (kilocalories) of energy. Highly ordered entities or situations have less entropy than disorganized ones. For example, a log cabin

FIGURE 2-23 When a forest burns, a great deal of entropy is created.

has less entropy than a pile of logs; living cells have less entropy than homogenized cells; a complex molecule such as a protein, which may be composed of hundreds of amino acids linked in a specific sequence, has less entropy than a solution of amino acids (Figure 2-23).

The tendency toward disorder or increasing randomness, as expressed in the Second Law, is a phenomenon of the natural world that is inescapable to us all. Automobiles eventually break down, houses fall into disrepair, our rooms become sloppy—all of these "randomizing" things can be expected to occur unless we expend energy, in one form or another, to reverse the process.

How do the Second Law of Thermodynamics and the concept of entropy explain the energy requirement of living systems? Although the application of these concepts to living organisms may seem obtuse, it is, in fact, reasonably simple. Living organisms are no exception to the Second Law. In order to maintain their highly ordered structure—to counteract the natural tendency toward disorder that is ultimately expressed in death—living organisms must constantly expend energy. Hence, all living organisms require a continual source of energy, not only to grow and reproduce, but for the general "housekeeping" functions of maintenance and repair.

## The Use of Chemical Energy by Cells

Where does the energy required by living systems come from, and how is it channeled to drive energy-requiring processes? We have already provided an answer to the first part of this question. Living systems obtain energy from catabolic reactions that involve the breakdown of complex, highly ordered substances into less ordered ones. Thus, *living systems use primarily chemical energy*—the potential energy inherent in chemical compounds—to drive energy-requiring metabolic processes. Even photosynthetic organisms, whose ultimate source of energy is solar radiation, convert solar energy into chemical energy before it can be effectively utilized.

One chemical process from which cells obtain much of their energy may be summarized by the following overall reaction:

$$C_6H_{12}O_6 \; + \; 6O_2 \; \rightarrow \; 6CO_2 \; + \; 6H_2O \; + \text{Energy released}$$
(glucose)　(molecular oxygen)　(carbon dioxide)　(water)

in which glucose is broken down with the aid of molecular oxygen into carbon dioxide and water. Because such catabolic reactions release energy, rather than requiring energy in order to occur, they are termed **spontaneous reactions.**

## The ATP-ADP Energy Transfer System

In practice, energy released from the breakdown of glucose or in other spontaneous reactions is not used directly to drive some energy-requiring chemical process. Such a situation would be extremely inflexible, as it would require the energy-yielding reaction to occur at the same time and in the same place in the cell as the energy-requiring reaction. As we shall see when we study cell structure, this clearly cannot be the case, since the energy-yielding breakdown of nutrients occurs primarily in one part of the cell, while the biosynthesis of cellular constituents occurs primarily in other parts.

Instead of being locked in to an impossibly rigid situation of energy transfer, cells use an intermediary chemical compound to store the energy released in energy-yielding reactions and shuttle it to the site of energy-requiring reactions. This compound is ATP (adenosine triphosphate), a substance we mentioned earlier. When energy is released in catabolic reactions, it is used to drive the synthesis of ATP from ADP (adenosine diphosphate) and phosphate, as follows:

$$ADP + Phosphate + Energy\ required \rightarrow ATP + H_2O$$

There is nothing complicated about this reaction. Adenosine diphosphate, containing two phosphate groups, adds a third such group, forming adenosine triphosphate.

When energy is needed at some specific site in the cell, ATP is shuttled to that site, and the energy stored in the ATP molecule may be released in the reverse of the reaction above:

$$ATP + H_2O \rightarrow ADP + Phosphate + Energy\ released$$

Such a reaction, in which water is used to break down a compound, is called **hydrolysis** (*hydro* = water; *lysis* = breakdown). The operation of the ATP-ADP system is shown in Figure 2–24.

The ATP-ADP system is the primary energy-transferring mechanism in all cells, and turns out to be an extremely flexible and efficient way of controlling energy flow in cells. To provide for their energy needs, cells store a reasonably constant supply of ATP. If the amount of ATP in the cell declines, energy-yielding reactions are speeded up until the deficit is made up; if ATP levels are adequate, then these same reactions are slowed down and excess nutrients may be stored in the form of glycogen or fats. The rate of ATP turnover within cells is

FIGURE 2–24 The ATP-ADP cyclic energy exchange system.

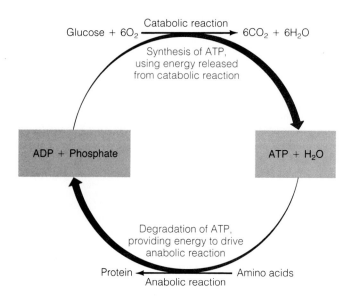

extremely high. It has been estimated that a human at rest converts approximately 40 kilograms (88 pounds) of ATP to ADP in 24 hours!

Because of its central role in providing energy for the metabolic reactions of the cell, ATP is often called the cell's "energy currency." As we shall see, most ATP is synthesized in intracellular structures called *mitochondria*.

## Metabolic Pathways that Lead to Energy Capture

Although we have already mentioned that cells use energy primarily in the form of ATP, and that ATP is produced using the energy released in catabolic reactions, we have not explained the mechanism—or means—by which cells extract energy from nutrient substances, nor have we discussed any of the metabolic pathways by which nutrient molecules are broken down and energy is released.

## Extraction of Energy by Moving Electrons

How do cells extract energy from nutrient substances? When we speak of the "breakdown" of nutrient molecules, we are, in fact, referring to a chemical process known as **oxidation.** Oxidation is the loss of electrons from a substance in a chemical reaction with another substance; the reverse of oxidation is called **reduction** and involves the gain of electrons in a chemical reaction. Since electrons cannot simply float around loose, electrons that are lost from one substance must be gained by another. Consequently, the oxidation of a substance, which loses electrons in the process, is always coupled to the reduction of another substance, which gains the electrons. In such reactions, called *oxidation-reduction reactions,* the substance that loses electrons (the substance that is oxidized) is often referred to as the **electron donor,** while the substance that gains electrons (the reduced substance) is termed the **electron acceptor** (Figure 2–25).

The spontaneous transfer of electrons from electron donors to electron acceptors is accompanied by the release of energy, because the electrons move to a state of lower energy. The amount of energy that is released by such an electron transfer depends upon the tendency of an electron donor to donate electrons to a particular electron acceptor; the greater the tendency to transfer electrons, the greater the amount of energy released. Compounds that have a strong tendency to donate electrons in a reaction with another substance are termed *strong reducing agents,* whereas compounds that have a strong tendency to accept electrons are called *strong oxidizing agents.*

In most biological oxidations and reductions, the loss or gain of an electron is accompanied by the simultaneous loss or gain of a hydrogen ion ($H^+$). Consequently, most biological oxidations involve the removal of hydrogen atoms (a hydrogen ion plus an electron) and are called **dehydrogenations;** most biological reductions involve the gain of hydrogen atoms and are called **hydrogenations.**

Energy release in such electron transfer reactions (oxidation-reduction reactions) is directly analogous to a situation in which water falls down a waterfall. At the top of the mountain, the water (analogous to the electron) has a certain amount of potential energy. This energy is released as kinetic energy when the water falls down the hill, because the water is moving to a state of lower potential energy. The greater the fall, the greater the amount of kinetic energy released.

In all metabolic pathways that lead to energy release and its capture in the form of ATP, electrons are transferred "downhill," or spontaneously, from one substance to another. For example, when glucose is oxidized ("broken down") in human cells, electrons are transferred spontaneously from glucose to molecular oxygen ($O_2$), a strong oxidizing agent. The energy released in these electron transfers is available for the synthesis of ATP, just as the kinetic energy of the falling water is available for doing useful work.

$$Cu^+ + Fe^{+++} \longrightarrow Cu^{++} + Fe^{++}$$

FIGURE 2–25 An example of an oxidation-reduction reaction. $Cu^+$ is oxidized to $Cu^{++}$ when it loses an electron (carrying one negative charge) to $Fe^{+++}$; $Fe^{+++}$ is reduced to $Fe^{++}$ when it gains an electron from $Cu^+$. $Cu^+$, then, is the electron donor and $Fe^{+++}$, the electron acceptor.

# Alcohol

**E**thyl alcohol, commonly known simply as alcohol, is the active ingredient of intoxicating beverages such as beer, wine, fortified wines, and distilled liquors. Its main effect is to tranquilize or sedate by depressing the reactions of the nervous system. Because alcohol's sedative effect may dull inhibitions, it commonly results in loud, active, or aggressive behavior, leading to the misconception that alcohol is a stimulant, which it is not.

$$H-\underset{\underset{H}{|}}{\overset{\overset{H}{|}}{C}}-\underset{\underset{H}{|}}{\overset{\overset{H}{|}}{C}}-OH$$

Ethyl alcohol

In addition to its well-known effects on behavior, alcohol has other short-term physiological effects that occur whenever you drink. One is to increase urination (act as a diuretic). Alcohol does this by inhibiting the secretion from the pituitary gland of a hormone (antidiuretic hormone), which normally acts to conserve the body's water by lessening urination. Another short-term effect of alcohol is to expand, or dilate, the blood vessels of the skin. This increases the flow of blood to the body's surface, giving the feeling of warmth, but in actuality it results in the loss of heat from the body. Particularly in a cold environment, such heat loss can lead to excessive cooling of the body (hypothermia). It is for this reason that giving alcohol to a chilled person in order to warm him or her up is misguided and can be dangerous.

Over and above alcohol's short-term, temporary effects, long-term heavy drinking can seriously damage many of the body's organs, as summarized in the accompanying figure. Moreover, heavy drinking frequently leads to severe social problems at home, at work, and at school.

When alcohol is consumed, it is absorbed from the digestive tract and enters the bloodstream, where it is distributed throughout the body's tissues. It is the level of alcohol in the blood that determines its effects on the body. Blood alcohol levels are, in turn, dependent upon a number of factors. One of these is body size. Because blood volume increases with body size, blood alcohol levels will rise more slowly and reach lower peak levels in large individuals than in smaller ones. Another important factor is the rate of alcohol consumption. The faster alcohol is consumed, the higher will be peak levels of blood alcohol. In this regard, it is a misconception that you can't get as drunk on beer as on whiskey; the crucial factor is how much alcohol is consumed in a given time period. Of course, it is necessary to drink a much greater volume of beer than of whiskey in order to ingest the same amount of alcohol. Eating before or during drinking will slow down the rate of alcohol absorption into the blood stream by as much as 50 percent, and will hence reduce peak levels of blood alcohol.

Counterbalancing the effects of alcohol ingestion on blood alcohol levels is the removal of alcohol from the bloodstream by the liver, where it is broken down. Very small amounts of alcohol are also excreted in the urine or are exhaled. The liver can remove from the blood each hour an amount of alcohol equivalent to about one ounce of 86-proof liquor. This means that it takes about 1½ hours to metabolize a 1½-ounce drink of liquor.

Alcoholic beverages differ in their alco-

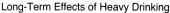

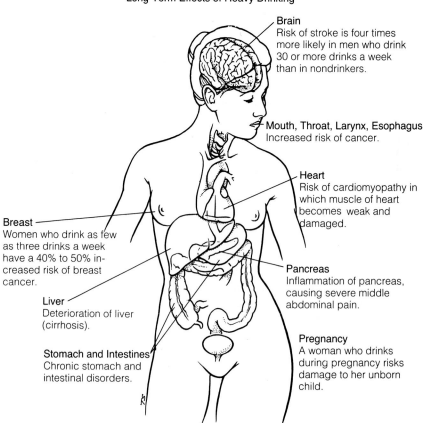

Long-Term Effects of Heavy Drinking

**Brain**
Risk of stroke is four times more likely in men who drink 30 or more drinks a week than in nondrinkers.

**Mouth, Throat, Larynx, Esophagus**
Increased risk of cancer.

**Heart**
Risk of cardiomyopathy in which muscle of heart becomes weak and damaged.

**Pancreas**
Inflammation of pancreas, causing severe middle abdominal pain.

**Pregnancy**
A woman who drinks during pregnancy risks damage to her unborn child.

**Breast**
Women who drink as few as three drinks a week have a 40% to 50% increased risk of breast cancer.

**Liver**
Deterioration of liver (cirrhosis).

**Stomach and Intestines**
Chronic stomach and intestinal disorders.

hol content, as illustrated in the accompanying figure. As a rough guide, 1½ ounces of distilled liquor is equivalent in alcohol content to 3 ounces of fortified wine (wine to which brandy has been added), 6 ounces of table wine, and 12 ounces of beer.

The effect of any particular blood alcohol level on mental alertness and physical coordination is roughly comparable for all individuals. In a person weighing 150 pounds and drinking in a single session of two or three hours, three bottles of beer (or any equivalent amount of another alcoholic drink) lead to a peak blood alcohol level of 50 mg% (read fifty milligrams percent, meaning 50 milligrams of alcohol per 100 milliliters of blood). At this level, judgment is noticeably impaired. After drinking five bottles of beer in several hours, the same person would reach a blood alcohol level of 80 mg% and show marked defects in muscle coordination, reaction time, and judgment. Ten bottles of beer produce a blood alcohol level of 150 mg%, which causes a pronounced loss of self-control and results in a 25-fold increased risk of having an automobile accident as compared to someone who has not been drinking. Blood alcohol levels defining legal intoxication vary considerably among the states, but range between 80 mg% and 150 mg%; 100 mg% constitutes legal intoxication in most states.

The ability to engage in long-term drinking without abusing alcohol is quite variable among individuals. Some people can drink moderately for their entire adult lives without ill effect, while others come to depend increasingly on alcohol to cope with everyday fears, pressures, and disappointments. At the farthest extreme is the alcoholic, whose overuse of alcohol is so frequent and intense as to overwhelm most of life's activities. Danger signs on the road to alcoholism are blackouts while drinking; drinking in the morning; drinking five or more drinks daily; and problems with family, friends, and co-workers over drinking.

The costs of alcohol consumption and alcoholism to the individual and to society are enormous. Alcohol has been implicated in over half of the fatal motor vehicle accidents that occur each year, and there is a direct association between the level of alcohol use and the seriousness of the crash. Government studies estimate that alcohol contributes overall to some 100,000 accidental or disease-related deaths each year in the United States and results in total economic costs of some 120 billion dollars. Many health professionals consider alcoholism our greatest health problem. Indeed, it has been estimated that between three and ten percent of the population will become alcoholics at one time in their lives and that one out of three families will be affected.

Considerable research has been done in the past decade on the cause of alcoholism. Although we do not as yet understand its cause, some intriguing results are accumulating that point to fundamental physiological and genetic differences between alcoholics and others. We now know that one primary symptom of alcoholism is a high tolerance for alcohol; that is, an alcoholic can function normally at blood alcohol levels that are significantly above those that make a normal individual feel somewhat "out of control." A high tolerance for alcohol may result from the way alcoholics metabolize alcohol. In a normal individual, blood alcohol is con-

verted in the liver to a substance called acetaldehyde; acetaldehyde is then converted to acetic acid, which in turn is broken down to carbon dioxide and water. In alcoholics, alcohol is converted to acetaldehyde at very high rates, much higher than those seen in normal individuals, but acetaldehyde is not converted to acetic acid at a comparable rate. Consequently, acetaldehyde builds up in the blood of alcoholics to reach levels much higher than in nonalcoholics. It is the increased conversion rate of alcohol into acetaldehyde that may explain the alcoholic's high tolerance for alcohol.

Other clinical studies have shown that a profile of 25 common blood tests can be used successfully and with great accuracy to identify alcoholics, in some cases even before an individual is aware of a drinking problem. Here again is a link between altered metabolic function and alcoholism or even potential alcoholism.

Researchers have also demonstrated a hereditary susceptibility to alcoholism. Studies with adopted children have shown that the children of alcoholics, as compared with those of nonalcoholics, have a four-fold increased risk of becoming alcoholic, even when they are raised by nonalcoholic adoptive parents. We do not as yet understand the physiological basis of this hereditary susceptibility to alcoholism, but further research is likely soon to unravel the mystery.

 =  =  =

12 oz     6 oz     3 oz     1½ oz

**Beer**
Most have about 5% alcohol by volume. Some ales have about 8%

**Table wine**
Alcohol content varies from about 10% to 13% by volume

**Fortified wine**
Includes sherry, port, and vermouth, to which alcohol is added to give about 20% alcohol by volume

**Distilled liquor**
Includes gin, vodka, whiskies, brandy, and liqueurs, all of which have between 40% to 50% alcohol by volume

## Energy-Capturing Metabolic Pathways

The pathways of energy capture in cells are designed to allow the transfer of electrons from a nutrient molecule, such as glucose, to an electron acceptor, such as $O_2$, *under the physiological conditions existing within the cell.* In order to maintain normal physiological conditions—a temperature of 98.6°F, a pH of about 7, and relatively constant intracellular concentrations of solutes—energy-capturing metabolic pathways must possess certain design features. One such feature is that metabolic pathways have many steps; another is that each step is catalyzed by a specific enzyme.

Enzymes are required to speed up reactions that otherwise would proceed at uselessly slow rates. Multistep pathways allow the release of energy in small amounts, rather than in one large amount that is difficult for the cell to handle and that would consequently result in a low efficiency of energy capture or, at the worst, a fatal rise in temperature as unused energy is dissipated as heat. For example, in the oxidation of glucose, electrons are not transferred from glucose *directly* to molecular oxygen. Instead, they are transferred in a series of enzyme-catalyzed reactions that permit a controlled release of energy and maximize the capture of energy in the form of ATP.

***Pathways of Glucose Oxidation.*** Human beings, as well as most other heterotrophs, use glucose as their primary source of energy. Consequently, the metabolic pathways of glucose oxidation are the most important energy-capturing pathways in human cells.

***Glycolysis and the Krebs Cycle.*** Glucose is oxidized in two consecutive metabolic pathways. In the first pathway, a nine-step sequence called **glycolysis,** glucose is partially oxidized to form two molecules of pyruvic acid (Figure 2–26). Some of the energy released in glycolysis is used to synthesize two

FIGURE 2–26 Diagrammatic summary of glycolysis, the first stage in the oxidation of glucose. Glucose (6-carbon) is oxidized to two molecules of pyruvic acid (3-carbon) with the net production of two molecules of ATP and two molecules of NADH for each glucose molecule oxidized.

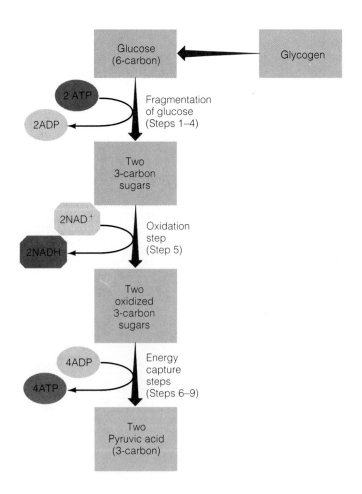

molecules of ATP for each molecule of glucose oxidized. Electrons lost from glucose in its oxidation during glycolysis are accepted by a special electron acceptor substance, called NAD (nicotinamide adenine dinucleotide), which consequently becomes reduced. The fate of this reduced NAD will be considered shortly.

Pyruvic acid, formed as the end-product of glycolysis, is then fully oxidized to carbon dioxide in another metabolic pathway known as the **Krebs cycle,** named for Hans Krebs, a biochemist who described it (Figure 2–27). The Krebs cycle results in the formation of the equivalent of two molecules of ATP for each original glucose molecule. In addition, electrons that are released during the oxidation of pyruvic acid in the Krebs cycle are accepted by NAD and by another electron acceptor substance, FAD (flavin adenine dinucleotide), both of which are consequently reduced.

Although glucose has now been fully oxidized to carbon dioxide, most of the energy released in its oxidation is stored in the reduced forms of NAD and FAD, or, more precisely, in the potential energy of the electrons accepted by these substances.

*The Respiratory Chain.*   The reduced forms of NAD and FAD are oxidized in a metabolic pathway called the **respiratory chain,** or the **electron transport system,** in which electrons are transferred "downhill" in a series of oxidation-reduction reactions to $O_2$ (Figure 2–28). The energy released in these reactions is captured in the form of 34 molecules of ATP per original glucose molecule. Hence, the net yield of ATP for each molecule of oxidized glucose is 38 (2 ATP molecules from glycolysis, 2 ATP molecules from the Krebs cycle, and 34 ATP molecules from the respiratory chain; see Table 2–2).

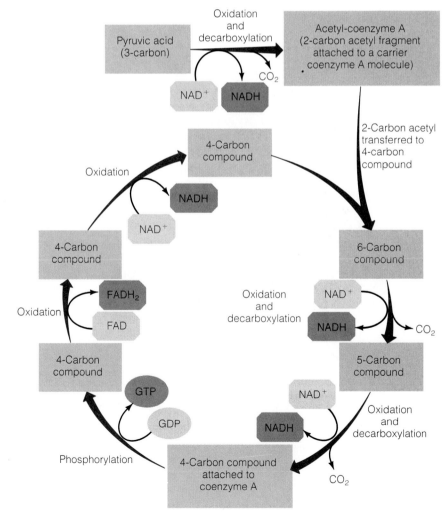

FIGURE 2–27 Diagram summarizing the Krebs cycle. Two molecules of pyruvic acid (3-carbon) are oxidized in the Krebs cycle for every original molecule of glucose (6-carbon). Notice that in one turn of the cycle, the three carbon atoms in pyruvic acid are released as $CO_2$ (a process termed *decarboxylation*). For each original molecule of glucose, 8 NADH, 2 $FADH_2$, and 2 GTP (equivalent of 2 ATP) are produced.

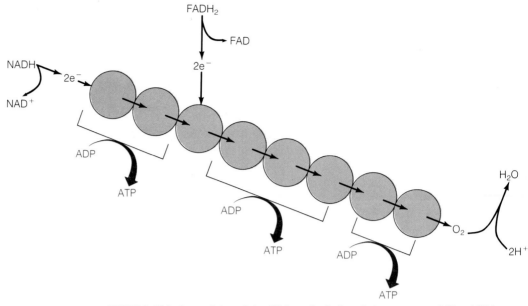

FIGURE 2–28 In the respiratory chain, ATP is synthesized as electrons are passed "downhill" from NADH and FADH$_2$ to O$_2$ through a series of electron carriers (represented by circles). Three ATP molecules are produced for each NADH, and two ATP molecules for each FADH$_2$.

***Oxidation of Other Substances as Energy Sources.*** Glycolysis, the Krebs cycle, and the respiratory chain serve a central role in the oxidation of other substances as well as glucose. Many different carbohydrates are oxidized in these pathways. In addition, we have mentioned previously that cells use fats, and sometimes proteins, as sources of energy. Indeed, fats serve as our primary energy reserve. The oxidation of fats and of the amino acids in proteins also occurs in the Krebs cycle and respiratory chain (Figure 2–29).

## The Use of Oxygen in Metabolism

Most people know that molecular oxygen (O$_2$) is required for human life. What most people do not understand is that the metabolic role of O$_2$ is solely that of final electron acceptor in the respiratory chain. Without O$_2$, the respiratory chain cannot function and consequently energy capture is greatly inhibited. Indeed, cyanide is a deadly poison because it inhibits energy capture by preventing the transfer of electrons to molecular oxygen in the respiratory chain.

Organisms that use O$_2$ as the ultimate electron acceptor in the oxidation of energy sources are called **aerobes.** Aerobes thus require O$_2$ in order to live. A few organisms, mostly bacteria, are **anaerobes;** that is, they can only live in environments where O$_2$ is absent or is present in low concentrations. Anaerobes use ultimate electron acceptors other than oxygen, including a wide range of organic and inorganic substances, in the oxidation of nutrients. One

TABLE 2–2 ATP Yield from the Aerobic Oxidation of One Glucose Molecule

| Pathway | Reduced coenzymes | ATP or equivalent |
|---|---|---|
| Glycolysis | 2NADH | 2 |
| Krebs Cycle | 8NADH + 2FADH$_2$ | 2 |
| Respiratory Chain | | |
|    Oxidation of 10 NADH from glycolysis and Krebs cycle | | 30 |
|    Oxidation of 2 FADH$_2$ from Krebs cycle | | 4 |
| Total Yield | | 38 |

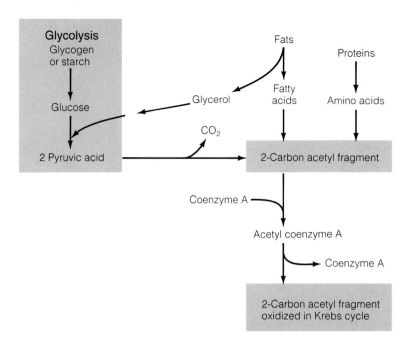

anaerobic bacterium, *Clostridium botulinum,* produces a toxin that causes human botulism, a type of food poisoning that is often fatal. Improperly canned food is the major cause of botulism.

## Study Questions

1. **a)** Describe the structure of an atom.
   **b)** How do atoms and elements differ?
2. Name the six major elements of which living things are composed.
3. Distinguish between covalent and ionic bonding.
4. **a)** What is a polar molecule?
   **b)** a nonpolar molecule?
5. Why are hydrogen bonds and hydrophobic bonds particularly important in the functioning of organic molecules?
6. Why are the solvent properties of water important to cellular metabolism?
7. **a)** What is an ion?
   **b)** List five ions important in living systems.
   **c)** Why is the ionic balance of body fluids important to living organisms?
8. **a)** What is pH?
   **b)** Why is the pH of body fluids important to living organisms?
9. What properties of water determine its central role in the regulation of body temperature?
10. List the six classes of small organic molecules present in living organisms and indicate the function of each.
11. List the four classes of organic macromolecules and indicate several functions of each.
12. Distinguish between catabolic and anabolic pathways of metabolism.
13. Distinguish between autotrophs and heterotrophs.
14. **a)** Define energy.
    **b)** What are the First and Second Laws of Thermodynamics?
    **c)** Explain why each of these laws is important to understanding the functioning of living organisms.
15. Diagram the ATP-ADP energy transfer system.
16. Describe the role of electrons in energy capture and release.
17. Describe in words the process by which glucose is oxidized in living cells to carbon dioxide and water, and indicate where energy is captured during the process.
18. Distinguish between aerobes and anaerobes.

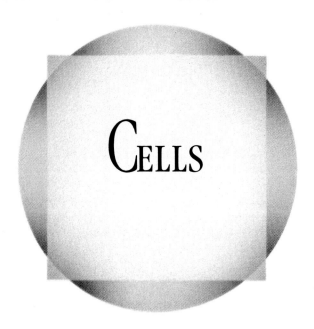

# CELLS

## CELL STRUCTURE

As we discussed in Chapter 1, the basic unit of organization of the human body is the cell. As with so many things in biology, our knowledge of cell structure has increased dramatically during the last two or three decades. Prior to that time, the light microscope was the only available instrument for viewing cell structure. Because the light microscope is only capable of resolving objects larger than about 0.2 $\mu$m (200 nm), much of the internal structure of cells was not visible (see Table 3–1). Consequently, cell biologists tended to view cells as sacks of cytoplasm in which floated a nucleus and a few subcellular structures.

TABLE 3–1 A Table of Measures

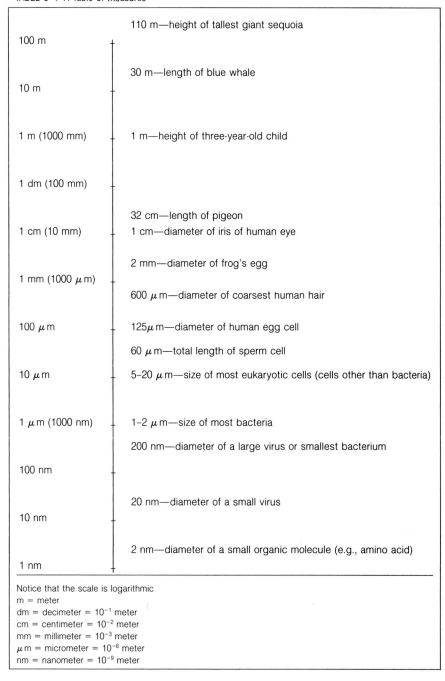

| | |
|---|---|
| | 110 m—height of tallest giant sequoia |
| 100 m | |
| | 30 m—length of blue whale |
| 10 m | |
| 1 m (1000 mm) | 1 m—height of three-year-old child |
| 1 dm (100 mm) | |
| 1 cm (10 mm) | 32 cm—length of pigeon |
| | 1 cm—diameter of iris of human eye |
| 1 mm (1000 $\mu$m) | 2 mm—diameter of frog's egg |
| | 600 $\mu$m—diameter of coarsest human hair |
| 100 $\mu$m | 125 $\mu$m—diameter of human egg cell |
| | 60 $\mu$m—total length of sperm cell |
| 10 $\mu$m | 5–20 $\mu$m—size of most eukaryotic cells (cells other than bacteria) |
| 1 $\mu$m (1000 nm) | 1–2 $\mu$m—size of most bacteria |
| | 200 nm—diameter of a large virus or smallest bacterium |
| 100 nm | |
| 10 nm | 20 nm—diameter of a small virus |
| 1 nm | 2 nm—diameter of a small organic molecule (e.g., amino acid) |

Notice that the scale is logarithmic
m = meter
dm = decimeter = $10^{-1}$ meter
cm = centimeter = $10^{-2}$ meter
mm = millimeter = $10^{-3}$ meter
$\mu$m = micrometer = $10^{-6}$ meter
nm = nanometer = $10^{-9}$ meter

The development of the electron microscope, capable of resolving structures as small as about 0.8 nm, radically changed our idea of cell structure. We came to realize not only that many subcellular structures exist that had not previously been seen, but that cells are *compartmentalized* in the sense that different subcellular structures constituted "little organs," or **organelles,** specialized for various metabolic functions. The idea that cells are complex entities in which specific metabolic activities, occurring in defined subcellular locations, are finely integrated with one another, was a major conceptual breakthrough in our understanding of cellular organization.

The structure of a typical animal cell is shown in Figure 3–1, which represents an artist's conception of the three-dimensional organization of the cell, based upon data from numerous electron micrographs, such as the one shown in Figure 3–2. Since each electron micrograph represents only one very thin section of a cell, the three-dimensional organization of any subcellular structure must be inferred by viewing many different sections.

## The Plasma Membrane

The membrane that surrounds an animal cell is called the **plasma membrane.** It selectively allows substances to pass into and out of the cell.

## Structure

Studies with the electron microscope have made it clear that all human cells are bounded by a plasma membrane, about 10 nm thick, which consists of two parallel dark lines separated by a light space. This three-part, dark-light-dark structure is characteristic of many biological membranes as viewed through the electron microscope, and has been termed a "unit membrane" (Figure 3–3).

FIGURE 3–1 A three-dimensional rendering of a typical animal cell sectioned in a number of planes. It illustrates the relationship between various cellular components likely to be found in eukaryotic cells.

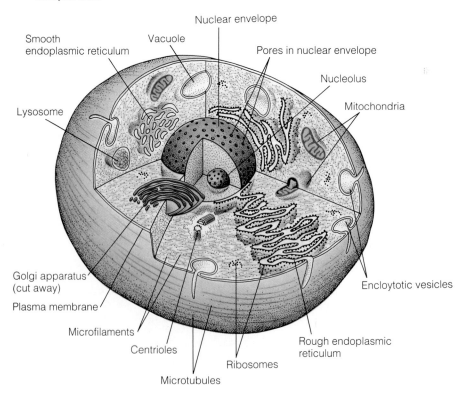

FIGURE 3-2 A transmission electron micrograph of a white blood cell.

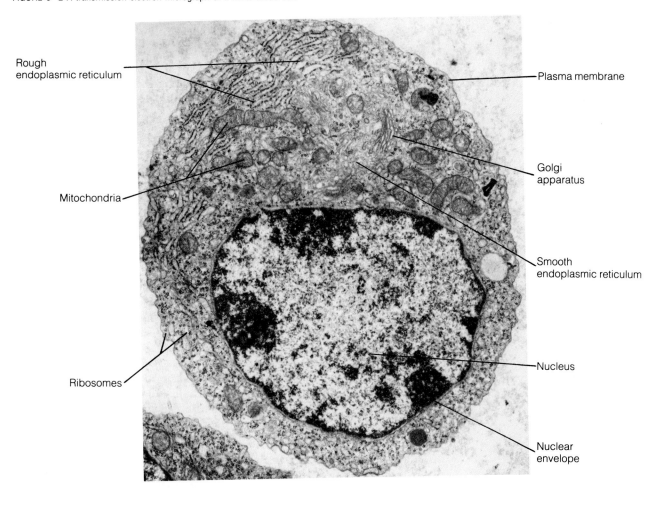

Rough
endoplasmic reticulum

Plasma membrane

Golgi
apparatus

Mitochondria

Smooth
endoplasmic reticulum

Nucleus

Ribosomes

Nuclear
envelope

FIGURE 3-3 The unit membrane structure of the plasma membranes of two apposed cells.

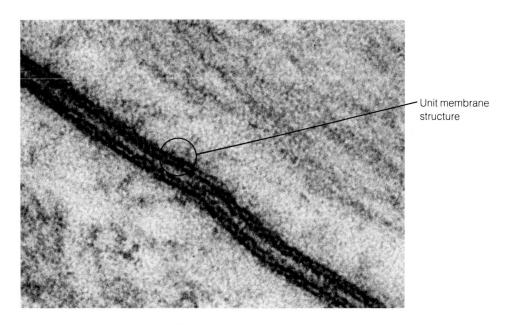

Unit membrane
structure

The plasma membrane of human cells is composed primarily of proteins and lipids, although carbohydrates may be present in association with these compounds, forming glycoproteins and glycolipids, respectively.

Although we do not know precisely how molecular components are organized to form a membrane, a number of models of membrane structure have been proposed. The most widely accepted of these models, called the **fluid mosaic model,** envisions membranes as mosaics of lipids and proteins (Figure 3–4). The lipids are thought to form the structural framework of the membrane and to be arranged primarily as a double layer, or *bilayer,* of molecules in which the membrane proteins are embedded to varying degrees. The proteins are envisioned to be able to diffuse laterally in the plane of the membrane, much as an iceberg may drift in the ocean. Thus, the entire structure, rather than being viewed as rigid, is dynamic or "fluid." While lipids seem to serve primarily a structural function, the proteins in membranes appear to have specific metabolic functions such as carriers of substances through membranes (*transport carriers*); *enzymes; receptor sites* for viruses, drugs and hormones; and *recognition molecules* (these functions are discussed below).

## Functions

The primary function of the plasma membrane is to maintain the physical and chemical integrity of the cell. By serving as a highly discriminating chemical barrier, the plasma membrane controls the passage of substances between the cell and its surroundings. As such, it is *selectively permeable;* that is, it permits some substances to pass through it while retarding or excluding the passage of others. For example, it allows the passage of nutrients and other essential substances into the cell while largely preventing the entry of harmful substances, and it permits the passage of waste products and secretory products out of the cell while preventing the loss of useful substances.

In addition to its function as a permeability barrier, the plasma membrane has a variety of other crucial functions as well. One of these is to attach cells to one another to form tissues. In most instances, cells in tissues are not tightly fused together. Instead, the plasma membranes of adjacent cells are separated by a narrow space, filled with an "extracellular cement," which is composed of polysaccharides and is secreted by the cells. The extracellular cement, as its name suggests, serves to glue the cells together, but with a seal through which

FIGURE 3–4 The fluid mosaic model of membrane structure. Protein and glycoprotein molecules are shown embedded in the lipid bilayer. Notice that the glycoproteins in the membrane are composed of a protein component (amino acid chain) to which a carbohydrate tail is attached.

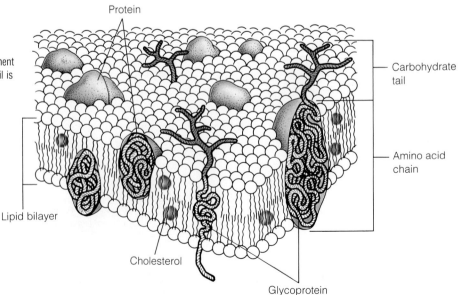

Protein

Carbohydrate tail

Amino acid chain

Lipid bilayer

Cholesterol

Glycoprotein

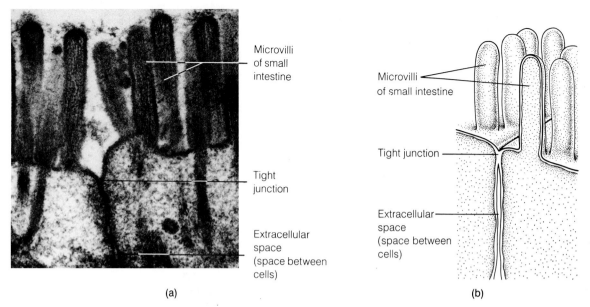

FIGURE 3–5 (a) Electron micrograph showing a tight junction between two cells that form the inner lining of the small intestine (intestinal epithelium). (b) Sketch showing how a tight junction forms a seal that prevents the leakage of substances between the cells. The fingerlike structures are microvilli, surface extensions of the plasma membrane that project into the cavity of the small intestine.

water and most soluble molecules may pass. In some cases, however, the plasma membranes of adjacent cells interact with one another to form particularly watertight or strong connections. For example, the plasma membranes of adjacent *epithelial cells* that form the outer surface of the skin, various glands, and the lining of some organs, among other structures, fuse with one another to form what is called a **tight junction,** which prevents water and other substances from leaking between the cells (Figure 3–5). Other junctional structures, called **desmosomes,** are often found also in epithelial tissues and apparently serve as particularly strong points of adhesion (Figure 3–6). Because they hold cells tightly together, desmosomes are particularly common in tissues such as skin that are subject to considerable mechanical stress.

The plasma membrane is also important in communication between cells of the body. Many cells have communicating junctions, termed **gap junctions,** which form pores or channels through which small substances such as ions, sugars, amino acids, nucleotides, vitamins, and some hormones can pass. Another type of communication involves specific glycoprotein or glycolipid molecules, called **recognition molecules,** that are components of plasma membranes. Recognition molecules allow a cell to recognize another cell as being of a similar or different type and are important in processes involving cell interactions. For example, cell recognition molecules help direct the extensive cell migrations that occur during the development of an embryo from a fertilized egg to a complex multicellular organism.

Plasma membranes also possess **receptor sites** to which viruses and certain hormones, drugs, and metabolic control compounds attach in order to influence the cell. Receptor sites are usually glycoproteins and are specific in the sense that each type of receptor site binds a particular substance or group of substances. This binding depends upon specific chemical interactions between the substance and its receptor site, much like a substrate fits an enzyme (lock and key mechanism of enzyme action; see Figure 2–18). Absence of receptors for a particular virus makes the cell resistant to infection by that virus, since a virus must attach to the plasma membrane before entering the cell. In a similar

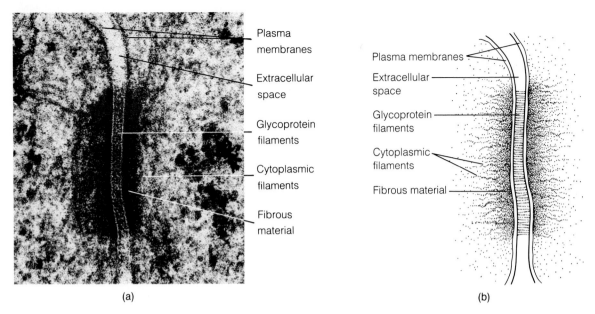

(a)

(b)

FIGURE 3–6 (a) A desmosome connecting the plasma membranes of two adjacent cells, as seen in the electron microscope. (b) Sketch of a desmosome. Notice that the space between the plasma membranes of the two cells is filled with glycoprotein filaments. Cytoplasmic filaments, extending from a dense fibrous material attached to the plasma membranes, extend into the cytoplasm of both cells.

way, hormones and drugs require attachment to plasma membrane receptor sites in order to elicit their effects. In some cases, attachment to a receptor site by a hormone or drug is followed by its entry into the cell, but in other cases, the hormone or drug may affect the cell merely by attaching to the receptor site, since attachment in itself triggers a metabolic response.

## Nucleus

The nucleus is usually the most prominent structure within a cell. It is bounded by a **nuclear envelope** consisting of two concentric membranes with openings called **nuclear pores,** through which various substances enter and leave the nucleus. Nuclear pores are visible in Figure 3–2 (arrows) as discontinuities in the nuclear envelope. The pores in the nuclear envelope are illustrated in Figure 3–1.

Most of the cell's hereditary information is contained in the nucleus, encoded in large molecules of DNA. Each molecule of DNA is associated with proteins and a little RNA to form a long *chromatin* strand. During cell division, chromatin strands condense by folding and coiling to form **chromosomes.** Human cells contain 23 different types of chromosomes. *Sperm* and *egg cells* (*germ cells*) contain one chromosome of each type; all other cells in the body, the *somatic cells,* contain two of each chromosome type, for a total of 46 chromosomes.

The nucleus of a nondividing cell contains, in addition to the diffuse chromatin strands, a dense region called the **nucleolus** ("little nucleus"), seen in Figure 3–1. The nucleolus is involved in the manufacture of **ribosomes,** small spherical structures that are transported to the cytoplasm and are the site of the synthesis of proteins in the cell. Ribosomes are evident in Figure 3–2 as numerous tiny, dark dots in the cytoplasm.

## Cytoplasm

Within the plasma membrane, but outside the nucleus lies the cytoplasm. This consists of an aqueous material containing a variety of subcellular structures that are responsible for carrying out specialized functions of the cell. There are three general groups of subcellular structures: membrane-bound organelles, internal membrane systems, and filamentous structures.

## Membrane-Bound Organelles

Some subcellular structures are bounded by membranes These include *mitochondria, lysosomes,* and *vacuoles.*

## Mitochondria

**Mitochondria** (sing: *mitochondrion*) generate most of the cell's ATP by the aerobic oxidation of nutrients. The oxidation of glucose to pyruvic acid occurs during glycolysis in the cell's cytoplasm; pyruvic acid then enters the mitochondria where reactions of the Krebs cycle and respiratory chain take place (see Chapter 2). Because mitochondria are the source of most of the cell's ATP, they are often called the "powerhouses" of the cell.

Although their size and shape can vary widely, mitochondria are typically rod-shaped. They are enclosed by two membranes: a smooth *outer membrane* that covers the surface, and an inner membrane that is convoluted to form folds, called **cristae** (Figure 3–7). Enzymes of the respiratory chain are present as components of the inner membrane, whereas most of the enzymes of the Krebs cycle are present in the aqueous material that is enclosed by the inner membrane.

## Lysosomes

**Lysosomes** are membrane-bound sacs or vesicles that contain enzymes capable of degrading proteins, nucleic acids, lipids and polysaccharides (see Figure 3–1). Consequently, lysosomes are important in digesting organic material brought into the cell and in degrading damaged or aging cells into their chemical components for use by other cells in the body. Because they often release their degradative enzymes into the cytoplasm previous to, or following death of the cell, lysosomes have been termed "suicide bags."

## Vacuoles

**Vacuoles** vary greatly in size and perform various functions. These include such things as storage of waste products, secretion products, and energy reserve compounds.

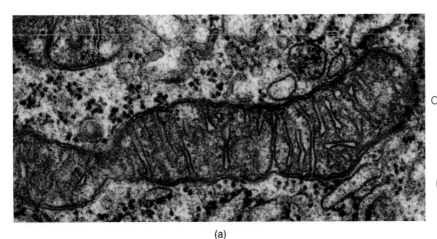

(a)

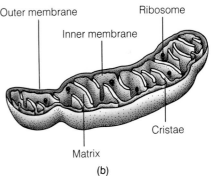

FIGURE 3–7 (a) Mitochondrion as revealed in the transmission electron microscope. (b) Sketch of the mitochondrion shown in (a), showing the arrangement of the inner and outer membranes.

Outer membrane    Ribosome
Inner membrane
Cristae
Matrix

(b)

## Internal Membrane Systems

In addition to membrane-bound structures, cells possess internal membrane systems that form an extensive maze of membranes within the cell. These serve as sites for chemical synthesis in the cell and as internal channels for the passage and storage of materials. There are two common types of membrane systems: *endoplasmic reticulum* and *Golgi complexes*.

### Endoplasmic Reticulum

The **endoplasmic reticulum** (or **ER**) consists of a series of large, flattened, membranous sacs that interconnect with one another, forming an extensive network of membrane-bound channels, the outer surface of which borders on the cytoplasm. The membranes of the endoplasmic reticulum have a "unit membrane" type of structure, but the width of the membrane is often somewhat narrower than that of the plasma membrane.

There are two distinct forms of endoplasmic reticulum present in most cell types. In one form, termed *rough endoplasmic reticulum* (or rough ER), the membranes have ribosomes attached to their outer surfaces, giving the

FIGURE 3-8 (a) An electron micrograph showing rough and smooth endoplasmic reticulum and a Golgi complex. (b) Sketch showing rough and smooth endoplasmic reticulum. (c) Sketch of a Golgi complex.

Rough endoplasmic reticulum

Smooth endoplasmic reticulum

Golgi complex

Golgi vesicle

(a)

(b)

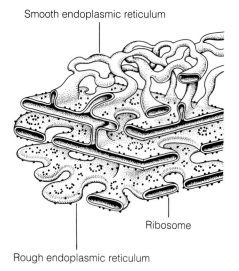

Smooth endoplasmic reticulum

Ribosome

Rough endoplasmic reticulum

(c)

Golgi vesicles

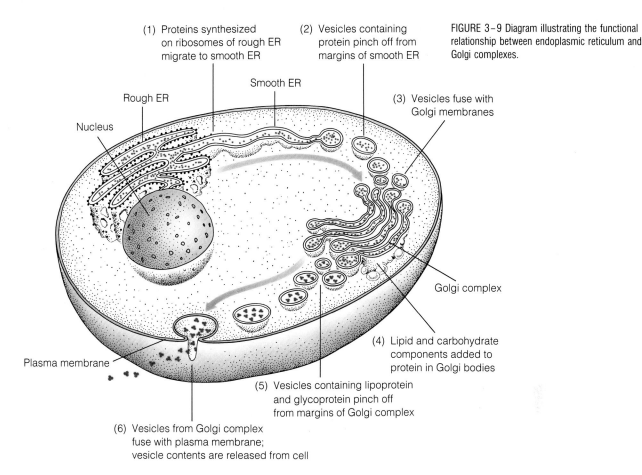

(1) Proteins synthesized on ribosomes of rough ER migrate to smooth ER

(2) Vesicles containing protein pinch off from margins of smooth ER

FIGURE 3-9 Diagram illustrating the functional relationship between endoplasmic reticulum and Golgi complexes.

Smooth ER

Rough ER

(3) Vesicles fuse with Golgi membranes

Nucleus

Golgi complex

(4) Lipid and carbohydrate components added to protein in Golgi bodies

Plasma membrane

(5) Vesicles containing lipoprotein and glycoprotein pinch off from margins of Golgi complex

(6) Vesicles from Golgi complex fuse with plasma membrane; vesicle contents are released from cell

membranes a granular or "rough" appearance when viewed in the electron microscope. Rough ER is involved in the synthesis of proteins, which may then be stored in the membranous channels or transported through the channels to other parts of the cell.

The other form of endoplasmic reticulum is called *smooth endoplasmic reticulum* (or smooth ER) because it is devoid of ribosomes. Smooth ER is involved in the synthesis, storage, and intracellular transport of lipids and lipid-containing substances such as glycolipids. Smooth and rough endoplasmic reticulum are shown in Figure 3-8.

## Golgi Complexes

**Golgi complexes** consist of small, spherical vesicles (membranous sacs) associated with stacks of larger vesicles collapsed in the form of discs (Figure 3-8). They are functionally associated with the ER, and their role within the cell can only be understood in that context.

Proteins synthesized on the ER are enclosed in vesicles that pinch off the edges of the ER, and these vesicles then migrate to Golgi complexes where they fuse with Golgi membranes. Inside the Golgi membranes, some of the proteins may be modified by the addition of carbohydrates to form glycoproteins. These and other proteins then accumulate at the edges of the discs of the Golgi complexes; here, they are packaged in vesicles that bud off the edges of the discs and enter the cytoplasm. Some of these vesicles contain products for secretion, which occurs when the vesicles, now called **secretory granules,** migrate to the plasma membrane and fuse with it, releasing their products at the cell surface. These products are discharged into the tissue fluid from where they may enter the circulation. The entire process is summarized in Figure 3-9.

Golgi complexes are also involved in the formation of lysosomes. This process is similar to the process by which secretory granules are formed; that is, enzymes produced in the ER are transported to the Golgi complex for packaging into vesicles. However, unlike secretory granules, lysosomes remain as organelles in the cytoplasm.

## Filamentous Structures

Typical human cells also contain a variety of structures that are formed of thin, rod-like filaments. **Microfilaments,** often found in bundles beneath the plasma membrane, are important in cell movement. For example, microfilaments in muscle cells are responsible for muscular contraction. **Microtubules** are cylindrical structures composed of an array of small filaments; they determine the shape of cells by providing internal support. In addition, microtubules form the **spindle fibers** that are involved in chromosome movements during cell division, and they form **centrioles,** two of which generally lie at right angles to one another near the nucleus of a nondividing cell (see Figure 3–1). Centrioles are important in the formation and organization of spindle fibers and will be discussed in a later section on cell division.

Some cells possess **cilia** or **flagella,** motile filamentous projections at the cell surface. Cilia, much shorter than flagella, serve to move substances past stationary cells, as, for example, when they clear dust and other small particles out of the respiratory passages (Figure 3–10). Flagella propel a whole cell such as a sperm cell (Figure 3–11). Both cilia and flagella are composed of an array of supporting microtubules covered by an outfolding of the plasma membrane.

## Structures of Some Representative Human Cell Types

Before leaving our discussion of cell structure, it would be well to emphasize that the foregoing description of a "typical" human cell represents a broad generalization. The human body is, in fact, composed of many different cell types, each of which has a specific structure that varies somewhat depending upon its function. Although many cells in the human body, such as liver cells, skin cells, and egg cells, look very much like our "typical" cell, others do not. As an example of the latter group, consider the sperm cell, shown in Figure 3–11a. It consists of a nucleus containing the cell's hereditary material, a prominent flagellum that propels the sperm in the reproductive tract of the female, and a very small amount of cytoplasm, in which are present closely packed mitochondria, which function to supply energy (in the form of ATP) for

FIGURE 3–10 Cilia on cells lining the trachea (windpipe).

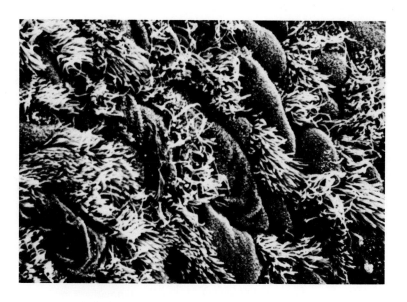

the movement of the flagellum. Other membrane-bound structures or internal membrane systems found in the "typical" human cell are absent.

The specialized structure of the sperm cell is consistent with its sole function, which is to transfer the hereditary material of the male to the female's egg during reproduction. To do this, the sperm cell must be motile and have a means for providing energy to sustain the continuous motion of its flagellum during its journey of many hours. It does not have a long life nor does it perform a variety of different functions. Consequently, it does not need the elaborate metabolic machinery, represented by ER, Golgi complexes, and the like, required to sustain a multipurpose metabolism or to keep it functioning for long periods of time.

The structures of several other human cell types are shown in Figure 3–11b and c. Although these cells form a structurally diverse group, it is well to remember that fundamentally, they operate by identical mechanisms.

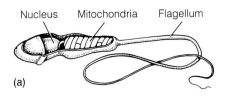
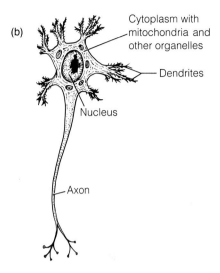

## MEMBRANE TRANSPORT: HOW THINGS GET INTO AND OUT OF CELLS

The plasma membrane has many important functions, but its role as a permeability barrier is clearly the most basic. We do not as yet fully understand all of the factors that account for the plasma membrane's **differential permeability**—its ability to determine which molecules will pass through the membrane, and at what rate—but we do know that there is nothing magic about it. Its selectivity depends upon the interaction between chemical components in the membrane and molecules that attempt to permeate it. Some of these interactions will be "favorable" and will allow permeation of the membrane; others will be "unfavorable" and will not allow permeation.

A molecule's size, shape, electrical charge, and solubility characteristics are all important in determining whether or not it will permeate the plasma membrane. In general, but by no means without exception, small molecules pass through more readily than large ones; uncharged molecules, more readily than charged ones; and molecules with high lipid solubility, more readily than those with low lipid solubility. The plasma membrane is most permeable to water, carbon dioxide, and oxygen, and considerably less so to the majority of solutes.

FIGURE 3–11 Some highly specialized human cell types. (a) Sperm cell; (b) nerve cell; and (c) red blood cell. Nerve cells possess long processes that conduct nerve impulses. Mature red blood cells contain the oxygen-carrying protein hemoglobin in the cytoplasm but have no nucleus or other significant intracellular structure. The three cell types are not drawn to scale.

The permeability of the plasma membrane is also related to its physiological state. Permeability increases with injury to the cell by heat, pH change, ion imbalance, radiation, and other means. Various drugs such as narcotics and anesthetics are also known to alter permeability. However, changes in permeability occur even in normal cells. For example, increased permeability of the plasma membrane to various ions is the basis for the excitability of muscle and nerve cells, as discussed in Chapters 7 and 13.

There are a number of different mechanisms, or means, by which substances pass through membranes. These include *passive diffusion, facilitated diffusion, active transport, endocytosis,* and *exocytosis.* Because the passage of substances into and out of cells is central to physiological processes in all systems, it is important to understand each of these mechanisms.

### Passive Diffusion

Water, $CO_2$, $O_2$, and some other substances pass through the plasma membrane by **passive diffusion.** This is the simplest type of transport mechanism, and it depends upon the process of diffusion—the natural tendency of molecules to distribute themselves evenly in any space by random motion. As a consequence of the phenomenon of diffusion, any single type of molecule in a solution, separated from other molecules of the same type by a membrane *to which they are permeable,* will tend to diffuse through the membrane from a region of high concentration to one of low concentration, until their concentrations are

## Sickle Cell Anemia and the Importance of Cell Shape

The human red blood cell does not have a structure or shape that is typical of other cell types. Its cytoplasm contains no intracellular organelles, not even a nucleus. Instead, it is filled with a dense solution composed of a watery medium in which is dissolved some 270 million hemoglobin molecules that function to transport oxygen to the tissues of the body. The shape of the human red blood cell, normally that of a biconcave disc, is maintained by a filamentous network of proteins underlying the plasma membrane. However, the red blood cell is extremely flexible, and its shape is temporarily distorted as it squeezes its way through the smallest blood vessels (capillaries) of the circulatory system, where it releases oxygen to the body's tissues. The pliable nature of the red blood cell is thus crucial to its function.

The relationship of the shape of red blood cells to their function has been clearly and dramatically demonstrated in studies of a blood disease called sickle cell anemia, first described in 1910. In sickle cell anemia, red blood cells become rigid and oddly shaped when their hemoglobin releases oxygen. Many of the cells assume a crescent, or sickle shape, from which the

disease gets its name (see accompanying figure). The rigid, sickle-shaped cells tend to get caught in capillaries and may clog them, thus preventing oxygen flow to the tissues. This causes severe tissue damage and pain. Because sickle-shaped cells are less flexible than normal ones, they have increased mechanical fragility and are frequently disrupted as they are propelled through the circulatory system. This leads to reduced numbers of red blood cells, which causes severe anemia.

The blockage of capillaries together with the severe anemia adversely affects virtually all organ systems of the body. Children with sickle cell anemia have impaired growth and development and an increased susceptibility to bacterial infection, particularly pneumonia. If untreated, sickle cell anemia is usually fatal in childhood, but with modern medical care, many affected individuals survive beyond the age of 50, albeit severely disabled. Mortality in childhood nearly always is a

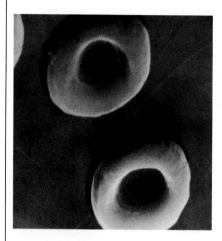

Normal red blood cells.

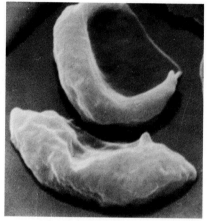

Sickle-shaped red blood cells.

equal on both sides of the membrane (Figure 3–12). The *rate* of diffusion is proportional to the concentration difference across the membrane. That is, the higher the concentration of a diffusible molecule on one side of the membrane, the faster will be its rate of diffusion to the other side.

Passive diffusion is termed "passive" because the movement of molecules does not require any special transport carrier molecules, as does facilitated diffusion. Because the molecular movements that are associated with diffusion are "driven" by the concentration gradient across the membrane, passive diffusion requires no energy expenditure by the cell.

### Osmosis

**Osmosis** refers to the passive diffusion of water across a membrane in response to a concentration gradient. Since it applies only to the diffusion of water molecules, osmosis is a special case of passive diffusion.

result of bacterial infection; mortality in adults occurs as a result of chronic organ damage.

Sickle cell anemia is an inherited disease and to date is not curable. It is prevalent in Africa and in some regions of Saudi Arabia, India, and the Mediterranean countries. It occurs in the United States primarily in people of African descent. One in about 600 black infants in the United States is afflicted with sickle cell anemia, and there are about 50,000 individuals in this country with the disease. There is also a milder form of the disease, called sickle cell trait, which rarely causes any physiological problems and is found in the same groups of people in which sickle cell anemia occurs. The patterns of inheritance of sickle cell anemia and sickle cell trait are discussed in Chapter 20.

Sickle cell anemia has been the subject of much basic research aimed at understanding the fundamental nature of the disease. As a result of this research, we now understand precisely why red blood cells in sickle cell anemia have an altered shape. It is because of the structure of the hemoglobin they contain. Normal hemoglobin consists of four amino acid chains:

two identical alpha chains, each consisting of a specific sequence of 141 amino acids, and two identical beta chains, each consisting of 146 amino acids. In addition, each chain has an attached heme group (a complex organic molecule) that can carry an oxygen molecule (see Chapter 8, Figure 8–6). Sickle cell hemoglobin is identical in structure to normal hemoglobin with the exception of one amino acid in the beta chains; that is, in the beta chain of sickle cell hemoglobin, a different amino acid substitutes for one found in normal hemoglobin.

This seemingly trivial change of only one amino acid out of a total of 287 in the alpha and beta chains greatly alters the biological function of the sickle cell hemoglobin. When sickle cell hemoglobin is deoxygenated, as it is when it releases oxygen to the tissues, its solubility in the cytoplasm of the red blood cell is decreased. This causes the hemoglobin molecules to pack closely together into fibrous aggregates. Because the aggregated hemoglobin molecules take up less space than do soluble hemoglobin molecules dispersed in the cytoplasm, the normal biconcave disc shape of the red blood

cell collapses, forming the more irregular and rigid sickle shape.

Other research has shown that people with sickle cell anemia have a permanent change, or mutation, in a region of their DNA (hereditary material) that determines hemoglobin structure; the mutation results in altered hemoglobin structure. We thus have a situation in which altered DNA causes the production of an altered hemoglobin protein, which, in turn, brings about a change in the shape and rigidity of the red blood cell, such that its function is drastically affected. The Nobel prize winning scientist Linus Pauling, who worked on sickle cell anemia, has called it a "molecular disease" because it derives from a defect in the structure of the hemoglobin molecule.

Our understanding of the molecular basis of sickle cell anemia has led to new ideas for its treatment. One of these is to treat people suffering from sickle cell anemia with chemicals that will enter the red blood cells and inhibit the aggregation of sickle hemoglobin molecules. A number of such chemicals are currently being tested, some of which show quite promising results.

Osmosis occurs because the plasma membrane allows water to pass freely through it, but retards to a certain degree and sometimes prevents the passage of solutes (those substances dissolved in the water). Consequently, if a cell is placed in an aqueous medium with a solute concentration equal to the solute concentration in its own cytoplasm, water will diffuse into and out of the cell, but the same amount of water will leave the cell as will enter it. Consequently, the volume of the cell is unaffected. Such an aqueous medium is termed an **iso-osmotic solution.** However, if a cell is placed in a solution that contains a solute concentration greater than that in its own cytoplasm **(hyperosmotic solution),** more water will leave the cell than will enter it. This occurs because the concentration of water, relative to that of solute, is initially greater inside the cell than outside. The cell will consequently shrink in volume due to the loss of water. On the other hand, a cell will swell in volume due to the net gain of water if it is placed in a **hypo-osmotic solution,** in which the solute concentration is less than that in the cell's cytoplasm (Figure 3–13).

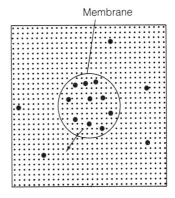

Membrane

FIGURE 3–12 Passive diffusion of a solute (large dots) dissolved in water (small dots). Solute molecules will move through the membrane until their concentration is identical on both sides of the membrane. Net flow of solute molecules is in the direction of the arrow.

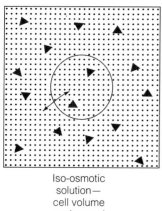

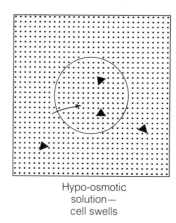

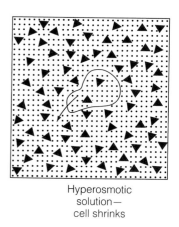

| Iso-osmotic solution— cell volume unchanged | Hypo-osmotic solution— cell swells | Hyperosmotic solution— cell shrinks |

FIGURE 3–13 Diagram illustrating osmosis, a special case of passive diffusion. Net flow of water is in the direction of the arrow. Small dots represent water molecules, and triangles represent solute molecules.

## Facilitated Diffusion

Some substances, such as glucose and certain amino acids, enter cells by a process called **facilitated diffusion.** In facilitated diffusion, substances diffuse through the membrane only if they are combined with specific transport proteins in the membrane, which aid, or facilitate, their diffusion through the membrane by acting as carriers or shuttles. Like passive diffusion, facilitated diffusion is driven by a concentration gradient, and consequently no energy expenditure by the cell is required.

## Active Transport

In passive diffusion (including osmosis) and facilitated diffusion, substances are transported in response to a concentration gradient, and hence, no energy expenditure by the cell is required. However, in another type of transport, called **active transport,** substances cross the plasma membrane against a concentration gradient (from one region to another of greater concentration), and the process requires cellular energy, usually in the form of ATP (Figure 3–14). As in the case of facilitated diffusion, active transport requires the use of transport proteins that shuttle the substance from one side of the membrane to the other.

## The Sodium-Potassium Pump

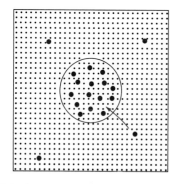

FIGURE 3–14 In active transport, solute molecules (large dots) are transported against a concentration gradient, as illustrated here. Net movement of solute is in the direction of the arrow. Compare this figure with Figure 3–12.

Cells regulate their intracellular concentrations of sodium ions (Na$^+$) and potassium ions (K$^+$) by the use of an active transport system called the **sodium-potassium pump.** In general, cells maintain a high internal concentration of K$^+$ and a low internal concentration of Na$^+$ while the extracellular fluids that bathe them are just the opposite: low in K$^+$ and high in Na$^+$. For example, in the human red blood cell, the concentration of Na$^+$ is about 12 times greater in the extracellular fluid than in the intracellular fluid, while the concentration of K$^+$ is some 21 times greater in the intracellular than in the extracellular fluid. Although it is not precisely clear why cells maintain low Na$^+$ and high K$^+$ levels, it must have significant physiological importance, since over 30% of the ATP consumed by a typical "resting" cell is used for this purpose! Furthermore, we know that imbalances of blood sodium and potassium levels can lead to severe metabolic disturbances that may be fatal.

The sodium-potassium pump is mediated by an enzyme, *Na$^+$-K$^+$ATPase,* that is located in the cell's plasma membrane. Na$^+$-K$^+$ATPase is capable of hydrolyzing ATP to ADP and phosphate, and using the released energy to transport $K^+$ into the cell and $Na^+$ out of the cell. The $Na^+$-$K^+$ATPase thus has a dual role: it acts as a transport carrier for $Na^+$ and $K^+$, and it provides the energy needed for the active transport of the two ions. A mechanism to explain how this process might occur is shown in Figure 3–15.

## Endocytosis and Exocytosis

Very large molecules and certain particles can sometimes enter cells by the processes of *endocytosis* and leave them by *exocytosis*. Unlike the other transport mechanisms discussed above, endocytosis and exocytosis do not actually involve the passage of substances *through* the plasma membrane.

In **endocytosis,** material outside the cell becomes trapped in small depressions, called **endocytotic vesicles,** formed by the plasma membrane on the outer cell surface (Figure 3–16). The depression then closes by a gradual fusion of the plasma membrane over the open end, trapping the external material in a membrane-bound vesicle. The vesicle then detaches from the inner surface of the plasma membrane and enters the cytoplasm. In the cytoplasm, the vesicle usually fuses with a lysosome to form a **digestive vacuole;** the lysosomal enzymes then digest the material into smaller molecules that diffuse through the membrane of the digestive vacuole to be used as energy sources or in other cell functions.

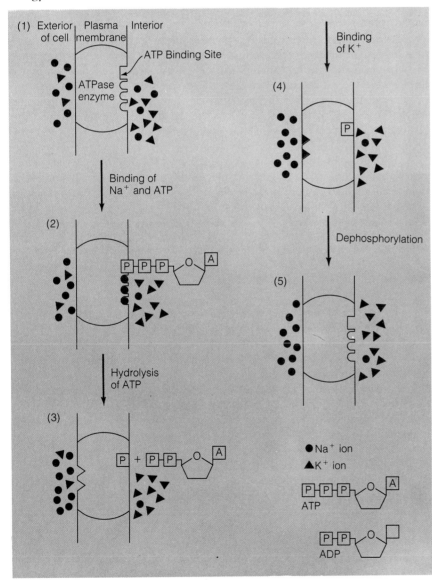

FIGURE 3–15 Diagram illustrating how $Na^+$-$K^+$ ATPase may act to transport $Na^+$ and $K^+$. (1) The $Na^+$-$K^+$ ATPase enzyme has binding sites for $Na^+$ and ATP. (2) $Na^+$ and ATP bind to the enzyme. (3) ATP is hydrolyzed by the enzyme to ADP and phosphate, and the phosphate binds to (phosphorylates) the enzyme. Phosphorylation of the enzyme causes a change in its shape, which carries the $Na^+$ ions to the cell exterior and exposes binding sites for $K^+$. (4) $K^+$ ions can now bind to the phosphorylated enzyme. (5) The phosphate group leaves the enzyme, returning the enzyme to its original shape (as in step 1), transporting $K^+$ ions to the cell exterior and again exposing binding sites for $Na^+$ and ATP.

FIGURE 3–16 Endocytosis in an amoeba. The indentation in the plasma membrane will eventually close around the two food particles.

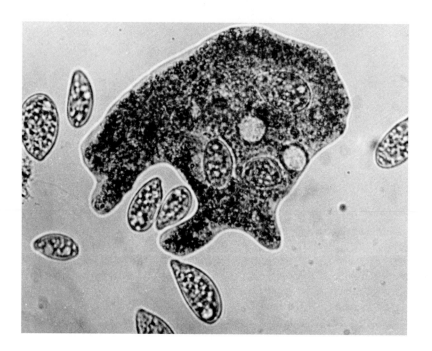

Endocytosis is used extensively by **scavenger cells** in the body (these include certain *white blood cells* and cells that derive from them) to remove foreign substances, bacteria and cellular debris. The scavenger cells will ingest foreign cells after recognizing certain specific recognition molecules that are present on the surface of the foreign cell. Interestingly, some pathogenic (disease-causing) bacteria are not attacked by scavenger cells because the bacterial cells are covered by a surface polysaccharide material (capsule) that scavenger cells do not recognize as foreign.

Although lysosomal enzymes can digest a wide range of materials, some materials are undigestible. These are released from the cell by the fusion of the digestive vacuole with the plasma membrane, in a process that is the reverse of endocytosis and is called **exocytosis**. Exocytosis is also used by cells in the secretion of hormones and other substances that are packaged in Golgi vesicles (see Figure 3–9).

## CELL DIVISION IN SOMATIC CELLS

The process of cell division leading to the formation of new somatic cells (all body cells other than germ cells) is fundamentally different from that leading to the formation of germ cells (sperm and egg cells). Here, we will consider cell division in somatic cells; we will discuss the formation of germ cells in Chapter 17.

Somatic cells show various patterns of division. Once formed, some somatic cells in our bodies do not divide. This is true, for example, of nerve cells and circulating red blood cells. Many other somatic cells, however, regularly proliferate by division; some of these divide only infrequently, while others, such as those of the intestinal lining or of the bone marrow, divide at an exceptionally rapid rate, replacing millions of dead or damaged cells every day.

When a somatic cell divides, each of the two new cells, termed **daughter cells,** must receive an exact copy of the original cell's hereditary material (DNA). If it did not, genetic abnormality would result—a condition that might lead to impaired cellular function, to death of the cell, and ultimately, even to death of the organism. The logistical problem of replicating (duplicating) a cell's DNA and distributing it equally among the daughter cells is more difficult than it might at first appear. As you will recall, the DNA in a human cell is *not* present as one giant molecule, but is broken up into 46 different molecules; each of these is associated with protein and RNA to form 46 chromatin strands (or, when coiled during cell division, 46 chromosomes). Thus, cell division requires not only the replication of all of these strands (to form 92 strands), but their distribution so that each daughter cell receives one copy of each of the 46 different chromatin strands, not just *any* 46 strands.

## The Cell Cycle

The sequence of events that occurs during somatic cell division is termed the **cell cycle.** The cell cycle consists of three main phases. In the first phase, called **interphase,** chromatin is replicated, other macromolecules are synthesized, and the cell usually increases substantially in mass. Interphase is followed by **mitosis,** during which the replicated chromatin strands are separated and enclosed in two daughter nuclei; mitosis itself occurs in five phases, as you will see. The third phase of the cell cycle, **cytokinesis,** involves the actual division of the cytoplasm; cytokinesis begins toward the end of mitosis and is completed shortly afterward. The cell cycle is diagrammed in Figure 3–17. For a typical human cell growing in tissue culture, the entire cycle lasts about 24 hours.

## Interphase

As shown in Figure 3–17, interphase is further divided into three stages: $G_1$ (first gap in DNA synthesis, or the stage preceding DNA synthesis), $S$ (DNA synthesis), and $G_2$ (second gap in DNA synthesis, the stage following DNA synthesis). DNA synthesis occurs only in the S stage, but the synthesis of proteins and other macromolecules, usually accompanied by a substantial increase in cell mass, occurs during all three stages. The length of $G_1$ varies greatly among different cell types, whereas the S and $G_2$ stages are relatively uniform in duration. Thus, a slowly dividing cell has a prolonged $G_1$, and a rapidly dividing cell has a shortened $G_1$, but once any cell enters the S stage, the rest of the cycle usually proceeds without interruption within similar time frames. By the end of interphase, the cell is committed to divide and the DNA has been replicated.

## Stages of Mitosis

The stages of mitosis are summarized in Figure 3–18. At some point, an interphase cell in $G_2$ becomes committed to enter mitosis. By this time, in the nucleus, each of the chromatin strands has been exactly replicated, but the two strands (the original and the new replica) remain attached to one another at a specific point on each strand called the **centromere.** In addition, by the end of interphase, a second pair of centrioles has been assembled near the nuclear envelope and next to the pair already present.

## Prophase

**Prophase** is the first visible stage of mitosis. During prophase, the doubled chromatin strands, which have been threadlike until this point (although not visibly so), condense by coiling and folding to form chromosomes. At this stage, each chromosome is composed of two identical parts, called **sister chromatids,** corresponding to the doubled chromatin strands attached at their centromeres.

One pair of centrioles migrates to the opposite side of the nucleus so that the two pairs eventually come to lie directly opposite one another with the nucleus between them. As centriole migration occurs, spindle fibers are seen radiating from each centriole pair, forming starlike structures called **asters;** this

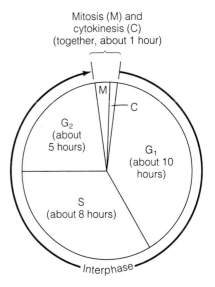

FIGURE 3–17 The three phases of the cell cycle of a typical human cell grown in tissue culture: Interphase, mitosis, and cytokinesis.

FIGURE 3–18 Mitosis and accompanying cytokinesis in an animal cell.

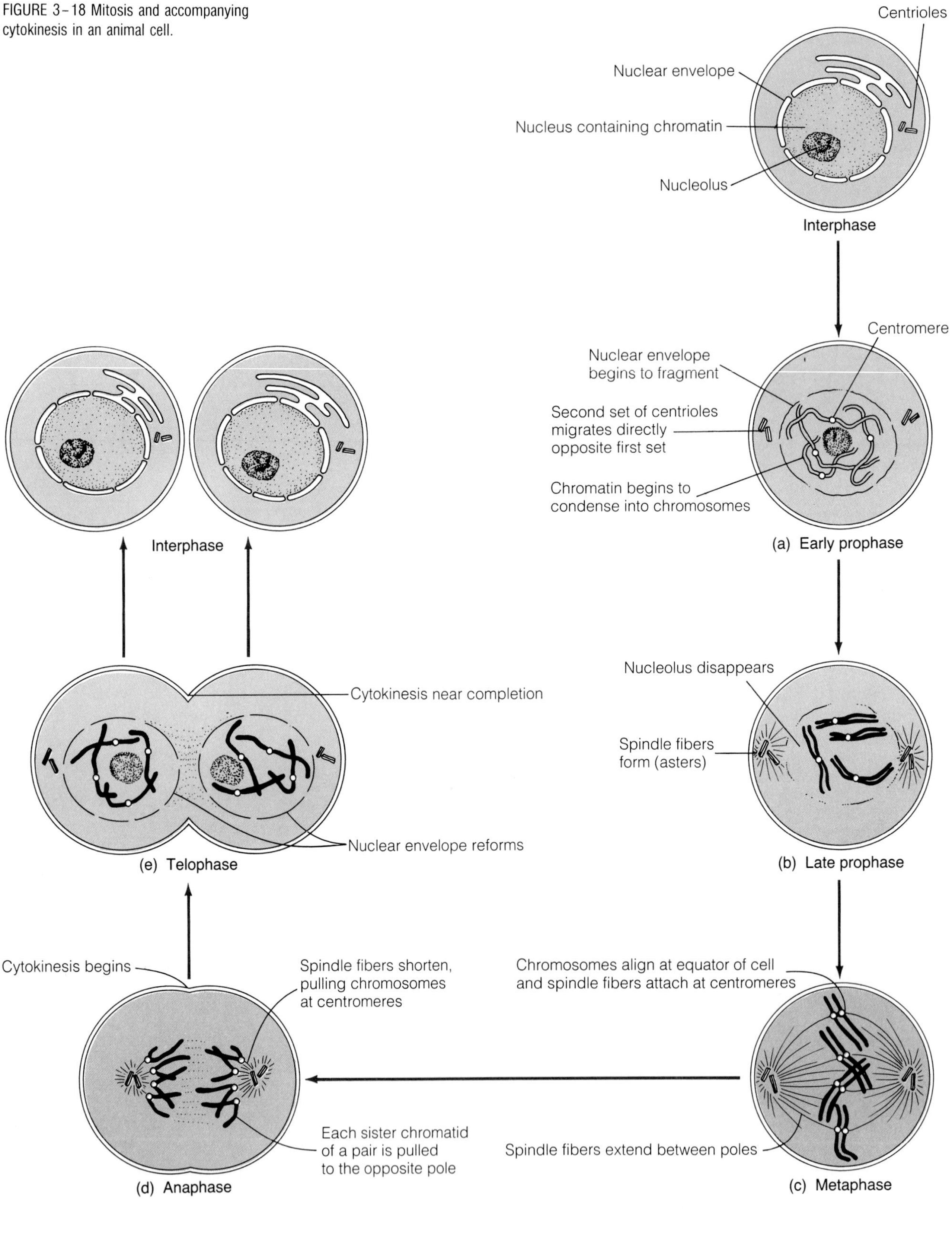

Centrioles

Nuclear envelope

Nucleus containing chromatin

Nucleolus

Interphase

Centromere

Nuclear envelope begins to fragment

Second set of centrioles migrates directly opposite first set

Chromatin begins to condense into chromosomes

(a) Early prophase

Nucleolus disappears

Spindle fibers form (asters)

(b) Late prophase

Interphase

Cytokinesis near completion

Nuclear envelope reforms

(e) Telophase

Cytokinesis begins

Spindle fibers shorten, pulling chromosomes at centromeres

Each sister chromatid of a pair is pulled to the opposite pole

(d) Anaphase

Chromosomes align at equator of cell and spindle fibers attach at centromeres

Spindle fibers extend between poles

(c) Metaphase

is the beginning of the formation of the **spindle apparatus** that is used to pull the duplicated chromosomes apart. Also at this stage the nuclear envelope fragments.

## Metaphase

The spindle apparatus is completed and consists of spindle fibers that extend from pole to pole. The doubled chromosomes migrate to a plane that is perpendicular to the middle of the spindle apparatus, and spindle fibers attach to each chromatid at the centromere. This attachment occurs in such a way that one of the sister chromatids of each chromosome is attached by spindle fibers to one pole, and the other sister chromatid is attached to the other pole.

## Anaphase

The sister chromatids of each chromosome are pulled toward opposite poles of the cell by shortening of the spindle fibers. As the chromatids are pulled apart, the centromeres of each chromatid are separated.

## Telophase

The separated chromatids, now single chromosomes, aggregate at the two poles, and the spindle apparatus is disassembled. By the end of telophase, the chromosomes have unfolded into chromatin threads, and a new nuclear envelope has formed to enclose each nucleus.

## Cytokinesis

As the events of mitosis are occurring, cytokinesis commences with the formation of a narrow groove, or *furrow,* which extends in a ring around the surface of the cell and deepens progressively until the cell is separated into two parts. The cytoplasm of the dividing cell, including organelles it may contain, is distributed roughly equally between the two cells.

## Summary of the Events of the Cell Cycle

The net result of the cell cycle is that one cell divides into two cells, each of which has exactly the same DNA and chromosome (or chromatin) content as the original cell, as well as a similar (although not necessarily identical) array of cytoplasmic organelles.

## Study Questions

1. Indicate the structure and function of each of the following: plasma membrane, tight junction, desmosome, gap junction, nucleus, chromatin, chromosomes, mitochondria, lysosome, smooth endoplasmic reticulum, rough endoplasmic reticulum, Golgi complex, microtubules, centrioles, spindle fibers, cilia, and flagella.
2. Indicate some of the factors that are important in determining a molecule's ability to permeate the plasma membrane.
3. Describe each of the following processes and indicate how they differ from one another: passive diffusion, osmosis, facilitated diffusion, active transport, endocytosis, and exocytosis.
4. A cell is placed in a solution that is hyperosmotic to its own cytoplasm. What will happen to the volume of the cell and why?
5. Describe the operation of the sodium-potassium pump.
6. Diagram the cell cycle.
7. **a)** Describe the phases of mitosis.
   **b)** What is the purpose of mitosis?

### Chapter Four

# Tissues, Organs, and Systems

# INTEGRATION OF CELLULAR ACTIVITY

Within the human body, cells function on essentially two levels: as individual cells and as integral parts of the organism which they compose. This means that individual cells must function not only to ensure their own survival, but to ensure the survival of the entire organism of which they are a part. Although such a statement may seem obvious, it serves to emphasize the relationship between cell function at the cellular level and at the level of the organism.

Individual cells perform functions that the organism itself performs. Individual cells must obtain oxygen, water, inorganic materials and organic nutrients from their environment; they must metabolize organic nutrients to release energy needed for energy-requiring metabolic activities, or to provide raw materials for the manufacture of cell constituents; they must eliminate metabolic wastes, and they usually must reproduce, among other functions. But as parts of a multicellular organism, individual cells are isolated from the external environment and consequently cannot exchange materials directly with their environment. A heart cell, for example, buried deep within the organism, has no way of obtaining oxygen or other essential materials directly from the environment of the organism nor can it eliminate wastes directly into the environment.

These functions are provided by the activities of the organism itself. The ingestion of food and water, and the digestion of food by an organism supplies essential nutrients and other required materials to all the cells of the body; breathing by an organism supplies oxygen to all the cells of the body; excretion by an organism serves to eliminate the cumulative wastes of all the body's cells; maintenance of body temperature and proper acidity of body fluids ensures optimum conditions at which cellular metabolic activities occur, and so forth. Thus, *the cells of the body, operating as a unit, sustain themselves.*

## Cells, Tissues, Organs, and Systems

The coordination between cells in the human body may occur on several levels. Cells with a specific structure and function are organized into **tissues.** Examples of tissues are the cells composing the muscle of the heart or those forming the lining of the stomach. One or more tissues may, in turn, be coordinated as a structural and functional unit termed an **organ.** The heart, for example, is an organ whose function is to pump blood; the stomach is an organ important in the digestion of food. A group of organs may form a **system** responsible for carrying out one general function, such as circulation or digestion. For example, the circulatory system consists of the heart, the blood vessels, and the blood. Together, these serve to transport materials between the various cells in the body. As another example, the digestive system includes a variety of organs that serve to supply the cells of an organism with nutrients, water, and other materials from the external environment (Figure 4–1). The activities of all the systems of the body are integrated in such a way as to maintain homeostasis—the steady-state internal conditions required for life (see Chapter 1).

# TISSUE FLUID

As we have just pointed out, the exchange of nutrients, oxygen, and metabolic wastes between the cells of the body and the organism's environment is a central function of metabolism.

The blood vessels of the circulatory system (which we shall discuss in due course) provide the passageway by which essential transfers of materials are made; the smallest blood vessels of the circulatory system—the capillaries— are the structural elements through which materials are actually transferred. It

FIGURE 4–1 The relationship between cells, tissues, organs, and systems.

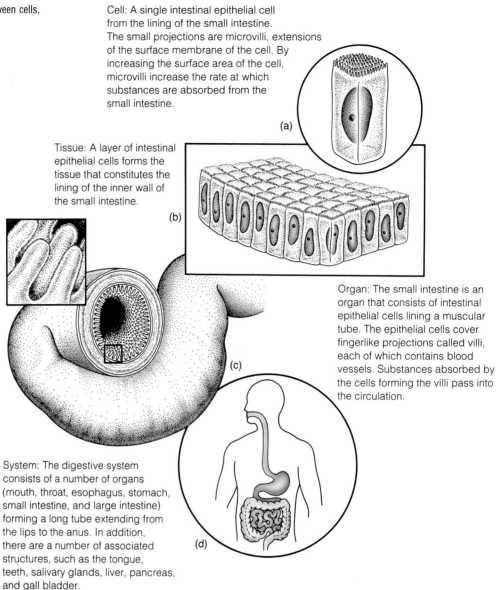

Cell: A single intestinal epithelial cell from the lining of the small intestine. The small projections are microvilli, extensions of the surface membrane of the cell. By increasing the surface area of the cell, microvilli increase the rate at which substances are absorbed from the small intestine.

(a)

Tissue: A layer of intestinal epithelial cells forms the tissue that constitutes the lining of the inner wall of the small intestine.

(b)

Organ: The small intestine is an organ that consists of intestinal epithelial cells lining a muscular tube. The epithelial cells cover fingerlike projections called villi, each of which contains blood vessels. Substances absorbed by the cells forming the villi pass into the circulation.

(c)

System: The digestive system consists of a number of organs (mouth, throat, esophagus, stomach, small intestine, and large intestine) forming a long tube extending from the lips to the anus. In addition, there are a number of associated structures, such as the tongue, teeth, salivary glands, liver, pancreas, and gall bladder.

(d)

may therefore be surprising to learn that the vast majority of cells in the human body do not actually come in physical contact with a capillary. How, then, are materials transferred between these cells and the capillaries of the circulatory system?

Cells that form tissues of the body not only contain an aqueous cytoplasm, but their plasma membranes are surrounded by an aqueous, ion-containing fluid, called **tissue fluid** or **interstitial fluid.** The tissue fluid serves as the medium through which substances are exchanged between the cells and the circulatory system. Nutrients and oxygen, dissolved in the blood, pass from capillaries that infiltrate a tissue into the tissue fluid; from there, these materials pass into the cells themselves through their plasma membranes. Conversely, cellular waste materials cross the plasma membrane and enter the tissue fluid, from which they pass into capillaries and are ultimately eliminated from the body (Figure 4–2). *The tissue fluid thus plays an indispensable role as intermediary in the exchange of materials between cells and the capillaries* (and ultimately, of course, between the cells and the external environment of the organism). The formation of tissue fluid is discussed in Chapter 8.

## TISSUE TYPES

The human body contains four basic types of tissues—*epithelial tissues, connective tissues, muscle tissues,* and *nervous tissues*—that combine in various ways to form the organs of the body. A variety of tissues, specialized to form particular functions, are derived from each of the basic tissue types.

## Epithelial Tissues

**Epithelial tissues** form sheets of cells that cover the outer surface of the body and line the interior surfaces of the body cavities and of hollow body organs. To better understand the function of epithelial tissues, it may be helpful to clarify use of the terms "covering" and "lining" in reference to anatomical structures. The outer surface of any structure, whether solid or hollow, may be covered by something, much as a tennis ball is covered with fuzzy material or an apple with a red "skin". The interior surface of a hollow structure is said to be lined, in the sense that a tennis ball is lined with rubber.

The individual cells of epithelial tissues fit very closely together, and, as mentioned previously, the plasma membranes of adjacent epithelial cells are fused by tight junctions that prevent leakage of substances between the cells (see Figure 3–5). Such an arrangement is appropriate for tissues that form coverings and linings since it minimizes the loss of vital materials from underlying tissues and substantially prevents the infiltration of substances from the environment. Sheets of epithelial cells, while not particularly strong themselves, are supported by underlying connective tissue to which the epithelial tissue is attached by a somewhat fibrous **basement membrane** (Figure 4–3). Epithelial tissues are not themselves infiltrated by blood vessels, and consequently are dependent on blood vessels in the underlying connective tissue for supplying essential substances and removing waste products.

Because epithelial cells are present on surfaces, they are continually worn off and must be replaced. Consequently, epithelial cells are among the most rapidly dividing of all cell types. Some rates of cell division in epithelial tissue are truly astounding. It has been estimated, for example, that the lining of the small intestine replaces billions of cells every hour.

## Types of Epithelial Tissues

Epithelial tissues are classified according to two criteria: (1) the shape of the component cells and (2) the number of cell layers. The shape of epithelial cells may be **squamous** (flat and thin), **cuboidal,** or **columnar.** In addition, cells in some epithelial tissues may vary in shape depending upon whether the tissue is in a stretched or relaxed state; such epithelial tissue is called **transitional.**

With regard to the number of cell layers present, epithelial tissues may be composed of a single cell layer **(simple epithelium)** or more than one layer **(stratified epithelium).** Another type of epithelial tissue, termed **pseudo-stratified,** also consists of one layer of cells. However, it gives the inaccurate impression that it is formed of more than one layer, since all cells do not extend from one surface of the tissue to the other, but all cells are attached to the basement membrane.

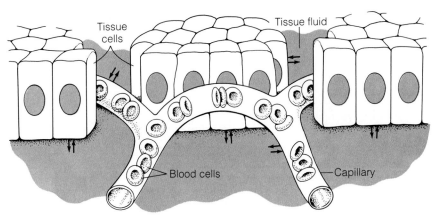

Tissue cells

Tissue fluid

Blood cells

Capillary

FIGURE 4–2 Tissue, or interstitial fluid, fills all spaces between cells and serves as the intermediary for the transfer of substances between the cells and the circulatory system.

Hollow structures that transfer substances through their walls, such as blood capillaries, the air sacs of the lung, and the small intestine, are lined with simple epithelium, the thinness of which is appropriate for this function. Stratified epithelium, on the other hand, makes up the outer layer (epidermis) of the skin, the lining of the mouth, and other structures that receive much abrasion. Often, the outermost layers of cells in stratified epithelium are altered in structure to form a tough surface layer that is particularly resistant to abrasion.

The structure of various types of epithelial tissues is shown in Figure 4–3. Also listed in that figure are examples of where each type of tissue may be found in the body.

### Specialized Epithelial Tissues

In addition to their protective function and their ability to transfer substances, epithelial tissues may be modified to perform other tasks. For example, the epithelium lining the seminiferous tubules of the testes, called **germinal epithelium,** gives rise to sperm cells. **Neuroepithelium,** in which nerve cells form part of the epithelium, is specialized for the reception of stimuli and is present in the eye, ear, nose, and tongue. Cells in epithelial tissue may also form glands **(glandular epithelium);** these are masses of cells that synthesize and release *(secrete)* specialized products. Glands are of two types, depending upon the relationship of the secreting cells to the epithelial tissue from which they are derived. **Exocrine glands,** including mammary glands, sweat glands, oil glands, and a number of glands that secrete digestive enzymes, retain a connection with the epithelial surface through one or more ducts, which convey secretions to the surface. **Endocrine glands,** which

FIGURE 4–3 Types of epithelial tissues.

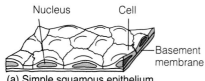

(a) Simple squamous epithelium
Walls of blood capillaries;
airsacs of lungs

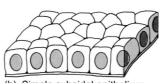

(b) Simple cuboidal epithelium
Lining of small ducts of salivary glands

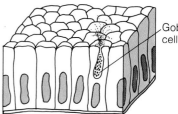

(c) Simple columnar epithelium
Lining of small intestine;
lining of stomach

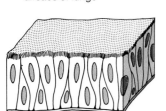

(d) Pseudostratified columnar epithelium (cilliated)
Lining of larger airways of the respiratory system

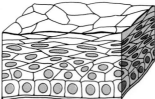

(e) Stratified squamous epithelium
Epidermis (outer layer of skin);
lining of oral cavity, throat, esophagus
lining of vagina

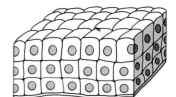

(f) Stratified cuboidal epithelium
Ducts of sweat glands

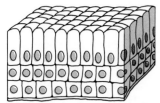

(g) Stratified columnar epithelium
Lining of short portions of ducts
of salivary glands

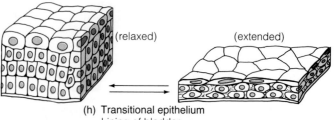

(h) Transitional epithelium
Lining of bladder

A scanning electron micrograph of the microvilli on cells lining the small intestine.
From TISSUES AND ORGANS: A text-atlas of scanning electron microscopy, by Richard G. Kessel and Randy H. Kardon. W.H. Freeman and Company, copyright © 1979.

secrete hormones, lose their connection with the epithelial surface and are consequently ductless; they secrete their products directly into the bloodstream (Figure 4–4).

Epithelial cells may show a variety of structural specializations. Some epithelial cells contain cilia (see Figure 3–10). Cilia move eggs along the Fallopian (uterine) tubes, or mucus and particulate matter up and out of the respiratory tubes. In other cells that are modified for absorption, such as those lining the small intestine, the plasma membrane is folded to produce numerous fingerlike projections, known as **microvilli,** that increase the surface area through which absorption can occur (see Figure 3–5). Still other epithelial cells, termed **goblet cells,** are modified to secrete mucus that moistens the epithelium and protects it from the action of digestive enzymes, or serves as a trap for dust and pollen (Figure 4–3c).

## Connective Tissues

**Connective tissues** are present in all parts of the body, where their primary function, as their name suggests, is to bind together, support, and protect other tissues. Indeed, without bone and a variety of other connective tissues of a

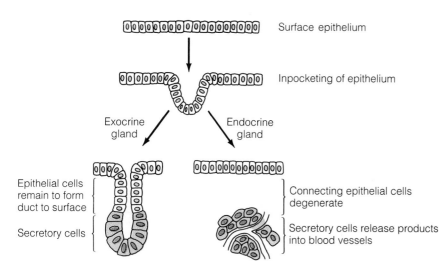

Surface epithelium

Inpocketing of epithelium

Exocrine gland

Endocrine gland

Epithelial cells remain to form duct to surface

Secretory cells

Connecting epithelial cells degenerate

Secretory cells release products into blood vessels

FIGURE 4–4 The development of exocrine and endocrine glands.

fibrous nature, the human body would lack its characteristic form. In addition to holding the cells of the body together as an organized structure, connective tissues store fat, form blood, and protect the body against disease, among other functions.

Unlike other tissues, connective tissues are not composed primarily of closely packed cells. Instead, connective tissues usually consist of relatively few cells, spaced apart from one another and surrounded by a large amount of nonliving *intercellular substance* that is secreted by some of the connective tissue cells. The intercellular substance is an amorphous material that, depending upon the particular tissue, can be a soft and jellylike material (as in the loose connective tissue underlying the skin), a very firm gel (as in cartilage), or an extremely hard substance (as in bone), and in which proteinaceous fibers may be embedded. The fibers may be of three types: **collagenous fibers,** which are composed of the protein *collagen* and are strong, flexible, and nonextensible; **elastic fibers,** which are made up of the protein *elastin* and are not unusually strong, but are resilient and can be stretched; and **reticular fibers,** which are delicate fibers, containing some collagen and polysaccharide, that form an extensive, intermeshing network.

### Types of Connective Tissue

Connective tissues can be broadly classified into three subgroups: (1) loose connective tissue, (2) strong, supporting types of connective tissue, and (3) blood, lymph and hemopoietic (blood cell-forming) tissue. The cells present in each kind of connective tissue are of various types, each specialized for particular functions, as discussed below.

***Loose Connective Tissue.*** Loose connective tissue is distributed widely throughout the body; in fact, with the exception of blood, it is the most pervasive of all tissues. It is termed "loose" because it is soft and pliable. These qualities are imparted to it by its intercellular substance which consists of loosely interwoven collagenous, elastic, and reticular fibers that are embedded within a gelatinous, semifluid material. Because loose connective tissue cannot withstand much strain without tearing, it is often supported by, and continuous with, other stronger types of connective tissue.

Loose connective tissue acts as a packing and binding material throughout the body. It is the connective tissue that underlies epithelial tissue and is associated with much muscle tissue. It surrounds, and extends into, many body organs, and serves as a loose sheath around glands and along the course of blood vessels and peripheral nerves.

In addition to its binding function, loose connective tissue plays an extremely important role in the transfer of materials between blood capillaries and other tissue cells of the body. The blood capillaries that bring oxygen and essential nutrients to the cells of all parts of the body and remove cellular waste materials course through the intercellular substance of loose connective tissue. As described earlier, these materials enter and leave the cells of the body by way of the tissue fluid that bathes the cells; the tissue fluid, in turn, permeates and bathes the intercellular substance of loose connective tissue, and hence is in contact with the capillaries in that tissue.

Loose connective tissue may contain a variety of cell types, and these particular types may vary in loose connective tissue from different parts of the body. The most numerous cells are **fibroblasts** (*fibro* = fiber; *blast* = forming), which secrete the fibrous and amorphous materials that constitute the intercellular substance of loose connective tissue. Other cell types include **macrophages,** which are large ameboid cells that ingest bacteria, viruses, and other foreign material, as well as dead or damaged cells; **plasma cells,** which produce antibodies that destroy disease-producing organisms; **mast cells,** which secrete the inflammatory substance *histamine* in response to tissue injury, and have a role in allergic responses; and certain other types of

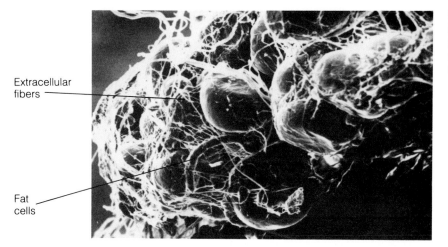

Extracellular
fibers

Fat
cells

FIGURE 4-5 Loose fatty connective tissue, also known as adipose tissue, as revealed in the scanning electron microscope. Note that the fat cells are embedded in an extensive network of extracellular fibers.

**leukocytes** (white blood cells), which, in response to tissue injury, may squeeze through the walls of blood capillaries and wander into surrounding loose connective tissue. (The function of all of these cell types will be discussed in more detail in later chapters).

**Fat cells,** capable of storing large amounts of fat in their cytoplasm, may also be present in loose connective tissue. In some cases, loose connective tissue consists almost entirely of fat cells; this specialized form of loose connective tissue is called **adipose tissue** (Figure 4-5). Adipose tissue, which is found under the skin and around various body organs, serves as a reserve food supply, as an insulator against heat loss, and as a protective cushion for the organs it surrounds. One place where adipose tissue normally exists is between the rear of the eye and the bony wall of the orbit (eye socket of the skull). In times of metabolic stress, this fat is utilized as a readily available source of energy, causing the eyes to sink further into their orbits and giving rise to the sunken-eyed appearance of an exhausted individual.

*Strong, Supportive Types of Connective Tissue.* There are three main types of strong, supporting connective tissues: (1) dense fibrous connective tissue, (2) cartilage, and (3) bone. These connective tissues consist of only a few cells, but a great deal of intercellular substance. This is the source of the tissues' strength. Thus, they have a particularly important function in supporting various parts of the body.

1. Dense fibrous connective tissue. **Dense fibrous connective tissue** consists primarily of collagenous fibers, although in some cases a good number of elastic fibers may be present. The few cells that are present in dense fibrous connective tissue receive their blood supply from capillaries that are embedded in loose connective tissue that is associated primarily near the periphery of the fibrous tissue mass. The nonliving intercellular substance, of course, requires no blood supply.

    Two common types of dense fibrous connective tissue are **tendons,** which join muscles to bones, and **ligaments,** which hold bones together, or bones and cartilage together, at joints. Tendons and ligaments have a glistening white color, imparted by the collagenous fibers that are arranged in compact parallel bundles that run more or less in the same direction and plane (Figure 4-6). This structural arrangement can withstand tremendous pulling forces exerted in the direction and plane of the fibers, but at the same time, the collagenous fibers impart an inelasticity to tendons and ligaments that makes them liable to be torn when stress is applied in other directions. The few cells in tendons and ligaments are virtually all fibroblasts and are located between the parallel bundles of collagenous fibers.

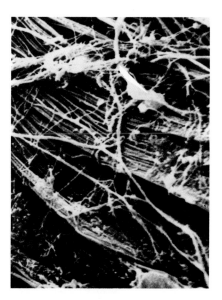

FIGURE 4-6 Tendons, which connect muscle to bone, consist of a dense, regular array of collagenous fibers, shown here lying beneath a small amount of loose connective tissue.

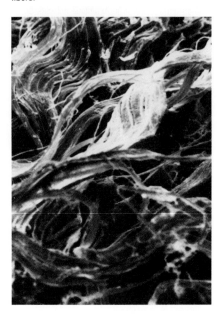

FIGURE 4–7 Irregular dense connective tissue, showing the meshwork of irregular interwoven fibers.

There is a second type of dense fibrous connective tissue, known as **irregular dense connective tissue,** in which the collagenous fibers, rather than being arranged in parallel bundles disposed in essentially one direction, are interwoven, sometimes in different directions or in different planes, or both (Figure 4–7). Unlike tendons and ligaments that can withstand strong unidirectional stretching, dense connective tissue with irregularly interwoven fibers can resist moderately strong stretching in multiple directions. Such sheets of tissue, called the **deep fascia,** form the membranous structures (capsules) that surround many internal organs; compose the sheaths in which the brain, spinal cord, and blood vessels are enclosed; envelop the body beneath the skin; and surround the muscles, often separating large masses of muscle tissue into bundles of separate muscles.

2. Cartilage. The second major category of strong, supporting connective tissue is **cartilage.** The cells of cartilage, called **chondrocytes,** are sparsely interspersed in little cavities within the material of the cartilage, and produce an intercellular substance that consists of an extremely dense network of collagenous fibers embedded in a firm gelatinous material. The intercellular substance of cartilage has some of the attributes of a firm, flexible plastic, strong enough to bear considerable weight. Furthermore, the cartilage surface that covers the ends of long bones that articulate with one another to form movable joints can, with natural lubrication, be very smooth and slippery, thus minimizing friction and wear.

Cartilage, with the exception of that which covers the articular surfaces of bones, is surrounded by a connective tissue envelope, called the **perichondrium,** composed of an outer fibrous layer and an inner vascularized layer capable of producing additional chondrocytes. Because cartilage does not contain blood vessels, nourishment and wastes must diffuse through the intercellular substance between the chondrocytes and the blood vessels in the inner layer of the perichondrium.

Cartilage grows in one of two ways. In **interstitial growth,** new chondrocytes proliferate within the cartilage from preexisting ones, and these produce additional intercellular substance. In effect, growth of the cartilage occurs from within its preexisting mass. A second type of cartilage growth, called **appositional growth,** occurs when chondrocytes proliferate in the vascularized layer of the perichondrium and synthesize intercellular substance that is deposited on the surface of existing cartilage, thus increasing its thickness. This is not unlike the process by which a tree increases in diameter as rings of cellulose are added beneath the bark.

Three types of cartilage, characterized by their fibrous organization, are present in the human body. These are *hyaline cartilage, fibrous cartilage,* and *elastic cartilage.*

**Hyaline cartilage,** the most abundant type, is composed of a network of delicate collagenous fibers embedded in a firm gel-like material. It is so named because of its frosty white appearance (*hyalos* = glass), which is familiar to anyone who has seen a fresh soup bone. Hyaline cartilage covers the articular surfaces of long bones at joints such as the knees and elbows. In addition, it makes up the rib cartilages and provides support for some structures of the upper respiratory passages such as the nose, the trachea (windpipe), and larynx (voice box) (Figure 4–8). In prenatal life, miniature models of most of the future bones are first formed from hyaline cartilage, which becomes mineralized as it grows during bone formation.

**Fibrous cartilage,** often found where tendons join some other cartilaginous structure, consists of bundles of collagenous fibers interspersed with sparse rows of very small chondrocytes. Fibrous cartilage forms the spongy, shock-absorbing discs between vertebrae, parts of the knee joint, areas where some of the pelvic bones fuse, and various other structures.

**Elastic cartilage** differs significantly in structure from other types of cartilage, since it contains some collagenous fibers embedded in a dense network of elastic fibers. Structures that must be stiff, yet elastic, are composed of elastic cartilage. These include the external ear, some parts of the larynx, and the epiglottis (a flap of cartilage that covers the entrance to the windpipe during swallowing).

3. Bone. The third primary type of strong, supporting connective tissue is **bone.** Its constituent cells, called **osteocytes,** secrete a collagenous intercellular substance that becomes mineralized, and hence, becomes extremely hard and rigid soon after it is formed. Like most cartilage, the outer surface of bone is covered with a connective tissue membrane called the *periosteum.* Further details of bone structure and the human skeleton will be discussed in Chapter 6.

*Blood Cells and Hemopoietic Tissues.* **Blood cells,** present in the circulatory system (including the related lymphatic system, to be discussed later) are a type of free connective tissue — free in the sense that the cells, except in the early stages of their development and proliferation, are not usually attached to one another, nor are they held in place by intercellular substance. Rather, blood (and related lymph) consists of cells suspended in an aqueous fluid that has been likened to the intercellular substance of other connective tissues, although the analogy is a poor one since the suspending fluid is not produced by the blood cells themselves. There are two types of blood cells: *red blood cells (erythrocytes: erythro* = red) and *white blood cells (leukocytes: leuko* = white). *Platelets,* important in blood clotting, are also present in blood, but are in fact not cells, only fragments of cytoplasm.

Blood cells are formed by the *hemopoietic (hemo* = blood; *poietic* = forming) *connective tissues,* of which there are two types: *myeloid tissue,* present in the bone marrow of human adults, and *lymphatic tissue,* present in lymph nodes, the thymus, and the spleen. These tissues will be discussed extensively in other chapters.

## Overview of Connective Tissue Function

Our discussion of connective tissues should make clear their pervasive nature and the host of functions they perform. But the most obvious function — their "connective" one — is exceedingly difficult for the beginning student to

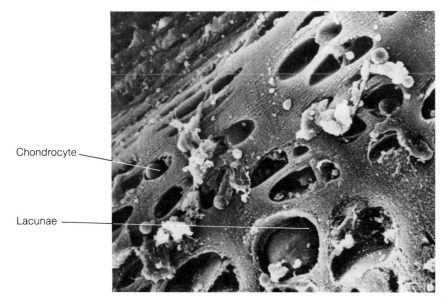

Chondrocyte

Lacunae

FIGURE 4–8 Hyaline cartilage showing the small cavities (lacunae) containing chondrocytes.

# Connective Tissue Diseases

**B**ecause connective tissue is found in every part of the body, diseases involving connective tissue can have widespread effects on a variety of physiological systems. Two well-known connective tissue diseases are *rheumatoid arthritis* and *systemic lupus erythematosus.*

Rheumatoid arthritis is a chronic inflammation of the connective tissue membrane that covers a joint (the synovial membrane, described in Chapter 6). As the inflammation progresses, it leads to irreversible damage to other parts of the joints, including the bones. The progressive destruction of the joints is accompanied by swelling and severe pain. Rheumatoid arthritis first affects the smaller joints in the hands and feet, but it usually spreads later to the larger joints of the body, such as the ankles, wrists, elbows, knees, and shoulders. The disease occurs more frequently in women than in men and shows an increasing incidence between the ages of 40 and 60. Children older than four can also be affected. It has been estimated that some 80 percent of the population experiences some discomfort from rheumatoid arthritis at some time in their lives.

We do not know the ultimate cause of rheumatoid arthritis, but we do know it is autoimmune in nature; that is, it occurs

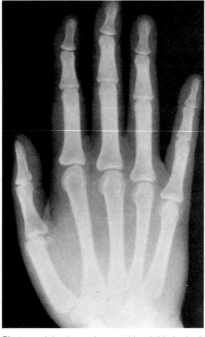
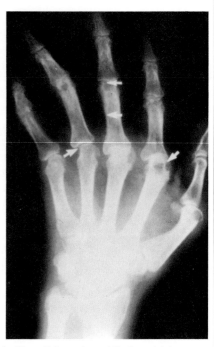

Photo on right shows rheumatoid arthritis in the hand, with multiple erosions of the finger joints (arrows), as compared to normal hand (left).

when the body's immune system, designed to recognize and destroy foreign invaders, begins to function abnormally, attacking tissues of the body itself. Recent evidence suggests that rheumatoid arthritis may occur in genetically predisposed individuals when a virus or some other environmental agent enters the joints, eliciting an immune response not only against the foreign agent, but against the joint tissue itself.

A related form of rheumatoid arthritis is rheumatic fever, which involves painful inflammation of the joints as well as of other body tissues, most notably the heart valves. Rheumatic fever begins several

conceptualize. The organization of the human body's connective tissues is a formidable topological problem. The human body is held together, *not* by a variety of separate, independently disposed connective tissues, but by different types of connective tissues that join together, interweave, and strengthen and nourish one another to form a fibrous framework that supports, and is essential to the functioning of, the entire body. For example, an extensive fibrous membrane system, composed of ordinary loose and adipose connective tissue, forms a *continuous sheet* throughout the body. This connective tissue membrane, called the **subcutaneous tissue,** or **superficial fascia,** attaches the skin to underlying structures such as the connective tissue that forms the deep fascia (whose myriad functions we discussed above), the perichondrium of cartilages, and the periosteum of bones. In order to appreciate how the human body is held together into a functional, three-dimensional structure, we must come to understand the organization of its connective tissues; this topic will consequently be emphasized in future discussions. For information on diseases of connective tissue, which can have profound effects on the body, please read *Focus: Connective Tissue Diseases.*

weeks after a sore throat or other infection caused by a particular strain of *Streptococcus* bacterium has seemed to clear up on its own. The *Streptococcus* bacteria that bring on rheumatic fever have molecules as part of their cell surface that are similar in structure to molecules in tissues of the joints, heart, and kidney, among others. Consequently, when the body's immune system gears up to destroy these bacteria by recognizing and attacking the foreign molecules on the bacterial cell surface, it will also attack the similar molecules present in the body's tissues, causing joint inflammation and other symptoms. Rheumatic fever is treated with antibiotics to eliminate *Streptococcus* and with anti-inflammatory drugs to reduce the inappropriate reactions of the immune system.

Another connective tissue disease of an autoimmune nature is systemic lupus erythematosus, also called SLE or lupus. In SLE, connective tissue in any part of the body may become inflamed and damaged. Although pain and inflammation in the joints are the most frequent symptoms, skin rashes are common as is inflammation of the kidneys, heart, and lungs, resulting in impairment of their function. SLE is considerably more common in females than in males, and attacks people most often in their twenties and thirties. Although SLE is a serious disease, its progression may be slowed or halted by drugs that suppress the activity of the immune system.

Other connective tissue diseases have been shown specifically to involve the collagen fibers that are present in virtually every type of connective tissue. One group of heritable disorders, known collectively as Ehlers-Danlos syndrome, is caused by defects in the synthesis of collagen fibers and results in extreme flexibility of the joints and hyperelasticity of the skin. Another heritable collagen disease, termed Marfan syndrome, is of special historical interest because Abraham Lincoln is thought to have had a mild form of it. We know that Marfan syndrome ran in Lincoln's family because one of his brothers died of severe manifestations of this disease. In addition, Lincoln's long limbs, fingers, and toes, and his loose-jointedness are characteristic of Marfan syndrome. Furthermore, there are reports that Lincoln had chronic heart trouble. This is also consistent with Marfan syndrome in which defects of the heart and blood vessels, especially weakening of the heart valves and walls of major arteries, are common.

The structural abnormalities associated with Marfan syndrome are due to a reduction in the tensile strength of collagenous connective tissue that supports affected structures. Some recent research has suggested that this reduced tensile strength may occur because the covalent crosslinkages that normally hold together and strengthen individual collagen molecules in a collagen fiber are defective.

Abraham Lincoln may have suffered from Marfan syndrome.

## Muscle Tissues

The third primary tissue is **muscle,** which is composed of specialized, contractile cells that cause movement or a change in the shape of some body part. Single muscle cells are elongated structures that are generally referred to as muscle fibers. What we normally think of as a functional muscle (e.g., the biceps muscle) is in fact groups of muscle fibers, tied together with connective tissue, supplied by blood vessels, and controlled by nerves.

The human body possesses three types of muscle tissues. These are *skeletal muscle, cardiac muscle,* and *smooth muscle.*

### Skeletal Muscle

**Skeletal muscle** occurs in bundles of fibers and is attached to bones, to facial skin, and to cartilage by tendons or other connective tissue. Skeletal muscle produces rapid, powerful contractions sustainable for short times. Normally, its activity is under conscious nervous control.

### Cardiac Muscle

**Cardiac muscle,** microscopically similar in many respects to skeletal muscle, is found in the walls of the heart and the pulmonary veins (veins that carry oxygenated blood from the lungs to the heart). It contracts spontaneously and rhythmically, and is adapted for steady activity, although it can sustain periods of very rapid and powerful contractions.

### Smooth Muscle

Found primarily in the walls of abdominal organs **(viscera)** and of blood vessels, **smooth muscle** is rather different in structure from the other two types. Its contraction, which is slow and wavelike, is not normally under conscious control; rather its activity is regulated by involuntary nervous control (although an individual can sometimes learn to control its contraction—the basis for the phenomenon of biofeedback). The contraction of smooth muscle moves material down the digestive tract in a sequential squeezing movement called **peristalsis;** smooth muscle in the walls of blood vessels is also involved in blood pressure maintenance.

### Nervous Tissues

**Nervous tissues** are composed of specialized, electrically excitable cells called **neurons.** Neurons are organized into extensive structural networks that receive, coordinate, integrate, interpret, and react to changes in the body's internal and external environment. As such, nervous tissues form the primary communication system of the body.

## ORGANS OF THE BODY

Tissues are building materials that combine to form various *organs* of the body. Composed of more than one tissue type, organs may have a comparatively simple structure, as in the case of some membranes, or they may be extremely complex, as in the case of internal organs.

### Membranes

The simplest organs of the body are membranes that are combinations of epithelial tissue and underlying connective tissue. Two such common membranes are *mucous membranes* and *serous membranes.*

### Mucous Membranes

Internal passageways of the body that open to the outside, such as the digestive, reproductive, and urinary tracts, and the respiratory passages, are lined by wet **mucous membranes.** The mucous membranes extend so as to cover the exposed openings of these passageways on the body surface (including the mouth, anus, nose, vagina, and so forth). The mucus, which is secreted by goblet cells in the epithelium or by glands that lie in the connective tissue layer of the membrane, helps protect the more delicate inner epithelial layer against drying, abrasion, and damage by noxious chemicals and other substances.

### Serous Membranes

**Serous membranes,** also composed of epithelium on the free surface and connective tissue below, cover various organs of the body and line body cavities that do not open to the outside (these cavities are described below). Epithelial cells at the surface of these membranes secrete a *serous* (watery and somewhat slippery) *fluid* that protects against friction when organs rub against each other or against the internal surface of a body cavity. For example, serous membranes surrounding the heart and others lining the cavity in which the heart resides permit the heart to move freely during its cycles of contraction and relaxation.

## Organs

In addition to membranes, tissues combine to form internal organs. The specialized cells that carry out the specific function, or functions, of the organ form what is called the organ's **parenchyma.** The parenchymal cells are supported by connective tissues that collectively are called the **stroma** of the organ. The organ is supplied by blood vessels (embedded, of course, in loose connective tissue) that provide nourishment and oxygen, and remove wastes; its activity is controlled by nerves, or sometimes by hormones, or both.

## HUMAN FORM AND BODY SYSTEMS

Externally, the human body has bilateral symmetry, which means that a plane running vertically through the midline of the body divides the body into identical right and left sides. The internal organs, many but not all of which are bilaterally arranged, can be grouped into functional units called **body systems.**

### External Form

The external form of the human body is shown in Figure 4–9, which indicates the various parts and regions of the body as a whole. The body is drawn in the so-called *anatomical position,* which serves, by convention, as the standard position of reference. A point of great confusion to the beginning student is the anatomical use of the terms "left" and "right." Notice in Figure 4–9 that handedness is defined with reference to the body itself, not to that of the observer. Thus, the body's left arm is on the observer's right side.

In future discussions of human structure and function, we will use repeatedly a number of anatomical terms to describe location and position. The most frequently used of these are shown in Table 4–1. It is important to note that these terms describe the position of one structure relative to another and do not indicate precise locations.

### Body Cavities

The human body, like the bodies of all vertebrates, contains two cavities—a *dorsal body cavity* and a *ventral body cavity*—in which many organs of the

TABLE 4–1 Definitions of Commonly Used Anatomical Terms

| | |
|---|---|
| Anterior (ventral) | Situated toward the front or abdominal side of the body; the nose is anterior to the ears. |
| Posterior (dorsal) | Situated toward the back side of the body; the ears are dorsal to the nose. |
| Superior | Situated in a higher position, toward the head end; the eyes are superior to the mouth. |
| Inferior | Situated in a lower position, toward the tail end; the mouth is inferior to the eyes. |
| Peripheral (external) | Situated toward the body surface; hair on the scalp is peripheral to the skull. |
| Internal | Situated in the interior, away from the body surface; the skull is internal to the hair on the scalp. |
| Proximal | Situated nearest the trunk or point of attachment; the upper arm is proximal to the hand. |
| Distal | Situated farthest from the trunk or point of attachment; the hand is distal to the upper arm. |

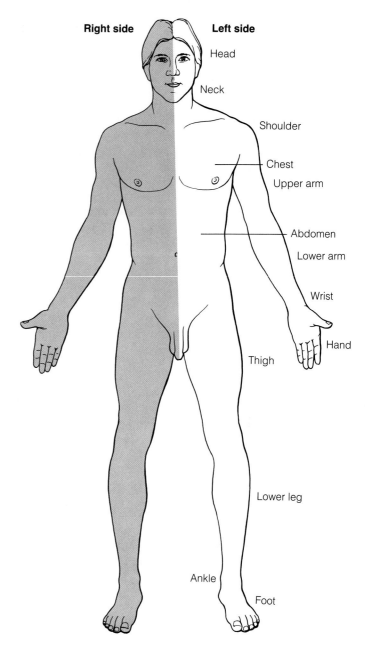

FIGURE 4–9 The external form of the human body, drawn in the standard anatomical position. Handedness is defined with reference to the body itself, not to that of the observer.

**Right side**     **Left side**

Head
Neck
Shoulder
Chest
Upper arm
Abdomen
Lower arm
Wrist
Hand
Thigh
Lower leg
Ankle
Foot

body are located and to which they are attached by connective tissues and other membranes (Figure 4–10). The **dorsal cavity,** located in the head region and extending down the back as a hollow tube, includes the cranial cavity, which contains the brain, and the spinal cavity, which houses the spinal cord and roots of the nerves extending from these organs.

The **ventral cavity** is subdivided by the muscular diaphragm into an upper *thoracic cavity,* or *chest cavity,* and a lower *abdominal cavity.* The thoracic cavity is subdivided into three compartments, separated by serous membranes. The central compartment, called the **pericardial cavity,** contains the heart, while two lateral compartments (the **pleural cavities**), each contains a lung. The abdominal cavity contains the organs of digestion, excretion, and reproduction. The lower most portion of the abdominal cavity, in which the end of the large intestine, the urinary bladder, and some reproductive organs are located, is sometimes called the **pelvic cavity.**

## A Brief Overview of the Systems of the Body

A system, as we have mentioned before, is a group of organs specialized to perform a particular function. It is commonly agreed that the body has 11 systems. We will discuss all of them in detail in later parts of the book, but the following will provide a brief overview to assist in understanding how the various systems integrate with one another.

## Systems

The **integumentary system** consists of the skin and its so-called appendages or accessory structures, which include hair, nails, and cutaneous glands such as sweat glands and sebaceous (oil-producing) glands. The integumentary system protects the body from loss of water and other critical materials, and from invasion by bacteria and viruses. It plays a significant role in temperature regulation and in sensing changes in the external environment.

The **skeletal system,** consisting of bones, cartilage, and other dense fibrous connective tissues, provides support for the body and protection of critical parts. The skull, for example, protects the brain and its related sense organs. Closely associated with the skeletal system is the **muscular system,** formed by the skeletal muscles that attach to movable parts of the skeletal system, allowing coordinated movements of the entire body.

The **circulatory system** consists of the heart and all vessels that circulate blood and lymph fluids through the entire body. We have discussed the role of the circulatory system in supplying nutrients and oxygen to cells of the body, and in removing wastes from them. Cellular elements in both the blood and lymph are also important in protecting the body from disease, a function performed by the **immune system.**

The **respiratory system,** including the lungs and the hollow passages through which air reaches the lungs, is involved in supplying the blood with oxygen and removing carbon dioxide and other volatile wastes from the blood.

Food and water is made available to the body through the **digestive system.** The digestive system consists of a long muscular tube, beginning at the lips and ending at the anus, in which food is digested and through which nutrient materials are absorbed into (enter) the blood. The digestive system is usually defined to include a number of associated glandular structures, such as the salivary glands, pancreas, liver, and gall bladder, that empty their secretions into the digestive tube proper.

The kidneys, along with the bladder and tubes involved in the evacuation of urine from the body, constitute the **urinary system.**\* Although the main

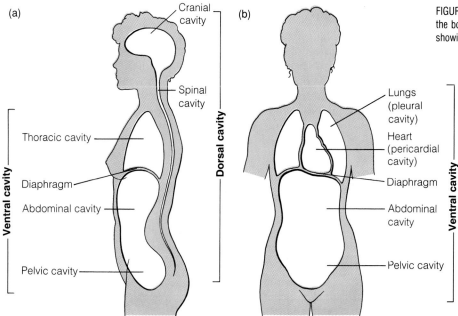

(a)

(b)

FIGURE 4–10 (a) Ventral and dorsal cavities of the body in lateral view. (b) Anterior view showing subcompartments of the thoracic cavity.

function of the urinary system is to filter water-soluble metabolic waste products out of the blood, the kidneys also play an important role in maintaining water and salt balance in the body.

The primary means by which bodily activities are controlled is through the **nervous system,** which consists of the brain, spinal cord, peripheral nerves, and special sense organs. The role of the nervous system and its importance in the intellectual development of humans has been mentioned previously.

All the endocrine glands of the body compose the **endocrine system.** There are many endocrine glands, and their secretions, which enter the bloodstream to be carried to target organs or tissues on which they have their effect, act to coordinate the metabolic activities of a wide variety of organs, and are intimately involved in the growth and development of the body.

The **reproductive system** is involved with the production of offspring. In the male, the primary reproductive organs (which are also endocrine glands) are the testes, which produce sperm cells. Accessory exocrine glands, ducts, and an external penis are involved in transferring viable sperm to the reproductive tract of the female. Female reproductive organs are the ovaries (which are also endocrine glands), which produce egg cells, the uterine (Fallopian) tubes, the uterus, and other external structures such as the mammary glands, which are important in providing nourishment for the young.

## Interdependence of the Systems of the Body

It is important to emphasize that the activities of all systems of the body are integrated in a complicated manner designed to optimize survival and eventual reproduction of the individual. As such, the activities of each system depend upon the activities of every other system, and we therefore cannot hope to understand the functioning of any one system in isolation from the others. As an illustration of this point, let us consider the interrelationship of the circulatory and digestive systems. The cells forming the tissues and organs of the circulatory system are completely dependent on the digestive system for obtaining water and essential nutrients from the external environment of the organism. In a similar way, the digestive system cannot operate independently of the circulatory system, which distributes materials supplied by the digestive system to all cells of the body, *including those of the digestive system itself.* Such an example, of course, does not begin to describe the integration of all of the 11 systems of the body. Indeed, as we study each system, it is imperative that we understand how it integrates with all other systems to form a functioning individual.

## Study Questions

1. Explain what is meant by the integration of cellular activities.
2. Describe the role of tissue fluid (interstitial fluid).
3. List the four basic tissue types of the human body.
4. What is the difference between simple and stratified epithelium?
5. Compare the origin of exocrine and endocrine glands.
6. Describe the basic structure of connective tissue.
7. a) List the three general categories of connective tissue.
   b) Give examples of each type and describe its structure.
8. Distinguish between interstitial and appositional growth of cartilage.
9. List the three types of cartilage in the human body and indicate one structure where each occurs.
10. Describe the primary function of connective tissue.
11. List two types of membranes and indicate their functions.
12. Describe the locations of each of the body cavities.
13. List the systems of the body and indicate the primary function of each.

*The urinary system is sometimes called the excretory system, although this latter designation is misleading because this system has nothing to do with the production or excretion of fecal wastes, a function of the digestive system.

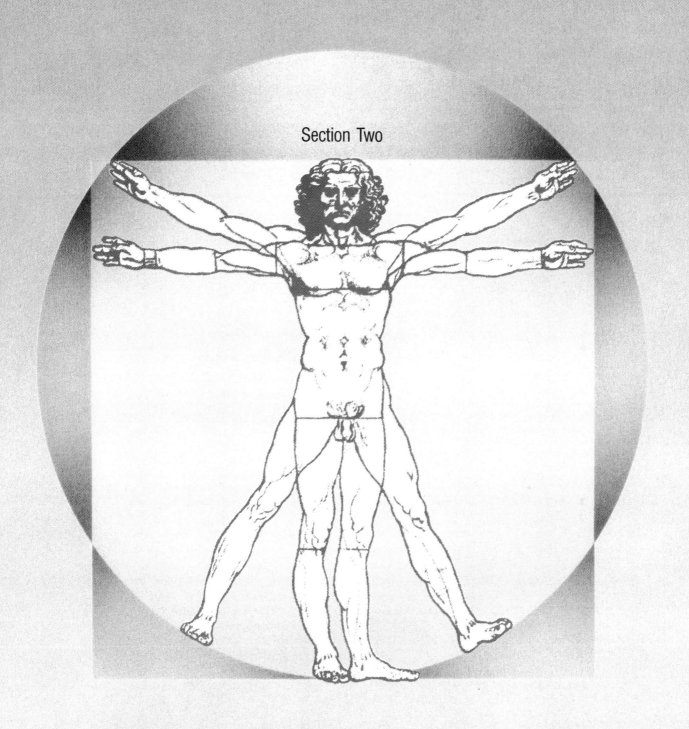

Section Two

# LINKING FORM AND FUNCTION

### Chapter Five

# THE INTEGUMENTARY SYSTEM

## STRUCTURE OF THE INTEGUMENTARY SYSTEM

The external surface of the human body is covered by **skin,** or **integument,** that consists of two layers of tissue: an outer layer of epithelial tissue, termed the **epidermis,** and an inner layer of connective tissue, called the **dermis.** During embryonic development, some of the cells that give rise to the epithelial tissue of the skin grow down into the developing dermal layer and give rise to exocrine glands (sweat glands and sebaceous glands), hair, and nails. These accessory structures that are derived from the skin are usually referred to as *appendages* of the skin, while the skin and its appendages grouped together form the *integumentary system.*

A diagram illustrating the structure of the integumentary system, including a typical cross-section of the skin and representative appendages, is shown in Figure 5–1.

## THE SKIN

The skin is the largest organ of the body. It functions as the body's protective covering, and although it is not hard or exceptionally tough, it forms an extremely effective barrier against penetration by disease-causing microbes. In addition, the skin possesses many nerve endings that are capable of eliciting a variety of sensations, including pain, temperature, touch, and pressure.

### Epidermis

The epidermis, forming the outer tissue layer of the skin, consists of stratified squamous **keratinizing epithelium;** "keratinizing" simply means that the outer layers of cells synthesize a fibrous protein called **keratin.** Keratin

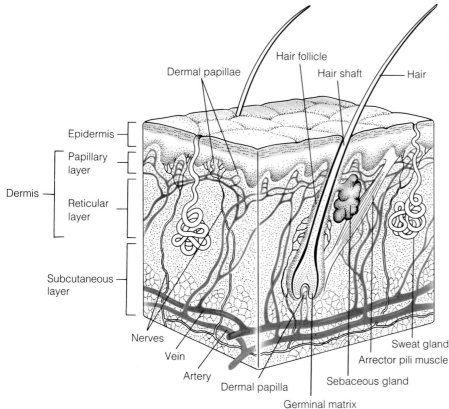

FIGURE 5–1 Diagram showing representative structures in hairy skin. The skin (epidermis and dermis) is attached to underlying subcutaneous tissue.

eventually fills the cells, turning groups of them into tough scales that form a nonliving surface layer firmly attached to the underlying living cells.

The cells of the epidermis must continuously proliferate to maintain the keratin layer as it is worn away or shed from the surface. This involves a process in which the division of cells in the deeper layers of the epidermis pushes overlying cells toward the surface. Because the epidermis contains no blood vessels and is nourished solely by tissue fluid from the dermal layer beneath it, cells nearest to the dermis receive substantial nourishment, while those in more external layers are removed from a source of nourishment. Consequently, as epidermal cells approach the surface of the skin, they lose a source of nourishment and become transformed into keratinized scales. Although these scales are shed from all skin surfaces, they are sometimes shed excessively from the scalp, where they are known as *dandruff*. Although the cause of dandruff is unknown, it usually can be controlled by the use of medicated shampoos.

It is largely the keratinized cells of the epidermal layer that give skin its protective qualities. They form a virtually germ-proof covering that protects the body from invasion by bacteria and other disease-producing organisms. In addition, keratinized cells form a waterproof barrier that prevents the loss of body fluids even in very dry atmospheres. Indeed, because keratin makes the skin impervious to water, it is possible for an individual to swim in fresh water without swelling or in salt water without shrinking! The keratinized cells also prevent the entry of some environmental chemicals, although recent research has shown that there are many chemicals that keratinized cells do not exclude, particularly organic ones that are likely to be carcinogens or have other toxic effects.

The thickness of the epidermis can vary considerably. The palms of the hands and soles of the feet are covered with an epidermis that has more cellular layers and a thicker layer of keratin than skin on the remainder of the body. The thickness and toughness of the epidermal layers of the palms and soles are, of course, appropriate to the increased abrasion that these surfaces must withstand.

Cells **(melanocytes)** in the deeper layers of the epidermis produce a pigment called melanin that may range in color from yellow, through various shades of brown, to black. The color and amount of melanin normally produced in the epidermis is responsible for differences in the skin color of various races. In some individuals with white skin, patches of cells normally produce melanin; these patches are called *freckles*. Exposure to ultraviolet light increases the amount of melanin synthesized in epidermal cells, giving rise to a "sun tan" that protects the skin from burning by ultraviolet light. The increase in melanin production is more pronounced in individuals with lighter than with darker skin. In all races, there is a hereditary condition in which melanin cannot be synthesized because of a missing enzyme, resulting in skin, hair, and eyes that lack pigment; such individuals are termed **albinos** (Figure 5–2).

Ultraviolet light induces the synthesis not only of melanin in white skin, but also of vitamin D in skin of any color. Children kept out of sunlight who do not have an adequate dietary source of vitamin D will develop *rickets,* a condition in which bone tissue is not properly calcified (see Chapter 6). Rickets is most common in cold climates. Excessive exposure to ultraviolet light from the sun and other sources may cause skin cancer, as discussed in Chapter 20.

## Dermis

The epidermis is firmly attached to the underlying *dermis,* which consists of two layers of collagenous loose connective tissue that merge with one another without clear delineation. The outer layer is much the thinner of the two and is called the **papillary layer** because its surface forms small, fingerlike projections **(papillae).** On the palms of the hands and soles of the feet, the papillary layer has particularly numerous papillae; the pattern that the papillae form is reflected on the surface of the epidermis as it follows their contours, giving rise

FIGURE 5–2 This albino boy lacks the heavy skin pigmentation characteristic of blacks.

(a)

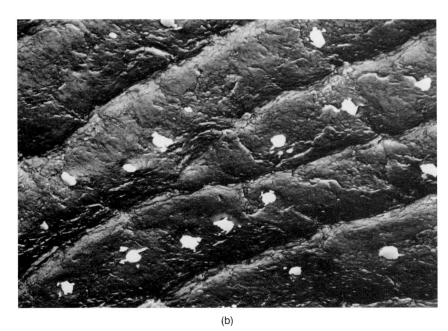

(b)

FIGURE 5-3 (a) Although no two individuals, even identical twins, have identical fingerprints, there are three basic fingerprint patterns, as shown here. From left to right: whorls, loops, and arches. (b) Scanning electron micrograph showing fingerprint ridges, formed by the epidermis following the contours of the dermal papillae. The white areas are openings of sweat glands.

to fingerprints and other ridges on the palms and soles (Figure 5–3). Beneath the papillary layer is the **reticular layer,** formed of a loose connective tissue that contains a somewhat more dense network of collagenous fibers than does the papillary layer.

The papillary and reticular layers vary greatly with regard to their blood supply. Arterioles (small arteries) in the subcutaneous tissue immediately beneath the dermis send branches that course through the reticular layer, supply hair follicles and cutaneous glands, and break up into an extensive capillary network in the papillary layer (see Figure 5–1). Thus, the dermis contains capillaries only in those regions that are in close proximity to epidermal cells, which require extensive nourishment for their rapid growth and replacement. The dermis itself, being largely composed of nonliving intercellular substance with a few fibroblasts and fat cells, does not require an extensive capillary supply. In addition to providing nourishment for the epidermis, the capillaries in the papillary layer function in temperature regulation, as will be discussed later in this chapter.

The skin is an important sense organ, and the dermis contains numerous nerve endings that carry sensations of touch, pain, heat, cold, and pressure *to* the **central nervous system** (the brain and the spinal cord). In addition,

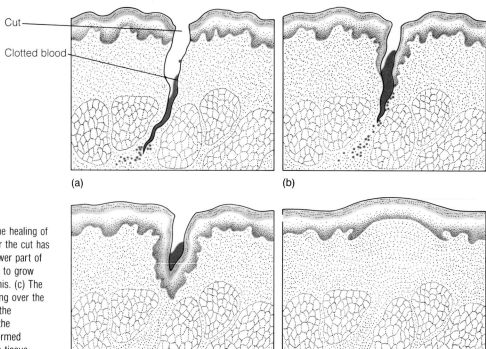

Cut

Clotted blood

(a)

(b)

FIGURE 5-4 Successive stages in the healing of a cut in human skin. (a) Shortly after the cut has been made, clotted blood fills the lower part of the wound. (b) The epidermis begins to grow down over the cut edges of the dermis. (c) The epidermis forms a continuous covering over the dermis. Meanwhile, fibroblasts from the subcutaneous tissue begin to repair the connective tissue below the newly formed epidermis. (d) As the new connective tissue grows, it pushes the epithelium upward toward the surface.

(c)

(d)

nerves in the dermis carry impulses *from* the central nervous system to arterioles, sweat glands, and smooth muscle associated with the hair follicle.

The epidermis and the dermis are firmly attached to one another by the basement membrane (an epithelial-derived structure, mentioned in Chapter 4). Excessive rubbing of the surface of the skin may mechanically separate the basement membrane of the epidermis from the dermis; in some cases, tissue fluid may collect between the two separated layers, forming a surface bubble known as a *blister*.

## Subcutaneous Tissue

The skin (epidermis and dermis) is attached to underlying **subcutaneous tissue** (also called the superficial fascia; see Chapter 4). The dermis is anchored to the subcutaneous tissue by irregularly spaced bundles of collagenous fibers that extend between the two layers. The subcutaneous tissue can accumulate considerable fat, which serves to insulate the body from excessive heat loss.

The skin and the subcutaneous tissue to which it is attached provides a strong, elastic and flexible covering for the body. These properties are especially apparent during pregnancy, obesity, or when swelling of tissue occurs. Sometimes excessive stretching will tear the skin, forming little silvery scars, evident particularly as the "stretch marks" of pregnancy. When an individual ages, the skin becomes less elastic and fat tends to disappear from the subcutaneous tissue, causing wrinkled skin. Recent research has suggested that the loss of skin flexibility during aging may be due to an increase in the amount of collagen present within the skin as well as to the formation of chemical bonds between collagen molecules, which prevent stretching of collagenous fibers.

## Healing of a Skin Wound

Following an accidental cut or surgical incision, a gap *(wound)* is left in the skin. The first step in healing is the formation of a blood clot, which prevents further loss of blood and tissue fluid from the wound. This is followed by the proliferation of the epidermis at the edge of the cut and its growth down the sides of the wound; eventually, new epidermis becomes continuous over the wound's surface. Fibroblasts and capillaries from the subcutaneous tissue—not from the dermis, since it has few cells and blood vessels—begin repairing the connective tissue by invading the area at the bottom of the epithelial-lined cleft.

As new connective tissue is laid down, the bottom of the epithelial-lined cleft is pushed toward the surface, and the area previously occupied by the cleft is filled in with connective tissue. This is covered by the epidermis, which had previously lined the cleft. Because the epidermis is increased in area, a bulge is formed; moreover, the surface of the bulge is smooth because the epidermis is not underlaid by papillae, which normally endow it with an uneven surface. This smooth, bulging tissue is called a *scar.* Scarring can be diminished by suturing the edges of the epidermis together, thus preventing the unrestricted growth of underlying connective tissue. The process of wound healing is illustrated in Figure 5–4.

## APPENDAGES OF THE SKIN

The appendages of the skin include hair, nails, and several types of exocrine glands, referred to as cutaneous glands. A brief discussion of each follows.

### Hair

**Hair follicles,** from which hair grows, develop from down-growths of epidermis into the underlying dermis or subcutaneous tissue. They are present on most parts of the body except the soles of the feet, palms of the hands, external genitalia, and nipples. In the adult, many of these follicles produce very fine hairs that are barely visible, while others give rise to the coarser hairs that are present in characteristic patterns on the body.

The structure of a hair and hair follicle is shown in Figure 5–1. The down-growth of epidermis forms a sheath of epithelial cells that lines the follicle; the deepest part of the down-growth forms a cluster of cells called the **germinal matrix,** because it gives rise to (germinates) the hair. The germinal matrix covers a small papilla of connective tissue, which is supplied by capillaries that bring nourishment to the cells of the germinal matrix.

Growth of hair is a process analogous to that of the proliferation of keratinizing epidermis of the skin. As the cells of the germinal matrix proliferate, the uppermost cells are pushed up the follicle sheath. As they move away from their nourishment source in the papilla, they become keratinized. The continuation of this process of cell proliferation and keratinization gives rise to hair growth. Contrary to popular belief, shaving or cutting the hair has no effect upon its growth, nor does it cause *baldness* (see *Focus: Baldness*).

The structure of hair depends upon the configuration and internal diameter of the hair follicle. Hair is curly if the follicle is bent, and straight if the follicle is straight. Coarse hair has a shaft diameter that is considerably greater than that of fine hair.

The color of hair depends on the quality and quantity of melanin, and sometimes other pigments, it contains. These pigments are produced by cells located at the tip of the dermal papilla that donate their pigment to the cells forming the outer shaft of the hair. Pigment synthesis often diminishes with age, giving rise to white hairs that contain no pigment. When white hairs intermingle with pigmented ones, the hair appears to be gray.

Most hair follicles are not perpendicular to the surface of the skin, but slant somewhat (see Figure 5–1). A small bundle of smooth muscle fibers, forming the **arrector pili muscle** (erector of the hair), is attached to the connective tissue on the side of the follicle toward which the hair slants; the other end of the muscle is attached to the papillary layer of the dermis, a short distance away

# Baldness

"**U**gly is a field without grass, a plant without leaves or a head without hair," wrote Ovid, the Roman poet. Many bald or balding men who spend millions of dollars every year on tonics and other worthless treatments for baldness would apparently agree.

Baldness may occur in both men and women. However, it is quite rare in women; when it occurs, it is usually caused by scalp infections, malnourishment, anemia, and various drugs; it is often reversible when the underlying cause is corrected. These factors may also cause some cases of baldness in males, but over 90 percent of male baldness is an inevitable result of heredity. In hereditary baldness, the hair follicles undergo progressive degenerative changes, shrinking in diameter and producing successively thinner hairs. The degeneration of the hair follicles is a preprogrammed event, intrinsic to the follicles themselves and not caused by an unfavorable environment in the scalp. This fact has been demonstrated by transplantation experiments. Skin transplanted from the scalp to other areas of the body loses its hair at the same time as does the scalp, whereas skin transplanted from other sites to the scalp never loses its hair.

The factors that initiate the degenerative changes in hair follicles are not well understood, but it is clear that male sex hormones (androgens) are involved. Castrated men, hereditarily disposed to baldness, do not become bald, but will become so if they are injected with androgens. It thus appears that baldness occurs as a result of a complex interaction between an individual's heredity and the hormones he produces.

To date, treatments for baldness have been largely unsuccessful. Estrogens can sometimes stimulate hair growth, but they can result in feminization and have been linked to an increased risk of cancer. The drug Minoxidil, used to treat severe cases of high blood pressure, can stimulate hair regrowth in some people when rubbed into the scalp. However, the stimulatory effect occurs only in men who are not already completely bald, and even then it is unpredictable. Moreover, new hair is usually thin and fragile, and it falls out when the treatment is stopped.

Hair transplants can be done to replace some of the hair in the bald area of the scalp, but the procedure does not correct the underlying cause of baldness. In hair transplantation, small plugs of skin containing healthy hair follicles are removed from the side of the head with a cookie cutter-like device several millimeters in diameter; the plugs are then reinserted into the skin of the bald region by placing them into small depressions cut with the same device. The results of this procedure are not always satisfying, since the new hair grows in tufts only in the area of the transplant. The cosmetic effect is improved as greater numbers of plugs are transplanted, but this can become very expensive. For now, at least, perhaps the best treatment for baldness is to face up graciously to the inevitable.

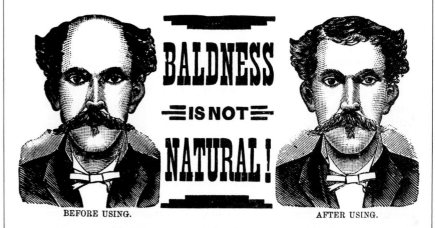

BEFORE USING.          AFTER USING.

Nature will restore the Hair if you will help it, even as a worn-out soil will grow good crops if you feed it with a proper fertilizer. If a farmer was to apply lime where stable manure was needed, and failed to get a crop, was it the fault of the soil or of knowledge on his part? If we have hitherto worked from a wrong principle and failed, is it any reason why success should not be reached? BENTON'S HAIR GROWER will **GROW HAIR, CURE DANDRUFF, and STOP FALLING HAIR. Price, $1.00 per Bottle, by mail free.**
*Address,* **BENTON HAIR GROWER CO., Brainard Block, Cleveland, O.**

from the point where the hair follicle opens to the surface. Contraction of the arrector pili muscle makes the follicle more perpendicular, and the hair "stands on end." In addition, the skin is pulled in over the site of the attachment of the muscle to the dermis, causing "goose flesh." The action of these muscles is controlled by the involuntary nervous system, and their contraction is stimulated by cold, fear, or apprehension.

Hair is sparse in humans and its function is limited. On the head, hair prevents sunburn and acts as an insulator, retarding excessive heat loss and heat gain. Hair on the eyelids and eyebrows protects the eyes from excessive light glare. Although these physiological roles of hair are not critical, it is nevertheless of crucial importance that skin possesses hair, since hair follicles provide an essential source of epidermis for repair of skin injured by burns and abrasion, and they also make certain types of skin-grafting possible (see *Focus: New Treatments for Burns*).

## Nails

Plates of hard keratin, called **nails,** cover the dorsal surface of the end of the fingers and toes. The initial step in their formation is the invasion of epidermis into the dermis near the area that will become the root of the nail. As the epidermal cells in the area of the invasion proliferate, the upper cells become keratinized into nail substance. The lower epidermal cells, on which the nail rests, form the **nail bed.** The nail grows as cells in the nail bed proliferate, pushing the nail forward as they become keratinized near the surface (Figure 5–5).

The nail is pink because of blood showing through underlying capillaries. A crescent-shaped white area ("moon") near the root of the nail is most prominent on the thumb and is usually absent on the little finger. This represents a region of the nail where the underlying epidermal cells form a particularly thick layer, preventing blood from showing through.

Fingernails grow about 0.5 mm per week, and toenails, considerably less. Both grow more rapidly in hot weather than in cold weather. Sometimes toenails become excessively curved, usually when an individual wears shoes that fit poorly, and grow into the skin at their edges. This condition is called an *ingrown toenail,* and is corrected by surgically removing the edges of the nail and underlying epidermal cells. This prevents the growth of new nail substance at the edges.

## Cutaneous Glands

Cutaneous glands include sebaceous glands (oil-producing glands), sweat glands, and ceruminous glands (wax-producing glands). All are derived from ingrowths of epidermal cells into the underlying dermis.

FIGURE 5–5 Diagram of a cross-section through a nail, nail bed, and underlying dermis.

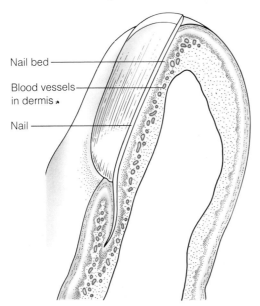

Nail bed

Blood vessels in dermis

Nail

## New Treatments for Burns

**B**urns are the third leading cause of accidental death in the United States, and the leading cause of accidental death among persons under 40. Of an estimated 130,000 Americans hospitalized each year for burns, some 10,000 die; patients with severe burns over two-thirds of their bodies have only a 50 percent chance of surviving. As grim as these statistics are, they do not even begin to speak of the severe pain, disfigurement, and psychological turmoil experienced by severely burned individuals.

A severe burn may destroy both layers of the skin and damage underlying tissue. If such burns are extensive, the most immediate dangers to life are fluid loss and infection. To minimize these dangers and to initiate the healing process, the wounds are covered when possible with grafts of the patient's own skin. This is done by removing thin slices of skin, including the epidermis and part of the underlying dermis, from unburned areas and suturing them over the wounds. Such grafts usually grow well in their new location; skin taken from the unburned areas regenerates a new epidermis as epidermal cells spread over the wound from its edges and from the epithelial lining of the remaining portions of hair follicles (see accompanying figure).

Unfortunately, severely burned patients do not usually have enough unburned skin to provide sufficient graft material. Temporary grafts of animal skin can be used, but these are soon rejected and must be removed in about a week. The most promising solution to this dilemma is the use of artificial skin.

Ioannis Yannas of MIT and John Burke of Massachusetts General Hospital have developed an artificial skin and pioneered its use in burn victims. The artificial skin is composed of collagen fibers extracted from cowhide that are covalently cross-linked into a meshwork with a polysaccharide extracted from shark cartilage. The

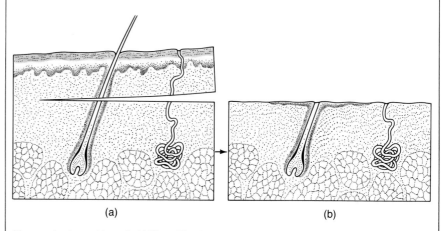

(a)　　　　　(b)

Diagram showing a skin graft. (a) The epidermis and part of the dermis are removed and transplanted over a burned site. (b) A new epidermis that grows up from hair follicles spreads over the site from where the graft material was removed.

### Sebaceous Glands

As the hair follicle develops, some cells forming the upper third of the follicle grow into the dermis and develop into **sebaceous glands** whose ducts open into the follicle at the site from which the outgrowth occurred. The sebaceous glands produce a lipid secretion called **sebum** that oils the hair and prevents the skin from becoming dry, chapped, or cracked.

The sebaceous glands are situated inside the triangle formed by the hair follicle, the surface of the skin, and the arrector pili muscle (see Figure 5–1). Consequently, contraction of the arrector pili muscle squeezes the sebum out of the gland, into the neck of the hair follicle, and onto the surface of the skin. Since cold is a stimulus for contraction of the arrector pili, additional sebum is expressed onto the skin in cold weather when the air tends to be driest.

Excess secretion of sebum at puberty can cause a blockage of the delivery of sebum to the skin surface and an accumulation of sebum within the skin, leading to inflammation and infection. This condition is known as *acne.*

meshwork, covered with a silicone sheet that helps retard fluid loss and prevent infection, is sutured into place over the wound. Fibroblasts from the subcutaneous tissue migrate into the collagenous network, forming a new dermis; as this occurs, the meshwork is slowly degraded by the growing tissue. In about three weeks, the silicone layer is removed and slivers of epidermis, obtained from unburned areas, are grafted over the dermis. The epidermis soon becomes confluent over both the grafted and source areas. This procedure has been used successfully in Boston and in other burn centers on hundreds of victims of severe burns.

In a recent experimental modification of this treatment, basal cells, the progenitors of the epithelial cells of the skin, are isolated from a small piece of skin and seeded into the meshwork. The dermis repairs as before, but in addition, the basal cells generate a new epidermis. When the silicone layer is removed in about two weeks' time, the wound is covered by a confluent epidermis and no epidermal grafts are required.

Another type of artificial skin has been developed by Howard Green of Harvard Medical School. By growing a patient's own epidermal cells in tissue culture, Green has been able to obtain sheets of epidermis several inches square that can be sewn directly over wounded areas. Although the skin lacks a natural dermis, it is fully functional and durable.

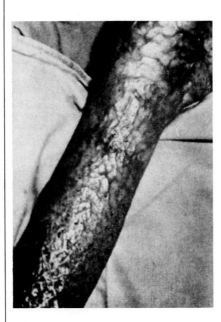 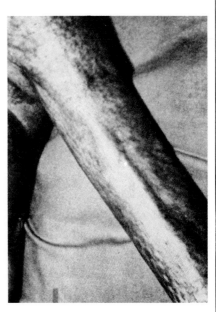

Photos of the shoulders and upper arms of a patient receiving conventional grafting to the right arm and shoulder (left photo), and artificial skin grafting to the left arm and shoulder (right photo), followed by epidermal grafting.

### Sweat Glands

These are simple tubular glands distributed over most of the body. The secretory part of the gland forms a tight coil, and the duct is a simple, unbranched tube that follows a spiral course through the dermis and epidermis (see Figure 5-1). There are about three million sweat glands in human skin; they are most numerous in the armpits and on the forehead and palms of the hands. Sweat glands produce a watery secretion, called sweat, which is important in the regulation of body temperature (see "Temperature Regulation by the Integumentary System").

### Ceruminous Glands

These are modified sweat glands that produce a waxy substance, termed **cerumen** (*cera* = wax), in the external canal of the ear. Cerumen protects the ear drum, but may impair hearing when it accumulates in great quantity.

## TEMPERATURE REGULATION BY THE INTEGUMENTARY SYSTEM

In addition to its other functions, the integumentary system has an important role in the regulation of the body's temperature. In humans, most heat is lost through the skin. Consequently, in order to maintain a constant body temperature, there must be mechanisms to minimize heat loss in a cold environment and to minimize heat gain in a hot environment or as a result of metabolic activity.

When body temperature begins to fall, the flow of blood through the capillaries in the papillary layer of the skin is reduced. If this mechanism fails to conserve enough heat to return the body temperature to normal, then shivering will eventually begin. The muscular activity involved in shivering will generate heat.

If body temperature begins to rise, blood flow through the skin is increased. However, in very hot environments or during strenuous activity, sufficient heat may not be lost by this mechanism. In these cases, sweat glands flood the surface of the skin with sweat. Since the evaporation of water requires relatively large amounts of heat (see Chapter 2), the evaporation of sweat removes additional heat from the body. Sweating, shivering, and the regulation of blood flow through the skin are all controlled by heat regulating centers in the hypothalamus of the brain.

## SKIN AND DISEASE

The condition and color of the skin are often an indication of an individual's general health. For example, the skin may become yellowed in a person with *jaundice*. Jaundice occurs because of the accumulation of bile pigments in the skin due to their improper breakdown; it is often an indication of liver malfunction or of certain endocrine diseases. Another symptom of ill health is *cyanosis,* a bluish or purplish coloration of the skin and mucous membranes. It occurs when blood does not possess normal levels of oxygen or when the passage of blood through the capillaries is greatly reduced. Consequently, cyanosis is an indication of circulatory or respiratory disease.

Because the skin is the most exposed part of the body, it is particularly susceptible to such injuries as *cuts, burns,* and *frostbite.* We have previously discussed burns and their treatment (see *Focus: New Treatments for Burns*), and the healing of cuts. Frostbite occurs when tissue is injured or destroyed by prolonged exposure to freezing temperature. Most commonly, fingers, toes, ears, and the nose become frostbitten, although other parts of the body are

FIGURE 5-6 Rash caused by allergy to poison ivy.

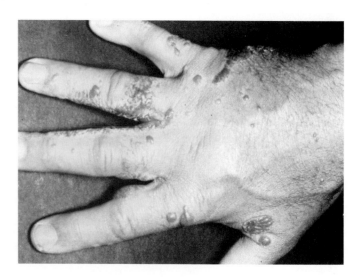

susceptible in severe cases. Although the treatment of frostbite has involved much controversy, it is now thought that gentle warming of the frostbitten area by immersing it in lukewarm water will result in the least tissue damage.

Many infections, diseases, and allergies affect the skin. Infections by staphylococcus, streptococcus, and pseudomonas bacteria commonly cause *boils* and other skin lesions, while *excessive dandruff production* and *athlete's foot,* characterized by cracks in the skin, are caused by fungal infections. The excessive production of sebum causes *seborrhea,* which results in oily crusts and scabs on the skin. Allergic responses to chemicals in foods, in medicines, and in the environment commonly result in characteristic rashes, such as *"hives"* (large raised, reddish welts on the skin), or the swelling, redness, itching, and blistering of the skin caused by contact with toxins in poison ivy and other plants (Figure 5–6). Characteristic rashes occur with measles, chicken pox, scarlet fever, herpes virus, and a wide range of other infectious diseases. Two chronic skin irritations of unknown cause are *eczema,* resulting in redness, burning, and itching; and *psoriasis,* characterized by silvery scales on elbows, knees, scalp, and trunk.

## Study Questions

1. **a)** Name the two tissue layers of skin.
   **b)** From which layer do exocrine glands, hair, and nails arise?
2. Summarize the process by which keratinized epithelial cells of the skin are formed.
3. Name several functions of the epidermal layer of the skin.
4. **a)** Name the two layers of the dermis.
   **b)** In which layer are capillaries present?
   **c)** Describe two functions of the capillaries in the dermis.
5. Summarize the process of wound healing.
6. Explain why hair follicles are crucial in the repair of damaged skin.
7. **a)** What is the function of sebaceous glands?
   **b)** Of sweat glands?

# THE SKELETAL SYSTEM AND JOINTS

The skeletal system consists of bones and cartilages that form the body's skeleton. The skeleton is a strong structural framework that supports the soft tissues of the body and, in the case of such structures as the skull and the rib cage, serves to protect certain critical internal organs. Muscles, ligaments, and tendons attach to the skeleton and by their actions, allow the body to move.

In addition to functioning in the support, protection, and movement of the body, the skeletal system has several other important physiological roles. For example, red blood cells, white blood cells and platelets are formed in the internal cavities of some bones. Also, bones contain large amounts of a mineral of calcium phosphate, which may be solubilized under certain conditions to form $Ca^{++}$ and $PO_4^{=}$ ions. Consequently, bones serve as a storehouse for these ions, which are required for many metabolic activities and may be released from bone to supply other tissues.

## BONE TISSUE

Mature bone is a connective tissue consisting of bone cells surrounded by large amounts of a hard intercellular substance. In fact, it is the intercellular substance that imparts to bones their characteristic properties of hardness, flexibility, and strength. The bone cells, called **osteocytes** (*osteo* = bone; *cytes* = cells), are responsible for maintaining the intercellular substance.

### Composition of Bone

The intercellular substance, or *matrix,* of bone has two components: (1) an organic material, primarily collagen, which forms the fibrous framework of bone and composes about 25 percent of a bone's dry weight, and (2) an inorganic material, chiefly a mineral of calcium phosphate, which is impregnated in the collagenous framework. By removing from bone one or the other of the two materials that compose its intercellular substance, it is possible to determine the function of each. If, for example, calcium phosphate is dissolved away by soaking a long bone in dilute mineral acid, the remaining collagen, which retains the shape of the original bone, can be tied in a knot; if, on the other hand, the collagenous material is burned away, the remaining mineral is hard, but extremely brittle. Thus, the organic material provides bone with its flexibility while the inorganic material contributes hardness and rigidity. Together, the two substances form a material which, for its weight, is comparable in strength to steel or reinforced concrete, but is considerably more flexible than both.

### The Formation of Bone Tissue

Bone is made by special bone-building cells called **osteoblasts,** which synthesize and secrete the organic portion of the intercellular substance. During the formation of bone tissue, a number of osteoblasts are connected to one another by thin cytoplasmic processes, or arms, that extend from the bodies of the cells. As the osteoblasts secrete intercellular substance, they eventually become completely surrounded by it. At this point, the osteoblasts are called osteocytes, cells specialized for bone maintenance. The organic intercellular substance then becomes impregnated with calcium phosphate, a process called *calcification.* Calcification turns the intercellular substance into a hard, stonelike material that entombs each osteocyte in a small chamber or space called a **lacuna** (pl: *lacunae*). Little channels, termed **canaliculi,** extend between lacunae; these correspond to tunnels formed by the deposition of intercellular substance around the cytoplasmic processes of the osteoblasts. By passing through the canaliculi, tissue fluid (see Chapter 4) can diffuse into

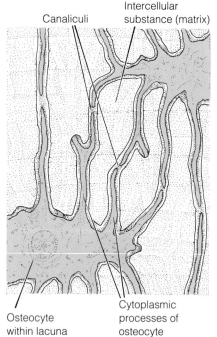

Canaliculi | Intercellular substance (matrix)

Osteocyte within lacuna | Cytoplasmic processes of osteocyte

FIGURE 6-1 Calcified bone is riddled with interconnecting cavities, called lacunae, in which bone cells (osteocytes) lie. Thus, bone is very much a living tissue.

the lacunae from regions near blood vessels that permeate the bone. The canaliculi thus serve an important function in allowing the transfer of materials between the circulatory system and the osteocytes. A simplified diagram showing a network of osteocytes in mineralized bone is shown in Figure 6-1.

## Bone's Dynamic Structure: Remodeling

Perhaps because of their strength and durability, it is a popular misconception that bones are static, unchanging structures. Nothing could be further from the truth. Bone is a living tissue and an extremely active one at that. During an individual's lifetime, skeletal bones undergo more or less continuous structural and metabolic changes, termed *remodeling.* Remodeling results from two processes: *resorption,* or dissolution of bone, and *deposition,* or laying down of new bone.

Structural remodeling of bone occurs not only during skeletal growth in youngsters, but also in individuals of all ages in response to the stress that bones must bear. Thus, for example, repeated movement that causes mechanical stress to a particular bone will result in the deposition by osteoblasts of intercellular substance in areas of the bone that will strengthen its structure in performing the movement. This deposition of new bone material may be accompanied by the resorption of intercellular substance in other areas of the bone that are under less stress. Such resorption of intercellular substance is mediated by bone cells called **osteoclasts,** which invade the bone and etch away the intercellular material. The repair of a bone fracture, which involves the resorption by osteoclasts of damaged bone and the deposition by osteoblasts of new bone material in the area of the break, is merely a special case of these remodeling phenomena.

In addition to the remodeling of bone in accordance with structural needs, bone is also remodeled in response to metabolic needs of the organism. Blood levels of calcium and phosphate ions must be maintained within narrow limits. There is a homeostatic mechanism, regulated by several hormones, that assures relatively constant blood levels of $Ca^{++}$ and $PO_4^{\equiv}$ by stimulating the deposition of these ions into bone when $Ca^{++}$ and $PO_4^{\equiv}$ blood levels begin to rise above acceptable levels, and bringing about bone resorption with the release of these ions into the circulation when blood levels begin to fall.

## Diseases Affecting Mineral Deposition and Resorption in Bones

A number of bone diseases result from disturbances in the normal processes of bone growth and bone remodeling. One such disease that affects children is *rickets,* in which the bones remain soft due to improper calcification. Weight-bearing bones, particularly those in the legs, often bow as a result. Rickets is caused by a lack of vitamin D, which is necessary for efficient absorption of minerals from the small intestine. Without vitamin D, the levels of calcium and phosphate ions in the blood remain low, retarding calcification. Normally, ample supplies of vitamin D are obtained by the conversion of a form of cholesterol in the skin into vitamin D by ultraviolet light from the sun. However, in northern climates where exposure to the sun may be minimal, particularly in winter, children often need supplementary vitamin D in their diet. In order to meet this need, milk is routinely supplemented with vitamin D.

Another disease, called *osteoporosis,* is characterized by abnormal resorption of bone. Osteoporosis is very common in postmenopausal women (see *Focus: Osteoporosis).*

## TYPES OF BONES

Based upon their shapes, the bones of the human skeleton have been classified into four groups: *long bones, short bones, flat bones,* and *irregular bones.* The long bones have a central shaft, called the **diaphysis,** with expanded ends, termed **epiphyses** (sing: *epiphysis*). The bones of legs, arms, fingers, and toes

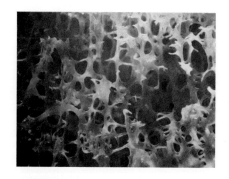

Spongy bone.

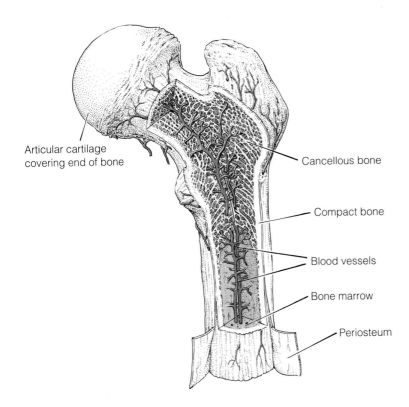

FIGURE 6-2 Structural features of a typical bone.

Articular cartilage covering end of bone

Cancellous bone

Compact bone

Blood vessels

Bone marrow

Periosteum

are long bones. Short bones, such as those in the wrist and ankles, have lengths that are similar to their widths. The bones that form the skull, ribs, and shoulder blades, among others, are flat bones. All other bones are irregular bones. Typical examples of irregular bones are the vertebrae that form the spinal column (see Figure 6-11).

## STRUCTURE OF BONE

The structural features of a typical bone are illustrated in Figure 6-2. A discussion of these features follows.

### Periosteum

In areas at their extremities where bones articulate with one another (come together at a joint), they are coated with articular cartilage. All other surfaces of bones are covered by a membrane called the **periosteum,** which is the counterpart of the perichondrium that covers cartilage (see Chapter 4). The periosteum consists of two main layers: an outer fibrous layer composed of dense, irregularly arranged connective tissue, and an inner layer containing *osteogenic cells*. Osteogenic cells give rise to osteoblasts and osteoclasts, which invade the bone tissue during the remodeling of bones. Blood vessels in the periosteum enter the bone at numerous places. Indeed, the blood supply to bone is so substantial that if the periosteum is stripped from a living bone, small bleeding points, marking the point of entry of blood vessels, cover the surface of the bone in all areas. The periosteum is also richly supplied with nerves; in fact, damage to the periosteum is responsible for most of the pain associated with fractures and bone bruises.

### Types of Bone Tissue

Every bone of the body is made up of two types of bone tissue. These tissues are *compact bone* and *cancellous,* or *spongy, bone,* which is more porous than compact bone.

# Osteoporosis

The elderly have a high incidence of broken bones as a result of relatively minor falls. Each year in the United States about 200,000 elderly individuals break their hips. Indeed, among people who live to the age of 90, about one-third of the women and one-sixth of the men will suffer a fracture of the hip. It has been estimated that between 15 and 20 percent of hip fracture patients die within months of their injury as a result of complications, and only about 25 percent recover without disability. There are also about 100,000 wrist fractures each year among the elderly. In addition, millions of people over the age of 45 suffer what are known as vertebral "crush fractures," a progressive compression of structurally weakened vertebrae that is responsible for the height loss and rounded shoulders often observed in aging individuals.

Most of the elderly who suffer from broken bones have a condition known as *osteoporosis*. In osteoporosis, bones decrease progressively in density and strength as a result of excessive bone resorption, with consequent demineralization and loss of matrix material.

Although everyone's bones begin to lose density after about the age of 40, the marked decreases in density associated with osteoporosis are most often seen in elderly women and bedridden patients. Furthermore, among those elderly women with osteoporosis, white and oriental women are the most likely to suffer from fractures since they have lighter bones, on average, than Black and American Indian women, and hence any bone loss is more serious.

Older women are more likely to develop osteoporosis than older men because the rate of bone loss among women accelerates after menopause to about double the rate seen in men. The bone loss in postmenopausal women has been traced to the lack of estrogen. Estrogen normally controls the rate of bone remodeling by maintaining a net balance between the rate of bone formation and resorption. However, in the absence of estrogen, the balance between these two processes is upset and resorption of bone predominates.

The majority of research indicates that bone weakened by osteoporosis cannot be significantly strengthened, and consequently, treatment of osteoporosis is aimed at minimizing further bone loss. High amounts of dietary calcium are usually prescribed to retard further demineralization of bone and to compensate for the reduced ability of postmenopausal women to absorb calcium from dietary sources. However, recent research has shown no correlation between calcium intake and bone loss when women consuming less than 500 mg of calcium per day are compared to women consuming more than 1400 mg per day.

Bone loss from osteoporosis in postmenopausal women can be halted, and sometimes reversed to a slight degree, by administering small doses of estrogen in conjunction with calcium supplements. Until recently, many physicians were reluctant to use estrogen to treat osteoporosis because of the association between estrogen use in postmenopausal women and an increased risk of uterine cancer. However, in the last few years evidence has accumulated that cancer risk is exceedingly low for postmenopausal women treated in cycles with small amounts of estrogen for three weeks followed by one week of progesterone. Furthermore, it is becoming clear that the severe crippling effects of osteoporosis are much more of a potential health problem than is uterine cancer, which is fully curable by hysterectomy (surgical removal of the uterus) if detected early by regular monitoring.

Because the effects of osteoporosis are, to date, largely irreversible, there is a growing emphasis on prevention. A good deal of evidence now indicates that peak bone mass in adults is directly related to the amount of calcium consumed during childhood and adolescence. Consequently, it is recommended that adolescents consume 1500 mg of calcium a day, the equivalent of drinking one and one-third

## Compact Bone

**Compact bone** is an extremely dense bone type, much like ivory in texture, that forms the hard exterior surface of all bones and most of the shaft of long bones. Its strength makes it particularly well adapted for providing support, and it can resist high levels of stress.

The microscopic structure of compact bone is shown in Figure 6-3. It consists of cylindrical units of bone tissue, known as **Haversian systems,** that run parallel to each other and to the long axis of the bone. Each Haversian system is composed of layers, or **lamellae,** that form concentric rings of bone tissue around a central channel, called a **Haversian canal.** The Haversian canals, which communicate with one another through cross-channels, serve as

quarts of whole or skim milk or eating seven and one-half ounces of hard cheese. In spite of the evidence showing a lack of correlation between bone loss and calcium consumption in women, the Food and Drug Administration (FDA) recommends that men and premenopausal women ingest about 1000 mg of calcium a day and that postmenopausal women should increase their calcium intake to 1500 mg a day.

A 12-year study conducted at the Mayo Clinic in Rochester, Minnesota, has suggested that sodium fluoride, given in conjunction with calcium supplements, may reverse the effects of osteoporosis by stimulating bone formation. Not only does osteoporotic bone become more dense after prolonged fluoride treatment, but some women developed low fracture rates after the first year of therapy. If these results are confirmed in further study, fluo-

ride may become an important treatment in osteoporosis.

There is also evidence that regular exercise can slow bone loss, presumably because stressing the bones leads to bone formation. Both smoking and drinking several alcoholic drinks a day can double the risk of osteoporosis, so that stopping these activities is also beneficial.

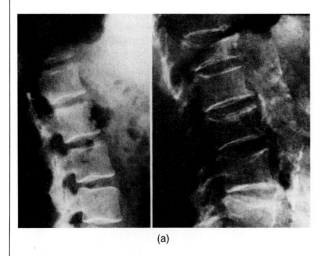

(a)

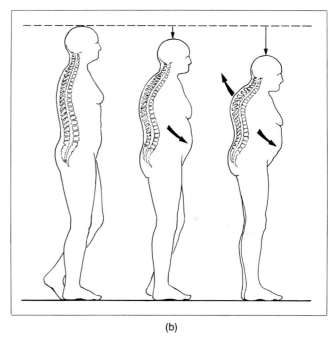

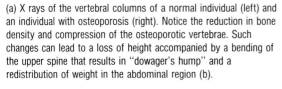
(a) X rays of the vertebral columns of a normal individual (left) and an individual with osteoporosis (right). Notice the reduction in bone density and compression of the osteoporotic vertebrae. Such changes can lead to a loss of height accompanied by a bending of the upper spine that results in "dowager's hump" and a redistribution of weight in the abdominal region (b).

(b)

conduits for blood vessels and nerves from the overlying periosteum. Osteocytes are embedded at more or less regular intervals in lacunae between the lamellae, and the lacunae are connected with each other and with the central Haversian canal by canaliculi, which allow tissue fluid to reach the osteocytes. The irregular spaces that are left between Haversian systems are filled in with curved lamellae.

Cancellous Bone

**Cancellous bone** (*cancellous* = chambered) is not nearly as dense as compact bone. It consists of a latticework of bone tissue that forms a porous, coral-like structure (see Figure 6-3). The interconnecting struts of bone that

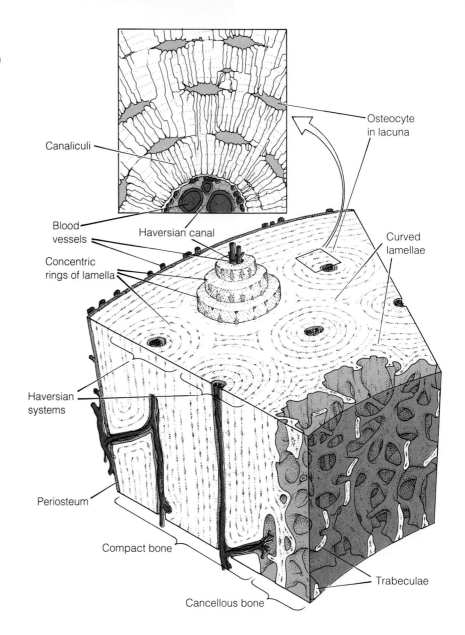

FIGURE 6-3 Diagram showing the microscopic structure of compact and cancellous bone. Osteocytes in lacunae (as shown in Figure 6-1) lie between lamellae in compact bone and are dispersed irregularly in the trabeculae of cancellous bone.

Canaliculi

Osteocyte in lacuna

Blood vessels

Haversian canal

Curved lamellae

Concentric rings of lamella

Haversian systems

Periosteum

Compact bone

Trabeculae

Cancellous bone

form the latticework are called **trabeculae.** Osteocytes are dispersed irregularly in lacunae within the trabeculae, and the space surrounding the trabeculae is filled with bone marrow. Blood vessels pass into the cancellous bone through minute orifices in the compact bone and run between the trabeculae. Tissue fluid derived from these vessels enters the canalicular network connecting the lacunae of the osteocytes.

Cancellous bone is not as strong as compact bone, but its porous structure is ideally suited to withstand compression and tension (stretching). Consequently, cancellous bone cushions forces of impact generated in such activities as walking, and resists effectively the downward drag of muscles and organs, which would otherwise crack less resilient bones.

## Bone Marrow

The cavities of bones are filled with bone marrow. In the fetus, most of the bones contain **red marrow,** which is the site for the production of red blood cells, white blood cells, and platelets. During the growing period following birth, some of the red marrow is invaded by fat cells, giving rise to **yellow marrow,** which functions as a fat reserve. In the adult, red marrow is present in the cancellous bone of the ribs, the sternum (breast bone), the skull bones, the bodies of the vertebrae, the epiphyses of long bones, and some of the short bones; all other bone cavities are filled with yellow marrow. In times of prolonged and serious anemia requiring increased red blood cell production, some of the yellow marrow can be converted back to red marrow.

## THE DEVELOPMENT OF BONE DURING PRENATAL LIFE

During prenatal development, bone is formed from connective tissue in one of two ways. *Intramembranous ossification* (ossification is the process of bone formation) occurs when a pre-existing fibrous membrane is changed into bone. In *intracartilaginous* (or *endochondral*) *ossification*, cartilage becomes changed into bone. It is important to remember, however, that these terms refer only to the method by which the bone starts to develop; the actual bone tissue that is formed is the same in both cases.

### Intramembranous Ossification

The bones that form the vault of the skull and some of the bones of the face are formed by **intramembranous ossification.** First, fibrous membranes appear in the region where the bones will form. Clusters of cells within the membrane turn into osteoblasts, which lay down cancellous bone by the bone-building process described previously (Figure 6–4). On its outer surfaces, the cancellous bone is converted into compact bone. This occurs by the deposition of new layers of bone material on the trabeculae, a process that gradually fills the spaces between the trabeculae with bone. The final result of this conversion is the formation of a bone that has an interior of cancellous bone and an exterior surface of compact bone.

The fibrous membranes that will become ossified to form the bones of the vault of the skull continue to expand and to grow together while ossification is proceeding during prenatal life. In fact, these bones are still not ossified at their edges when the baby is born. The unossified membranes that connect the bones forming the skull vault in the newborn infant are called **fontanelles** (Figure 6–5). Once all the fontanelles become ossified by the middle of the second year of life, the skull bones are connected by immovable joints, termed **sutures,** formed by thin layers of connective tissue. It is not known whether the vault of the skull enlarges during postnatal life by adding bone at the sutures or by adding bone on the outer surfaces (and resorbing bone on the

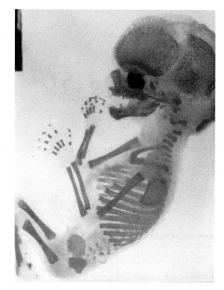

FIGURE 6–4 The formation of bone in a human embryo is revealed by a dye that specifically stains bone. Intramembranous ossification is apparent as lacy areas in the skull region. Most other bones are formed by the ossification of cartilage.

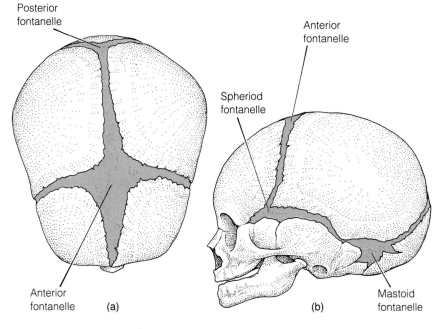

Posterior fontanelle

Anterior fontanelle

Spheriod fontanelle

Anterior fontanelle  **(a)**

**(b)**

Mastoid fontanelle

FIGURE 6–5 The fontanelles of a newborn infant. (a) Superior and (b) lateral view.

inner surfaces) of the bones that form the vault. In any case, the sutures do not become completely ossified until old age.

## Intracartilaginous Ossification

Most bones of the body are formed by **intracartilaginous ossification.** In this process, miniature cartilaginous models of what will become the bones are formed first, and the cartilage is then replaced by bone. Most of the models have developed by the end of the seventh week of prenatal life, when the embryo is approximately 20 mm long.

Intracartilaginous ossification is best studied in long bones. The process begins by the third month of fetal life when chondrocytes (cartilage cells) in the central region of the shaft of the model degenerate. The degeneration of the chondrocytes brings about the break up of intercellular substance surrounding the chondrocytes, which leaves extensive cavities in the model. At about the same time that the breakdown of cartilaginous substance is occurring, cells in the perichondrium (the membrane that covers the cartilage) turn into osteoblasts, which lay down bone tissue in a collar around the central area of the shaft. At this point, the perichondrium is called the periosteum, since it is now capable of producing bone tissue. Osteoblasts and blood vessels in the periosteum invade the degenerating area of the cartilage, and bone is laid down in this area through the process described at the beginning of this chapter.

A continuation of these processes results in a gradual replacement of cartilage by bone toward both ends of the shaft. Notice in Figure 6–4 that the bones of the arm have not yet ossified at their ends. Near the time of birth, the degeneration of chondrocytes begins in both epiphyses, and blood vessels and osteoblasts invade these areas, forming cancellous bone. However, a region of cartilage, called the **epiphyseal plate,** remains between the bone that has formed in the shaft at the so-called *primary ossification center* and at both ends, the *secondary ossification centers.* The epiphyseal plate is a temporary union between parts of the growing bone and is the region where bone elongation occurs as the child grows, as described in the following section (Figure 6–6).

## Growth of Long Bones

During prenatal life and in the postnatal period until the attainment of adult size, long bones are growing more or less continuously in both width and length (see Figure 6–6). Long bones grow in width by the deposition of

FIGURE 6–6 Intracartilaginous ossification and growth of long bone. Pink areas represent cartilage, and white areas represent bone. (a) Primary ossification center in cartilage model; (b) and (c) ossification proceeds toward both ends; marrow cavity begins to be hollowed out; (d) ossification centers appear in epiphyses; (e) entire bone is ossified, except for epiphyseal plates and articular surfaces.

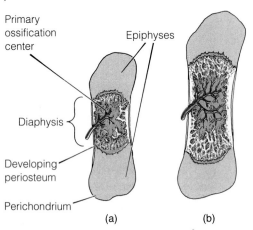

Primary ossification center

Epiphyses

Diaphysis

Developing periosteum

Perichondrium

(a)

(b)

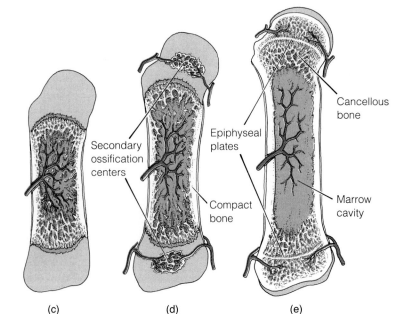

Secondary ossification centers

Epiphyseal plates

Compact bone

Cancellous bone

Marrow cavity

(c)

(d)

(e)

successive layers of compact bone beneath the periosteum. While this new bone material is being laid down by osteoblasts, osteoclasts are resorbing bone tissue in the central area of the shaft, creating the marrow cavity. These two processes, occurring simultaneously, bring about a widening of the bone and an increase in the size of the marrow cavity. The tubular shape that is thus attained has maximum strength with minimum weight.

Long bones increase in length by expanding in the areas of the two epiphyseal plates. As cartilage cells proliferate near the epiphyseal surface of the plates, other cartilage cells degenerate, and bone is laid down near the diaphyseal surface of the plates. The proliferation of cartilage cells occurs at a rate that is comparable to their degeneration and ossification, resulting in lengthening of the bone. Long bones can continue to grow in length until the epiphyseal plate becomes ossified—an event that occurs at the time that adult size is attained, at about the age of sixteen or seventeen in women and eighteen in men.

## THE HUMAN SKELETON

The adult human skeleton is composed of 206 bones. For purposes of study, the bones of the skeleton are usually divided into two groups: those bones, situated along the axis of the body, that form the *axial skeleton;* and bones of the appendages and their supporting structures that collectively are known as the *appendicular skeleton.* The bones of the axial and appendicular skeletons are listed in Table 6–1, and the positions of many of them are shown in Figure 6–7.

TABLE 6-1 Bones of the Human Skeleton

| Axial Skeleton | Number of Bones | Appendicular Skeleton | Number of Bones |
|---|---|---|---|
| Skull | | Upper Extremity | |
| Cranium | 8 | Pectoral girdle | 4 |
| Frontal (1) | | Clavicle (2) | |
| Occipital (1) | | Scapula (2) | |
| Parietal (2) | | Arm | 60 |
| Temporal (2) | | Humerus (2) | |
| Sphenoid (1) | | Ulna (2) | |
| Ethmoid (1) | | Radius (2) | |
| Face | 14 | Carpal (16) | |
| Palatine (2) | | Metacarpal (10) | |
| Zygomatic (2) | | Phalanges (28) | |
| Maxilla (2) | | Lower Extremity | |
| Lacrimal (2) | | Pelvic girdle | 2 |
| Nasal (2) | | Hip bones (2) | |
| Vomer (1) | | Leg | 60 |
| Inferior nasal concha (2) | | Femur (2) | |
| Mandible (1) | | Fibula (2) | |
| Others | 7 | Tibia (2) | |
| Hyoid (1) | | Patella (2) | |
| Auditory ossicles | | Tarsal (14) | |
| Incus (2) | | Metatarsal (10) | |
| Malleus (2) | | Phalanges (28) | |
| Stapes (2) | | Total Appendicular Skeleton | 126 |
| Vertebral Column | 26 | | |
| Cervical vertebra (7) | | | |
| Thoracic vertebra (12) | | | |
| Lumbar vertebra (5) | | | |
| Sacral vertebra (1) | | | |
| Coccygeal vertebra (1) | | | |
| Thorax | 25 | | |
| Sternum (1) | | | |
| Rib (24) | | | |
| Total Axial Skeleton | 80 | | |

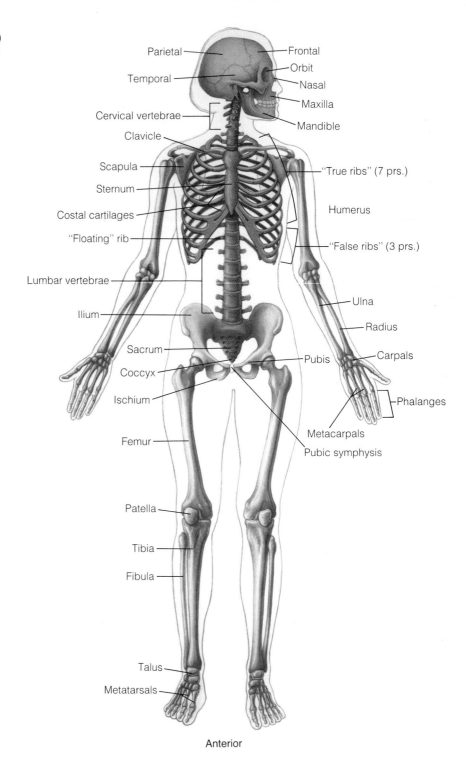

Anterior

## The Axial Skeleton

The **axial skeleton** consists of the bones of the skull; the hyoid bone tucked beneath the mandible, or lower jaw bone; the bones that form the vertebral column; the ribs; and the sternum, or breastbone. The ribs and the sternum together form the thorax, or rib cage. The bones of the axial skeleton support and protect organs of the head, neck, and trunk.

## Skull and Associated Bones

The skull proper (Figure 6–8) is composed of 22 bones, including those that form the bony vault, or **cranium,** immediately around the brain, and those that

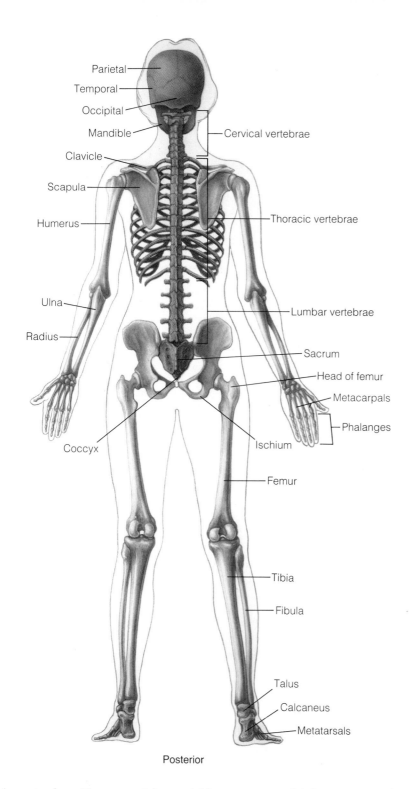

Parietal
Temporal
Occipital
Mandible
Clavicle
Scapula
Humerus
Ulna
Radius
Coccyx

Cervical vertebrae
Thoracic vertebrae
Lumbar vertebrae
Sacrum
Head of femur
Metacarpals
Phalanges
Ischium
Femur
Tibia
Fibula
Talus
Calcaneus
Metatarsals

Posterior

form the face. There are eight cranial bones: two *parietal,* two *temporal,* one *frontal,* one *occipital,* one *sphenoid,* and one *ethmoid.* The fourteen facial bones include two *palatine,* two *zygomatic,* two *maxilla,* two *lacrimal,* two *nasal,* two *inferior nasal concha,* one *vomer* and one *mandible.*The **mandible,** or lower jaw, is freely movable, but the other bones of the cranium and face are fused together by sutures.

Seven other bones are associated with the skull but are not a part of the framework of the cranium or face. These are the *middle ear bones (auditory ossicles),* of which there are three in each ear (two *incus,* two *malleus,* and two *stapes*), and the *hyoid bone.* The role of the middle ear bones in the transmission of sound will be discussed in Chapter 15. The U-shaped hyoid bone is tucked beneath the mandible and is suspended by ligaments from the

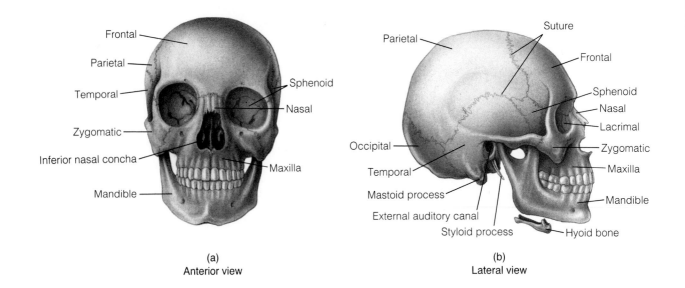

(a)
Anterior view

(b)
Lateral view

FIGURE 6–8 (a) anterior; (b) lateral; and (c) inferior view of the skull.

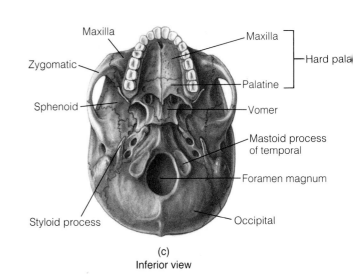

(c)
Inferior view

styloid processes of the temporal bones (see Figure 6–8). It serves as the attachment site for muscles that move the tongue and are involved in swallowing and speaking.

The primary function of the skull is to protect the brain and its associated sense organs—the eyes, ears, and nose. Not only is the vault of the skull spherical in shape, which adds to its strength, but the bones that form the vault have some elasticity, which allows it to withstand surprisingly severe blows without fracturing.

Some of the bones of the skull possess hollow chambers called **air sinuses.** These lighten the weight of the skull and serve as resonance chambers for the voice. Four pairs of sinuses communicate with the nasal cavity and are lined with mucous membranes that are continuous with it (Figure 6–9). On occasion, the lining of these sinuses becomes inflamed—a condition known as *sinusitis.* Swelling associated with sinusitis may result in *sinus headaches.* Another pair of sinuses, the mastoid sinuses, are located in each of the temporal bones and open into the middle ear. Inflammation of these sinuses is called *mastoiditis,* which occurs as a complication of middle ear infections.

***Development of the Face.*** The formation of the bones of the face is one of the most complicated processes that occurs during embryonic development. It involves the coordinated growth and fusion of a number of separate membranous tissues, which eventually become ossified to form the membranous bones of the face. Any interference with the precise coordination of events that occur during this process is likely to result in developmental abnormalities (birth defects).

One such birth defect that is relatively common is a cleft palate. A cleft palate is characterized by a fissure in the roof of the mouth, or **palate,** which separates the nasal cavity from the mouth cavity. Most of the *hard palate*—the anterior bony portion of the palate—is normally formed by two shelves of tissue that grow toward the midline, eventually fuse, and become ossified. However, if the growth of these tissues is slowed so that fusion does not occur before they reach the end of their normal growth period, a cleft palate results. Frequently, a cleft palate is associated with a fissure in the upper lip, termed "harelip". Babies that are born with a cleft palate have difficulty nursing, because they cannot generate suction between their tongue and hard palate. Although at one time a cleft palate was difficult to repair, surgical techniques have been developed recently with promising results.

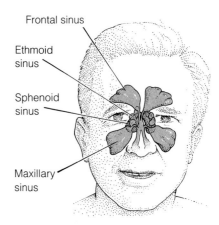

FIGURE 6-9 Locations of the four pairs of air sinuses that communicate with the nasal cavity. The mastoid sinuses, which open into the middle ear, are not shown.

***Growth of the Skull.*** At birth, the skull is large relative to other parts of the skeleton; however, the face takes up a considerably smaller portion of the skull than it does in an adult. The reduced size of the newborn's face is due to the small size of the upper and lower jaws, nasal cavity, and air sinuses. The face enlarges first with the eruption of the deciduous teeth ("baby" teeth) and then further as the permanent teeth appear.

The skull grows very rapidly between birth and the seventh year. Very little growth occurs between the seventh year and puberty, at which time the development of the air sinuses causes a marked expansion in the frontal (forehead) and facial regions of the skull.

## Vertebral Column

The **vertebral column,** or backbone, extends from the base of the skull down the dorsal midline of the body. It consists of a number of segments, called **vertebrae,** which are arranged in series to form a hollow bony column that surrounds and protects the spinal cord. A large opening at the base of the skull, the **foramen magnum,** allows the brain stem to pass into the vertebral column as the spinal cord.

Children and young adults have 33 vertebrae that are classified on the basis of the regions of the body they occupy (Figure 6-10). There are seven *cervical vertebrae* in the neck region, twelve *thoracic vertebrae* in the thorax, five *lumbar vertebrae* in the small of the back, five *sacral vertebrae* associated with the pelvis, and four *coccygeal vertebrae* posterior to the pelvis. By the age of about twenty-five or thirty, the sacral vertebrae have fused to form the *sacrum,* and the coccygeal vertebrae have fused to form the *coccyx* or terminal bone. Because they are fused in the adult, the sacral and coccygeal vertebrae are termed *false,* or *fixed vertebrae,* while all other vertebrae are referred to as *true,* or *movable vertebrae.*

***Structure of Vertebrae.*** The structure of the first two cervical vertebrae differs substantially from that of the other vertebrae. The uppermost cervical vertebra is a bony ring through which the spinal cord passes and which articulates with two rounded projections, one on either side, on the back of the skull. Because it supports the skull, this vertebra is called the **atlas,** named for the mythological Titan who was doomed to carry the heavens and earth on his shoulders. The articulation of the atlas and skull forms a joint, stabilized by strong ligaments, which permits nodding of the head. Beneath the atlas is the **axis,** the second cervical vertebra. A projection on the axis articulates with the anterior part of

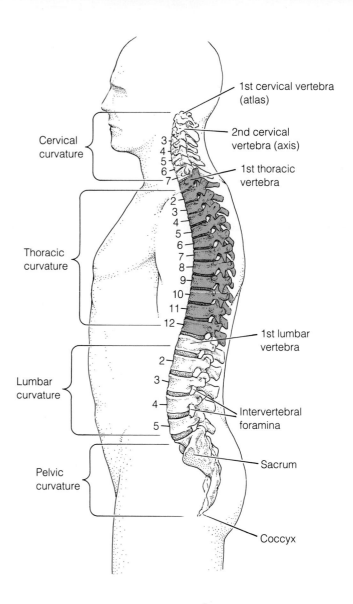

FIGURE 6–10 Side view of the vertebral column showing how it is positioned within the body. Note the four curvatures.

1st cervical vertebra (atlas)

2nd cervical vertebra (axis)

1st thoracic vertebra

Cervical curvature

Thoracic curvature

Lumbar curvature

Pelvic curvature

1st lumbar vertebra

Intervertebral foramina

Sacrum

Coccyx

the atlas, forming an axis of rotation (hence the name) that allows turning the head from side to side as the atlas moves with the skull.

Though there are variations in structure according to their position in the vertebral column, the other movable vertebrae generally share common structural features. A typical vertebra from the thoracic region is shown in Figure 6–11. The largest part of the vertebra is the *body,* shaped like a somewhat flattened solid cylinder. The body of the vertebra is the weight-bearing region where successive vertebrae are stacked on one another to form the vertebral column. Between the bodies of each vertebra is an **intervertebral disk** (see Figure 6–10) of fibrous cartilage, possessing a tough fibrous exterior and a highly elastic, soft center (the *nucleus pulposus*). The intervertebral disk serves as a cushion between vertebrae to absorb vertical shock and to prevent the vertebrae from grinding together when the column bends.

Two short, thick projections, one on either side, extend dorsally from the body; these are called **pedicles.** Extending from the pedicles are two broad bony plates, the *laminae,* that fuse in the midline to form the dorsal part of the vertebral arch. The large opening formed by the body, pedicles, and laminae is the **vertebral foramen,** through which the spinal cord passes. Two pairs of *articular processes,* one pair extending anteriorly and the other posteriorly, project as short nubs from the junctions of the pedicles and laminae. These are smooth surfaces, coated with hyaline cartilage, where each vertebra articulates with the vertebra below and above it. Several other bony processes (projections) serve for the attachment of ligaments and muscles that hold the vertebrae in place. These are the *spinous process,* projecting from the dorsal surface of the laminae, and the *transverse processes,* which project at either side laterally from the points where the pedicle and laminae join.

***Structure of the Vertebral Column.*** The compressible intervertebral disks, the cartilage-coated articular processes, and the elastic muscles and ligaments that hold the vertebrae in register, are structural features that allow the vertebral column to bend easily and smoothly, and to withstand considerable vertical shock. In addition, the vertebral column in an adult has four distinct curvatures that distribute the weight of the body over its center of gravity (see Figure 6–10). This arrangement provides much more strength and stability for supporting the head, trunk, and upper extremities than would a straight column. Moreover, the curvatures provide resilience to the vertebral column, acting as a kind of spring during walking, running, and jumping.

The newborn infant has only one spinal curvature—an anteriorly concave curve along the entire length of the vertebral column. When the child is able to hold up its head at three to four months, the anteriorly convex cervical curve develops; the lumbar curve, also anteriorly convex, appears at twelve to eighteen months when the child begins to walk.

***The Vertebral Column in Injury and Disease.*** Damage to the vertebral column or its associated structures can have serious consequences. Because the spinal cord carries nerves that stimulate the contraction of skeletal muscles, injury to the spinal cord as a result of a vertebral fracture can result in paralysis or even death. Sometimes, either because of undue strain on the vertebral column or for no obvious reason, the soft central portion of an intervertebral disk may rupture through the enclosing fibrous layer. The rupture of material may cause severe pain by pressing either against the spinal cord itself or on roots of nerves that arise from the spinal cord and exit between the vertebrae to supply various parts of the body. This condition is called a "slipped disk" (see *Focus: Back Pain*).

Birth defects or disease may cause structural abnormalities of the vertebral column. In the embryo, failure of the vertebral laminae to meet and fuse properly causes a birth defect known as *spina bifida,* in which the spinal cord or its surrounding membranes, or both, may protrude from the vertebral canal. Spina bifida may involve only one or two vertebrae, causing a condition with few symptoms that is easily repaired surgically. If the defect is large and involves many vertebrae, damage to the unprotected spinal cord often results in paralysis.

Abnormalities in the spinal curvatures are also not uncommon and may be severe enough to cause crippling. Exaggeration of the curvature in the thoracic region is called *kyphosis* and results in a hunchback posture, whereas exaggeration of the lumbar curvature is termed *lordosis,* causing a swayback. In a condition known as *scoliosis,* the vertebral column deviates laterally, disrupting posture, distorting the position of the ribs, and reducing chest capacity; these effects, in turn, may restrict normal functioning of the lungs and heart.

## Skeleton of the Thorax

The skeleton of the thorax, also known as the rib cage, is formed by the *sternum* (breast bone), twelve pairs of *ribs,* and the *costal cartilages.* The **sternum** is a flat, narrow bone shaped like an elongated arrow head. It forms the front of the rib cage at the midline.

The **ribs** are flexible arches of bone (see Figure 6–7). Each pair of ribs is attached dorsally by ligaments to either side of a separate thoracic vertebra. The first seven pairs of ribs are called *true ribs* because they are attached anteriorly to the sternum by their own costal cartilages that extend from the ends of the ribs. The remaining five pairs are called *false ribs.* The costal cartilages of the upper three pairs of false ribs are attached to the cartilage of the rib above, not

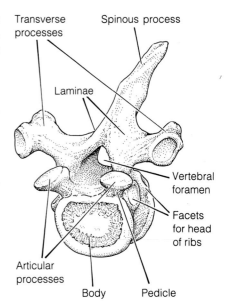

FIGURE 6–11 Structure of a thoracic vertebra.

# Back Pain

**A**lthough severe back pain is not life threatening, it is a major source of disability among adults in the United States and elsewhere. Indeed, it has been estimated that 60 to 80 percent of all Americans have at least one episode of severe, disabling back pain.

Back pain can be brought on by a variety of conditions. Some of it is a result of underlying disease, such as arthritis, a spinal cord tumor, or abdominal cancer. Most, however, is caused by injury to the back that occurs when excessive strain is placed on the vertebral column and the muscles, ligaments, and tendons that support it. Overworking is perhaps the greatest source of back injury, but other conditions can injure the back in a less obvious way. Pregnancy, for example, can put great strain on the vertebral column by increasing its curvature as the buttocks are forced backward in an attempt to compensate for the additional weight in the front of the body. Overweight individuals, particularly those with a pot belly, strain the spine in a similar way. Poor posture may also cause or contribute to back injury.

Whatever the causes of back injury, it usually takes one of three forms. Most common are spasms of the back muscles. These sustained contractions compress blood vessels and thus deprive muscle cells of oxygen and essential nutrients, causing tissue damage and pain. Another source of back pain, generally more severe than muscle spasms, results from a so-called "slipped disc," more accurately termed a *ruptured,* or *herniated,* disc. In this condition, the soft central portion of an intervertebral disc ruptures through the enclosing fibrous layer. The ruptured material may cause severe pain by pressing either against the spinal cord itself or on roots of nerves that arise from the spinal cord and exit between the vertebrae to supply various parts of the body (see accompanying figure). Most commonly involved are the disks between the last several lumbar vertebrae, or the one between the final lumbar vertebra and the sacrum.

A third form of back pain occurs when the vertebral column is twisted, and the articular processes of one vertebra be-come misaligned with those of an adjacent vertebra. Pain results when the articular processes rub together and put pressure on the nerves that supply them.

Frequently accompanying all three of these conditions is *sciatica,* pain and inflammation of one or both of the sciatic nerves that supply the legs. Because the sciatic nerve has roots that arise from the spinal cord in the lumbar and sacral regions of the vertebral column, any injury in these regions may affect the sciatic nerve.

Most severe back pain is relieved by two or three weeks of bed rest. Aspirin to reduce inflammation and muscle relaxants may also be prescribed. If the pain is not relieved by rest in bed and continues for a number of weeks at an intolerable level, a myelogram (X ray of the spinal cord following injection of a radiopaque dye into the space surrounding the spinal cord) or a CAT (computerized axial tomography) scan may be performed to aid the diagnosis. If the existence of a herniated disc is confirmed, surgery to remove the offending material may be appropriate. An alter-

directly to the sternum. The last two pairs of false ribs, termed *floating ribs,* are not attached to the sternum in any way.

The bones of the thorax are covered by muscles, and the floor of the thorax is separated from the abdominal cavity by a dome-shaped muscle called the diaphragm. The thorax thus forms a muscle-bound cavity that protects the lungs, heart, and major blood vessels. In addition, the liver and stomach, lying just beneath the diaphragm, are also protected by the lower ribs.

## The Appendicular Skeleton

The appendicular skeleton consists of the bones that form the upper and lower extremities of the body. These include the arms and legs and their respective supporting structures, the shoulders and hips.

### Upper Extremity

***Shoulder Girdle.*** The **shoulder girdle,** or **pectoral girdle,** provides the supporting framework for the attachment of the bones of the upper extremities to the axial skeleton. Two *collarbones,* or *clavicles,* one on each side, form the anterior portion of the shoulder girdle; two *shoulder blades,* or *scapulae,* also

native to surgery is *chemonucleolysis,* in which a digestive enzyme called *chymopapain* is injected into the center of the herniated disc. The central portion of the disc and the material extruded from it are digested by the enzyme, eliminating pressure on spinal nerves and relieving pain. Chemonucleolysis alone relieves the pain of a herniated disc in about 75 percent of patients, but it is not a risk-free procedure since some people may have life-threatening allergic reactions to the enzyme. If back pain is caused by two vertebrae slipping relative to one another and becoming misaligned, the vertebrae may be fused together surgically.

Back pain is best prevented before it occurs. Regular exercise that does not strain or twist the back excessively has been shown to be the best preventive regimen, both for individuals who have never had back pain and for those who have chronic back problems. Swimming and walking are ideal, as are special back-strengthening exercises recommended by orthopedists.

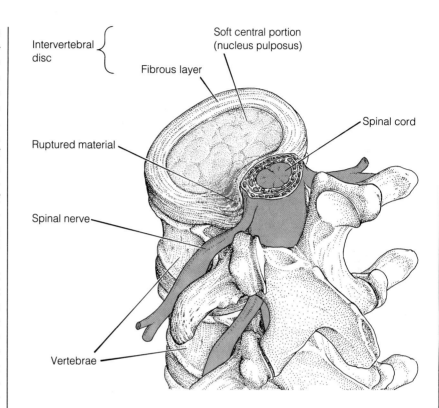

A herniated disc, commonly known as a "slipped disc," can put pressure on the root of a spinal nerve.

one on each side, form the dorsal part (see Figure 6–7). One end of each clavicle articulates in the midline with the upper part of the sternum, while the other end of each clavicle articulates dorsally with one of the scapulae. The scapulae are anchored to the back only by muscles. The connections of the bones of the shoulder blade to one another and to the axial skeleton are loose, allowing free movement of the shoulder.

*Arm Bones.* The upper arm is supported by one large long bone, the *humerus.* The humerus has a smoothly rounded head that fits into a shallow socket in the scapula. The shallowness of the socket allows the arm to move in virtually all directions, but does not provide much stability at the joint, leaving the humerus more prone to dislocation than any other bone. Such a dislocation is inaccurately referred to as a *dislocated shoulder.* A prominent projection on the distal end of the humerus can be easily felt on the inside of the elbow joint. The ulnar nerve, which passes below the projection, can be pinched when the elbow is bumped, causing an unpleasant tingling sensation known as hitting the "funny bone"—a pun on humerus.

The humerus articulates distally, at the elbow, with the two lower arm bones, the *radius* and *ulna.* The ulna is the thinner of the two, and its head is easily seen as the bony projection forming the tip of the elbow.

FIGURE 6-12 An X ray of the bones of the wrist, palm, and fingers. Notice how the radius and ulna (of the lower arm) articulate with the carpals of the wrist.

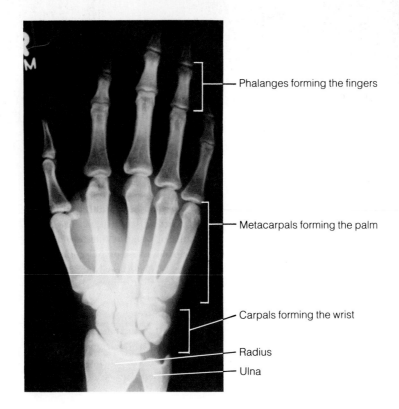

— Phalanges forming the fingers

— Metacarpals forming the palm

— Carpals forming the wrist

— Radius
— Ulna

***Bones of the Hand.*** The bones of the hand include those of the wrist, palm, and fingers (Figure 6-12). The wrist has two rows of four bones called *carpals.* The radius and ulna articulate with the proximal row of carpals. The bones of the distal row of carpals articulate with five *metacarpals* that compose the palm. The metacarpal of the thumb is placed lower than the metacarpals of the digits, allowing the thumb to move across the palm—an arrangement that is responsible for the great manual dexterity of humans. The metacarpals articulate with the first row of *phalanges,* of which there are three for each finger and two for each thumb.

***Fractures of the Upper Extremity.*** The clavicles and the arm bones are fractured quite commonly. Since the clavicles brace the shoulder, a fall on the shoulder to one side is liable to snap the clavicle on that side. The humerus is most often fractured at the shaft near its proximal end. One of the more frequently encountered of all fractures is that of the radius at a point about a third of the way from its distal end; this is caused by falling on the palm of the hand.

## Lower Extremity

***Pelvic Girdle.*** The **pelvic girdle,** or pelvis, is a sturdy, bony ring that supports the trunk and attaches the lower limb bones, on which it rests, to the body (see Figure 6-7). In addition, the pelvis protects the bladder, some reproductive organs, and lower parts of the large intestine.

The pelvis is composed of four bones: two hipbones, which form the lateral and anterior parts of the ring, and the sacrum and coccyx, which complete the ring dorsally (Figure 6-13). Each hipbone consists of three bony parts, the *ilium, ischium,* and *pubis,* which are distinct from each other in youngsters; however, areas between the three parts become ossified to form a single bone in the adult. The two ilia form the lateral and upper, flared portions of the pelvis; the ischia form its bottom; and the bones of the pubis curve forward to form the anterior portion of the pelvis. The pubic regions of the two

hipbones are firmly attached to one another by a disk of fibrous cartilage and ligaments that form a connecting structure known as the *pubic symphysis.* Fibrous cartilage also attaches the sacrum to the two hipbones, and consequently the pelvis is, in a sense, an integral part of the vertebral column. This arrangement allows for great stability and strength in supporting the body, but does not permit the freedom of movement that characterizes the shoulder girdle.

Males and females show differences in their pelvic construction. The pelvis of the female is broader and shallower, and possesses a substantially larger outlet than the pelvis of the male. The breadth of the female's pelvis is adaptive for supporting the fetus during pregnancy, and the large outlet aids in childbirth. In fact, the diameter of the pelvic outlet is increased during childbirth as estrogens soften and loosen the pubic symphysis.

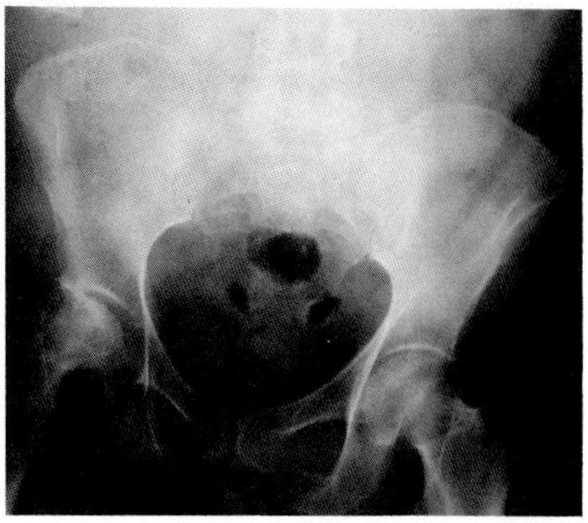

FIGURE 6–13 An X ray of the pelvis, showing the articulation between it and the femur (thigh bone). Notice the marked arthritic changes of the right hip joint.

***Leg Bones.*** Each hipbone articulates with a *femur,* or thigh bone, which is the longest and strongest bone of the body (see Figure 6–13). The femur possesses a rounded head connected nearly at a right angle to the cylindrical shaft by a short stretch of bone, termed the *neck.* The head of the femur fits into a deeply rounded socket in the hipbone. Because the head of the femur is offset from the shaft, the femur has considerable movement even though the joint is a weight-bearing one.

Although the lower leg possesses two bones—the *tibia* and the *fibula*—the femur articulates only with the tibia, the larger of the two. The tibia is prominent in the front of the lower leg as the shinbone. The articulation between the femur and tibia forms the knee joint, which is protected by a flat, triangular bone, the *patella* or knee cap. The patella is held in place by the tendon that attaches a group of anterior thigh muscles to the tibia. The exposed position of the patella makes it liable to fracture and dislocation. The other lower leg bone, the *fibula,* articulates at either end with the tibia; the fibula is not a weight-bearing bone, but allows twisting of the lower leg. Two bony lumps at either side of the ankle—often referred to as ankle bones, which they are not—correspond to the end of the tibia, on the inside, and to the end of the fibula, on the outside.

Leg bones are often fractured by serious falls. Falls onto the hip are apt to break the femur at its neck—an occurrence that is particularly common in older people whose bones tend to be brittle. Fracture of the femur at the neck is called a *broken hip.* In another common fracture, the fibula is broken a few inches above its distal end as a result of a sudden twist of the ankle.

***Bones of the Foot.*** The weight of the body is transferred from the tibia to the *talus,* one of the seven *tarsals* or ankle bones. From the talus, the body's weight is distributed about equally to the heel bone (a tarsal, called the *calcaneus*)

## Sliding Filament Theory of Muscle Contraction

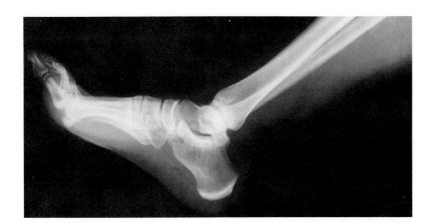

FIGURE 6–14 X ray showing a lateral view of the ankle bones and the articulation of the talus with the tibia and fibula.

and to the other tarsals and the five *metatarsals,* all of which form the arch. There are two toe bones or *phalanges* in the big toe and three in each of the other toes (Figure 6–14, page 113).

The construction of the foot reflects its weight-bearing and supporting function. The foot possesses two arches—one along the length of the foot and one across the ball of the foot. These are normally maintained by powerful ligaments, tendons, and muscles that hold the bones in place. The arches distribute weight throughout the foot and provide a springy cushion that absorbs shock generated by locomotion. The absence of arches results in *flat feet,* caused by a congenital defect in the supporting ligaments or by the stretching of the ligaments, which can occur in individuals who must stand on their feet much of the day.

The toes, shorter and less mobile than fingers, bear weight well. In addition, they are important in maintaining balance and in pushing the foot off the ground during walking, running, and jumping.

## BONE INJURY AND DISEASE

The most common injury to bones are fractures, but these usually repair completely with proper treatment. Diseases affecting bones, although often serious, are rare.

### Bone Fractures

There are various types of bone fractures. In a *simple fracture,* the bone is broken into two separate fragments, but the fragments do not break through the overlying skin or mucous membrane. A *compound fracture,* on the other hand, is one in which the broken bone breaks through the skin or mucous membrane that overlies it. In other types of fractures, the bone may be fragmented into many smaller pieces (*comminuted fracture*), the bone may be cracked but not completely severed (*greenstick fracture*), or the bone may be compressed (*crush fracture*). All of these fractures are accompanied by various degrees of damage to the periosteum and blood vessels in the bone and to the tissue surrounding the area of the break.

### Repair of a Fractured Bone

Before the biological processes of repair begins, it is usually necessary to align the broken ends of the bone and to immobilize the bone with a cast or some other device while healing occurs.

Repair of a simple fracture, summarized in Figure 6–15, begins with the proliferation of osteogenic cells in the healthy periosteum adjacent to the fracture site and in the marrow cavity. This proliferation gives rise eventually to a collar of new tissue, called a **callus,** that bridges the area between the broken ends of the bone and bulges somewhat at the surface. The callus looks much like a glob of glue that might be used to hold the ends of a broken stick together.

The osteogenic cells in the region of the callus that lies close to the bone have a good blood supply and give rise to osteoblasts, whereas those in the outer area of the callus are not supplied by blood vessels and develop into chondroblasts. The osteoblasts lay down bony trabeculae that are attached to the broken ends of the bone and eventually extend across the break. The chondroblasts form cartilage that exists only temporarily before it is replaced by cancellous bone in a process similar to that seen during the formation of bones by intracartilaginous ossification.

The callus, which comes to consist entirely of cancellous bone, is extensively remodeled. Cancellous bone between the broken ends of the original bone is converted into compact bone (as described in an earlier

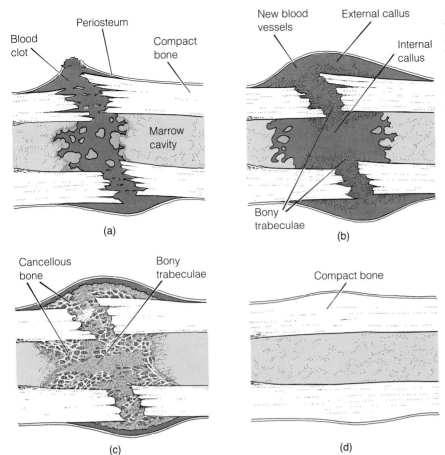

Periosteum

Blood clot

Compact bone

Marrow cavity

(a)

New blood vessels

External callus

Internal callus

Bony trabeculae

(b)

Cancellous bone

Bony trabeculae

(c)

Compact bone

(d)

FIGURE 6–15 Stages in the repair of a simple fracture. (a) The bone shortly after the break; (b) a callus has formed and bony trabeculae connect the ends of the broken fragments; (c) the callus is replaced entirely by cancellous bone; (d) the cancellous bone between the broken ends is remodeled into compact bone and the remainder of the cancellous bone is resorbed.

section), and the remaining cancellous bone in the bulge of the callus is gradually resorbed. When the remodeling process is complete, the bone may be so well restored to its original shape that the site of the break cannot be detected by X ray.

## Diseases of Bones

In addition to those mentioned earlier, there are a few other bone diseases that are sometimes encountered. *Osteomyelitis* is an infection of the bone and marrow cavity that results in bone destruction. *Benign bone tumors* consisting of hard bony material sometimes develop, most frequently at the end of a long bone. They may require surgical removal if they cause pain or make the bone susceptible to fracture. *Cancerous bone tumors* are rare, but are often dangerous since cancer cells can spread throughout the body at an early stage of tumor development.

## JOINTS (ARTICULATIONS)

Bones of the skeleton are connected with one another at **joints,** or **articulations.** Joints may be immovable or provide for various degrees of movement.

## Types of Joints

Some joints, such as those between the bones of the skull, are immovable and extremely strong. Others provide slight movement combined with great strength because they are united by fibrous cartilage; joints between vertebral bodies and in the pubis are of this type.

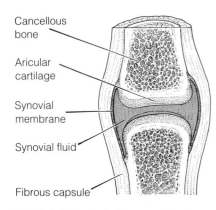

Cancellous bone

Aricular cartilage

Synovial membrane

Synovial fluid

Fibrous capsule

FIGURE 6–16 A sketch of one type of freely movable joint.

Freely movable joints have a more complex structure than either of the first two types (Figure 6–16). The expanded ends of the bones involved in the joint are coated with cartilage to prevent them from grinding together. The bones are held together by a strong capsule formed by ligaments that consist of dense fibrous connective tissue. The inner lining of the joint capsule forms a membrane, the **synovial membrane,** that is composed entirely of connective tissue and secretes a slippery lubricating fluid, **synovial fluid.** Joints may be reinforced by muscles that overlie the fibrous capsule.

Freely movable joints include, among others: (1) **ball-and-socket joints,** such as those at the hip, in which the rounded head of one bone articulates with a cuplike cavity in the other, allowing angular movement in all directions; (2) **hinge joints,** such as those at the elbow and knee, which allow movement in one plane; and (3) **pivot joints,** such as those between the radius and humerus or between the atlas and axis, which permit rotation as one bone turns about another.

## Joint Injury and Disease

Injury to joints is common. A *sprain* occurs when joint attachments are stretched or torn, but the bones remain essentially in place. In more severe injuries, bones at joints may be *dislocated* such that the bones are no longer properly aligned and must be physically manipulated to be realigned. Sometimes, injury to joints is accompanied by excessive secretion of synovial fluid, causing such conditions as *water on the knee.*

Joint inflammation caused by disease, rather than by injury, is called *arthritis,* of which there are several types. The most common form of arthritis is *rheumatoid arthritis.* Although its cause is unknown, it is associated with inflammation of the synovial membrane, which may eventually result in irreversible damage to the joint capsule. Rheumatoid arthritis is characterized by a painful swelling of the joints. *Osteoarthritis* is a chronic degenerative joint disease, usually of older people. It results in destruction of articular cartilage, particularly in weight-bearing joints and in the terminal joints of the fingers. An abnormal growth of bone at affected joints may cause distortion of the form of the bone (Figure 6–17). Pain associated with osteoarthritis usually occurs only during joint movement. *Gout* or *gouty arthritis* is caused by an inherited condition in which nitrogen-containing compounds such as proteins are not broken down properly because of the lack of a particular enzyme. As a consequence, uric acid, an abnormal metabolic by-product, builds up in the blood and eventually precipitates as crystals in the joints. This brings about an extremely painful swelling of affected joints. Gout may be prevented by drugs that encourage the excretion of uric acid.

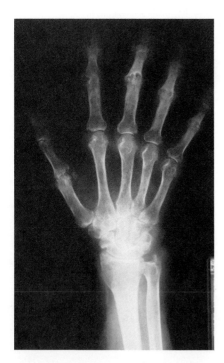

FIGURE 6–17 Abnormal growth of bone in the terminal (distal) and proximal joints of the fingers are degenerative changes typical of osteoarthritis. Compare this figure with Figure 6–12.

## Study Questions

1. Describe several functions of the skeletal system.
2. a) What characteristics of bone are determined by its organic constituents?
   b) By its inorganic constituents?
3. Distinguish between the role of osteocytes, osteoblasts, and osteoclasts in the formation and remodeling of bone.
4. a) What are the cause and symptoms of rickets?
   b) Of osteoporosis?
5. Describe the role of the periosteum in the repair of bone fractures.
6. Distinguish between the structure of compact and cancellous bone.
7. a) Indicate the two ways in which bones form during embryonic development.
   b) Summarize each of these processes.
8. Explain why the skulls of newborn babies are more resilient to impact than those of adults.

9. Diagram the process of long bone growth.
10. a) Name the five general types of vertebrae.
    b) Indicate which types of vertebrae are movable in adults.
11. a) Indicate the function of intervertebral disks.
    b) What is a "slipped disk?"
12. Distinguish between true and false ribs.
13. Explain why the upper extremities can move freely at the shoulder while the lower extremities have less freedom of movement at the hip.
14. Draw a labeled diagram of a cross-section through a freely movable joint.
15. Indicate three types of freely movable joints and give an example of each type.
16. Name two types of arthritis and indicate the symptoms of each.

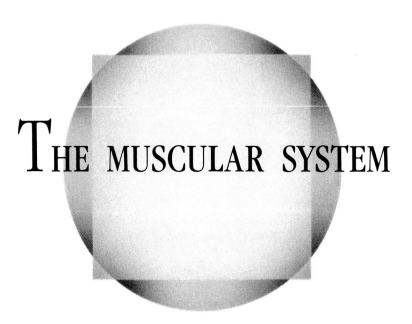

# THE MUSCULAR SYSTEM

As we noted in Chapter 4, muscle tissue contains contractile muscle cells that function to bring about the movement of some body part. Although the ability to contract is the basis for the action of all muscle tissue, the human body contains three different types of muscle tissue, each specialized for particular tasks. These are *skeletal muscle, cardiac muscle,* and *smooth muscle;* the distribution and general properties of each of these muscle types was summarized in Chapter 4. This chapter will consider the anatomy and physiology of skeletal muscle and smooth muscle; cardiac muscle will be considered in the context of discussing the circulatory system (Chapter 9).

## OVERVIEW OF THE STRUCTURE AND FUNCTION OF SKELETAL MUSCLE

Skeletal muscle, as its name implies, is connected by tendons or other connective tissues to the bones and cartilage of the skeleton and to some of the body's fleshy parts, such as the face and the tongue. The action of skeletal muscle maintains body posture, brings about movement of the skeleton by exerting pull on bones across joints, and allows us to speak and change our facial expression. These are complicated tasks and account for the fact that skeletal muscle comprises about 40 percent of total body weight in men and 25 percent in women. Skeletal muscle is sometimes referred to as *voluntary muscle* because it controls our voluntary actions and is hence under our conscious control.

Skeletal muscle tissue forms what people commonly refer to as their muscles. A typical skeletal muscle is composed of *muscle cells,* also called *muscle fibers,* and noncontractile connective tissue elements that serve to organize the muscle cells into an effective mechanical unit.

The organization of connective tissues in a skeletal muscle occurs on a number of levels (Figure 7–1). A single muscle is covered by a tough, smooth outer sheath of connective tissue, called the **epimysium,** which allows the muscle to slide freely over nearby muscles and other structures. Connective tissue extends inward from the epimysium to divide the muscle into bundles of muscle cells. These bundles are called **fasciculi** (sing: *fasciculus*), and the connective tissue that surrounds each fasciculus is referred to as the **perimysium.** At the lowest level of organization, each muscle cell is surrounded by a thin connective tissue sheath, the **endomysium.**

The connective tissues of a muscle serve to connect the muscle with other types of connective tissue to which the muscle is attached. For example, the connective tissues of skeletal muscles that move bones are continuous with tendons at the ends of the muscles; the tendons, in turn, attach to bones by merging with the periosteum (see Chapter 4).

In addition to muscle cells and connective tissues, skeletal muscle tissue contains nerves that stimulate it to contract and capillaries that provide it with oxygen and necessary nutrients.

## SKELETAL MUSCLE CELLS

Skeletal muscle cells are the functional units of skeletal muscle. The contraction of the individual cells that form a skeletal muscle results in the contraction of the muscle as a whole.

### Cell Structure

Muscle cells have the shape of elongated cylinders (Figure 7–2). Because of the way muscle cells are formed during embryonic development, muscle cells

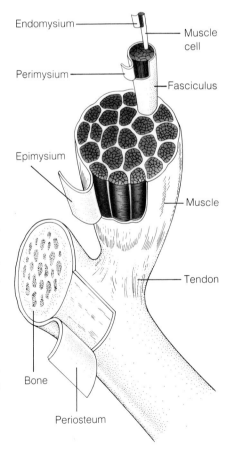

FIGURE 7–1 The connective tissue components of skeletal muscle.

are multinucleate. The muscle cell is surrounded by a plasma membrane, termed the **sarcolemma,** which encloses the cell's cytoplasm or **sarcoplasm.** In the sarcoplasm are numerous fine fibrils, called *myofibrils,* which take up most of the volume of the cell. In addition to the myofibrils, the sarcoplasm contains numerous mitochondria and two prominent membrane systems: the *transverse tubules* and the *sarcoplasmic reticulum.* The **transverse tubules** are tubular structures formed by the invagination (infolding) of the sarcolemma. They lead into the interior of the cell encircling the myofibrils. Between the transverse tubules runs the **sarcoplasmic reticulum,** an extensive system of smooth endoplasmic reticulum consisting of connected vesicles and channels that surround each myofibril.

## Fine Structure of Myofibrils

Each myofibril possesses a repeating pattern of stripes, called *striations.* Because the myofibrils in a cell are arranged in such a way that their striations are usually aligned, or almost so, the entire cell appears striated. This is why skeletal muscle is also referred to as *striated* muscle.

Various regions in the striation pattern can be distinguished; these have been designated by letters, as shown in Figure 7–2. From Z line to Z line is one functional unit, called a **sarcomere,** which is repeated down the length of the myofibril and which shortens in length upon contraction of the myofibril. Since

FIGURE 7–2 A single muscle cell (muscle fiber), showing details of internal structure. The endomysium, a connective tissue sheath, surrounds the sarcolemma in intact muscle. Inset: patterns of striations in the myofibrils and a diagrammatic interpretation of these patterns.

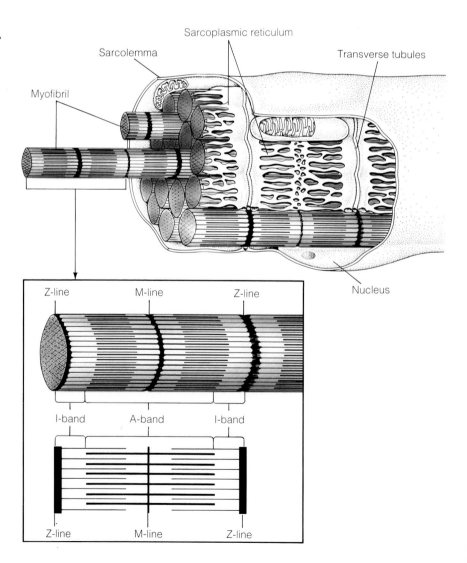

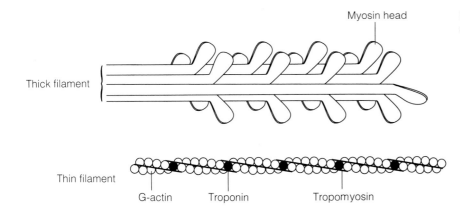

FIGURE 7-3 The structure of a thick filament and a thin filament.

Thick filament

Myosin head

Thin filament

G-actin     Troponin     Tropomyosin

all the sarcomeres in all the myofibrils of a single muscle cell shorten simultaneously, the result is a shortening, or contraction, of the muscle cell itself.

In each sarcomere are two types of filaments: *thick filaments* and *thin filaments.* Thick filaments are composed primarily of a bundle of *myosin* molecules. **Myosin** is a rod-shaped protein molecule that has an enlarged "head" at one end; the rod portion has two areas of structural flexibility that are thought to allow the molecule to bend. In forming the thick filament, myosin molecules are arranged so that their heads project from the surface of the filament, as shown in Figure 7-3.

The thin filament (Figure 7-3) is formed of more or less spherically shaped protein molecules, called *G-actin* (*G* for globular), that line up in two helically wound rows to produce an *F-actin* (*F* for fibrous) strand. Two additional proteins are associated with the F-actin strand. One of these is **tropomyosin,** an elongated molecule that runs in the groove between the two F-actin strands; the other is **troponin,** a globular protein that occurs at regular intervals along the strand in association with both the actin and tropomyosin molecules.

The arrangement of thick and thin filaments in the myofibril creates dark and light areas, referred to as bands and shown diagrammatically in Figure 7-2 (inset). The thick filaments, which extend throughout the A-band, are anchored together at the vertical M-band. The region of the sarcomere that contains only thin filaments is designated the I-band, but the thin filaments also extend into the A-band where they interdigitate with the thick filaments. On their other ends, some of the thin filaments are anchored to the Z-line; others pass through the Z-line, extending into the adjacent sarcomere.

The thick and thin filaments do not change significantly in length during contraction of a sarcomere. Instead, contraction of a sarcomere results in an increased overlap of thick and thin filaments, thus giving rise to the *sliding filament theory* of muscle contraction. The **sliding filament theory** proposes that muscle contraction results when thin filaments slide past the thick filaments toward the center of the sarcomere. This action pulls the Z-lines, to which thin filaments are attached, closer together, shortening the sarcomere (Figure 7-4).

What events lead to the initiation of contraction, and how is the force generated that causes the thick and thin filaments to slide past one another during contraction as proposed by the sliding filament theory? Contraction is initiated when a nerve impulse reaches the sarcolemma, the plasma membrane surrounding a muscle cell. The nerve impulse initiates a wave of electrical activity *(depolarization)* that spreads throughout the sarcolemma and is transmitted over the T-tubules to the sarcoplasmic reticulum that surrounds

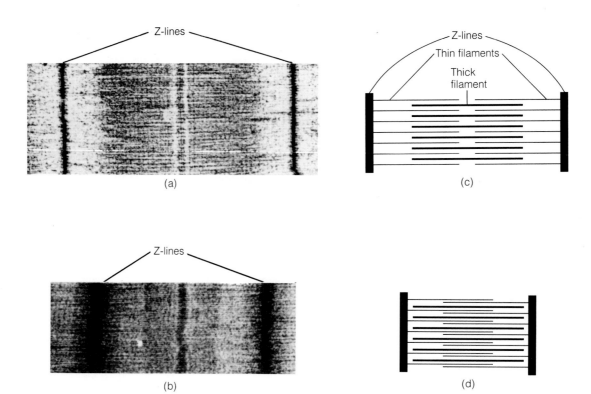

Z-lines

Z-lines — Thin filaments

Thick filament

(a)

(c)

Z-lines

(b)

(d)

FIGURE 7-4 Left: Electron micrographs of skeletal muscle in (a) partly contracted and (b) fully contracted state. Notice that the Z-lines are pulled closer together as a result of contraction; the sarcomere thus shortens. Right: Diagrams interpreting the events in (a) and (b). The thin filaments (thin lines) are displaced relative to the thick filaments (thick lines) during shortening of the sarcomere.

each myofibril. This transmission of electrical activity to the sarcoplasmic reticulum causes the release of calcium ions ($Ca^{++}$) that are normally stored within the sarcoplasmic reticulum of resting muscle cells. The calcium ions diffuse among the myofibrils, coming in contact with the thick and thin filaments.

Once contraction is initiated, resulting in the release of calcium ions and their diffusion among the myofibrils, the sliding filament theory proposes that tension is generated as the thick and thin filaments pull against (slide past) one another (Figure 7-5). The generation of tension begins when calcium ions bind to troponin molecules on the thin filaments, causing a change in shape of the troponin molecules. The change in shape in turn is thought to cause the tropomyosin of the thin filaments to shift position, exposing G-actin molecules in the thin filaments to which the heads of myosin molecules of the thick filaments can bind. When myosin heads bend to G-actin molecules in the thin filaments, structures are formed called *crossbridges* that are apparent in electron micrographs of contracted myofibrils.

The binding of myosin heads to G-actin molecules induces a bending of the myosin molecules that pulls the thin filament a short distance past the thick filament. The myosin molecules then straighten, the heads bind to G-actin molecules further along the filament, and the myosin molecule bends again, repeating the cycle. Hence, the force required to move the actin filaments, and thus to shorten the sarcomere, comes from the alternate straightening and bending of myosin molecules. To prevent the thin filament from backsliding, only some of the myosin heads that form crossbridges to a particular filament are unbound at any one time. When relaxation occurs, all the crossbridges are broken so that the filaments are free to slide past each other.

Contraction, resulting in repeated cycles of straightening and bending of myosin molecules, will continue as long as calcium remains outside the sarcoplasmic reticulum. When nerve impulses no longer reach the muscle cell, an active transport mechanism carries $Ca^{++}$ into the sarcoplasmic reticulum. Lack of $Ca^{++}$ near the myofibril prevents the structural changes in the troponin molecule that ultimately result in the exposure of sites on the actin strand to which myosin heads can bind.

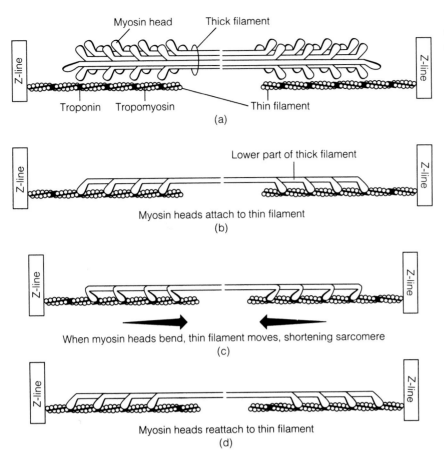

Myosin head    Thick filament

Troponin    Tropomyosin    Thin filament

(a)

Lower part of thick filament

Myosin heads attach to thin filament

(b)

When myosin heads bend, thin filament moves, shortening sarcomere

(c)

Myosin heads reattach to thin filament

(d)

FIGURE 7–5 Diagram summarizing the sliding filament theory of skeletal muscle. (a) Structure of thick and thin filaments in relaxed muscle; (b) relaxed myosin heads bind to G-actin molecules on the actin (thin) filament; (c) bending of myosin heads pulls thin filament toward center of sarcomere, shortening the sarcomere; (d) myosin heads relax and bind to G-actin molecules further down the filament, repeating the cycle. To prevent backsliding of the filament, not all myosin heads relax simultaneously.

## Source of Energy for Muscle Contraction

The energy required for muscle contraction is supplied by ATP, the energy currency of the cell (see Chapter 2). ATP binds to the myosin head, which has the enzymatic activity to catalyze the breakdown of ATP to ADP and phosphate. The energy that is released in that reaction is used to extend the myosin molecule, the head of which then binds to a G-actin molecule on the thin filament. The subsequent bending of myosin that brings about the sliding of the filament does not in itself require energy. Thus ATP may be thought of as providing the energy to "cock" the myosin molecule and to break the previously formed crossbridge, much as energy is used to cock the hammer of a revolver.

The available energy in one ATP molecule is required for the straightening of one myosin molecule, and consequently large amounts of ATP are needed for muscle contraction. Indeed, without ATP, the crossbridges are locked in place, and the muscle is unable to contract further or to relax. The lack of ATP in muscle cells after death causes just such an event, resulting in muscle stiffness known as *rigor mortis.*

## Aerobic and Anaerobic Metabolism

Very little ATP is stored in muscle, and consequently the ATP necessary for contraction must be supplied by metabolic processes occurring within muscle cells. During periods of minimal muscle activity ("resting" muscle), ATP is produced in muscle cells by the aerobic oxidation of glucose and fatty acids (see Chapter 2). Aerobic oxidation, which requires $O_2$, is the most efficient way of making ATP, but it is not the fastest.

When strenuous muscle activity is begun, the oxygen demands of aerobic metabolism increase substantially in muscle cells. Because there is a lag between the time that more oxygen is needed and the time when it can be

# Anabolic Steroids

Testosterone and related male sex hormones, collectively known as androgens, have well-known effects on the development of male secondary sexual characteristics. These include deepening of the voice as a result of growth of the larynx (voice box), growth of body hair, and development of a roughened skin texture. In addition, androgens also directly influence the growth of skeletal muscle by stimulating protein synthesis in skeletal muscle cells. Indeed, this synthetic, or *anabolic,* effect of androgens is the primary factor responsible for the normally observed difference in body structure between males and females. We know this because castrated males acquire a pattern of muscle and fat distribution similar to that of females, whereas females treated with androgens (for example, during breast cancer therapy) acquire the male pattern.

A number of synthetic testosterone derivatives have been developed in the laboratory that stimulate protein synthesis in muscle without having marked masculinizing effects. These substances are termed *anabolic steroids* and include drugs with the tradenames Dianabol, Androyd, Winstrol, Anavar, and Proviron. They have been used effectively to stimulate muscle growth during convalescence following severe illnesses. Unfortunately, they are also widely used by male and female athletes, particularly weight lifters, wrestlers, field eventers, football players, and the like, who feel that additional muscle growth provides them with added strength and a competitive edge. They are also used by some young men who simply wish to look more muscular.

Although growth of skeletal muscle and resultant weight increases unquestionably occur in individuals who take anabolic steroids, it has never been proven that the added muscle growth contributes to an overall increase in strength or, more significantly, in competence. Some studies, for example, show substantial strength gains, amounting to 17–22 pounds in bench press and squat lifts; other studies show little or no increases in strength. There seem to be no consistent effects, whatever the type of anabolic steroid used, the dosage, or the accompanying training program, and it therefore seems likely that there are great differences in individual response to anabolic steroids.

The use of anabolic steroids has been barred by both the International Olympic Committee and the International Amateur Athletic Association, and athletes participating in events sponsored by these organizations must submit to tests for anabolic steroids. Such tests are effective in detecting anabolic steroids for two weeks to two months after their use has been discontinued, depending on the user's metabolism and the dosage. Within a few days after anabolic steroids have been discontinued, their effects begin to wear off, but a temporary increase in muscle mass may still remain at the time that steroids can no longer be detected.

Because muscle mass is lost when the usage of anabolic steroids is discontinued, they must be taken more or less regularly during an athlete's training program in order to have an effect. In fact, many athletes who use anabolic steroids have developed bizarre and dangerous self-treatment programs. In a recent study, published in the *Journal of the American Medical Association,* of ten weight-trained female athletes who use anabolic steroids, the athletes admitted taking, on average, three different types of anabolic steroids together (a practice called "stacking") for a period lasting about nine weeks prior to an important contest. Injectable as well as oral types of anabolic steroids were used. Dosages were as much as nine times the manufacturers' recommended dose. On average, participants in the study used the "stacking" method for about three nine-week periods per year, and 40 percent of the women continued to take smaller amounts of anabolic steroids between the "stacking" periods.

Aside from the obvious ethical questions involving fairness in competition, the use of anabolic steroids is associated with serious long-term side effects. These include liver damage, liver cancer, blood and circulatory disorders, behavioral changes, and psychotic episodes. High doses may have masculinizing effects on females such as deepening of the voice, increased body and facial hair, and decreased breast size, and ovarian cycles may be disrupted, leading to infertility. In males, sperm production is lowered and infertility may result. Considering the delicate balance that must be maintained between the levels of individual hormones in the body, it is likely that additional serious side effects will be discovered.

Use of anabolic steroids may or may not contribute to an overall increase in strength.

supplied as a result of increased heart and breathing rates, there is a period immediately following the beginning of strenuous activity in which oxygen levels are insufficient to maintain high levels of aerobic oxidation. During this period, the muscle must obtain additional ATP from anaerobic processes.

There are two anaerobic processes that are important in supplying ATP to muscles. The first process to be utilized by oxygen-deficient muscle cells involves the breakdown of an energy-rich compound called *phosphocreatine* that is synthesized and stored in resting muscle cells. Phosphocreatine is broken down anaerobically with the transfer of its phosphate group to ADP to form ATP, as follows:

$$\text{Phosphocreatine} + \text{ADP} \rightarrow \text{creatine} + \text{ATP}$$

However, phosphocreatine levels are rapidly depleted in active muscle, and ATP must then be supplied by an alternative anaerobic process. This process is glycolysis (see Chapter 2). In glycolysis, glucose is oxidized to pyruvic acid; however, in the absence of $O_2$, pyruvic acid cannot enter the Krebs cycle, and instead is converted to lactic acid. The energy released in anaerobic glycolysis is used to synthesize ATP.

As a consequence of anaerobic glycolysis, lactic acid accumulates in muscle cells, causing an increase in intracellular acidity. This in turn inhibits the enzymes of glycolysis, so anaerobic glycolysis cannot proceed for very long. Indeed, the accumulation of lactic acid after several minutes of intense muscle activity without ample oxygen results in *muscle fatigue,* a state in which the strength of contraction becomes increasingly weak and inefficient. As muscle fatigue progresses and becomes more severe, an individual will be unable to continue the activity.

Within a short time after beginning strenuous activity, increased circulatory and respiratory rates are translated into the increased delivery of oxygen to muscle tissue, and aerobic oxidation of glucose can resume. In addition, accumulated lactic acid can also be oxidized aerobically to produce ATP. As long as muscular activity is not too strenuous, aerobic oxidation can proceed at a *steady state level,* providing sufficient ATP and preventing muscle fatigue from lactic acid buildup. In fact, one of the purposes of training in athletes is to increase lung capacity and heart efficiency so that high levels of oxygen are supplied to muscles (see *Focus: Anabolic Steroids*).

## Diseases Affecting Skeletal Muscle Cells

The *muscular dystrophies* are a related group of severely debilitating diseases of skeletal muscle cells. They are characterized by a chronic and progressive degeneration of muscle cells, which are gradually replaced by adipose and other connective tissues. The loss of muscle cells is accompanied by progressive weakening of muscles. The severity and course of these diseases depends upon the particular form. They are known to be hereditary, but their cause is unknown, and there is no cure.

## EXCITATION OF SKELETAL MUSCLES BY NERVES

Skeletal muscle cells are supplied by nerve cells (neurons) called **motor neurons.** As we noted above, the contraction of a skeletal muscle cell is initiated when an electrical impulse from a motor neuron reaches it. Each motor neuron supplies a varying number of muscle cells. In muscles that bring about fine, delicate movements, such as those that move the eyeball, a single neuron may supply one muscle cell or only a few. On the other hand, in muscles that control less precise movements, such as the muscles of the leg, one motor neuron may branch extensively to supply a hundred or so muscle cells. A single neuron and all the muscle cells it supplies form a **motor unit.** A motor unit is a functional unit in the sense that stimulation of the nerve will initiate simultaneous contraction in all the muscle cells that the nerve supplies.

## Structure of a Neuromuscular Junction

The place where the ends of neurons approach muscle cells is called a **neuromuscular junction** (Figure 7–6). The motor neurons terminate in enlarged bulbs, called *terminal bulbs,* that contain many mitochondria and numerous small vesicles. The terminal bulbs lie in slight depressions of the sarcolemma, called *motor end plates,* and the sarcolemma in this area is thrown into numerous folds, termed *junctional folds.* These effectively expand the surface area of the sarcolemma that comes in contact with the nerve cell ending. The terminal bulbs of the motor neuron do not actually touch the peaks of the junctional folds; rather, there is a space some 200–300 Å wide that separates the plasma membranes of the two cells. This space is called a **synaptic cleft.** The entire neuromuscular junction forms what is usually referred to as a **synapse** (juncture) between a nerve and muscle cell.

## Molecular Events at the Neuromuscular Junction

The stimulation of a muscle cell by a motor neuron occurs as follows. An electrical impulse, reaching the terminal bulb of the nerve fiber, causes the membrane surrounding the bulb to change its permeability to $Ca^{++}$, such that calcium enters the bulb. The calcium ions cause some of the vesicles, which are filled with the chemical *acetylcholine,* to release their contents by exocytosis into the synaptic cleft. **Acetylcholine,** which acts as a chemical mediator between the nerve and muscle cells, diffuses across the synaptic cleft to the junctional folds, where it attaches to receptor sites on the sarcolemma.

In a resting muscle cell, the sarcolemma normally is *polarized;* that is, the outer surface of the sarcolemma carries a positive charge relative to the inner surface. This charge differential is due to an excess of positively charged ions in the extracellular fluid. Binding of acetylcholine to the sarcolemma at the junctional folds makes the sarcolemma more permeable to sodium ions ($Na^+$) in the extracellular fluid. As $Na^+$ diffuses into the muscle cell at the junctional folds, the polarization of the membrane disappears; that is, it becomes

**FIGURE 7–6** (a) Scanning electron micrograph showing connections between a neuron and skeletal muscle cells; (b) sketch of a neuromuscular junction as it would appear at high power under the electron microscope. The motor end plate includes the entire area containing junctional folds.

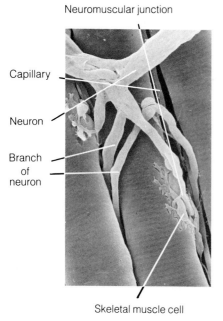

Neuromuscular junction

Capillary

Neuron

Branch of neuron

Skeletal muscle cell

(a)

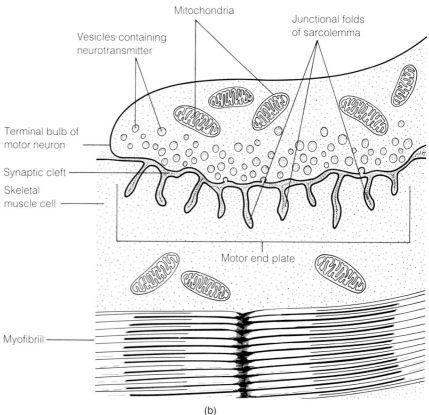

Mitochondria

Vesicles containing neurotransmitter

Junctional folds of sarcolemma

Terminal bulb of motor neuron

Synaptic cleft

Skeletal muscle cell

Motor end plate

Myofibriil

(b)

*depolarized.* Within a fraction of a second after depolarization occurs, the membrane again becomes relatively impermeable to Na$^+$; this event is associated with a rapid diffusion of potassium ions (K$^+$) to the exterior of the cell, which restores the normal resting potential, or *repolarizes* the membrane.

This cycle of instantaneous depolarization and repolarization constitutes what is known as an **action potential.** In a muscle cell, the action potential is propagated as an electrical current along the membrane. This occurs as depolarization, followed immediately by repolarization, spreads from the point of stimulation to adjacent areas of the membrane and so on over the sarcolemma and into the muscle cell by way of the transverse tubules.

After binding to the junctional folds, the acetylcholine is broken down within 2 to 3 milliseconds (thousandths of a second) by **acetylcholinesterase,** an enzyme present within the junctional folds. The breakdown of acetylcholine restores the original permeability characteristics of the sarcolemma, and consequently no further depolarization of the sarcolemma occurs until another nerve impulse arrives. The entire process of neuromuscular transmission is summarized in Figure 7–7.

## Diseases and Disorders Affecting Motor Pathways

The functioning of skeletal muscle is adversely affected by a variety of conditions resulting from disease that affects the brain, spinal cord, and motor neurons. These conditions, including Parkinson's disease, chorea, cerebral palsy, poliomyelitis, Lou Gehrig's disease, multiple sclerosis, and many others are discussed in Chapters 13 and 14. Each of these diseases causes a malfunction of the nervous system that prevents the normal stimulation of muscle cells by motor neurons.

Certain other diseases interfere with the transmission of nerve impulses at the neuromuscular junction. One of these is *myasthenia gravis,* in which the body's immune system damages and destroys acetylcholine receptors on the muscle cell sarcolemma, preventing membrane permeability changes that lead to depolarization and muscular contraction. All the voluntary muscles of the body are affected. In the early stages of the disease, the main symptom is muscle fatigue. In more advanced cases, loss of voluntary muscle function leads to crippling and death.

A number of chemicals also impede or block impulses at neuromuscular junctions, causing muscle paralysis. These include black widow spider toxin, nerve gases used in chemical warfare, and curare used on blowgun darts by South American Indians.

## THE SKELETAL MUSCLES

The skeletal muscles, formed of individual muscle cells with associated connective tissue, blood vessels, and nerves, produce movement of body parts. The structural and metabolic features of individual skeletal muscles vary considerably depending upon the type of movement they elicit.

## Cell Types Composing Skeletal Muscles

Most skeletal muscles are composed of a mixture of two types of cells: *red muscle cells* and *white muscle cells.* Red muscle cells are richly supplied by capillaries and contain in their sarcoplasm many mitochondria and large amounts of the reddish protein myoglobin. White muscle cells, on the other hand, have fewer mitochondria, contain only small amounts of myoglobin and are supplied by fewer capillaries. Red muscle cells and white muscle cells are frequently referred to as red fibers and white fibers, respectively, because of the fibrous appearance of muscle cells.

The two cell types function differently. Muscles composed primarily of red muscle cells are ideally suited for sustained activity supported by aerobic

FIGURE 7-7 Summary of the cellular and molecular events occurring during the transmission of a nerve impulse to a muscle cell.

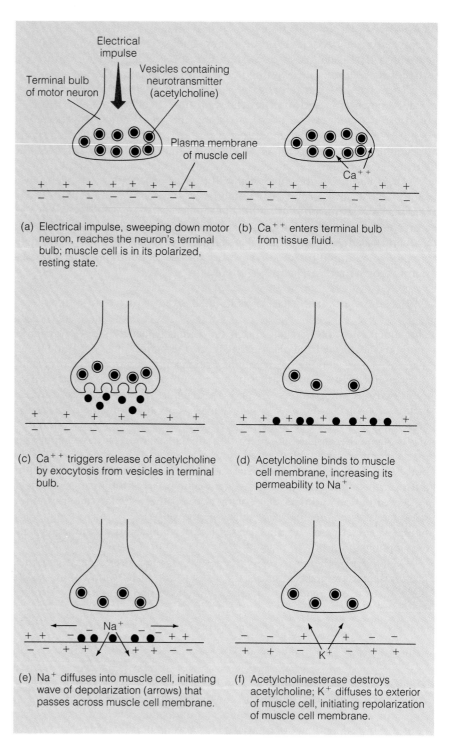

(a) Electrical impulse, sweeping down motor neuron, reaches the neuron's terminal bulb; muscle cell is in its polarized, resting state.

(b) Ca$^{++}$ enters terminal bulb from tissue fluid.

(c) Ca$^{++}$ triggers release of acetylcholine by exocytosis from vesicles in terminal bulb.

(d) Acetylcholine binds to muscle cell membrane, increasing its permeability to Na$^{+}$.

(e) Na$^{+}$ diffuses into muscle cell, initiating wave of depolarization (arrows) that passes across muscle cell membrane.

(f) Acetylcholinesterase destroys acetylcholine; K$^{+}$ diffuses to exterior of muscle cell, initiating repolarization of muscle cell membrane.

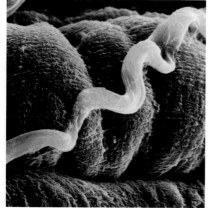

Nerve (light gray) along which electrical nerve impulses pass to appropriate muscle fibers (dark gray).

metabolism. Their numerous mitochondria provide the biochemical machinery to support high levels of aerobic ATP production, and oxygen needed for aerobic metabolism is supplied not only by the ample capillaries, but by use of the myoglobin molecule. Myoglobin is an oxygen-carrying protein similar in structure to the hemoglobin present in red blood cells. The oxygen stored by myoglobin in red muscle cells is released as it is needed by the cell, and consequently, red muscle cells have an alternative source of oxygen when blood oxygen levels are low. As a result, such muscles fatigue fairly slowly. Muscles composed primarily of white muscle cells are suited for short bursts of intense activity, but fatigue rapidly. They have large carbohydrate stores in the form of glycogen and are able to obtain energy by breaking down these carbohydrates through anaerobic metabolism.

## Slow Twitch and Fast Twitch Fibers

Exercise physiologists classify muscle cells according to the speed of their response to an electrical stimulus. On this basis, they refer to red muscle cells (red fibers) as *slow twitch fibers* and to white muscle cells (white fibers) as *fast twitch fibers*. Although there is a large variability among individuals in the proportion of slow and fast twitch fibers in a given muscle, some generalizations are possible. For example, muscles that maintain posture and thus must be well adapted for sustained activity, consist primarily of slow twitch fibers. Muscles that perform powerful movements of short duration, such as the biceps and triceps muscles of the upper arm, have a preponderance of fast twitch fibers.

Because slow twitch fibers are better for activities that require endurance, and fast twitch fibers are more appropriate to power-related activities, exercise physiologists have done a variety of studies to determine whether athletic performance in various events is influenced by the proportion of fiber types in an athlete's muscles. Such studies have shown that, on average, highly successful distance runners have leg muscles with about 80 percent slow twitch fibers as compared to about 45 percent in nonathletes. Successful sprinters, on the other hand, have leg muscles with about 75 percent fast twitch fibers. It is important to remember, however, that although a particular fiber composition may be advantageous in certain events, it is merely one factor among many important to success.

Fiber composition of muscles appears to be hereditary. Much evidence suggests that training cannot alter the proportion of fiber types in a muscle, but it can make both types of fibers respond more efficiently.

## Skeletal Muscle Parts

Typically, the two ends of a muscle are attached to different bones, and contraction of the muscle moves one bone relative to the other. The end of the

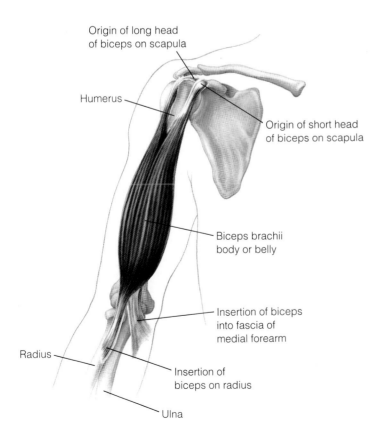

Origin of long head
of biceps on scapula

Humerus

Origin of short head
of biceps on scapula

Biceps brachii
body or belly

Insertion of biceps
into fascia of
medial forearm

Radius

Insertion of
biceps on radius

Ulna

FIGURE 7-8 Structure and attachment of the biceps muscle of the upper arm.

FIGURE 7-9 (a) Major muscles of the body, anterior view; (b) major muscles of the body; posterior view.

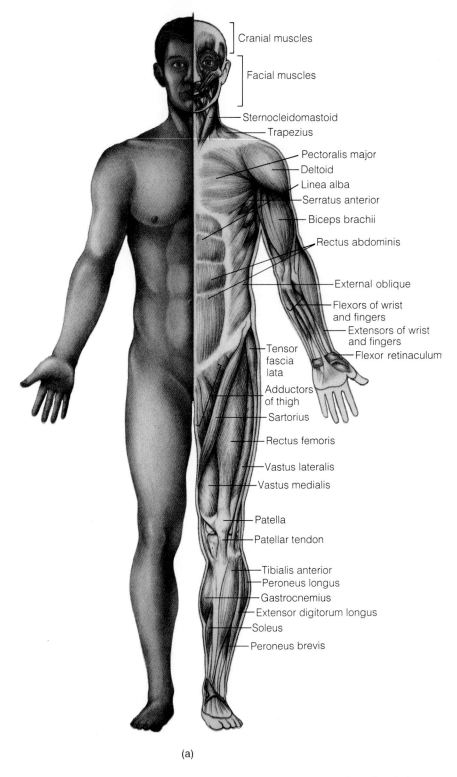

Cranial muscles

Facial muscles

Sternocleidomastoid
Trapezius
Pectoralis major
Deltoid
Linea alba
Serratus anterior
Biceps brachii
Rectus abdominis
External oblique
Flexors of wrist and fingers
Extensors of wrist and fingers
Tensor fascia lata
Flexor retinaculum
Adductors of thigh
Sartorius
Rectus femoris
Vastus lateralis
Vastus medialis
Patella
Patellar tendon
Tibialis anterior
Peroneus longus
Gastrocnemius
Extensor digitorum longus
Soleus
Peroneus brevis

(a)

muscle that is attached to the bone that remains more or less fixed during contraction is called the *origin;* the other end, the *insertion,* is attached to the bone that moves as a result of contraction. Many muscles separate into bundles near their origin or insertion ends; when this occurs, the bundles will attach to more than one bone or to different regions on the same bone. The *body,* or *belly,* of the muscle is the thick region between the two ends.

The parts of the *biceps* muscle are shown in Figure 7-8 on page 129. The **biceps** is the large muscle, lying anterior to the humerus, that bends, or *flexes,* the forearm. At its origin, the biceps (*bi* = two; *ceps* = heads) separates into two bundles, each of which attaches by tendons to different points on the

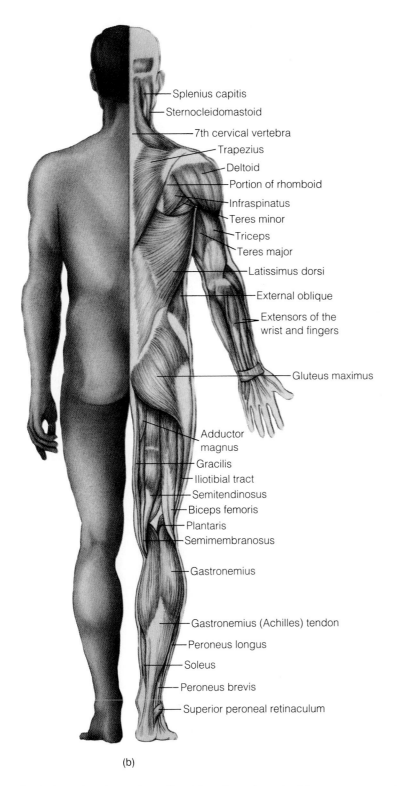

- Splenius capitis
- Sternocleidomastoid
- 7th cervical vertebra
- Trapezius
- Deltoid
- Portion of rhomboid
- Infraspinatus
- Teres minor
- Triceps
- Teres major
- Latissimus dorsi
- External oblique
- Extensors of the wrist and fingers
- Gluteus maximus
- Adductor magnus
- Gracilis
- Iliotibial tract
- Semitendinosus
- Biceps femoris
- Plantaris
- Semimembranosus
- Gastronemius
- Gastronemius (Achilles) tendon
- Peroneus longus
- Soleus
- Peroneus brevis
- Superior peroneal retinaculum

(b)

scapula; its insertion is on the radius. Note that when the biceps contracts, the radius moves at the elbow and the shoulder remains fixed.

## Individual Skeletal Muscles

Some of the more prominent of the skeletal muscles are shown in Figure 7–9. However, because there are more than 600 skeletal muscles, we will not attempt to describe individual ones. Students interested in a comprehensive description of the skeletal muscles should consult an advanced textbook of human anatomy.

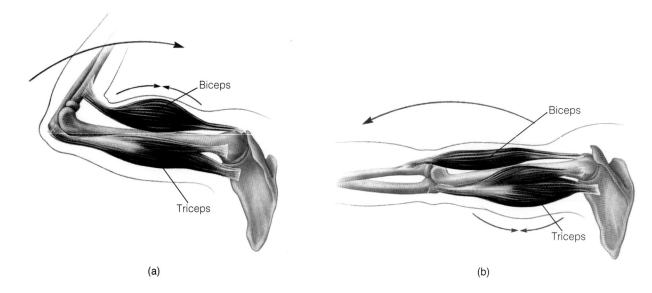

(a)                                                                      (b)

FIGURE 7–10 Role of antagonistic muscles in muscle action. (a) Contraction of the biceps muscle and relaxation of the triceps muscle brings about flexion of the forearm. (b) Contraction of the triceps and relaxation of the biceps causes forearm extension.

## Muscles in Action

Because muscles only generate tension by contracting and not by lengthening, they can exert a pull on bones but not a push. Consequently, in order to move a bone in one direction and then back to its original position, two muscles, operating as an **antagonistic pair,** are usually required.

### Agonists, Antagonists, and Synergists

The biceps muscle and the *triceps* muscle make up an antagonistic pair (Figure 7–10). In order to bend the forearm, the biceps muscle must contract and the triceps muscle, located at the back of the upper arm, must relax. Extension of the forearm requires contraction of the triceps and relaxation of the biceps.

In an antagonistic pair, the muscle that, by contracting, brings about a particular movement is called the *agonist* or *prime mover,* while the other member of the pair is called the *antagonist;* the antagonist relaxes when the agonist contracts. For example, in the bending of the forearm, the biceps operates as the agonist and the triceps, as the antagonist. However, in the extension of the forearm, the triceps is the agonist and the biceps becomes the antagonist.

Other muscles in addition to those forming the antagonistic pair may aid the agonist in bringing about a particular motion. These muscles are called *synergists,* and they are essential for smooth and efficient movement. For example, when the forearm is flexed, the pectoralis major and deltoid muscles (see Figure 7–9a) contract, maintaining the shoulder and upper arm in an appropriate position for the flexing movement.

### Muscle Tone

Healthy muscles at "rest" are in a state of partial contraction called **muscle tone,** or **tonus.** To maintain tone, groups of motor units operate in relays, such that some muscle cells are contracted at all times. Muscle tone is important in supporting the body during sitting and standing; without it our bodies would collapse. Patients requiring long periods of bed rest have a reduction in muscle tone that is responsible for the weakness they experience when first getting out of bed. Indeed, in some cases, such weakened muscles will not support the body, and the patient cannot stand without assistance.

Muscle tone is also responsible for maintaining our normal facial expression. The loss of tone in paralyzed facial muscles accounts for the characteristic "sagging" appearance of one side of the face in stroke victims.

## Strength of Stimulus and the All-or-None Response

We have seen that each skeletal muscle is composed of a number of motor units, and that all the muscle cells of a single motor unit contract simultaneously in response to stimulation by the nerve cell that supplies them. However, the muscle cells in different motor units require different levels of nervous stimulation in order to contract; that is, the muscle cells in some motor units will be depolarized at a stimulus strength that is inadequate to depolarize the muscle cells in other units. In conjunction with these observations, it has been shown that when the muscle cells in any motor unit receive a stimulus of a particular strength, they will contract either fully or not at all. This property of muscle cells has been termed the **all-or-none response.**

Although all muscle cells display the all-or-none response, it is important to point out that the strength of their contraction depends upon the overall physiological state of the muscle. Thus, muscles weakened by malnutrition or fatigue cannot contract as strongly as healthy muscles.

## Matching Force to Work

Flexing the forearm while holding an ordinary construction brick in the hand requires much less muscle strength than performing the same movement while holding a lead brick. This means, of course, that the contraction of fewer motor units in the biceps muscle is required to perform the former movement than the latter. Because different motor units are activated at different stimulus strengths, more and more motor units will be activated—and hence the force of contraction will increase—as the strength of the stimulus increases. This situation provides a mechanism by which the force of muscle contraction can be graded to match the requirements of the work performed.

As we develop muscular coordination, we learn how to judge the force required for particular movements and to bring about the appropriate degree of muscle contraction. On occasion, such judgments may be incorrect, as when we think an object is heavier than it is, and practically "throw it through the roof."

## Maintaining Smooth Contractions

If a muscle receives a single electrical stimulus of sufficient strength, contraction will occur, followed by relaxation. There is a very short period following contraction, called the **refractory period,** during which the muscle will not respond to a second stimulus. However, if a muscle receives a series of stimuli at intervals just longer than the refractory period, it will remain contracted and show no apparent relaxation. This phenomenon is called **tetanic contraction** or **tetanus** (not to be confused with the disease called tetanus discussed later). Tetanic contraction results in the smooth sustained contractions that are required for most movements performed by skeletal muscle.

## Isotonic and Isometric Contractions

When a muscle contracts, tension always develops in the muscle, but the muscle does not always shorten. Contraction accompanied by shortening of the muscle is called **isotonic contraction** and occurs whenever work is done. For example, when an individual lifts an object, muscles contract, develop tension, and shorten. However, in situations in which an individual attempts to lift an object that is beyond his or her capacity, tension develops but the involved muscles cannot shorten, resulting in what is called **isometric contraction** (Figure 7–11).

A few years ago, exercises using isometric contractions were touted for developing muscles without the risk of muscle strain. Typical isometric exercises involve pressing against or pulling an immovable object. Although

FIGURE 7–11 Isometric contractions occur as this individual attempts to lift an automobile.

isometric exercises can increase muscle size significantly, it is now known that the strength of the muscle is not significantly increased except at the one angle at which the tension is applied.

## Disorders of Skeletal Muscles

### Spasms and Cramps

A sudden involuntary contraction of a muscle or group of muscles is called a *spasm.* For example, a "stiff" neck is caused by protracted spasms in muscles of the neck and shoulder. If a spasm is unusually severe and painful, it is called a *cramp.* Cramps usually last only a few minutes before relaxation of the muscle occurs. Muscle spasms can usually be relieved by heat and vigorous massage; in severe cases, skeletal muscle relaxants may be prescribed.

### Strains

Vigorous exercise to which an individual is unaccustomed may overstretch and tear some of the muscle cells, resulting in a *pulled* or *strained muscle.* Strained muscles become stiff, painful, and tender and are best treated by applying ice to reduce swelling and resting the muscle for several days.

### Hernias

When muscles are weakened because of a strain or a congenital condition, underlying soft tissues may push through or between the muscles. This condition is called a *hernia.* Hernias occur most commonly in the abdominal wall, which is composed of sheets of muscle and associated connective tissues that enclose the abdominal organs. If a part of the abdominal wall is weakened, it is not uncommon for a small section of intestinal tissue to push through the abdominal wall, causing a bulge (hernia) clearly visible beneath the skin. Such hernias tend to worsen gradually. Complications include *obstruction,* in which constriction of a loop of intestine may prevent passage of intestinal contents, and *strangulation,* in which constriction cuts off blood supply to the protruding tissue. Most abdominal hernias are corrected surgically by pushing in the protruding tissue and sewing up the weak area in the muscle.

One type of abdominal hernia that usually heals without treatment is an umbilical hernia in newborns. These are quite common, appearing as bulges around the navel. The risk of obstruction or strangulation is small since the opening in the abdominal wall is quite broad.

Another common type of hernia is a *hiatus hernia,* caused by weakness of muscles surrounding the hiatus, the hole in the diaphragm through which the esophagus passes. Hiatus hernias are discussed in Chapter 10.

### Tetanus

Tetanus is a serious, often fatal disease caused by the soil bacterium *Clostridium tetani* that invades the body through wounds. Infection by *Clostridium tetani* occurs most frequently in puncture wounds that do not bleed profusely, and hence the bacteria are not washed away. A toxin produced by the bacteria attacks motor neurons in the spinal cord, causing severe spasmodic contractions that do not abate. Suffocation brought about by spasms in the throat and respiratory muscles causes death in about 60 percent of all cases. Tetanus may be prevented by immunization. Tetanus is also called *lockjaw* because spasms of the jaw muscles clamp the jaw tightly shut, but this is somewhat of a misnomer since most skeletal muscles, not just those of the jaw, are involved. Contrary to popular belief, rusty nails are no more likely to be contaminated with the tetanus bacillus than other sharp objects in or near the soil.

# SMOOTH MUSCLE

Smooth muscle is found primarily in the walls of many internal organs, where it performs a variety of functions. For example, in the walls of the digestive tube, the contraction of smooth muscle in rhythmic, descending waves (peristalsis) serves to propel food through the digestive tract, while other rhythmic contractions bring about the mixing of food with digestive juices. Evacuation of the bladder and ejection of the fetus from the uterus during labor occur as a result of smooth muscle contraction. The presence of smooth muscle in the walls of blood vessels and of the respiratory passages to the lungs helps control, respectively, the flow of blood and the flow of air through these structures.

In addition to its presence in the walls of hollow organs, smooth muscle occurs in the iris of the eye, where it determines constriction and dilation of the pupil. Smooth muscle also forms the ciliary muscles that shape the eye's lens and the arrector pili muscles in skin that cause hairs to stand on end (see Chapter 4).

## Structure of Smooth Muscle Cells

Smooth muscle cells are thin and spindle shaped (Figure 7-12). Within the cell, which is bounded by a plasma membrane or sarcolemma, is a centrally located nucleus surrounded by sarcoplasm containing numerous mitochondria and a small amount of sarcoplasmic reticulum. Thick and thin filaments, oriented parallel to the long axis of the cell are present in the sarcoplasm, but these are not neatly organized into the striated myofibrils typical of skeletal and cardiac muscle. Hence, smooth muscle cells are nonstriated in appearance.

## Smooth Muscle Tissue

The arrangement of smooth muscle cells varies greatly from tissue to tissue. They may occur singly or, with their tapered ends overlapping, they may be arranged roughly parallel to one another to form bundles or layers of smooth muscle tissue. For example, the arrector pili muscles consist of small bundles of smooth muscle cells. In the walls of small arterioles, isolated smooth muscle cells are oriented in a circular fashion around the vessels; in larger vessels, spirally arranged smooth muscle cells form a continuous layer. In most hollow internal organs, there are two layers of smooth muscle: an inner layer in which the cells are arranged circularly around the organ, and an outer one of longitudinally arranged cells. Sheaths of connective tissue on the surface of such smooth muscle layers penetrate into the layers and separate groups of cells into bundles. Spaces between the bundles are richly supplied by blood capillaries.

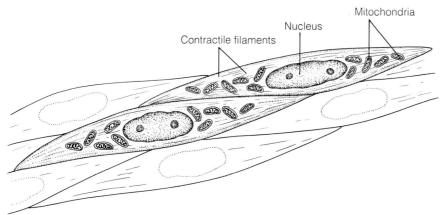

Contractile filaments    Nucleus    Mitochondria

FIGURE 7-12 Structure of smooth muscle cells.

## Contraction of Smooth Muscle

Smooth muscle, like cardiac muscle, is *involuntary* in the sense that its contraction is not generally subject to our conscious control. However, recent work involving the techniques of biofeedback has shown that it is possible to teach humans and other animals to alter contractions of smooth muscle, as discussed in Chapter 13.

### Initiation of Contraction

In general, two types of smooth muscles can be distinguished on the basis of how their contraction is initiated: *single-unit smooth muscles* and *multiunit smooth muscles.* Single-unit smooth muscles are the most common. They are present in the walls of most of the body's hollow internal organs and in the walls of small blood vessels. Contraction of a single-unit smooth muscle is initiated at more or less regular intervals when a small, localized group of its muscle cells spontaneously depolarize. Because the individual muscle cells composing single-unit muscles are bound firmly together, a wave of depolarization is rapidly conducted from cell to cell throughout the muscle, triggering a wave of contraction. Hence, the muscle contracts as if it were a single unit. Although the contraction of single-unit smooth muscles is triggered spontaneously, the rate of contraction may be modified by nerves and hormones. Moreover, stretching of single-unit muscles can also be a stimulus to contraction. This phenomenon is particularly important in organs like the bladder and large intestine that are evacuated by contractions.

In contrast to single-unit smooth muscles, the contraction of multiunit smooth muscles requires nerve stimulation. Each muscle cell of a multiunit smooth muscle is supplied by nerve endings. This is similar to the situation in skeletal muscle, although the nerves that supply smooth muscle are from the autonomic (involuntary or "automatic") division of the nervous system. The contraction of multiunit smooth muscle can be modified by hormones. Multiunit smooth muscle is found in the walls of larger blood vessels and in the iris of the eye, and forms the ciliary muscles and arrector pili muscles. Nervous stimulation of these muscles can generate relatively fast and graded contractions as compared to contractions of single-unit smooth muscles, which occur at a slower and uniform rate.

### Mechanism of Contraction

Although smooth muscle cells are not striated and the myofibrils are not organized into sarcomeres, it is generally accepted that the contraction of smooth muscle occurs by a sliding filament mechanism similar to that postulated for skeletal and cardiac muscle. This view is supported by the fact that tropomyosin and troponin have been found in association with the thick and thin filaments and by the detection of crossbridges between thick and thin filaments in smooth muscle cells. However, the precise details of the mechanism of contraction are not known.

### Characteristics of Smooth Muscle Contraction

Smooth muscle contractions develop more slowly than that of other muscle types, but forceful contractions of smooth muscle can be sustained for longer periods of time. These characteristics are particularly evident during the contraction of smooth muscle in the uterus at the time of labor.

Like skeletal muscle, smooth muscle can exist in a continuous state of partial contractions (tonus) that helps maintain the shape of hollow internal organs. Increased tonus of smooth muscle in the walls of the respiratory tubes causes a narrowing of the air passages and labored breathing, a condition known as *asthma.* Asthma is usually caused by allergic sensitivities to inhaled or ingested substances, and it is treated by chemicals such as epinephrine that relax smooth muscle.

## Study Questions

1. Describe the anatomy of a skeletal muscle, including the organization of its connective tissues.
2. Describe the structure of a myofibril of a skeletal muscle cell.
3. In your own words, summarize the sliding filament theory of skeletal muscle contraction.
4. Describe the sequence of metabolic events that occurs when strenuous muscular activity is begun and sustained.
5. a) Diagram a neuromuscular junction.
   b) Explain how nervous impulses are conducted across the neuromuscular junction.
6. Indicate the functional differences between red and white muscle cells (fibers).
7. Explain how the actions of antagonistic pairs of muscles bring about movement.
8. a) What is muscle tone (tonus)?
   b) Why is it important?
9. Explain why muscle strain cannot occur when a muscle contracts isometrically.
10. Give examples of the arrangement of smooth muscle cells in several different tissue types.
11. Distinguish between multiunit and single-unit smooth muscle.

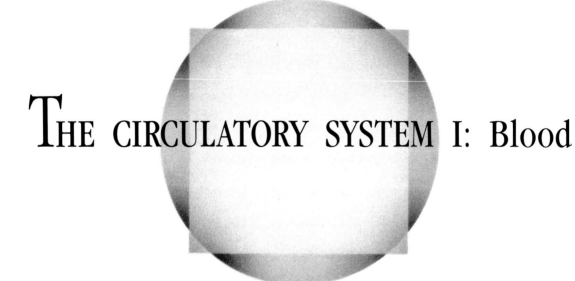

# THE CIRCULATORY SYSTEM I: Blood

The circulatory system, consisting of the heart, blood vessels, and blood, is the body's transportation system. By the pumping action of the heart, blood is circulated throughout the tissues of the body in a massive network of continuous, interconnected tubes, the blood vessels. Since all cells have access to the blood by way of the tissue fluid (see Chapter 4), materials can be effectively transported between various specialized parts of the body. For example, as blood courses through the lungs it picks up oxygen that is transported to all the body's cells, and it releases carbon dioxide, a metabolic waste product that is discharged by cells into the blood. In a similar manner, blood flowing through vessels near the lining of the intestine absorbs water and nutrients, which are then carried to all cells. Urea and other organic waste products of cellular metabolism, discharged into the blood by the cells, are carried to the kidneys where they are eliminated as urine.

In addition to transporting water, nutrients, and oxygen to cells, and waste products from cells, the circulatory system carries out a variety of other functions as well. Circulating blood distributes heat throughout the body, thus playing an important role in temperature regulation. Blood also serves as a buffer that maintains the body's proper acid-base balance. Hormones and other chemical substances that regulate and coordinate the metabolic activities of the body are carried by the circulatory system from specialized cells that synthesize them to target cells. And lastly, but crucially, white blood cells carried in blood protect the body from invasion by pathogenic organisms and other harmful substances.

## BASIC DESIGN OF THE CIRCULATORY SYSTEM

The functional relationship between the heart and the blood vessels is shown diagrammatically in Figure 8–1. The heart pumps blood into major vessels called **arteries.** These divide and redivide, eventually giving rise to medium-sized vessels, the **arterioles.** Each arteriole branches into a number of still smaller vessels, termed **capillaries,** that form vast networks that permeate the body's tissues. Blood, flowing from the arterioles through the capillaries, enters slightly larger collecting vessels, the **venules,** and then drains into larger **veins.** Major veins return blood to the heart, completing the circuit.

It has been estimated that there are over 60,000 miles of blood vessels in the human body. That most of this distance is covered by capillaries demonstrates the fantastic extent of the capillary networks.

FIGURE 8–1 A simplified diagram showing the relationship between the heart and blood vessels. Blood flow occurs in the direction of the arrows.

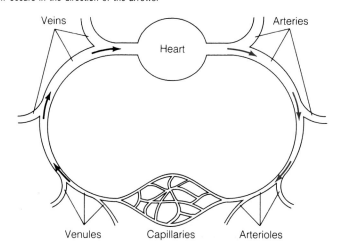

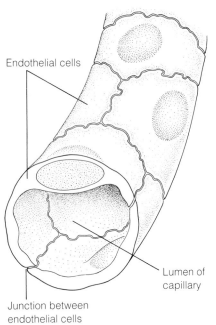

Endothelial cells

Lumen of capillary

Junction between endothelial cells

FIGURE 8-2 Structure of a blood capillary, showing the arrangement of the endothelial cells that form the capillary wall. Notice that the spaces between adjacent endothelial cells provide channels through which substances can pass.

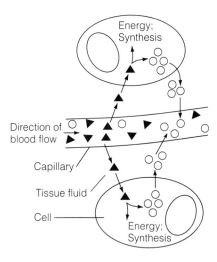

Energy; Synthesis

Direction of blood flow

Capillary

Tissue fluid

Cell

Energy; Synthesis

FIGURE 8-3 Diagram summarizing the exchange of materials between cells and capillaries. Nutrients and oxygen (solid triangles) diffuse from the capillaries to the tissue fluid and then into the cells. As a result of cellular metabolism, these substances are converted to waste products (open circles) that diffuse in the opposite direction.

## Function of the Capillary Networks

The transfer of heat and materials between the bloodstream and the cells of the body occurs at the capillaries. You will recall that all cells are bathed in tissue fluid, which serves as an intermediary in the transfer of substances between the cells and the bloodstream (in both directions). Essential materials pass through the walls of capillaries into the tissue fluid, and from there, they enter cells by permeating the cells' plasma membranes; waste products leave the cell and enter the bloodstream by the reverse route. Thus, the transfer of any substance between the bloodstream and the cells is regulated by two events: the passage of the substance through the plasma membrane and the passage of the substance through the capillary wall.

### Transfers through the Plasma Membrane

In Chapter 3, we discussed the mechanisms that control the passage of substances through the selectively permeable plasma membrane. These mechanisms, which include diffusion (both passive and facilitated), osmosis, active transport, exocytosis and endocytosis, are responsible for the passage of essential materials from the tissue fluid into the cell and of waste products from the cell into the tissue fluid.

### Transfers through the Capillary Wall

Exchanges of substances between the bloodstream and the tissue fluid occur through the walls of the capillaries. Capillary walls consist of a single layer of specialized epithelial cells, called *endothelial cells,* with an associated basement membrane. The plasma membranes of adjacent endothelial cells are not fused tightly together, leaving intercellular spaces that serve as channels through which substances can enter and leave the capillaries (Figure 8-2). The capillary walls thus operate as a kind of sieve, permitting the passage of oxygen, carbon dioxide, water, and organic molecules that are small enough to penetrate the channels, but excluding, or at least substantially retarding, the passage of proteins, other macromolecules and cells.

Although a variety of substances are small enough to permeate the walls of capillaries, it is clear that there must be mechanisms that regulate the direction of their flow. Why, for instance, do waste materials pass from the tissue fluid into the bloodstream, while nutrients and other essential materials move in the opposite direction? The direction of flow through capillary walls is determined by a combination of processes, including diffusion, hydrostatic pressure, and osmosis.

*Diffusion through Capillaries.*   As you remember, molecules of any type will tend to move from a region where they are highly concentrated to one in which they are less concentrated. This tendency is called diffusion, and it is an important driving force in determining the direction that any substance will move through the capillary wall. Substances such as oxygen and nutrients that are essential to cells will diffuse out of the bloodstream and into the tissue fluid because they are more concentrated in the bloodstream than they are in the tissue fluid, and they will continue to diffuse out of the bloodstream because their concentrations in the tissue fluid are kept low as a result of the movement of these substances into the cells. The reverse is true, of course, of waste materials: their concentrations build up in the tissue fluid as they are eliminated by cells, and they consequently pass by diffusion into the bloodstream (Figure 8-3).

*Effects of Hydrostatic Pressure and Osmosis on Diffusion.*   The pumping action of the heart generates hydrostatic pressure (pressure exerted by liquids) in the bloodstream; this pressure is greatest in the arteries and drops to near zero as blood returns to the heart. At the arteriole ends of capillaries, the pressure of blood in the bloodstream is greater than the pressure exerted by the surrounding tissue fluid. Consequently, some water and small molecules dissolved in it are pushed through the intercellular spaces in the capillary wall. Large molecules, including most proteins, are unable to penetrate the intercellular spaces, and therefore remain in the bloodstream. Near the venule ends of the capillaries, the pressure of the blood is reduced significantly, which largely eliminates further filtration of water out of the bloodstream; at the same time, the proteins in the blood tend to pull water by osmosis from the tissue fluid back into the bloodstream. The net effect, however, is that more water is pushed out of the bloodstream at the capillaries than is returned to it. The excess water, containing dissolved substances, drains into *lymphatic capillaries* that permeate the tissues. This fluid, now called **lymph,** is transported from the lymphatic capillaries into larger lymphatic vessels that ultimately dump into the bloodstream.

Tissue fluid, then, is derived from fluid filtered out of the bloodstream. Tissue fluid volume is maintained relatively constant by a delicate balance between processes leading to the accumulation of tissue fluid and those responsible for its removal. This fluid balance may be upset on occasion as a result of tissue injury or disease, leading to the accumulation of excess tissue fluid, known as *swelling* or *edema,* or the net loss of tissue fluid, termed *dehydration.*

## Blood Flow through Capillaries

The flow of blood through capillaries is controlled by the arterioles. The walls of arterioles contain smooth muscle, contraction of which narrows the diameter of arterioles, reducing blood flow through the capillaries they supply. The ability to alter the flow of blood through capillaries allows the body to meet the metabolic requirements of different tissues by shunting blood to areas where it is needed. For example, during strenuous activity, blood flow through particular skeletal muscles can be increased up to about fourfold. Regulation of blood flow through capillaries near the surface of the skin is also an important mechanism in temperature regulation, as mentioned in Chapter 5.

## BLOOD

Blood constitutes about seven percent of the body's weight. This percentage translates into a volume of about 3 to 3.5 liters of blood for the average woman, and on the order of 4.5 to 5 liters for the average man.

Circulating blood is a red-colored viscous liquid that clots readily when removed from blood vessels. If, however, blood is treated with an anticoagulant to prevent clotting and is then allowed to settle or is spun in a centrifuge, it separates into three distinct layers, as shown in Figure 8–4. The upper layer, constituting about 55 percent of the total blood volume, is a straw-colored fluid called **plasma;** the middle layer, a thin whitish band, consists of *white blood cells* and *platelets;* the bottom layer consists of *red blood cells.* Whole blood is thus a suspension of various cellular elements in plasma.

## Plasma

Plasma consists, by weight, of about 92 percent water, 6 to 8 percent dissolved proteins, and small amounts of ions and dissolved organic compounds such as glucose, amino acids, fats, urea, hormones, and vitamins. The proteins, termed *plasma proteins,* are of three main types: *albumins, globulins,* and *fibrinogen.* Albumins are the most plentiful. They serve an important function by pulling water from the tissue fluid into the blood by osmosis, thus helping to maintain plasma volume. The globulins in plasma can be divided into *alpha* ($\alpha$), *beta* ($\beta$), and *gamma* ($\gamma$) types. The $\alpha$- and $\beta$-globulins serve primarily to transport hormones and other substances by combining with them. Gamma-

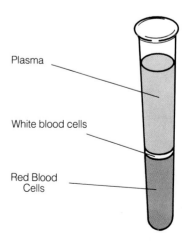

Plasma

White blood cells

Red Blood Cells

FIGURE 8–4 A diagram showing normal human blood, treated with anticoagulant and centrifuged.

globulins, or *antibodies,* are the only plasma proteins not manufactured in the liver; instead, they are synthesized by specialized cells derived from lymphocytes, a type of white blood cell. **Antibodies** protect the body by providing immunity to disease. *Fibrinogen* is important in blood clotting. At the site of a break in a blood vessel, soluble plasma fibrinogen is converted into insoluble strands of *fibrin,* forming a mesh which traps blood cells and helps plug the break.

The ions in plasma are similar to those found in sea water and are present in the same relative proportions, but not in the same amounts. The similarity in the ionic balance of sea water and plasma has been explained as an evolutionary legacy of our ancestral relationship to organisms that evolved in the sea.

Because tissue fluid is a filtrate of plasma, both have similar amounts of ions and small organic compounds. However, the plasma proteins, with the exception of a small amount of albumin, are not present in tissue fluid.

## Red Blood Cells

Red blood cells, also known as *red blood corpuscles* or **erythrocytes** (*erythro* = red; *cytes* = cells) form about 45 percent of the volume of whole blood. The average adult has approximately 5 million red blood cells per cubic millimeter of blood, or an astounding 5 trillion per liter.

### Shape and Structure

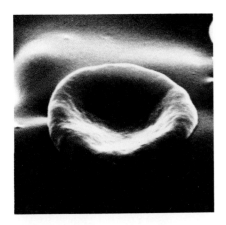

FIGURE 8-5 Shape of a red blood cell as revealed in the scanning electron microscope.

Erythrocytes have the shape of biconcave discs and are quite flexible—a characteristic that allows them to squeeze single file through the smallest capillaries whose diameters are less than their own (Figure 8-5). Structurally, red blood cells are not typical cells. Although they possess a limiting plasma membrane, they do not contain a nucleus, and their cytoplasm, virtually devoid of intracellular structure, consists of little else than a large number of *hemoglobin* molecules dissolved in the intracellular fluid.

### Hemoglobin

**Hemoglobin** is a protein with a somewhat complicated molecular structure. It consists of four amino acid chains—two identical alpha chains and two identical beta chains—each of which has associated with it a *heme group,* a complex iron-containing organic molecule. The four amino acid chains and their heme groups are arranged in a specific three-dimensional configuration relative to one another to form the hemoglobin molecule (Figure 8-6). The iron atom in each heme group is capable of reversibly binding a molecule of oxygen, and hence one hemoglobin molecule can carry four oxygen molecules. Because of its high affinity for oxygen, the hemoglobin in red blood cells allows the blood to carry vastly more oxygen than it could if the oxygen were simply dissolved in the plasma.

Hemoglobin picks up oxygen in the lungs where oxygen is plentiful and releases it at the tissues where oxygen is scarce. After discharging its oxygen at the tissues, hemoglobin can bind some carbon dioxide, which it releases at the lungs. However, most of the carbon dioxide transported in the blood is *not* bound to hemoglobin. (In Chapter 10, we will discuss in detail the dynamics of the transport and exchange of respiratory gases between the lungs and the blood, and between the tissues and the blood.)

Oxygenated hemoglobin is bright scarlet and gives blood its characteristic color. Indeed, as we mentioned previously, it is oxygenated hemoglobin in erythrocytes flowing through capillaries that imparts a pinkish cast to whitish skin and a somewhat darker pink to mucous membranes. When hemoglobin is completely deoxygenated it is a bluish color, but circulating erythrocytes usually contain enough oxygenated hemoglobin, even in oxygen-poor blood, to give the blood a dark red color instead. There are, however, certain circumstances where large amounts of deoxygenated hemoglobin are present in blood cells. In extreme cold, lips turn blue because the local capillary

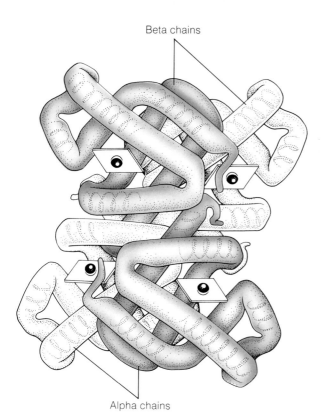

Beta chains

Alpha chains

FIGURE 8-6 A model of hemoglobin structure. Bound to each of the four amino acid chains (two alpha chains and two beta chains) is a heme group, represented here as a rectangular plate, containing a central iron atom (sphere) that binds $O_2$.

circulation is sufficiently shut down to conserve heat. Consequently, hemoglobin in this region becomes abnormally deoxygenated, and the blue color becomes visible in the superficial capillaries. In diseases or pathological conditions in which the oxygenation of blood in the lungs is impaired, blood containing a substantial amount of deoxygenated hemoglobin is delivered to all the capillaries of the body, giving the skin a bluish color. This condition is called *cyanosis*.

### Formation of Red Blood Cells

The life span of an erythrocyte is about 100 to 120 days. This means that about one percent of the total circulating red blood cells must be replaced every day in order to keep the number of red blood cells in the circulation constant. Aging or damaged erythrocytes are removed from the blood stream by scavenger cells located in the liver, spleen, and bone marrow; the scavenger cells ingest these cells and break down the hemoglobin to its raw materials for recycling. Red blood cells that are removed from the circulation are normally replaced by new red blood cells produced, in the child, in the red marrow of most bones, and in the adult, in the red marrow of the ribs, sternum, cranial bones, vertebrae, and epiphyses of long bones. The bone marrow houses blood cell precursors called *stem cells* that proliferate to produce immature nucleated erythrocytes (as well as other blood cell types). The immature erythrocytes subsequently synthesize hemoglobin and extrude their nuclei, becoming mature erythrocytes that enter the circulation. By this process, the bone marrow of a typical adult produces something on the order of a half ton of erythrocytes in a lifetime.

### Diseases Involving Red Blood Cells

A number of diseases and abnormal conditions result in *anemia*. This term is used to describe a reduction in the number or hemoglobin content of circulating erythrocytes—a condition that causes a decrease in the oxygen-carrying capacity of the blood. Reduction in oxygen supply to the tissues

# Blood Transfusions and Artificial Blood

Throughout the ages, blood has symbolized the very essence of life. Nowadays, the transfusion of blood to the terminally ill or injured is one of the simplest and most effective procedures in the physician's lexicon of life-saving technologies. So essential is an ample supply of blood to modern medicine that the Red Cross Bloodmobile has become a fixture on the American landscape. Recently, however, "the gift of life" has sometimes caused unintended tragedy. For many blood recipients, transfusions have been vectors for the transmission of serious, even fatal, infectious diseases.

Prior to 1984 as many as 12,000 Americans may have received blood contaminated with the AIDS virus. The routine use of blood screening since that time has reduced the chances of receiving AIDS-contaminated blood to an estimated 1 in 100,000. These highly favorable odds still mean that every year some 350 people in the United States are transfused with blood that is tainted with the AIDS virus. AIDS-infected blood slips into the blood supply because the existing screening tests detect the presence of antibodies to the AIDS virus, not the AIDS virus itself. Since an individual does not make antibodies for several weeks after initial infection with the AIDS virus, blood donated during that time period will test negative even though it contains the virus.

Far greater than the danger of contacting AIDS from contaminated blood is the possibility of developing viral hepatitis. Although donated blood is routinely screened for hepatitis, current techniques do not detect the hepatitis virus of the so-called non-A, non-B type. This form of hepatitis often leads to chronic liver disease and is sometimes fatal, particularly in the seriously ill. In some cities, as many as one in ten blood recipients have contracted the disease.

Aside from risking infectious disease, people who receive donated blood some-times have severe, even fatal, allergic reactions as a result of errors in blood type matching. Furthermore, some people have allergic reactions to blood that is not their own, even if it is of the proper type.

The fear of receiving contaminated blood or of suffering allergic reactions has drawn attention to new ways of ensuring the safety of blood transfusions. At a recent meeting at the National Institutes of Health, a panel of physicians and blood bank administrators recommended that patients anticipating elective surgery donate some of their own blood to be stored in case they need it. This procedure is now available at many hospitals. Because blood cannot be stored indefinitely, stockpiling

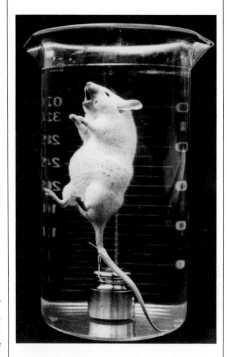

Liquid breathing mouse. Mouse is totally immersed in fluorocarbon liquid FC 75 (perfluorobutyltetrahydrofuran; 3M Company) which has been saturated with oxygen by bubbling at room temperature. A mouse can survive breathing such a liquid for many hours and when withdrawn and drained, head down, recovers in good health.

is not a practical solution to the need for blood in unanticipated, emergency situations.

Problems with the natural blood supply have also led to renewed interest in an old idea — developing an artificial blood substitute. Two major approaches have been followed. One approach involves the use of organic compounds, primarily a group of liquid carbon compounds containing fluorine (fluorocarbons), which can dissolve large amounts of oxygen. One mixture of fluorocarbons, called Fluosol DA, had shown some promise as a blood substitute in experimental animals, but recent clinical trials on several dozen severely anemic patients who had refused blood transfusions on religious grounds were unsuccessful. Although other fluorocarbons have been developed and are awaiting testing, questions have been raised about the short-term and long-term toxicity of fluorocarbon compounds.

A second approach has involved the use of purified hemoglobin, either encapsulated in microscopic membranous vesicles made of phospholipids and cholesterol ("neohemocytes"), or modified chemically and allowed to circulate free in the bloodstream. These preparations are able to carry and release oxygen. However, in tests with experimental animals, the neohemocytes are recognized as foreign and are destroyed by the immune system in several hours; the free-flowing hemoglobin, while not attacked by the immune system, is cleared from the circulation within a day or two. Another serious problem with artificial blood that uses hemoglobin is that the hemoglobin must be isolated from blood and hence carries the risk of transmitting viral infections.

Although the technical problems in developing artificial blood are daunting, progress to date has been substantial. Many researchers in the field are predicting that an artificial blood solution will be clinically available within a few years.

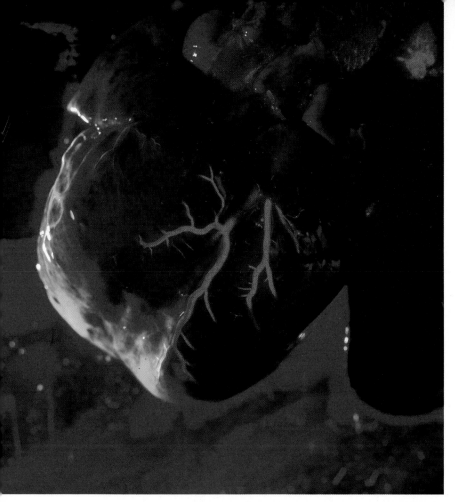

# THE HUMAN HEART

The heart is life's incredible pump. Weighing between 10 and 16 ounces in an adult, the heart beats some 70 times a minute for a lifetime. It pumps five liters of blood every minute through 60,000 miles of blood vessels, sustaining life by supplying oxygen and other nutrients to all cells of the body and removing their accumulated waste products.

To work ceaselessly and efficiently, the cardiac muscle that generates the heart's pumping pressure requires substantial amounts of oxygen and nourishment. Indeed, in a person at rest, only the brain requires more oxygen for its weight. The heart receives blood from two small arteries, the left and right coronary arteries, that branch off from the aorta just as it leaves the heart. The coronary arteries run along the surface of the heart, eventually breaking up into an extensive capillary network that permeates every part of the heart muscle. Highlighted with ultraviolet light in this photograph of the heart, the coronary arteries are clearly visible as delicate red branches. About the diameter of drinking straws, the coronary arteries are especially prone to blockage by the fat deposits of atherosclerosis.

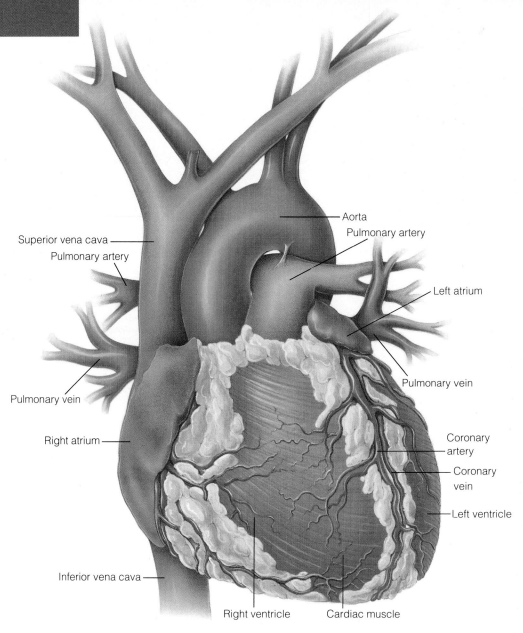

Superior vena cava

Pulmonary artery

Aorta

Pulmonary artery

Left atrium

Pulmonary vein

Pulmonary vein

Right atrium

Coronary artery

Coronary vein

Left ventricle

Inferior vena cava

Right ventricle

Cardiac muscle

The external shape of the heart is somewhat conical. It is positioned in the chest cavity with its apex pointing downward and somewhat to the left. Blood vessels that connect to the heart enter and leave at its top. When the heart contracts, its apex strikes the inner chest wall a few inches to the left of the sternum, causing a pulsation that is visible or can be felt at the surface of the chest.

The heart is composed of four chambers: two thin-walled atria that receive blood returning to the heart, and two thick-walled ventricles that pump blood from the heart. Most of the thickness of the atrial and ventricular walls is cardiac muscle. The wall of the left ventricle is the thickest, since it does most of the work of the heart, pumping

blood to all parts of the body except the lungs.

The force required for the pumping action of the heart is generated by alternate cycles of contraction and relaxation of cardiac muscle. All cardiac muscle cells have the inherent ability to contract simultaneously and rhythmically, but the coordinated contraction of all the muscle cells that compose the heart is controlled by a group of specialized cardiac muscle cells, called the pacemaker. Located in the upper region of the right atrium, the pacemaker initiates contraction of the heart by generating an electrical impulse about 70 times every minute. Each impulse spreads as a wave of electrical activity through both atria, stimulating their contraction. Near the base of the right

atrium, the spreading electrical wave reaches another group of specialized cardiac muscle cells, constituting the atrioventricular node (AV node). The impulse is conducted from the AV node along specialized fibers that run down both sides of the walls separating the two ventricles and spread into the cardiac muscle of the right and left ventricles. The wave of electrical activity conducted from these specialized fibers causes essentially simultaneous contraction of both ventricles following contraction of the atria. Although the "resting" heart rate is determined by the spontaneous activity of the pacemaker, the rate of the heartbeat can be altered by nerves and hormones depending upon the oxygen demands of the body.

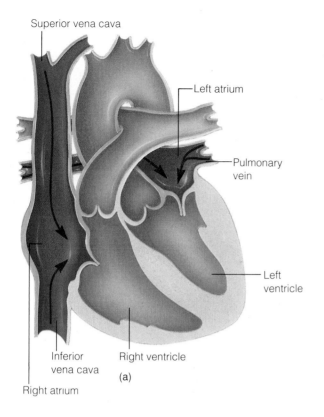

Superior vena cava

Left atrium

Pulmonary vein

Left ventricle

Inferior vena cava

Right ventricle

Right atrium

(a)

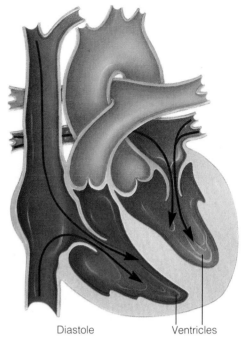

Diastole

Ventricles

(b)

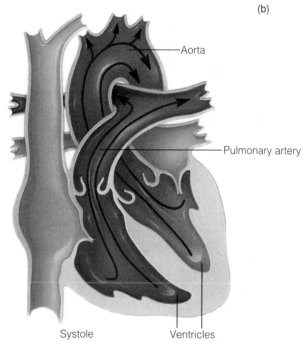

Aorta

Pulmonary artery

Systole

Ventricles

(c)

These three diagrams explain the pumping action of the heart. The heart is really two interconnected pumps that operate simultaneously. One pump consists of the right atrium and the right ventricle. The right atrium receives blood low in oxygen (blue color) from the two major veins of the body, the superior vena cava, which drains the head, neck, arms, and upper part of the trunk, and the inferior vena cava, which drains the middle and lower trunk, and the lower extremities. The other pump consists of the left atrium, which recieves oxygenated blood (red color) from the lungs by way of the pulmonary veins, and the left ventricle.

As the heart relaxes, blood flows into the two atria. The contraction of the atria, which precedes that of the

ventricles, forces blood through one-way valves into the two ventricles. When the ventricles contract, oxygenated blood is pumped from the left ventricle into the aorta, and deoxygenated blood is pumped from the right ventricle into the pulmonary artery. The entrance to the

aorta and to the pulmonary artery is guarded by one-way valves that prevent the backflow of blood into the heart during relaxation. The two pumps of the heart are linked together in the sense that blood pumped out of either pump is returned to the other.

The Human Heart

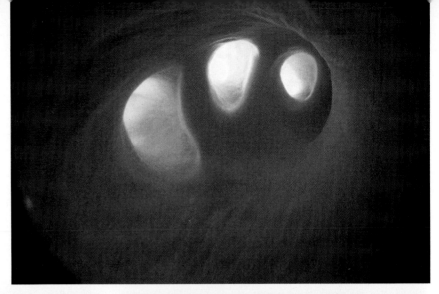

This internal view of the aortic arch—the first part of the aorta that rises from the heart and then bends downward behind the heart—gives the appearance of windows in a vaulted ceiling. The openings are the three major arteries that leave the aorta to supply the head, neck, and arms.

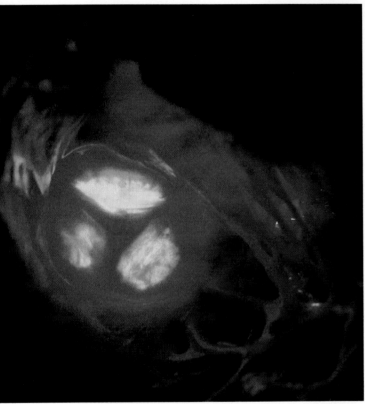

The delicate appearance of the closed aortic valve belies its strength and durability. When the left ventricle contracts, the leaflets are separated and pressed against the wall of the aorta as blood is pumped out of the ventricle.

The right atrium is separated from the right ventricle by the tricuspid valve. The tricuspid valve opens during atrial contraction to allow blood to enter the ventricle, but slams tightly shut during ventricular contraction to prevent the backflow of blood into the atrium. As shown in this dramatic photograph from inside the right ventricle, the leaflets of the tricuspid valve are reinforced by strong fibrous cords that run from the ventricular surfaces of the valve to thick columns of muscle that attach to the inner walls of the ventricle. Like the spokes of an umbrella, this structural reinforcement of the tricuspid valve prevents it from turning inside out.

interferes with the efficient operation of metabolic pathways of oxidation and energy capture because these pathways require oxygen. Depending upon the severity of the condition, individuals with anemia may show pallor in the palms of the hands, under the fingernails, and in the lining of the eyelids, and they may experience fatigue, shortness of breath, and heart palpitations. In severe cases, anemic individuals may be bedridden, able to sustain only vital activities such as heart beat and respiration.

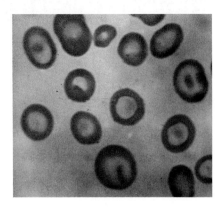

Anemic red blood cells contain less hemoglobin than normal red blood cells.

Anemia has many causes. The most common type of anemia is *iron-deficiency anemia,* which afflicts about 20 percent of the world's population. Because iron is necessary for the synthesis of hemoglobin, but not for the production of red blood cells themselves, individuals with an iron deficiency possess erythrocytes that contain less hemoglobin than usual. Iron salvaged from hemoglobin in aging erythrocytes is normally recycled by the body, and consequently dietary requirements for iron are reasonably low. However, growing children, who must make new erythrocytes in addition to those required for replacing aging ones, and women of reproductive age, who lose blood during menstruation, need additional dietary supplies of iron.

Some anemias are caused by a decrease in the normal rate of red blood cell production. Vitamin $B_{12}$ and the vitamin folic acid are required for normal development and maturation of erythrocytes. Deficiency of either of these vitamins leads to a situation in which the production of new erythrocytes does not occur fast enough to replace all of the aging erythrocytes that are removed from the circulation. Folic acid deficiency is common in alcoholics, who usually do not eat enough of the green vegetables that are the dietary source of the vitamin. Deficiency of vitamin $B_{12}$ produces a severe anemia, called *pernicious anemia,* that can lead to spinal cord lesions and death. Pernicious anemia occurs when the stomach, for some unknown reason, stops producing a substance called *intrinsic factor,* which is necessary as a carrier substance for the absorption of vitamin $B_{12}$ from food in the intestinal tract. The condition is treated by injections of vitamin $B_{12}$, which must be continued indefinitely.

Other anemias occur when there is an increased rate of erythrocyte loss or destruction that is not accompanied by a compensatory increase in erythrocyte production. This condition is characteristic of the short-term effects of severe hemorrhage or of a variety of so-called *hemolytic diseases,* often of a hereditary nature, that bring about the abnormal disruption of erythrocytes. One type of hereditary hemolytic anemia is *sickle cell anemia,* common in Blacks and some people of Mediterranean origin. Sickle cell anemia is discussed extensively in Chapter 3 *(Focus: Sickle Cell Anemia and the Importance of Cell Shape)* and in Chapter 20.

The opposite of anemia is a pathological condition called *erythrocytosis,* in which there is an overproduction of red blood cells. As the disease progresses, blood vessels become clogged with erythrocytes and clots may form, thus increasing the chance for heart attacks and stroke. Treatment involves venesection (bloodletting) to remove blood cells and the administration of drugs that inhibit the proliferation of red cells in bone marrow.

## White Blood Cells (Leukocytes)

White blood cells, or **leukocytes** (*leuko* = white; *cytes* = cells) protect the body against disease and against injury by foreign substances. Unlike erythrocytes, which remain within the bloodstream while performing their function, most leukocytes function outside the bloodstream. They leave the bloodstream by passing through the walls of blood vessels and entering the tissues at the site of a diseased or damaged area, where they exert their effects. Leukocytes are, in a sense, transitory blood cells that are present in the bloodstream while they are being transported from their sites of manufacture to their sites of action. The blood of a healthy individual contains approximately 5000 to 10,000 leukocytes per cubic millimeter, but during an acute infection these numbers may increase dramatically—an occurrence that is important in diagnosis.

## Types of White Blood Cells

There are two main categories of leukocytes: *granular leukocytes* and *nongranular leukocytes* (Figure 8–7). The granular leukocytes are so named because they have small granules in their cytoplasm, while nongranular leukocytes generally lack granules. Three types of granular leukocytes have been distinguished on the basis of their staining reactions with various dyes. These are *neutrophils, eosinophils,* and *basophils.* There are two types of nongranular leukocytes. The smaller ones are called *lymphocytes* because they are found in lymph as well as in blood, while the larger ones are termed *monocytes.* The proportion of the various leukocyte types in normal blood is also shown in Figure 8–7.

## Formation of White Blood Cells

All types of white blood cells are formed in the red bone marrow from the same stem cells that also give rise to erythrocytes. In addition, certain types of lymphocytes migrate from the bone marrow to the thymus gland, where they proliferate and become functionally mature. Lymphocytes from the bone marrow and thymus seed the lymph nodes and other lymphatic tissues where they multiply.

## Function of White Blood Cells

Lymphocytes are the cellular elements of the body's immune system; they are involved in the production of antibodies that recognize infectious agents, foreign cells, and other foreign substances. Neutrophils and monocytes are *phagocytes* (*phago* = eat) that ingest invading microorganisms, bits of dead tissue, and other cellular debris by endocytosis. Eosinophils are also phagocytic and, in addition, are involved in the allergic response. Basophils release chemical substances that attract leukocytes to sites where microorganisms or other foreign substances have infiltrated the tissues. The specific function of

FIGURE 8–7 Types of leukocytes and their frequencies in normal blood.

| Granular leukocytes | Morphology | Approximate percentage in normal blood |
|---|---|---|
| Neutrophil |  | 55 |
| Eosinophil |  | 3 |
| Basophil |  | 0.5 |
| Nongranular leukocytes | | |
| Lymphocyte | 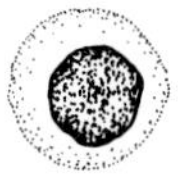 | 35 |
| Monocyte | 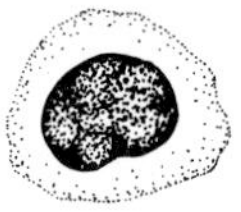 | 6.5 |

each leukocyte type in protecting the body from disease will be discussed in Chapter 21, "Disease and the Immune System."

## Diseases Affecting White Blood Cells

*Infectious mononucleosis* is a disease common in adolescents and young adults and is caused by a virus (Barr-Epstein virus). Because the virus is transferred through mucous membranes, particularly the lips, mononucleosis has been referred to as the "kissing disease." It is characterized by the appearance in the blood of large numbers of abnormally formed nongranular leukocytes. Individuals with mononucleosis experience fatigue, sore throat, swollen lymph nodes, and fever for two to three weeks.

*Leukemia* is a cancerous disease of the blood-forming tissues in the bone marrow. It results in the uncontrolled production of large numbers of functionally abnormal leukocytes and a diminished production of red blood cells. There are different types of leukemia that vary in severity and in the kinds of leukocytes present in the bloodstream. Two severe forms of leukemia are *acute lymphocytic leukemia* and *acute myelocytic leukemia.* Acute lymphocytic leukemia occurs most frequently in children, with two-thirds of its victims below the age of 15. It is characterized by increased numbers of immature lymphocytes in the circulation. Considerable progress has been made in the treatment of acute lymphocytic leukemia in children. Once a disease that was usually fatal in about six months, the average survival rate with chemotherapy (treatment with anticancer drugs) is now three to five years, with many children surviving for much longer times and some apparently being cured. Acute myelocytic leukemia is one of the most intractable of the leukemias. It occurs with nearly equal frequency in individuals of all ages and is associated with the production of large numbers of mature and immature granulocytes, which appear in the circulation.

In addition to the two acute forms of leukemia, each type occurs in a less severe, chronic form. The chronic forms are not curable, but the course of the disease is prolonged.

The cause of leukemia is not clearly understood. Exposure to radiation can cause acute leukemia, as it did in some survivors of the atomic bomb blasts at Hiroshima and Nagasaki, Japan. Other evidence has shown that some leukemia is of viral origin, as discussed in Chapter 22. If treated early by radiation and chemotherapy, remission of symptoms may occur for extended periods of time.

*Hodgkin's disease* is a chronic cancer of the lymphoid tissues (lymphoma) that is marked by the enlargement of lymph nodes, spleen, and liver. Giant, bizarre leukocyte forms appear in the lymph nodes. It occurs most often in adults between the ages of 18 and 38, and more frequently in men than women. The treatment of Hodgkin's disease is similar to that for leukemia, and it is not curable. Its cause is unknown.

## Platelets

Suspended in the blood plasma along with red blood cells and leukocytes are *platelets,* which play an essential role in mending tears in blood vessels and in promoting the clotting of blood. Platelets are not actually cells, but are membrane-bound fragments of cytoplasm that pinch off from large cells, called *megakaryocytes.* Megakaryocytes reside in the bone marrow and are formed from the same stem cell that gives rise to erythrocytes and leukocytes. Platelets are smaller than red or white blood cells, and normal blood contains about 250,000 platelets per cubic millimeter.

### Function of Platelets in Stopping Flow of Blood

Platelets are involved in a sequence of events that operates to stem the flow of blood from broken blood vessels. This sequence involves the following steps:

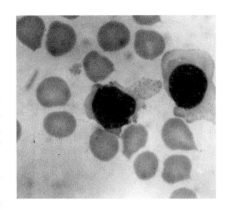

Abnormal nongranular leukocytes typical of mononucleosis.

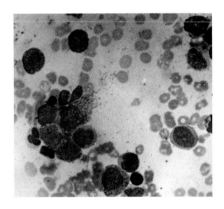

Functionally abnormal leukocytes typical of leukemia.

1. When a blood vessel wall is cut, damage to the vessel immediately causes an intense spasm of the smooth muscle surrounding the vessel. (This does not occur in capillaries since they have no muscular tissue.) The spasm, which may last up to 30 minutes, narrows the diameter of the blood vessels in the injured area, thus reducing blood loss.

2. Platelets will normally adhere to rough or foreign surfaces, but not to the smooth inner surface of blood vessels. Consequently, platelets in blood flowing out of the damaged vessel adhere to its cut edges and to one another, forming a loose plug that may completely seal the break, especially in small vessels. In fact, clumped platelets stop the flow of blood in the thousands of small vessels that routinely rupture every day as a result of normal wear and tear.

   In addition to plugging breaks in small vessels, aggregated platelets release serotonin and epinephrine—two chemicals that act as powerful constrictors of blood vessels (vasoconstrictors). The effect of these chemicals is to prolong the initial spasm induced by the injury.

3. The clumping of platelets usually leads to the formation of a blood clot. The process of blood clotting is initiated by a protein liberated from aggregated platelets and damaged tissue cells. This protein reacts, in a sequence of steps, with at least a dozen *clotting factors,* most of which are plasma proteins, to bring about the formation of fibrin strands from the plasma protein fibrinogen, as mentioned earlier. A blood clot forms when blood cells are trapped in a fibrin mesh (Figure 8–8).

   After a gelatinlike clot forms, a process called *clot retraction* occurs. This involves the contraction of a contractile protein within the platelets and results in the formation of a clot that is denser and stronger than before. When clot retraction occurs, a clear yellowish fluid is extruded from the clot. This fluid is called *serum* and consists of plasma without fibrinogen and other clotting factors that are used up in the clotting process.

### Abnormal Clotting

***Thrombus Formation.*** On occasion, a clot may develop *within* an intact blood vessel, producing what is known as a **thrombus** (pl: *thrombi*). Usually forming on and adhering to blood vessel walls, a thrombus may reduce blood flow through the vessel. Alternatively, it may become dislodged and carried to smaller vessels where it blocks blood flow completely, causing tissue death. Such a thrombus is called an **embolism.** An embolism can cause a heart attack if it blocks blood flow to the heart or a stroke if it blocks a vessel in the brain.

FIGURE 8–8 A scanning electron micrograph of a blood clot, showing erythrocytes and platelets enmeshed in fibrin.

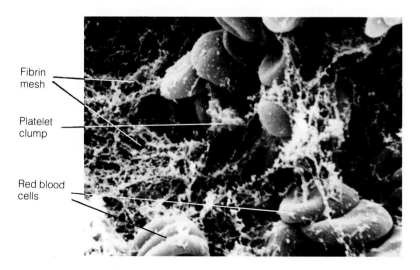

Fibrin mesh

Platelet clump

Red blood cells

It is not clear why thrombi form, although a number of ideas have been advanced. Some researchers have suggested that the blood of certain individuals may be more prone to clot than that of others because of the presence of excess clotting factors. Although this explanation is undoubtedly true in some cases, a more generally accepted idea is that damage to the lining of blood vessels may provide roughened areas that trigger platelet aggregation and the formation of thrombi. For example, bacterial infection, toxic substances, and allergic reactions may inflame the lining of veins and thus damage them. Vein damage could also occur when a sluggish flow of blood through veins, brought on by inactivity or heart insufficiency, deprives vessel cells of necessary oxygen. In arteries, the deposition of lipids causes thickening and roughening of the arterial walls *(arteriosclerosis)* and may induce clot formation.

*Anticoagulant Therapy.* Patients prone to develop thrombi may be treated with anticoagulant drugs that discourage the formation of new thrombi. These drugs include *heparin,* which prevents the formation of fibrin from fibrinogen; *aspirin* and related agents, which prevent platelets from sticking together; and a class of chemicals that blocks the action of vitamin K—a vitamin required for the synthesis of plasma clotting factors. One of these latter chemicals is *dicoumarol,* which was originally isolated from spoiled sweet clover and shown to be the agent that causes fatal hemorrhaging in cattle. Indeed, one complication of anticoagulant therapy is severe bleeding from ulcers and other lesions that may be present in the body.

*Hemophilia.* There is a class of genetic diseases, called *hemophilia,* in which one or another of the clotting factors is missing or has reduced activity. Blood of individuals with hemophilia clots slowly, if at all, and consequently a relatively minor wound may cause fatal hemorrhaging. In addition, internal bleeding may occur in skin, muscles, and joints as a result of relatively minor injuries. The accumulation of blood in joints may lead to their gradual deterioration. Bleeding episodes may be treated by transfusion of whole blood. As a preventive measure, hemophiliacs may receive regular injections of missing clotting factor isolated from human blood. Tragically, however, many hemophiliacs have contracted AIDS when treated with clotting factor unknowingly contaminated with the AIDS virus. Blood is now routinely screened for the AIDS virus before it is used to isolate clotting factors.

## Platelet Deficiencies

Diseases or conditions that suppress the production of platelets lead to small hemorrhages under the skin or in mucous membranes. For example, the proliferation of tumor cells in bone marrow may crowd out megakaryocytes, reducing platelet production. Megakaryocyte production is also suppressed by radiation and chemotherapy treatment for cancer.

## Study Questions

1. Explain the role of tissue fluid in the exchange of materials between the cells and the circulatory system.
2. What is the origin of lymph?
3. Describe the cellular and fluid components of blood.
4. a) What is the primary function of red blood cells?
   b) How do the chemical composition and structure of red blood cells facilitate this function?
5. List three types of anemia and describe the cause of each type.
6. List the various types of leukocytes and indicate the function of each type.
7. a) What is the origin of platelets?
   b) What is their function?
8. a) What is a thrombus?
   b) Why are thrombi dangerous?

# THE CIRCULATORY SYSTEM II: The Heart, Vessels, and Circulation of Blood and Lymph

## THE HEART

The heart is a living pump, roughly the size of a clenched fist, that circulates blood through the blood vessels. The force required for the pumping action of the heart is generated by alternate cycles of contraction and relaxation of cardiac muscle that forms most of the heart's structure. The contraction of the heart forces out blood into the major arteries leaving the heart, while relaxation allows the heart to fill with blood that enters from the major veins.

### Pericardium

The heart is located in the pericardial compartment in the central region of the thorax (rib cage). The pericardial compartment is bounded by a tough membrane called the *fibrous pericardium,* which holds the heart in place. The fibrous pericardium attaches to the major vessels at the upper part of the heart and, by ligaments, to structures that surround the heart, including the diaphragm, sternum, and ribs. Lining the interior of the fibrous pericardium is a slippery membrane, called the *serous pericardium;* at the bases of the major blood vessels the serous pericardium folds back upon itself to cover the surface of the heart. The inner layer of the serous pericardium thus forms the outer layer of the heart, also referred to as the **epicardium.** (The inner and outer layers of the serous pericardium are also referred to as the *visceral* and *parietal pericardium,* respectively.) Between the two layers of the serous pericardium is a small amount of slippery fluid that allows the heart to move easily during contraction and relaxation (Figure 9–1).

Inflammation of the pericardium, called *pericarditis,* may be caused by viruses or other agents. Although pericarditis is usually a benign disease, it is associated with chest pain that may be mistaken for the pain of a heart attack. Pericarditis is diagnosed by a characteristic creaking sound, audible in the stethoscope, that results from the rubbing of the two layers of the serous pericardium against one another. This sound is called a pericardial rub.

### Chambers, Valves, and Major Blood Vessels of the Heart

The internal structure of the heart is shown in Figure 9–2. It consists of four chambers: two **atria** (sing: *atrium*) with thin muscular walls and two larger

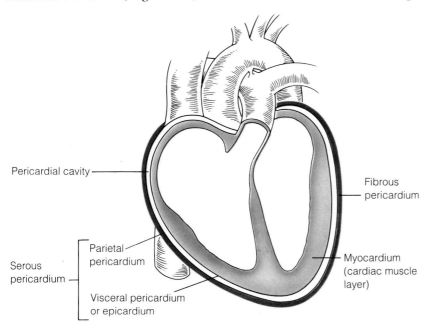

Pericardial cavity

Serous pericardium

Parietal pericardium

Visceral pericardium or epicardium

Fibrous pericardium

Myocardium (cardiac muscle layer)

FIGURE 9–1 Section through the heart, showing its outer coverings. The inner layer of the serous pericardium, attached firmly to the outer surface of the myocardium, is also referred to as the epicardium.

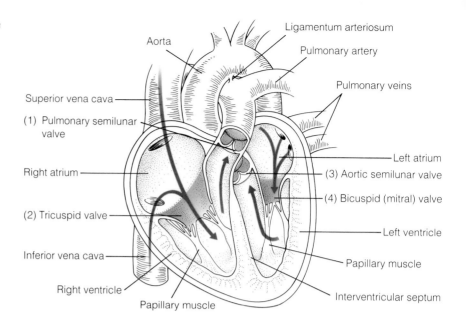

ventricles with thick muscular walls. The walls of the heart, whether forming the atria or ventricles, are composed of three distinct layers. The outermost layer is the *epicardium,* mentioned above. The innermost layer, the **endocardium,** lines the chambers of the heart, forms the valves, and is continuous with the endothelial lining of blood vessels. Between the epicardium and the endocardium is the **myocardium** (*myo* = muscle), which makes up most of the thickness of the heart walls. The myocardium consists of bundles of cardiac muscle fibers that are richly supplied with blood vessels.

The heart possesses a number of valves that assure a one-way flow of blood through the heart chambers and through the major blood vessels that leave and enter the heart. Valves are formed of dense connective tissue covered by endothelium; their structure is illustrated in Figure 9-3. The valves and the major blood vessels of the heart are discussed functionally in the following section.

FIGURE 9-3 A photograph of the pulmonary valve at the base of the pulmonary artery. In (a) the valve is partly open and in (b) it is almost closed. Blood can flow through the valve toward the viewer, but not in the opposite direction; the cusps of tissue forming the valve flatten against the inside of the vessel wall when blood flows through the valve.

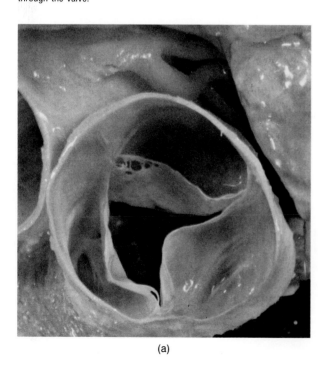

(a)

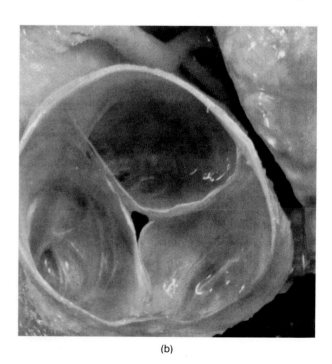

(b)

# PULMONARY AND SYSTEMIC CIRCULATION

The heart is really two pumps or functional units that operate simultaneously. However, the two pumps are linked to one another in the sense that blood pumped out of either pump returns to the other (Figure 9–4). One pump consists of the right atrium and right ventricle. The right atrium receives blood, low in oxygen and high in carbon dioxide, from two major veins: the **superior vena cava,** which drains the head, arms, and upper part of the trunk; and the **inferior vena cava,** which drains the rest of the trunk and lower extremities. The right atrium is separated from the right ventricle by the **tricuspid valve,** so named because it consists of three cusps of tissue. The tricuspid valve allows blood to flow from the atrium to the ventricle, but prevents flow in the opposite direction. The leaflets that form the tricuspid are reinforced by strong fibrous cords that run from the ventricular surfaces of the valve to conical projections of muscle, forming the **papillary muscles,** which attach to the muscular walls of the ventricle and keep the valves from turning inside out.

When the heart contracts, contraction of the atria precedes that of the ventricles. Consequently, blood that has flowed into the right atrium from the two venae cavae is squeezed into the right ventricle. Contraction of the right ventricle then forces the blood into the *pulmonary artery* to the lungs for oxygenation. At the base of the pulmonary artery is the **pulmonary semilunar valve,** a one-way valve that is forced open during the contraction of the right ventricle, but is closed when flow from the heart decreases, thus preventing the back-flow of blood from the pulmonary artery to the right ventricle.

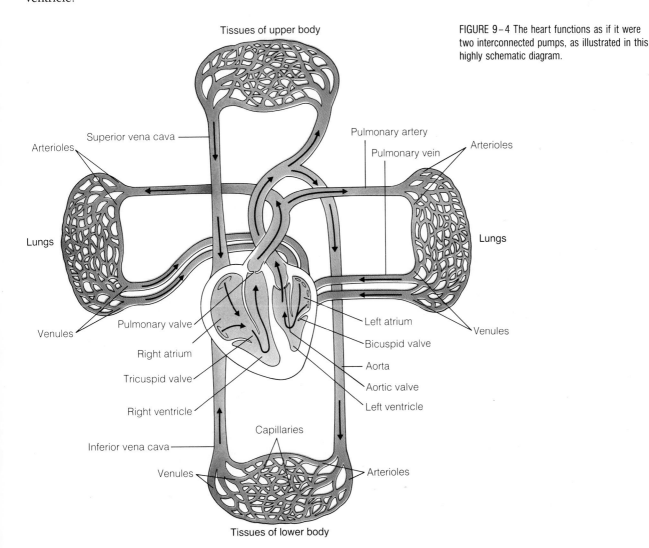

Tissues of upper body

FIGURE 9–4 The heart functions as if it were two interconnected pumps, as illustrated in this highly schematic diagram.

Superior vena cava

Pulmonary artery

Pulmonary vein

Arterioles

Arterioles

Lungs

Lungs

Venules

Pulmonary valve

Left atrium

Right atrium

Bicuspid valve

Tricuspid valve

Aorta

Aortic valve

Right ventricle

Left ventricle

Inferior vena cava

Venules

Capillaries

Arterioles

Venules

Tissues of lower body

Blood passing through the capillary network in the lungs discharges carbon dioxide and picks up oxygen, returning to the left atrium through four pulmonary veins. This circulatory loop, by which blood is pumped from the right ventricle through right and left branches of the pulmonary artery to each lung and then back to the left atrium by way of the pulmonary veins is called the **pulmonary circulation.**

Contraction of the left atrium (which occurs at the same time as contraction of the right atrium) forces blood into the left ventricle through the **bicuspid,** or **mitral valve,** so called because it is formed of two cusps of tissue that somewhat resemble a bishop's mitre. The mitral valve, like the tricuspid valve, is strengthened by fibrous cords that attach to papillary muscles, but the papillary muscles and the flaps of tissue forming the mitral valve are heavier and stronger than the counterparts of these structures on the heart's right side. The mitral valve is reinforced structurally because it is subject to greater blood pressures than the tricuspid valve.

The wall of the left ventricle is approximately three times thicker than that of the right ventricle and is capable of generating considerably more pressure upon contraction. This additional strength is needed because oxygenated blood leaving the left ventricle is pumped through the thousands of miles of blood vessels in the body (with the exception, of course, of those leading to and from the lungs).

Blood leaving the left ventricle begins a second circulatory loop, called the **systemic circulation,** by which oxygenated blood from the lungs is delivered to all other tissues of the body. From the left ventricle, blood flows through the *aortic semilunar valve* into the largest blood vessel of the body, the **aorta.** The aorta branches into major arteries that carry blood to the body's tissues. In the tissues, these arteries branch further into arterioles and finally into capillary networks that deliver oxygen and nutrients and pick up carbon dioxide and waste materials. From the capillaries, blood passes into venules and finally into the two venae cavae. These return blood to the right atrium, completing the circuit of the systemic circulation.

## DETAILS OF THE SYSTEMIC CIRCULATION

The major arteries of the systemic circulation include the aorta and vessels that branch from it (Figure 9–5a, p. 156). The major veins of the systemic circulation include the superior vena cava and the inferior vena cava, and the vessels that drain into them (Figure 9–5b, p.157).

### Coronary Circulation

Although the heart pumps blood to all organs of the body, it needs its own blood supply to do this. Indeed, the oxygen demands of the heart are so great that only the brain requires more oxygen per unit weight. Because the tissues of the heart cannot obtain oxygen and nutrients directly from the blood within the heart's chambers, the heart is supplied with its own set of arteries and veins. Blockage of arteries that supply the heart can lead to the death of cardiac muscle cells (heart attack), as discussed in a later section.

The heart is supplied by two small arteries that branch off from the aorta just beyond the aortic semilunar valve. These vessels are the *left* and *right coronary arteries,* so named because they encircle the upper region of the heart somewhat like a crown (Figure 9–6a, p. 158). The coronary arteries, about as large in diameter as a soda straw, branch extensively toward the tip of the heart, forming an interconnecting network of small vessels that ultimately break up into vast capillary beds that supply every region of the myocardium with oxygen and nutrients.

Veins that run roughly parallel to the coronary arteries drain the capillary networks of the myocardium and carry blood to the **coronary sinus,** a wide venous channel that empties into the right atrium below the opening of the inferior vena cava (Figure 9–6b, p. 158).

## Major Arteries of the Head, Neck, Trunk, and Extremities

Shortly after leaving the heart, the aorta bends to descend behind the heart; this bend is called the *arch of the aorta.* Three major arteries branch off from the arch. The first of these, the *brachiocephalic artery,* branches into the *right common carotid artery,* which sends blood to the right side of the head and neck, and into the *right subclavian artery,* which sends blood to the right arm. The other two arteries that arise from the aortic arch are the *left common carotid artery,* serving the left head and neck, and the *left subclavian artery,* serving the left arm.

As the aorta descends into the thoracic and abdominal regions, it sends branches to the walls of the thorax and to the abdominal organs. The aorta terminates in the lumbar region, branching into the *right* and *left common iliac arteries* that send branches to supply the pelvic organs and legs.

## Major Veins of the Head, Neck, Trunk, and Extremities

These veins drain the blood carried to the tissues by the systemic arteries and run approximately parallel to their corresponding arteries (see Figure 9–5b). The *right* and *left internal jugular veins,* and the *right* and *left external jugular veins* drain the capillary networks and venous sinuses of the head and neck. These join on either side with the *subclavian veins,* forming the *left* and *right brachiocephalic veins,* which unite in the *superior vena cava.* Veins draining the thoracic region also empty into the superior vena cava.

The *right* and *left common iliac veins,* receiving blood from the legs and pelvic organs, fuse to form the *inferior vena cava.* This receives blood from veins that drain the skin and muscles of the back, the vertebral column, the spinal cord, the reproductive organs, the kidneys, the diaphragm, and the liver, among other organs.

The veins draining the abdominal portion of the digestive tract (stomach and intestines) and the spleen, pancreas, and gall bladder do not empty separately into the inferior vena cava. Instead, they come together to form the *hepatic portal vein* to the liver (Figure 9–7, p. 158). The portal vein empties into spaces (sinusoids) between the liver cells, and its contents are mixed with the oxygen-rich blood that reaches the liver by the hepatic artery (arising from the aorta). The *hepatic veins,* formed by vessels that drain the liver sinusoids, carry blood to the inferior vena cava. This entire system of veins that carries blood from abdominal organs to the liver and then to the inferior vena cava is referred to as the **hepatic portal system.** The hepatic portal system is important because it allows toxic substances that may have been absorbed into the blood from food in the digestive tract to be detoxified by the liver before they enter the general circulation.

## CONTRACTION OF THE HEART

The work of the heart is done by contraction of the heart muscle. The cardiac muscle cells that form the heart muscle possess an inherent ability to contract in a rhythmic fashion. Contractions of these cells are coordinated by mechanisms that assure efficient operation of the heart.

### Structure of Cardiac Muscle

Cardiac muscle cells are elongated, branching cells with a centrally located nucleus and surrounding sarcoplasm. Within the sarcoplasm are myofibrils possessing cross-striations similar to those seen in skeletal muscle and giving

FIGURE 9–5 (a) Major arteries of the systemic circulation; (b) major veins of the systemic circulation.

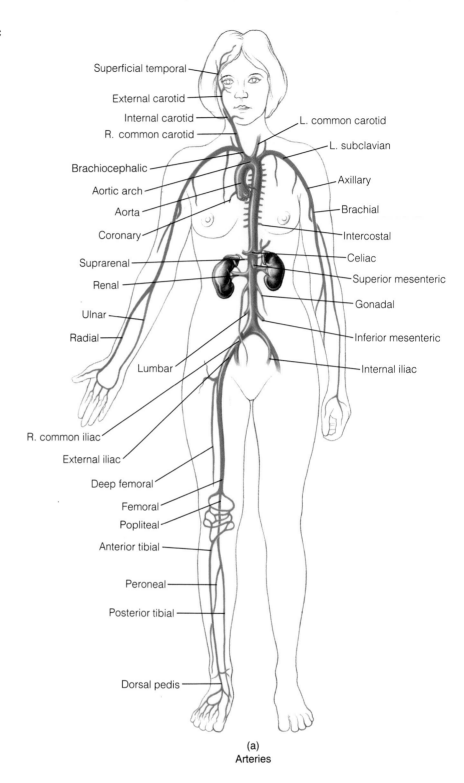

Superficial temporal

External carotid

Internal carotid

R. common carotid

Brachiocephalic

Aortic arch

Aorta

Coronary

Suprarenal

Renal

Ulnar

Radial

Lumbar

R. common iliac

External iliac

Deep femoral

Femoral

Popliteal

Anterior tibial

Peroneal

Posterior tibial

Dorsal pedis

L. common carotid

L. subclavian

Axillary

Brachial

Intercostal

Celiac

Superior mesenteric

Gonadal

Inferior mesenteric

Internal iliac

(a)
Arteries

cardiac muscle cells a striated appearance. In addition to myofibrils, the sarcoplasm contains sarcoplasmic reticulum, a system of transverse tubules, and numerous mitochondria.

In cardiac muscle tissue, the muscle cells are arranged roughly parallel to one another to form a branching network (Figure 9–8, p. 159). Spaces between the cells are abundantly supplied with blood vessels, lymphatic vessels, and nerves. Individual cardiac muscle cells are joined together at their plasma membranes, forming structures called *intercalated discs* where contact occurs. The intercalated discs keep the plasma membranes of adjacent cells fused tightly together, preventing them from being pulled apart and allowing electrical impulses to pass from cell to cell essentially unimpeded. Conse-

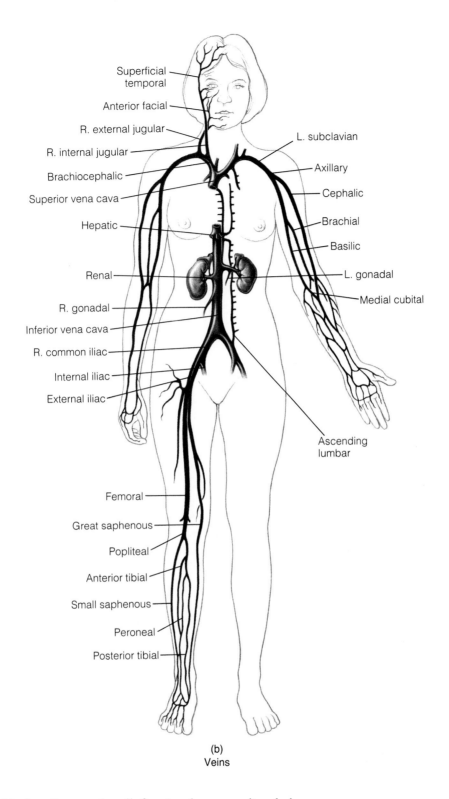

Superficial temporal
Anterior facial
R. external jugular
R. internal jugular
Brachiocephalic
Superior vena cava
Hepatic
Renal
R. gonadal
Inferior vena cava
R. common iliac
Internal iliac
External iliac
Femoral
Great saphenous
Popliteal
Anterior tibial
Small saphenous
Peroneal
Posterior tibial

L. subclavian
Axillary
Cephalic
Brachial
Basilic
L. gonadal
Medial cubital
Ascending lumbar

(b)
Veins

quently, the individual cardiac muscle cells forming the myocardium behave as if they were a single cell.

## Mechanism of Cardiac Muscle Contraction

The basic cellular events involved in cardiac muscle contraction are essentially similar to those seen in skeletal muscle. A wave of depolarization, followed by repolarization, is propagated throughout the sarcolemma (plasma membrane) and transmitted to the sarcoplasmic reticulum by the transverse tubules, causing contraction of the myofibrils. Presumably, contraction occurs by the sliding filament model (see Chapter 7).

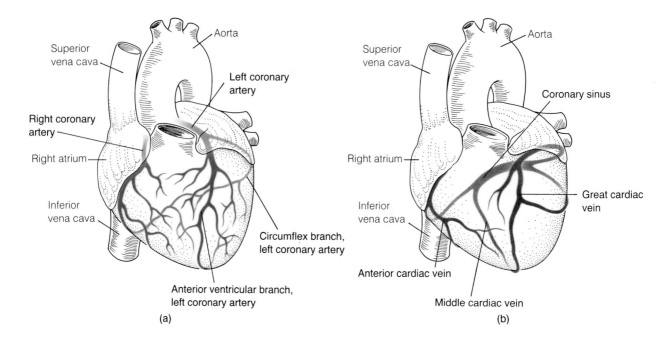

FIGURE 9–6 Coronary circulation. Anterior view of (a) coronary arteries and (b) venous drainage of the heart. Coronary vessels are shown in bold type.

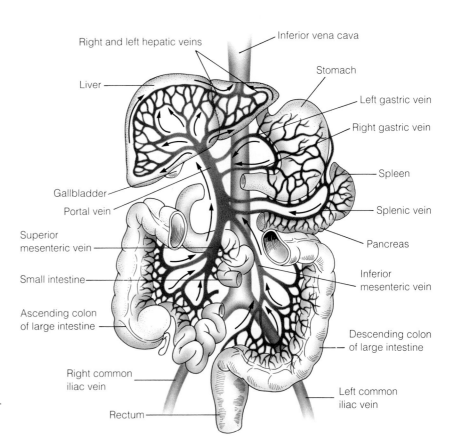

FIGURE 9–7 The hepatic portal system that provides venous drainage of the digestive organs. Arrows indicate direction of blood flow.

## Properties of Cardiac Muscle

Cardiac muscle, like skeletal muscle, follows the all-or-none response. However, there are some important differences in the action of the two types of muscle. Cardiac muscle cells have an inherent ability to contract rhythmically; that is, they depolarize spontaneously at regular intervals without receiving external stimuli. In addition, cardiac muscle repolarizes more slowly than skeletal muscle. Consequently, cardiac muscle has a long refractory period during which contraction cannot occur. The long refractory period gives the heart time to fill with blood before the next contraction, and it prevents tetanic contractions—the sustained contractions seen in skeletal muscle that would be lethal if they occurred in the myocardium.

## Initiation of Contraction and Propagation of Impulses

Efficient operation of the heart requires that cycles of contraction and relaxation of the atria and ventricles occur in an orderly sequence. For example, contraction of the atria, which forces blood into the ventricles, must precede the contraction of the ventricles; before contracting again, the atria must relax for a sufficient amount of time to allow them to fill with blood. By what mechanisms are these different events synchronized in the normal heart?

Although cardiac muscle cells have an intrinsic ability to undergo rhythmic contraction, there are specialized groups of these cells within the myocardium that normally initiate the impulse for contraction and conduct the impulse throughout much of the heart (Figure 9–9). One such group of cardiac muscle cells, called the **sinoatrial (SA) node** or **pacemaker,** is located in the region where the superior vena cava joins the right atrium. Cells that form the sinoatrial node have the greatest tendency for inherent rhythmical contraction of any cardiac muscle cells; that is, sinoatrial node cells in the "resting" heart depolarize spontaneously about 70 times per minute, considerably more frequently than do other cardiac muscle cells. As the wave of depolarization from the sinoatrial node spreads through the muscles of both atria, it stimulates their contraction.

At the base of the right atrium near the wall between the two atria, the wave of depolarization spreading through the atria encounters a second group of specialized cardiac muscle cells, forming the **atrioventricular (AV) node.** After reaching the AV node, the wave of depolarization passes along specialized conducting fibers (the *AV bundle*) that run down both sides of the wall between the ventricles and branch extensively to form *Purkinje fibers* that spread into much of the myocardium of the right and left ventricles. From these specialized conducting fibers, the depolarization wave is conducted rapidly from cell to cell throughout the ventricles, causing essentially simultaneous contraction of the two ventricles following the simultaneous contraction of the two atria.

Because the atria and the ventricles are separated by nonconducting connective tissue, the wave of excitation passing through the atria can reach the ventricles *only* through the AV node and the conducting fibers that run from it.

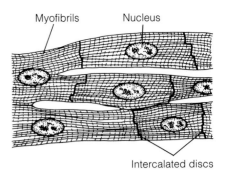

FIGURE 9–8 Drawing of cardiac muscle cells. Although not shown here, connective tissue and blood vessels fill the spaces between muscle cells.

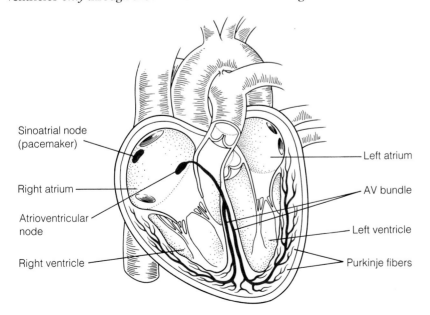

FIGURE 9–9 Structures responsible for the conduction of electrical impulses within the heart.

FIGURE 9-10 The cardiac cycle.

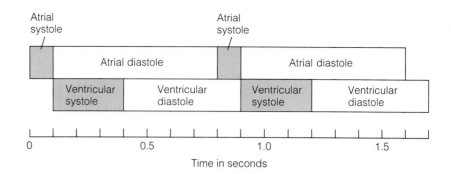

This is important in the synchronization of the contraction and relaxation cycles of the atria and ventricles because the AV node delays the wave of excitation until the atria have had time to contract and empty their contents into the ventricles.

In certain cases, the rate of depolarization of the SA node may be altered as a result of the influence of hormones, drugs, or ingested chemicals. For example, an elevated heart rate, called *tachycardia,* may be caused by such agents as caffeine and/or the hormone epinephrine, while a decreased heart rate, or *bradycardia,* may be caused by small amounts of alcohol.

## The Cardiac Cycle

Physiologists refer to rhythmically recurring contraction of the heart by the term **systole,** and rhythmically recurring relaxation by the term **diastole.** In an individual at rest, one complete cycle of systole and diastole lasts about 0.8 seconds. As shown in Figure 9-10, contraction of the atria, or atrial systole, precedes ventricular systole and lasts about 0.1 seconds, whereas atrial diastole lasts about 0.7 seconds. Ventricular systole, lasting 0.3 seconds, is longer than atrial systole and is followed by 0.5 seconds of ventricular diastole. If the heart rate increases, the length of atrial and ventricular diastole decreases, but not that of systole.

## Electrocardiograms

Waves of depolarization passing through the cardiac muscle cause current flows through the body fluids that may be detected at the surface of the body. Consequently, to determine electrical activity in the heart, metal plates attached to wire leads are placed on the chest and a number of other points on the body. The leads are connected to an instrument, called an *electrocardiograph,* that is equipped with a current-sensing stylus, and a graphic tracing of the electrical activity of the heart is obtained. This tracing is called an **electrocardiogram,** or EKG.

As shown in Figure 9-11, a normal electrocardiogram is characterized by five deflections; the origin of each is described in the figure inset. Any pathological process that disturbs the normal electrical responses of the heart will produce specific changes in one or more of these deflections. Indeed, the changes are so specific that they can be used for diagnostic purposes as discussed in the next section.

It is important to point out that although an EKG traces the heart's electrical activity during the cardiac cycle, it tells nothing about the actual physical state of the organ. A heart that is diseased, even severely so, but with no electrical disturbances will appear normal in an electrocardiogram.

## Rhythm Disturbances

Any change in the heart's normal rhythm — an irregularity of the heart beat — is referred to as *cardiac arrhythmia.* Cardiac arrhythmia may take a number of forms (see Figure 9-11).

## Ectopic Beats

The refractory period of cardiac muscle, during which it will not respond to a stimulus by contracting, lasts throughout atrial and ventricular systole. As we have mentioned previously, the length of the refractory period prevents the heart from exhibiting tetanic contractions, which would be fatal. In early ventricular diastole, the heart begins to recover excitability and will respond to a stimulus that arrives before the next regular impulse from the SA node. Such a *premature beat,* or *ectopic beat* (*ectopic* = misplaced), is weaker than a normal beat. When the next regular SA node impulse is generated, the heart is still in systole from the ectopic beat and consequently will not respond to the stimulus. Because the heart will not contract again until the next regular SA node impulse arrives, there is a long diastole, called a *compensatory pause,* between the end of the premature systole and the next regular beat (see Figure 9–11b).

Occasional ectopic beats are a frequent occurrence of no medical importance. They occur when localized areas of the heart depolarize spontaneously, and they are often brought on by anxiety, fatigue, and stimulants such as caffeine. Usually, an ectopic beat is not noticed, but the compensatory pause sometimes is, causing an individual to think that his heart has "skipped a beat."

## Heart Block

Disturbances in the conduction of electrical impulses through the AV node and the fibers that pass from it are called **heart block.** In nrst degree heart block, the mildest form, the time required for the impulses from the SA node to reach the ventricles is simply increased somewhat. More severe is second degree heart block, in which the AV node fails to conduct all of the impulses reaching it from the SA node; this results in a situation in which the rate of atrial contraction is greater than that of ventricular contraction. In third degree heart

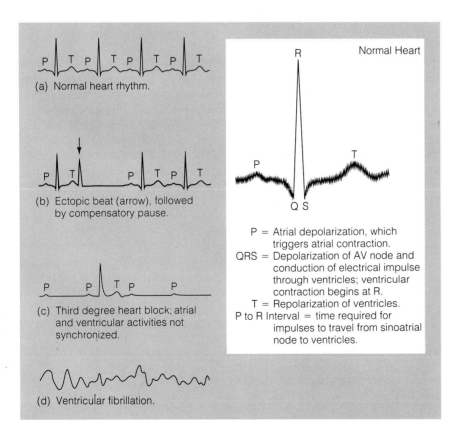

(a) Normal heart rhythm.

(b) Ectopic beat (arrow), followed by compensatory pause.

(c) Third degree heart block; atrial and ventricular activities not synchronized.

(d) Ventricular fibrillation.

Normal Heart

P = Atrial depolarization, which triggers atrial contraction.
QRS = Depolarization of AV node and conduction of electrical impulse through ventricles; ventricular contraction begins at R.
T = Repolarization of ventricles.
P to R Interval = time required for impulses to travel from sinoatrial node to ventricles.

FIGURE 9–11 Electrocardiograms, showing a normal heart rhythm and some abnormal heart rhythms (cardiac arrhythmias).

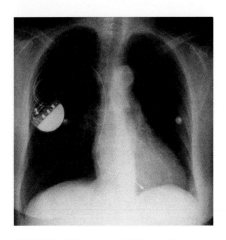

FIGURE 9-12 X ray of an individual with an artificial pacemaker, showing positioning of the device. The electrode is implanted in the wall of the right ventricle.

block, none of the impulses from the SA node are conducted through the AV node. In this case, the atria will contract at the rate set by the SA node, and the ventricles will contract at a much slower rate determined by the spontaneous depolarization of the AV node or of ventricular tissue itself. Hence, the contractions of the atria are not synchronized with those of the ventricles (Figure 9-11c). As a consequence of third degree block, the ventricles may not beat fast enough to meet the physiological needs of the body, a condition referred to as **heart failure.**

An *artificial pacemaker* may be implanted to overcome the effects of the more severe forms of heart block. Pacemakers in use today are capable of monitoring the heartbeat and delivering electric shocks to the heart when the heart rate falls below some preset rate. Electrodes that deliver the shock are usually implanted in the inside wall of the right ventricle, with wires leading through the right atrium and into a vein in the chest. The wires pierce the wall of the vein and connect to a small battery and timer that is implanted beneath the skin of the chest (Figure 9-12). The batteries last a number of years before they must be replaced.

## Flutter and Fibrillation

Under certain conditions, localized areas of the heart may repeatedly depolarize spontaneously, causing large numbers of sequential ectopic beats. One frequent trigger for this occurrence is oxygen deprivation of cardiac muscle, usually as a result of coronary artery disease or a heart attack (discussions of both conditions follow). If this situation causes the atria or the ventricles to contract exceedingly rapidly, but in a regular rhythm, the condition is called **flutter.** Flutter often degenerates into **fibrillation,** a rapid, disorganized twitching of the myocardium (Figure 9-11d).

Atrial flutter and fibrillation in the absence of ventricular fibrillation is not life threatening since the ventricles, even without contraction of the atria, will fill with substantial amounts of blood. For example, atrial fibrillation reduces the filling ability of the ventricles by only about 25 percent. Digitalis, a drug extracted from the foxglove plant, is used to improve myocardial function in atrial fibrillation and other forms of heart disease. It increases the force of each beat and decreases the heart rate by slowing conduction between the atria and ventricles.

Ventricular fibrillation, however, prevents the pumping of blood entirely and often leads to *cardiac arrest*—a condition in which the heart stops beating in any way. Treatment for ventricular fibrillation, which must be initiated within a minute or so of onset, involves use of an electrical defibrillator. Electrodes attached to the defibrillator are placed on the surface of the chest and are activated to send a strong electrical impulse through all cardiac muscle cells simultaneously. This throws all of the cells into a refractory state and thus halts electrical activity in the heart for a few seconds. If the defibrillation treatment is successful, the SA node will begin to fire rhythmically at its normal rate and the heart will resume a regular beat.

## Heart Sounds

Contraction and relaxation of the heart are accompanied by distinctive sounds that may be heard with a stethoscope, an instrument that consists of little more than flexible hollow tubing.

## Normal Sounds

During a normal cardiac cycle, two sounds are heard in quick succession, followed by a pause. The first sound, generally described as a "lub" sound, is the louder, longer, and lower-pitched of the two and is caused by ventricular systole and closure of the tricuspid and mitral valves. The second sound, a shorter and higher pitched "dup," is due to closure of the pulmonary and aortic semilunar valves.

## Valve Disease

Since heart sounds are due primarily to vibrations caused by valve closure, abnormal sloshing sounds will occur if the heart valves are not functioning properly and hence disrupt the smooth flow of blood. These sounds are called **heart murmurs.** So-called *functional murmurs* are essentially benign, occurring most frequently in childhood when the heart is beating forcibly. *Resting murmurs,* on the other hand, usually indicate valve damage.

Resting murmurs may be caused by improper development of the valves during the embryonic period or injury to the valves due to disease. In some cases, the valves do not open completely because they are thick and stiff, or the leaflets are partly fused together. Such valves are said to be *stenosed* (narrowed) and impede the flow of blood. In other cases, the edges of the valves may be eroded so that the valve cannot close completely. Valves with this condition are said to be *insufficient* because blood flows back in the wrong direction, a condition known as *regurgitation.* Inflammation of the endocardium (endocarditis) occurring as a result of rheumatic fever caused by a streptococcus infection may cause both conditions. Valves on the left side of the heart that do the heaviest work are the most commonly affected.

The timing of a murmur relative to the normal sounds of the heart gives clues to its cause. For example, stenosis of the mitral valve is heard just prior to ventricular systole as blood is forced through the constricted valve, whereas mitral insufficiency, causing blood to regurgitate into the left atrium during contraction of the left ventricle, is heard during ventricular systole. In some cases, a murmur may be due not to valve malfunctioning, but to congenital defects. This is the case when a small hole occurs in the wall between the two sides of the heart, causing blood to slosh between them.

Valve disease may vary in severity, but it tends to worsen over time. As it progresses, the left ventricle may be unable to pump normal amounts of blood, reducing blood flow through the coronary arteries and to other vital organs, or causing blood to back up in the pulmonary veins. In severe conditions, valves may be replaced by artificial ones.

## Cardiac Output

The amount of blood pumped by *each* ventricle in one minute is called the cardiac output. The cardiac output obviously depends upon two factors: the **heart rate,** or the number of beats per minute, and the **stroke volume,** or the amount of blood pumped by each ventricle during each contraction. Thus, for example, if the heart rate at rest is 72 beats per minute and the stroke volume is about 70 milliliters, then

$$\text{Cardiac output} = 72 \text{ beats/min} \times 70 \text{ mL/beat}$$

or about 5000 milliliters (5 liters) of blood per minute.

Cardiac output can increase substantially during strenuous exercise as the heart increases both its beat rate and stroke volume, giving a cardiac output of about five times the resting level in an average individual (see *Focus: The Physiology of Exercise and the Importance of Exercise to Health*). Although heart rate, as we mentioned above, is controlled by nerves and hormones, we still do not understand the mechanism by which increased oxygen demands of the tissues trigger increased heart rate.

Stroke volume is determined to a large extent by the amount of blood that returns to the heart from the body's tissues. As the inflow of blood to the right atrium and ultimately into other parts of the heart increases during exercise, contraction of the heart becomes stronger, increasing stroke volume. This

# The Physiology of Exercise and the Importance of Exercise to Health

**W**hether in a usually sedentary individual or in an endurance-trained athlete, strenuous aerobic exercise—exercise such as running or swimming that relies heavily on oxygen for energy production—triggers a set pattern of short-term physiological responses. As skeletal muscles begin to use increased amounts of oxygen, heart rate and stroke volume start to increase, bringing about a corresponding increase in cardiac output. During peak levels of aerobic exercise, oxygen may be consumed by working muscles at a rate ten to twenty times greater than when they are at rest, and cardiac output must increase substantially to help meet the increased oxygen demands. For example, in a sedentary but otherwise healthy college student, cardiac output increases from 4 to 5 liters of blood per minute at rest to about 22 liters per minute at peak exercise levels, about a fourfold increase. In an endurance-trained athlete, cardiac output may reach peak levels of about seven times the resting level.

An increased cardiac output during strenuous exercise would not, by itself, increase blood flow to the working muscles were it not for accompanying changes in the vascular system (blood vessels). First, increased cardiac output raises systolic blood pressure substantially in both sedentary individuals and endurance-trained athletes (see accompanying table). The elevated systolic blood pressure, in turn, increases the rate at which blood is circulated through the tissues. Second, arterioles that supply the working muscles dilate, increasing the flow of blood to the muscles. At the same time, the distribution of blood to other tissues such as the liver, kidney, and digestive organs is reduced by constriction of arterioles that supply these tissues, although the constriction of blood vessels does not occur in the brain or the heart. The net result is a diversion of oxygen-rich blood to the working muscles.

The physiological changes in the circulatory system that result in an increased flow of blood to skeletal muscle during exercise also increase blood flow to the heart. The coronary arteries dilate, and the increased systolic blood pressure forces more blood into the coronary arteries, increasing blood flow through the coronary arteries to as much as five times the resting value.

Accompanying the exercise-induced changes in the circulatory system are changes in the functioning of the respiratory system. The elevated demand for supplying oxygen to the muscles and removing excess carbon dioxide requires an accelerated exchange of oxygen and carbon dioxide between the blood and the lungs. This is accomplished by increases in the rate and depth of breathing and in the flow of blood through the lungs, up to a maximum value of about fivefold over resting levels.

It is important to note that healthy individuals, even if sedentary, are not usually limited in their capacity for aerobic exercise by inadequacies of the respiratory system. Instead, the limiting factor is the pumping capacity of the heart.

One of the most important benefits of regular aerobic exercise is an improvement in the function of the circulatory and respiratory systems, so that oxygen is supplied more efficiently to the cells of the body. At least five major population studies have shown that men who exercise regularly have about half as many heart attacks as those who do not. In addition, many studies have shown that exercise can help reduce cholesterol levels, hypertension, and obesity—three risk factors for heart disease.

Like any other muscle of the body, the heart gets stronger and operates more efficiently if it must work strenuously. If a substantial workload is placed on the heart on a regular basis, as would occur with regular aerobic exercise, the heart muscle enlarges as cardiac muscle cells increase in size and functional capacity. As a consequence, each heartbeat becomes stronger, increasing stroke volume. This means that an individual with a well-conditioned heart not only can increase

Physiological Values for Normal, Healthy Males as Compared to Endurance-Trained Male Athletes

| Physiological Parameter | Normal Healthy Males | | | | Male Endurance Athlete | |
| --- | --- | --- | --- | --- | --- | --- |
| | Preconditioning | | Postconditioning | | | |
| | Resting | Peak levels | Resting | Peak levels | Resting | Peak levels |
| Heart rate/min | 75 | 185 | 60 | 185 | 45 | 180 |
| Strole volume (mL) | 60 | 120 | 80 | 160 | 110 | 200 |
| Cardiac output (liters/min) | 4.5 | 22.2 | 4.8 | 29.6 | 4.9 | 36 |
| Maximal oxygen uptake (liters/min) | 0.27 | 3.1 | 0.28 | 4.4 | 0.29 | 5.8 |
| Breathing rate/min | 14 | 40 | 12 | 45 | 10 | 55 |
| Tidal volume (liters) | 0.4 | 2.75 | 0.5 | 3 | 0.6 | 3.5 |
| Systolic blood pressure (mm Hg) | 135 | 210 | 130 | 205 | 115 | 210 |
| Diastolic blood pressure (mm Hg) | 83 | 85 | 80 | 82 | 70 | 72 |

With regular aerobic exercise, the heart muscle enlarges and pumps blood more efficiently.

cardiac output during exercise beyond that of an average individual, but can pump blood more efficiently when at rest, lowering the resting heart rate substantially. Indeed, a decrease in the resting heart rate is a characteristic sign of physical training. Also accompanying physical training are decreases in resting systolic and diastolic blood pressures.

As the capacity of the heart increases with regular aerobic exercise, the heart muscle itself requires more oxygen. This oxygen requirement in turn stimulates an increase in the size and complexity of the capillary network that supplies the heart.

Yet another benefit of regular aerobic exercise is a slight increase in numbers of red blood cells and hence in total blood hemoglobin. Consequently, more oxygen can be transported to the tissues of the body in a given volume of blood.

Regular exercise can also have some minor effects on the respiratory system. Improving the strength of the skeletal muscles that control breathing can in-crease tidal volume (the volume of air inhaled or exhaled in a single breath). This results in decreases in breathing rate both during exercise and at rest.

The beneficial effects of exercise are not limited to improved fitness of the circulatory and respiratory systems. A recent study of Harvard alumni has shown that regular exercise increases longevity (length of life) by lowering the risk of death, not just from heart disease, but from all natural causes.

The Harvard study, published in 1986, correlates exercise habits with death rates of almost 17,000 men who entered Harvard between 1916 and 1950. The results of the study show that men who engaged in regular physical activity three or four times a week had lower death rates than those who did not exercise. Furthermore, the reduction in death rates was roughly correlated with the number of calories expended in physical activity, up to an optimum level of about 3500 calories per week. At this optimum level, equivalent to walking briskly for 30 to 35 miles or bicycling strenuously for 6 to 8 hours, overall death rates were reduced 28 percent as compared to alumni who did not exercise. Even moderate exercise lowered death rates significantly. Interestingly, those who expended more than 3500 calories a week in vigorous sports like squash had increased death rates as compared to those who exercised more moderately.

The study suggests that men who exercised vigorously and regularly would gain, on average, an extra year or two of life. This gain, although seemingly small because it is a population-wide average, is approximately equivalent to the years that would be gained if all cancer were eradicated.

Because the Harvard study includes only men who are primarily white, the authors caution against extrapolating the results strictly to the entire population. It nevertheless seems reasonable to conclude that regular vigorous exercise would be beneficial to all groups.

observation is known as **Starling's law of the heart;** it states that, within limits, the more cardiac muscle is stretched, the more forceful is its contraction. Starling's law of the heart thus provides an automatic mechanism whereby elevated blood flow to the heart increases the amount of blood it pumps.

## PROPERTIES AND DISEASES OF BLOOD VESSELS

The human body possesses some 60,000 miles of blood vessels. These form a network of tubes that convey the blood throughout the tissues of the body.

### Arteries and Arterioles

#### Structure

The walls of arteries consist of three layers (Figure 9–13). The inner layer (*tunica intima*) is formed of endothelium, the outer surface of which is coated with elastic connective tissue. The middle layer (*tunica media*) is the thickest of the three layers. In large arteries, it consists primarily of elastic connective tissue and, in medium-sized and small arteries and in arterioles, it consists of more or less spirally arranged smooth muscle cells. Contractions or relaxation of the smooth muscle in the smaller arteries and arterioles causes, respectively, constriction or dilation of the vessel. The outer layer in all arteries (*tunica adventitia*) consists of elastic and collagenous connective tissue that gives strength to the wall.

#### Function

Arteries and arterioles carry blood to the capillary networks that supply all the tissues of the body. Because they are strong elastic and muscular tubes, they are able to withstand high blood pressures generated by the pumping action of the heart. The expansion of elastic arteries during systole prevents an excessive rise in blood pressure, while elastic recoil of distended arteries during diastole maintains arterial blood pressure, keeping blood moving throughout diastole. As we noted before, the ability of arterioles to dilate or constrict permits the shunting of blood to tissues that need it.

FIGURE 9–13 (a) Scanning electron micrograph of a typical medium-sized artery and larger vein with surrounding connective tissue. (b) Diagram showing the layers that form the walls of arteries and veins. Although arteries and veins both have the same three layers, the layers in veins are thinner; consequently, veins are more distensible than arteries.
From TISSUES AND ORGANS: A text-atlas of scanning electron microscopy, by Richard G. Kessel and Randy H. Kardon. W.H. Freeman and Company, copyright © 1979.

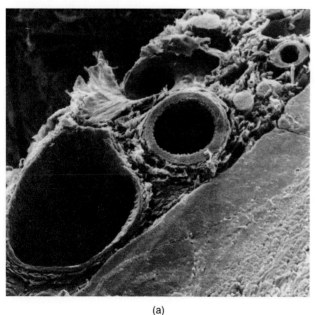

(a)

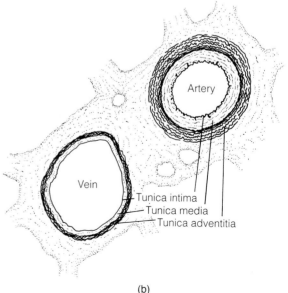

(b)

## Arterial Disease

***Arteriosclerosis and Atherosclerosis.*** Most arterial disease results from *arteriosclerosis,* also called hardening of the arteries. Arteriosclerosis is a general term used to describe a group of pathological conditions that result in a loss of elasticity of the arterial wall. Arteries weakened in this way do not function properly in the normal maintenance of blood pressure, and they may rupture as a result of the pressure of blood within them.

One particularly common form of arteriosclerosis is *atherosclerosis*. In atherosclerosis, fatty deposits known as *plaque* accumulate on the inner wall of arteries, leading to a hardening and thickening of the walls and a narrowing of the internal diameter of the vessel which results in a diminished blood flow (Figure 9–14a). As the accumulation of plaque continues, an artery may become completely blocked (Figure 9–14b). In addition, platelets may adhere to the roughened surface of the blood vessel at the site of plaque formation, producing a clot that may block the affected vessel. Alternatively, such a clot may become dislodged and be carried to a smaller vessel, which it then blocks.

Clot formation is particularly dangerous because it occurs so suddenly, blocking off blood supply without warning and causing death of the tissue supplied by the blocked artery. On the other hand, if the buildup of plaque leads to only gradual occlusion of a vessel, accessory vessels may develop over time to supply the affected area.

Although the cause of atherosclerosis is unknown, it is associated with aging. In addition, high blood pressure, lack of exercise, cigarette smoking, and diets high in cholesterol and saturated fatty acids appear to accelerate the occurrence of atherosclerosis (see *Focus: Cholesterol and Heart Disease*).

Atherosclerosis can occur, and does occur, in all arteries of the body. However, it is particularly dangerous in the coronary arteries that supply the heart and in arteries that supply the brain.

***Coronary Artery Disease.*** Contraction of the heart depends upon an adequate blood supply through the coronary arteries, the only vessels that supply the myocardium. When the development of atherosclerosis occurs gradually so that blood flow is restricted through the coronary arteries, the heart may not receive ample oxygen, particularly during physical exertion or emotional stress when the heart beats more rapidly. This condition in which cardiac muscle cells have an oxygen debt is called *ischemia,* and it results in a severe constricting pain in the chest, often radiating to the left shoulder and down the arm. The pain is called *angina pectoris* (*angina* = pain; *pectoris* = chest). Attacks of angina pectoris may be relieved by nitroglycerin and amyl nitrite, two substances that dilate the coronary arteries.

When blood supply to an area of the myocardium is completely blocked, sudden death of the affected tissue may occur. This condition is called **myocardial infarction** (*infarct* = death) or **heart attack.** A heart attack is most frequently caused by the blockage of a coronary artery by a blood clot that develops in atherosclerotic coronary arteries. Such a clot is called a *coronary thrombosis*. In many individuals, a heart attack is preceded by recurrent episodes of angina pectoris, lasting often for years and signaling pathological changes in the coronary arteries.

Heart attack is associated with pain that is more severe and prolonged than angina pectoris and is not relieved by nitroglycerin. The death of a large area of myocardium may cause heart function to cease, often as a result of electrical disturbances such as ventricular fibrillation. However, as much as a quarter of the tissue of the left ventricle—the hardest working chamber of the heart—can be lost without affecting the ability of the heart to pump adequately. After a heart attack, dead myocardial tissue is replaced by inelastic scar tissue that is unable to contract.

The extent of damage caused by a heart attack may be judged by determining the amount of a particular enzyme present in blood serum. This enzyme, serum glutamic oxaloacetic transaminase (SGOT), is released into the

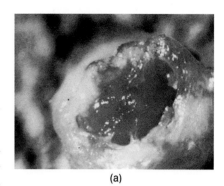

(a)

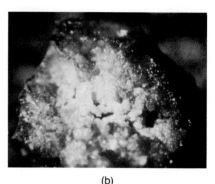

(b)

FIGURE 9–14 The effects of atherosclerosis. A coronary artery (a) somewhat thickened and (b) almost completely obstructed by fatty deposits of atherosclerosis.

# Cholesterol and Heart Disease

As the cause of most heart attacks and strokes, atherosclerosis is responsible for over half of the deaths in the United States. Consequently, understanding the factors that lead to atherosclerosis is of great importance. It has been known for at least several decades that the development of atherosclerosis correlates with high blood levels of cholesterol. For example, in the rare genetic disease called *familial hypercholesterolemia* (FH), blood cholesterol is six to eight times the normal level. Atherosclerosis develops at an early age, with heart attacks usually occurring in childhood, followed by death at an early age from severe heart disease.

Several population studies have shown a direct correlation between blood cholesterol levels and the severity of atherosclerosis. In one particularly large study, published in 1986, blood cholesterol levels and deaths from coronary artery disease (a result of atherosclerosis) were recorded for some 360,000 men between the ages of 35 and 57 who were followed for six years. The results of the study are summarized in the accompanying graph. As is clearly evident, the relative risk of mortality from coronary artery disease increases as blood cholesterol levels rise, but at high levels of blood cholesterol, the risk of mortality is dramatically increased over that seen at lower levels.

The major population studies also show that the severity of atherosclerosis, at any cholesterol level, is increased substantially if other major risk factors—smoking and hypertension—are present. Thus, for example, in the absence of other risk factors, a blood cholesterol level of 200 mg/dL may bring about a substantially increased risk of coronary artery disease (from severe atherosclerosis) by age 70. If smoking is added as a risk factor, the stage of increased risk is reached by age 60; and with the third risk factor, hypertension, the same stage is reached at age 50.

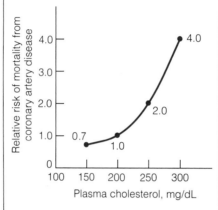

Because of the clear relationship between blood cholesterol levels and the development of atherosclerosis, many researchers feel that blood cholesterol levels for the population should be as low as practical. The National Institutes of Health and the American Heart Association recommend that individuals should strive to maintain cholesterol levels below 200 mg/dL, although some physicians are now suggesting even lower levels of about 185 mg/dL.

The simplest and usually most effective way to lower blood cholesterol levels is by diet. Cholesterol in the diet should be restricted, as should saturated fatty acids (those derived from animal fats and from fats of palm and coconut), which increase cholesterol levels. (The various types of fatty acids are described in Figure 2–14). Unsaturated fatty acids (those derived from most plant fats) seem to have no effect, either way, on cholesterol levels, and hence should be used in the diet, when possible, to substitute for saturated fatty acids. One class of polyunsaturated fatty acids, the so-called omega-3 fatty acids found in fish oils, have recently been touted as lowering blood cholesterol levels. There is currently no good evidence that they do so, but there are suggestions that they may reduce the incidence of heart attack by inhibiting the formation of blood clots, as does aspirin. In any case, substituting fish for red meat or cheese in the diet is generally a good practice, as it lowers the intake of cholesterol and saturated fatty acids.

If diet is ineffective in lowering blood cholesterol to acceptable levels, drugs may be useful. Cholestyramine, a cholesterol-lowering drug that has been used for a number of years, can reduce blood cholesterol levels in people whose levels are not very high. Lovastatin, a new cholesterol-lowering drug developed by Merck and Company and marketed under the trade-name Mevacor, has just received FDA approval. By inhibiting a key enzyme re-

blood by injured cardiac muscle cells; the amount of SGOT in the serum is proportional to the damage to cardiac muscle.

Angina pectoris and heart attack are not always brought on as a result of atherosclerosis. Recent data have shown that blood flow in the coronary arteries may be restricted or even interrupted by unexplained spasms of the coronary arteries that constrict their diameter.

There are a number of treatments that are currently used for coronary artery disease. Two new classes of drugs—*beta blockers* and *calcium blockers*—are

quired for cholesterol synthesis in the liver, lovastatin can cause a dramatic lowering of blood cholesterol levels, from 19 to 39 percent. It should be stressed, however, that drugs are no substitute for diet. All drugs have side effects, and adverse reactions often don't show up immediately. There are already suggestions that long-term lovastatin use may in some cases cause cataracts or liver problems.

Our understanding of the relationship between blood cholesterol levels and heart disease has been greatly aided by recent advances in our knowledge of the regulation of cholesterol levels in the body. Cholesterol that gives rise to atherosclerotic plaques is derived from particles in the blood termed low density lipoprotein (LDL). An LDL particle contains hundreds of cholesterol, long chain fatty acid, and phospholipid molecules.

LDL particles bind specifically to LDL receptors that are present in the plasma membranes of cells. Once attached to a receptor, LDL is taken into the cell by endocytosis. It is then broken down, its cholesterol being used for the formation of plasma membranes in all cells, for the synthesis of steroid hormones in the adrenal glands, ovaries, and testes, and for the synthesis of bile acids in the liver.

LDL receptors on cells thus remove LDL, needed as a raw material for biosynthesis, from the bloodstream. This is important from the standpoint of atherosclerosis, since the development of atherosclerosis has been shown to accelerate when blood levels of LDL rise. Furthermore, there is a feedback mechanism involved in the production of LDL receptors by a cell: as cholesterol is accumulated within a cell, the cell makes fewer LDL receptors and consequently less LDL is removed from the circulation.

A great deal of evidence suggests that atherosclerosis develops when high blood levels of LDL accumulate due to an inadequate number of LDL receptors on cells. Supporting this idea is the finding that LDL receptors are absent on the cells of persons with familial hypercholesterolemia; it is this defect that accounts for the high blood levels of LDL in those individuals.

In most normal individuals, a high dietary intake of cholesterol and saturated fatty acids presumably suppresses the manufacture of LDL receptors by causing an accumulation of cholesterol within cells. In some people who maintain low LDL levels while consuming high fat diets, other mechanisms for the regulation of LDL levels must be operative.

Regular exercise has been shown to decrease blood cholesterol by increasing the blood levels of another type of lipoprotein particle, called high density lipoprotein (HDL). HDL acts to scavenge cholesterol from the blood and hence to reduce its deposition in arteries.

Total Fat, Saturated Fatty Acids, and Cholesterol in Foods

| Food | Total fat (grams) | Saturated fatty acids (grams) | Cholesterol (milligrams) |
| --- | --- | --- | --- |
| Grilled lean ground beef (3 ounces) | 15 | 6 | 80 |
| Fried beef liver (3 ounces) | 9 | 2 | 372 |
| Roasted pork rib (3 ounces) | 20 | 7 | 69 |
| Roasted chicken, with skin (3 ounces) | 12 | 3 | 75 |
| Broiled halibut fillet (3 ounces) | 6 | 1 | 48 |
| Boiled large egg | 6 | 2 | 274 |
| Whole milk (1 cup) | 8 | 5 | 33 |
| Skim milk (1 cup) | 1 | trace | 5 |
| Creamed cottage cheese (1 cup) | 9 | 6 | 31 |
| Natural cheddar cheese (1 ounce) | 9 | 6 | 30 |
| Vanilla ice cream (½ cup) | 7 | 4 | 30 |
| Butter (1 tablespoon) | 11 | 7 | 31 |
| Stick margarine (1 tablespoon) | 11 | 2 | 0 |
| Mayonnaise (1 tablespoon) | 11 | 2 | 8 |
| Peanut butter (2 tablespoons) | 16 | 2 | 0 |

Vegetables and fruits are very low in fat and none contains cholesterol. However, fat may be added in cooking.

now in wide use. Beta-blockers slow the heart rate under stressful situations, thus lowering blood pressure and helping to control angina and cardiac arrhythmias. Calcium blockers inhibit the flow of calcium into cardiac muscle cells, decreasing their contractility. They slow the heart rate and dilate coronary and peripheral arteries. They are particularly useful in reducing angina by preventing coronary artery spasms and in quieting cardiac arrhythmias. Coronary arteries that have deteriorated severely and narrowed due to plaque buildup may be treated by *coronary bypass surgery*. In this operation, dispensable veins in the leg are used as grafts for bypassing obstructions in coronary arteries. Another treatment, still somewhat experimental but very promising, involves threading a tiny deflated balloon into a partially blocked coronary artery; inflating the balloon compresses atherosclerotic plaques against the arterial wall, opening up the channel of the vessel. In another newly developed technique, victims of a heart attack caused by coronary thrombosis can be treated in the early stages of the attack by the injection into the blocked coronary artery of an enzyme that dissolves the thrombus.

***Stroke.*** Blood flow to the brain can be interrupted by the formation of a thrombus within an artery supplying the brain. It can also be interrupted by the rupture of a brain artery weakened by atherosclerosis. Either of these events may cause **stroke,** or *cerebral vascular accident,* which is characterized by damage to brain tissue and often results in paralysis, depending upon the site of the lesion.

***Aneurysm.*** Weakness in the wall of an artery may cause the wall to balloon out under the pressure of blood flowing through it. Such a swelling is called an **aneurysm.** An aneurysm may be due to a congenital defect in the structure of the arterial wall, to inflammation that weakens the wall, or to the degenerative effects of arteriosclerosis, often in conjunction with high blood pressure. Aneurysms occur most frequently in the aorta and in arteries that supply the brain. The danger of an aneurysm is that the arterial wall will become so weakened that it will burst, resulting in fatal hemorrhaging.

## Veins

### Structure and Function

The walls of veins, like those of arteries, consist of three layers, but these are generally thinner and the walls more distensible (expandable) than those of arteries. The pressure of blood in veins is low, and the flow of blood is bolstered by the massaging of veins by skeletal muscles during movement. Also aiding the movement of blood is the presence in many veins of one-way valves that permit blood to flow toward the heart but not in the other direction. Valves are particularly abundant in veins of the extremities in which blood must flow a considerable distance against the force of gravity.

### Diseases of Veins

***Varicose Veins.*** Veins that course near the surface of the skin are relatively unsupported by surrounding tissue. These veins, particularly those in the legs where blood tends to accumulate due to gravity, have a tendency to be stretched. Under conditions in which the walls of the veins are weakened, the veins tend to become greatly distended, and the valves incompetent. This leads to further stretching, ultimately causing the development of dilated torturous vessels, called *varicose veins,* that can be very painful.

Varicose veins occur most frequently in veins weakened by heredity or disease, or in individuals whose professions force them to stand a lot without moving. In addition, varicose veins often develop during pregnancy when pressure of the fetus on the large veins in the pelvis interferes with the return of blood from the legs.

The development of varicose veins may be slowed by wearing elastic support hose that prevents blood from pooling in surface veins. Large and painful varicose veins may be removed surgically.

***Phlebitis.*** An inflammation of veins leading to the formation of thrombi that adhere to the walls of the veins is called **phlebitis.** Individuals with varicose veins are more likely than others to get phlebitis since blood pooled in varicose veins may clot due to the slowness of its flow. Phlebitis usually clears up quickly when the source of the inflammation is treated, but there is a slight chance that a blood clot may become dislodged and block a more important vessel.

## Capillaries

The structure and function of capillaries were discussed extensively in an earlier section of this chapter.

## BLOOD PRESSURE

As blood is pumped by the heart through the circulatory system, the flow of blood exerts a force against the walls of the blood vessels. This force is called **blood pressure.** The maintenance of adequate blood pressure is critical to the body, for a substantial pressure is required to distribute blood throughout the capillary networks of the body, particularly to those of the brain where blood flow must overcome the force of gravity. Blood pressure is highest in the arteries, less in the capillaries, and drops to near zero in the large veins that return blood to the heart.

### Measurement of Blood Pressure

Blood pressure is measured using an instrument called a sphygmomanometer. A sphygmomanometer consists of an inflatable rubber cuff, a bulb to pump air into the cuff, and a manometer that measures the pressure within the cuff in millimeters of mercury (Figure 9–15). To measure blood pressure, the cuff is placed snugly around the upper arm, and sufficient air is pumped into the cuff to collapse the brachial artery that runs down the arm. The bell (round part) of a stethoscope is placed over the brachial artery just below the cuff, and air is released slowly from the cuff. As the pressure in the cuff falls, a point will be reached when blood can just force its way through the collapsed artery, and a distinct tapping sound, corresponding to the first turbulence of blood flow, can be heard in the stethoscope. The pressure at the point when the tapping sound is first heard is the **systolic blood pressure,** the pressure generated during ventricular systole. As air continues to be released from the cuff, the tapping sound will increase in intensity, then become suddenly muffled just before disappearing. The pressure at the point of the muffling is the **diastolic blood**

FIGURE 9–15 Measurement of arterial blood pressure using a sphygmomanometer.

**pressure,** the pressure maintained in the arteries during relaxation of the heart.

A blood pressure reading consists of two numbers. The normal average blood pressure for an individual at rest is 120/80 (read 120 over 80), where 120 refers to the systolic pressure and 80 to the diastolic pressure. In actuality, normal healthy individuals show a range of blood pressure readings, something on the order of 90–130 for systolic pressure and 60–85 for diastolic pressure.

## Factors Determining Blood Pressure

In a healthy person, arterial blood pressure is determined mainly by two factors: *cardiac output* and *peripheral resistance.* **Peripheral resistance** refers to the degree to which the flow of blood is impeded by muscular blood vessels, primarily the arterioles. The more the arterioles are constricted and hence the smaller their diameter, the greater the arterial pressure must be to force blood through them. With respect to cardiac output, the greater the amount of blood forced into the arterial system by contraction of the heart, the higher the pressure in the arteries.

In the regulation of blood pressure, cardiac output and peripheral resistance are controlled in tandem, mainly by the part of the autonomic nervous system that controls unconscious or "autonomic" functions (autonomic nervous system). Blood pressure changes are registered by specialized nerve cells called **baroreceptors,** or **pressoreceptors,** that are present in the walls of the aortic arch and of the carotid sinuses. (The carotid sinuses are localized enlargements, or dilatations, at the lower ends of the internal carotid arteries, see Figure 9–5a). Stretching of the baroreceptors due to increased blood pressure within these arteries sends nerve impulses to cardiac control centers at the base of the brain; impulses carried by nerves from the brain inhibit cardiac output and dilate peripheral arterioles, resulting in a compensatory drop in blood pressure. Conversely, a fall in blood pressure, relayed to the cardiac control centers by the baroreceptors, causes impulses carried by other nerves to increase cardiac output and constrict arterioles. The net effect is a compensatory rise in blood pressure.

Higher brain centers may also affect peripheral resistance. For example, blushing due to embarrassment is caused by dilation of the arterioles in the face and neck, which shunts blood into the capillary networks of these regions. Blanching brought on by fear is caused by constriction of these same arterioles.

## Hypertension

High blood pressure, or **hypertension,** is defined as a sustained pressure of 140/90 or higher in a person at rest, although diastolic pressure is usually a more accurate indicator of hypertension than systolic pressure. Hypertension is harmful because increased pressure in the circulatory system strains the heart by making it work harder in order to keep the blood circulating. In addition, the increased pressure may damage the walls of arteries, leading to development of arteriosclerosis and the possible rupture of weakened arteries. Heart attack and stroke are two likely consequences of these conditions. Prolonged severe hypertension may also damage the kidneys.

Most cases of hypertension are caused by increased peripheral resistance due to inappropriate constriction of the arterioles. In the majority of individuals with hypertension, we do not know what brings about arteriolar constriction. Such individuals are said to have *essential hypertension.* The tendency to develop essential hypertension appears to have a strong hereditary component, but other factors are also important. For example, people who are overweight when they are young and those under prolonged psychological stress are more likely than others to have essential hypertension in middle age. In some cases of hypertension, arteriolar constriction is known to be caused by endocrine (hormonal) problems or diseases that damage the kidneys. Hypertension for which we know the cause is referred to as *secondary hypertension.*

We do not completely understand why the baroreceptors in hypertensive individuals do not initiate events that return the blood pressure to normal. There is some evidence to suggest that in sustained hypertension the baroreceptors are reset so that they regulate blood pressure at a higher level than normal.

## Treatment of Hypertension

In individuals with borderline hypertension, weight loss may lower blood pressure. In addition, reduced sodium consumption is almost universally advised, although very recently some evidence has suggested that lack of calcium—and not excessive sodium consumption—may cause hypertension. In those with more severe hypertension, drugs may be required. One group of drugs widely used in treating chronic hypertension is the beta blockers that act to slow the rate of the heartbeat. Another group of drugs acts as diuretics that increase the excretion of sodium and water through the kidneys. This treatment reduces blood volume and hence tends to lower blood pressure.

## Heart Failure and Congestive Heart Failure

Individuals with high blood pressure are six times as likely as others to develop **heart failure,** a condition in which the heart does not pump efficiently due to a decrease in strength of the heart's contraction. As heart failure progresses, the heart eventually becomes unable to generate an adequate pumping pressure, and blood backs up in the veins. The resultant increase in venous and capillary pressure causes an increased filtration out of the capillaries, leading to the accumulation of large amounts of tissue fluid in the intercellular spaces of all the tissues. The reduction of blood flow through the kidneys leads to an increase in the retention of water, contributing to the accumulation of tissue fluid.

As the lungs become congested with tissue fluid, the patient has difficulty breathing and, if the condition remains untreated, may eventually drown in the fluid. Congestion of the lungs is generally somewhat alleviated during the day when the patient is upright and the fluid accumulates in the legs due to the force of gravity.

The accumulation of tissue fluid as a result of heart failure is known as *congestive heart failure.* Treatment for congestive heart failure involves taking diuretics, substances that increase the volume of urine excreted, to remove excess fluid and digitalis to strengthen the heartbeat.

## Shock

When the flow of blood throughout the body becomes inadequate due to a sudden drop in blood pressure, vital functions are disrupted and the individual is said to be in **shock.** Shock may be caused by failure of the heart to pump an adequate supply of blood; by severe hemorrhaging or burns, which reduce blood volume to the extent that sufficient blood pressure cannot be maintained; and by dilation of blood vessels brought on by a severe allergic reaction (*anaphylactic shock;* see Chapter 21) or severe infection.

The initial treatment for shock involves restoring the blood pressure to normal. This is usually done by transfusing plasma or whole blood into a vein to increase blood volume. After stabilizing the patient, the cause of the shock can be treated.

## THE LYMPHATIC SYSTEM

The lymphatic system consists of an extensive network of vessels that drain excess tissue fluid from the intercellular spaces and return it to the circulatory system. In addition, several lymphatic organs, including lymph nodes, thymus, and spleen, play an important role in defending the body against disease.

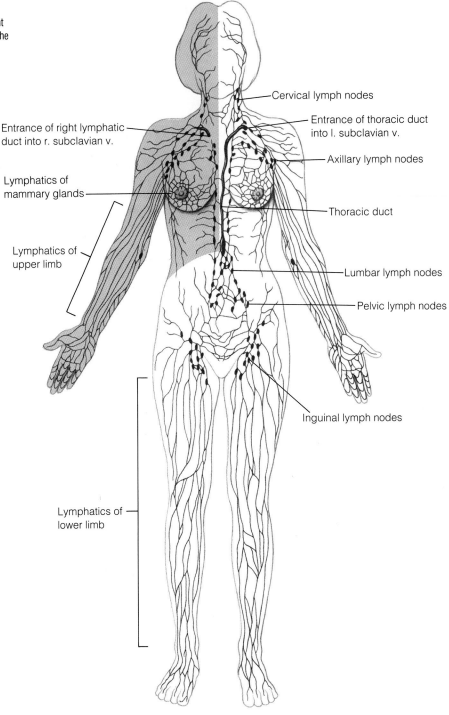

FIGURE 9–16 Major vessels of the lymphatic system. The shaded area is drained by the right lymphatic duct; all other areas are drained by the thoracic duct.

Cervical lymph nodes

Entrance of right lymphatic duct into r. subclavian v.

Entrance of thoracic duct into l. subclavian v.

Axillary lymph nodes

Lymphatics of mammary glands

Thoracic duct

Lymphatics of upper limb

Lumbar lymph nodes

Pelvic lymph nodes

Inguinal lymph nodes

Lymphatics of lower limb

## Lymphatic Vessels

The smallest lymphatic vessels are the *lymphatic capillaries* that arise in the intercellular spaces as blind-ended tubes that join one another to form a vast network throughout the body. The walls of lymphatic capillaries are thin like those of blood vessels, but are considerably more permeable. Excess tissue fluid containing dissolved substances enters lymphatic capillaries; once the tissue fluid has entered a lymphatic vessel it is referred to as **lymph.** From the lymphatic capillaries, lymph drains into larger vessels called *lymphatics.* These in turn ultimately drain into one of the two major lymphatic ducts, the *thoracic* *duct* and the *right lymphatic duct,* which empty into the left and right subclavian veins, respectively (Figure 9–16).

Lymph is moved through the lymphatic vessels by the contraction of smooth muscle in their walls and by contraction of skeletal muscles that massage the vessels. The flow of lymph toward the veins of the circulatory system is maintained by one-way valves similar to those in veins.

## Lymphatic Organs

Along the course of the lymphatics are numerous *lymph nodes* through which lymph must flow on its way to the circulatory system. Lymph nodes help protect the body against disease by filtering out bacteria and other foreign substances. This function will be discussed further in Chapter 21. The roles of the thymus and the spleen in the production and maturation of lymphocytes will also be considered in that chapter.

## Study Questions

1. Indicate the arrangement of the membranes that cover the heart.
2. Draw a diagram of the internal structure of the heart, labeling chambers, valves, and vessels. Indicate the direction of blood flow with arrows.
3. Outline the path of blood in the pulmonary and systemic circulations.
4. Describe the hepatic portal system.
5. How does the structure of cardiac muscle compare to that of skeletal muscle?
6. Describe the electrical events that lead to the contraction of the heart.
7. a) What is an ectopic beat?
   b) Explain how an ectopic beat can occur.
8. Describe the three types of heart block.
9. What is the difference between a functional and resting heart murmur?
10. Describe Starling's law of the heart and indicate its importance in the functioning of the heart.
11. a) What is atherosclerosis?
    b) Why is it a dangerous condition?
12. What do the two numbers in a blood pressure reading indicate?
13. Explain how blood pressure is maintained.
14. Distinguish between essential and secondary hypertension.
15. Describe the function of the lymphatic system.

# THE RESPIRATORY SYSTEM

With few exceptions, the cells of the human body obtain most of their energy by the oxidation of glucose and other nutrients in metabolic pathways that require oxygen gas ($O_2$) and that release carbon dioxide ($CO_2$) as a major end product (see Chapter 2). However, most of the body's cells cannot exchange $O_2$ and $CO_2$ directly with the environment. Not only do the vast majority of cells lie deep within the body, far removed from the external environment, but the dead keratinized skin cells that cover the outer surface of the body prevent the diffusion of $O_2$ and $CO_2$ through the skin. Supplying the body's cells with $O_2$ and removing $CO_2$ requires coordination between two specialized systems— the respiratory system and the circulatory system. The respiratory system functions in the actual *exchange* of $O_2$ and $CO_2$ with the environment, a process called *respiration*. The circulatory system transports $O_2$ and $CO_2$ between the cells of the body and the part of the respiratory system where gas exchange occurs.

## OVERVIEW OF THE RESPIRATORY SYSTEM

In essence, the respiratory system consists of a highly branched hollow tube, forming the *air passages*, the smallest branches of which terminate in clusters of microscopic inflatable sacs, called *alveoli*. The air passages, which serve merely to conduct air into and out of the alveoli, include in descending order the *nasal cavities, pharynx* (throat), *larynx* (voice box), *trachea* (wind pipe), *bronchi*, and *bronchioles*. The alveoli, bronchi, and bronchioles are organized into two paired organs, the *lungs*, that are surrounded by a two-layered membrane.

The alveoli, the structures where actual gas exchange occurs, are formed of delicate elastic membranes that are covered by vast capillary networks of the pulmonary circulation. As blood from the pulmonary arteries enters the capillaries, it carries relatively high concentrations of $CO_2$ and low concentrations of $O_2$.

During inspiration (breathing in), fresh air is drawn into the alveoli through the air passages. Oxygen diffuses from within the alveoli into the capillaries, and, simultaneously, carbon dioxide diffuses from the capillaries into the alveoli. The air within the alveoli, now containing high amounts of $CO_2$ and low amounts of $O_2$, is exhaled. The net result is that $CO_2$ in the blood is exchanged for $O_2$ in the alveoli. The organization of the respiratory system is shown in Figure 10–1.

## STRUCTURES OF THE RESPIRATORY SYSTEM

### Air Passages

#### Nose and Nasal Cavities

The nose is considered the normal route through which air enters the respiratory passages. However, air may also enter through the mouth, since the nasal cavities and the mouth are both connected to the throat region, which serves as a common passageway for the respiratory and digestive systems (see Figure 10–1).

The two external openings of the nose are the *nostrils*, which lead into the right and left nasal cavities. The anterior portion of the nasal cavities just behind the nostrils is lined with skin possessing coarse hairs that strain large particles from the air. Farther back, the lining of the nasal cavities changes to a highly vascular, ciliated mucous membrane, called the *nasal mucosa*. Within the skull, three plates of bone, covered with nasal mucosa and lying one on top of the other, project from the lateral walls into each nasal cavity. By expanding the surface area of the nasal mucosa, the bony plates facilitate the conditioning of air as it passes through the nasal cavities. Blood vessels that permeate the nasal mucosa act as radiators, cooling hot air and warming cold air. The mucus secreted by the nasal mucosa adds moisture to dry air and traps fine dust particles. The ciliated cells of the mucosa constantly move dust-containing mucus into the throat, where it is swallowed.

The nasal cavity is associated with other functions as well. For example, part of the nasal mucosa at the roof of the nasal cavities is specialized to form the *olfactory region*—the organ of smell (see Chapter 15). In addition, four paired air sinuses in the bones of the skull (see Chapter 6) communicate with the nasal cavities on each side by small openings. Inflammation of the mucous membrane that lines the sinuses is called *sinusitis,* mentioned in Chapter 6.

Inflammation of the membrane lining the nasal cavities is called *rhinitis* (*rhin* = nose; *itis* = inflammation) and occurs in the *common cold*. The common cold is caused by a group of related viruses that infect the nasal mucosa. Sometimes, a cold is associated with a bacterial infection that may worsen the cold's symptoms and may require treatment with antibiotics.

Allergies to pollen, feathers, mites in housedust, and dandruff on animal fur may cause *allergic rhinitis,* commonly known as *hayfever.* Symptoms include sneezing, a runny nose, and eyes that become red, watery, and itchy. Antihistamines are often effective in reducing or eliminating the symptoms of hayfever.

FIGURE 10–1 Organs of the respiratory system.

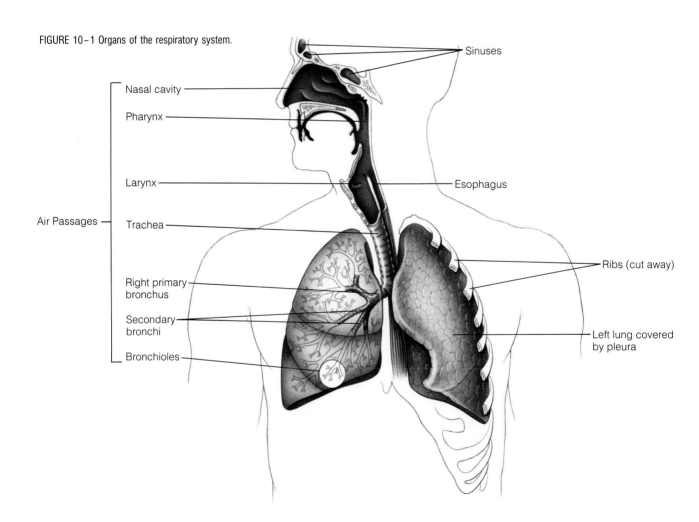

Sinuses

Nasal cavity

Pharynx

Larynx

Esophagus

Air Passages

Trachea

Ribs (cut away)

Right primary bronchus

Secondary bronchi

Left lung covered by pleura

Bronchioles

## Pharynx

The **pharynx,** or throat, extends from the nasal cavities to the larynx, or voice box. The portion of the pharynx behind the nasal cavities and above the soft palate is referred to as the *nasopharynx.* During swallowing, food is prevented from entering the nasopharynx by muscles that elevate the soft palate.

The Eustachian tubes from the middle ear open into the lateral walls of the nasopharynx on either side. As a result of this anatomical arrangement, infection can spread from the nasopharynx along the Eustachian tube to the middle ear. Such infections are most common in children, since their Eustachian tubes are wider, shorter, and straighter than those of adults.

In the medial region of the posterior wall of the nasopharynx lies a mass of lymphatic tissue called the *pharyngeal tonsil,* which functions to protect the respiratory system from airborne infectious agents. The pharyngeal tonsil often becomes enlarged during early childhood, and a child with this condition is said to have *adenoids.* Adenoids may obstruct the passage of air through the nasopharynx so that the child is forced to breathe through his/her mouth. In addition, adenoids may become more or less permanently infected. For these reasons, adenoids are commonly removed surgically.

Oval masses of lymphatic tissue on either side of the pharynx behind the mouth are called the *palatine tonsils.* Their function is similar to that of the pharyngeal tonsil. Inflammation of the palatine tonsils is called *tonsillitis.* If tonsillitis is unusually severe or frequent, surgical removal of the tonsils, called *tonsillectomy,* may be required.

The pharynx branches at its lower end into two tubes: the esophagus, through which food passes from the pharynx to the stomach, and the larynx, through which air passes to the lungs (see Figure 10–1).

## Larynx

The **larynx** is a cartilaginous structure connecting the pharynx and the trachea (Figure 10–2). The thyroid cartilage, one of the cartilages that forms the larynx, has a projection, called *Adam's apple,* that is visible in the front of the throat. Adam's apple is much more pronounced in adult males than it is in children and adult females, reflecting the considerable increase in size of the larynx that occurs in males at the time of puberty. Most of the larynx is lined by a ciliated mucous membrane that moves foreign particles up toward the pharynx.

In addition to serving as an air passageway, the larynx functions in the production of sound and prevents substances other than air from entering the lower air passages. The cavity within the larynx is narrowed to a slit by two sets of heavy, membranous folds that project from either side of the lateral walls of the larynx (see Figure 10–2). The upper pair of folds, called *false vocal cords* or vestibular folds, do not function in the production of sound. The lower pair, the *vocal cords,* contain bands of connective tissue, the *vocal ligaments,* at their free edges. Vibration of the free edge of the vocal cords by air exhaled from the lungs produces sounds that can be modified into words by muscles of the neck, lips, tongue, and cheeks. Because the length of the vocal cords is one factor that determines pitch, adult females and children, who have shorter vocal cords than mature males, have a voice of higher pitch.

Not only are the vocal cords involved in the production of sounds, but their closure across the larynx prevents food, water, and other foreign substances from entering the lower respiratory passages. Also serving to prevent the entrance of foreign substances is the fact that during swallowing, the larynx is brought upward and forward so that the upper end of its opening is pressed against the **epiglottis,** a cartilaginous flap projecting above the rest of the larynx on its ventral side. The entrance to the larynx is thus effectively sealed off. If, in spite of the protection offered by this maneuver, food or other substances should enter the larynx, a violent cough reflex is immediately initiated. This coughing usually evokes the comment that "food went down the

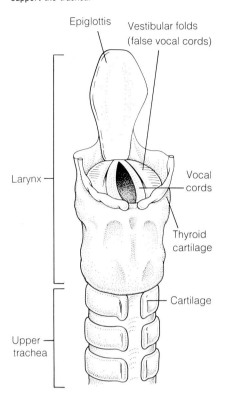

FIGURE 10–2 A posterior view of the larynx and upper trachea. Note the C-shaped cartilages that support the trachea.

Epiglottis

Vestibular folds (false vocal cords)

Larynx

Vocal cords

Thyroid cartilage

Cartilage

Upper trachea

# The Heimlich Maneuver

In 1974, Dr. Henry Heimlich, a Cincinnati surgeon, introduced a first-aid procedure for clearing food or other objects from the air passageways. Referred to as the Heimlich maneuver, it has since been used effectively on thousands of people who otherwise might have died of choking. It is one of the simplest of life-saving procedures, and it can be performed easily by any adult.

The Heimlich maneuver is illustrated in the accompanying figure. The rescuer clasps her arms around the victim from behind, making a fist of one hand and holding the fist with the other hand. The fist is placed just above the victim's navel and well below the lower end of the sternum, and a quick thrust is made upward and toward the chest. The thrust pushes against the diaphragm, decreasing the volume of the thoracic compartments and increasing air pressure within the lungs. If done successfully, the maneuver should cause a forceful expiration that expels the object. If the object is not dislodged, several additional thrusts may be necessary.

The Heimlich maneuver can be performed in the same manner on choking victims who are sitting down. A victim who has collapsed should be turned on his back on the floor. The rescuer places her two hands, one on top of the other, slightly above the navel and thrusts downward and forward with the heel of the hand. It is even possible for a choking victim to perform the Heimlich maneuver on himself by raising the chin to straighten the respiratory passages and administering sudden pressure to the abdomen by pushing forcefully against a chair back.

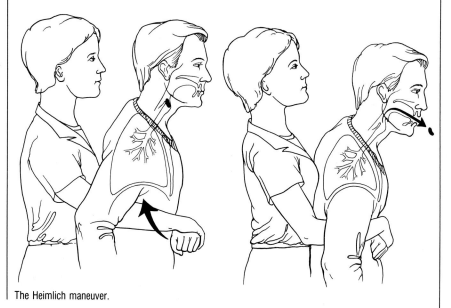

The Heimlich maneuver.

wrong way." Sometimes, food may enter the larynx and get caught in the area of the vocal cords, preventing the passage of air. This event is called *choking,* and it is best treated by the *Heimlich maneuver* (see *Focus: The Heimlich Maneuver*).

Bacterial or viral infections of the larynx or irritation of the larynx by smoking, alcohol, or excessive speaking may cause *laryngitis,* inflammation of the larynx. The vocal cords become inflamed and swollen, reducing the size of the laryngeal opening and the flow of sound through it. The result is a distortion in sound known as hoarseness.

Cancer of the larynx may require a *laryngectomy,* removal of the larynx. Because the vocal cords are absent, sounds cannot be made in the usual way. In order to regain speech, an individual must learn to produce sounds by belching air through the esophagus and shaping the sounds into speech. Alternatively, a valve may be constructed surgically between the trachea and esophagus. Air passing up the trachea and through the valve causes the valve to vibrate, producing sounds used in speech.

### Trachea, Bronchi, and Bronchioles

The larynx opens into a 4- to 5-inch long rigid tube, called the **trachea,** or windpipe, which is present in the midline of the neck and extends into the thorax. The trachea is supported and held open by a series of C-shaped cartilages stacked one upon the other and open in the back (see Figure 10–2). The gaps between adjacent cartilages and between the ends of the cartilages are filled with smooth muscle and connective tissue.

The trachea branches into two **primary bronchi** (sing: *bronchus*) and these, in turn, branch repeatedly in a treelike fashion to give progressively smaller bronchi (see Figure 10–1). The walls of the primary bronchi have a structure the same as that of the trachea, but as these divide into smaller bronchi, the walls contain progressively less cartilage and more smooth muscle.

At the point where bronchi lose the cartilage in their walls and their diameter is reduced to about 1 millimeter, they become known as **bronchioles.** Bronchioles are constructed entirely of smooth muscle supported by connective tissue. The bronchioles continue to divide until they form the smallest air passageway, the *terminal bronchioles.*

The trachea, bronchi, and larger bronchioles are lined with ciliated mucous membrane; in the smaller bronchioles, nonciliated cells are present. As in the other respiratory passages, the mucus secreted by cells of the membrane helps moisten air and trap small particles, including dust, bacteria and viruses; the cilia constantly move the mucus to the pharynx, where it is swallowed. In fact, a major cause of respiratory infections is the paralysis of cilia by noxious agents such as cigarette smoke and other chemical pollutants in the air. The early morning cough of smokers is an attempt to clear accumulated mucus from the air passageways.

Inflammation of the bronchi is called *bronchitis* and is caused by the same viruses that cause colds. It is marked by a deep cough that brings up gray or yellow phlegm, pain in the upper chest, breathlessness, wheezing, and fever. No treatment is generally necessary unless a bacterial infection develops. Another condition affecting the bronchi and bronchioles is *asthma.* Asthma is characterized by periodic attacks of wheezing and difficulty in breathing, and it is caused by spasms of smooth muscle in the walls of bronchi and bronchioles. Asthma attacks are usually triggered by an allergy, but may be brought on by infection, psychological stress, and the effects of certain drugs. In severe attacks, the face and lips may turn blue due to a lack of oxygen. Treatment involves removal of the suspected irritant and the use of bronchodilator drugs such as epinephrine that can relieve attacks by relaxing smooth muscle.

## The Alveoli: Site of Gas Exchange

The terminal bronchioles give rise to *respiratory bronchioles,* so named because their walls contain small, cup-shaped outpocketings, the *alveoli,* through which gas exchange occurs. The respiratory bronchioles terminate in clusters of alveoli that open into a central space continuous with the lumen (inside space) of the respiratory bronchioles (Figure 10–3a).

## The Respiratory Membrane

The walls of alveoli are lined by a layer of epithelial cells that are attached proximally to a basement membrane. Numerous capillaries cover the outside of the alveolar wall (Figure 10–3b). The walls of the capillaries are formed by endothelial cells, also attached proximally to a basement membrane. The basement membranes of the alveolar epithelium and the capillary endothelium are usually separated from one another by a thin interstitial space that contains some elastic or reticular fibers embedded in amorphous intercellular substance. However, in some areas, the interstitial space may be absent altogether, and the basement membrane of the alveolar epithelium may be fused to the basement membrane of the capillary endothelium.

The entire membranous structure that separates air in the alveoli from blood in the capillaries is termed the **respiratory membrane,** and it constitutes an extremely thin barrier that permits rapid exchange of gases (Figure 10–4). Gas exchange is also facilitated by enormous numbers of alveoli, estimated in excess of 300 million in the average adult. These alveoli

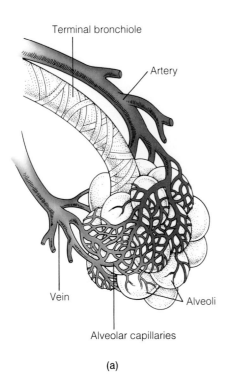

Terminal bronchiole

Artery

Vein

Alveoli

Alveolar capillaries

(a)

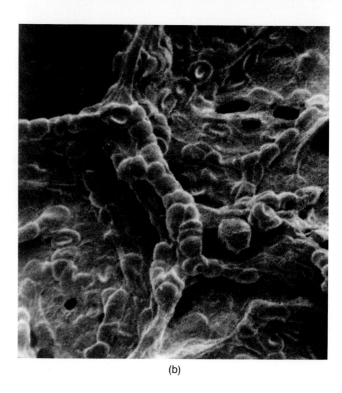

(b)

FIGURE 10-3 (a) Sketch of a cluster of alveoli with surrounding blood capillaries. Respiratory gases are exchanged by diffusing, in both directions, between the air within the alveoli and the blood. (b) Scanning electron micrograph showing capillaries located on the walls of alveoli. Red blood cells are clearly seen within capillaries. The dark holes are openings in the walls of the alveoli.

provide a surface area in contact with capillaries of about 80 square meters or more.

## The Lungs

While the air passageways provide a conduit for air to pass into and out of the alveoli, the site where gas exchange occurs, there must be some mechanism by which fresh air is drawn into the alveoli and stale air is exhaled. In order to provide for such ventilation—which we call *breathing*—the air passageways and the alveoli are organized into two functional and structural units, the *lungs*.

### Structure of Lungs and Pleura

The lungs lie in separate compartments on either side of the thoracic cavity; the two compartments are formed in the thoracic cavity by a median dividing wall that contains the heart and other thoracic organs embedded in connective tissue. (You will recall from Chapter 4 that the thoracic cavity is a closed compartment formed by the chest wall and bounded at its upper end by muscles and connective tissue of the neck and at its lower end by a large dome-shaped sheet of muscle, the diaphragm.) The lungs consist of all parts of the respiratory system beyond the primary bronchi, which branch shortly after they enter the lungs on each side. The lungs, somewhat conical-shaped, have an elastic, spongy texture that derives from the millions of alveoli they contain. The right lung is divided by deep fissures into three *lobes,* and the left lung, into two. Each lobe, in turn, is subdivided into a number of smaller *lobules* that are supplied by the largest bronchioles.

Each lung is surrounded by a two-layered serous membrane called the *pleura* (Figure 10–5). The inner layer, the *visceral pleura,* covers the surface of each lung and dips into the fissures between the lobes. The outer layer of the pleura, the *parietal pleura,* lines the compartments of the thoracic cavity in which the lungs lie. The visceral and parietal pleura are continuous with one another at the point, termed the *root* of the lung, where the primary bronchus, blood vessels, and nerves enter each lung. Thus, the two layers of the pleura

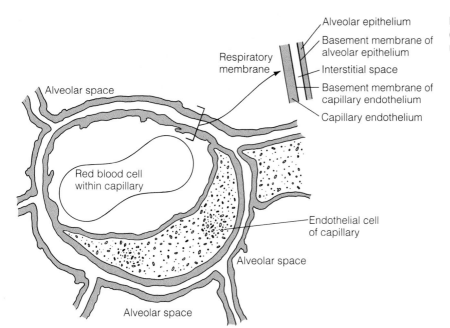

FIGURE 10-4 Cross-section through a blood capillary and surrounding alveoli, showing the respiratory membrane.

Labels: Alveolar epithelium; Basement membrane of alveolar epithelium; Respiratory membrane; Interstitial space; Basement membrane of capillary endothelium; Capillary endothelium; Alveolar space; Red blood cell within capillary; Endothelial cell of capillary; Alveolar space; Alveolar space

form a collapsed sac. The area within the sac—between the visceral and parietal layers—is called the **pleural cavity,** although it is not really a cavity at all. It is filled with a small amount of serous fluid that holds the two layers in close contact, but allows them to glide smoothly over one another, much as a layer of water holds two glass plates together, but lets them slide relative to each other. As we shall see, cohesion between the two layers of the pleura is of great importance for breathing.

### Blood Supply to the Lungs

The pulmonary arteries give rise to the pulmonary capillaries, which are closely apposed to the alveolar wall and through which gas exchange takes place. Other structures of the lung, including the walls of the air passageways, the walls of blood vessels, the pleura, and nerves, are supplied by the bronchial arteries that arise from the thoracic region of the aorta.

### Disease of the Lungs

Pulmonary emphysema (*emphysema* = bodily inflation) occurs when the epithelium lining the bronchioles grows abnormally and partially obstructs the air passages within the bronchioles. Air is trapped in the alveoli, which become increasingly distended with the rupture of adjacent alveolar walls (Figure 10–6). The lungs cannot deflate properly, causing a permanently expanded chest (barrel chest) from which the disease gets its name.

Severe cases of emphysema occur mostly in heavy cigarette smokers, although the disease is also associated with conditions such as asthma and chronic bronchitis, which cause narrowing of bronchioles, resulting in forceful breathing that weakens the alveoli.

*Pneumonia* is a general term for inflammation of lung tissue. It is usually caused by a viral or bacterial infection. This disease is characterized by fever, cough, and the accumulation of fluid within alveoli. In a weakened individual, pneumonia can develop very quickly and is a frequent cause of death. Bacterial pneumonia is treated with antibiotics, but there is no specific treatment for viral pneumonia.

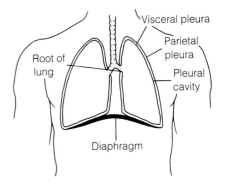

FIGURE 10-5 A greatly simplified diagram showing the arrangement of the pleura covering the lungs. The visceral and parietal pleura are continuous with one another at the root of the lung.

Labels: Visceral pleura; Parietal pleura; Pleural cavity; Root of lung; Diaphragm

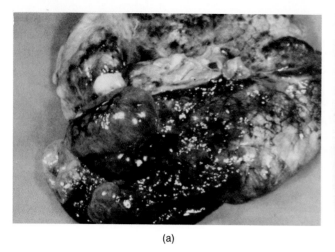

(a)

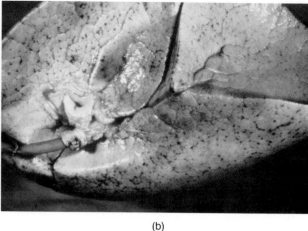

(b)

FIGURE 10-6 (a) Surface view of a lung from an individual with emphysema. Notice the balloonlike distensions resulting from rupture of adjacent alveolar walls. The dark color of the lung tissue is caused by carbon particles deposited from smoking. (b) Surface of a normal lung with small spots of carbon deposits resulting from breathing smog or mildly polluted city air.

One disease that is widespread in underdeveloped parts of the world, but is increasingly rare in the United States, is *tuberculosis,* or TB. Tuberculosis is caused by a bacterium that often attacks and destroys lung tissue, but the bacteria are usually virulent only in individuals who are weak, ill, or undernourished. The disease may be treated with special antibiotics.

*Lung cancer* is caused primarily by tobacco smoking (Figure 10-7). It is the most frequent cause of cancer death in males in the United States. Until recently, it was the second most common cause of cancer death, behind breast cancer, in American females, but in 1985, breast cancer was overtaken by lung cancer as the number one cancer killer in females. Lung cancer is difficult to cure, since it has often spread by the time it is discovered. Treatment involves surgical removal of the affected lung or lobe of the lung, followed by radiation or chemotherapy. Lung cancer is discussed extensively in Chapter 22.

Inflammation of the pleura, known as *pleurisy,* is usually a symptom of an underlying disease such as pneumonia or tuberculosis, although it may be caused by such things as chest injury. It usually causes pain during breathing because the pleural layers do not glide smoothly over one another.

## TRANSFER OF GASES

The respiratory system functions to oxygenate the blood and to remove carbon dioxide from it. This transfer of gases occurs at the lungs during breathing.

### Mechanism of Breathing

The process of breathing moves air into and out of the lungs, and hence the alveoli. It takes advantage of the principle that air, like any gas, will flow from a region of high pressure to one of low pressure. Consequently, air will enter the lungs through the air passageways if the air within the lungs is at a lower pressure than that of atmospheric air, and air will leave the lungs if the air pressure within them is higher than that of atmospheric air. During breathing, pressure changes are generated within the lungs by the action of a number of skeletal muscles, known collectively as the *respiratory muscles.*

### Generation of Pressure Changes

The generation of pressure changes by the respiratory muscles depends upon the elasticity of lung tissue and upon the anatomical relationship of the pleura to the lungs and to the closed thoracic compartments in which the lungs lie. As we noted above, the outer layer of the pleura (parietal pleura) lines, and is firmly attached to, the structures that form the wall of each thoracic compartment; the visceral pleura covers, and adheres firmly to, the outer surface of each lung. In addition, serous fluid between the two layers of the pleura hold the layers securely together.

During inspiration, contraction of a set of muscles between the ribs (the external intercostal muscles) raises the ribs upward and outward; the diaphragm moves downward when it contracts. The end result is that contraction of the respiratory muscles greatly increases the volume of the two thoracic compartments. As the volume of the thoracic compartments increases, the elastic lungs, attached to the walls of their thoracic compartments by the pleura, are stretched; that is, the lungs expand as the thoracic compartments expand. Expansion of the lungs increases the volume of the air passageways and the alveoli. Consequently, the pressure of air within these structures is *decreased,* causing atmospheric air to flow into the lungs. (Remember that the pressure of a given amount of gas is inversely proportional to the volume it occupies.)

During expiration, the respiratory muscles relax, and the thoracic compartments return to their original volume. This results in an *increase* in air pressure within the lungs such that air flows out of them. Expiration is thus normally a passive process, although both expiration and inspiration can be forced with heavy breathing.

It is a common misconception that lung tissue itself contains muscles that cause inflation and deflation of the lungs. This clearly is not the case. Lung tissue is elastic but not muscular (with the exception of smooth muscle in the walls of the airways), and the inflation and deflation of the lungs is entirely a result of pressure changes elicited by the respiratory skeletal muscles. This fact, and the importance of the pleura to breathing, is obvious when one side of the thoracic wall (including the parietal pleura) is punctured, but the lung itself is not; the lung on the side of the puncture will nevertheless collapse immediately. Collapse of the lung occurs as recoil of the elastic tissue of the lung pulls air into the pleural cavity, separating the two layers of the pleura. This is what happens when someone is stabbed in the chest. However, the collapse of one lung does not cause the collapse of the other, since the pleura of each lung are entirely separate from one another.

### Artificial Respiration

If an individual should stop breathing as a result of drowning, shock, or some other trauma, artificial means can be used to attempt resuscitation. The technique of mouth-to-mouth resuscitation is now considered the most efficient method of forcing air into and out of the lungs.

Mechanical devices have been designed for individuals who need extended assistance in breathing. *Resuscitators* are portable instruments used primari-

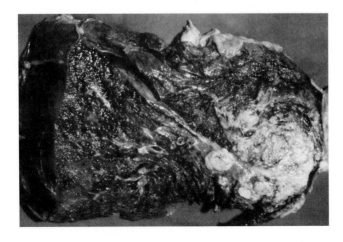

FIGURE 10–7 Lung cancer is evident as the whitish area in a lung blackened by carbon deposits from heavy cigarette smoking. Compare this to the lung of a nonsmoker in Figure 10–6b.

ly to assist respiration. They blow air into the lungs through a mask placed over the nose and mouth, and then allow the air to escape passively from the lungs. Individuals whose respiratory muscles are paralyzed can receive long-term artificial respiration by use of an *iron lung.* This device consists of a large air-tight chamber that surrounds the entire body up to and around the neck and performs the function of the respiratory muscles. The generation of alternating negative and positive pressure changes within the chamber move air into and out of the lungs by expanding and reducing the volume of the thoracic cavity—a function normally performed by the respiratory muscles.

## The Work of Breathing

The work required to inflate the lungs is done by the respiratory muscles. The amount of work that the respiratory muscles must perform depends to a large extent on the ease with which the lungs can be stretched or expanded. This, in turn, depends upon the physical properties of the fluid that coats the inner surface of the alveoli. Under normal circumstances, water coating the alveoli is intermingled with phospholipoprotein substances, called *surfactants,* which are secreted by specialized alveolar cells. The surfactants reduce water's high surface tension, an attractive force between water molecules that resists stretching of the molecules and that must be overcome in order to expand the alveoli. By lessening the surface tension of water, surfactants allow the alveoli to expand with considerably less work than would otherwise be required.

The importance of surfactants to breathing is demonstrated by the condition called *respiratory distress syndrome,* seen in some very small, premature newborns. In this condition, the specialized alveolar cells that normally secrete surfactant have not matured sufficiently to produce adequate amounts of surfactant. Consequently, inspiration requires strenuous effort that may ultimately lead to exhaustion and death. If it is known that a premature birth may occur, the mother may be injected with a steroid hormone that speeds up the maturation of surfactant-producing cells. However, this treatment must be done at least 24 hours before birth in order to be effective. A baby that develops respiratory distress syndrome must be given artificial respiration and other supportive treatment in an intensive care unit.

Breathing does not normally require a great deal of work. Even during strenuous exercise when inspiration and expiration are forced, only about 3 percent of the energy required by the body is used for breathing.

## Volume of the Lungs

During normal breathing under resting conditions, about 500 mL of air enters and leaves the lungs. This volume of air is referred to as the **resting tidal volume.** After normal expiration, about 2500 mL of air remains in the lungs. Even after forced expiration, the lungs still contain about 1000 mL of air (residual volume). Thus the lungs are never completely deflated. Forced inspiration can draw into the lungs 2500 to 3500 mL of air in addition to the 500 mL inspired during normal breathing. These figures, summarized in Figure 10–8, mean that the maximum amount of air that can be inhaled and exhaled, termed the *vital capacity* of the lungs, is about 5000 mL (total volume of the lungs is 6000 mL, but 1000 mL is left in the lungs following forced expiration). In practice, however, only about half of this volume is normally moved into and out of the lungs even during strenuous activity, since maximum inspiration and expiration requires exhausting muscular effort.

## Transport and Exchange of Respiratory Gases

In order to understand how the respiratory system functions, it is necessary to explore the mechanisms by which $O_2$ and $CO_2$ are transported in the blood and exchanged at the blood-tissue and at the blood-lung interfaces.

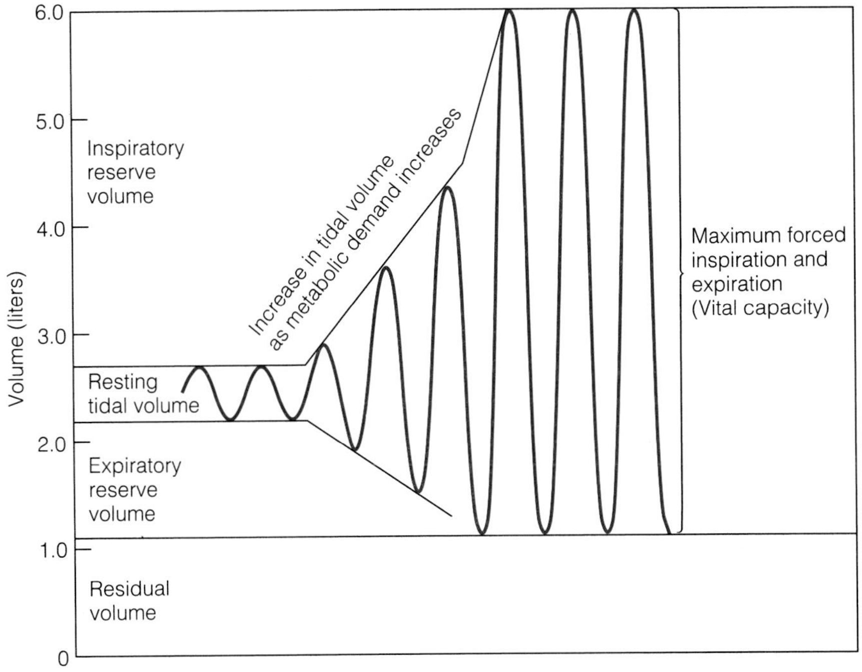

FIGURE 10-8 Volume of the lungs during various breathing states.

Chart labels (left axis): Volume (liters) — 6.0, 5.0, 4.0, 3.0, 2.0, 1.0, 0

Inspiratory reserve volume

Increase in tidal volume as metabolic demand increases

Resting tidal volume

Expiratory reserve volume

Residual volume

Maximum forced inspiration and expiration (Vital capacity)

## Transport of Oxygen and Carbon Dioxide in the Blood

$O_2$ and $CO_2$ are transported in the blood by somewhat different mechanisms. About 2 percent of the $O_2$ in blood is dissolved in the plasma or in the intracellular fluid of red blood cells; the rest is carried by some 300 million hemoglobin molecules present within each red blood cell. As discussed in Chapter 8, the molecular structure of hemoglobin is such that each molecule is capable of reversibly binding four molecules of $O_2$. When oxygen is bound to hemoglobin, the complex is referred to as *oxyhemoglobin,* whereas free hemoglobin is termed *reduced hemoglobin.*

As in the case of $O_2$ transport, only a small fraction of the $CO_2$ in the blood is carried as dissolved gas in the plasma and intracellular fluid of red blood cells. Additional $CO_2$ is carried in combination with reduced hemoglobin — hemoglobin that has released its oxygen at the tissues. Together, these two mechanisms account for between a quarter and a third of the $CO_2$ transported by blood. The remainder of the $CO_2$ that diffuses into the blood is converted into bicarbonate ions ($HCO_3^-$), formed primarily in the red blood cells in a two-step chemical reaction. In the first step, $CO_2$ combines with water to form carbonic acid:

$$CO_2 + H_2O \rightleftarrows H_2CO_3$$

This reaction is catalyzed by an enzyme, *carbonic anhydrase,* present in red blood cells. Once carbonic acid is formed, it readily dissociates into $HCO_3^-$ and hydrogen ions ($H^+$), as follows:

$$H_2CO_3 \rightleftarrows HCO_3^- + H^+$$

The sum of these two reactions is

$$CO_2 + H_2O \rightleftarrows H_2CO_3 \rightleftarrows HCO_3^- + H^+ \qquad \text{(Reaction 10-1)}$$

As bicarbonate ions are produced, most of them diffuse out of the red blood cell into the plasma. Because $HCO_3^-$ is very soluble in the aqueous components

FIGURE 10–9 Exchange of $O_2$ and $CO_2$ between the blood and the tissues of the body.

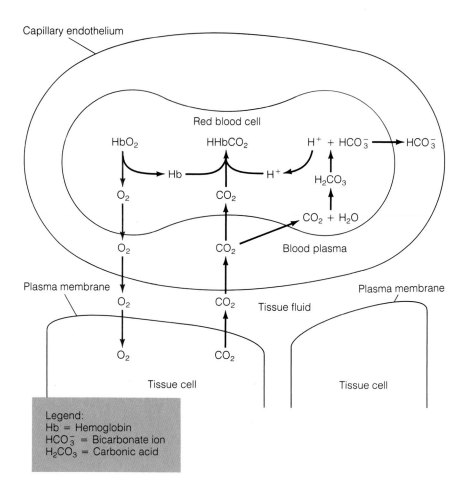

Legend:
Hb = Hemoglobin
$HCO_3^-$ = Bicarbonate ion
$H_2CO_3$ = Carbonic acid

of the blood, and $CO_2$ has only limited solubility, the conversion of $CO_2$ into bicarbonate allows the blood to carry far greater amounts of $CO_2$ than it could if it were merely carrying dissolved $CO_2$.

### Exchange at the Blood-Tissue Interface

At the tissues, $O_2$ in the blood is exchanged for $CO_2$ from the tissues (Figure 10–9). This exchange, as previously noted, occurs by diffusion. As the concentration of dissolved $CO_2$ increases within cells as a result of metabolic activity, $CO_2$ diffuses into the tissue fluid. Increased concentrations of dissolved $CO_2$ in the tissue fluid in turn bring about a diffusion of dissolved $CO_2$ into the blood plasma and from there into the intracellular fluid of red blood cells. Simultaneously, $O_2$ diffuses from the plasma into the tissue fluid. The resultant decrease in dissolved $O_2$ in the plasma causes a corresponding reduction in dissolved $O_2$ in the intracellular fluid of red blood cells. The lowered concentration of $O_2$ within the red blood cells, coupled with the concomitant increase in $CO_2$ levels, decreases the binding of $O_2$ to hemoglobin, causing the release of $O_2$ and its diffusion into the tissue fluid. Other factors besides low $O_2$ and high $CO_2$ levels bring about decreased binding of $O_2$ to hemoglobin and thus encourage the release of $O_2$ at the tissues. For example, an increase in hydrogen ion concentration (increase in acidity) and in temperature—both of which occur in actively metabolizing tissue such as skeletal muscle—also reduce binding of $O_2$ to hemoglobin.

Some of the $CO_2$ that diffuses into the red blood cells becomes bound to hemoglobin that has released its $O_2$, but most of the $CO_2$ is converted into bicarbonate ions, as summarized in Reaction 10–1 earlier. Most of the bicarbonate ions, as noted previously, diffuse into the blood plasma.

## Exchange at the Blood-Lung Interface

At the alveoli, $O_2$ in the air is exchanged for $CO_2$ in the blood (Figure 10–10). Again, the driving force is diffusion. As blood passes through the alveoli, $O_2$ diffuses into the plasma, increasing the amount of dissolved $O_2$ the plasma contains. This increase in plasma $O_2$ results in the diffusion of $O_2$ into the intracellular fluid of red blood cells. The increased levels of dissolved $O_2$ within red blood cells promote the binding of $O_2$ to hemoglobin; simultaneously, high $O_2$ levels decrease the binding of $CO_2$ to hemoglobin. The released $CO_2$ diffuses into the alveoli.

Recall that additional $CO_2$ is present in the blood plasma as bicarbonate ions. The diffusion of dissolved $CO_2$ from the plasma into the alveoli decreases the concentration of dissolved $CO_2$ in all components of the blood, driving Reaction 10–1 to the left and producing more dissolved $CO_2$. The net result is that bicarbonate, formed from $CO_2$ at the tissues, is converted back to $CO_2$ and exhaled at the lungs.

The exchange of gases between the blood and the air is so rapid that arterial blood leaving the lungs has virtually the same proportions of $O_2$ and $CO_2$ as does alveolar air. This fact is illustrated in Table 10–1, which shows the gaseous composition of inspired (tracheal) air, alveolar air, arterial blood, venous blood, and expired air. Two points are important in interpreting the significance of the data shown in Table 10–1. First, the composition of alveolar air does not vary significantly during the normal respiratory cycle. This is a result of the fact that inspired air is diluted by a large volume of air remaining in the lungs after normal expiration. Second, the composition of expired air is not the same as that of alveolar air. Instead, expired air has more $O_2$ and less $CO_2$ than alveolar air. This is because expired air consists of a mixture of alveolar air and air that was present in the respiratory passageways and that does not exchange gases with the blood.

*Carbon Monoxide Poisoning.* Although $O_2$ has a high affinity for hemoglobin, carbon monoxide (CO) binds to hemoglobin several hundred times as

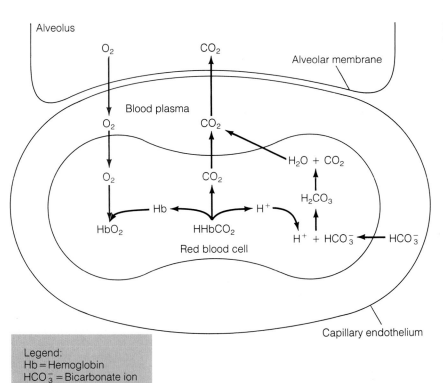

FIGURE 10–10 Exchange of $O_2$ and $CO_2$ between the lungs and the blood.

Legend:
Hb = Hemoglobin
$HCO_3^-$ = Bicarbonate ion
$H_2CO_3$ = Carbonic acid

TABLE 10-1 Partial Pressures (and Percentages) of Respiratory Gases[a]

| Gases | Atmosphere[b] | Tracheal Air[c] | Alveolar Air | Arterial Blood | Venous Blood | Expired Air |
|---|---|---|---|---|---|---|
| $O_2$ | 158.0 (20.8) | 149.0 (19.6) | 100 (13.1) | 95 (12.6) | 40 (5.7) | 116 (15.3) |
| $CO_2$ | 0.3 (0.04) | 0.3 (0.04) | 40 (5.3) | 40 (5.3) | 46 (6.5) | 32 (4.2) |
| $H_2O$ | 5.7 (0.8) | 47.0 (6.2) | 47 (6.2) | 47 (6.2) | 47 (6.6) | 47 (6.2) |
| $N_2$ | 596.0 (78.4) | 563.7 (74.2) | 573 (75.4) | 573 (75.9) | 573 (81.2) | 565 (74.3) |
| Total | 760 (100) | 760 (100) | 760 (100) | 755 (100) | 706 (100) | 760 (100) |

a) Data from Mountcastle, V.B. (1980). *Medical Physiology,* Fourteenth Edition, C. V. Mosby, St. Louis, p 1693.

b) Atmospheric air with average water content.

c) Tracheal air is atmospheric air that has been humidified while passing through the nasal cavities and pharynx.

strongly. In fact, the association between carbon monoxide and hemoglobin is so strong that it is essentially irreversible, and hence hemoglobin bound with carbon monoxide cannot function as an oxygen carrier. If an individual breathes air containing only 0.5 percent carbon monoxide, so much hemoglobin becomes bound with carbon monoxide in 30 minutes as to prove fatal. The combination of carbon monoxide with hemoglobin produces a cherry red color that is visible in the superficial capillaries of victims of carbon monoxide poisoning and belies the oxygen-starved condition of their cells.

## The Heart-Lung Machine

During open-heart surgery, when the heart and lungs cannot perform their normal functions, the tasks of pumping blood through the body, oxygenating blood, and eliminating carbon dioxide can be performed by a heart-lung machine. A heart-lung machine is shown in Figure 10–11.

## Control of Breathing

Unlike heart muscle that has inherent contractility, the respiratory muscles are skeletal muscles that contract only when stimulated by nerves. Furthermore, the strength and rate of stimulation must be coordinated with the physiological requirements of the body for oxygen and for the removal of carbon dioxide.

### Respiratory Control Center

The electrical activity of the nerves that supply the respiratory muscles is controlled by the *respiratory control center,* located in the lower portion of the brainstem in the same region that contains the cardiac control center. During normal, quiet breathing, rhythmic impulses generated by the respiratory control center activate the nerves supplying the diaphragm and external intercostal muscles, causing inspiration. Expiration occurs when the diaphragm and intercostal muscles relax as a result of cessation of nervous stimuli.

The rate and depth (force) of breathing vary, depending upon the metabolic demands of the body. As we have seen, the cellular increase in $O_2$ utilization and $CO_2$ production that occurs as a result of metabolic activity brings about corresponding changes in the relative amounts of $O_2$ and $CO_2$ in the tissue fluid and in the blood. Specialized receptors located in various parts of the body monitor respiratory conditions in body fluids and continually send impulses along nerves to the respiratory control center, which processes this information and makes appropriate adjustments in breathing rate and depth. The rate of breathing depends upon the frequency at which the respiratory control center dispatches stimuli to the respiratory muscles; the depth of breathing depends upon stimulus strength, which determines the number of motor units that are stimulated to contract in the respiratory muscles.

## Respiratory Receptors

Where are the respiratory receptors located, and how do they monitor the oxygen requirements of the body? The most important respiratory receptors are called **central chemoreceptors** that are located in the brainstem. These monitor $CO_2$ levels. As the $CO_2$ concentration builds up in arterial blood, $CO_2$ readily diffuses into the cerebrospinal fluid that bathes the central chemoreceptors. Increased levels of $CO_2$ in the cerebrospinal fluid cause a corresponding increase in the concentration of hydrogen ions, by the chemical reaction shown in Reaction 10–1. Heightened levels of $H^+$ activate the central chemoreceptors, which send nerve impulses to the respiratory control center. In response to these impulses, the respiratory control center increases the frequency and strength of stimuli reaching the respiratory muscles, resulting in an increase in breathing rate and depth. As $CO_2$ is exhaled at the lungs, the $H^+$ concentration drops in the blood, causing a resultant decrease in the $H^+$ concentration of the cerebrospinal fluid. Reduced $H^+$ levels in the cerebrospinal fluid inhibit the activation of the central chemoreceptors.

Other, less important respiratory receptors are the *peripheral chemoreceptors,* located in the *carotid bodies* at the bifurcation of the common carotid arteries, and in the *aortic bodies* in the arch of the aorta, near the baroreceptors that regulate blood pressure (see Figure 9–5a). The peripheral chemoreceptors operate by monitoring dissolved $O_2$ in the blood. The peripheral chemoreceptors are stimulated by low amounts of $O_2$ and send nerve impulses to the respiratory control center, which increases the rate and depth of breathing.

It should be emphasized that the central chemoreceptors are far more important in the control of respiration than the peripheral receptors. Consequently, the body is much more sensitive to rises in $H^+$ levels (as a result of increased $CO_2$ production) than it is to falls in $O_2$ levels.

## Voluntary Breathing

The actions of the central and peripheral chemoreceptors represent involuntary mechanisms that control respiratory function by regulating the activity of

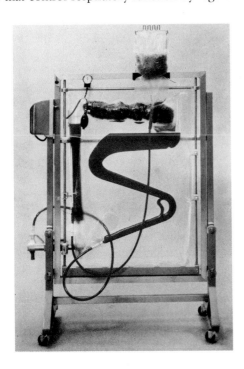

FIGURE 10–11 A heart-lung machine.

the respiratory control center. However, these mechanisms can be voluntarily overridden, but only up to a point, by nerve impulses to the respiratory control center originating in the cerebral cortex. Such voluntary control of breathing allows us to hold our breath during swimming or for other reasons, and to learn to inhibit inhalation during speaking and singing. However, as the lack of ventilation causes higher and higher levels of $CO_2$ and $H^+$ in body fluids, impulses from the central chemoreceptors will ultimately overcome the voluntary inhibition of breathing, forcing us to take a breath. This is why it is impossible to commit suicide by holding your breath — a fact that is comforting to parents confronted with a toddler who holds his or her breath during a tantrum.

Voluntary deep and rapid breathing performed prior to holding your breath prolongs the length of time you are able to hold your breath, and is a practice that can be dangerous or even fatal. Such *hyperventilation* reduces $CO_2$, and hence $H^+$, in the blood to levels far lower than normal. Consequently, the central chemoreceptors will not override the voluntary inhibition of breathing until an extended time has passed — a time sufficient to build up high blood levels of $H^+$. It is for this reason that underwater swimmers who hyperventilate prior to submerging are courting disaster. Since the body is not as sensitive to $O_2$ depletion as it is to $H^+$ buildup, $O_2$ levels may drop sufficiently in blood supplying the brain to cause underwater swimmers to faint before they have any urge to breathe.

### Drug Inhibition of Respiratory Control Center

The activity of the respiratory control center may be modified by various drugs. One class of drugs that is particularly potent in inhibiting the respiratory control center is the opiates. Deaths from overdoses of opiates such as heroin and morphine are usually the result of respiratory failure; that is, breathing simply stops due to lack of stimuli emanating from the respiratory control center. The administration of Demerol (a synthetic opiate) to mothers during labor may also cause a dangerous depression of respiratory activity in newborns.

### Breathing at High and Low Pressures

As shown in Table 10–1, air contains about 79 percent nitrogen gas ($N_2$). Because $N_2$ is somewhat soluble in aqueous solutions, blood and other body fluids contain dissolved $N_2$, just as they contain dissolved $O_2$ and $CO_2$. Under normal conditions, the dissolved $N_2$ has no physiological effect, but under conditions of high and low environmental pressures, the existence of dissolved $N_2$ in body fluids may pose a serious problem. For example, underwater scuba divers are subject to increased water pressures, and the air they breathe is also of increased pressure. In fact, the pressures of gases increase about 1 atmosphere for each 33 feet of depth. As air pressure increases, greater amounts of $N_2$ become dissolved in the body fluids. In deep dives of 100 feet or more, the increased amounts of dissolved $N_2$ in body fluids may cause an anesthetic effect called *nitrogen narcosis*, characterized by giddiness and stupor.

During ascent to the surface following diving, the body returns to a reduced pressure, and the body's fluids cannot hold as much dissolved $N_2$ as they could at higher pressures. If ascent is slow, the excess $N_2$ can be eliminated through the lungs. However, in a rapid ascent, excess $N_2$ is released as bubbles that are trapped in the tissues, causing pain in the muscles and joints. This condition is referred to as the *bends, decompression sickness,* or *caisson disease.* In severe cases of the bends, $N_2$ bubbles may lodge in blood vessels, causing fatal circulatory blockages.

Symptoms of the bends develop within 24 hours of the dive. They may be relieved or prevented by placing the diver in a *decompression chamber.* The diver is kept at increased pressure in the decompression chamber until the $N_2$

bubbles are redissolved, and then the pressure in the chamber is reduced slowly so that excess $N_2$ is expelled by the lungs without forming bubbles.

Rapid ascent to very high altitudes in an airplane causes a similar condition to the bends, usually referred to as *aviation sickness*. The pressurization of airplane cabins to ground-level pressures prevents the development of aviation sickness.

## Study Questions

1. Distinguish between the roles of the respiratory and circulatory systems in supplying the cells of the body with $O_2$ and removing $CO_2$ from them.
2. List the air passages in descending order starting from the nose.
3. **a)** Describe three functions of the nasal cavities.
   **b)** Distinguish between sinusitis and rhinitis.
4. Indicate the function of the pharyngeal and palatine tonsils.
5. **a)** Describe the structure of the larynx.
   **b)** Why do foreign objects tend to get lodged at the larynx?
6. Explain why cigarette smoking may lead to respiratory infections.
7. Describe the structure and function of the respiratory membrane.
8. Describe the organization of the lungs and the membranes that surround them.
9. **a)** Explain how breathing occurs.
   **b)** Indicate the role of surfactants in breathing.
10. **a)** What is the vital capacity of the lungs?
    **b)** What is the resting tidal volume?
11. **a)** Explain how $O_2$ and $CO_2$ are transported in the blood.
    **b)** Indicate the chemical basis of carbon monoxide poisoning.
12. **a)** Explain how the rate and depth of breathing are regulated in the body.
    **b)** Indicate why hyperventilating prior to swimming underwater can be dangerous.
13. **a)** Would you say that breathing is an automatic or voluntary process?
    **b)** Explain your answer.

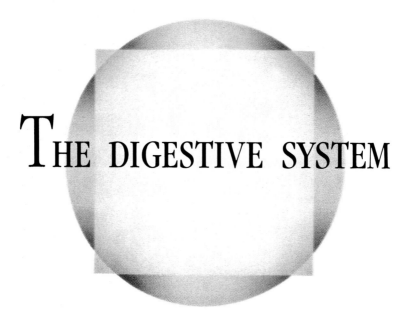

# THE DIGESTIVE SYSTEM

Ingested food and water is processed by the digestive system for distribution to the cells of the body. The digestive system consists of two groups of organs (Figure 11–1). One group forms a long muscular tube, the **digestive tube** or **tract,** that extends from the lips to the anus and is composed, in descending order, of the *mouth, pharynx, esophagus, stomach, small intestine,* and *large intestine.* (The pharynx serves both respiratory and digestive functions and hence is considered an organ of both systems.) Associated with the digestive tract is a second group of organs, including the *salivary glands, liver, pancreas,* and *gallbladder,* that lie outside of the digestive tract and that empty their secretions, essential to the digestive process, into the digestive tube by way of ducts.

In order to provide the body's cells with nutrients and water, the digestive system must transfer these substances from within the digestive tube into the circulatory system, through which they are transported to the cells. In this regard, it is important to realize that food and water inside the digestive tube are external to the body in the sense that they are totally unavailable to the body's cells until they have been transferred into the circulatory system. The initial step in this transfer is the absorption of nutrients and water by the epithelial cells that line the small and large intestine. However, most food is not in a form that can be absorbed. It usually consists of large particles of organic matter that contain, in addition to water, ions, and a few small organic molecules, relatively high amounts of the large organic molecules that compose living tissue: proteins, polysaccharides, lipids, and nucleic acids. These large organic molecules must be broken down into their constituent parts before absorption can occur.

This process, by which large food particles are broken down in the digestive tube to smaller substances that can be absorbed, is termed **digestion.** Although digestion involves mechanical processes such as chewing, most digestion is a result of the degradative action of digestive juices secreted by glands that lie in the wall of the digestive tube or that empty into the tube.

## GENERAL STRUCTURE OF THE DIGESTIVE TUBE

With some minor variations noted below, the digestive tube has a similar structure throughout much of its length. As shown in Figure 11–2, it consists of four main layers. From the inside of the tube out, these are (1) *mucosa* (*mucous membrane*); (2) *submucosa;* (3) *muscularis externa* (external muscle coat); and (4) an *external layer.*

### Mucosa

The **mucosa** lining the digestive tube consists of three layers: (1) an inner, single layer of epithelial cells; (2) a supporting layer of connective tissue, the *lamina propria;* and (3) a thin, usually double layer of smooth muscle, the *muscularis mucosae.* The thin epithelial lining is lubricated extensively with mucus, which helps protect it from damage by digestive juices and helps prevent penetration by microbes present in the digestive tract. Mucus, hormones, or digestive chemicals may be secreted by cells scattered in the epithelial layer or by glands that develop from invaginations of the epithelial layer and lie in the lamina propria or submucosa. The liver, pancreas, and salivary glands—glands that lie outside of the digestive tube—also develop from cells of the epithelial layer, and their products drain into the digestive tube through ducts that mark their site of origin in the embryo.

The lamina propria is sprinkled with nodules of lymphatic tissue and acts as an important line of defense against microbes that manage to invade the surface layer. Such defense is vital since the digestive tube is in contact with the

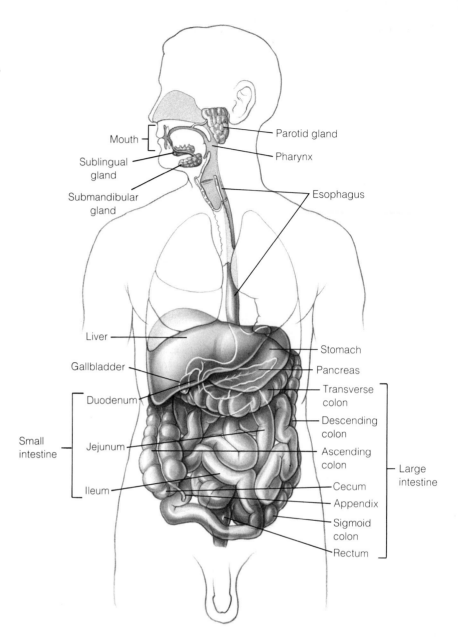

FIGURE 11–1 Organs of the digestive system. The small intestine consists of the duodenum, jejunum, and ileum; the large intestine consists of the cecum, ascending colon, transverse colon, descending colon, sigmoid colon, rectum, and anal canal.

external surroundings and receives food contaminated with many types of microbes. The lamina propria also contains blood vessels, lymphatic vessels, and nerves that run close to the epithelial layer. This arrangement facilitates the passage of absorbed substances into blood and lymphatic vessels, since these substances do not have to diffuse for any great distance through the tissue fluid of the lamina propria.

The **muscularis mucosae,** the outermost layer of the mucosa, consists of two layers of smooth muscle cells. In the inner layer, the muscle fibers are arranged circularly, and in the outer layer, longitudinally. These muscle layers probably allow some localized movement of the mucosa, although their function is not fully understood.

### Submucosa

The **submucosa** connects the overlying mucous membrane to the muscularis externa. It is composed of loose connective tissue in which course the larger blood vessels and lymphatics that send branches into the mucous membrane and the surrounding smooth muscle layers of the muscularis externa.

## Muscularis Externa

The **muscularis externa** is formed of two substantial layers of smooth muscle. In the inner layer, the muscle fibers are arranged in a more or less circular fashion around the circumference of the wall; the outer layer has muscle fibers that are arranged longitudinally along the wall. The smooth muscle cells of the muscularis externa are innervated by a rather extensive network of nerve cells, termed the *myenteric plexus,* lying between these two layers.

Contraction of the muscularis externa serves to mix the contents of the digestive tube and to propel the contents along the tube toward the anus. Because smooth muscle cells undergo spontaneous contractions that can spread from cell to cell, the smooth muscle of the muscularis externa can undergo rhythmic contractions over short distances. However, the waves of contraction that progress down the tube and propel the contents within, termed *peristaltic movements,* require the assistance of nervous coordination from the nerves of the myenteric plexus. During rest or recovery from metabolic stress, nerves of the autonomic nervous system tend to increase the tone and frequency of peristaltic movements of the muscularis externa, during stressful situations, autonomic nerves inhibit tone and peristaltic movements. Emotional disturbances can affect the function of the digestive system by altering the contractile activity of the smooth muscle layer. Although there are individual differences, frustration and hostility tend to increase the motility of the digestive tract, while fear tends to decrease it.

## External Layer and Peritoneum

In those parts of the digestive tube that lie above the diaphragm and that merge with adjacent tissues, the external layer of the digestive tube is fibrous connective tissue. However, the external layer of those organs that lie in the

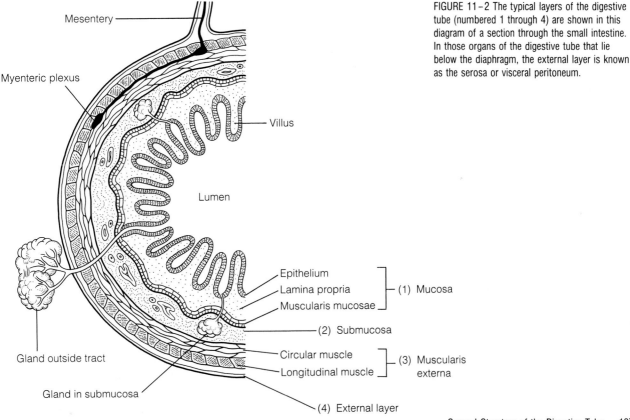

FIGURE 11-2 The typical layers of the digestive tube (numbered 1 through 4) are shown in this diagram of a section through the small intestine. In those organs of the digestive tube that lie below the diaphragm, the external layer is known as the serosa or visceral peritoneum.

Mesentery

Myenteric plexus

Villus

Lumen

Epithelium
Lamina propria — (1) Mucosa
Muscularis mucosae

(2) Submucosa

Circular muscle — (3) Muscularis
Longitudinal muscle — externa

Gland outside tract

Gland in submucosa

(4) External layer

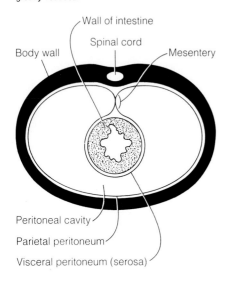

FIGURE 11–3 Greatly simplified diagrams of a cross-section through the body at the level of the intestines, showing the arrangement of the peritoneum. In actuality, the body cavity is completely filled with folds of the intestine, and hence the volume of the peritoneal cavity is greatly reduced.

Wall of intestine

Spinal cord

Body wall

Mesentery

Peritoneal cavity

Parietal peritoneum

Visceral peritoneum (serosa)

abdominal cavity below the diaphragm (i.e., stomach, small intestine, and large intestine) is a serous membrane called the **serosa** or *visceral peritoneum* (Figure 11–3).

The visceral peritoneum not only covers the stomach and intestines, but two layers of it fuse together to form membranous sheets that are continuous with another serous membrane, the *parietal peritoneum,* that lines the wall of the abdominal cavity. The membranous sheets, which connect the stomach and intestines to the abdominal wall, are referred to as *mesenteries.* Connective tissue within the mesenteries contains varying amounts of fat and carries blood vessels, lymphatic vessels, and nerves to the digestive tube.

The visceral and parietal peritoneum collectively form the **peritoneum,** a closed sac that is fully analagous to the pericardium and to the pleura. The slippery secretions of these serous membranes allow the outer walls of the abdominal digestive organs to slide past each other and past the walls of the abdominal cavity. The area between the parietal and visceral layers of peritoneum, filled with a small amount of slippery fluid, is a potential space referred to as the **peritoneal cavity.**

Inflammation of the peritoneum is called *peritonitis.* It is almost always the result of an underlying disease or an injury. For example, peritonitis may be caused by a ruptured ulcer or appendix, or by a stabbing wound that pierces the abdominal wall. These events would introduce bacteria into the peritoneal cavity. Symptoms of peritonitis include severe abdominal pains that may become less severe after several hours due to paralysis of the intestines. Death will occur unless emergency surgery is initiated to repair or remove the organ that is causing the problem. With the help of antibiotics, few patients now die of peritonitis.

## ORGANS OF THE DIGESTIVE TUBE

### Mouth

The **mouth,** the first part of the digestive tube, is the cavity that lies between the lips and the pharynx (Figure 11–4). The opening of the mouth is formed by the lips; its roof, by the hard and soft palates; its sides, by the cheeks; and its floor, by the tongue. The mouth serves to store, moisten, and chew food in preparation for swallowing. In addition, enzymes secreted by the salivary glands begin the digestion of starch.

### Lips

The lips are muscular folds, covered on their outer surface with skin and on their inner surface with mucous membrane that lines the mouth and is continuous with that of the pharynx. The red margins of the lips are covered with modified skin, possessing relatively transparent epidermal cells through which can be seen underlying blood capillaries that give the lips their pinkish color. Because the epidermis in this area does not have a heavy protective layer of keratin, it will become dry unless it is wetted frequently with the tongue, which is why under dry climatic conditions, the lips may become chapped or cracked.

Lesions on the lips or in the lining of the mouth may be caused by viral infection or by injury to the mucous membranes. Cold sores on the lips result from infection by herpes simplex virus.

### Hard and Soft Palates

The hard palate is formed by the fusion during embryonic development of several bones at the base of the skull (see Chapter 6). It is covered by stratified squamous keratinizing epithelium that can withstand much wear and tear as food is scraped along its surface by the tongue. The soft palate, which extends posteriorly from the hard palate, is covered on both sides by mucous membrane and is supported by a dense network of collagenous fibers and skeletal muscle. The little flap of tissue hanging down at the back of the soft palate is called the *uvula.*

Because the soft palate extends from the hard palate into the pharynx, its upper surface forms the floor of the nasopharynx (see Chapter 10). During swallowing, sucking, blowing, and the formation of some speech sounds, the soft palate is drawn upward to close off the nasopharynx.

## Tongue

The tongue is important in swallowing, chewing, speech, and taste. It is composed primarily of skeletal muscle covered by mucous membrane. The anterior two-thirds of the tongue lies in the mouth, while the posterior one-third, the *root of the tongue,* lies in the pharynx. The mucous membrane on the underside of the tongue is thin and smooth, and blood vessels run close to the surface. This arrangement allows certain drugs such as nitroglycerin, used to dilate the coronary arteries during an angina attack, to be absorbed directly into the bloodstream if a pill is placed under the tongue.

The thick mucous membrane covering the dorsal side of the oral part of the tongue is covered by little projections of mucous membrane called *papillae.* These give the tongue its rough appearance and allow us to lick certain soft substances such as ice cream. Many of these papillae contain taste buds along the sides of their walls. Specialized receptors in taste buds are stimulated by certain basic tastes, including sweet, sour, salty, and bitter. The enormous variety of different tastes that we experience results from a combination of these basic tastes.

The mucous membrane covering the root of the tongue has no papillae, but beneath the mucous membrane are nodules of lymphatic tissue. These collectively are called the **linguil tonsil.**

## Teeth

The primary function of teeth is to tear off chunks of food and grind them into smaller pieces that can be swallowed. Although prolonged chewing of most food is not necessary for subsequent digestion, chewing helps prevent choking.

*Structure.*    Teeth are the hardest of all tissues of the body. The structure of a typical tooth is shown in Figure 11–5. Most of a tooth is composed of a calcified connective tissue called *dentin.* The *anatomical crown* of the tooth, the part

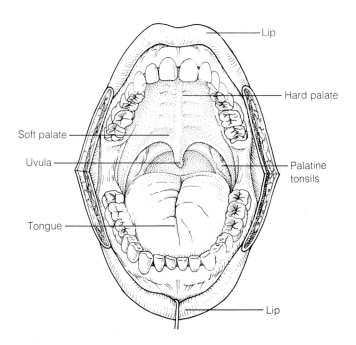

FIGURE 11–4 The mouth.

Lip

Hard palate

Soft palate

Uvula

Palatine tonsils

Tongue

Lip

FIGURE 11-5 Longitudinal section through a tooth and surrounding tissue.

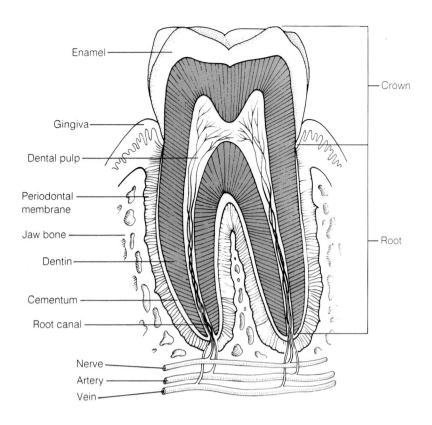

that projects outward from slightly below the gum line, is covered by a very hard cap of whitish *enamel.* The rest of the tooth, the *root,* is covered with bonelike, calcified connective tissue termed *cementum.*

Within the center of each tooth is a cavity, the *pulp cavity,* that is expanded toward the crown of the tooth and narrowed toward its root. The narrowed part of the cavity that passes through the tip of the root is called the *root canal.* The pulp cavity is filled with *dental pulp,* consisting of connective tissue that contains many small blood vessels and nerve fibers. These provide the vascular and nerve supply of *odontoblasts,* a layer of cells that surround the pulp cavity and are involved in the production of dentin. Odontoblasts maintain the dentin and are analogous to the osteoblasts of bone.

The teeth are set in sockets present in bony ridges that project upward from the lower jaw bone (mandible) and downward from the upper jaw bone (maxilla). Bundles of connective tissue fibers, forming what is called the *periodontal membrane* or *ligament,* anchor each tooth to its socket by running between the bone of the socket wall and the cementum covering the root of the tooth. This attachment allows slight movement of the tooth within the socket during chewing.

***Periodontal Disease.*** The bony ridges in which teeth are embedded are covered by mucous membrane referred to as the *gums,* or *gingiva.* The gums surround each tooth like a collar. Next to the tooth, the epithelium of the gums extends a short way down from the crest of the collar (see Figure 11-5). This epithelium is only weakly attached to the enamel, but firmly attached to the more porous cementum of the root just below the enamel. This situation creates a crevice between the gum and the tooth surface. *Tartar,* a calcium-containing precipitate from saliva, may accumulate in the crevice, eventually expanding to the point that it tends to separate the epithelial attachment from the tooth. If this occurs, bacteria can invade the region. Bacterial infection may result in the detachment of the periodontal ligament from the cementum, loosening the tooth. *Periodontal disease* of this type is common among middle-aged and older individuals, and is the leading cause of tooth loss.

***Deciduous and Permanent Teeth.*** Humans develop two sets of teeth during their lifetime: a temporary set of small teeth, called *deciduous* or *baby teeth,* and a set of larger, *permanent teeth* (Figure 11–6). There are 20 deciduous teeth, 10 in each jaw. The two deciduous teeth on either side of the midline of both jaws are called *incisors;* they are thin, flat teeth adapted for cutting. Next to the incisor is a single *canine,* or *cuspid,* tooth, and next to this are two *molars.* The canine teeth, sharp and conical in shape, are adapted for tearing, whereas the wider and flatter molars are modified for grinding. The first incisor usually appears in infants at about six months of age, and an additional tooth is added about every month, although the timing of eruption varies greatly among different individuals.

Beginning at about the age of six, the deciduous teeth begin to be replaced by permanent teeth. The incisors and canine teeth are replaced by permanent teeth of the same shape, but of larger size. The two deciduous molars are replaced by the first and second *bicuspids* or *premolars.* Three molars are added behind the second bicuspid on each side of each jaw. The first of these, nearest the bicuspids, erupts at about the age of 6 (the *6-year molars*); the second, at about the age of 12 (the *12-year molars*); and the third (*"wisdom teeth"*), in early adulthood, if they erupt at all. Not infrequently, the wisdom teeth become impacted—wedged between the jawbone and another molar—and must be removed surgically.

***Dental Caries.*** Specialized cells that are involved in the deposition of enamel during tooth development degenerate once all the enamel has been formed. Consequently, there is no way that enamel can be repaired if it is chipped or injured by decay. On the other hand, the life of the tooth depends upon the health of the dental pulp, which supplies the odontoblasts with necessary materials.

Localized areas of decay, called *dental caries,* cause cavities to develop on exposed tooth surfaces. This most common of all diseases begins when bacteria attack food trapped between teeth or in tiny crevices of the enamel. Bacterial metabolism is believed to result in the formation of acid products that decalcify the enamel locally and destroy it. Cavities in the enamel, left untreated, will eventually reach the dentin and the underlying pulp. Since

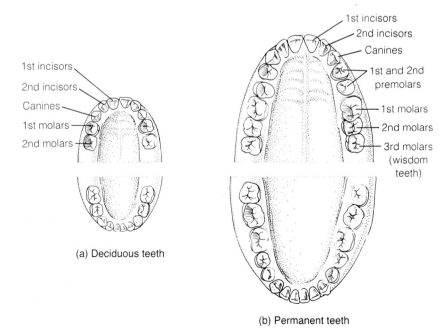

1st incisors
2nd incisors
Canines
1st molars
2nd molars

(a) Deciduous teeth

1st incisors
2nd incisors
Canines
1st and 2nd premolars
1st molars
2nd molars
3rd molars (wisdom teeth)

(b) Permanent teeth

FIGURE 11–6 (a) Deciduous and (b) permanent teeth. The deciduous incisors and canine teeth are replaced by permanent incisors and canines, respectively; the deciduous molars are replaced by permanent premolars; and three permanent molars (6-year, 12-year, and wisdom teeth) are added sequentially at the back of the jaws.

cavities that are developing in enamel cause no pain, and may or may not cause pain when in the dentin, the best way to prevent severe damage from cavities is to undergo regular dental checkups. To treat cavities that do not involve the pulp, the surrounding enamel and dentin are drilled away, and the hole is filled with metal or plastic material. If the pulp is killed as a result of inflammation or infection, it can often be removed and the space filled with an inert material, in a procedure termed *root canal treatment*. Such a tooth is dead, but usually can remain functional if it has not decayed badly.

### Salivary Glands

The lining of the oral cavity contains numerous small glands that secrete saliva. In addition to these, there are three pairs of large glands, for which the term **salivary glands** is commonly reserved. These are the *parotid glands,* the *submandibular glands,* and the *sublingual glands.* Their locations are shown in Figure 11–1.

Saliva is the mixed secretions of all the salivary glands, large and small. It is a watery secretion that, depending upon the stimuli received by the glands, contains variable amounts of *mucus,* a protein that gives saliva its viscous, slippery texture. In addition to water and mucus, saliva contains ions and the enzyme *salivary amylase.* Some one to two liters of saliva are secreted in 24 hours.

Saliva has a number of functions. It lubricates the lips and the lining of the mouth, thus preventing these mucosal surfaces from drying out. It serves to moisten food so that it can be more easily swallowed. Moreover, the moistening of food allows it to be tasted, since only dissolved substances will activate the taste buds. Furthermore, the more or less continuous secretion of saliva serves to wash bacteria and food debris out of the mouth.

Saliva is also involved to a limited extent in the digestion of starch. Starch, as you recall from Chapter 2, is the primary storage polysaccharide in plants and consists of long chains of glucose molecules. *Salivary amylase* in saliva begins the digestion of starch to *dextrins,* fragments smaller than starch, and finally to *maltose,* a sugar consisting of two glucose molecules linked together. However, food is not usually in the mouth long enough for starch to be digested to any significant extent, and once the food becomes mixed with the stomach acid, salivary amylase is inactivated.

Secretion of saliva is controlled by the autonomic nervous system. The presence of food in the mouth stimulates taste buds, which send impulses directly to the salivary center in the base of the brain; impulses reaching the salivary glands from the salivary center initiate salivation. Psychological factors may also be important in regulating salivation. For example, the thought, sight, and smell of food may stimulate secretion by way of cerebral pathways to the salivary center. Unpleasant associations, on the other hand, inhibit the secretion of saliva.

### Pharynx

The **pharynx,** a common passageway for the respiratory and digestive systems, is a short tubelike chamber that lies behind the nasal cavity, mouth, and larynx. Its mucous membrane lining covers skeletal muscle that composes most of its wall. The pharynx is divided into three regions. The *nasopharynx,* lying above the soft palate, was described in Chapter 10. The portion of the pharynx behind the tongue is referred to as the *oropharynx;* the *laryngopharynx* connects the oropharynx to the esophagus (see Figure 11–7a).

When food is swallowed, it must be propelled from the back of the mouth through the oropharynx and laryngopharynx to the esophagus. In addition, food must be prevented from entering the larynx. The events that occur during swallowing are summarized in Figure 11–7. When the tongue forces food into the pharynx, a sequence of involuntary events is initiated which, once begun,

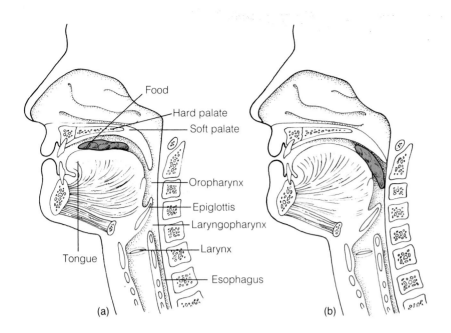

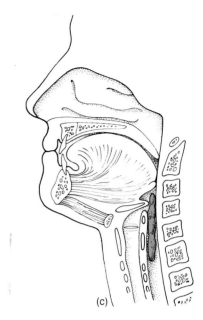

FIGURE 11-7 Sequence of events during swallowing. (a) Food is in the mouth, just prior to swallowing. (b) As swallowing begins, the tongue is pressed against the roof of the mouth; the soft palate is elevated, sealing off the nasopharynx; the larynx rises, pressing against the epiglottis; and (not shown) the glottis closes to seal off the entrance to the larynx. (c) As the food passes through the oropharynx and into the laryngopharynx, the epiglottis is tipped backward.

cannot be stopped. These events are controlled by the autonomic nervous system in the swallowing center in the base of the brain. First, breathing is inhibited. Then the soft palate is elevated, sealing off the nasopharynx and preventing food from entering the nasal cavity. The larynx is also elevated so that its upper end is pressed against the epiglottis, and the entrance to the larynx is closed off by closure of the glottis (the opening between the vocal cords). Involuntary contraction of skeletal muscles in the wall of the pharynx moves the food past the epiglottis (which tips backward) and into the laryngopharynx. The passage of food into the esophagus requires the relaxation of the *pharyngoesophageal sphincter,* an area of increased muscular tone in the wall of the esophagus just after the pharynx joins it. Once food enters the esophagus, the larynx drops, the glottis opens, and breathing resumes.

## Esophagus

The **esophagus** serves merely to conduct food between the pharynx and stomach. Like the pharynx, it has no digestive functions.

### Structure and Function

The esophagus is a straight tube about ten inches long, lying behind the trachea and connecting the pharynx with the stomach. It travels through the thoracic cavity and passes through the diaphragm, extending an inch or two below the diaphragm before joining the stomach. The section of the esophagus below the diaphragm is referred to as the *abdominal portion* of the esophagus.

The wall of the esophagus has the four layers typical of the remainder of the digestive tube. However, in the upper third of the esophagus the outer muscular layer (muscularis externa) consists of skeletal muscle that changes over gradually to smooth muscle in the middle third, such that the lower third is entirely smooth muscle. This skeletal muscle, as with that in the wall of the pharynx, is not under voluntary control.

The esophagus is normally relaxed and collapsed upon itself except when distended with food. In addition to the pharyngoesophageal sphincter, mentioned earlier, there is another region of the esophagus just as it joins the stomach that is tonically contracted (contracted more or less continuously). This area, forming the *gastroesophageal sphincter,* prevents the regurgitation of stomach contents into the esophagus. Weakness of the pharyngoesophageal and gastroesophageal sphincters in newborns accounts for their tendency to "spit up" after a meal.

As food enters the esophagus from the pharynx, local distension of the esophagus initiates the peristaltic wave that moves down the esophagus toward the stomach. Such a peristaltic wave takes about six seconds to pass over the esophagus. By narrowing the lumen of the tube, the peristaltic wave propels the mass of food in front of it. The gastroesophageal sphincter relaxes briefly just before the peristaltic wave passes over it, allowing the food to enter the stomach. If the food does not reach the stomach during passage of the first peristaltic wave, repeated waves are initiated in the region where the esophagus remains distended by food.

### Hiatus Hernia

A common condition affecting the esophagus is *hiatus hernia.* As noted above, the esophagus passes through the diaphragm; the hole through which it passes is called the *hiatus.* If the diaphragm's muscular tissue surrounding the hiatus weakens, as it often does in overweight, elderly people, the abdominal portion of the esophagus, and sometimes part of the upper stomach, herniates (ruptures) upward through the hiatus into the thoracic cavity. This results in an expanded region at the base of the esophagus (Figure 11–8).

Many people have no symptoms from a hiatus hernia, but in some cases, the function of the gastroesophageal sphincter is impeded, resulting in the regurgitation of stomach acid into the lower esophagus. This process, called *acid reflux,* produces a burning sensation known as *heartburn* because it occurs in the general area where the heart is located. Persistent acid reflux may be treated with antacids that neutralize stomach acid and hence help relieve symptoms. If severe, a hiatus hernia may need to be repaired surgically.

Heartburn need not be caused by hiatus hernia. It may occur after a large meal if distension of the stomach forces some stomach contents into the esophagus. It is also common during the last few months of pregnancy, as pressure on the stomach from growth of the fetus causes regurgitation of stomach contents into the esophagus.

### Stomach

The esophagus empties into the **stomach,** an expanded portion of the digestive tube that serves to store and to partially digest food. The stomach lies just below the diaphragm and leads into the upper part of the small intestine.

### Structure and Function

The major parts of the stomach are shown in Figure 11–9. The area where the esophagus opens into the stomach is called the *cardiac orifice,* because of its proximity to the heart. Lying above and to the left of the cardiac orifice is the *fundus.* The *body* of the stomach is its central and major region. The body narrows down significantly to form the *pyloric antrum,* which communicates with the small intestine through the *pylorus,* a circular opening surrounded by a ring of smooth muscle, the *pyloric sphincter.*

The wall of the stomach has the four characteristic layers, but the muscularis externa consists of three layers of smooth muscle instead of two. Each of the three smooth muscle layers runs in a different direction, as shown in Figure 11–9. The external layer of the stomach corresponds to visceral peritoneum (serosa).

The mucous membrane lining the stomach is relatively thick. Its epithelial surface is covered by a single layer of columnar cells that secrete mucus; the mucus forms a coat that serves to protect the stomach wall from digestion by stomach acid and digestive enzymes. Another protective feature is the replacement of epithelial cells every few days, which mediates against damage to the

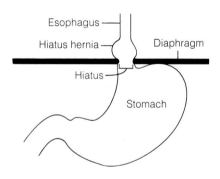

FIGURE 11–8 Hiatus hernia. Notice that a part of the lower esophagus and upper stomach has ruptured upward through the hiatus.

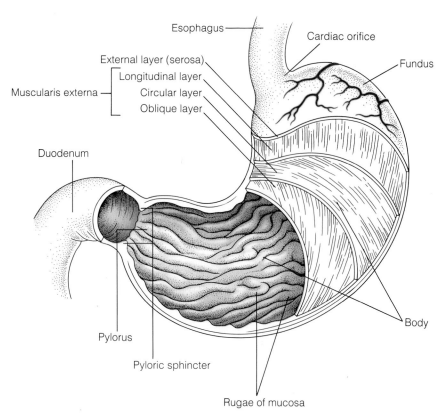

FIGURE 11–9 Parts of the stomach.

Esophagus

Cardiac orifice

External layer (serosa)

Fundus

Longitudinal layer

Muscularis externa

Circular layer

Oblique layer

Duodenum

Pylorus

Pyloric sphincter

Body

Rugae of mucosa

cells as a result of prolonged exposure to digestive juices. Nevertheless, lesions may develop in the mucous membrane, causing *ulcers.*

The surface epithelium descends into the mucous membrane to line numerous little crevices called *gastric pits.* These serve as openings to tubular *gastric glands,* located in the lamina propria, that deliver their secretions into the gastric pits from where they are conducted to the mucosal surface. The gastric glands secrete *gastric juice* containing hydrochloric acid, digestive enzymes, and mucus. They are really compound glands composed of several cell types, each specialized for secreting individual products. Thus, *parietal cells* of the gastric glands secrete hydrochloric acid; *zymogen,* or *chief cells* produce *pepsinogen,* which is degraded (and hence activated) by acid in the stomach to *pepsin,* a protein-digesting enzyme; and *mucous neck cells* secrete mucus. Also secreted by parietal cells is the *intrinsic factor,* a protein required for the absorption from the intestine of vitamin $B_{12}$, the lack of which causes pernicious anemia (see Chapter 8). In addition to the presence of gastric glands, some of the epithelial cells in the pyloric antrum are endocrine cells that secrete the hormone *gastrin.*

The empty stomach has a small lumen (central cavity), and its mucous membrane lining is crumpled into folds called *rugae* (Figure 11–10). The rugae disappear as the stomach fills with food and the stomach wall becomes stretched.

## Digestion in the Stomach

***Phases of Gastric Secretion.*** Although small amounts of gastric juice are secreted more or less continuously, secretion is greatly stimulated immediately preceding and during a meal. Stimulation of gastric juice secretion is said to occur in three phases, but they may overlap considerably. Together, the three phases of secretion produce about two liters of gastric juice per day.

The initial phase, the *cephalic phase,* occurs as a result of the thought, sight, smell, and taste of food, and is mediated by nerves of the autonomic nervous system that stimulate the gastric glands. The *gastric phase* of secretion

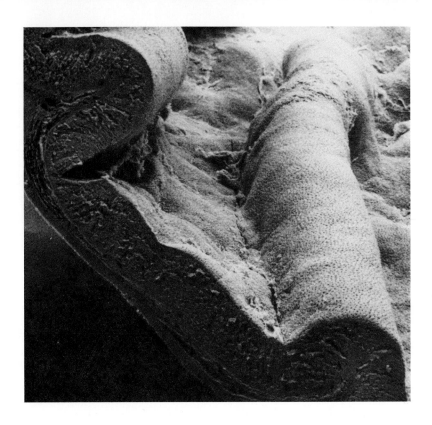

FIGURE 11–10 Scanning electron micrograph of the inner surface of the stomach, showing rugae. From TISSUES AND ORGANS: A text-atlas of scanning electron microscopy, by Richard G. Kessel and Randy H. Kardon. W.H. Freeman and Company, copyright © 1979.

occurs when food reaches the stomach. It is triggered by the mechanical distension of the stomach wall and by the presence in the stomach of certain foods, notably partially digested proteins and amino acids. These foods stimulate the release of gastrin from cells in the pyloric antrum. Gastrin is carried by the bloodstream to cells of the fundus and body of the stomach, stimulating increased secretion of gastric juice. In the *intestinal phase* of gastric secretion, the presence of partially digested proteins and amino acids in the small intestine promotes the secretion of a hormone, probably *cholecystokinin,* by the intestinal mucosa that acts on the gastric glands to stimulate gastric juice secretion. On the other hand, the presence of fat and acid in the small intestine serves to inhibit gastric secretion.

The hydrochloric acid in gastric juice breaks down large food particles into smaller ones, but it does not produce products small enough to be absorbed. Similarly, the enzyme pepsin in gastric juice catalyzes the degradation of proteins into small fragments, but not into their constituent amino acids. By the time food leaves the stomach and enters the small intestine, it is partially digested in the form of a soupy, highly acidic liquid known as *chyme.*

***Gastric Motility.*** Digestion in the stomach requires that the food be mixed with gastric juice. In addition, of course, the stomach contents must be moved into the small intestine once digestion in the stomach is completed. These functions are carried out by contractions of the stomach musculature. As food enters the stomach, peristaltic waves, beginning near the stomach's upper end, pass over the stomach wall, gathering strength as they proceed. By the time the contractions reach the pyloric antrum, they are powerful enough to mix the stomach contents. Pressure against the pyloric sphincter allows well-digested, fluid food to be squeezed in small amounts into the small intestine, but more solid, undigested food is held back.

Gastric motility is influenced by a variety of factors, including psychological ones. For example, pain, sadness, and fear tend to decrease gastric motility. Aggression tends to increase gastric motility and this, in turn, accelerates the rate of gastric emptying; distension of the first part of the small intestine (the duodenum), or presence of fat in the duodenum, inhibits gastric motility.

## Absorption in the Stomach

Only a very small amount of absorption occurs in the stomach. Among the substances absorbed to some degree are alcohol, a few other drugs such as aspirin and acetaminophen (Tylenol), and water. Because alcohol is absorbed more rapidly from the small intestine than from the stomach, the rate of alcohol absorption—and hence the effects of ingested alcohol—can be delayed by inhibiting the emptying of the stomach. The ingestion of fatty foods before or while drinking alcoholic beverages has such an inhibiting effect.

## Vomiting

Upsets or infections of the digestive system, morning sickness during pregnancy, motion sickness, and many other conditions may cause vomiting. Vomiting, like swallowing, involves a coordinated sequence of events and is controlled by the autonomic nervous system. Vomiting is usually preceded by feelings of nausea, which are accompanied by salivation, perspiration, aversion to food, and weakness. This is followed by inhibition of breathing, closure of the glottis, elevation of the soft palate, and violent contractions of the abdominal and thoracic muscles. These contractions raise the abdominal pressure, compressing the stomach. As the gastroesophageal sphincter relaxes, the contents of the stomach are ejected into the esophagus and forced past the pharyngoesophageal sphincter into the mouth. Bile, produced in the liver and secreted into the small intestine, may be present in the vomit when strong contractions in the small intestine cause its contents to be forced into the stomach.

Repeated vomiting is dangerous because it can lead to large losses of fluids and ions, and lower the acidity of the blood. However, when poisonous substances have been ingested, it is often helpful to induce vomiting. This is done by administering substances known as *emetics.* Syrup of ipecac and concentrated solutions of sodium chloride (table salt) are examples of emetics.

## Small Intestine

The **small intestine** extends from the stomach to the large intestine (see Figure 11–1). It is a convoluted tube about 20 feet long and an inch or so in diameter. The first 10 to 12 inches of the small intestine is the **duodenum,** which loops around the head of the pancreas. The next two-fifths of the small intestine is called the **jejunum,** and the remainder is the **ileum.** The ileum is separated from the large intestine by the *ileocecal valve,* which is formed from the terminal musculature of the ileum and prevents material from flowing back into the small intestine from the large intestine. The jejunum and ileum are attached to the posterior abdominal wall by folds of mesentery.

The function of the small intestine is twofold: to complete the digestion of the material that enters from the stomach and to absorb the final products into the blood and lymph. In fact, *most digestion and absorption occur in the small intestine,* and consequently, conditions that impair its function may be life threatening.

### The Absorptive Surface of the Small Intestine

The wall of the small intestine has the typical four structural layers, but because the small intestine is specialized for absorption, it differs structurally in significant ways from other portions of the gastrointestinal tract. In addition to its length, which provides a significant amount of absorptive surface, the mucosa of the small intestine has three structural features that greatly increase its surface area and hence facilitate absorption. These features are the *plicae circulares, villi,* and *microvilli.* Together, they provide a surface area approximately equivalent to that of a tennis court (300 square meters).

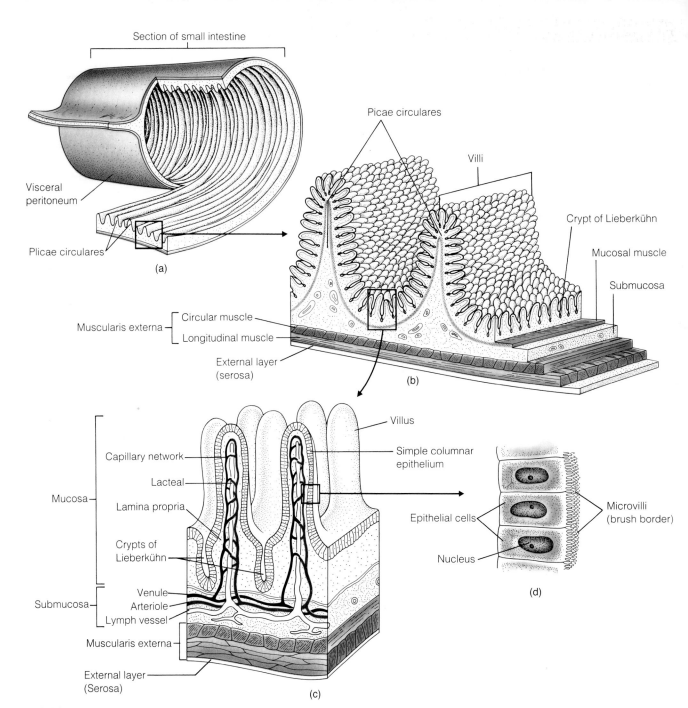

FIGURE 11–11 Structural features of the small intestine. (a) Section of the small intestine, showing the *plicae circulares*. (b) Structure of the *plicae circulares*, showing villi. (c) Detailed structure of villi. (d) Ultrastructure of surface epithelial cells of villi, showing microvilli.

***Plicae Circulares.*** The mucous membrane lining the small intestine is arranged in circularly or spirally disposed folds, called **plicae circulares,** that project into the lumen (Figure 11–11a). These are permanent folds having cores of submucosa, and, unlike the rugae of the stomach, they are not flattened out when the intestine is distended. The plicae are most prominent in the duodenum; they become smaller and farther apart down the length of the small intestine, disappearing entirely in the middle or lower part of the ileum.

***Villi.*** The mucous membrane lining the entire inner surface of the small intestine, both over the plicae and between them, forms numerous tongue- or finger-shaped projections called **villi** (Figure 11–11b). These are packed so closely together as to give the mucosal surface the appearance of velvet. The villi decrease in number, length, and surface area along the course of the small intestine; they are about 1.5 mm high in the duodenum and about one-third that size in the lower ileum.

The internal structure of a villus is shown in Figure 11–11c. Beneath the single layer of epithelial cells is the lamina propria, consisting of loose connective tissue supported by reticular fibers. A single arteriole arises from the arterial network in the submucosa and ascends into each villus, where it breaks up into an extensive capillary network that lies just beneath the epithelial cells. The capillaries are drained by a venule that empties into the venous network in the submucosa. Also projecting into each villus is a single blind-end lymphatic vessel, called a *lacteal,* that is surrounded by smooth muscle fibers. Nerve fibers enter the villus from the myenteric plexus in the submucosa.

Not only do villi increase surface area, but their structure is such that blood and lymphatic vessels are brought close to the absorptive epithelial surface. This feature greatly aids the movement of substances into the circulatory system.

*Microvilli.*    The surface epithelial cells of the small intestine are of two types: about 90 percent are tall columnar cells and the rest, scattered among the columnar cells, are mucus-secreting goblet cells (there are, in addition, a very few endocrine cells, mentioned later). The columnar cells are responsible for absorption and are usually referred to as *absorptive cells.* On the surface that projects into the lumen of the small intestine, the plasma membrane of the absorptive cells forms fingerlike processes, termed **microvilli.** They collectively form the *brush border* of the absorptive cells (Figure 11–11d).

The microvilli are a third means by which the absorptive surface of the lining cells is increased. Moreover, the microvilli contain several enzymes that are important for digestion, as discussed below.

## Glands Associated with the Small Intestine

The liver and pancreas are two glands that arise developmentally from the epithelial layer of the small intestine. They come to lie outside the small intestine, but are connected to it by ducts, through which their secretion products drain into the duodenum. The liver and pancreas synthesize and secrete *bile* and *pancreatic juice,* respectively. Both of these secretions are essential to the normal digestive process. The structure and function of the pancreas and liver, and of the gallbladder, which stores bile, will be discussed in detail in a later section.

The *submucosal glands* lie in the submucosa of the small intestine, primarily in the duodenum. These glands secrete mucus in addition to that supplied by the mucus-secreting goblet cells of the intestinal epithelium. Mucus secreted by the submucosal glands provides extra protection to the duodenum, which receives both the highly acidic chyme from the stomach and the digestive juices from the pancreas and liver.

The ducts of the submucosal glands empty into the bottom of a third type of gland, present in the lamina of the mucous membrane, called *crypts of Lieberkühn,* or *mucosal glands* (see Figure 11–11c). The crypts open into the small intestine at the base of the villi. The epithelial lining of the crypts, continuous with the epithelium that covers the villi and the area between them, contains some mucus-secreting goblet cells as well as other cells that secrete *intestinal juice.*

In addition to their secreting function, the crypts play an important role in the replacement of epithelial cells. Some of the epithelial cells lining the crypts are capable of rapid division. As these cells proliferate, they push adjacent epithelial cells toward the tips of the villi, from which epithelial cells are shed into the lumen of the small intestine. This continuous process of division, upper displacement, and shedding of epithelial cells causes the replacement of the entire intestinal epithelium about every five days. Literally billions of aging cells, amounting to some 250 grams (more than half a pound), are shed every day from the epithelium of the small intestine. Because radiation therapy and drugs used for cancer treatment are designed to kill rapidly dividing cells, these agents often damage the rapidly proliferating intestinal epithelial cells, causing diarrhea and malfunctioning of the digestive tract.

## Digestion in the Small Intestine

***Digestive Secretions of the Pancreas.*** Some one to two liters of pancreatic juice is secreted into the duodenum every 24 hours. Pancreatic juice is a watery fluid containing high concentrations of sodium bicarbonate as well as a number of digestive enzymes. By neutralizing the highly acidic chyme from the stomach, the sodium bicarbonate in pancreatic juice functions to protect the intestinal lining and to create the proper pH environment for functioning of the digestive enzymes.

Pancreatic juice contains a wide spectrum of enzymes. These include *proteases, lipase, amylase,* and *nucleases.* The proteases, including the enzymes *trypsin, chymotrypsin,* and *carboxypeptidase,* split proteins into smaller fragments such as peptides (several amino acids hooked together) and amino acids. Pancreatic lipase breaks down fats (triglycerides) to fatty acids and glycerol; pancreatic amylase converts starch and dextrins to maltose; and nucleases split DNA and RNA into their constituent nucleotides.

***Control of Pancreatic Juice Secretion.*** The secretion of pancreatic juice is controlled by hormonal mechanisms. In response to the presence of chyme in the duodenum, the duodenal mucosa secretes the hormone *secretin,* which stimulates the pancreas to secrete alkaline pancreatic juice, but without digestive enzymes. The presence of fats or partially digested proteins in the duodenum causes the release of *cholecystokinin,* previously known as *pancreozymin,* from the duodenal mucosa. Cholecystokinin stimulates the pancreas to secrete pancreatic juice rich in digestive enzymes.

***Production of Bile by the Liver.*** The liver secretes bile, which is stored and concentrated in the gallbladder prior to its release into the small intestine. Bile is a mixture of a number of substances, including bile salts, lecithin, cholesterol, and bile pigments. Bile salts, a major breakdown product of cholesterol, are essential for the digestion of fats. In the chyme, dietary lipids form large aggregates that are not easily attacked by lipase. Bile salts act as highly effective detergents that emulsify the aggregates by breaking them up into a suspension of small droplets. The resultant increase in the surface area of the lipids greatly promotes their digestion by lipase. In addition, bile salts facilitate the absorption of fatty acids and other lipid molecules by combining with them to form water soluble droplets called *micelles.*

***Control of Bile Secretion.*** Secretin and cholecystokinin, the two duodenal hormones that stimulate the pancreas to secrete pancreatic juice, also control bile secretion. Secretin stimulates the liver to secrete bile, whereas cholecystokinin causes the gallbladder to contract, which releases stored bile into the duodenum.

***Digestive Secretions of the Small Intestine.*** Intestinal juice, secreted by the crypts of Lieberkühn, is a watery, neutral fluid. It contains small amounts of amylase as well as *enterokinase,* an enzyme that activates trypsin, one of the proteases present in pancreatic juice.

A variety of digestive enzymes have been found attached to, or as a membrane component of, the brush border of the absorptive cells covering the villi. These include the following:

1. *Disaccharidases,* enzymes that split sugars composed of two carbohydrate units (disaccharides) into their individual units, called monosaccharides. Of these, *maltase* breaks down maltose to two molecules of glucose, *lactase* breaks down lactose to glucose and galactose, and *sucrase* breaks down sucrose to glucose and fructose.

TABLE 11-1 Major Digestive Enzymes and Substances

| Enzyme or substance | Action | Source | Organ where digestion occurs |
|---|---|---|---|
| Salivary amylase | Starch → dextrins + maltose | Salivary glands | Mouth |
| Pepsin | Proteins → peptides | Zymogen cells of stomach gastric glands | Stomach |
| Hydrochloric acid | Large food particles → smaller particles | Parietal cells of stomach gastric glands | |
| Bile | Emulsifies fats | Liver | Small intestine |
| Proteases | Proteins → peptides, some amino acids | Pancreatic juice secreted by pancreas | |
| Pancreatic lipase | Triglycerides → fatty acids + glycerol | | |
| Pancreatic amylase | Starch → dextrins + maltose | | |
| Nucleases | DNA and RNA → nucleotides | | |
| Amylase | Starch → dextrins + maltose | Crypts of Lieberkühn of small intestine | |
| Enterokinase | Activates the protease trypsin in pancreatic juice | | |
| Maltase | Maltose → glucose | Component of brush border of epithelial cells of small intestine | |
| Lactase | Lactose → glucose + galactose | | |
| Sucrase | Sucrose → glucose + fructose | | |
| Peptidases | Peptides → amino acids | | |
| Enteric lipase | Tryglycerides → fatty acids + glycerol | | |
| Nucleotidases and phosphorylases | Nucleotides from DNA and RNA → pentose sugar (ribose and deoxyribose) + nitrogenous bases + phosphate | | |

2. *Peptidases,* which split peptides into amino acids.
3. *Enteric lipase,* similar in activity to pancreatic lipase.
4. *Nucleotidases* and *phosphorylases,* which split nucleotides into their component parts, including monosaccharides, nitrogenous bases, and phosphorate acid.

*Products of Digestion.* A list of digestive enzymes and other digestive substances and their action is shown in Table 11–1. By the time digestion is complete in the small intestine, virtually all usable nutrients have been digested to yield substances that can be readily absorbed. Polysaccharides, mainly in the form of starch, and other carbohydrates are broken down into monosaccharides, including glucose, fructose, and galactose, among others. However, there are no enzymes in humans capable of degrading cellulose (fiber), the structural polysaccharide of plants, and most cellulose is excreted in the feces except for a small amount that is digested by bacteria in the large intestine. Most proteins are digested to their constituent amino acids. Fat in the diet, mostly in the form of triglycerides, is digested to fatty acids and glycerol, while nucleic acids yield monosaccharides, nitrogenous bases, and phosphate acid.

## Absorption in the Small Intestine

Although there is obviously enormous variation among individuals, an average adult consumes something on the order of 1000 g of food and 1000 mL of water in 24 hours. In addition, about 7000 mL of digestive juices, composed mostly of water, are secreted into the digestive tract from glands that empty into the mouth, stomach, and small intestine. Of this 9000 mL of soupy fluid, only about 500 mL passes into the large intestine; the remainder—almost 95 percent by volume—is reabsorbed in the small intestine.

Substances entering the capillaries and lacteals of the intestinal villi must first pass through the epithelial layer of the intestinal mucosa and into the tissue fluid surrounding the capillaries and lacteals; from the tissue fluid, they enter the capillaries and lacteals by diffusion. Movement across the epithelial layer—the unique aspect of nutrient absorption—occurs by a variety of mechanisms (see Chapter 3 for a review of cellular transport mechanisms). Monosaccharides produced by carbohydrate digestion, amino acids produced by protein digestion, and nitrogenous bases and monosaccharides resulting from the digestion of nucleic acids are transported across the intestinal epithelium by active transport mechanisms that use carrier molecules located in the plasma membranes of these cells. Small amounts of some undegraded proteins are also absorbed at the luminal side of the absorptive cells by endocytosis and released into the tissue fluid by exocytosis. Indeed, the ability to absorb intact proteins is greater in newborns than in adults and permits absorption from mother's milk of antibody molecules that provide temporary immunity (passive immunity) to disease.

The absorption of lipids differs from the absorption of other substances. Fatty acids, glycerol, and other lipids, solubilized by bile salts in micelles, enter the absorptive cells by simple diffusion. Once inside the absorptive cells, fatty acids and glycerol enter the endoplasmic reticulum, where they are resynthesized into triglycerides (fats). These are packaged along with small amounts of cholesterol, fatty acids, and some other lipids into protein-covered droplets, called *chylomicrons.* The chylomicrons are then released into the tissue fluid by exocytosis. Because the lacteals are more permeable to the chylomicrons than are capillaries, the chylomicrons enter the lacteals, forming in the lymph a milky suspension that gives the lacteals their name. After entering the lacteals, the chylomicrons are transported through the lymphatic system to the circulatory system.

Vitamins present in food are absorbed in undigested form. The water-soluble vitamins, including vitamin C and vitamins of the B group, are absorbed by diffusion or carrier-mediated transport, with the exception of vitamin $B_{12}$. In order to be absorbed, vitamin $B_{12}$ must combine with the proteinaceous *intrinsic factor,* secreted by the parietal cells of the stomach. By attaching to binding sites on the luminal surface of the absorptive cells in the lower ileum, the intrinsic factor, carrying vitamin $B_{12}$ with it, triggers these cells to absorb the complex by endocytosis. The fat soluble vitamins, dissolved in micelles along with other fats, are absorbed by diffusion.

Large volumes of water are also absorbed in the small intestine. As nutrient molecules enter the absorptive cells by active transport, water follows by osmosis. The loss of water concentrates solutes in the intestinal contents and produces a favorable concentration gradient for the absorption of substances that enter by diffusion.

A variety of ions are also absorbed. These include large amounts of $Na^+$ and $Cl^-$ and lesser amounts of $K^+$, $Mg^{++}$, $Ca^{++}$, $PO_4^{\equiv}$, and $SO_4^{=}$, as well as a number of trace elements.

## Motility of the Small Intestine

Motility of the small intestine is important for mixing the contents so that they are exposed to digestive juices and for propelling food along the digestive tract. Mixing occurs by *segmentation,* in which ringlike areas of smooth muscle contract and relax rhythmically and spontaneously. During segmentation, the small intestine comes to look somewhat like a chain of sausages, the "links" between the "sausages" corresponding to areas of contraction, and the "sausages" formed in regions where the smooth muscle remains relaxed. Because the sites of the contraction rings vary, the small intestine is segmented in one way and then resegmented in another, forcing the intestinal contents up and down the tract (see Figure 11–12).

Peristaltic waves, beginning after most absorption has occurred, move the intestinal contents toward the large intestine. They start at the duodenum and sweep down the small intestine for variable distances; the site at which the wave is initiated slowly moves down the intestine.

## Disorders of the Small Intestine

*Ulcers.* Erosions in the mucosal surface, or **ulcers,** may develop in the stomach, but more commonly they occur in the duodenum (Figure 11–13). *Stomach* or *gastric ulcers* are commonly characterized by an intermittent burning or gnawing pain in the upper abdomen that may last several hours, then recede until a new bout of pain begins a short time thereafter. There is generally no consistent relationship between the pain and the ingestion of food. In *duodenal ulcers,* on the other hand, pain in the upper middle abdomen typically occurs several hours after a meal. The pain is similar to severe hunger pangs and is relieved by eating or by drinking a glass of milk.

Serious complications of both types of ulcer include slow bleeding that may lead to anemia and severe hemorrhage that can be fatal. Rarely, the stomach or duodenal wall at the site of the ulcer may break through, or *perforate,* releasing contents of the digestive tract into the peritoneal cavity and causing *peritonitis.* Ulcers are treated with drugs that partially inhibit the secretion of gastric juice and with antacids to neutralize excess acid. In intractable cases, surgery may be warranted.

Ulcers are a very common disorder. Autopsies show that 10 percent of Americans have ulcers. Although we do not know what causes ulcers, stress, excessive secretion of gastric juice, and overuse of alcohol and certain drugs such as aspirin and steroids have been implicated.

*Malabsorption.* Injury to the intestinal lining or lack of certain digestive enzymes may impair digestion and absorption in the small intestine. In *celiac disease,* usually discovered during infancy or early childhood, an allergic reaction to gluten protein in wheat and other grains may destroy the intestinal villi and seriously interfere with absorption. Avoidance of all foods containing gluten repairs the defect.

*Lactose Intolerance.* A condition known as *lactose intolerance* occurs when the intestinal epithelial cells synthesize little or no lactase. Consequently lactose, the sugar in cow's milk and mother's milk, cannot be digested. However, lactose is metabolized by bacteria in the large intestine, causing the production of large amounts of gas and organic by-products that interfere with water absorption from the large intestine. Diarrhea and abdominal pain result. If the disease is present at birth, lactose intolerant infants must drink soybean-based formulas instead of milk. Not infrequently, lactase levels may diminish during childhood or adolescence, leading to the onset of symptoms that may be mild or severe. Some degree of lactose intolerance occurs in most blacks, Orientals, and American Indians, and in about 25 percent of whites.

Sometimes, a temporary lactose intolerance will develop as a result of a viral gastrointestinal infection. Damage to the cells of the intestinal lining reduces lactase levels, which return to normal only when the affected cells are repaired or replaced.

## Large Intestine

The **large intestine** is the last section of the digestive tube. It extends from the end of the small intestine to the exterior.

### Structure

The large intestine is about four to five feet long and some two and one-half inches in diameter. Unlike the small intestine, the large intestine is not coiled. The large intestine consists of three regions: (1) cecum, (2) colon, and (3) rectum and anal canal (see Figure 11–1).

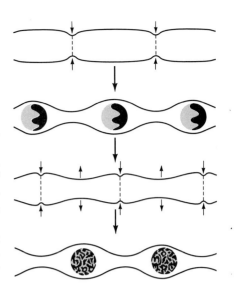

FIGURE 11–12 Segmentation in the small intestine. Dashed lines indicate positions of future contraction rings.

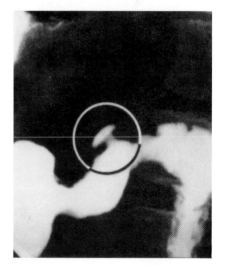

FIGURE 11–13 This X ray was taken after the patient had swallowed barium salts, which are opaque to X rays. The mushroom-shaped projection is a typical benign duodenal ulcer, caused by erosion of tissue part way into the wall of the duodenum.

The **cecum**, beginning at the ileocecal valve, forms a pouch from which extends a narrow blind tube, about the size of the little finger, called the *vermiform appendix*. The appendix has no proven role in humans, but it has been suggested that it may have some function in the immune system. It is commonly twisted and coiled, and may become severely inflamed if bacteria infect its walls, a condition known as *appendicitis*. The chief symptoms of appendicitis are abdominal rigidity and pain in the lower, right-hand portion of the abdomen. An inflamed appendix must be removed surgically or it may rupture, releasing the intestinal contents and the bacteria they contain into the peritoneal cavity—an occurence that may prove fatal.

The **colon** has four sections. Where it ascends on the right side of the body from the cecum toward the undersurface of the liver, it is called the *ascending colon*. From there, the *transverse colon* crosses the abdominal cavity from right to left at about the level of the kidneys. It then turns downward as the *descending colon*. At its end, the descending colon forms an S-shaped region, the *sigmoid colon,* which empties into the rectum.

The **rectum** is a tube, five to six inches long, that may become greatly distended with feces. It empties into the **anal canal**, which is an inch or so in length. The anal canal passes through the muscles of the pelvic diaphragm, and terminates in the **anus,** the posterior opening of the digestive tube. The anus is surrounded by strong muscles, including a thickened region of circular smooth muscle, called the *internal anal sphincter,* and an outer layer of voluntary, skeletal muscle that forms the *external anal sphincter.* These muscles hold the walls of the anus tightly shut.

The mucous membrane of the large intestine is thicker than that of the small intestine, and it contains no villi. Numerous mucus-secreting goblet cells are scattered throughout the epithelium. Crypts of Lieberkühn, set deeply in the mucosa, also contain goblet cells as well as cells that give rise to new epithelial cells. In the anal canal, the mucous membrane is folded longitudinally to form *anal columns* that fuse together at their lower ends (Figure 11–14). Small veins present in this area may become dilated and bulge inward, forming *internal hemorrhoids;* dilated veins near the anus form *external hemorrhoids.*

FIGURE 11–14 The end of the large intestine, showing sigmoid colon, rectum, and anal canal.

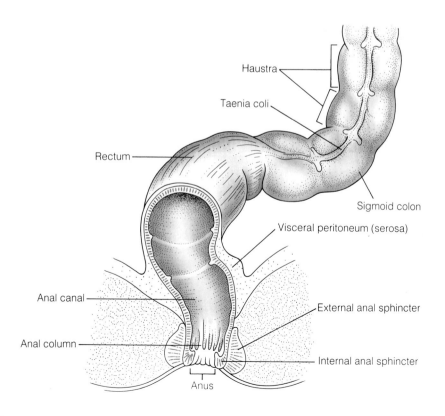

Haustra

Taenia coli

Rectum

Sigmoid colon

Visceral peritoneum (serosa)

Anal canal

External anal sphincter

Anal column

Internal anal sphincter

Anus

Hemorrhoids are very common and are not dangerous. However, they may bleed occasionally and may cause painful bowel movements. External hemorrhoids sometimes *prolapse,* or protrude outside the anal orifice, causing a mucousy discharge, itching, and pain. Assuring soft bowel movements by eating a diet high in fiber will help keep hemorrhoids under control. Astringents in the form of rectal suppositories will often shrink hemorrhoids, or they may be removed surgically.

In the cecum and colon, the longitudinal smooth muscle fibers are not arranged in a continuous layer that surrounds the circular muscle. Instead, most longitudinal fibers are arranged in three narrow bands, forming the *taenia coli,* that extend along the course of the large intestine. Because these bands are not as long as the large intestine, they pucker its walls, forming outpocketings called *haustra.* The puckering effect of the taenia coli is much the same as that obtained by sewing elastic bands into fabric.

## Role of the Large Intestine

The main function of the large intestine is the active absorption of sodium ions, with the accompanying osmotic reabsorption of water. The large intestine secretes no digestive enzymes. However, enzymes from the small intestine are carried into the large intestine and may be responsible for a limited amount of digestion.

Bacteria, predominantly *Escherichia coli* (*E. coli*), are permanent, beneficial residents of the large intestine. They break down cellulose to some degree. More importantly, they synthesize vitamins that may then be absorbed by the large intestine. This source of vitamins may be important if dietary vitamin intake is low. Most vitamin K is produced by bacteria in the large intestine, and consequently, prolonged antibiotic therapy that eliminates intestinal bacteria can provoke symptoms of vitamin K deficiency. As a result of their metabolic processes, bacteria may produce considerable amounts of gas, consisting mostly of $N_2$, $CO_2$, and small amounts of methane ($CH_4$), hydrogen ($H_2$), and hydrogen sulfide ($H_2S$).

## Feces

The solid matter emptied from the colon into the rectum is termed *feces.* Feces are 65 to 80 percent water, the water content varying inversely with the length of time the fecal material remains in the colon. In addition to water, feces consist of bacteria, undigested cellulose, cellular debris from the intestinal epithelium, and modified bile pigments that give feces their brown color.

## Motility of the Large Intestine

Segmentation, similar to that in the small intestine but occurring at a much slower rate, mixes the intestinal contents; peristaltic waves move the contents along the intestinal tract. Strong contractions of large sections of the colon occur several times a day, usually after meals, and empty most of the contents of the large intestine into the rectum.

## Defecation

The process of defecation, by which feces are emptied from the digestive tract, has an involuntary component, controlled by the autonomic nervous system, that can be overridden by voluntary mechanisms.

Activation of stretch receptors in the wall of the rectum, occurring as a result of the accumulation of feces in the rectum, results in a conscious urge to defecate. Defecation is initiated when voluntary inhibition is removed by relaxation of the external anal sphincter. Reflex contraction of the rectum, accompanied by relaxation of the internal anal sphincter and increased peristaltic activity, propels the fecal material through the anus. The movement of fecal material may be assisted by contractions of abdominal and diaphragm muscles, which increase pressure within the abdomen. Before the age of about eighteen months when voluntary inhibition can be learned, defecation is entirely involuntary.

### Disorders of the Large Intestine

***Irritable Colon.***   A condition characterized by irregular, spasmodic contractions of the colon is called *irritable* or *spastic colon*. These uncoordinated contractions interfere with the normal movement of material through the colon, and may result in diarrhea or constipation as well as cramplike pains, usually relieved by defecation. The cause is unknown, but it is prevalent in people under stress. It is twice as common in women as in men. The use of antispasmodic drugs and avoidance of certain foods will generally control the condition.

***Diverticulitis.***   Small saclike swellings, called diverticula, occasionally develop in the sigmoid colon, particularly in elderly people. The diverticula may become inflamed, causing *diverticulitis*. Diverticulitis causes severe abdominal pain and, if left untreated, may lead to complications, including the development of an abcess in the colon or even perforation. Treatment with antibiotics may help clear up the inflammation. In severe cases, surgery may be required to remove the affected section of the colon.

***Colon and Rectal Cancers.***   Cancer of the large intestine, including colon cancer and rectal cancer, is the second most common type of cancer in males, behind lung cancer, and the third most common in females, behind lung and breast cancer. It occurs most frequently in people over the age of 40, especially those in their sixties and seventies. Early warning signs are persistent, abnormal constipation or diarrhea, and bloody stools (Figure 11–15).

Surgery is the best treatment if the cancer has not spread too far. A section of intestine containing the tumor is removed, and the cut ends are sewn together. However, if the tumor occurs in the rectum close to the anus, the entire rectum and anal canal must be removed, and a new opening is constructed in the abdominal wall for the elimination of feces. This operation is called a *colostomy*.

***Diarrhea and Constipation.***   Most people have one bowel movement a day, but some have more, and others, fewer. When bowel movements become less frequent than is normal for a particular individual, the condition is called *constipation*. The main causes of constipation are diets low in roughage and overuse of laxatives, which disrupt the normal pattern of defecation. In addition, certain drugs such as opiates may cause constipation. Contrary to popular opinion, there is no evidence that constipation is toxic. Loss of appetite, headache, and nausea associated with constipation appear to result from distension of the rectum.

*Diarrhea,* characterized by unusually frequent, watery bowel movements, can be caused by certain foods (such as prunes), food sensitivities, food poisoning, overuse of alcohol, stress, and drugs (such as antibiotics and antacids). Severe diarrhea may also result from infections of the intestinal tract by bacteria (as in typhoid fever and cholera), by viruses (influenza), and by protozoa (amoebic dysentery). Diarrhea can become life threatening if it results in severe dehydration or in the loss of certain ions such as potassium and bicarbonate.

## ACCESSORY DIGESTIVE ORGANS

The accessory digestive organs include the salivary glands, liver, gallbladder, and pancreas. The salivary glands, discussed in an earlier section, secrete saliva

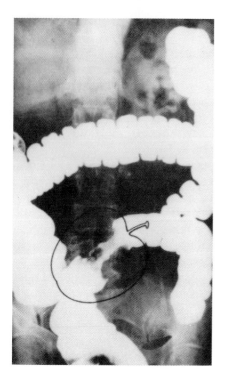

FIGURE 11–15 A barium X ray of the colon, showing a typical constricting lesion caused by cancer of the colon. The appearance of such a lesion is often likened to an apple core.

into the mouth. The liver produces bile, important in the digestion of fats; bile is then stored in the gallbladder prior to its release into the duodenum of the small intestine. The pancreas secretes a variety of important digestive enzymes into the duodenum. The liver and pancreas also serve other functions that are not directly related to digestion.

## Liver

The **liver** is the largest gland of the body. It lies in the upper part of the abdominal cavity just beneath the diaphragm (see Figure 11–1).

### Structure

The liver is composed of two main lobes: the *right lobe,* which is the largest and accounts for more than 80 percent of the substance of the liver, and the *left lobe.* Its upper surface is convex and its lower, concave; a deep fissure extends across the undersurface of the right lobe. The liver is covered with visceral peritoneum, which in certain areas forms folds, or ligaments, that connect the liver to the diaphragm and the anterior abdominal wall. Beneath the visceral peritoneum, the liver is covered with a fibrous connective tissue capsule. Connective tissue, continuous with this capsule, extends into the substance of the liver at the deep fissure, and branches extensively, providing internal support. Accompanying the connective tissue into the liver are the portal vein and hepatic artery; lymphatic vessels and the right and left hepatic ducts (bile ducts) leave the liver in the same area (Figure 11–16).

You will recall from Chapter 9 that the liver receives blood from two sources. The hepatic artery, a branch of the aorta, carries oxygenated blood to the liver tissue. In addition, the portal vein collects blood from the visceral organs and carries it to the liver; this blood is rich in absorbed food but low in oxygen. Blood from these two sources mixes together within the liver and then leaves the liver through the hepatic veins. These empty into the inferior vena cava.

The basic functional unit of liver tissue is the liver lobule (Figure 11–17). Each lobule, separated from adjacent lobules by small amounts of connective tissue, consists of thin plates of *hepatocytes,* or liver cells, that radiate out from a

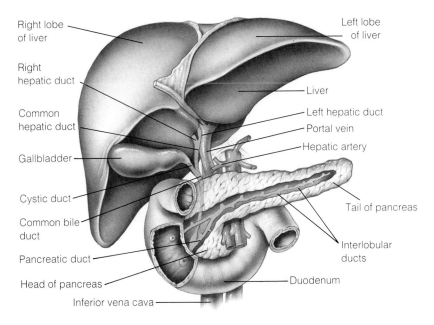

Right lobe of liver

Right hepatic duct

Common hepatic duct

Gallbladder

Cystic duct

Common bile duct

Pancreatic duct

Head of pancreas

Inferior vena cava

Left lobe of liver

Liver

Left hepatic duct

Portal vein

Hepatic artery

Tail of pancreas

Interlobular ducts

Duodenum

FIGURE 11–16 The liver and associated organs.

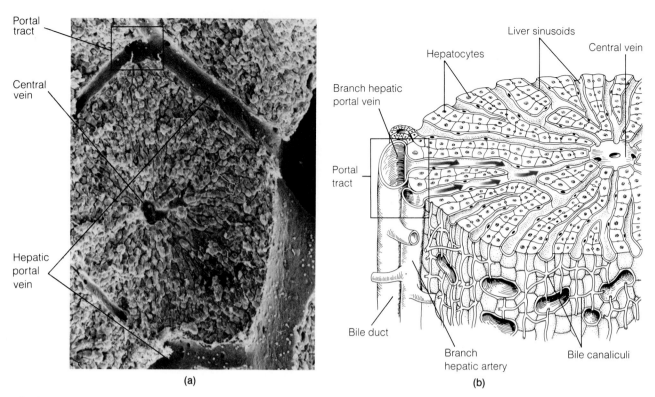

Portal tract

Central vein

Hepatic portal vein

(a)

Branch hepatic portal vein

Portal tract

Bile duct

Hepatocytes

Liver sinusoids

Central vein

Branch hepatic artery

Bile canaliculi

(b)

FIGURE 11–17 (a) Scanning electron micrograph of a liver lobule as seen in cross-section. (b) Diagram interpreting the organization shown in (a).

*central vein;* the central vein is a small branch of one of the hepatic veins. A lobule is typically hexagonally shaped. At each corner of the hexagon is a *portal tract,* consisting of a branch of the hepatic artery, a branch of the hepatic portal vein, and a small bile duct, all sheathed in connective tissue.

The branches of the hepatic artery and hepatic portal vein in the portal tract dump blood into *liver sinusoids.* These are microscopic channels that are wider and more permeable than ordinary capillaries, and hence allow plasma to come in contact with, and nourish, the hepatocytes. The sinusoids are also lined with phagocytic cells that engulf bacteria, other microbes, and foreign particles that may have entered the blood stream. Blood flows from the sinusoidal spaces into the central vein of the lobule.

While one surface of a hepatocyte borders on a sinusoid, the other surface abuts another microscopic channel, a *bile canaliculus* (see Figure 11–17). Bile produced by the hepatocytes accumulates in the bile canaliculi and drains into bile ducts in the portal tracts. Because the plasma membranes of adjacent hepatocytes are fused tightly together, there is no mixing of plasma and bile. Ultimately, all bile is collected into one large duct from each lobe. These ducts, the *right* and *left hepatic ducts,* fuse to form the *common hepatic duct* that conveys bile from the liver.

### Function

The liver is essential for life because it has many varied and important functions. A central function is the production of bile, which contains bile salts necessary in the digestion of fats. Also present in bile are bile pigments, primarily *bilirubin.* Bilirubin, produced from the breakdown of hemoglobin by phagocytes in the spleen and bone marrow, is transported to the liver by plasma proteins and excreted in the bile. Upon entering the small intestine, the bilirubin in bile is broken down by digestive enzymes and bacteria, and is eliminated in the feces.

In addition to bile production, other metabolic functions of the liver include the following:

1. The liver is central to the maintenance of relatively constant blood glucose levels. When blood glucose levels rise after meals, the liver, under the influence of the hormone *insulin,* converts excess blood glucose into glycogen, which it stores. As glucose levels fall between meals, the liver breaks down stored glycogen to glucose and releases it into the blood. This latter function is mediated by another hormone, *glucagon.* Both insulin and glucagon are secreted by the pancreas, discussed later in this chapter. The liver can also synthesize glucose from amino acids and other noncarbohydrate sources during fasting or starvation.

2. The liver is important in the synthesis and degradation of amino acids and proteins. All plasma proteins except immunoglobulins are synthesized in the liver.

3. It is responsible for the formation of *urea* from ammonia and other nitrogen-containing waste products; urea is then excreted in the urine.

4. The liver synthesizes a variety of lipids such as cholesterol, lipoproteins (see Chapter 8), phospholipids, and triglycerides. It is also important in the degradation of lipids into molecules that can be metabolized for energy or excreted.

5. The liver detoxifies many harmful compounds, including alcohol and drugs, which are eliminated in the bile and urine. In this respect, the portal blood flow to the liver is particularly significant since potentially harmful substances, absorbed from the digestive tract, are sent first to the liver for detoxification before they reach other cells of the body. One inadvertent—and potentially dangerous—consequence of the liver's detoxification function is that sometimes harmless substances are converted by the liver into harmful ones. For example, it is now known that some chemicals become carcinogenic only after their molecular structure is altered during their "detoxification" in the liver (see Chapter 22).

6. The liver inactivates hormones, particularly steroid hormones, by converting them into substances that are excreted in the bile and urine.

7. The liver stores a number of necessary substances, including iron and vitamins A, D, and B$_{12}$. It also synthesizes vitamin A from precursor compounds obtained in the diet. Important precursors for vitamin A synthesis are the carotenes found in carrots and dark-green leafy vegetables.

## Diseases of the Liver

*Jaundice.*   Presence of abnormal levels of bilirubin in the bloodstream causes **jaundice,** in which the skin and whites of the eye take on a yellow color. Because the liver is involved in the elimination of bilirubin from the body, any condition that impairs liver function may lead to jaundice. Furthermore, obstruction of the ducts that transport bile into the small intestine, as occurs when gallstones become lodged in the ducts, also causes jaundice. Jaundice will disappear with successful treatment of the underlying cause.

*Hepatitis.*   **Hepatitis** is an inflammation of the liver that is most commonly caused by a viral infection or by a reaction to certain drugs. Among several types of viral hepatitis, the most common are *hepatitis A* and *hepatitis B,* caused by different viruses. In hepatitis A, viral infection of the liver causes it to swell and become tender. Bilirubin may accumulate in the blood, causing jaundice. Other symptoms include weakness, nausea, and whitish feces. The virus is present in the blood and feces for several weeks before and after symptoms develop, and may be transmitted to healthy individuals who come in contact with these substances. There is no specific treatment other than bed rest for hepatitis A, and recovery is usually complete in several weeks.

Hepatitis B, also called *serum hepatitis,* causes symptoms similar to those of hepatitis A, but they are more severe and of longer duration. Because the virus is present in saliva, blood, semen, and nasal discharge of infected individuals, it is frequently transmitted by blood transfusions, sexual activity, and contaminated hypodermic needles. Although most people recover completely from hepatitis B with prolonged bed rest, some develop a chronic form of the disease. Still others can appear to recover but continue to harbor the virus, which may be transmitted to uninfected individuals.

*Cirrhosis.* Alcoholism, malnutrition, hepatitis, infection by liver parasites, and exposure to toxic chemicals are common causes of a chronic deterioration of the liver, termed **cirrhosis.** The decline in liver function resulting from cirrhosis may eventually lead to liver failure. Cirrhosis is particularly common among alcoholics who do not eat properly. If alcoholics with cirrhosis do not stop drinking, they eventually die of liver failure.

## Gallbladder

The **gallbladder** is a pear-shaped sac, lying under the right lobe of the liver (see Figure 11–16). It functions to concentrate and store bile secreted by the liver and to release bile when it is required for digestion.

### Structure and Function

After bile has been synthesized by the liver, it travels to the gallbladder for storage. From there it is released into the small intestine through ducts that connect the gallbladder and small intestine. As shown in Figure 11–16, the common hepatic duct from the liver fuses with the *cystic duct* from the gallbladder to form the *common bile duct.* The common bile duct joins the pancreatic duct from the pancreas before emptying into the duodenum. A sphincter formed of smooth muscle surrounds the common bile duct just before it joins the pancreatic duct. When fatty foods are not present in the duodenum and hence bile is not required for digestion, the sphincter is closed, and bile, secreted more or less continuously from the liver, backs up through the cystic duct into the gallbladder. Another sphincter surrounds the duct formed by the fusion of the common bile duct and the pancreatic duct.

Like the digestive tube, the gallbladder is lined with mucous membrane. Its epithelial cells, resembling those of the small intestine, are absorptive cells possessing microvilli. The mucous membrane is wrinkled when the gallbladder is empty, but the folds are ironed out when the gallbladder is full. External to the mucous membrane of the gallbladder is a thin layer of smooth muscle, which in turn is covered by loose connective tissue that conveys blood vessels, lymphatic vessels, and nerves. Visceral peritoneum covers the gallbladder except where it is in contact with the liver.

As bile from the liver enters the gallbladder, the epithelial cells that line the gallbladder absorb sodium ions and water from the bile. The loss of water results in a five- to tenfold concentration of the bile.

The presence of fatty acids in the duodenum causes the hormone cholecystokinin to be released into the bloodstream from the cells of the intestinal mucosa. Cholecystokinin causes contraction of the gallbladder, which is accompanied by the relaxation of the sphincters mentioned earlier. Bile thus flows from the gallbladder into the duodenum.

### Gallstones

Increased concentrations of cholesterol relative to other components of bile may lead to the precipitation of solid cholesterol and the formation of one or more hard, stonelike objects, called **gallstones,** within the gallbladder. Once formed, gallstones grow in size by accretion. Gallstones may remain in the gallbladder, usually without consequence, or they may pass out of the gallbladder in the bile. In this latter case, a gallstone may sometimes lodge in the cystic duct or in the common bile duct, causing intense pain, referred to as *biliary colic,* in the upper right-hand side of the abdomen. If the lodged gallstone eventually passes into the small intestine, the pain will cease. If it remains lodged, it must be removed surgically, usually along with the gallbladder to prevent future problems. Removal of the gallbladder, termed *cholecystectomy,* does not affect digestive functions since bile from the liver can drain directly into the small intestine.

It has been estimated that about 20 million Americans have gallstones. For unknown reasons, they are about twice as frequent in men as in women, and they occur far more commonly in older people than in younger ones.

## Pancreas

The **pancreas** is a moderately large gland, about 100 grams in weight. In color, it is white with a pinkish tinge. It is somewhat tadpole-shaped, with its head resting in the concave portion of the duodenum and its tail touching the spleen (see Figures 11–1 and 11–16). The pancreas is surrounded by a capsule of very thin connective tissue, and this is covered by visceral peritoneum. Internally, the tissue of the pancreas is divided into lobules by thin partitions of connective tissue that carry blood vessels, lymphatic vessels, secretory ducts, and nerves.

The pancreas is an organ with both digestive and endocrine functions, but these two functions are not carried out in separate anatomical regions of the organ. Instead, tissues that perform each of the functions are intermingled in a lobule.

The digestive function of the pancreas is performed by exocrine tissue, which makes up the bulk of each lobule. It is organized into sac-like structures called *acini* (sing: *acinus*). These are formed of a single layer of *acinar cells* surrounding a central space, into which the acinar cells secrete pancreatic juice. Each acinus is drained by a small duct. Ducts from a number of acini empty into *interlobular ducts* that in turn drain into the *pancreatic duct,* which runs longitudinally through the pancreas.

Scattered throughout the exocrine tissue are clusters of endocrine cells, called **islets of Langerhans.** The endocrine function of the pancreas is discussed extensively in Chapter 16.

## NUTRITION

Within the last decade or so, newly discovered correlations between diet and various disease conditions, such as the linkage between atherosclerosis and high blood cholesterol levels, have created a heightened awareness among many Americans of the importance of nutrition to human health. Unfortunately, this knowledge has not translated, for the most part, into changes in our dietary habits, and the diet of the average American today, while providing most essential nutrients, is still far from optimal.

Before beginning a discussion of nutrition, it is perhaps well to stress several points. First, nutrition is by no means an exact science, and the nutritional requirements of humans are not precisely known. This is particularly true with regard to substances that might be needed in only trace amounts, since we may have difficulty detecting their presence, let alone establishing a requirement for them or understanding their function. Second, the nutritional requirements described here are those necessary for long-term optimal health, not for short-term survival. Clearly, people can survive, sometimes even for long periods of time, with various nutritional deficiencies that cause metabolic abnormalities or disease conditions. These compromise health, usually in significant ways, but they may not always lead to immediate death or even, if the deficiency is eventually eliminated, to permanent disability. For example, the effects of severe starvation, in which an individual receives little or no food for many days (but receives water), are often completely reversible in adults.

### Essential Nutrients

For optimal health, about 50 specific nutrients, and some general categories of nutrients, are required on a regular basis. These so-called *essential nutrients,* which must be supplied in the diet because they cannot be manufactured by

TABLE 11–2 Essential Nutrients*

| Energy foods | Fatty acids | Amino acids | Fat-soluble vitamins | Water-soluble vitamins | Bulk minerals | Trace elements |
|---|---|---|---|---|---|---|
| Carbohydrates | Linoleic acid | Isoleucine | Vitamin A | Vitamin B$_1$ (thiamin) | Calcium | Iron |
| Fats and fatty acids | | Leucine | Vitamin D | Vitamin B$_2$ (riboflavin) | Phosphorus | Iodine |
| Proteins | | Lysine | Vitamin E | Niacin | Magnesium | Zinc |
| | | Methionine | Vitamin K | Folic acid | Sodium | Chromium |
| | | Phenylalanine | | Vitamin B$_6$ (pyridoxine) | Potassium | Copper |
| | | Threonine | | Vitamin B$_{12}$ (cobalamin) | Chlorine | Fluorine |
| | | Tryptophan | | Vitamin C (ascorbic acid) | Sulfur | Manganese |
| | | Valine | | Biotin | | Molybdenum |
| | | Arginine** | | Pantothenic acid | | Selenium |
| | | Histidine** | | | | Cobalt |
| | | | | | | Perhaps others*** |

*Excluding water.

**Histidine and arginine are considered essential in infants to meet their growth requirements, and there is some evidence that they may be essential in adults, since deficiency disorders may occur as a result of long periods of limited availability.

***Aluminum, arsenic, silicon, vanadium, nickel, tin, and cadmium.

the cells of the body, are listed in Table 11–2 and discussed shortly. Not included in Table 11–2 is water, which is essential for life, but which is not strictly classified as a nutrient.

## Energy Foods

A great deal of energy is required to support the body's work, including the synthesis of cellular components, cellular transport, muscular contraction, and so forth. Indeed, most of the food we ingest goes to supplying the energy requirements of the body; a much smaller proportion provides the minerals or organic compounds we cannot synthesize and which we need either in unaltered form or as raw materials for biosynthesis.

*Metabolism of Energy Foods.*    The energy we need is supplied by carbohydrates, fats, and proteins, obtained from animal and plant sources. The main natural dietary sources of these foods are shown in Table 11–3.

As we discussed in Chapter 2, energy is released within the cells by the metabolic breakdown of organic substances, and it is used to manufacture ATP, the energy currency of the cell. It is important to remember, however, that the fate of various types of energy foods may vary once they are ingested. Some,

TABLE 11–3 Main Natural Sources of Energy Foods

| | |
|---|---|
| Carbohydrates | Cereal grains (wheat, corn, rice) |
| | Tubers (potatoes, yams) |
| | Legumes (soybeans, peanuts, peas, and beans) |
| Fats | Triglycerides (as solids in meat, fish, butter; as liquids in vegetable oils) |
| Proteins | Animal sources |
| | Meat |
| | Fish |
| | Eggs |
| | Milk and other dairy products |
| | Plant sources |
| | Cereal grains (wheat, corn, millet) |
| | Legumes (soybeans, beans, peanuts) |

such as glucose in grapes, is absorbed unchanged from the digestive tract and may be metabolized directly by cells. Other more complex molecules, such as starch in potatoes or protein in meat, must be broken down by enzymes in the digestive tract into their smaller organic subunits, which are then absorbed and metabolized.

It is also true that large organic molecules in the diet are not necessarily used exclusively as energy sources. If the body's energy supply is adequate, the small organic subunits that result from the digestion of macromolecules can also be used as essential raw materials for the synthesis of cellular components. For example, the breakdown of dietary proteins by digestive enzymes yields amino acids that may be used not only as energy sources, but also as essential raw materials for the manufacture of "human" proteins.

*Weight Gain.*    Roughly 40 percent of the energy content of food is captured as utilizable energy in the form of ATP. The remainder is lost as heat due to the inefficiency of the body's metabolic machinery; this heat is used to maintain our body temperature, and the excess is dissipated in the environment. The energy captured as ATP is used to meet the body's energy requirements, as noted earlier. However, if the amount of food consumed, and hence the capture of energy as ATP, exceeds the energy needs of the body, the excess ATP, which cannot be stored in significant amounts, is used to synthesize fats that are stored in adipose tissue. The result is a weight gain, the magnitude of which is directly related to the amount of excess food consumption.

The storage of fat makes sense from the standpoint of survival because the fat can be broken down to yield ATP when energy foods are scarce. In affluent societies, however, scarcity of food is not generally a problem, and consequently, the normal biological mechanisms that promote weight gain in the face of nutrient abundance tend to be counterproductive, leading to overweight and obesity.

The energy content of food and the energy requirements of the body are stated in units of heat called *kilocalories* (kcal), also known as *Calories,* written with a capital C. About 35 to 40 kcal are required per day by reasonably active adults for each kilogram (2.2 pounds) of ideal body weight. On this basis, an individual with an ideal body weight of 55 kilograms (120 pounds) requires between 1925 and 2200 kcal per day, whereas a 70 kilogram (154 pound) individual needs between 2450 and 2800 kcal (see *Focus: Weight Gain and Eating Disorders*).

## Essential Fatty Acids

The only fatty acid that cannot be synthesized by the body and hence is required in the diet is linoleic acid, an 18-carbon polyunsaturated fatty acid found in most common animal fats and vegetable oils. It is used as a precursor for the synthesis of other, more complex fatty acids, and for the synthesis of a group of specialized hormones, the prostaglandins, which will be discussed in Chapter 16, *The Endocrine System.*

## Essential Amino Acids

As shown in Table 11–2, at least eight amino acids are essential beyond infancy. Another amino acid, histidine, is essential for infants and may be so for adults, since a deficiency syndrome may appear after long periods of limited intake. There is some evidence that the amino acid arginine may also be required in infancy.

Amino acids needed for the synthesis of "human" proteins are obtained from the breakdown of dietary proteins from animal and plant sources. All readily available animal proteins, with the exception of gelatin (collagen), provide adequate sources of all the essential amino acids. This is not true of plant proteins. Any one type of plant protein does not supply adequate quantities of all of the essential amino acids. For example, the protein of

# Weight Gain and Eating Disorders

With the relative affluence of Western societies has come a growing health problem: large numbers of overweight individuals. Indeed, a quarter of all Americans are classified as obese, meaning that their weights are 20 percent or more greater than their ideal weights, as determined by life insurance tables.

A great deal of epidemiological research has demonstrated that obesity is a contributing factor to the development of a number of serious disease conditions. These include high blood pressure and atherosclerosis, both of which may lead to strokes and heart attacks; some forms of cancer; adult-onset diabetes; and gallbladder disease. In addition, weight-bearing bones and joints are liable to be damaged from excessive strain, causing or aggravating a host of orthopedic problems such as arthritis and back pain. Statistics show that obese individuals are three times as likely as individuals of normal weight to develop high blood pressure and adult-onset diabetes. Highly obese men (40 percent or more overweight) have increased risks of colon, rectal, and prostate cancers, and highly obese women show a fivefold increase in the incidence of endometrial cancer as compared with normal weight women.

Although the problems of obesity have been well documented, its cause is poorly understood. Clearly, many factors are involved. It has long been known that obese parents tend to have fat children who become obese adults. Until recently, most evidence suggested that this correlation was primarily due to social factors; that is, that children pick up bad eating habits from their parents. Several studies supported this conclusion. One study showed that obese parents tend to have fat adopted children as well as fat pets. Another showed that obesity, particularly among women, is correlated with social class, with 30 percent of women in the lower social class being obese, 18 percent in the middle class, and 5 percent in the upper class. However, several recent studies on large groups of adopted children in Denmark and in Iowa show that an individual's relative adult weight is strongly correlated with the relative weights of the biological parents and not with the weights of the adoptive parents, suggesting that the tendency toward obesity is inherited. Most researchers now feel that obesity results from a complex interplay of hereditary and social factors.

Recent research by several groups of scientists at Rockefeller University has uncovered some striking biochemical differences between fat cells of obese individuals and those of normal weight individuals—differences that provide some clues to the nature of obesity. Not only is it known that obese, or formerly obese, people have up to three times as many fat cells as normal individuals, but the Rockefeller researchers have shown that individual fat cells in obese people are two to two and one-half times as large as the fat cells in people of normal weight.

Further evidence suggests that physiological mechanisms designed to maintain the expanded fat cell size are operative in obese individuals and give rise to signals for overeating in formerly obese people. For example, the researchers have shown that the metabolic processes by which triglycerides are degraded and resynthesized in the fat cells of formerly obese individuals are similar to the processes that occur in the fat cells of normal weight people who have fasted for four or five days; these processes are distinctly different than fat cell metabolism in normal weight individuals or in obese individuals who are not trying to lose weight. The researchers propose that the altered metabolism in the fat cells of formerly obese people somehow triggers a signal for eating that is similar to that in a normal, fasting individual. Related research has shown that formerly obese individuals need 25 percent fewer calories to maintain their weight than do individuals who were never obese—another contributing factor to obesity.

These metabolic differences suggest that there is a physiological basis for obesity, and they may account for the fact that two-thirds of all obese individuals and 95 percent of those who are extremely obese cannot lose their excess weight without regaining it. However, this research does not explain why obesity develops in the first place.

As much of a problem to health as is excessive weight gain, excessive weight

legumes (beans, peas, and lentils) does not contain sufficient amounts of the sulfur-containing amino acid methionine, and many legumes are also deficient in tryptophan. On the other hand, the cereal grains (including rice, oats, rye, corn, barley, and wheat) lack adequate amounts of lysine, and some also are deficient in isoleucine. In order to overcome the amino acid deficiencies of individual plant proteins, vegetarians construct their diets to combine cereal grains with legumes.

Lack of an adequate supply of protein that supplies all of the essential amino acids is the major nutritional problem throughout the world, particularly in developing countries. Because high quality animal protein from livestock and fish is usually not available in these countries, cereal grains are often used

loss can be even more devastating. An estimated 400,000 Americans, most of them women and many in the 12 to 25 age group, suffer from *anorexia nervosa,* self-induced starvation that results in progressive and sometimes fatal malnutrition. Accompanying the avoidance of food is a striking increase in physical activity and a delusion of fatness that continues even when the body has become emaciated. Anorectics show a host of abnormal physiological symptoms, including hypothermia, low blood sugar levels, and low blood pressure, that are a direct consequence of starvation. Their malnourished condition makes anorectics vulnerable to other illnesses, and their mortality rate is some 20 percent above that of the overall population.

The cause of anorexia is unknown, but it is thought to result from a complex interaction between psychological, social, and biological factors. It typically strikes young women who are bright, energetic, and perfectionist, but who have, beneath an outward facade of emotional stability, a low self-esteem, severe emotional conflicts, and a tendency to depression. Because the self-induced starvation of anorexia is thought to be a symptom of these underlying problems, treatment is primarily psychiatric.

A related eating disorder is *bulimia,* which involves recurrent episodes of gorging on food, followed by self-induced vomiting and administration of laxatives and diuretics. Studies have shown that as many as 20 percent of college women are bulimic at one time or another. Unlike anorectics, individuals with bulimia usually are normal weight or overweight. Bulimia can result in serious medical complications including chronic diarrhea, dehydration, and electrolyte imbalances. A common sign of bulimia is the erosion of tooth enamel by stomach acid as a consequence of frequent vomiting.

Research has shown that individuals with bulimia feel guiltier after eating a large meal, count calories more often, diet more frequently, and exercise less than their nonbulimic peers. Many workers feel that the undue emphasis placed in our society on the virtues of thinness may lead psychologically vulnerable individuals into weight control efforts that result in the vicious binge-purge cycles of bulimia.

This anorexic patient is trying to overcome her problem by observing her body in a mirror.

as the sole protein sources. Not only do cereal grains have a low protein content relative to their total mass, but reliance on a single cereal crop does not supply all essential amino acids in adequate amounts. Protein deficiency especially affects infants and children, whose growth requires high levels of protein synthesis.

## Vitamins

On the basis of their solubility characteristics, vitamins are grouped into two categories: the *fat-soluble vitamins* and the *water-soluble vitamins* (see Table 11–2). The fat-soluble vitamins are usually associated with lipids in foods.

**TABLE 11–4** Characteristics of Vitamins

| Fat-soluble vitamins | | | |
| --- | --- | --- | --- |
| Vitamin | Major functions | Source | Deficiency disease |
| Vitamin A | Essential for vision; growth of bones and teeth. | Liver, eggs, butter. Synthesized in the body from carotene present in green and yellow vegetables and tomatoes. | Night blindness, loss of sight; stunting of growth; spermatogenesis impaired. |
| Vitamin D | Promotes intestinal absorption of $Ca^{++}$ and $PO_4^{\equiv}$; good bone and tooth structure. | Liver, oily fish, eggs, butter. Can be synthesized in the skin in the presence of sunlight. | Rickets in children; deficiency in bone matrix formation in adults. |
| Vitamin E | Protection of tissues from oxidation by toxic substances (antioxidant); normal fertility. | Animal fats and vegetable oils; most foods. | Deficiency unlikely. |
| Vitamin K | Synthesis of clotting factors in liver. | Synthesized by intestinal bacteria; present in leafy vegetables. | Impaired blood coagulation. |

| Water-soluble vitamins | | | |
| --- | --- | --- | --- |
| Vitamin | Major functions | Source | Deficiency disease |
| Vitamin $B_1$ (thiamin) | Cellular metabolism of energy sources. | Whole grains, nuts, legumes, liver, dairy products, eggs. | Beriberi |
| Vitamin $B_2$ (riboflavin) | Involved in oxidation-reduction reactions. | Liver, eggs, dairy products. | Ariboflavinosis |
| Niacin | Involved in oxidation-reduction reactions. | Whole grains, nuts, legumes, liver, meat, fish, poultry. | Pellagra |
| Folic acid | Synthesis and breakdown of many compounds; needed for red blood cell production. | Leafy vegetables, liver, poultry, fish. | Anemia |
| Vitamin $B_6$ (pyridoxine) | Cellular metabolism of energy sources; functioning of red blood cells, nerves. | Meat, whole grain cereals, legumes. | Nerve disorders; skin irritations. |
| Vitamin $B_{12}$ | Synthesis and breakdown of many compounds; red blood cell production; functioning of nervous system. | Eggs, meat, dairy products. | Pernicious anemia |
| Vitamin C (ascorbic acid) | Involved in oxidation-reduction reactions; repair of tissues; many roles in growth. | Citrus fruits, potatoes, tomatoes, peppers, leafy vegetables. | Scurvy |
| Biotin | Cellular metabolism of energy sources. | Liver, egg yolk, whole grain cereals. | Ill health, but no specific symptoms that characterize deficiency. |
| Pantothenic acid | Cellular metabolism of energy sources. | Meat, egg yolk, whole grain cereals, legumes. | Ill health, but no specific symptoms that characterize deficiency. |

They are stored in the body, primarily in the liver and in adipose tissue, and consequently are not required on a daily basis. On the other hand, water-soluble vitamins (with the exception of vitamin $B_{12}$) are not stored in appreciable amounts and are required more or less daily in the diet. Most water-soluble vitamins are essential components of enzymes.

The sources, functions, and deficiency diseases of the vitamins are listed in Table 11–4.

## Bulk Minerals

The bulk minerals (see Table 11–2) are required daily in relatively large quantities. They exist in foods and water primarily as ions, and in that form they

TABLE 11–5 Bulk Minerals Required by Humans*

| Mineral (ionic form) | Major functions |
|---|---|
| Sodium ($Na^+$) | Principal cation in extracellular fluids. Important in the propagation of action potentials in excitable cells. |
| Potassium ($K^+$) | Principal intracellular cation. Important in the propagation of action potentials in excitable cells. Necessary for the activity of some enzymes. |
| Calcium ($Ca^{++}$) | Composes, with phosphate, the mineral of bones and teeth. Necessary for muscular contraction and for the activity of some enzymes. Involved in the control of many cellular processes. |
| Phosphorus ($HPO_4^=$, $H_2PO_4^-$, $PO_4^{\equiv}$) | Intracellular and extracellular buffer. Constituent of phospholipids in cell membranes and of high energy phosphate compounds used in biosynthesis and energy transfer. |
| Magnesium ($Mg^{++}$) | Required for the activity of many enzymes and for normal protein synthesis. |
| Chlorine ($Cl^-$) | Principal intracellular and extracellular anion. |
| Sulfur ($SO_4^=$) | Constituent of the amino acids methionine and cysteine, of some high energy compounds, and of some vitamins. |

*All are present in most foods, with the exception of calcium, which is available in dairy products and dark-green leafy vegetables.

are indispensable constituents of body fluids. Their concentrations in blood and other body fluids are precisely regulated, primarily by the kidney. The functions of the bulk minerals are summarized in Table 11–5.

## Trace Elements

At least ten trace elements are required in microgram to milligram amounts. Their functions are listed in Table 11–6. In addition to these ten, seven other trace elements including aluminum, silicon, vanadium, nickel, arsenic, tin, and cadmium are present consistently in human tissues, but their functions, if any, are not known.

TABLE 11–6 Trace Elements Required by Humans

| Element | Function | Food sources |
|---|---|---|
| Iron | Necessary component of hemoglobin (in red blood cells), myoglobin (in muscle cells), and many enzymes. | Liver, egg yolk, red meat, dark-green leafy vegetables, prunes, raisins. |
| Iodine | Necessary for the synthesis of thyroid hormone. | Seafood, iodized table salt. |
| Zinc | Component of many enzymes. | Seafood, meats, legumes, whole grain cereals, nuts. |
| Chromium | Involved in the activation of insulin. | Liver, red meat, nuts. |
| Copper | Component of many enzymes. | Liver, meat, whole grain cereals, nuts; water in copper piping. |
| Fluorine | Increases hardness of teeth, protecting them from decay; bone mineralization. | Seafood, tea, fluoridated water. |
| Manganese | Component of several important enzymes. | Whole grain cereals, legumes, nuts, dark-green leafy vegetables. |
| Molybdenum | Component of several important enzymes. | Legumes, liver, leafy vegetables. |
| Selenium | Component of peroxidase enzyme, which degrades potentially harmful peroxide. | Whole grain cereals, seafood. |
| Cobalt | Component of vitamin $B_{12}$. | Meats, seafood. |

## Dietary Guidelines for Americans

**E**at a variety of foods: To assure yourself an adequate diet, eat a variety of foods daily, including selections of (a) fruits; (b) vegetables; (c) whole grain and enriched breads, cereals, and grain products; (d) milk, cheese, and yogurt; (e) meats, poultry, fish, and eggs; and (f) legumes (dry peas and beans).

*Maintain ideal weight:* To improve eating habits, eat slowly, prepare smaller portions, and avoid "seconds." To lose weight, increase physical activity, eat less fat and fatty foods, eat less sugar and sweets, and avoid too much alcohol.

*Avoid too much fat, saturated fat, and cholesterol:* To avoid too much fat, saturated fat, and cholesterol choose lean meat, fish, poultry, dry peas and beans as your protein sources. Moderate your use of eggs and organ meats (such as liver). Limit your intake of butter, cream, hydrogenated margarines, shortenings, and coconut oil, and foods made from such products. Trim excess fat off meats. Broil, bake, or boil rather than fry. Read labels carefully to determine both amount and types of fat contained in foods.

*Eat foods with adequate starch and fiber:* To eat more complex carbohydrates daily substitute starches for fats and sugars. Select foods that are good sources of fiber and starch, such as whole grain breads and cereals, fruits and vegetables, beans, peas, and nuts.

*Avoid too much sugar:* To avoid excessive sugars use less of all sugars, including white sugar, brown sugar, raw sugar, honey, and syrups. Eat less of foods containing these sugars, such as candy, soft drinks, ice cream, cakes, and cookies. Select fresh fruits canned without sugar or light syrup rather than heavy syrup. Read food labels for clues on sugar content—if the names sucrose, glucose, maltose, dextrose, lactose, fructose, or syrups appear first, then there is a large amount of sugar. Remember, how often you eat sugar is as important as how much sugar you eat.

*Avoid too much sodium:* To avoid too much sodium learn to enjoy the unsalted flavors of foods. Cook with only small amounts of added salt. Add little or no salt to food at the table. Limit your intake of salty foods, such as potato chips, pretzels, salted nuts, and popcorn, condiments (soy sauce, steak sauce, garlic salt), cheese, pickled foods, and cured meats. Read food labels carefully to determine the amounts of sodium in processed foods and snack items.

*If you drink alcohol, do so in moderation.*

### Dietary Fiber

Dietary fiber, primarily cellulose and other nondigestible components of plant tissues, is not an essential nutrient, but it is necessary for optimal health of the large intestine. Because it is indigestible and retains water, fiber contributes to the formation of soft, bulky feces, which move rapidly through the large intestine. In addition, fiber may absorb toxic byproducts of bacterial metabolism as well as other potentially dangerous environmental chemicals, and thus hinder their absorption into the bloodstream. Because regions of the world with high fiber diets have a lowered rate of colon and rectal cancers and of diverticulosis, nutritionists have increasingly stressed the importance of fiber in the diet. This, in turn, has generated some consumer interest in whole grain breads and breakfast cereals.

### The American Diet

During the first half of this century, the diet of Americans emphasized meats, eggs, and dairy products with cereal grains, vegetables, and fruit providing fiber and additional energy. Our high consumption of animal protein was once assumed to be nutritionally ideal. In the last two decades, however, nutritionists have increasingly questioned this assumption, based upon the realization that meats, eggs, and dairy products are high in saturated fats and cholesterol, which have been linked to the development of atherosclerosis. Although our consumption of animal fats has in fact declined steadily in the last two decades—in large part because of fears of cardiovascular disease—our overall fat consumption, including unsaturated vegetable oils, has been increasing. Accompanying increased fat consumption has been a decline in the consumption of whole grain cereals, vegetables, and fruit; this has resulted in a significant decrease in dietary fiber, a situation that may contribute to the formation of colon and rectal cancers. In addition, sugar consumption in the form of candy, soft drinks, and pastries has been increasing dramatically. These sweet foods, along with other "junk" foods and alcoholic beverages, are high in calories but low in nutrients, and excessive consumption of these leads to overweight and obesity. The ubiquitous "fast" foods, which are becoming an integral part of the American diet, tend to be too high in animal fats, sugar, and calories, and too low in fiber and iron, reflecting the nutritional inadequacies in our diet as a whole.

## Dietary Guidelines for Americans

In 1980, a joint committee of the U.S. Department of Agriculture and the Department of Health, Education, and Welfare, expressed concern about the American diet and published dietary guidelines in a booklet entitled *Nutrition and Your Health.* These guidelines are summarized in *Focus: Dietary Guidelines for Americans.*

## Study Questions

1. Diagram the general structure of the digestive tube, labeling the layers that compose it.
2. List the parts of the digestive tube in descending order.
3. In what sense are the contents of the digestive tube considered to lie outside the body?
4. Indicate how the structure of the peritoneum and peritoneal cavity compares with that of similar membranes and cavities that surround the lungs and the heart.
5. **a)** Diagram the structure of a tooth, labeling pertinent structures.
   **b)** What is the function of odontoblasts?
   **c)** Why are odontoblasts important to the health of a tooth?
6. **a)** Identify the three salivary glands.
   **b)** Indicate the functions of their secretions.
7. Summarize the events that occur during swallowing.
8. Why is the stomach not digested by the strong hydrochloric acid solution it contains?
9. Describe the three structural features of the small intestine that increase the surface area available for absorption.
10. List the functions of the digestive enzymes that are secreted by the small intestine, the pancreas, and the liver.
11. Describe the process by which lipids are absorbed by the small intestine.
12. Describe the cause and symptoms of lactose intolerance.
13. Describe the structure and function of the large intestine.
14. Indicate how the structure of a liver lobule relates to its function.
15. List eight important functions of the liver.
16. Trace the path of bile, beginning with its secretion from the liver.
17. **a)** List the seven classes of essential nutrients.
    **b)** What other substance not listed in a) is indispensable to life?
18. Explain how weight gain occurs.
19. What is the function of most water-soluble vitamins?
20. Why is dietary fiber important to health?

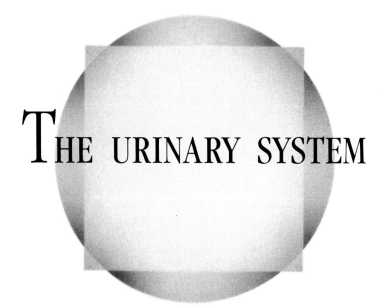

# THE URINARY SYSTEM

Cellular metabolism results in the production of a variety of waste products that diffuse from the cells into the tissue fluid and from there into the bloodstream. One of these waste products is carbon dioxide which, as we discussed in Chapter 10, is eliminated from the blood at the lungs. Other metabolic waste products, such as nitrogen-containing compounds resulting from the degradation of proteins and nucleic acids, are removed from the blood by the kidneys, the most important organs of the urinary system.

In addition to cleansing the blood of various waste products, the kidneys have other important functions as well. They maintain the water balance of the body by controlling the amount of water lost in the urine. For example, if the body contains more water than it needs, the excess is excreted by the kidneys in the urine; on the other hand, dehydration of the tissues will cause the kidneys to conserve water by excreting into the urine only as much water as is necessary to eliminate soluble waste products. Similarly, the kidneys can vary the kinds and amounts of inorganic ions that they excrete, thus maintaining the proper ionic balance of the blood and other fluids of the body. The kidneys also play a central role in maintaining the acid-base balance of the body's fluids by regulating the excretion of hydrogen ions and of compounds that buffer their effects (see Chapter 2). The net result of these activities is that the kidneys—and the urinary system of which the kidneys are a part—maintain a hospitable fluid environment for the cells of the body. The kidneys are thus intimately involved with the maintenance of homeostasis—the constancy of the body's internal environment.

## OVERVIEW OF STRUCTURE AND FUNCTION

The components of the urinary system are shown in Figure 12–1. The urinary system consists of paired organs, the **kidneys,** which filter the blood and produce urine. Two tubes called *ureters,* one from each kidney, convey urine to the urinary *bladder.* There, urine is stored prior to excretion through another tube, the *urethra,* which extends from the bladder to an external orifice. It is important to note that the kidneys perform all of the specialized work of the urinary system, while the urinary bladder and the tubes that run to and from it function merely to store and convey urine.

The kidneys act as highly selective filtering devices. Although many of the details of kidney function are complex and not yet fully understood, the basic mechanism is simple. As blood under pressure passes through capillaries in the kidneys, large amounts of tissue fluid are produced. You will recall that tissue fluid, derived from blood plasma, consists of water and virtually all dissolved, low molecular weight substances, but it does not contain blood cells or plasma proteins, since these are too large to pass through the intercellular spaces in the capillary walls. The kidney processes the tissue fluid by retaining useful substances, which are returned to the bloodstream, and eliminating waste products and other useless or toxic substances as urine. The operation of the kidney has been described as similar to cleaning a room by removing everything from it (the tissue fluid), and then returning to the room only those things you wish to keep (useful substances).

## NEPHRONS: FUNCTIONAL UNITS OF THE KIDNEYS

Each kidney contains a million or more microscopic structural and functional units called **nephrons,** which serve as miniature filtering devices. Nephrons form the bulk of the tissue of the kidney; they are packed closely together in connective tissue through which blood vessels, nerves, and lymphatic vessels course. Urine, a watery solution of inorganic ions, nitrogenous wastes, and

FIGURE 12–1 Organs of the urinary system. The adrenal glands, located on top of the kidneys, are not part of the urinary system.

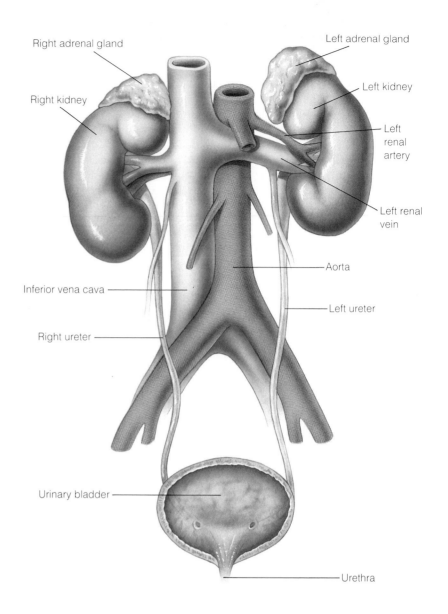

Right adrenal gland

Left adrenal gland

Right kidney

Left kidney

Left renal artery

Left renal vein

Aorta

Inferior vena cava

Left ureter

Right ureter

Urinary bladder

Urethra

various other useless or toxic substances, is the collective product of all the nephrons of the kidneys.

## Microscopic Structure

The structure of a nephron is shown in Figure 12–2. It consists of two parts: a tuft of capillaries, called the **glomerulus,** pushed, as it were, into the blind bulbous end of a hollow tubular structure, the **renal tubule.** Starting from the region near the glomerulus, the four parts of the renal tubule are (1) Bowman's capsule, (2) the proximal convoluted tubule, (3) the loop of Henle, and (4) the distal convoluted tubule. The renal tubule is composed of a single layer of epithelial cells with an associated basement membrane; the structure and function of the epithelial cells forming the tubule vary in different regions of the tubule.

The part of the renal tubule that surrounds the glomerulus is called **Bowman's capsule.** Because it is formed by the invagination of the blind end of the renal tubule, Bowman's capsule is, of course, a double-walled structure. The inner wall, forming the concave surface of the capsule, covers and closely apposes the endothelial membranes that form the walls of the glomerular capillaries. The outer wall of Bowman's capsule is continuous with the next region of the renal tubule, called the **proximal convoluted tubule.** The inner and outer walls of Bowman's capsule enclose a space that connects with the lumen of the rest of the tubule.

The proximal convoluted tubule, as its name suggests, winds tortuously as it leaves Bowman's capsule and then straightens and descends to form a thin, U-shaped region known as the **loop of Henle.** As the loop of Henle ascends from the hairpin turn, it joins a thickened segment of tubule, at first straight and then convoluted, called the **distal convoluted tubule.** The distal convoluted tubule is considered the last part of the nephron unit.

At its end, the distal convoluted tubule merges with a straight, descending **collecting tubule,** into which the distal convoluted tubules of many nephrons empty. Collecting tubules merge into still larger vessels called **papillary ducts.** The hollow tubular portions of the nephron (renal tubule), together with the collecting tubules and papillary ducts, serve to convey tissue fluid through the kidney while processing it into urine.

## Blood Supply to Nephrons

In the upper abdominal region, the aorta gives rise to right and left *renal arteries,* supplying the right and left kidneys, respectively (see Figure 12–1). Within each kidney, the renal artery branches extensively to form a large number of arterioles, termed *afferent arterioles* (*afferent* = leading to), one of which supplies the capillary beds of each glomerulus with blood (see Figure 12–2). As they leave the glomerulus, the glomerular capillaries merge together forming another arteriole, called the *efferent arteriole* (*efferent* = leading from), which has a diameter less than that of the afferent arteriole. The efferent arteriole subsequently breaks up into a second set of capillaries, the *peritubular capillaries,* which form an extensive network of vessels that surround the tubules of one or more nephrons. The peritubular capillaries merge to form venules; and these in turn merge into larger veins, finally to form the *renal vein.* Right and left renal veins leave the right and left kidney, respectively, carrying blood to the inferior vena cava.

The anatomical arrangement of the blood vessels that supply the nephrons is unique. The capillary network of the glomerulus does not drain directly into

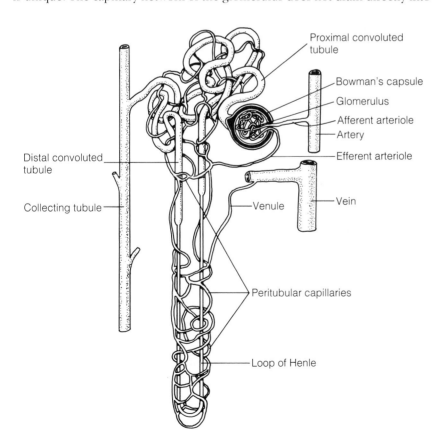

Proximal convoluted tubule

Bowman's capsule

Glomerulus

Afferent arteriole

Artery

Efferent arteriole

Distal convoluted tubule

Collecting tubule

Venule

Vein

Peritubular capillaries

Loop of Henle

FIGURE 12–2 Structure of a nephron with surrounding blood vessels. The nephron consists of the glomerulus and the renal tubule; the renal tubule is composed, in turn, of Bowman's capsule, the proximal convoluted tubule, the loop of Henle, and the distal convoluted tubule.

A capillary tuft of a glomerulus.

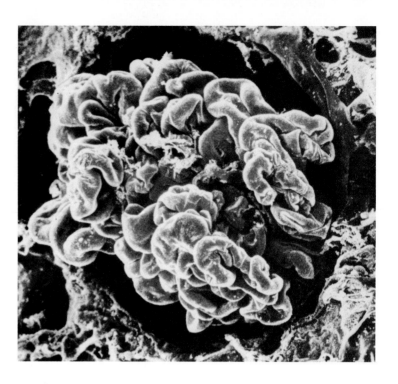

a venule as do capillaries in other parts of the body. Instead, it drains into an arteriole (the efferent arteriole). Because the diameter of the efferent arteriole is smaller than the diameter of the afferent one, and because arterioles are not as distensible as venules, the blood pressure within the glomerular capillaries is considerably higher than in other capillaries of the body. As will become apparent, the increased pressure within the glomerular capillaries favors the production of large amounts of tissue fluid that is cleansed of wastes and harmful substances within the kidneys.

## THE KIDNEYS

The two kidneys lie on either side of the vertebral column, surrounded by adipose connective tissue and embedded high in the posterior abdominal wall directly beneath the diaphragm. External to the adipose tissue is a supporting layer of tough fibrous tissue, the *renal fascia.* The kidneys are held loosely in place by the renal fascia and the renal arteries and veins.

The kidneys are shaped like kidney beans. The indented region of the surface is directed toward the vertebral column and forms the *hilus,* through which blood vessels and nerves enter and leave the kidney and from which the ureter descends to the bladder. The surface of each kidney is covered by a strong, smooth fibrous capsule.

### Internal Structure

Details of the internal structure of the kidney are shown in Figure 12–3. The dark, more or less solid tissue of the kidneys is divided into two zones: the **renal cortex** and the **renal medulla.** The outermost zone, the renal cortex, is granular in appearance because it contains the glomeruli and most of the proximal and distal tubules cut in various cross and oblique sections. The medulla consists of striated, cone-shaped structures called *renal pyramids,* which vary in number from 6 to 18. The renal pyramids of the medulla contain the straight loops of Henle as well as the collecting tubules and papillary ducts, giving the pyramids a raylike appearance as they are cut in longitudinal sections. The apex of each of the renal pyramids, called a *renal papilla,* contains a variable number of openings of the papillary ducts. Between adjacent renal pyramids are areas of typical cortical substance, called *renal columns.*

Within the indented region of the kidney, the upper end of the ureter is expanded into a funnel-shaped cavity called the *renal pelvis.* Hollow, tubelike extensions of the renal pelvis, the *renal calyces* (sing: *calyx*), surround the end of each renal papilla. The renal calyces serve as collecting chambers that receive urine from the papillary ducts; from the calyces, urine passes into the renal pelvis and down the ureter to the bladder.

The arrangement of several nephrons within the kidney is shown diagrammatically in Figure 12–4. It is important to emphasize that the structure of the kidney is a consequence of the specific arrangement of nephrons and associated collecting vessels within the kidney. The arrangement of these structures is more than a matter of anatomical interest, for it determines several important aspects of kidney function, as described shortly.

## Kidney Function

As mentioned earlier, the nephrons of the kidneys produce large amounts of tissue fluid from the blood. This is then processed so that useful substances are retained by the body and waste products are eliminated.

## Glomerular Filtration

The first step in the formation of urine by the kidneys is the production of tissue fluid. This process occurs at the glomeruli and is referred to as **glomerular filtration.**

Blood passing through the glomerular capillaries is separated from the space within Bowman's capsule by two membranes—the endothelial membrane that forms the wall of the glomerular capillaries and the overlying

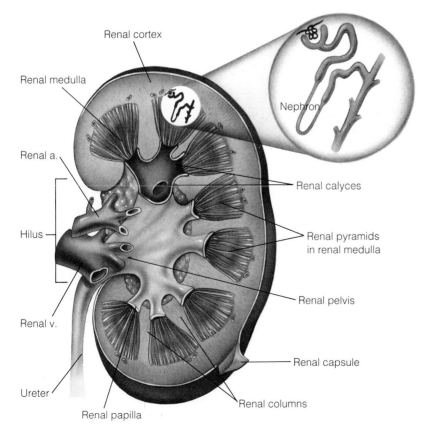

FIGURE 12–3 Structure of the kidney.

Renal cortex

Renal medulla

Nephron

Renal a.

Renal calyces

Hilus

Renal pyramids in renal medulla

Renal v.

Renal pelvis

Renal capsule

Ureter

Renal papilla

Renal columns

FIGURE 12-4 Diagram showing how nephrons are arranged within the kidney. A single nephron is also shown in Figure 12-3 in order to indicate its relationship to other structures in the kidney.

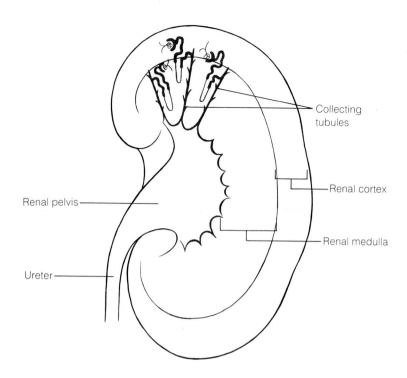

Collecting tubules

Renal cortex

Renal medulla

Renal pelvis

Ureter

epithelial membrane of the inner wall of Bowman's capsule (see Figure 12-2). Although the cellular structure of this double membrane is complex, it forms a barrier possessing permeability characteristics similar to that of other capillaries in the body. Because the blood pressure within the glomerular capillaries is relatively high and the pressure of fluid within Bowman's capsule is low, water and small solutes are pushed out of the blood stream, through the double membrane and into the space within Bowman's capsule. This space, of course, is the beginning of the renal tubule. Particles such as red blood cells and most plasma proteins are too large to pass through the double membrane and remain in the bloodstream as the blood passes out of the glomerulus. The fluid that enters Bowman's capsule after being filtered from the blood plasma is referred to as *glomerular filtrate*. The glomerular filtrate is virtually identical in composition to tissue fluid that is extruded through capillaries in other parts of the body.

Although the rate of glomerular filtration varies somewhat with physiological conditions, it results in the production of enormous volumes of filtrate—between 170 and 200 liters per day on the average. From the amount of glomerular filtrate produced, it can be calculated that about one-fifth of the blood plasma entering the glomeruli is filtered through them, and the rest remains in the blood stream. Since the total blood plasma volume of an average individual is something on the order of three liters, the production of 170 to 200 liters of filtrate also means that the entire plasma volume is filtered through the kidneys about 60 times a day.

The high output of glomerular filtrate is due primarily to two factors. First, the flow of blood to the kidneys—and hence to the glomeruli—is extremely high, amounting to between one-fifth and one-quarter of the total cardiac output. Thus the two kidneys together receive about 1.1 liters of blood per minute, compared to a total cardiac output of about 5 liters per minute. For a body at rest, the flow of blood to the two kidneys is greater than that to any other organ. Second, the total filtration surface of all glomeruli is very large, estimated at well over one square meter. It is the ability to filter such huge volumes of plasma that allows the kidneys to purify the body fluids and to regulate their composition so precisely.

The average individual excretes between one and two liters of urine in 24 hours. Consequently, it is apparent that of the large volume of glomerular filtrate that enters the renal tubule, only a small proportion (about 1 percent) leaves the kidneys as urine. The rest is conserved by the body.

### Processing of Glomerular Filtrate into Urine

*Tubular Reabsorption.* As the glomerular filtrate passes along the renal tubule, useful substances are reabsorbed by cells of the tubular epithelium. The reabsorbed substances then pass into the tissue fluid that surrounds the outer surface of the renal tubule and return to the circulation by diffusing into the blood as it flows through the peritubular capillaries (Figure 12–5).

The reabsorption of substances by the tubule cells occurs by a variety of mechanisms. Some substances are reabsorbed by passive transport processes such as simple diffusion, facilitated diffusion, and osmosis. Others are reabsorbed by complex active transport mechanisms that require energy. (See Chapter 3 for a review of transport mechanisms.)

Although tubular reabsorption occurs to some extent in all parts of the renal tubule, a great deal takes place in the proximal convoluted tubule. Cells that form the proximal convoluted tubule are particularly well suited for reabsorption because of the presence of microvilli on surfaces that face the lumen of the tubule. Like the microvilli of cells lining the small intestine, these greatly expand the surface area available for reabsorption. It has been estimated that the total surface area of all the proximal convoluted tubules of both kidneys is between 50 and 60 square meters.

As glomerular filtrate passes through the proximal convoluted tubule, almost 90 percent of the sodium ions ($Na^+$) and the water are reabsorbed. The reabsorption of $Na^+$ occurs by active transport, using specific enzymes that are present in the cells of the proximal convoluted tubule and that function as sodium pumps (see Chapter 3). As $Na^+$ is pumped through the cells of the proximal convoluted tubule and into the surrounding tissue fluid, negatively charged chloride ($Cl^-$) and bicarbonate ($HCO_3^-$) ions follow by electrical attraction. The increase in the concentration of ions in the tissue fluid brings about a corresponding movement of water by osmosis from the renal tubule into the tissue fluid. Thus, although the reabsorption of $Na^+$ in the proximal tubule, like all active transport, requires energy, the reabsorption of water does not.

In addition to reabsorbing a large proportion of the $Na^+$ and the water in the glomerular filtrate, the proximal convoluted tubule is responsible for most

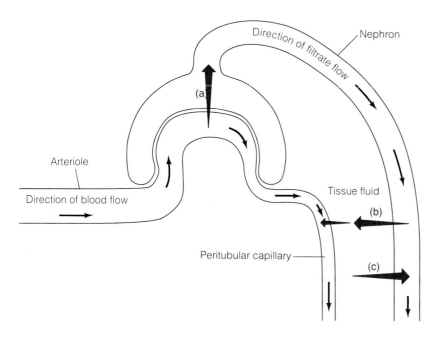

FIGURE 12–5 A highly diagrammatic summary of kidney function. (a) Glomerular filtration; (b) tubular resorption; and (c) tubular secretion.

of the reabsorption of other useful solutes. These include ions such as potassium ($K^+$) and calcium ($Ca^{++}$); important nutrients such as glucose, amino acids, and vitamins; and any small plasma proteins that may have leaked through the glomerulus. Most substances are reabsorbed by active transport, although one exception is the reabsorption of proteins, which occurs by endocytosis (see Chapter 3). As in the case of $Na^+$ reabsorption mentioned above, the accumulation of solutes in the tissue fluid acts to pull additional water out of the tubule by osmosis. From the tissue fluid, solutes diffuse into the peritubular capillaries, followed by water, which enters by osmosis.

Different substances are reabsorbed from the proximal tubule at different rates and consequently to varying degrees. Some substances are reabsorbed completely under normal dietary conditions. One of these is glucose. However, if large amounts of candy or other glucose-containing foods are eaten, the concentration of glucose in blood plasma—and hence in the glomerular filtrate—may become so high that the upper limit for reabsorption is exceeded. In this event, some glucose will appear in the urine, a condition called *glucosuria.* Although glucosuria occurs normally in response to a meal very high in glucose, fasting glucosuria is abnormal and is a symptom of *diabetes mellitus* (see Chapter 16). In this disease, plasma glucose concentrations are not properly regulated by hormones and reach such high levels that the renal tubules, although normal in function, are incapable of reabsorbing all of the glucose. The presence of excess glucose in the renal tubules also contributes to the osmotic retention of water in the tubules, leading to tissue dehydration and thirst.

Urea, the chief nitrogenous waste product in the blood, and hence in the glomerular filtrate, is an example of a substance that must be eliminated from the body. As water and useful solutes are reabsorbed from the glomerular filtrate, the concentration of urea within the renal tubules increases. Eventually, the concentration of urea becomes high enough that some begins to diffuse through the tubular cells and into the surrounding tissue fluid, from where it diffuses into the peritubular capillaries, returning to the bloodstream. Although the reabsorption of urea may seem like an odd occurrence since urea is a useless and ultimately toxic waste product, the renal tubules have no mechanism for retaining all of the filtered urea within them. However, only 40 to 60 percent of the urea in the glomerular filtrate is normally reabsorbed in this manner, while the rest is excreted in the urine. The result is a net removal of urea from the blood. The efficiency of reabsorption of various substances by the kidney is shown in Table 12-1.

*Tubular Secretion.*    The renal tubule is also capable of **tubular secretion,** a process which is the reverse of tubular reabsorption (see Figure 12-5). Tubular secretion results in the transport of substances from the tissue fluid surrounding the tubules to the lumen of the tubules. Tubular secretion may occur by passive or active transport mechanisms, and it constitutes a second way that substances can enter urine. Substances that are transported by tubular secretion are usually derived from blood within the peritubular capillaries, but in some instances, substances reabsorbed in one part of the tubule are secreted in another. Although tubular secretion occurs in both the proximal and distal convoluted tubules, the latter are the major site for this activity.

TABLE 12-1 Reabsorption by the Kidney

| Substance | Amount filtered in 24 hours | Percent reabsorbed |
|---|---|---|
| Glucose | 140 g | 100 |
| Sodium | 540 g | 99 |
| Potassium | 28 g | 86 |
| Bicarbonate | 300 g | 100 |
| Urea | 53 g | 52 |
| Water | 180 L | 99 |

Tubular secretion is particularly important in the regulation of the acid-base balance of the body (see Chapter 2). Since many acidic substances are by-products of normal metabolism, hydrogen ions ($H^+$) derived from these substances tend to accumulate in the body's fluids. The effect of these hydrogen ions must be counteracted in order to prevent a fatal rise in acidity. Buffers in the blood and other fluids play an important role in this respect, as does the respiratory system in which the elimination of carbon dioxide is coupled to a reduction in $H^+$ levels. In addition to these mechanisms, excess hydrogen ions are added to the urine by tubular secretion. In order to prevent a resultant increase in urine acidity, which might harm the tubules or collecting structures, some tubular cells produce ammonia ($NH_3$), which diffuses into the renal tubule and combines with $H^+$ to form ammonium ion ($NH_4^+$).

Tubular secretion is also involved in the regulation of $K^+$ concentrations in the extracellular fluids. Since potassium ions play a central role in the excitability of nerve and muscle, $K^+$ concentrations must be regulated at precise levels in order to prevent abnormalities in the functioning of these tissues. As mentioned above, virtually all filtered $K^+$ is reabsorbed in the proximal tubule. Consequently, the regulation of $K^+$ levels occurs primarily in the distal convoluted tubule, which is capable of eliminating excess $K^+$ by tubular secretion.

The renal tubule is also able to secrete a large number of foreign chemicals, such as penicillin and certain dyes. In fact, the ability of a healthy tubule to secrete diagnostic dyes is the basis of several tests of kidney function.

*Concentration of Urine.*   The fluid leaving the proximal tubule and entering the loop of Henle is roughly iso-osmotic to the blood and has about one-eighth the volume of the glomerular filtrate, thus amounting to some 25 liters per day. Before excretion as urine, this tubular fluid must somehow be concentrated to a hyperosmotic fluid with a volume of one to two liters.

The concentration of urine depends upon events that occur in the loop of Henle, but it does not actually take place in the loop of Henle. Instead, some water is selectively reabsorbed from the tubular fluid in the distal convoluted tubule, but most is reabsorbed as fluid passes down the collecting vessels. The loops of Henle function to create a hyperosmotic tissue fluid in the renal medulla—the region of the kidney that contains the loops of Henle and the collecting tubules (see Figure 12–4). As the collecting tubules pass through this fluid, water is drawn out of them by osmosis. This water eventually enters the peritubular capillaries, thus returning to the circulation.

The creation of a hyperosmotic tissue fluid by the loops of Henle is summarized in Figure 12–6. The process is best understood by examining first the events that occur in the ascending limb of the loop of Henle (the portion of the loop of Henle that connects with the distal convoluted tubule). The cells that form the wall of the ascending limb actively transport chloride ion ($Cl^-$) from the tubular fluid into the surrounding tissue fluid. Because $Na^+$ follows the $Cl^-$ passively as a result of electrical attraction, we can think of the ascending limb as transporting sodium chloride ($NaCl$) into the tissue fluid. In addition, the ascending limb is largely impermeable to water, so that the transport of $NaCl$ is not followed by the osmotic movement of water. Because $Na^+$ is pulled out all along the course of the ascending loop, and water is retained inside, the tubular fluid becomes progressively hypo-osmotic, relative to the filtrate, as it ascends toward the distal tubule. Concomitantly, the tissue fluid in the surrounding area becomes hyperosmotic.

The descending limb of the loop of Henle is permeable to water and allows the passive diffusion of $NaCl$, but it does not actively transport $NaCl$. As tubular fluid moves down the descending limb, the extra $NaCl$ in the surrounding tissue fluid withdraws water by osmosis. At the same time, $NaCl$ diffuses into the descending limb from the more concentrated tissue fluid. As a consequence, the fluid in the descending limb becomes increasingly hyper-osmotic as it nears the bottom of the loop. However, as this fluid passes into the ascending limb, $NaCl$ begins to be pumped out, as described earlier.

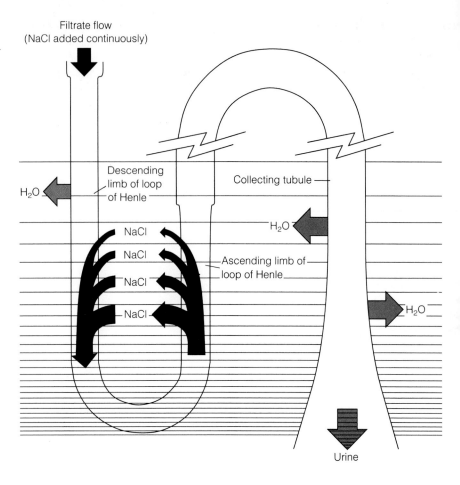

FIGURE 12-6 Events occurring at the loop of Henle and in the collecting tubule. The horizontal lines depict the NaCl gradient; the closer together the lines are drawn, the greater the concentration of NaCl.

Filtrate flow
(NaCl added continuously)

H₂O

Descending limb of loop of Henle

Collecting tubule

H₂O

NaCl

NaCl

Ascending limb of loop of Henle

NaCl

NaCl

H₂O

Urine

The net result of these activities is the establishment of a concentration gradient of NaCl in the tissue fluid surrounding the loops of Henle (in the medulla of the kidney). This gradient is such that the highest concentrations of NaCl are near the bottom of the loops of Henle, the lowest levels are at the top, and a gradual change in concentration occurs between these two extremes. The opposing flow of fluid in the ascending and descending limbs of the loop of Henle is called a countercurrent flow. This, in conjunction with the characteristic permeabilities of the ascending and descending limbs to NaCl and water, results in a multiplier effect that allows the buildup of hyperosmotic concentrations of NaCl in the tissue fluid of the medulla. For these reasons, the activities occurring in the loop of Henle are termed a *countercurrent multiplier* system.

Since the collecting tubules pass through the hyperosmotic tissue fluid generated by the loops of Henle, water is drawn by osmosis from the fluid within them, thus concentrating the urine (see Figure 12-6). The hyperosmotic gradient can be maintained, even in the face of dilution by water entering the tissue fluid from the collecting tubules, because NaCl is constantly added to the loop of Henle as new filtrate flows into it.

### Control of Kidney Function

***Regulation by Nerves.*** Stimulation of autonomic nerves that supply the renal arterioles causes severe constriction of these vessels, greatly reducing blood flow through the glomeruli, lowering the rate of glomerular filtration, and decreasing kidney function. Since the kidneys normally receive more than 20 percent of the resting cardiac output, stimulation of these nerves allows large amounts of blood to be shunted to other organs of the body such as the brain, heart, and skeletal muscles during stressful situations requiring mobilization of body resources (the "fight or flight" mechanism).

***Renin, Angiotensin, and Aldosterone.*** Embedded in the walls of the afferent arterioles are specialized cells called *juxtaglomerular cells* (JG cells). The juxtaglomerular cells are involved in the maintenance of blood pressure—a function crucial to the operation of the kidney, since adequate blood pressure is essential for glomerular filtration. JG cells act as a type of baroreceptor that, if not stretched by normal blood pressure, secrete an enzyme called *renin.* Renin catalyzes the cleavage of a large plasma protein, *angiotensinogen,* to yield a small molecule *angiotensin I.* Angiotensin I, which is inactive, is converted into *angiotensin II* by further enzymatic cleavage (Figure 12–7).

Angiotensin II elevates blood pressure in two ways. One, it causes contraction of smooth muscle in the walls of arterioles, leading to constriction of arterioles throughout the body. Because all arterioles constrict, rather than just the afferent arterioles as when the autonomic nerves are stimulated, systemic blood pressure increases, resulting in increased glomerular filtration. Secondly, angiotensin II stimulates the cortex of the adrenal gland to secrete a hormone, *aldosterone.* Aldosterone acts on the renal tubules to stimulate sodium reabsorption, which has the effect of increasing the reabsorption of water by osmosis. The net effect is to increase the volume of blood within the circulatory system and hence to elevate blood pressure.

The secretion of renin by the JG cells is under feedback control. Thus, as the blood pressure within the afferent arterioles rises to acceptable levels, the secretion of renin by the JG cells is inhibited. Normally, these responses are important in the overall regulation of blood pressure, but the system may function inappropriately in certain diseased states. For example, the thickening of arterial walls due to atherosclerosis may reduce blood flow and pressure to the kidneys, causing an essentially continuous secretion of renin and a consequent abnormal elevation of blood pressure. Such a condition is termed *renal hypertension.* In *congestive heart failure,* the inability of a failing heart to maintain adequate blood pressure may lead to increased aldosterone secretion, resulting in high levels of $Na^+$ reabsorption and a greatly increased blood volume. The increased blood volume causes an elevation of blood pressure within capillaries throughout the body, leading to excessive production of tissue fluid and resultant swelling (edema). In another disease, *Addison's disease,* there is a deficiency in aldosterone secretion that results in excessive loss of $Na^+$ and water from the body, causing fatal dehydration if untreated.

***Antidiuretic Hormone.*** The amount of water that is withdrawn from the tubular fluid as it passes through the distal convoluted tubules and the

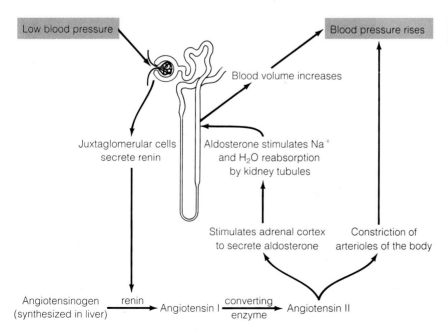

FIGURE 12–7 The regulation of blood pressure by the kidney.

collecting tubules depends upon the permeability of these structures to water. This property is regulated by **antidiuretic hormone (ADH),** also known as *vasopressin*. (*Diuresis* refers to an increased excretion of urine; antidiuretic hormone counteracts diuresis.) ADH is secreted from the posterior lobe of the pituitary gland (Chapter 16) in response to a high solute concentration in the blood; that is, when the body, and hence the blood, loses too much water. ADH acts by increasing the permeability of the distal convoluted tubules and the collecting tubules to water, and so accelerating the rate of water reabsorption. This conserves water for the body because the amount of water lost in the urine is reduced. On the other hand, ADH secretion is inhibited when the blood becomes too dilute because of the retention of too much water. Lack of ADH reduces the water permeability of the distal tubules and collecting tubules, minimizing water reabsorption and resulting in the excretion of excess water in the urine.

Under normal situations, ADH secretion closely regulates the water balance of the body. However, certain drugs may interfere with this activity. For example, ethyl alcohol inhibits ADH release, causing diuresis with loss of inappropriate amounts of water. The resultant dehydration accounts for the thirst experienced the morning after overindulgence in alcohol. In addition to drugs, certain diseases may affect ADH secretion. In one such disease, called *diabetes insipidus,* ADH secretion is chronically reduced, often due to damage to the pituitary from severe head injuries. Vast quantities of water are lost in the urine, as much as 20 liters in 24 hours. Such a fluid loss results in unquenchable thirst, and patients must drink large volumes of liquids to prevent fatal dehydration. In severe cases, diabetes insipidus is usually treated with a synthetic form of ADH, administered in nose drops or by injection.

The human kidney has a limit beyond which it cannot concentrate urine. Even if no food or water is ingested, a minimum of about 500 mL of urine must be produced each day in order to eliminate ions and toxic waste products that continually build up as a result of metabolic processes occurring within cells. Thus, in times of water deprivation, the water conservation function of the kidney is overridden by the necessity to eliminate waste products and to maintain ionic balance.

### Properties of Urine

Urine is a clear, yellowish solution, usually acidic in pH. On average, 100 mL of urine contains about 2.5 g of urea and 1.2 g of NaCl; these two substances together account for about 90 percent of the solutes in urine. In addition small amounts of a number of other organic waste products and ions are also present. It is important to note that the composition of urine varies, depending upon the types and concentrations of substances that the kidneys must remove from the blood in order to maintain the normal composition, volume, and pH of body fluids.

The composition of urine can be determined clinically in a procedure known as *urinalysis*. The presence of certain abnormal constituents in urine is indicative of various pathological conditions.

## THE EXCRETORY PASSAGES

The excretory passages include the ureters, which convey urine to the bladder; the bladder, which stores urine; and the urethra, which carries urine from the bladder to the outside of the body.

### Ureters

The **ureters,** each about 10 to 12 inches long, carry urine from the pelvis of each kidney to the urinary bladder (see Figure 12–1). They enter the bladder on either side of its posterior wall, passing obliquely through the wall for about an inch, and opening through slitlike apertures into the cavity of the bladder.

Although there are no valves at the openings of the ureters into the bladder, the terminal portions of the ureters become compressed when the bladder begins to distend with urine, preventing the backflow of urine from the bladder.

The walls of the ureters are composed of three coats: a mucous membrane that lines the cavity of the ureters and is continuous with the mucosal lining of the renal pelvis and urinary bladder; a middle muscular coat, consisting of two layers of smooth muscle, one with fibers arranged longitudinally and one, circularly; and an external coat of fibrous connective tissue.

Peristaltic contractions of the smooth muscle layers, sweeping down the ureters at a frequency of several times a minute, squirt urine into the bladder where it is stored prior to elimination.

## Urinary Bladder

The **urinary bladder** is a thick muscular sac lying in the pelvic cavity. Its walls are composed of an inner mucous membrane surrounded by consecutive coats of loose connective tissue, smooth muscle, and serous membrane. The mucous membrane lining is thrown into folds when the bladder is empty, but these are stretched out when the bladder is full (Figure 12–8). In addition, the epithelium of the mucous membrane is of the so-called *transitional* type, in which the cells are more or less cuboidal in shape when the bladder is empty, and flat in shape when the bladder is stretched (see Figure 4–3). The smooth muscle coat, termed the *detrusor muscle,* consists of a layer of circularly disposed fibers sandwiched between two layers of longitudinally arranged ones. In the region where the bladder opens into the urethra, the smooth muscle coat forms the *internal sphincter of the bladder.* Although technically not a separate muscle, contraction of the internal sphincter helps retain urine in the bladder.

## Urethra

In females, the **urethra** has a purely excretory function, while in males it serves as a common passageway for both urine and semen. The female urethra, some one to two inches long, is lined with mucous membrane that is surrounded by a layer of circularly arranged smooth muscle fibers. At its external orifice, striated muscle fibers join the smooth ones to form a *urethral sphincter.* The male urethra consists of a mucous membrane tube covered by connective tissue rich in elastic fibers; some smooth muscle is also present. There are three distinct sections of the male urethra. The first section,

FIGURE 12–8 Scanning electron micrograph showing folds in the inner lining of the undistended bladder.

beginning at the bladder and running for about an inch, is surrounded by the prostate gland. When the prostate gland is enlarged, as it often is in elderly men, it presses on this section of the urethra and interferes with, or even stops, the flow of urine. The next section of the male urethra, somewhat shorter than the first, extends from the prostate gland through the pelvic floor of the body wall. The smooth muscle in this region is surrounded by striated muscle, forming the urethral sphincter. The last region, five to six inches in length, lies entirely within the penis.

## Urination (Micturition)

When the adult bladder contains 200 to 300 mL of urine, stretch receptors in the bladder wall, stimulated by distension, send impulses through the spinal cord to a center in the base of the brain. These impulses arouse a desire to empty the bladder. The process of urination is initiated by removal of voluntary control; that is, by voluntary relaxation of the urethral sphincter and of the *pelvic diaphragm* formed by skeletal muscles that fill the outlet of the pelvis and help support some of the pelvic organs. Relaxation of these muscles allows the neck of the bladder to move downward, further stretching the walls of the bladder. Strengthened impulses from the stretch receptors reach the spinal cord, where they stimulate autonomic nerves to the detrusor muscle of the bladder. This stimulation leads to the reflex contraction of the detrusor muscle, accompanied by relaxation of the internal sphincter of the bladder. Because removal of voluntary control is necessary for the initiation of urination, it is possible to inhibit urination for a considerable period of time after the initial urge to urinate. However, when the amount of urine in the bladder increases to about a pint, autonomic stimulation of the detrusor muscle will eventually override voluntary inhibition, and the bladder will empty.

In infants, urination occurs by a reflex pathway not involving higher brain centers that allow voluntary control. Filling of the bladder causes a simple reflex contraction of the detrusor muscle and relaxation of the internal sphincter, emptying the bladder.

## COMMON DISEASES AND DISORDERS OF THE URINARY SYSTEM

### Cystitis

**Cystitis,** or inflammation of the bladder, is usually caused by an infection that results when bacteria enter the urethra from the urethral opening and spread to the bladder. Cystitis is common in females because the female urethra is short and the urethral opening is not far from the anus, allowing bacteria that contaminate the skin after a bowel movement to spread relatively easily to the bladder. Cystitis rarely occurs in males unless an enlarged prostate, a bladder tumor, or some other urinary tract abnormality interferes with the normal flow of urine. In these cases, incomplete emptying of the bladder allows small numbers of bacteria that may enter the bladder, and that normally would be washed out in the urine, to flourish in the bladder.

In both sexes, the major symptoms of cystitis are a frequent urge to urinate accompanied by stinging as the urine passes through the urethra. Sometimes the urine is strong smelling and contains blood. Treatment of cystitis involves increased fluid intake in conjunction with antibiotic therapy.

### Pyelonephritis

*Pyelonephritis,* or bacterial infection of the kidneys, may be either acute or chronic. It usually begins as an infection in the bladder, which then spreads to the kidneys through the ureters. Pyelonephritis is characterized by chills, fever, and flank pain, and is treated with antibiotics. Acute pyelonephritis is a reasonably common condition in women because of the increased frequency of bladder infection.

## Glomerulonephritis

*Glomerulonephritis* is a disease that usually follows a throat infection by certain strains of streptococcus. Toxins, produced by the bacteria, convert a chemical component of the basement membranes of the glomerular capsule into another substance that the body's immune system recognizes as foreign. Antibodies, produced against the new substance, react with it and in the process cause damage to the basement membranes. This damage allows red blood cells and plasma proteins to pass through the glomeruli and into the glomerular filtrate and to be excreted in the urine. Because plasma proteins are diminished in the blood, water is not properly drawn by osmosis back into the circulation, resulting in the accumulation of tissue fluid and consequent edema. Many forms of glomerulonephritis require no specific treatment, while others can be treated with steroid drugs to reduce the body's immune reaction.

## Kidney Stones

As the renal tubules concentrate the glomerular filtrate, some substances in the tubular fluid may reach high enough concentrations to precipitate out of solution. These tiny granules, deposited in the renal pelvis, may serve as a nucleus for the deposition of additional material, forming hard, stonelike objects termed *kidney stones.* Kidney stones are usually a result of infections or metabolic disorders that cause the excretion of high levels of certain substances such as calcium or uric acid (as in gout; see Chapter 6).

Kidney stones less than about 5 mm in diameter are often carried into a ureter with the urine and pass into the bladder; from the bladder, they usually pass out of the body without difficulty, since the urethra is of larger diameter than the ureters. Kidney stones that are large enough just to enter the ureter may cause intense pain as they pass through the ureter. Sometimes, a stone may get stuck in the ureter and block the flow of urine. In that case, the stone must be shattered ultrasonically or removed surgically. If a kidney stone does not happen to leave the kidney before it is too large to enter the ureter, it will usually remain harmlessly in the kidney.

## Bladder Stones

Stones sometimes form within the bladder itself and may become large enough to be trapped in the bladder. These must be removed to prevent troublesome symptoms such as an overfrequent urge to urinate, pain during urination, and bloody urine. Bladder stones can be shattered ultrasonically or removed under general anesthesia by grasping them with an instrument called a *cystoscope,* which is inserted through the urethra and into the bladder. Alternatively, the bladder can be opened surgically and the stone removed.

## Tumors of the Urinary System

Cancerous and noncancerous tumors of the bladder and kidneys are quite rare. Because there are usually no clear cut early symptoms of kidney cancer, it is often not discovered until after it has spread to other organs. If it is detected early, kidney cancer is best treated by surgical removal of the organ.

Bladder cancer has been linked to the use of aniline dyes by workers in the textile industry. Cancerous bladder tumors or cancerous sections of the bladder must be removed surgically.

## Kidney Failure

Kidney disease, physical damage to the kidneys, certain drugs, and some hormonal conditions may lead to *kidney failure.* In *acute kidney failure,* which may be caused by such things as acute glomerulonephritis, a sudden drop in blood pressure due to shock, or blockage of the flow of urine through the excretory passages, the kidneys stop functioning abruptly. Treatment of the

A patient undergoes kidney dialysis.

# The Artificial Kidney (Hemodialysis)

The artificial kidney is an apparatus that is used to remove waste products, excess ions, and other potentially dangerous substances from the blood of individuals whose kidneys no longer function. It operates by an extremely simple principle. As shown in the accompanying figure, blood is pumped from one of the patient's arteries through Teflon tubing that is connected to coils of cellophane tubing within the artificial kidney; blood circulates through the cellophane coils and is then returned to a vein. The cellophane tubing is immersed in a fluid that has the same approximate chemical composition and concentrations of substances as normal tissue fluid.

The process of filtering the blood artificially is called *hemodialysis,* because the artificial kidney works on the principle of *dialysis,* the separation of substances in solution by means of their diffusion through a selectively permeable membrane (in this case, cellophane). Because cellophane has permeability characteristics much like those of blood capillaries, it allows ions and other small molecules to diffuse through it, but is impermeable to blood cells and to large molecules such as proteins. As blood circulates through the cellophane tubing, waste products, ions, and other chemicals that are present in higher concentrations within the blood than in the bath fluid will diffuse through the cellophane into the bath fluid in response to the concentration gradient (see Chapter 3). To maintain the concentration gradient that is necessary for diffusion to occur, fresh bath fluid is added to the chamber continuously as used bath fluid, containing waste products and other dialyzed substances, is removed.

An individual with complete lack of kidney function must be dialyzed for an average of about four hours, three times a week. With this treatment, otherwise healthy individuals can lead surprisingly normal lives.

In some patients with kidney failure, a new hemodialysis technique known as *continuous ambulatory peritoneal dialysis* (CAPD) may be used, which is more convenient than conventional dialysis. In CAPD, dialyzing fluid from a plastic bag attached to the body flows through a catheter into the patient's peritoneal cavity. Waste products pass from blood vessels in the abdominal cavity, through the peritoneum, and into the dialyzing fluid. When dialysis is complete, the fluid containing waste products is drained into the plastic bag and discarded.

Before 1972, the artificial kidney was made available only to comparatively young individuals who suffered from kidney failure but had no other significant health problems. In that year, the U.S. Congress extended Medicare coverage to provide federal funds for dialysis or kidney transplants to all patients with kidney failure. As a result of this federal program, the use of the artificial kidney has expanded dramatically from approximately 6000 patients in 1971 to some 60,000 patients in 1985. Many of these patients are terminally ill, mentally incompetent or suffering from a variety of other serious diseases. Use of dialysis for these patients has been criticized by many health professionals, who feel that it merely sustains life without offering any hope for a normal existence.

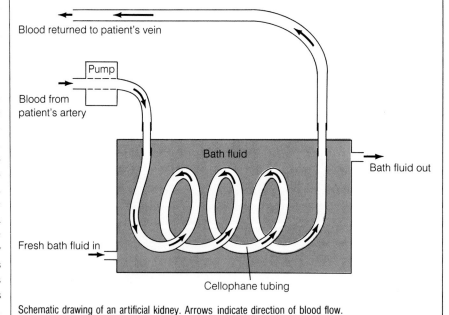

Schematic drawing of an artificial kidney. Arrows indicate direction of blood flow.

underlying cause returns the kidneys to normal function, but use of an *artificial kidney* may be needed to substitute for normal kidney functions during recovery [see *Focus: The Artificial Kidney (Hemodialysis)*].

In chronic kidney failure, repeated mild attacks of inflammation injure and scar the kidneys, gradually reducing their effectiveness. Chronic kidney failure is caused by a number of conditions, including chronic pyelonephritis, chronic glomerulonephritis, and hypertension. As the efficiency of kidney function slowly diminishes, waste products and other chemicals build up in the blood, and an abnormal amount of water is excreted.

*End-stage kidney failure* usually occurs as a result of the progression of chronic kidney failure to a point where the kidneys can no longer sustain life. Symptoms include lethargy, weakness, headache, nausea, vomiting, diarrhea, and edema of the lungs—a syndrome termed *uremic poisoning.* End-stage kidney failure results because of irreversible kidney damage. The only treatments are permanent use of an artificial kidney or *kidney transplantation,* the surgical implantation of a donated, healthy kidney. Unfortunately, some patients with end-stage kidney failure are too ill for treatment by either of these methods.

## Study Questions

1. Indicate several functions of the urinary system.
2. a) Draw a diagram of a nephron and surrounding blood vessels, labeling as many structures as you can.
   b) Draw another labeled diagram showing how nephrons are arranged in the kidney.
3. a) Describe the process of glomerular filtration.
   b) Why is the composition of glomerular filtrate similar to that of tissue fluid?
4. Distinguish between tubular reabsorption and tubular secretion.
5. a) Explain how the loop of Henle functions to establish a hyperosmotic fluid in the medulla of the kidney.
   b) What is the significance of this hyperosmotic fluid to the functioning of the kidney?
6. Describe the role of the kidneys in the regulation of blood pressure.
7. How does ADH function in the maintenance of water balance by the kidney?
8. It is not possible to survive by drinking sea water. From what you know about the functioning of the kidney, why do you suppose this is so?
9. Describe the structure and functional properties of the epithelium lining the bladder.
10. a) Describe the physiological events that trigger urination in a newborn baby.
    b) In an adult.
11. Explain how the artificial kidney functions.

### Chapter Thirteen

# THE NERVOUS SYSTEM I:
## Structure, Function, and Organization

Foregoing chapters have discussed various physiological systems of the body, but in order to understand how the human body functions, it is essential to remember that these systems do not operate in isolation of one another. Instead, the activities of all systems are integrated and interdependent.

The coordination of bodily activities is the function of two closely interrelated systems of the body—the nervous system and the endocrine system. Both of these systems function by accumulating information on the state of the internal and external environment, analyzing this information to determine an appropriate response, and finally, initiating a response. However, because the nervous and endocrine systems carry out their regulatory functions by fundamentally different mechanisms, their activities complement one another. The nervous system, composed of a vast network of nerve cells that conduct information in the form of electrical impulses, is designed for rapid communication between cells of the body, allowing for fast, short-term physiological adjustments. On the other hand, the endocrine system, which secretes regulatory chemicals that are transported throughout the body in the bloodstream, provides adjustments that occur more slowly, but are of longer duration than those brought about by the nervous system.

This chapter and Chapters 14 and 15 deal with physiological regulation mediated by the nervous system. Chapter 16 discusses the endocrine control of bodily activities.

## OVERVIEW OF THE NERVOUS SYSTEM

The nervous system has often been likened to a telecommunication system, with its single lines, trunk lines, and switching centers. This analogy is a gross oversimplification, particularly with respect to the decision-making capacity of the nervous system, which has no real counterpart in a telecommunication system. However, it is a useful analogy because it emphasizes several important characteristics of the nervous system: information is relayed along tracts formed of elongated nerve cells analogous to telephone wires, and messages are sent, as over telephone wires, in electrical form at high speeds.

### Structure

As shown in Figure 13–1, the nervous system consists of the *brain* and the *spinal cord* and a network of nerves, called *peripheral nerves,* that connect the brain and the spinal cord with other parts of the body.

The brain and the spinal cord together compose the **central nervous system,** or **CNS.** The brain is protected by the skull, and the spinal cord, by the vertebral column through which it passes. Both the brain and the spinal cord are composed of vast numbers of **nerve cells,** or **neurons,** that actually transmit electrical impulses, and some ten times as many **glial cells** that support the neurons, protect them and help sustain them metabolically.

The **peripheral nerves,** of which there are 43 pairs, are composed of bundles of neurons and some supporting cells. Collectively, these nerves form the **peripheral nervous system (PNS).** The members of each pair of peripheral nerves serve opposite sides of the body. Twelve of the 43 pairs connect with the brain through small openings in the bones of the skull; these nerves, called **cranial nerves,** supply the head and neck, with the exception of one of them, which supplies various internal organs. The remaining 31 pairs of peripheral nerves are called **spinal nerves.** The most anterior pair of spinal nerves emerges between the atlas and the occipital bone of the skull; all others run through canals, the **intervertebral foramina,** that are present between adjacent vertebrae (see Figure 13–1b). The spinal nerves connect the body below the neck with the spinal cord.

FIGURE 13-1 (a) Components of the nervous system. (b) Posterior view of the spinal column and spinal nerves.

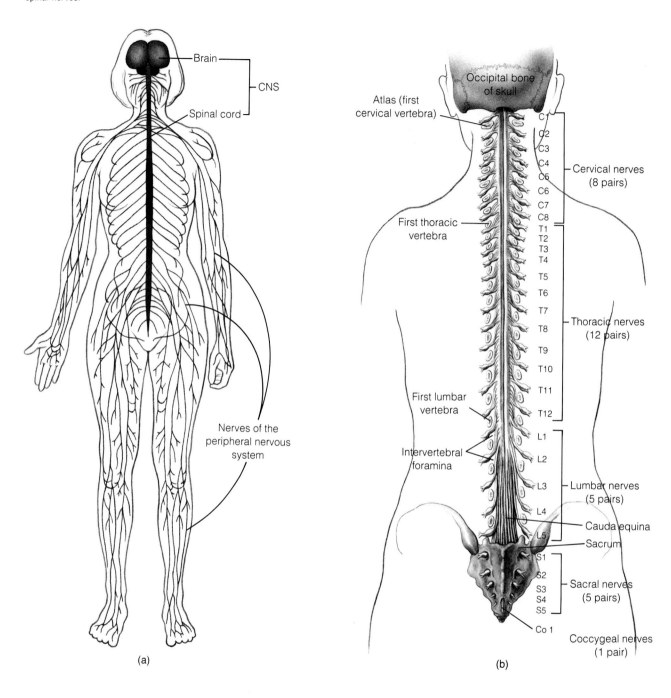

(a)

(b)

It is important to understand that the division of the nervous system into central and peripheral components is largely artificial, and hence misleading. The central nervous system and the peripheral nervous system are connected through neurons both anatomically and functionally, and both divisions of the nervous system are required for most nervous function, with the exception of certain thought processes.

## Function

Input to the nervous system is provided by a variety of *sense receptors* that collect information about the body's internal and external environment. This information is converted into electrical impulses that are transmitted by neurons in peripheral nerves to the spinal cord and brain. The spinal cord serves as a vast trunk line of neurons that carries impulses received over spinal nerves to the brain for processing. In addition, the spinal cord possesses preprogrammed circuitry that allows some of these impulses to initiate directly certain responses independently of the brain; such responses are termed **reflexes.**

Once an impulse reaches the brain, either by way of the spinal cord or directly over cranial nerves, the brain must interpret its significance and make decisions as to what responses, if any, are required.

The processing in the brain of incoming nerve impulses results in the generation of new impulses that are transmitted directly over cranial nerves, or down the spinal cord and then over spinal nerves, to muscles and glands, which carry out the appropriate response.

Many of the functions that are controlled by the nervous system are conscious ones, such as the movement of skeletal muscles, while others are not. Those that operate below our level of consciousness serve to integrate the activities of internal organs. Thus, for example, control centers in the brain regulate such vital functions as heartbeat, blood pressure, breathing, and internal temperature; other areas of the brain regulate the synthesis and secretion of hormones that control water balance, growth, sexual function, and other processes.

The brain is responsible not only for maintaining homeostasis and hence assuring survival, but for higher thought processes as well, such as abstract reasoning, creativity, insight, imagination, and language.

## NEURONS

Neurons are the functional cells of the nervous system. They transmit electrical impulses from one area of the body to another.

### Shape, Size and Structure

Neurons vary quite considerably in size, some being less than a millimeter in length and others, longer than a meter. They also have a variety of shapes, as shown in Figure 13–2. Nevertheless, all neurons have similarities in structure. The more or less spherical *cell body* contains the nucleus and most of the cell's cytoplasm, in which are suspended various cytoplasmic organelles. The metabolic activities that occur in the cell body are responsible for sustaining the entire neuron. Extending from the cell body are elongated cytoplasmic processes. These are of two types: **axons** (also called nerve fibers) and **dendrites.** Dendrites carry impulses from their distal ends toward the cell body; from there, impulses are transmitted down the axon to its termination. Although each neuron has only one axon, it may have a side branch or two, called **axon collaterals.** Toward its end, the axon typically divides into a number of branches, each of which terminates in a swelling called the **terminal bulb** or **synaptic knob.** Nerve impulses that reach the terminal bulb may be transmitted to another neuron, or to a muscle or gland where they produce their effect.

In most nerve cells, nerve fibers are surrounded by a cellular sheath. The sheath serves to support the fiber and to protect it. Sheath cells are types of glial cells. Those that surround fibers of neurons in peripheral nerves are called **Schwann cells,** and those that surround fibers of neurons in the central nervous system are termed **oligodendrocytes.** Nerve fibers that are covered by a simple sheath of Schwann cells or oligodendrocytes are called *unmyelinated* (Figures 13–2a and 13–3a and b). Another type of fiber, said to be *myelinated,* is covered by flattened sheath cells that wrap repeatedly around the fiber, investing it with concentric rings of the sheath cells' plasma membranes (Figure 13–3c to e). Because the plasma membranes of the sheath cells are

These two photographs are of different neurons found in the human brain.

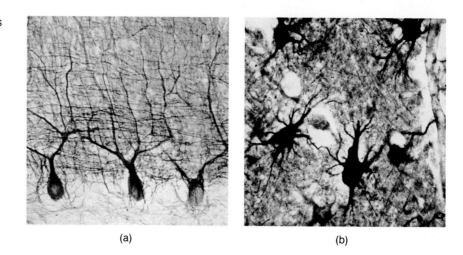

(a)                    (b)

FIGURE 13-2 Two types of neurons: (a) is unmyelinated and (b) is myelinated.

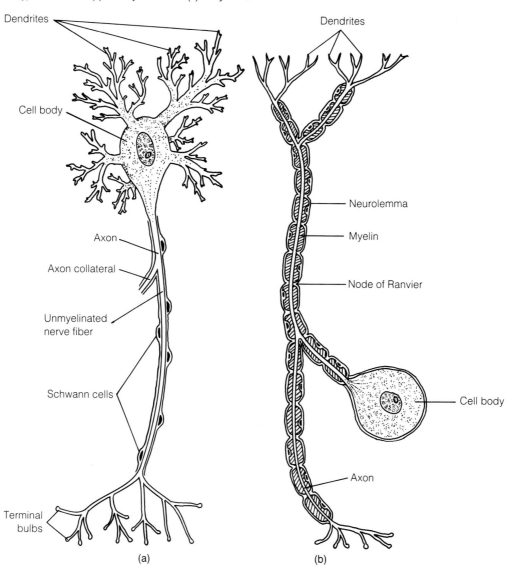

Dendrites

Cell body

Axon

Axon collateral

Unmyelinated
nerve fiber

Schwann cells

Terminal
bulbs

(a)

Dendrites

Neurolemma

Myelin

Node of Ranvier

Cell body

Axon

(b)

composed of about 80 percent lipid, the multiple layers form a thick lipid material. This material is called **myelin.** The myelin is *not* continuous along a myelinated nerve fiber, since a single sheath cell only covers about a millimeter or so of the length of the fiber. Between successive sheath cells on the fiber are short, unmyelinated gaps, termed **nodes of Ranvier.** As we shall see shortly, the nodes of Ranvier play an important role in the conduction of electrical impulses along myelinated fibers.

Because myelin is a fatty material, myelinated fibers appear white. These whitish fibers form the *white matter* of the brain and spinal cord. Unmyelinated fibers, on the other hand, appear gray, and together with cell bodies and dendrites form the *gray matter* of these two structures. Peripheral nerve trunks contain a mixture of both myelinated and unmyelinated fibers, but appear a glistening white.

## Generation of Action Potentials in Neurons

The generation of **action potentials,** or electrical impulses, in neurons is much the same as in muscle cells. Normally, the neuron's plasma membrane is polarized such that its outer surface is positively charged with respect to its inner surface; this charge differential is due to an excess of positively charged ions in the extracellular fluid surrounding the neuron. An action potential is initiated when the plasma membrane becomes depolarized. Depolarization occurs as a result of changes in the permeability of the membrane to sodium ions ($Na^+$) and potassium ions ($K^+$). At the site of stimulation, $Na^+$ ions diffuse into the cell from the extracellular fluid, and the polarization of the plasma membrane disappears. Immediately after depolarization, the membrane repolarizes when it becomes relatively impermeable to $Na^+$, but permeable to $K^+$, which then diffuses out of the cell. This depolarization followed by repolarization constitutes an action potential.

Regardless of the strength of the stimulus, a nerve cell will depolarize fully or not at all; this is the so-called *all-or-none response.* However, an increased strength of stimulus can be signaled by an increased frequency in the

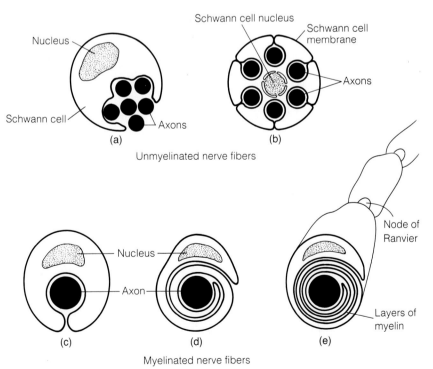

(a) Unmyelinated nerve fibers

Nucleus

Schwann cell

Axons

Schwann cell nucleus

Schwann cell membrane

Axons

(b)

(c) Nucleus

Axon

(d)

(e)

Node of Ranvier

Layers of myelin

Myelinated nerve fibers

FIGURE 13–3 Development of the sheaths of unmyelinated and myelinated nerve fibers. Unmyelinated axons lie in grooves of a single Schwann cell, whereas myelinated neurons are wrapped singly by concentric layers of a Schwann cell.

generation of action potentials in a single neuron or by the stimulation of additional neurons.

## Propagation of Nerve Impulses along Neurons

Action potentials are propagated along neurons as electrical currents (nerve impulses) in one of two ways. The mechanism of propagation depends on the type of fiber along which the action potential is traveling.

### Unmyelinated Fibers

In unmyelinated fibers, the action potential travels as it does in muscle cells. Briefly stated, local current flow, generated between the depolarized area of the membrane and the polarized region immediately adjacent to it, causes the adjacent site to change its membrane permeability to Na$^+$ and hence become depolarized. In this way, a wave of depolarization sweeps down the nerve fiber (Figure 13–4a).

FIGURE 13–4 Propagation of nerve impulses along (a) unmyelinated nerve fibers as compared to (b) myelinated nerve fibers.

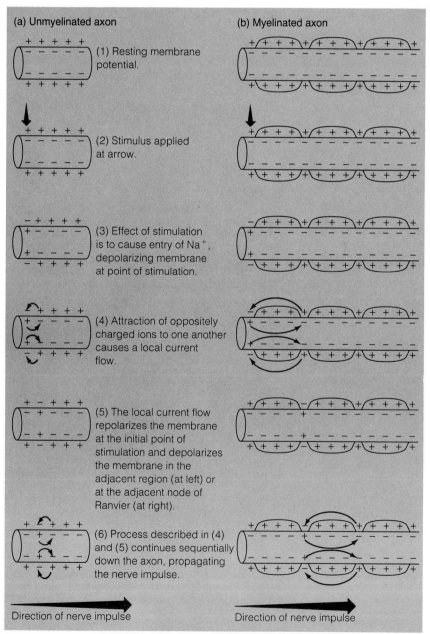

(a) Unmyelinated axon    (b) Myelinated axon

(1) Resting membrane potential.

(2) Stimulus applied at arrow.

(3) Effect of stimulation is to cause entry of Na$^+$, depolarizing membrane at point of stimulation.

(4) Attraction of oppositely charged ions to one another causes a local current flow.

(5) The local current flow repolarizes the membrane at the initial point of stimulation and depolarizes the membrane in the adjacent region (at left) or at the adjacent node of Ranvier (at right).

(6) Process described in (4) and (5) continues sequentially down the axon, propagating the nerve impulse.

Direction of nerve impulse    Direction of nerve impulse

## Myelinated Fibers

In myelinated fibers, the action potential is propagated by a different mechanism (Figure 13–4b). Between the nodes of Ranvier on a myelinated fiber, the myelin forms a thick lipid coat that effectively insulates the axon from the ion-containing tissue fluid that surrounds the neuron (and that ultimately determines the membrane potential). However, at the nodes of Ranvier there is no myelin, and the nerve fiber is exposed to tissue fluid. Consequently, action potentials can be generated at the nodes, and the wave of depolarization jumps from one node to the next, rather than passing over the entire fiber. This mechanism allows for the conduction of impulses at very rapid velocities, and it accounts for the fact that myelinated fibers conduct impulses at a much faster rate than do unmyelinated fibers.

Myelination is thus of obvious importance in the transmission of nerve impulses, and diseases that damage the myelin sheaths of nerve fibers can have a marked effect on the function of the nervous system. Destruction of myelin may be caused by exposure to toxic substances, viral infection, trauma, and autoimmune diseases (diseases in which the body's immune system attacks the body's components). Much evidence suggests that *multiple sclerosis* (MS), in which myelin undergoes patchy degeneration, may result from an autoimmune response to myelin that may follow certain viral infections, particularly measles (Figure 13–5). Affecting primarily young adults between the ages of 18 and 40, multiple sclerosis causes increasingly serious muscular weakness, problems with coordination, various degrees of paralysis, and impaired vision and speech; in some cases, it may progress to mental impairment and death. The disease does not progress at an even rate, but is instead characterized by repeated attacks, during which symptoms worsen, interspersed with periods of remission that may last for long periods.

## Fiber Diameter

In addition to an increase in conduction velocity due to myelination, nerve fibers with large diameters conduct impulses faster than those with small diameters. For instance, a large diameter myelinated fiber can conduct impulses as fast as 130 meters per second. This type of fiber generally relays information that involves the detection of danger in the external environment, and hence a high speed of conduction is distinctly advantageous. Myelinated fibers with medium-sized diameters conduct impulses at a speed of about 10 meters per second, whereas the smallest unmyelinated fibers conduct at a rate of about 0.5 meters per second. Both of these fiber types transfer information in situations where an instantaneous response is not crucial to survival.

## Transmission of Nerve Impulses between Neurons

### Synapses

Nerve impulses may be transmitted from one neuron to another. However, unlike a telecommunication system in which a relay is used to provide contact between separate wires, two neurons that transmit impulses between them do not actually come in physical contact with one another. Instead, impulses are transmitted over spaces, termed **synapses.** At a synapse between neurons, the axon of one neuron typically terminates on the dendrites or cell body, or both, of another neuron.

The neuron that transmits an impulse to another neuron at a synapse is called the **presynaptic neuron;** the neuron that receives the impulse is called the **postsynaptic neuron** (Figure 13–6a). At the termination of an axon branch at a synapse, the axon is expanded into a *terminal bulb,* as mentioned previously. Between the terminal bulb and the postsynaptic neuron is a space, about 200 Å wide, termed the **synaptic cleft.** Stored within small vesicles in the terminal bulb of the presynaptic neuron are neurotransmitters, chemicals that transmit an electrical impulse from the presynaptic to the postsynaptic neuron. The structure of a synapse is shown in Figure 13–6.

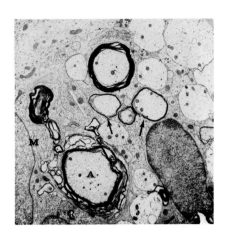

FIGURE 13–5 Electron micrograph showing degeneration of the myelin sheath around an axon (A), caused by the action of a macrophage (M). A macrophage is a type of white blood cell that normally ingests foreign invaders; in this case, the body's immune system has gone awry, and myelin is interpreted by macrophages as foreign matter. Arrows point to thin layers of myelin undergoing repair.

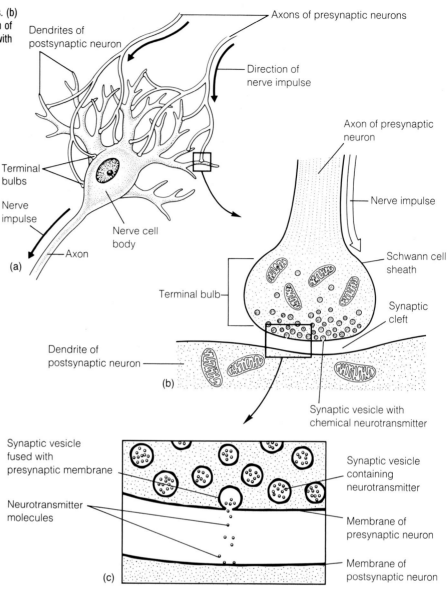

FIGURE 13-6 (a) Synapses between neurons. (b) Details of synapse structure. (c) Magnification of (b) showing the fusion of a synaptic vesicle with the membrane of the presynaptic neuron and release of neurotransmitter.

Dendrites of postsynaptic neuron

Axons of presynaptic neurons

Direction of nerve impulse

Axon of presynaptic neuron

Nerve impulse

Terminal bulbs

Nerve impulse

Nerve cell body

Axon

(a)

Terminal bulb

Schwann cell sheath

Synaptic cleft

Dendrite of postsynaptic neuron

(b)

Synaptic vesicle with chemical neurotransmitter

Synaptic vesicle fused with presynaptic membrane

Synaptic vesicle containing neurotransmitter

Neurotransmitter molecules

Membrane of presynaptic neuron

Membrane of postsynaptic neuron

(c)

## Neurotransmission

The transmission of impulses between neurons at synapses, a process termed **neurotransmission,** occurs precisely in the same way as it does at neuromuscular synapses (see Figure 7–7). When an impulse reaches the terminal bulb, depolarization of the membrane triggers an influx of $Ca^{++}$ into the bulb. This, in turn, causes synaptic vesicles to fuse with the membrane of the terminal bulb, releasing neurotransmitter molecules into the synaptic cleft (Figure 13–6c). The neurotransmitter molecules diffuse across the cleft and bind to receptor sites on the membrane of the postsynaptic neuron. This binding initiates a localized depolarization in the postsynaptic neuron by bringing about a transient change in the permeability of its plasma membrane. Once depolarization occurs, an action potential is propagated along the fibers of the postsynaptic neuron.

Although chemically mediated transmission of this type, involving the diffusion of neurotransmitters across a synapse, would seem to take a long time, this is not the case. In fact, impulses are transmitted across synapses in less than one thousandth of a second.

The events that trigger depolarization in the postsynaptic neuron are terminated in various ways, depending upon the particular synapse and the type of neurotransmitter involved. These include such mechanisms as the enzymatic degradation of the neurotransmitter substance, uptake of the neurotransmitter by the presynaptic cell, and diffusion of the neurotransmitter away from the receptors.

It is important to point out that synapses may be one of two types. At an **excitatory synapse,** of the type described earlier, changes in the permeability of the postsynaptic membrane brought about as a result of neurotransmission initiate depolarization, triggering an action potential. However, at another type of synapse, termed an **inhibitory synapse,** the permeability changes that occur in the postsynaptic membrane *inhibit* depolarization. At synapses of the inhibitory type, the transmission of an impulse to a postsynaptic neuron is blocked. The inhibitory synapse provides an additional type of nervous control of bodily activities, for it allows stimulation of a presynaptic neuron to *inhibit* the firing of a postsynaptic neuron. For example, inhibitory synapses can prevent the stimulation of those neurons that activate antagonistic muscles in an agonist-antagonist pair.

Because a single axon may branch extensively at its end and thus have a reasonably large number of terminal bulbs, one neuron may transmit impulses to a number of other neurons with which it synapses. Moreover, one neuron may receive impulses from many other neurons (see Figure 13–6a). For example, it has been estimated that certain neurons in the brain may receive impulses from more than 100,000 other neurons. The scope of such interactions between neurons accounts for the complexity of neural integration.

## Neurotransmitters

One important neurotransmitter released by many neurons is *acetylcholine.* Other neurotransmitters include *norepinephrine, epinephrine (adrenalin), dopamine, serotonin, gamma-aminobutyric acid (GABA),* and *endorphins,* as well as many others. The use of different types of neurotransmitters by various neurons may allow specialization of function or prevent interference between adjacent neurons.

## Drugs and Neurotransmission

Drugs that act on the nervous system often alter the transfer of impulses at synapses. *Anesthetics,* for example, inhibit synaptic transmission. *Caffeine,* present in coffee, tea, and many soft drinks, stimulates activity of the nervous system by stimulating synaptic transmission across certain neurons. *Curare,* used by South American Indians on blowgun tips and arrowheads, competes with acetylcholine for receptors on the postsynaptic membrane, thus preventing nerve transmission and leading to respiratory paralysis and death. The deadly *botulism toxin* produced by the bacterium *Clostridium botulinum* blocks the release of acetylcholine, causing a physiological effect similar to that of curare. *Black widow spider toxin* enhances the release of acetylcholine at synapses, causing skeletal muscles to go into spasm. The *organophosphate pesticides* (organophosphates are organic compounds containing phosphorus) such as diisopropylfluorophosphate, parathion, and malathion kill insects by interfering with the action of acetylcholinesterase, an enzyme that normally degrades acetylcholine at synapses, halting transmission of impulses. Because acetylcholine is not degraded in the presence of these pesticides, impulses are transmitted continuously across many synapses, causing skeletal muscles to go into spasm and resulting in death from lack of respiratory function. *Cocaine* acts on the brain by inhibiting neurotransmitters from being taken up by the presynaptic neurons after their release at synapses, effectively preventing the termination of neurotransmission across the synapse (see *Focus: Cocaine*).

# Cocaine

Cocaine is the drug of the 1980s. Ten percent of Americans have tried it, and four to eight million Americans use it at least once a month. Many users abuse it so severely that their entire lives are dominated by the drug.

The widespread use and abuse of cocaine has come on so suddenly as to make cocaine seem like the drug culture's newest discovery. In fact, cocaine has a long history as a "recreational" drug. Leaves from the coca shrub, the natural source of cocaine, have been chewed for at least two thousand years by the inhabitants of the Andes Mountains of South America, where the shrub is indigenous (see accompanying figure). It was not until the mid-nineteenth century that coca leaves were imported in quantity by Europe. In 1863, a coca leaf extract was incorporated into a wine called "Vin Mariani" (Mariani Wine), that soon became a highly popular beverage all over Europe and in the United States. Advertisements for Mariani Wine, "The World Famous Tonic for Body and Brain," touted its delicious, refreshing taste, and recommended it "for overworked men, delicate women, and sickly children" and as a treatment for "General Debility, Overwork, Profound Depression and Exhaustion, Throat and Lung Disease, Consumption and Malaria."

In 1866, John Pemberton, a Georgia pharmacist, used Mariani Wine as a model to concoct Coca-Cola. Pemberton's original Coca-Cola contained a coca leaf extract in an alcoholic wine, but the wine was subsequently removed and replaced with

The coca shrub, the source of cocaine.

an extract of kola nut, which contained caffeine. When, in 1888, Pemberton substituted soda water for the plain water base, the drink became identical to present-day "Classic" Coca-Cola, except that it contained cocaine. In 1903, cocaine was removed from Coca-Cola when its addictive property became appreciated (see accompanying figure).

When coca leaves are chewed, small amounts of cocaine are swallowed and then slowly absorbed into the bloodstream, bringing about a mild sense of well-being and alleviating hunger. Chewing coca leaves does not appear to lead to addiction, and most authorities consider the practice less harmful than the regular use of tobacco or alcohol. In fact, cocaine abuse and addiction was not a significant problem until cocaine became available in purified chemical form around 1860.

Sigmund Freud used purified cocaine and published a paper on its effects in 1884. He noted the anesthetizing effect of cocaine on the skin and mucous membrane, and mentioned that

*Cocaine brings about an exhilaration and lasting euphoria which in no way differs from the normal euphoria of a healthy person. . . . You perceive an increase of self-control and possess more vitality and capacity for work. . . . This result is enjoyed without any of the unpleasant after-effects that follow exhilaration brought about by alcohol.*

With Freud's endorsement, cocaine was widely prescribed for anxiety and depression in the late 1800s. However, when many people became psychotic as a result of cocaine addiction, physicians began to realize its danger, and Freud was soundly criticized for his endorsement.

Today, cocaine is no longer used for medical purposes and, except for scientific research, is available only illegally. When cocaine is inhaled, it is absorbed through the mucous membrane of the nasal cavity into the bloodstream, reaching the brain within a few minutes. The "high," or pleasurable effect, of cocaine lasts for 20 to 40 minutes, and is followed by a "crash" or

A 1914 advertisement for Coca Cola.

"low," a period of depression and anxiety. Because the high lasts for such a short period of time, frequent doses are needed to maintain it. To produce a more intense high, some users inject cocaine intravenously or smoke it. In order to smoke cocaine, the crystalline powder, which cannot be ignited, must be converted into another form. Sometimes, it is converted into the pure alkaloid, the form that is present in coca leaves, by using ether. The dried alkaloid, or "free base," can then be smoked. Freebasing is exceedingly dangerous because of the extreme flammability of ether.

Cocaine can also be smoked in the form of "crack" or "rock," small chunks of cocaine produced by heating and then drying a mixture of cocaine, baking soda, and water. Crack produces a short, intense, rushing high, followed in minutes by a crashing low. Both the high and low are far more extreme than that obtained by inhaling cocaine, often driving the user to repeat the experience. Crack's high is very dangerous, since the user obtains a large dose of drug over a short period of time. It causes dramatically increased blood pressure and heart rate, which can lead to a heart attack or a stroke, and its effect on the brain can trigger convulsions.

The principal mental effects of cocaine are feelings of euphoria and alertness, suppression of appetite, and psychosis (serious mental derangement). Cocaine psychosis, brought on by heavy use of the drug and addiction, resembles paranoid schizophrenia. Drug users first experience feelings of a vague suspiciousness that grow into overt delusions that everyone is plotting against them. In this state, many users become violent, lashing out at their presumed enemies. Another symptom of cocaine psychosis is an itching sensation that the user often interprets as bugs crawling underneath the skin. Additional effects of cocaine are summarized in the accompanying figure.

We do not fully understand cocaine's mode of action on the nervous system. We know that cocaine inhibits the normal uptake of the neurotransmitters norepinephrine and dopamine into presynaptic neurons after they have been released at synapses. Consequently, neurotransmitter levels at the synapse do not fall as they normally would after stimulation of a presynaptic neuron; transmission across the synapse is not terminated, and the postsynaptic neuron continues to be stimulated. The feeling of euphoria brought on by cocaine is probably a result of an increase in norepinephrine levels at synapses in neurons of the limbic system, a region of the brain responsible for emotional feelings (the limbic system is discussed in Chapter 14). Cocaine psychosis, with its symptoms similar to schizophrenic paranoia, is thought to be due to an increase in brain dopamine levels, since antischizo-phrenic drugs act by blocking dopamine receptors on postsynaptic neurons.

Treatment of cocaine addiction usually involves psychotherapy. This is sometimes done in conjunction with the administration of antidepressant tricyclic drugs, such as desipramine and imipramine, that help overcome the craving for cocaine by acting as a nonaddictive replacement.

Cocaine's impact on society has not been uniformly negative. The discovery that cocaine can act as a local anesthetic by blocking conduction in nerve fibers must be considered one of the most important advancements of modern medicine. It led to the development of a group of local anesthetics, including procaine (Novocaine) and lidocaine, that are routinely used for dental and eye surgery as well as for other types of minor surgery for which general anesthesia is inappropriate.

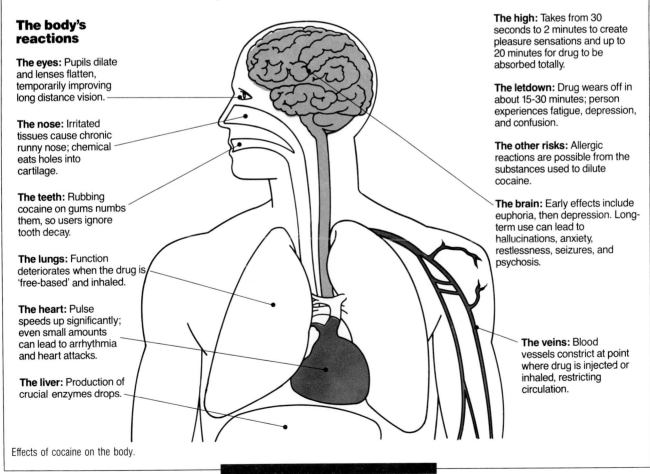

## The body's reactions

**The eyes:** Pupils dilate and lenses flatten, temporarily improving long distance vision.

**The nose:** Irritated tissues cause chronic runny nose; chemical eats holes into cartilage.

**The teeth:** Rubbing cocaine on gums numbs them, so users ignore tooth decay.

**The lungs:** Function deteriorates when the drug is 'free-based' and inhaled.

**The heart:** Pulse speeds up significantly; even small amounts can lead to arrhythmia and heart attacks.

**The liver:** Production of crucial enzymes drops.

**The high:** Takes from 30 seconds to 2 minutes to create pleasure sensations and up to 20 minutes for drug to be absorbed totally.

**The letdown:** Drug wears off in about 15-30 minutes; person experiences fatigue, depression, and confusion.

**The other risks:** Allergic reactions are possible from the substances used to dilute cocaine.

**The brain:** Early effects include euphoria, then depression. Long-term use can lead to hallucinations, anxiety, restlessness, seizures, and psychosis.

**The veins:** Blood vessels constrict at point where drug is injected or inhaled, restricting circulation.

Effects of cocaine on the body.

### Diseases Affecting Neurotransmission

Certain disease conditions may also affect the transmission of nervous impulses. In *myasthenia gravis* (see Chapter 7), an autoimmune disease, an individual's immune system destroys acetylcholine receptors, blocking neuromuscular function. *Parkinson's disease,* discussed extensively in Chapter 14, is a result of degeneration of neurons in the brain that secrete the neurotransmitter dopamine.

### Types of Neurons

On the basis of their function, neurons can be divided into three classes: (1) *afferent,* or *sensory neurons;* (2) *efferent,* or *motor neurons;* and (3) *interneurons,* also called *association neurons.* The functional relationship between these neuronal types is illustrated in Figure 13–7 and discussed in the paragraphs that follow.

**Afferent neurons** (*affere* = to bring to), also referred to as **sensory neurons,** carry impulses from other parts of the body to the central nervous system. At the peripheral termination of an afferent neuron or chain of afferent neurons are nerve endings or specialized sensory receptors that are activated by specific stimuli, including light, sound, odor, touch, pressure, position, pain, heat, cold, and movement. When impulses from afferent neurons are transmitted to the CNS, they may be perceived as conscious sensations, but this is not always the case.

**Efferent neurons** (*effere* = to carry away), or **motor neurons,** carry impulses from the CNS to muscles or glands, the so-called *effector* organs; that is, the organs whose activity is influenced by the impulse. The juncture between efferent neurons and an effector organ is called a **neuroeffector junction;** the neuromuscular junction, discussed in Chapter 7, is one type of neuroeffector junction.

The peripheral nerves are composed of bundles of afferent or efferent neurons or both. Afferent neurons terminate in the brain or spinal cord; efferent neurons originate in the brain or spinal cord and pass into the periphery of the body. In some cases, the afferent neuron synapses directly with an efferent neuron, and nervous impulses traveling along the afferent neuron are therefore relayed directly to the efferent neuron. In most cases, however, the connection between the two types of neurons is not so direct. Instead, an afferent neuron generally synapses with one or more **interneurons,** or **association neurons,** which transmit impulses from one part of the CNS to another before

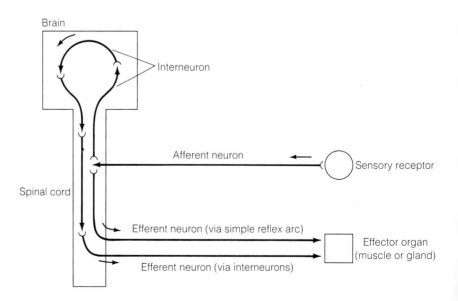

FIGURE 13–7 Functional relationship between afferent neurons, efferent neurons, and interneurons.

Brain

Interneuron

Afferent neuron

Sensory receptor

Spinal cord

Efferent neuron (via simple reflex arc)

Effector organ (muscle or gland)

Efferent neuron (via interneurons)

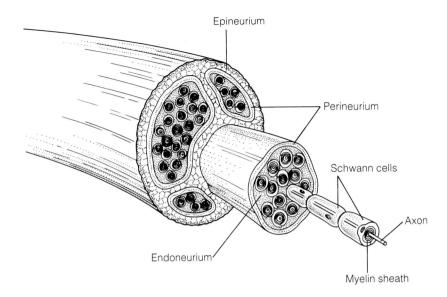

Epineurium

Perineurium

Schwann cells

Axon

Endoneurium

Myelin sheath

FIGURE 13-8 Structural organization of a nerve.

synapsing with an efferent neuron. In fact, most neurons—more than 99 percent—are interneurons that compose the brain and spinal cord, and they are involved in complex movements, regulatory phenomena, and higher thought processes. The number of interneurons in a pathway between afferent and efferent ones varies according to the complexity of the process that is being controlled.

## NERVES

Nerves are bundles of individual nerve fibers (neurons). They may contain afferent or efferent fibers, or both, and some of the fibers may be myelinated and others not.

### Structure

The structure of a nerve is shown in Figure 13–8. Individual nerve fibers are bound to one another by fine connective tissue called **endoneurium.** Bundles of nerve fibers, so joined, form cablelike structures termed **fascicles.** Each fascicle is surrounded by the **perineurium,** a relatively dense connective tissue sheath. Groups of fascicles are, in turn, bound together with a loose connective tissue, not nearly as strong as the perineurium, called the **epineurium,** through which course blood vessels and lymphatics that supply the nerve.

### Nerve Disorders

**Neuritis,** or inflammation of a nerve, may be caused by trauma, infection, and other conditions. One type of neuritis, *sciatica,* affects one or both of the sciatic nerves serving the legs. It is commonly caused by a "slipped" intervertebral disc that puts pressure on these nerves as they leave the spinal cord through the intervertebral foramina. Neuritis may be accompanied by *neuralgia* (nerve pain of a severe throbbing sort), numbness, paralysis, muscular atrophy, and lack of reflexes.

A number of diseases can also damage nerves and impair nerve function. In *shingles,* the herpes roster virus (which also causes chickenpox) infects sensory neurons, producing pain and skin rashes in areas served by the infected nerves. Another viral disease that damages nerves and can cause severe crippling, particularly of children, is *poliomyelitis* (*infantile paralysis* or *polio*), now largely eliminated from Western countries by use of vaccines developed by Jonas Salk and Albert Sabin in the early 1950s. Poliovirus infection is characterized by the destruction of motor neurons that control skeletal muscles, producing paralysis; death may occur if motor neurons controlling respiratory muscles are affected.

Vitamin deficiencies can also damage nerves. Dietary deficiency of vitamin B₁ causes *beriberi,* a neuritis that preferentially involves sensory neurons. Severe *pernicious anemia,* caused by vitamin B₁₂ deficiency, may result in damage to sensory fibers that convey information on touch and sense of position, leading to loss of touch sensation and poor muscle coordination.

### Regeneration of Peripheral Nerves

If a peripheral nerve is severed, portions of axons that are separated from their cell bodies will degenerate. For example, the severance of a spinal nerve, in which cell bodies lie in or near the spinal cord, will cause the degeneration of axon fragments distal to the cut. This leads to the loss of sensory and motor function in the area served by the nerve. Although the loss of function is often permanent, it is possible under favorable conditions for nerve fibers to regenerate, restoring at least partial function.

Regeneration of a nerve fiber requires continuity of its Schwann cell sheath, which provides a channel that guides a regenerating nerve fiber more or less along its old path. Although nerve fibers distal to the cut degenerate, the Schwann cells of the sheath do not. Furthermore, Schwann cells may grow to fill a small gap between them. In order to keep the gap small enough to allow the Schwann cells to reestablish continuity, the connective tissue wrappings of the two severed ends of the nerve must usually be sutured together surgically. Because peripheral nerves have the ability to regenerate, severed appendages that have been reunited surgically can often regain at least partial nervous function.

## FUNCTIONAL DIVISIONS OF THE NERVOUS SYSTEM

As mentioned previously, the nervous system consists anatomically of the CNS (brain and spinal cord) and the PNS (31 pairs of spinal nerves that connect with the spinal cord and 12 pairs of cranial nerves that connect directly with the brain). The nerves of the PNS contain afferent and efferent fibers that reach throughout most of the body.

From a functional standpoint, it is useful to think of the peripheral nervous system as having two divisions: **somatic** and **autonomic.** The somatic division (also called the **somatic nervous system**) controls voluntary functions. **Somatic afferent (sensory) neurons** carry sensory information from the skin, skeletal muscles, and joints to the CNS; **somatic efferent (motor) neurons** carry impulses from the CNS to the skeletal muscles to effect voluntary movement. The autonomic division (**autonomic nervous system**), on the other hand, controls involuntary functions ("automatic" functions, as it were) such as heartbeat, blood pressure, and temperature. The autonomic nervous system is composed of **visceral afferent (sensory) neurons** that convey sensory information from visceral organs to the CNS, and **visceral efferent (motor) neurons** that carry impulses from the CNS to smooth muscles, cardiac muscle, and glands.* Functions and components of the two divisions of the peripheral nervous system are summarized in Figure 13–9.

Although division of the PNS into somatic and autonomic divisions is helpful for purposes of discussion, it is crucial to emphasize that these are not

---

*Some schemes include in the autonomic nervous system only efferent fibers to smooth muscle, cardiac muscle, and glands, and relegate afferent (sensory) fibers from visceral organs to the somatic system. Many schemes now include both afferent and efferent fibers to and from visceral organs in the autonomic nervous system, a practice that is also followed here.

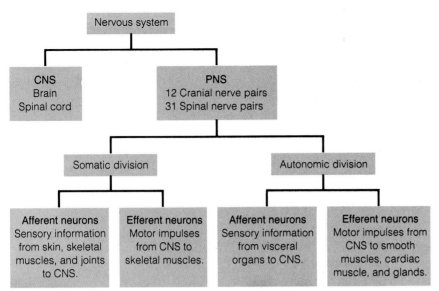

entirely separate subsystems, either anatomically or functionally. Instead, their activities overlap and are closely integrated with one another. Both somatic and autonomic functions are regulated by the CNS, and all spinal nerves and many cranial nerves contain both somatic and visceral fibers.

In order to explain how the somatic and autonomic systems function, we must first describe the structure of the spinal cord, spinal nerves, and cranial nerves.

## THE SPINAL CORD

The function of the spinal cord is to receive afferent impulses by way of spinal nerves from peripheral sense organs located below the neck, to transmit some of this information to the brain for further processing, and to carry efferent impulses from the brain to spinal nerves. In addition, some of the afferent neurons that enter the spinal cord synapse directly, or through interneurons, with efferent neurons (reflex arc), allowing rapid, automatic responses to sensory information.

The spinal cord is a nearly cylindrical, elongated structure about the diameter of the little finger. It begins at the foramen magnum at the base of the skull, where it is continuous with the brainstem, and it runs through the vertebral canal to the level of the second lumbar vertebra (see Figure 13–1b). From the end of the spinal cord extend all of the spinal nerves below the first lumbar pair, forming a structure called the *cauda equina,* likened to a horse's tail. Figure 13–1b also shows the position of the spinal cord in the vertebral column and the points of emergence of spinal nerves between vertebrae. The spinal nerves are numbered with reference to the vertebral segments through which they emerge.

### Spinal Meninges

The spinal cord is protected not only by the vertebral column, but by three connective tissue membranes, called the *meninges,* that are continuous with those that cover the brain. The meninges of the spinal cord are shown in Figure 13–10. The innermost membrane, the **pia mater** (*pia* = tender; *mater* = mother), is delicate and is surrounded by another membrane, the **arachnoid,** also of moderately delicate structure. Between the pia mater and the arachnoid is a space, the **subarachnoid space,** filled with **cerebrospinal fluid,** a modified tissue fluid that completely surrounds the spinal cord (and the brain).

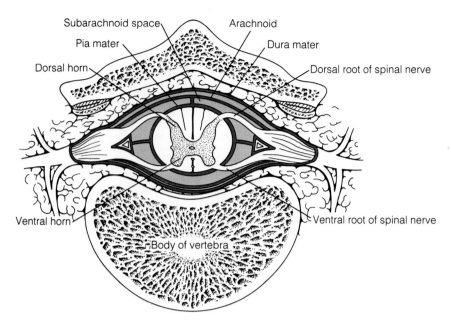

FIGURE 13-10 Cross-sectional view of the spinal cord, showing the meninges.

The cerebrospinal fluid acts as a watery shock absorber that protects the soft tissues of the CNS, and it serves as a medium through which nutrients and waste materials are exchanged between the bloodstream and the cells of the CNS. The outermost membrane, composed of dense connective tissue of a fibrous nature, is called the **dura mater** (*dura* = tough).

Because the cerebrospinal fluid surrounding the spinal cord is continuous with that surrounding the brain, analysis of cerebrospinal fluid can be extremely useful in diagnosing disease or injury in the CNS. Blood in the cerebrospinal fluid may indicate a skull fracture, accompanied by the rupture of blood vessels. In *bacterial meningitis,* the meninges become inflamed as a result of bacterial infection, and bacteria can be isolated from the cerebrospinal fluid. In order to obtain cerebrospinal fluid for analysis, a *spinal tap* or *lumbar puncture* is performed. In this procedure, cerebrospinal fluid is withdrawn with a hypodermic needle inserted between the third and fourth lumbar vertebrae and through the outer meninges. Because the meninges extend past the end of the spinal cord, insertion of a needle at this point will prevent inadvertent damage to the spinal cord.

The cerebrospinal fluid can also be used for the administration of anesthetics. Injections of anesthetics into the cerebrospinal fluid temporarily eliminate sensations in the lower part of the body, but the patient remains conscious. This can be a valuable procedure in situations where the use of a general anesthetic is dangerous.

## Internal Structure of the Spinal Cord

In a cross-section, the spinal cord has an elliptical shape with a central butterfly-shaped region of gray matter surrounded by white matter (see Figure 13-10). The white and gray matter, supported by glial cells, form columns that run up and down the spinal cord. The dorsal extensions of the butterfly's wings, one on each side, are termed the **dorsal,** or **posterior, horns,** and the ventral extensions are the **ventral,** or **anterior, horns.**

The gray matter of the spinal cord contains unmyelinated neurons, including the terminal portions of afferent neurons (both somatic and visceral), interneurons, and the cell bodies of efferent somatic and efferent visceral neurons. The gray matter of the spinal cord is thus an area where afferent and efferent neurons synapse, with or without one or more interneurons interceding between the two junctions.

The white matter consists of myelinated fibers. Fibers forming the white matter connect neurons in the gray matter at different levels of the spinal cord, and they also serve to connect the spinal cord with the brain. The cell bodies of these fibers are in the gray matter of the brain or spinal cord, or in certain spinal ganglia. (**Ganglia** [sing: *ganglion*] are clusters of nerve cell bodies lying outside the brain or spinal cord.) The axons in the white matter of the central nervous system are organized into bundles or *tracts*. **Ascending tracts,** which transmit information to the brain, have a sensory function; **descending tracts,** which carry impulses from the brain down the spinal cord, have a motor function. Some tracts carry only one type of information up or down the spinal cord, but others carry several. Thus, for example, there is an ascending tract that conveys information for subconscious *proprioception* (awareness of position and movement of body parts); another ascending tract carries sensations of pain and temperature.

During transmission of information in the CNS, much sensory information is relayed to the opposite side of the brain. This is because many afferent fibers cross over **(decussate)** as they ascend in the spinal cord; most descending motor fibers also decussate. The decussation of both afferent and efferent fibers results in motor information being delivered to the correct side of the body.

It has been possible to map the functions and locations of various tracts in the spinal cord. This knowledge allows physicians to pinpoint the area of the spinal cord that is damaged by trauma, tumorous growths, or other conditions, since injury to a particular tract or tracts will lead to impaired sensory or motor function. In some cases, this knowledge has also permitted the relief of intractable pain by surgically severing certain tracts without affecting other functions.

## Connection of Spinal Nerves to the Spinal Cord

The anatomical relationship between the spinal nerves and the spinal cord is illustrated in Figure 13–11. Remember that each pair of spinal nerves serves a different segment of the body; this arrangement is a remnant of the segmental development of the human embryo (as will be discussed in Chapter 18). Thus, the spinal nerves that pass out between the vertebrae from each side of the spinal cord contain fibers of the afferent and efferent neurons serving that body segment. If a large muscle develops from more than one body segment, it will have nerve fibers from each of those segments. This is true, for example, of the *rectus abdominis* muscle, which forms the anterior, central portion of the abdominal wall. In fact, the rectus abdominis retains some of its segmental character, which is clearly evident in people with well-developed abdominal muscles.

As a spinal nerve from the periphery, containing both afferent and efferent fibers, approaches the spinal cord, it splits into two branches: the **dorsal root** and the **ventral root,** each of which breaks into a number of smaller branches near the spinal cord (see Figure 13–11). The dorsal root enters the spinal cord

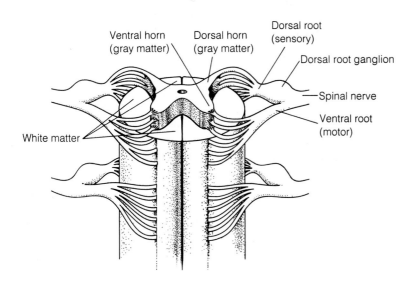

FIGURE 13–11 Diagram showing how the spinal nerves connect with the spinal cord. All sensory (afferent) fibers of a nerve enter the spinal cord in the nerve's dorsal root; motor (efferent) fibers leave in the ventral root.

Ventral horn (gray matter)

Dorsal horn (gray matter)

Dorsal root (sensory)

Dorsal root ganglion

Spinal nerve

Ventral root (motor)

White matter

at the back or dorsal aspect of the body, and it contains all the afferent (sensory) fibers, both somatic and visceral, of the spinal nerve. A swelling on the dorsal root, the **dorsal root ganglion,** contains all the cell bodies of these afferent fibers. Thus, in an afferent neuron, one process extends from a peripheral sense organ to the cell body in the dorsal root ganglion and another from the cell body into the spinal cord. These afferent fibers enter the gray matter of the spinal cord in the region of the dorsal horn. As mentioned previously, afferent neurons synapse with efferent neurons, either directly or by way of interneurons, in the gray matter of the spinal cord. Efferent neurons leave the gray matter of the spinal cord at its ventral horn and emerge by the ventral root of the spinal nerve.

## Spinal Cord Function

### Summary of Connections in the Somatic Nervous System

As summarized in Figure 13–12, somatic afferent neurons enter the spinal cord through the dorsal root of spinal nerves and terminate in the dorsal horn of the gray matter of the spinal cord. At their termini in the gray matter, the afferent neurons may synapse directly with somatic efferent neurons or with interneurons that relay information between segments of the spinal cord and between the spinal cord and the brain. In the latter case, the interneurons then synapse with somatic efferent neurons. The efferent neurons pass directly, without synapsing, from the ventral horn to the skeletal muscles that they innervate. At the neuromuscular junction, all somatic efferent neurons use acetylcholine as neurotransmitter.

### Spinal Reflex Arcs

The spinal cord is not only a conduction pathway to and from the brain, but it can act as a center of **reflex activities**—automatic stimulus-response reactions. Reflex activities are carried over anatomical pathways termed **reflex arcs,** which provide neural connections between receptor and effector organs. These preprogrammed connections allow automatic reflex activities that occur without conscious intervention; however, we may or may not be aware of them.

The simplest type of reflex arc is represented by the knee jerk. Indentation of the tendon just below the knee cap (patella) stretches the large muscle in the front of the thigh (the *quadriceps femoris*). This stretching stimulates afferent receptors in the muscle, sending impulses along an afferent neuron in a spinal nerve to the spinal cord. The afferent neuron synapses directly in the gray matter of the spinal cord with an efferent neuron that sends impulses through the same spinal nerve to the muscle, initiating its contraction (see Figure 13–12, left side). Contraction of the muscle causes the foot to jerk forward.

Other stretch-contraction reflexes of this type occur throughout the body, and these may be tested in a physical exam to indicate whether various parts of the nervous system function properly. Absence of a reflex may indicate nervous system damage involving a spinal nerve or the spinal cord.

FIGURE 13–12 Connections between neurons of the somatic nervous system and the CNS. Left side shows connections in the simplest type of reflex arc; right side shows more complicated connections involving interneurons. In the latter situation, interneurons may also connect with efferent neurons that leave the spinal cord in other spinal nerves.

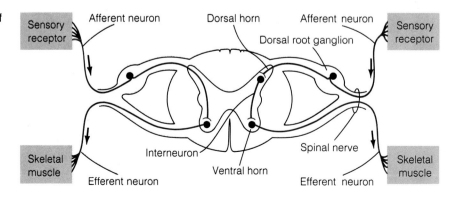

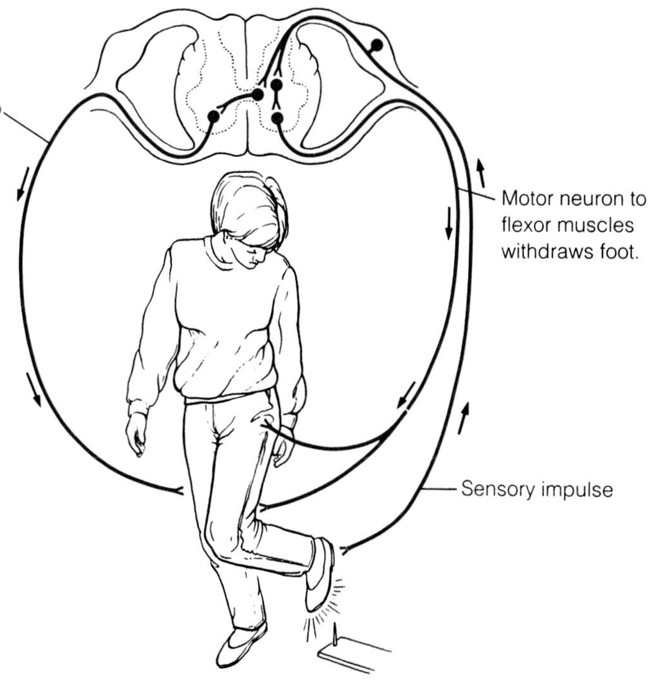

Motor neuron to extensor muscles to maintain balance and support weight.

Motor neuron to flexor muscles withdraws foot.

Sensory impulse

FIGURE 13-13 Example of the coordination of reflexes. The leg that receives a painful stimulus flexes, and the other leg extends.

The knee jerk and similar stretch-contraction reflexes are of the simplest type because they involve only the afferent and efferent neurons and no interneurons. However, most reflexes involve circuits that include many interneurons. In these cases, interneurons connect sensory and motor neurons and may synapse with neurons that are a part of other reflex arcs at different levels of the spinal cord. Thus, reflexes can be coordinated with one another to give an appropriate response. For example, if your right foot contacts a painful stimulus, the right leg flexes and the opposite leg extends, as diagrammed in Figure 13-13. In addition, other reflex pathways are activated that help to maintain balance by bringing about specific movements of the arms and trunk. Such action, in which a particular stimulus always causes a particular response, is of great value not only because it occurs automatically, but because it can occur much more rapidly than if the brain were involved in decision making.

It is important to realize that most reflex circuits are not fully isolated from our conscious control. For example, if you are walking barefoot and know you must step on a nail in order to walk across a board that spans a gorge, you can do so consciously without withdrawing your foot. In this case, the reflex arc is overridden by conscious decision making. Similarly, afferent neurons of a reflex arc may connect with interneurons that convey the sensory stimulus to the brain, where it becomes a conscious sensation, but in that situation, the reflex response will occur before you are aware of the sensation. Such an event occurs if you mistakenly place your hand in hot water: you will withdraw your hand before you become consciously aware of the pain.

Spinal reflexes are not just the province of the somatic nervous system; they also involve the autonomic nervous system. Autonomic reflexes help regulate many automatic body functions, including heartbeat, blood pressure, and breathing, among others. However, the activities of these reflex circuits are monitored, coordinated, and in other ways influenced by the brain, as will be explained shortly.

## Spinal Cord Injuries

Even minor damage to the spinal cord usually will result in impaired function in parts that are served by spinal nerves located below the point of injury. The amount and type of function lost depends upon the severity of the injury and its location. Spinal cord injury usually leads to some degree of permanent disability, since regeneration of neurons in the CNS does not normally occur.

Severance of the spinal cord by bullet wounds, severe vertebral fractures, or other trauma leads to loss of all sensory and motor function (other than some simple reflexes) below the level of the cut. This will result in paralysis of affected muscles. If the transection occurs above the fifth cervical segment, the phrenic nerves to the diaphragm will be involved, causing paralysis of respiratory muscles and death. Such is the result of the hangman's noose. Below this point, breathing is not affected. Paralysis occurring in the lower part of the body including the legs is referred to as *paraplegia.* If the upper as well as lower limbs are paralyzed, the condition is termed *quadriplegia.*

Loss of spinal reflexes below the level of transection or other site of severe spinal cord injury is known as *spinal shock.* Recovery from spinal shock, which may take months, is associated with the reappearance of simple somatic and autonomic reflexes (e.g., knee jerk and bladder evacuation) in the isolated section of the spinal cord.

Recent research has suggested that it may one day be possible to obtain functional regeneration of axons in the human central nervous system—a goal that has been unattainable to date. One promising line of research, so far tried only in experimental animals, is the use of grafts of a peripheral nerve, such as the sciatic nerve, to bridge injured areas of the CNS. For some unexplained reason, the proximal stumps of CNS axons, although normally unable to regenerate, will send new sprouts through a PNS graft and sometimes even make functional reconnections with target organs. Research is now focusing on the reasons why PNS tissue provides an environment conducive to regeneration that is not present in CNS tissue.

## THE CRANIAL NERVES

Twelve pairs of cranial nerves can be identified emerging from the underside of the brain, where they pass into the periphery through foramina (small openings) in the skull. The names of the cranial nerves, their distribution, fiber types, and functions are summarized in Figure 13–14 and Table 13–1. Notice that not all cranial nerves have both afferent and efferent components; some have both, and some have one or the other. Afferent fibers have their cell bodies in ganglia that lie outside the brain, whereas the cell bodies of efferent fibers are present in *nuclei* within the brain. (Nuclei are clusters of cell bodies within the central nervous system.) Four of the cranial nerves (III, VII, IX, and X) have both somatic and visceral components, whereas the others have only somatic fibers.

## AUTONOMIC NERVOUS SYSTEM

Peripheral nerve fibers other than those belonging to the somatic nervous system constitute the autonomic nervous system (ANS) (see Figure 13–9). Autonomic nerves innervate smooth muscle, cardiac muscle, and glands. The autonomic nervous system is so named because it regulates functions that are largely beyond our conscious control; that is, automatic functions. Organs that are served by the ANS include the eyes, salivary glands, heart and blood vessels, lungs, organs of the digestive tract, kidneys, bladder, and external genitals. In previous chapters, we have discussed the role of the ANS in regulating many activities involving these organs, including blood pressure maintenance, heartbeat, breathing, swallowing, vomiting, defecating, bladder evacuation, and ejaculation.

Although the autonomic nervous system, acting as it does to control involuntary functions, is to some degree independent of the rest of the nervous system, it is nevertheless regulated to a significant extent by the CNS. Not only are control centers for many autonomic functions located in the base of the

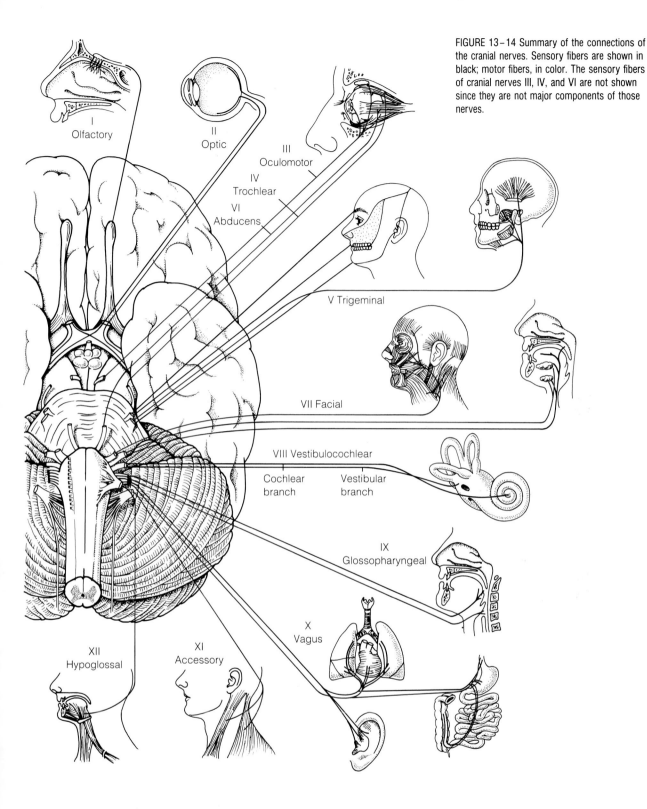

FIGURE 13-14 Summary of the connections of the cranial nerves. Sensory fibers are shown in black; motor fibers, in color. The sensory fibers of cranial nerves III, IV, and VI are not shown since they are not major components of those nerves.

brain, but these centers are modified by input from other parts of the brain. Thus, for example, anger, fear, worry, and other emotions generated in higher brain centers may bring about physiological changes that are mediated by the ANS. We also know that some autonomic functions can be altered by conscious will; for example, people can learn to lower their blood pressures. Moreover, the regulation of most bodily activities that invade adjustments to changes in the external and internal environment requires both the somatic and autonomic systems.

TABLE 13–1  The Cranial Nerves

| Nerve | No. | Fiber Type | Origin | Function |
|---|---|---|---|---|
| Olfactory | I | Sensory | Olfactory receptors in nasal mucosa. | Sense of smell. |
| Optic | II | Sensory | Retina of eye. | Sense of vision. |
| Oculomotor | III | Mixed | Motor: midbrain. | Motor: muscles that control movement of eyeball, eyelids, focusing of lens, pupil changes. |
| | | | Sensory: extrinsic eye muscles. | Sensory: focusing of lens, pupil changes, muscle sense (contraction state of extrinsic eye muscles). |
| Trochlear | IV | Mixed | Motor: midbrain. | Motor: eye muscle involved in movement of eyeball. |
| | | | Sensory: eye muscle. | Sensory: muscle sense (contraction state of eye muscle). |
| Trigeminal | V | Mixed | Motor: pons. | Motor: muscles involved in chewing. |
| | | | Sensory: surface of eye, scalp, mouth, and face. | Sensory: sensation of touch, pain, and temperature for surface of eye, scalp, mouth, and face. |
| Abducens | VI | Mixed | Motor: pons | Motor: eye muscle involved in movement of eyeball. |
| | | | Sensory: eye muscle. | Sensory: muscle sense. |
| Facial | VII | Mixed | Motor: pons. | Motor: muscles that control facial expression, salivary gland secretion, and tear gland secretion. |
| | | | Sensory: taste buds in anterior two-thirds of tongue. | Sensory: sense of taste in anterior two-thirds of tongue; muscle sense. |
| Vestibulocochlear | VIII | Sensory | Cochlear branch: inner ear, cochlea. | Sense of hearing. |
| | | | Vestibular branch: inner ear, semicircular canals, saccule, utricle. | Sense of equilibrium. |
| Glossopharyngeal | IX | Mixed | Motor: medulla oblongata. | Motor: muscles of pharynx that control swallowing; salivary gland secretion. |
| | | | Sensory: pharynx; taste buds in posterior one-third of tongue; carotid sinus. | Sensory: sensations from pharynx; sense of taste in posterior one-third of tongue; regulation of blood pressure. |
| Vagus | X | Mixed | Motor: medulla oblongata. | Motor: muscles of pharynx and larynx that control speech and swallowing; control of cardiac muscle; muscles that control movement of visceral organs. |
| | | | Sensory: visceral organs, heart. | Sensory: sensations from heart and visceral organs. |
| Accessory | XI | Motor | Medulla oblongata | Muscles of soft palate, pharynx, and larynx that control swallowing; muscles of neck and back that control movement of head. |
| Hypoglossal | XII | Motor | Medulla oblongata | Muscles of tongue involved in swallowing, and speech. |

## Connections Between the ANS and the CNS

The ANS is connected to the CNS by visceral fibers in cranial and spinal nerves. Receptors sensitive to pain, heat, cold, pressure, stretch, and the composition of blood gases, among others, send impulses over visceral afferent fibers that enter the brain or spinal cord together with somatic afferent fibers in cranial nerves or in the dorsal roots of spinal nerves. The cell bodies of visceral afferent fibers are located, together with the cell bodies of somatic afferent fibers, in the dorsal root ganglia of spinal nerves and in ganglia of cranial nerves.

The transmission of impulses from the CNS to effector organs (smooth muscle, cardiac muscle, and glands) occurs in visceral efferent neurons. Unlike somatic efferent fibers in which one neuron connects the CNS with the effector organ, two efferent neurons are always required to connect the CNS with visceral effector organs (Figure 13–15). The cell body of the first of these two neurons is located in the CNS, either in the gray matter of the spinal cord, or in the nuclei of cranial nerves. The axon of the first neuron extends from the cell body to a ganglion of the autonomic system, where it synapses with the dendrites and cell body of the second neuron. The axon of the second neuron then carries impulses to the visceral effector organ. The first neuron of the two is called a **preganglionic neuron,** and its axon is called a **preganglionic fiber;** the second neuron is called a **postganglionic neuron,** and its axon is a **postganglionic fiber.**

## Parasympathetic and Sympathetic Divisions of the ANS

Based upon their functions, visceral efferent neurons can be divided into two types: *parasympathetic* and *sympathetic.* Parasympathetic fibers collectively compose the **parasympathetic division** of the ANS, and sympathetic fibers compose the **sympathetic division.**

## Control of Visceral Activities

Most visceral organs are innervated by *both* parasympathetic and sympathetic fibers. Furthermore, the effect of stimulating sympathetic fibers to a particular organ is usually antagonistic to the effect of stimulating parasympathetic fibers to the same organ. This dual, antagonistic innervation of visceral organs permits a fine degree of control over their activities. It is, in effect, like controlling a vehicle with a brake and an accelerator rather than just with one or the other.

We have seen many examples of this type of dual, autonomic control in previous chapters. For instance, the stimulation of sympathetic fibers to the heart increases heart rate and stroke volume, while stimulation of parasympathetic fibers has the opposite effect, slowing and weakening the heartbeat. However, it should be noted that the sympathetic division does not necessarily increase activity, nor does the parasympathetic division decrease it, in all organs. In the small intestine, in contrast to the heart, sympathetic impulses reduce the strength and frequency of peristaltic movements, whereas parasympathetic stimulation increases them.

Although both systems are used continuously in the routine maintenance of homeostasis, there is a pronounced division of function among the two systems. The parasympathetic division controls those activities that are appropriate during rest or recovery from physiological stress. For example, parasympathetic stimulation reduces the frequency and strength of the heartbeat, and decreases blood pressure. This decreases the supply of $O_2$ and nutrients to the cells, since demand for these substances is less during rest. Parasympathetic stimulation increases salivary secretion and mobility of the digestive tract, and inhibits contraction of the anal sphincter, thus stimulating digestion and defecation. It stimulates contraction of the muscle of the bladder and inhibits

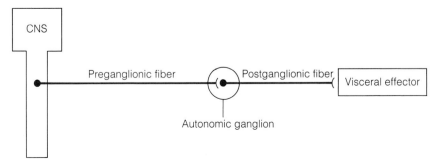

FIGURE 13–15 Diagram of connections between the CNS and neurons of the autonomic nervous system.

contraction of the internal sphincter of the bladder, effects that promote urination. It constricts smooth muscle surrounding the bronchioles and decreases breathing rate (reducing air flow into the lungs); promotes the formation of glycogen in the liver (conserving glucose); constricts the pupil (preventing excess light, that might otherwise be needed for optimal vision in times of stress, from entering the eye); and so forth.

The sympathetic division, on the other hand, aids the body during physiologically stressful situations and in situations involving strong emotions. It is also responsible for what has been called the "fight or flight" response to danger. Blood loss, severe pain, asphyxia, anger, fear, and many other such circumstances evoke sympathetic stimulation.

Activation of the sympathetic division mobilizes the resources of the body and involves responses that are usually antagonistic to those of the parasympathetic division. Thus, for example, heart rate and strength of contraction increase, as do cardiac output and blood pressure; digestion and elimination are inhibited; glycogen is broken down in the liver, raising blood sugar levels; breathing rate increases and bronchioles dilate; and pupils dilate. Stimulation of sympathetic fibers to the sweat glands (there is no parasympathetic innervation of sweat glands) increases sweat production. In addition, sympathetic fibers to the adrenal gland stimulate the secretion of the hormone *epinephrine,* which has the systemic effect of increasing and sustaining all of the above events.

### Neural Components of the Parasympathetic Division

The neural components of the parasympathetic division and their functions are summarized in Figure 13–16. As shown in that figure, the parasympathetic division is made up of two groups of efferent neurons. One group, composing the **cranial outflow,** includes preganglionic fibers that leave the brain in cranial nerves III, VII, IX, and X. In this group, note particularly the course of the **vagus nerve** (cranial nerve X; *vagus* = wandering), which sends parasympathetic fibers to nearly all thoracic and abdominal viscera. The other group of efferent parasympathetic fibers, constituting the **sacral outflow,** consists of preganglionic fibers that leave the spinal cord in the second, third, and fourth sacral spinal nerves. Because the parasympathetic division is composed of groups of fibers in the cranial nerves and in the sacral spinal nerves, it is also sometimes called the *craniosacral division.*

The preganglionic fibers of the parasympathetic division are generally long, and they usually reach all the way to their effector organs, where they synapse with short postganglionic fibers in ganglia that are located near or in the walls of the effector organs. Such ganglia, situated near effector organs, are called **terminal ganglia.**

### Neural Components of the Sympathetic Division

The sympathetic division, also called the *thoracolumbar division,* consists of visceral efferent preganglionic fibers of all thoracic spinal nerves and the first two lumbar nerves, as well as postganglionic fibers (see Figure 13–16). The cell bodies of these neurons, located in the gray matter of the spinal cord, have axons that leave the cord through the ventral nerve roots. Shortly after the spinal nerve roots fuse to form the spinal nerve, the efferent fibers leave the spinal nerve in a short nerve trunk called the **white ramus communicans** (*ramus* = branch; *communicans* = communicating) and synapse with postganglionic fibers in one of two types of ganglia. One type, termed the **sympathetic, lateral** or **vertebral ganglia,** forms chains of 22 ganglia that lie on either side of the vertebral column from the cervical to the sacral region. These two chains of ganglia are referred to as **sympathetic chains** (Figure 13–17). Preganglionic fibers may synapse in the sympathetic ganglia (either the one they first enter or one further up or down the chain), or they may pass through the sympathetic ganglia to synapse in **collateral,** or **prevertebral ganglia.** These ganglia lie in a group in front of the vertebral column, near the organs

| Parasympathetic division (craniosacral) | | Sympathetic division (thoracolumbar) | |
|---|---|---|---|
| Function | Neural components | Neural components | Function |
| Pupil constricts. | Iris | Iris | Pupil dilates. |
| Slight increase in secretion of tears. | Lacrimal gland | Lacrimal gland | Excessive secretion of tears. |
| Increased secretion of saliva. | Salivary gland | Salivary gland | Reduced secretion of saliva |
| Decreased rate and strength of contractions of heart; coronary vessels constrict. | Heart | Heart | Increased rate and strength of contractions of heart; coronary vessels dilate. |
| Constriction of bronchial tubes. | Lungs | Lungs | Dilation of bronchial tubes. |
| Increased motility of stomach and duodenum; stimulation of secretion of pancreatic juices and enzymes. | Stomach, duodenum, pancreas | Celiac g. Stomach, duodenum, pancreas | Decreased motility of stomach and duodenum; inhibition of secretion of pancreatic juices and enzymes. |
| No innervation to adrenal gland. | Adrenal | Adrenal | Increased secretion by adrenal gland. |
| Increased peristaltic actions of colon; relaxation of anal sphincter. | Colon | Superior mesenteric g. Colon | Decreased peristaltic actions of colon; contraction of anal sphincter. |
| Contraction of bladder wall; relaxation of internal urethral sphincter. | Urinary bladder | Inferior mesenteric g. Urinary bladder | Relaxation of bladder wall; contraction of internal urethral sphincter. |
| Erection of penis in males; erection of clitoris and increased vaginal secretions in female. | Gonads and sex accessories | Gonads and sex accessories | Ejaculation in male; stimulation of uterine peristalsis in female. |

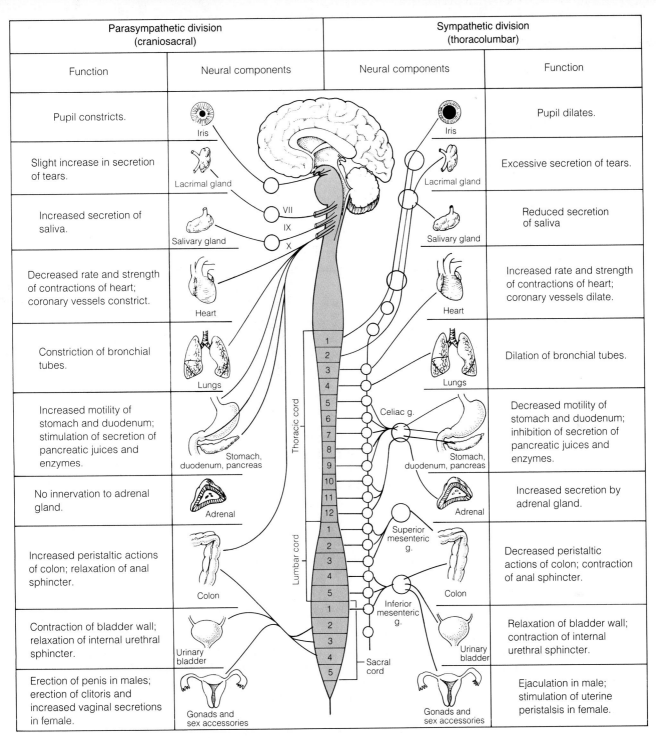

FIGURE 13-16 Neural components and functions of the parasympathetic division (left) and the sympathetic division (right) of the autonomic nervous system.

they serve. The collateral ganglia include the *celiac, superior mesenteric,* and *inferior mesenteric ganglia.* Those preganglionic fibers that synapse within the sympathetic ganglia may rejoin the spinal nerve by way of another short nerve trunk, the **gray ramus** (see Figure 13-16).

## Neurotransmitters in the ANS

All preganglionic fibers in the autonomic nervous system—both sympathetic and parasympathetic—release acetylcholine as a neurotransmitter. Postganglionic parasympathetic fibers also release acetylcholine. Fibers that use acetylcholine as neurotransmitters are said to be *cholinergic.* Most, but not all, sympathetic postganglionic fibers use *norepinephrine* as a neurotransmitter; such fibers are called *adrenergic* because the effects of norepinephrine resemble those of adrenalin (epinephrine). The antagonistic effects of sympa-

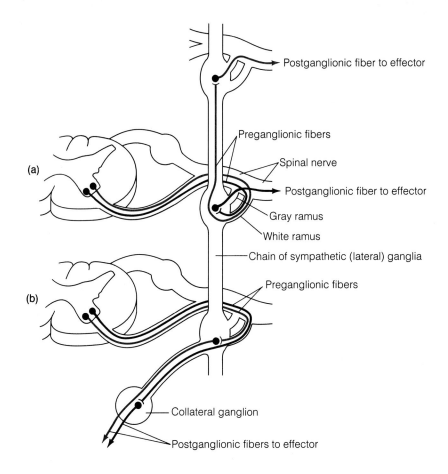

FIGURE 13–17 The ganglia of the sympathetic division and their relationship to the spinal nerves. (a) Preganglionic fiber synapses with postganglionic fiber either in a sympathetic ganglion at the same level (fibers shown in black lines) or in a sympathetic ganglion at a different level (fibers shown in colored lines), then rejoins spinal nerve via gray ramus. (b) Preganglionic fiber synapses with postganglionic fiber either in sympathetic ganglion (fibers shown in colored lines) or in collateral ganglion (fibers shown in black lines).

(a)

Postganglionic fiber to effector

Preganglionic fibers

Spinal nerve

Postganglionic fiber to effector

Gray ramus

White ramus

Chain of sympathetic (lateral) ganglia

Preganglionic fibers

(b)

Collateral ganglion

Postganglionic fibers to effector

thetic and parasympathetic stimulation on a particular organ are thus a result of the different types of transmitters that are released at the neuroeffector junction.

## Drugs That Affect the Autonomic Nervous System

Many drugs affect the ANS. Three well-known ones are beta blockers, atropine, and amphetamine.

### Beta Blockers

Hormones and neurotransmitters secreted by the sympathetic division of the ANS in response to stress and excitement bind to specific membrane receptors, called *beta-adrenergic receptors,* on cardiac muscle cells, blood vessels, and other tissues. The effect of the binding of these substances to their receptors is to increase heart rate and the force of the heart's contraction and to increase blood pressure by causing constriction of arterioles. There is a class of drugs, called *beta blockers,* that block the action of beta-adrenergic receptors and thus inhibit sympathetic stimulation of the heart and blood vessels. Consequently, beta blockers slow the action of the heart under stressful situations and lower blood pressure. They are used extensively in the treatment of high blood pressure and to control angina pains and cardiac arrhythmias. The beta blocker *propanolol,* manufactured by Ayerst Laboratories under the trade name Inderal, is the nation's second most prescribed drug (behind Valium).

In a recent study conducted by the National Heart, Lung, and Blood Institute, propanolol was also shown to be highly effective in preventing deaths among heart attack survivors. Patients in the study who received continuous propanolol treatment beginning shortly after a heart attack had 26 percent fewer deaths than those who received no treatment. Indeed, the early results of the study were so significant that it was stopped nine months early, since it was deemed unethical to deprive heart attack victims of propanolol.

## Atropine

*Atropine* blocks the action of the parasympathetic division by competing with acetylcholine for receptors on the postsynaptic membranes. When atropine is applied to the eyes, it causes a dilation of the pupils that is useful during eye examinations. Because it also inhibits the secretion of acid in the stomach, atropine is sometimes used in the treatment of ulcers.

Atropine has a colorful history. It is the active ingredient in *belladonna* (*bella* = beautiful; *donna* = lady), an extract of the deadly nightshade plant. It was used by Roman and Egyptian women to dilate their pupils, which was thought to make them more beautiful. Assassins of the Middle Ages frequently used deadly nightshade extracts as a source of poison. Even today, leaves of plants containing atropine are still used in many areas of the world to prepare intoxicating beverages that cause euphoria, drowsiness, and delirium.

## Amphetamine

*Amphetamine,* known colloquially as "speed," is structurally similar to norepinephrine. By mimicking the action of norepinephrine, amphetamine stimulates the activity of the sympathetic division. This effect is further enhanced because amphetamine also increases the release of norepinephrine at synapses.

For unknown reasons, amphetamine is useful in treating certain hyperkinetic disorders in children. It is widely abused both as an appetite suppressant and because it produces euphoria and a slight improvement in intellectual and athletic performance, most likely by augmenting alertness and the "fight or flight" reaction.

## Study Questions

1. How do the functions of the endocrine and nervous systems complement one another?
2. a) Describe the components of the central nervous system and the peripheral nervous system.
   b) Why is it a contrivance to divide the nervous system into central and peripheral components?
3. Diagram a typical myelinated neuron, labeling as many structures as you can.
4. Contrast the mechanism by which nerve impulses are propagated along unmyelinated and myelinated neurons.
5. a) Diagram the structure of a synapse between two neurons.
   b) Outline the process by which impulses are transmitted between neurons at synapses.
6. Indicate the functions of 1) afferent neurons, 2) interneurons, and 3) efferent neurons.
7. Diagram the structure of a nerve, labeling the various levels of its organization.
8. Compare the functions and outline the components of the somatic and autonomic divisions of the peripheral nervous system.
9. Explain why analysis of the cerebrospinal fluid is helpful in diagnosing disease and injury in the CNS.
10. a) Diagram the spinal cord in cross-section, labeling gray and white matter, dorsal and ventral horns, and dorsal and ventral roots of a spinal nerve.
    b) Indicate the functions of the gray and white matter of the spinal cord.
11. Describe how the knee-jerk reflex operates.
12. Why are the cranial nerves so named?
13. Compare the functions of the sympathetic and parasympathetic divisions of the autonomic nervous system.

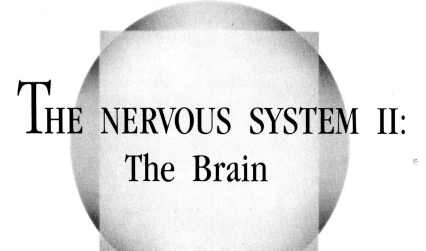

# THE NERVOUS SYSTEM II:
## The Brain

## THE BRAIN

The brain, enclosed within the skull, is the most complex organ of the nervous system, responsible for neural coordination and thought. It weighs about three pounds in the adult, and it is composed of some one hundred billion neurons and an equal number of cells that have supporting, protective, and nutritive functions (glial cells).

### General Organization

The anatomy of the brain is best understood by tracing the embryological development of the central nervous system (CNS). The CNS arises in the early embryo from a hollow, tubelike structure, the *neural tube.* The elongated caudal (tail) portion of the neural tube becomes the spinal cord; the brain develops from the cephalic (head) portion of the neural tube, which at an early stage has three pouchlike swellings, forming the *forebrain,* the *midbrain,* and the *hindbrain* (Figure 14–1a). Subsequently, the forebrain gives rise to the *thalamus, hypothalamus,* and *cerebrum* (*cerebral hemispheres*). The midbrain undergoes no marked modification in later development, but the hindbrain forms the *pons, cerebellum,* and *medulla oblongata* (Figure 14–1b).

In some anatomical schemes, the term *brain stem* is used to describe those portions of the brain that are continuous with the spinal cord and through which all afferent and efferent impulses must pass on their way to and from other brain regions. The brain stem consists of two parts of the hindbrain, the medulla oblongata and pons, and the midbrain (Figure 14–1b, insert).

### Brain Ventricles and Cerebrospinal Fluid

In the development of the hindbrain, midbrain, and forebrain from the neural tube, the central cavity persists within these structures to form four *ventricles* and the *cerebral aqueduct* (Figure 14–2). Each cerebral hemisphere contains one *lateral* ventricle. In their posterior regions, the two lateral ventricles are continuous with the *third ventricle,* which lies in the midline of the forebrain between the two sides of the thalamus and somewhat below it. The third ventricle connects with the *fourth ventricle* in the hindbrain by the cerebral

FIGURE 14–1 (a) Development of the brain from the neural tube in a four-week-old embryo. (b) The major subdivisions of the brain.

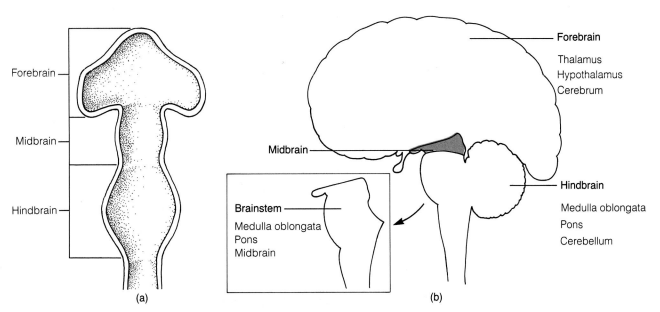

The Brain    277

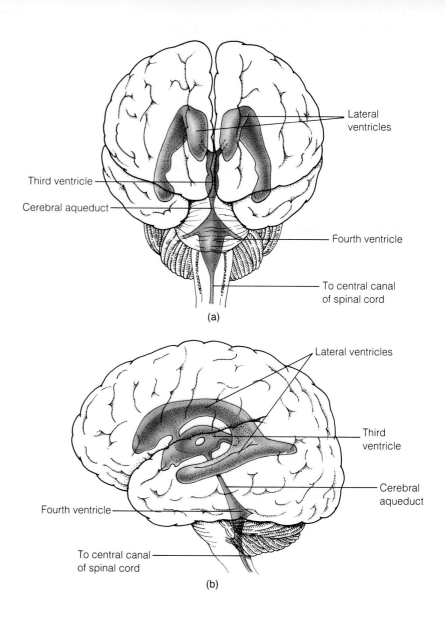

FIGURE 14–2 (a) Anterior view and (b) lateral view of the ventricles in the brain.

Lateral ventricles

Third ventricle

Cerebral aqueduct

Fourth ventricle

To central canal of spinal cord

(a)

Lateral ventricles

Third ventricle

Cerebral aqueduct

Fourth ventricle

To central canal of spinal cord

(b)

aqueduct, a narrow channel in the midbrain. The fourth ventricle is continuous with the central canal of the spinal cord, but this canal is generally obliterated in adults.

## Meninges of the Brain

Inside the skull, the brain is covered and protected by three membranes, the **cranial meninges,** that are continuous with the meninges that cover the spinal cord. The cranial meninges have the same basic structure as the spinal meninges and they have the same names: the *dura mater* (outermost), *arachnoid* (middle), and *pia mater* (innermost). The arrangement of the cranial meninges differs somewhat from that of the spinal meninges (Figure 14–3). First, the pia mater closely covers the surface contours of the brain, but the arachnoid does not. Consequently, there are expanded spaces between the two membranes for accommodating cerebrospinal fluid. Second, the outer surface of the dura mater is fused over most of its extent with the periosteum of the cranial bones. However, the dura mater extends into the deep fissures (clefts) on the surface of the brain and at those points is not continuous with the periosteum. Cavities thus formed between the periosteum and dura mater are called **dural sinuses.** They are lined with endothelium and form channels through which venous blood flows from the brain to vessels leading to the heart.

Blood vessels that serve the brain course between the arachnoid and pia mater (see Figure 14–3). Branches of these vessels pass through the pia mater into the brain. The capillaries of the brain are considerably less permeable than other capillaries in the body to many substances in the blood. Hence their walls form what has been termed the *blood-brain barrier* that excludes substances such as proteins, urea, sucrose, and most antibiotics and other drugs. The blood-brain barrier presumably helps protect the brain from potentially harmful substances that may be present in the blood.

## Cerebrospinal Fluid

**Cerebrospinal fluid** fills the space between the pia mater and the arachnoid, as well as the ventricles of the brain (Figure 14–4). It is produced by membranous structures, rich in capillaries, called **choroid plexuses,** that are rich in capillaries and that penetrate into the lumen of the four ventricles. Because cerebrospinal fluid is formed by filtration through the capillary walls of the choroid plexuses and their epithelia, it is a modified tissue fluid. The cerebrospinal fluid flows from the ventricles where it is produced, through the roof of the fourth ventricle to the space between the arachnoid and pia mater on the surface of the brain and spinal cord. It then enters the dural sinuses through projections of the arachnoid, called **arachnoid villi** (see Figure 14–3), from where it is reabsorbed into the bloodstream.

The cerebrospinal fluid functions as a shock absorber for the brain. It also serves as an intermediary for the exchange of some materials between the brain cells and the blood, but most exchange occurs at the capillaries.

Accumulation of cerebrospinal fluid within the ventricles as a result of blockage at any point before its reabsorption into the bloodstream causes *hydrocephalus,* or "water on the brain." Because the cerebrospinal fluid is produced but cannot drain, relatively high pressures of cerebrospinal fluid can result. In severe, untreated cases, this pressure causes destruction of brain tissue and mental retardation. Hydrocephalus may occur as a result of injury, disease, or congenital deformity. It is a special problem of newborns, in whom it is detected by a greater than normal head circumference at birth. It is treated by the surgical implantation of a one-way shunt between the ventricles and a major vein in the neck.

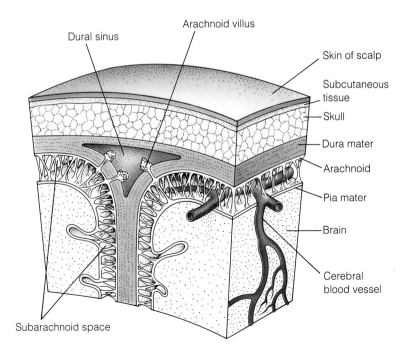

Dural sinus

Arachnoid villus

Skin of scalp

Subcutaneous tissue

Skull

Dura mater

Arachnoid

Pia mater

Brain

Cerebral blood vessel

Subarachnoid space

FIGURE 14–3 Arrangement of the meninges of the brain and associated structures.

FIGURE 14–4 Median section of the brain, spinal cord, and meninges, showing the circulation of cerebrospinal fluid (arrows).

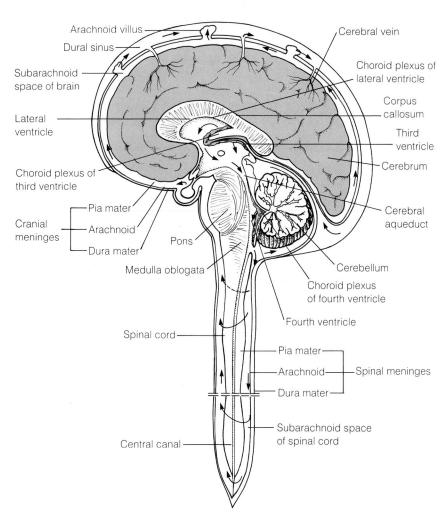

## The Hindbrain

### Medulla Oblongata

The **medulla oblongata,** or **medulla,** is a 3 cm long section of the hindbrain that is continuous with the spinal cord and lies below the pons (Figure 14–5). It functions as a conduction pathway, carrying all ascending and descending tracts that convey information between other parts of the brain and the spinal cord.

There are no distinct regions of gray and white matter in the medulla, as there are in the spinal cord; instead, gray and white areas are intermingled. Many of the descending motor neurons that pass through the medulla from the cerebral cortex *decussate* (cross over) in the medulla to run down the opposite side of the spinal cord. The decussation of these fibers accounts for the fact that skeletal muscles on one side of the body are controlled by the opposite side of the cerebral cortex. Many ascending tracts also cross over in the medulla.

In addition to its function as a conduction pathway, the medulla oblongata is also the site of three reflex centers that control vital functions. These are the **cardiac control center,** which regulates cardiac output; the **vasomotor control center,** which affects blood pressure by regulating the diameter of arterioles; and the **respiratory control center,** which regulates the rate and depth of breathing. The importance of these control centers is underlined by the fact that damage to the medulla, as a result of disease or physical injury, often proves fatal. Other reflex centers in the medulla control so-called nonvital functions including swallowing, coughing, sneezing, and vomiting.

A number of cranial nerves have their nuclei (cell bodies) located in the medulla, from where their axons emerge. These include the *glossopharyngeal nerve (IX)*, with motor fibers that affect swallowing and sensory fibers involved in taste sensation; the *vagus nerve (X)*, possessing motor fibers that influence visceral movement and sensory fibers for visceral sensation; the *accessory nerve (XI)*, possessing motor fibers that influence swallowing and head movement; and the *hypoglossal nerve (XII)*, with motor fibers that influence speech and swallowing.

## Pons

Above the medulla oblongata is a bulbous area called the *pons* (see Figure 14–5). The **pons,** which means *bridge* in Latin, is so named because it serves as an important relay center, or bridge as it were, between various regions of the brain. The pons has fibers that course principally in two directions. Transverse motor fibers connect to the two hemispheres of the cerebellum. Fibers running in a longitudinal direction form motor and sensory tracts that connect the spinal cord and the medulla with the midbrain and cerebral hemispheres.

The pons, like the medulla, contains the nuclei of several cranial nerves. These are the *trigeminal nerve (V)*, with motor fibers that are involved in chewing and sensory fibers that convey sensations from the scalp and face; the

FIGURE 14–5 Major parts of the brain as shown in a median section.

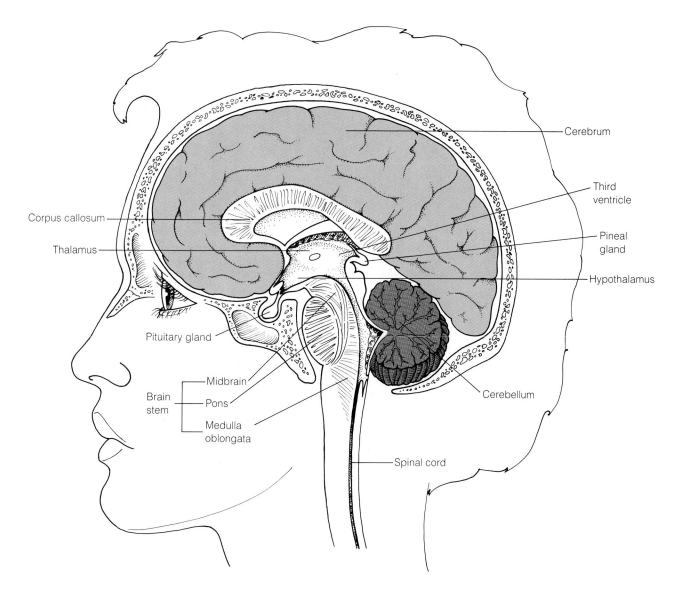

*abducens nerve (VI),* with motor and sensory fibers that are involved in eye movement and muscle sense; the *facial nerve (VII),* containing motor fibers that influence facial expression and sensory fibers for taste; and the *auditory nerve (VIII),* with sensory fibers for posture and hearing.

The pons is also the site of nuclei forming the *pneumotaxic center* and the *apneustic center,* which function in conjunction with the respiratory control center of the medulla to regulate breathing rhythm. Damage to these centers results in *apneustic breathing,* characterized by prolonged inspiration alternating with short, rapid expiration.

## Cerebellum

Above and behind the medulla and the pons is the **cerebellum** shown in Figure 14-5. It consists of two **cerebellar hemispheres** lying side by side, connected by a narrow region termed the *vermis* (= worm). Vast numbers of interneurons connect the parts of the cerebellum to one another and to other parts of the brain and spinal cord.

The cerebellum functions at the subconscious level to monitor, coordinate, and control voluntary and involuntary movement. It receives sensory impulses from peripheral receptors and coordinates these with motor impulses from the cerebral cortex and other centers, modifying impulses to skeletal muscle so as to provide fine coordination of muscular activity, especially in rapid, highly detailed movements. It influences such parameters as the rate, force, and direction of movement. The cerebellum also functions in the maintenance of posture and equilibrium.

Damage to the cerebellum from injury or disease results in deficiencies in coordination and fine muscular control, and defects in the maintenance of muscle tone and equilibrium. However, motor function is not lost; only the coordination of muscular activity is adversely affected. Because speech involves fine control of many skeletal muscles, it may also be impaired by cerebellar damage.

## The Midbrain

The **midbrain,** the upper portion of the brain stem, is a small area lying above the pons and below the thalamus (see Figure 14-5). It serves as a major conduction pathway between the forebrain and the hindbrain.

The ventral part of the midbrain contains two fiber bundles, called the **cerebral peduncles** (*peduncle* = stalk). These contain motor fibers from the cerebral hemispheres to the lower parts of the brain stem (pons and medulla) and the spinal cord. In addition, there are sensory fibers from the spinal cord to the thalamus.

The dorsal portion, or *roof,* of the midbrain contains four nuclei, known collectively as the *corpora quadrigemina.* These integrate visual and auditory sensory input to allow for specific motor reflex responses, including the movement of the eyeball in response to changes in the position of the head, the movement of the head toward sounds, and the avoidance of objects during walking and running.

The cerebral aqueduct, which connects the third and fourth ventricles, runs through the central gray matter of the midbrain. Also extending throughout the length of the midbrain is the *substantia nigra,* a large, darkly pigmented nucleus. The degeneration of the substantia nigra, which sometimes occurs inexplicably, results in Parkinson's disease, discussed later in this chapter.

## Reticular Formation

The **reticular formation** is a complex network of nuclei and nerve fibers extending from the uppermost part of the spinal cord through the brain stem to the lower part of the thalamus (Figure 14-6). It functions to arouse the brain to incoming sensory information and to keep it alert. Without arousal by the

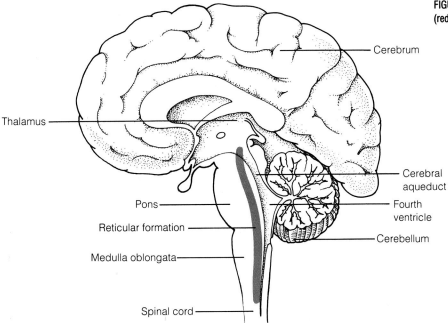

FIGURE 14-6 Location of the reticular formation (red).

Cerebrum

Thalamus

Cerebral aqueduct

Pons

Fourth ventricle

Reticular formation

Cerebellum

Medulla oblongata

Spinal cord

reticular formation, the higher brain centers are unaware of sensory stimuli and consequently cannot react to them.

The reticular formation receives information from most peripheral sensory organs and from a number of motor regions of the brain itself. It sorts through this information, disregarding some of it as unessential and conveying what it regards as important to the thalamus and cerebral cortex (the outermost layer of the cerebrum, which is the site of higher brain functions).

Cycles of wakefulness (consciousness) and sleep (during which consciousness is suspended) are also controlled by the reticular formation. Human beings normally have relatively constant periods of wakefulness and sleep in a 24-hour day. Feelings of sleepiness begin when, apparently as a result of normal fatigue, the reticular formation starts to become less active. Wakefulness and arousal diminish progressively, bringing about relaxation that gradually leads into a *deep sleep*. Deep sleep, also called *nonrapid eye movement sleep (NREM sleep)* or *slow-wave sleep* (because brain waves occur in slow patterns), lasts for about 90 minutes. During this period, body temperature and blood pressure fall, and respiration and heart rate decrease.

Deep sleep is followed by a short period, lasting about ten minutes, called *rapid eye movement sleep (REM sleep),* during which dreaming occurs. Respiration and heart rate increase and become irregular, and blood pressure fluctuates. REM sleep is then followed by another 90 minutes or so of deep sleep. The deep sleep-REM sleep cycle is repeated four or five times in eight hours of sleep, with REM sleep increasing in duration about ten minutes at each successive cycle. Toward the end of a night's sleep, the reticular formation increases in activity until awakening occurs. Of course, a loud noise like an alarm clock or some other stimulus can reactivate the reticular formation prematurely, causing awakening.

It is not clear why sleep is necessary. Although the brain shows electrical activity during sleep, it has been proposed that various areas of the brain may be quiescent at different times during sleep. Presumably, regenerative changes necessary for normal brain function occur during the quiescent phase.

Damage to the reticular formation can cause *coma,* a profound state of unconsciousness from which an individual cannot be aroused by sensory stimuli, no matter how intense. A number of drugs that alter consciousness, such as anesthetics and sedatives, do so in part by affecting the function of the reticular formation.

## The Forebrain

### Thalamus

The **thalamus** (see Figure 14–5) consists of two lobes of tissue, one on either side of the third ventricle, that are connected by a bridge of gray matter. The thalamus receives sensory fibers by way of the spinal cord or cranial nerves from all sensory receptors except those associated with smell. It sorts out and integrates this information in more than 25 nuclei and relays it to the cerebral cortex. There, the sensory information is perceived as conscious sensations of varying intensities and emanating from specific locations.

In addition to its function as a sorting and relay center, the thalamus itself is thought to be responsible for a general, crude type of sensory awareness. It recognizes pain, touch, and temperature change as well as the pleasantness or unpleasantness of sensations. However, it is the cerebral cortex that interprets in a precise way the sensory information received from the thalamus.

Fibers from the cerebral cortex to the thalamus can also modify the activity of the thalamus. Consequently, the functions of these two regions of the brain are closely coordinated.

### Hypothalamus

Lying below the thalamus is a small region of the forebrain called the **hypothalamus,** which has critical functions in maintaining homeostasis. It receives fibers from the spinal cord and most areas of the brain, and it sends important efferent tracts to the thalamus, posterior pituitary gland, and the control centers of the medulla.

Key functions of the hypothalamus include the following:

*Regulation of Visceral Activities.* The hypothalamus regulates the autonomic nervous system by sending fibers to visceral control centers in the brain stem and spinal cord. It controls heart rate, arterial blood pressure, mobility of the gastrointestinal tract, contraction of the urinary bladder, and secretion of many glands.

*Temperature Regulation.* The hypothalamus contains *temperature regulation centers* that operate to maintain body temperature as close as possible to a constant level of 98.6°F (37.5°C). Neurons in the posterior region of the hypothalamus, responsive to a fall in the temperature of the blood reaching them, act to elevate body temperature by stimulating the constriction of skin arterioles, thus reducing heat loss through the skin, and initiating shivering. The muscular contractions associated with shivering generate heat that raises body temperature. Other neurons in the anterior region of the hypothalamus, activated by a rise in the temperature of the blood, facilitate heat loss from the body by causing the dilation of skin arterioles and stimulating the secretion of sweat by sweat glands. Because heat is required for the evaporation of water, heat is removed from the body when sweat evaporates from the skin surface.

Malfunctioning of hypothalamic temperature-regulating mechanisms may cause an elevated body temperature (fever) or a low body temperature (hypothermia), depending upon which areas of the hypothalamus are affected. Many microorganisms produce chemicals called *pyrogens* that are released during infection; these "reset" the hypothalamic thermostat to a higher value, causing fever. Damage to the hypothalamus as a result of injuries or tumors may also cause disruption of temperature regulation.

*Control of Pituitary Function.* Nuclei in the hypothalamus control the function of the anterior pituitary gland by secreting chemicals that stimulate or inhibit the release of hormones from the anterior pituitary. Anterior pituitary hormones influence most body functions, either directly or indirectly, as discussed in Chapter 16. Hypothalamic nuclei also synthesize two hormones, antidiuretic hormone (ADH) and oxytocin, that are transported to the posterior pituitary gland and stored there before being released into the circulation.

*Regulation of Thirst.* Specialized neurons in the hypothalamus, termed **osmoreceptors,** are sensitive to osmotic changes in the blood. When the blood's osmotic pressure is too high (the blood's water content is too low), these neurons, which form the *thirst center,* generate the sensation of thirst. In addition, a high osmotic pressure of the blood stimulates the secretion of ADH from the posterior pituitary gland. As discussed in Chapter 12, ADH stimulates the reabsorption of water from the kidney tubules, lowering the blood's osmotic pressure.

*Control of Appetite.* Hypothalamic neurons, capable of responding to the level of glucose in the blood, form an *appetite center,* activated by low blood glucose, and a *satiety center,* activated by a high level of blood glucose. Research with experimental animals suggests that some obese individuals may have appetite and satiety centers that do not function properly.

*Emotions and Behavioral Responses.* The hypothalamus is part of the *limbic system* (discussed later in this chapter) that is responsible for the expression of many emotional reactions and behavioral responses.

*Generation of Sleep.* The hypothalamus appears to be involved along with the reticular formation in regulating normal patterns of wakefulness and sleep, since it controls temperature, heart rate, blood pressure, and other autonomic activities that change during the course of the awake/sleep cycle.

## Cerebrum

The **cerebrum** composes about four-fifths of the mass of the brain and fills most of the space in the cranial cavity. Its surface is creased and convoluted, possessing ridges called *gyri* (sing: *gyrus*) and creases termed *sulci* (sing: *sulcus*). There are also deep clefts, called *fissures.* One of these, the *longitudinal cerebral fissure,* separates the cerebrum into two portions, the *right* and *left cerebral hemispheres.* These are connected in their middle lower portions by myelinated fibers that form the **corpus callosum** through which the two cerebral hemispheres communicate with one another (see Figure 14–5).

Each cerebral hemisphere is roughly divided by fissures into four lobes: *frontal, parietal, temporal,* and *occipital* (Figure 14–7). The three fissures that divide these lobes, also shown in Figure 14–7, are the **central fissure (fissure of Rolando)** that separates the frontal from the parietal lobe; the **lateral fissure (fissure of Sylvius)** that separates the frontal lobe and part of

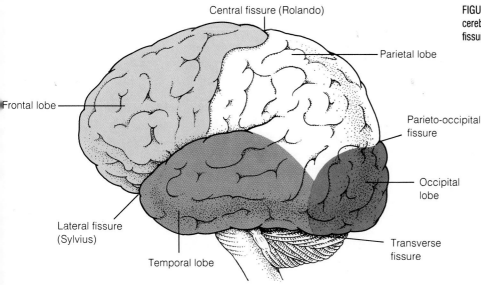

Central fissure (Rolando)

Parietal lobe

Frontal lobe

Parieto-occipital fissure

Occipital lobe

Lateral fissure (Sylvius)

Transverse fissure

Temporal lobe

FIGURE 14–7 Side view of the left cerebral and cerebellar hemispheres, showing the lobes and fissures of the cerebrum.

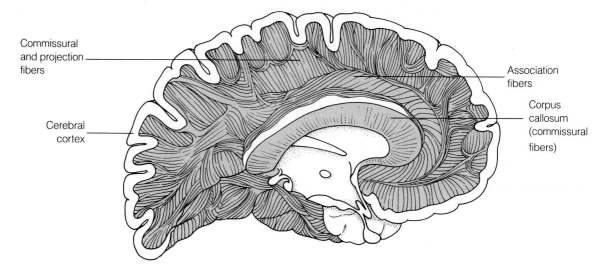

Commissural
and projection
fibers

Cerebral
cortex

Association
fibers

Corpus
callosum
(commissural
fibers)

FIGURE 14–8 White matter tracts forming the medullary body are shown in this drawing of a sagittal section through the brain.

the parietal lobe from the temporal lobe; and the **parieto-occipital fissure** that divides the parietal from the occipital lobe.

In cross-section, each cerebral hemisphere is seen to possess an outer, thin layer of gray matter, 2.5 to 4.0 mm thick, called the **cerebral cortex.** The neurons that compose the cerebral cortex are oriented roughly vertically, and they form six rather poorly defined layers. The function of the cerebral cortex is discussed in the next section.

Lying beneath the cerebral cortex is white matter, consisting of myelinated fibers that form what is called the **medullary body** (Figure 14–8). Impulses reaching and leaving the cerebral cortex travel over the fibers of the medullary body. These fibers are categorized into three groups, based upon the general direction in which they travel.

*Commissural fibers,* many of which form the corpus callosum, connect the two cerebral hemispheres. *Association fibers* connect different areas of the cerebral cortex within the same hemisphere, and *projection fibers* form ascending and descending tracts that connect the cerebral cortex with other regions of the brain.

Situated deep within the medullary body of each hemisphere are masses of gray matter, collectively known as the *basal ganglia* or *cerebral nuclei.* (They are not really ganglia since they lie within the CNS, but basal ganglia is the preferred designation.) The basal ganglia are connected to one another and to the cerebral cortex, thalamus, and brain stem. They have an important role, in conjunction with the cerebral cortex, cerebellum, and other brain regions, in regulating voluntary movement, as discussed in a later section.

The *right* and *left lateral ventricles* form elongated cavities within the right and left cerebral hemispheres, respectively. Although all of the ventricles of the brain produce cerebrospinal fluid, most of it is formed in the lateral ventricles.

## THE CEREBRAL CORTEX

The cerebral cortex is the brain's command center. It consists of many billions of neurons—estimates run as high as 50 billion—that integrate sensory information to form conscious sensations, help coordinate voluntary motor activity, provide ultimate regulatory control over all activities of the nervous system, are involved in emotional reactions, and determine higher brain functions such as speech, language, memory, creativity, judgment, abstract reasoning, and so forth.

## Functional Areas of the Cerebral Cortex

Neurons that control specific functions of the cerebral cortex are localized in various cortical regions. Consequently, it has been possible to construct rough maps of the functional areas of the cerebral cortex, as shown in Figure 14–9. The functional areas are divided into three groups: *motor areas,* which control the contraction of skeletal muscles; *sensory areas,* which receive sensory information from all over the body and interpret it as sensations; and *association areas,* which are involved in integrative activities and higher brain functions such as memory, learning, and language. All of these areas are present in both cerebral hemispheres with the exception of those controlling language functions which, as discussed in a later section, may be located in one or both cerebral hemispheres, depending upon the individual.

### Motor Areas

Located in the frontal lobe just anterior to the central fissure (fissure of Rolando) is the so-called **motor cortex** or **primary motor area,** from which descending pathways lead to the skeletal muscles of the body, controlling their contraction. The neurons of the motor cortex are grouped according to the location in the body of the muscles they control. This is illustrated in Figure 14–10a, which shows a cross-section of the cerebrum in the area of the motor cortex indicating the regions of the body that are innervated by motor neurons. It also shows, in cartoon form, the proportions of motor cortex that are devoted to various motor functions. The large amount of motor cortex devoted to the

FIGURE 14–9 Map of the functional areas of the cerebral cortex.

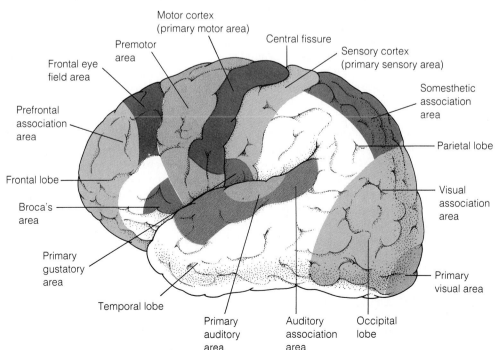

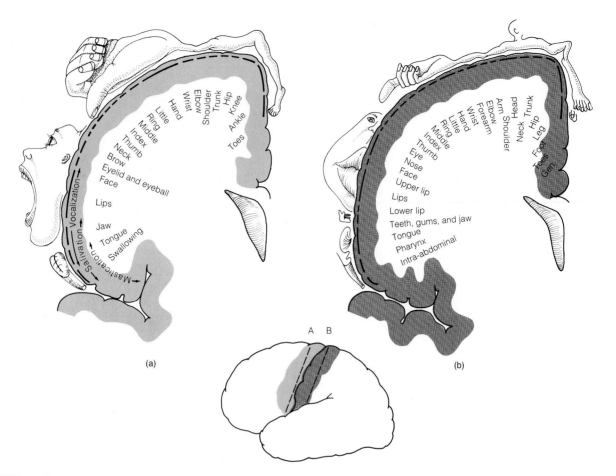

Labels in figure (a), motor cortex:
Hand, Little, Ring, Middle, Index, Thumb, Neck, Brow, Eyelid and eyeball, Face, Lips, Jaw, Tongue, Swallowing, Wrist, Elbow, Shoulder, Trunk, Hip, Knee, Ankle, Toes, Salivation, Vocalization, Mastication

Labels in figure (b), sensory cortex:
Head, Arm, Shoulder, Forearm, Elbow, Wrist, Hand, Little, Ring, Middle, Index, Thumb, Eye, Nose, Face, Upper lip, Lips, Lower lip, Teeth, gums, and jaw, Tongue, Pharynx, Intra-abdominal, Neck, Trunk, Hip, Leg, Foot, Toes, Gen.

(a)          A   B          (b)

FIGURE 14–10 A now classic diagram illustrating the proportion of (a) the motor cortex and (b) the sensory cortex devoted to various functions of the body. Notice that large areas of the motor and sensory cortices control functions of the hands and face, allowing fine manual dexterity and complex speech.

hands and face accounts for the high degree of motor control over these parts, allowing fine manipulations of the hand and sophisticated speech. In the motor cortex, the left cerebral hemisphere is mostly associated with motor function on the right side of the body, whereas the right cerebral hemisphere is associated with motor functions on the body's left side.

Immediately anterior to the motor cortex in the frontal lobe is the *premotor area.* This acts, in conjunction with the motor cortex and basal ganglia, to help maintain posture and balance by controlling movements that accompany voluntary activities, including such things as positioning of the head and swinging of the arms during walking. It also controls learned motor activities such as writing.

Two other important areas that are concerned with motor function are the *frontal eye field area* and the *motor speech area,* also called *Broca's area.* The frontal eye field area, which lies anterior to the premotor area, controls scanning types of eye movement. The motor speech area, inferior to the frontal eye field area, operates in concert with parts of the premotor area and motor cortex to control muscles of the mouth, larynx, and pharynx, allowing thoughts to be translated into speech.

### Sensory Areas

Just posterior to the central fissure, a band of tissues in the parietal lobe forms the **sensory cortex** or **primary sensory area.** It receives sensory information from receptors in the skin, muscles, and viscera, and generates sensations from this information. It determines not only the nature of the stimulus, such as pain, heat, cold, touch, and pressure, but its location in the body. Neurons of the sensory cortex are grouped according to the location in the body of the receptors that generate sensory impulses. Figure 14–10b shows a cross-section in the area of the sensory cortex, indicating those regions from which sensory information is received and the proportions of sensory cortex devoted to various sensory functions. As in the case of the motor cortex, large amounts of the sensory cortex are devoted to the hands and face. This accounts for the sensitivity of these parts and provides complex sensory feedback information needed for the intricate motor control of the hands and face during precise hand manipulations and speech. In a situation also similar to that of the motor cortex, the left cerebral hemisphere is primarily associated with sensory functions on the body's right side, and the right hemisphere is associated with those functions on the left side of the body.

The posterior part of the occipital lobe contains the *primary visual area.* This receives sensory impulses from the eyes, allowing vision. The *primary auditory area,* located in the temporal lobe, receives sensory fibers from the organs of hearing in the ears. At the base of the sensory cortex in the parietal lobe is the *primary gustatory area,* which interprets sensations related to taste. The *primary olfactory area,* involved with the sensation of smell, is located in the medial area of the temporal lobe.

### Association Areas

Association areas are present in regions of the cerebral cortex that are not occupied by motor and sensory areas. The *prefrontal association area,* lying in the frontal lobe anterior to the frontal eye field area, is involved in complex intellectual activities including problem solving, planning, concentration, and controlling behavior to conform to society's expectations.

The *somesthetic association area,* posterior to the sensory cortex in the parietal lobe, receives sensory information from the sensory cortex, thalamus, and other areas of the lower brain. It interprets the meaning of sensations and integrates sensory information. It also functions to store memories of past sensory experiences.

The *visual association area* lies just anterior to the primary visual area and receives impulses from it. It allows visual images to be compared to past visual experiences, permitting recognition. The *auditory association area,* just inferior to the primary auditory area, discriminates between various types of sounds and allows speech to be understood. Association areas in the temporal lobe are involved in the interpretation of language in its written and spoken forms, and with the memory of complex visual scenes and auditory patterns.

### Impaired Function of the Cerebral Cortex

Damage to motor and sensory areas of the cerebral cortex usually causes obvious deficits in the functions determined by these areas. Thus, the ability to move skeletal muscles is adversely affected when the motor cortex is damaged. This is the case in *cerebral palsy,* often caused by injury to the motor cortex before or during birth, or when a *stroke* injures some region of the motor cortex by cutting off its blood supply. Similarly, damage to the sensory cortex interferes with the ability to perceive sensations. Curiously, limited damage to an association area often produces only subtle effects, presumably because other areas can compensate in some manner for the function of the injured area.

### Control of Movement

Voluntary movements involve complex, highly integrated networks of neurons connecting the motor cortex, premotor cortex, basal ganglia, and cerebellum with sensory pathways. It is thought that voluntary movements are initiated by "command" neurons somewhere in the cerebral cortex, which send instruc-

# The Legacy of a Drug Pusher

In June, 1982, a 42-year-old drug addict was admitted to the Santa Clara Valley Medical Center in San Jose, California. Bent over, rigidly immobile, and unable to speak, he had the symptoms of advanced Parkinson's disease. His young age, however, seemed to rule out a diagnosis of Parkinson's disease, which occurs primarily in people over the age of sixty. A week later, the man's sister was admitted to the same hospital with a condition similar to her brother's but not as severe. She had a hand tremor, slow movement, and the expressionless face typical of Parkinson's disease.

The neurologists in charge of the two cases were eventually able to learn from the patients that their symptoms had come on suddenly, several days after they had purchased and used a "synthetic heroin" supplied by a local drug pusher. By a serendipitous turn of events, they also found out that another neurologist from Watsonville, a town 30 miles from San Jose, was treating two brothers, both drug addicts in their twenties, who also had symptoms of Parkinson's disease. Alarmed at the possibility that a highly toxic drug was on the streets, the San Jose neurologists made a public announcement warning of the danger. The announcement turned up three more cases of drug addicts suffering from stiffness and tremors.

Samples of the suspected drug, obtained from the patients, were sent for chemical analysis to several laboratories. However, its identity was difficult to determine, and no progress was made until help came from a toxicologist familiar with the case. He remembered reading a report, published in 1979 in a relatively unknown medical journal, that described the case of a 23-year-old Maryland graduate student who had developed neurological symptoms similar to Parkinson's disease after he had taken a Demerol-like drug, called MPPP, that he had synthesized himself. Analysis of the homemade drug had revealed that the MPPP was contaminated with a related by-product, MPTP. With this information to guide the analytical chemists, it was quickly determined that the "synthetic heroin" contained MPTP.

Subsequent experiments on animals have revealed that MPTP is not the cause of the neurological damage. Instead,

(a) The chemical structures of MPTP and its oxidation product MPP$^+$. Both are strikingly similar to the structure of the herbicide paraquat (b).

MPTP is oxidized in glial cells to form the toxic compound, MPP$^+$ (see accompanying figure). MPP$^+$ is then taken up by cells of the substantia nigra, which are rapidly and specifically destroyed. These are the same cells that degenerate in Parkinson's disease.

The discovery of a drug that specifically destroys cells of the substantia nigra, albeit under tragic circumstances, has had far-reaching implications for Parkinson's research. Since the late 1960s when L-dopa was first used for the treatment of Parkinson's disease, most research on Parkinson's disease had centered on the effects of L-dopa and other drug therapies. Now, MPTP treatment of animals can be used as a model for studying the fundamental mechanisms of Parkinson's disease. Furthermore, recognition of the dramatic effects of MPTP has led to the concept that environmental factors may be important in the cause of the disease.

An environmental cause for Parkinson's disease is supported by several lines of circumstantial evidence. First, MPTP is chemically similar to the herbicide paraquat (see accompanying figure) as well as to a number of other common industrial chemicals. Second, Parkinson's disease was apparently unknown prior to the industrial revolution, as its symptoms are not described in the medical literature, the Bible, or other sources predating that period. This suggests a link to industrialization and consequent pollution. Supporting this link are data that appear to indicate that more people are developing Parkinson's disease at an early age, perhaps because industrial pollution is be-

tions to the basal ganglia. The basal ganglia, in turn, activate neurons in the motor cortex, which sends impulses to motor units in skeletal muscles, causing them to contract. Once the movement commences, the basal ganglia receive continuous feedback from sensory neurons indicating the actions that are taking place. Reacting to this information, the basal ganglia signal instructions to the motor cortex to make adjustments needed to ensure smooth, coordinated movement. Whereas the basal ganglia coordinate slow, continuous move-

coming more extensive.

Other researchers have questioned the importance of environmental factors, citing evidence which shows that countries with different degrees of industrialization have no significant differences in the incidence of Parkinson's disease. Studies of identical twins do not support an environmental cause, but neither do they show a hereditary component to the disease. At the moment, the relevance of environmental factors to Parkinson's disease remains unresolved.

A five-year clinical trial, funded in 1987 by the National Institutes of Health, may help resolve the debate. The trial is designed to test the effect of two drugs on the progression of Parkinson's disease. One of the drugs, Deprenyl, has been shown in experimental animals to prevent the oxidation of MPTP to $MPP^+$ by blocking a cellular enzyme that is known to catalyze the reaction. Consequently, it is thought that Deprenyl might inhibit the production of a toxic chemical from an environmental chemical similar to MPTP. The other drug being tested is tocopherol, a vitamin E derivative. By inactivating chemicals that act as oxidizing agents within the body, tocopherol might also be protective against an MPTP-like chemical.

The renewed research interest in Parkinson's disease has stimulated the development of other approaches to the treatment of the disease. In 1987, a team of neurologists in Mexico City reported dramatic improvement of Parkinson's symptoms when they transplanted dopamine-secreting tissue from the medulla of a patient's adrenal gland into one of the lateral brain ventricles. It is not clear whether the improvement is due to the release of dopamine by the grafts or to some other effect produced by the surgery. Whatever the reason, the technique had been performed in about 80 patients in Mexico, the United States, and other countries by the end of 1987. About one-third of them show significant benefit.

Luck, serendipity, and bizarre coincidences have often played an important role in the advancement of scientific knowledge. The story of Parkinson's disease and a drug pusher trying to make Demerol is a case in point.

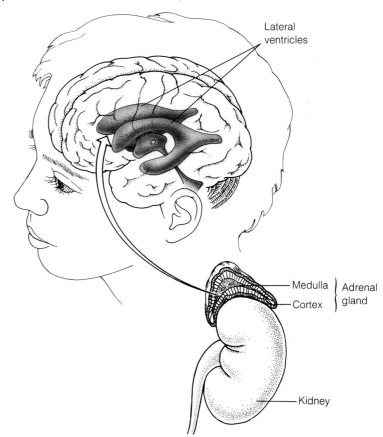

In surgery for Parkinson's disease, tissue from the medulla of the adrenal gland is transplanted into a lateral brain ventricle.

ments, the cerebellum is involved in a similar interaction with the motor cortex to coordinate fine, rapid movements once they have begun. In addition, the basal ganglia and premotor area cooperate to aid smooth movement by making changes in posture and initiating compensatory movements that help maintain balance.

Diseases or injuries that interfere with the operation of motor coordination pathways may result in motor deficiencies. For example, several conditions are known that affect the functioning of the basal ganglia, causing sluggish, stiff, poorly coordinated and purposeless movements. One such condition is Parkinson's disease, which affects about a million people in the United States, a third of whom are severely incapacitated. Most of those with Parkinson's disease are over the age of 60. In Parkinson's disease, the substantia nigra in the midbrain degenerates for unknown reasons. Axons with cell bodies in the substantia nigra normally synapse in the basal ganglia. Degeneration of the substantia nigra results in diminished levels of the neurotransmitter *dopamine* in the basal ganglia, a deficit that brings on symptoms of Parkinson's disease. These include tremors in the hand or head, muscular rigidity, and disturbances of posture and voluntary movement.

The symptoms in most Parkinson's patients can usually be controlled and often dramatically improved by treatment with the drug L-dopa, a precursor in the biosynthesis of dopamine. Apparently, L-dopa reduces the symptoms of the disease by increasing dopamine synthesis in the basal ganglia. Unfortunately, treatment with L-dopa becomes progressively ineffective over a period of five to ten years, and the therapy eventually has to be stopped in some patients, in whom it causes severe convulsions and frightening hallucinations (see *Focus: The Legacy of a Drug Pusher*).

Another disease affecting the basal ganglia is *Huntington's chorea.* It is a hereditary disorder in which neurons in the basal ganglia degenerate over a number of years. Symptoms of the disease become apparent between the ages of 35 and 50. Movements become increasingly jerky and uncontrollable until normal muscular activity is entirely lost. This is followed by mental deterioration and death, usually within 10 to 15 years of the onset of the disease.

## Higher Functions of the Cerebral Cortex

In addition to the control of motor function and the conscious perception of sensations, the cerebral cortex is concerned with other higher functions that involve complex interactions between the vast number of neurons that compose the cerebral cortex. These functions, including memory, learning, language, and the subjective aspects of emotion, are the basis for such uniquely human characteristics as self-awareness, abstract reasoning, judgment, moral sense, and creativity, among many others. Although we have accumulated considerable information about these functions, we are far from understanding the precise neurological mechanisms that mediate them.

## Memory and Learning

Memory, the remembrance of past experiences, is an essential component of all higher brain functions since all of the intellectual capacities of the brain are dependent on the ability to remember information on which further reasoning is based. Loss of memory is a primary symptom of Alzheimer's disease (see *Focus: Alzheimer's Disease*).

It is generally recognized that there are two forms of memory: *short term* and *long term.* Short-term memory allows us to remember past experiences for as short a time as a few seconds or for as long as several weeks. Memories that are short term—what you ate for breakfast yesterday—are usually lost for all practical purposes although some can be recalled under hypnosis. However, some short-term memories may be retained as long-term memories if the information is rehearsed, either consciously or subconsciously. The transfer of information from short-term to long-term memory is called **memory consolidation** and is essential to advanced types of learning and reasoning.

We do not understand the basis of memory, either short- or long-term, but several theories have been proposed. One such theory suggests that long-term memory results when the transmission of impulses between certain circuits of neurons becomes increasingly efficient as short-term experiences are practiced, to the point that the particular circuit becomes easily activated. Presumably, any such change that makes transmission among certain neurons more efficient must ultimately result from changes in the synthesis of molecules, such as neurotransmitters, proteins, and nucleic acids, in the neurons themselves. Indeed, the synthesis of new RNA and proteins is known to occur during learning in experimental animals.

## Conditioned Reflexes

One simple type of learning is a *conditioned reflex.* These reflexes are the basis of many of our behavior patterns. In a **conditioned reflex,** the repetition of a particular stimulus comes to evoke automatically a particular response. Conditioned reflexes are illustrated by Pavlov's classic experiments in dogs. By ringing a bell during feeding when salivation normally occurs, Pavlov was able to condition dogs to salivate in the absence of food whenever the bell was rung. Another example of a conditioned reflex involves increases in blood pressure in humans in response to specific stimuli. For example, many people become so worried about their blood pressure after obtaining several high readings that the mere sight of a physician with a sphygmomanometer will elevate their blood pressure.

## Language and Cerebral Dominance

Although both hemispheres of the cerebral cortex are involved in all functions, there is considerable specialization between the two hemispheres. Consequently, one side of the brain may have a significantly greater responsibility for one function than for another—a phenomenon termed **cerebral dominance.** The development of dominance is particularly extreme with regard to language, both written and spoken, including the ability to say and write what we want to and to understand what is said and written. In over 90 percent of individuals, these functions are mediated by the left cerebral hemisphere, as is fine motor control. These individuals are right handed, since the left hemisphere controls motor function in the right side of the body. In this case, the left hemisphere is said to be dominant over the right hemisphere with respect to handedness and language. However, for many left-handed people, language is controlled by the left hemisphere, and hence in these individuals, language and handedness are not controlled by the same cerebral hemisphere. For other left-handed people, language is controlled in the right hemisphere or bilaterally.

Other skills are also controlled by one or the other cerebral hemispheres. For example, the left hemisphere is dominant in most individuals for mathematical computation, and the right hemisphere for spatial recognition and musical creativity.

Whichever hemisphere governs language in a particular individual, different areas of that hemisphere control different language functions, and complex integration between the areas must occur for normal language abilities. Neurological disorders affecting particular areas will result in various types of deficiencies. These include disorders such as *alexia,* the inability to understand written language, and *aphasia,* the inability to use written and spoken language. One form of alexia, termed *dyslexia,* is an incomplete form of alexia in which letters or numbers are not perceived in their proper positions or orientations, resulting in reading difficulties.

## Intelligence and IQ

Intelligence, or mental acuity, is not easily defined. It is really a characteristic that is derived from a combination of many traits and abilities, including all higher brain functions such as memory, creativity, motivation, reasoning ability, mathematical ability, language ability, spatial sense, artistic aptitudes, and social facilities, among others. However, in any individual certain of these

# Alzheimer's Disease

**P**rogressive deterioration of memory and intellectual capacity, termed *dementia,* was once considered a normal consequence of aging. We now know that dementia results from various disease conditions that affect the brain, and that it is not inevitable.

*Senile dementia (senile* = occurring in the aged) in mild to severe forms affects some 10 to 15 percent of Americans over the age of 65. A rare form of dementia, called *presenile dementia,* can also occur earlier in life, typically in the fifth decade; clinically, presenile and senile dementia are indistinguishable.

Dementia may be caused by a number of conditions, including atherosclerosis of blood vessels in the brain ("hardening of the arteries"), brain tumors, and alcoholism, among others. However, most individuals with presenile dementia and about half of those with senile dementia suffer from a devastating condition, called Alzheimer's disease, named for the French physician who first described it in 1906 (see accompanying figure). Although Alzheimer's disease is difficult to distinguish from other types of severe dementia

on the basis of its clinical symptoms, it has recently been shown that the brains of all patients with Alzheimer's disease have characteristic pathological alterations that can be detected upon autopsy.

The relentless mental deterioration associated with Alzheimer's disease (and with other types of severe dementia) has a typical course. Initially, there is difficulty in remembering minor facts associated with recent events. Short-term memory continues to fail, even the memory of things that just happened, but memory of events in the distant past remains surprisingly detailed. As mental deterioration continues, disorientation grows. Long-term memories begin to fade, and with them are lost the ability to read, write, and even to talk. These losses are often accompanied by drastic swings in mood, hallucinations, paranoia, and disruptive behavior. Eventually, all higher mental functions and learned motor activities are lost. Severely demented patients can live up to 15 years, totally dependent on others to care for them.

Degenerative changes in the brains of individuals with Alzheimer's disease are mainly of two types. First, neurons with cell bodies that contain abnormal clumps of helical (spirally shaped) filaments, called *neurofibrillary tangles,* are scattered among normal cells in the cerebral cortex. Second, there are areas termed *neuritic plaques* that contain abnormal axons and axon terminals wrapped around extracellular deposits of *brain amyloid protein* (see accompanying figure). The density of neuritic plaques in the cerebral cortex of Alzheimer's patients correlates with the severity of their dementia at the time they died.

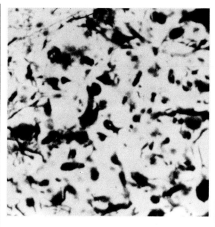

Amyloid-containing neuritic plaques in the brains of Alzheimer's victims.

Recent research shows that neuritic plaques and neurofibrillary tangles are a result of more profound abnormalities in the brains of Alzheimer's victims. Chemical studies of brain tissue have shown that in the cerebral cortex and hippocampus, the levels of choline acetyltransferase, an enzyme that synthesizes the neurotransmitter acetylcholine in the axon terminals of cholinergic neurons, are reduced by 60 to 90 percent in individuals who died of Alzheimer's disease as compared to age-matched controls who died of other causes. (The *hippocampus,* a component of the limbic system, is involved in learning and short-term memory.) This deficit in choline acetyltransferase, and hence presumably in acetylcholine levels, is significant, since it has long been known that drugs that block acetylcholine receptors cause a temporary short-term memory loss and impairment of cognitive functions.

### Total cases of Alzheimer's disease

| 1980 | 1985 | 1990 | 2000 | 2020 | 2030 |
|------|------|------|------|------|------|
| 1.9 mil. | 2 mil. | 2.4 mil. | 2.6 mil. | 3.8 mil. | 4.8 mil. |

Present and projected cases of Alzheimer's disease in the United States. The total number of cases will grow as the population grows.

characteristics may be developed considerably more than others, so that, for example, a person may have high mathematical aptitude but lack creativity or the ability to get along with others.

It is because of the difficulty in defining intelligence that IQ (intelligence quotient) tests have been criticized. They are known to measure mathematical

Another important recent finding is that the reduction in choline acetyltransferase levels in cortical and hippocampal neurons is associated with the degeneration of nerve cell bodies in the *nucleus basalis* (see accompanying figure) and in some adjacent nuclei of the basal forebrain, which lies in the region just above the optic chiasma (the optic chiasma is described in Chapter 15). Cell bodies composing the nucleus basalis send axons that innervate the entire cortex; cell bodies in adjacent nuclei are the source of neurons to the hippocampus. The selective degeneration of these specific neuronal populations presumably accounts for the deficit of choline acetyltransferase in the cortex and hippocampus; the ensuing lack of acetylcholine prevents these nerves from functioning properly in the integration of information relayed to the cortex and hippocampus. It has been suggested that neuritic plaques form as a result of the degeneration of axons that project into the cortex from cell bodies in the nucleus basalis.

Although lesions in nuclei of the basal forebrain are an important cause of Alzheimer's, we do not know why the cells degenerate. Some investigators have proposed that a "slow" virus—a virus with an incubation period of decades—may be involved, but as yet none has been found. Other evidence has suggested that aluminum may play some, as yet unidentified role, since neurofibrillar tangles in the brains of Alzheimer's victims contain high levels of the metal. At least one form of early onset Alzheimer's disease is hereditary, and many researchers now think that all Alzheimer's may be. One research group, for example, has evidence to suggest that a mother, father, sister, brother, or child of any Alzheimer's patient has a 50 percent probability of getting the disease if they live to age 90.

Recent research has shown that the genetic defect in hereditary early-onset Alzheimer's patients is located on chromosome 21. Interestingly, individuals with Down's syndrome, characterized by severe mental retardation, have inherited an extra copy of chromosome 21; upon autopsy, the brains of Down's patients who live beyond their twenties show pathological changes similar to those of Alzheimer's victims, suggesting that the cause of the two diseases may be similar. It has also been determined that the gene that codes for amyloid protein, a component of the neuritic plagues seen in the brains of Alzheimer's (and Down's) patients, is located on chromosome 21. However, it is not presently clear whether abnormal synthesis of amyloid protein is the cause of Alzheimer's disease or an effect of the disease.

There is no effective therapy for Alzheimer's disease. Treatment with chemicals that prevent the breakdown of acetylcholine is not effective, nor is the administration of chemicals that are used by the body as percursors in acetylcholine synthesis. As we come to understand more about the fundamental nature of Alzheimer's disease, we may eventually be able to develop a treatment.

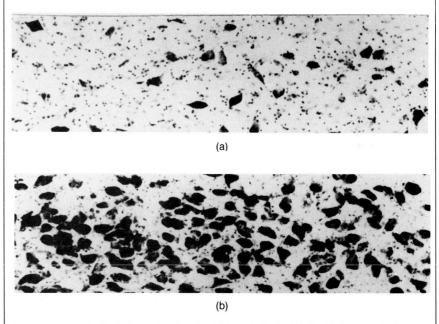

(a)

(b)

Stained nerve cell bodies in the nucleus basalis of (a) a patient with Alzheimer's disease and (b) a person of the same age without Alzheimer's. Notice the loss of neurons in the Alzheimer's patient.

and language abilities required by schools, but it is not clear what else they measure, or how well, nor do they predict success in life.

Much debate currently centers on the degree to which the normal range of intelligence is dependent on heredity or environment—a question that seems impossible to answer as long as we are unable to define or measure intelligence with reliability. The best we can say is that both heredity and environment are involved in intelligence.

### Emotions and the Limbic System

The limbic system is a functional complex of structures encircling the brain stem. It includes areas of the cerebral cortex in the frontal and temporal lobes as well as portions of the thalamus, hypothalamus and certain subcortical nuclei, along with the fibers that connect all of these areas.

The limbic system controls the emotional aspects of behavior, including sexual feelings and emotions such as anxiety, fear, anger, rage, sorrow, affection, and pleasure. It also regulates autonomic functions and body movements that are related to this behavior. Sweating, blushing, blanching, bristling of hair, and changes in heart rate and in blood pressure are types of autonomic responses that accompany strong emotions and are under the control of the limbic system. Laughing, crying, gasping, and cringing are examples of emotionally related activities that involve specific body movements. Because of its central role in behavioral responses, the limbic system is sometimes called the "emotional" brain.

## ASSESSING BRAIN FUNCTION

Normal functioning of the brain can be affected by a number of diseases and disorders as well as by a variety of drugs.

### Electroencephalogram (EEG)

An electroencephalogram is a recording of the electrical activity of the brain. We do not understand how the wavelike patterns obtained in an electroencephalogram are related to brain function, but we do know that characteristic patterns are normally obtained that correspond to various states of consciousness, as shown in Figure 14–11a through c. In addition, an electroencepha-

FIGURE 14–11 EEG patterns in several normal states and during epileptic seizures.

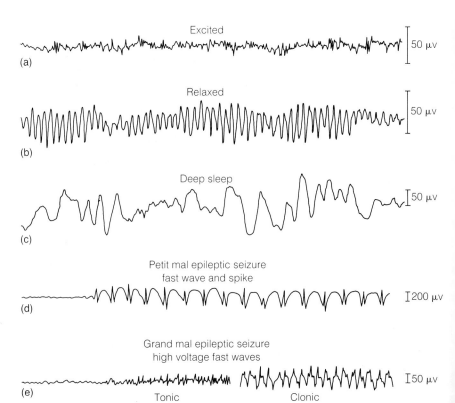

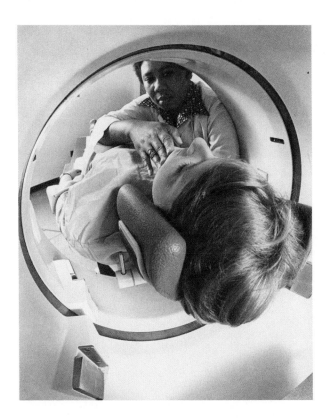

FIGURE 14-12 A person undergoing a CAT scan.

logram is useful as a diagnostic tool since changes in normal wave patterns occur over damaged or diseased brain areas.

In an electroencephalogram, potential changes occurring in neurons are conducted through the surrounding fluids of the brain and may be recorded by placing electrodes on the scalp in various regions. The signal picked up by the electrodes is amplified and then converted by a chart recorder into a visual record. The technology used for obtaining an EEG is identical to that used for an electrocardiogram (see Chapter 9).

Abnormal EEG patterns are obtained in a variety of neurological disorders, including injury, infection, brain tumors, and congenital abnormalities. One moderately common condition in which brain waves are altered is *epilepsy*, which affects about one percent of the population. Epilepsy is characterized by recurrent bursts of excessive electrical activity in neurons in the brain, causing sudden onset of symptoms known as a *seizure*. Seizures may be mild *(petit mal epilepsy)* lasting from a few seconds up to a minute and accompanied by few external signs other than a blank stare. In *grand mal epilepsy,* the seizure is severe. It may be preceded by a distorted sensation of smell, sight, or sound. The individual slumps to the ground unconscious, after which the skeletal muscles undergo sustained, involuntary contractions *(tonic phase),* followed by violent jerking of the body *(clonic phase).* The clonic phase is usually followed by deep sleep. EEG patterns during epileptic seizures are shown in Figure 14-11d and e.

Although epilepsy may be caused by brain injury at birth or in later life, and by brain tumors or infections, in about two-thirds of cases there is no obvious organic cause. Epileptic seizures may often be prevented or decreased in frequency and severity by the use of anticonvulsive drugs such as barbiturates (e.g., phenobarbitol) and bromides.

## CAT Scans

The CAT scanner, developed in the early 1970s, has revolutionized the diagnosis of brain disorders such as tumors, strokes, birth defects, and damage due to injury (Figure 14-12). *CAT* stands for *computerized axial tomography*

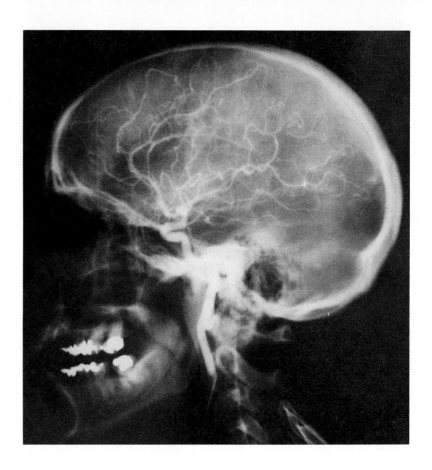

FIGURE 14–13 A cerebral angiogram taken from the left side of the head.

(*tomo* = section), a term that refers to a computer-enhanced technique for the visualization of sections through the brain (or other soft parts of the body).

In a CAT scan of the brain, a scanner rotating around the axis of the patient's head takes a series of X rays at many different angles using a pencil-thick beam. The X-ray pictures are then reconstructed by a computer to form visual images of brain sections.

## Cerebral Angiography

*Cerebral angiography* (*angio* = blood vessel) is a technique that makes blood vessels in the brain visible, allowing the detection of aneurysms, breaks, and other blood vessel abnormalities. It is done by injecting a dye that is opaque to X rays into a blood vessel and then taking an X ray of the brain. The opaque dye makes the blood vessels stand out as white lines against the black background of the brain tissue, which is transparent to X rays (Figure 14–13).

## Drug Effects

Many drugs can affect brain function. *Barbiturates* ("downers") such as phenobarbital depress the action of the central nervous system, causing, with increasing doses, sedation, hypnosis, anesthesia, coma, and death. Because alcohol augments the effects of barbiturates, taking barbiturates with alcohol can be particularly dangerous and may lead to death from overdose.

Marijuana and hashish, extracted from the hemp plant, produce a relaxation of the mind and body and heightened perceptual sensations. Cocaine, which is now widely abused, also heightens sensations (see *Focus: Cocaine* in Chapter 13).

Opiates and related synthetic drugs, including heroin, morphine, and Demerol, relieve pain and produce temporary euphoria. They lead to addiction, and users risk death from overdose due to respiratory arrest. The opiates block the transmission of pain by binding to specific opiate receptors in the brain. We now know that the body produces a class of substances, called *endogenous opiates,* that can bind to opiate receptors and are thus able to block pain perception. There is evidence, for example, that the feeling of well-being experienced by many joggers is due to the secretion of increased amounts of endogenous opiates. Two important types of endogenous opiates are the *endorphins* and *enkephalins.*

## Study Questions

1. Indicate the major functions of each of the following parts of the brain: (1) medulla oblongata, (2) pons, (3) cerebellum, (4) midbrain, (5) reticular formation, (6) thalamus, (7) hypothalamus, (8) cerebral cortex, (9) limbic system.
2. What brain defects are associated with (1) Parkinson's disease, (2) Huntington's chorea, and (3) Alzheimer's disease?
3. Explain how short-term memories are related to long-term memories.
4. Give several examples of the importance of conditioned reflexes to autonomic function.
5. Describe the phenomenon of cerebral dominance.
6. Why is intelligence impossible to measure?
7. Explain why a bacterial infection in the brain is particularly difficult to treat.
8. Describe the use of electroencephalograms, CAT scans, and cerebral angiograms in the diagnosis of brain injury and disease.

### Chapter Fifteen

# THE NERVOUS SYSTEM III: The Sense Organs

## SENSORY RECEPTORS AND SENSATIONS

The human nervous system has a large number of different kinds of **sensory receptors.** These allow human beings to perceive a variety of specific sensations.

### Types of Sensory Receptors

Sensory receptors are of five basic kinds, as determined by the type of stimuli they respond to. *Thermoreceptors* respond to heat and cold. *Pain receptors* respond to tissue damage. *Mechanoreceptors,* sensitive to mechanical changes that deform them, include receptors responsible for touch, pressure, hearing, and equilibrium (movement and position relative to gravity), and for the detection of changes in limb position and in the tension of muscles and tendons. *Photoreceptors* are sensitive to light and are responsible for vision. *Chemoreceptors* respond to changes in the concentration of chemical substances, and include receptors responsible for taste, smell, and for detecting chemical changes in the blood (such as the concentrations of $CO_2$, $O_2$, $H^+$ and glucose).

### Neural Mechanisms in the Perception of Sensations

Whatever the particular type of sensation, its perception involves similar neural mechanisms. First, each sensation occurs by activation of a particular type of receptor that responds more or less exclusively to a particular kind of stimulus. For example, photoreceptors respond to light but not to sound, taste, or other types of stimuli. Second, all sensory receptors convert the stimuli that activate them into nervous impulses. Thus, no matter what the stimulus, all receptors send the same kind of message—a nervous impulse—to the brain. Third, since all nerve impulses are qualitatively similar, the particular kind of sensation that is perceived depends upon the region of the brain to which the sensory receptor is connected by chains of neurons.

### Location of Sensory Receptors

Thermoreceptors, pain receptors, and mechanoreceptors for pressure and touch, and for the detection of changes in limb position and tension in muscles and tendons, are scattered in many places throughout the body. They consist either of naked nerve endings or nerve endings that are encapsulated in connective tissue or epithelial cells.

All other receptors are aggregated in localized sense organs. These include photoreceptors located in the eyes; chemoreceptors for taste, in the taste buds of the tongue; chemoreceptors for smell, in the olfactory organs in the nasal cavities; and mechanoreceptors for hearing and for equilibrium, both located in the ears.

## SENSORY RECEPTORS SCATTERED THROUGHOUT THE BODY

### Thermoreceptors

Several types of encapsulated receptors as well as free nerve endings that are sensitive to temperatures above body temperature (heat receptors) and below body temperature (cold receptors) are scattered unevenly throughout the dermis and subcutaneous tissue of the skin, various mucous membranes, and other localized areas of the body. The face and hands have the greatest number of thermoreceptors. Some areas of the body are more sensitive to heat and others to cold, but overall, cold receptors greatly outnumber heat receptors, suggesting perhaps that cold is usually a greater threat to homeostasis than heat.

Cold receptors are stimulated not only by temperatures below body temperature but also between 46°C (115°F) and 50°C (122°F). Stimulation of cold receptors at these latter temperatures is responsible for the sensation of *paradoxical cold,* so named because a temperature that is actually quite warm seems cold.

Thermoreceptors adapt to stimulation rather quickly, meaning that the sensation they elicit, in this case, of heat or cold, soon fades even though the receptors continue to be stimulated. This adaptation of thermoreceptors accounts for the common experience that bath water, which at first may feel too hot, becomes comfortable within a short time.

## Pain Receptors

Nerve endings that are sensitive to pain are highly branched and located in the deeper layers of the epidermis of the skin and in many connective tissues of the body. These nerve endings respond to stimuli of a variety of types (mechanical, chemical, and thermal) that are strong enough to cause tissue damage. It has been proposed that a chemical released from damaged tissue may initiate depolarization of the sensory nerve endings. There is some evidence that sharp, stabbing, localized pain is conveyed to the brain by different neuronal pathways than is dull, aching pain.

Two general types of pain are recognized: *somatic pain* and *visceral pain.* **Somatic pain** occurs when pain receptors in skin, muscles, tendons, and joints are stimulated; **visceral pain,** from the stimulation of pain receptors in the visceral organs. The site of somatic pain is usually identified quite precisely by the sensory cortex. However, visceral pain is often felt in the skin or just under the skin at a site distant from the organ from which the pain arises. Such pain is called **referred pain** (Figure 15–1). The cause of referred pain is not clear. Some investigators have suggested that pain fibers from visceral organs may synapse in the spinal cord with somatic pain fibers from the same spinal segment, causing the brain to confuse the source of the stimulus.

Amputees sometimes experience pain that appears to emanate from a limb that has been amputated. Termed **phantom pain,** it is thought to occur when the ends of sensory pain nerves that are cut (and hence damaged) during amputation continue to send pain impulses to the CNS; these are interpreted by past experience as coming from the limb.

Unlike thermoreceptors, pain receptors adapt to stimulation only slightly or not at all. This is important to survival, for it helps prevent continued tissue damage. One common type of pain is discussed in *Focus: Headache.*

## Pressure and Touch Receptors

Pressure receptors are located deep in subcutaneous connective tissue, in the connective tissue of skeletal muscles and joints, in serous and mucous membranes, and in the external genitals as well as in other areas. In pressure receptors, termed *pacinian corpuscles,* the nerve endings are wrapped in concentric layers of connective tissue and epithelial cells. Elongation of the pacinian corpuscle by pressure apparently triggers its depolarization, leading to the generation of a nerve impulse. In addition to detecting pressure, pacinian corpuscles are sensitive to high frequency vibration.

Touch receptors are of several types. Encapsulated *Meissner's corpuscles* are located just below the epidermis of the skin and are most numerous in hairless areas such as the palm of the hand, lips, nipples, and external genitals. Meissner's corpuscles also detect low frequency vibration. Other types of touch receptors are located in the tongue and in the cornea, and surround hair follicles; the latter are stimulated by the movement of hairs.

Both pressure and touch receptors adapt quite rapidly. Thus, for example, we notice the feel of clothes when we first put them on, or pressure on the

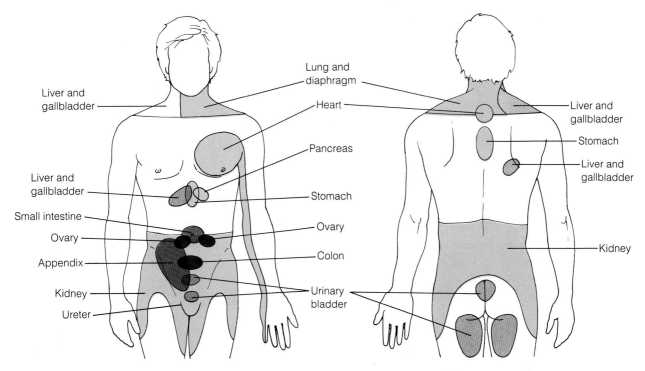

FIGURE 15-1 (a) Anterior view and (b) posterior view of the cutaneous areas of the body to which visceral pain is referred.

lower back and upper legs when we first sit down, but the sensation quickly fades, allowing us to concentrate on more immediate matters.

### Afferent Nerve Endings in Muscles, Tendons, and Joints

Afferent nerve endings in muscles, tendons, and joints provide sensory information about muscle tone and the state of tension in tendons. They also provide a sense of the position and rate of movement of different parts of the body relative to one another. Most of the time, impulses from these afferent nerve endings do not give rise to conscious sensations, but they are crucial in every movement, from simple stretch reflexes to complex activities.

## SENSE ORGANS

### The Eyes

The eyes are specialized organs for photoreception that provide us with vision. They are roughly spherical, about an inch in diameter, and located in bony sockets, the **orbits** of the skull, which provide substantial protection. Between the eye and the orbit is fatty connective tissue that cushions the eye and connects it to the orbit. Also located in this area are the six *extrinsic muscles* that are attached by ligaments to the outer surface of the eye and allow for its movement, and the tear glands that produce tears. Tears serve to wash and lubricate the surface of the eye and contain enzymes that are antibacterial. The tear glands empty by a number of small ducts at the inner, upper lateral surface of the eyelids, and excess tears are drained by ducts into the nose.

### Structure of the Eye

The structure of the eye is shown in Figure 15-2. The outer surface,* or *supporting layer,* of most of the eye is a membrane of dense connective tissue,

---

*With regard to the eye, the terms "inner" and "outer" refer to the center and exterior of the eye, respectively, and not to the body as a whole.

FIGURE 15-2 Median section through the eye.

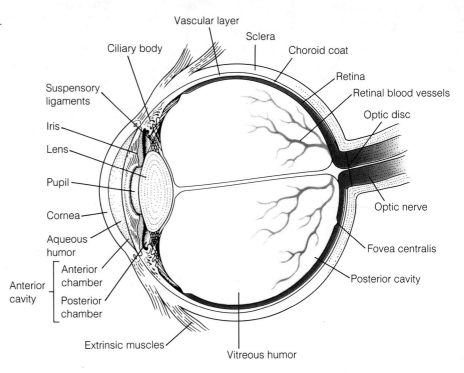

the **sclera,** which is white in color and forms the "white" of the eye. Covering the sclera and lining the eyelids is a thin, transparent mucous membrane, the **conjunctiva.** Anteriorly, the supporting layer is transparent and bulges out somewhat, forming the **cornea.**

Inside the supporting layer is the **vascular layer,** which, as its name suggests, contains many blood vessels and is heavily pigmented. The vascular layer consists of the **choroid** in the posterior two-thirds of the eye; anteriorly, the vascular layer forms the thickened **ciliary body** and the **iris.** The ciliary body contains the *suspensory ligaments* and the *ciliary muscles.* The suspensory ligaments support the lens; contraction or relaxation of the ciliary muscles changes the tension on the suspensory ligaments and hence changes the shape of the lens to permit focusing. The iris, the most anterior portion of the vascular layer, is the colored part of the eye that lies behind the cornea and in front of the lens. It forms a diaphragm, the aperture of which is termed the **pupil.** A ring of smooth muscle cells in the iris contracts or expands the iris causing, respectively, dilation or constriction of the pupil. The iris regulates the amount of light entering the eye, since light enters the eye only through the pupil.

The innermost layer of the eye, the **retina,** consists in turn of two layers (Figure 15-3). The outer layer, lying next to the vascular layer, contains large amounts of black pigment and is called the pigment layer. The inner, or *neural* layer, of the retina, and lines somewhat more than half of the posterior part of the eye, consists of various types of neurons. Closest to the pigment layer are the **rods** and **cones.** These are specialized neurons that function as photoreceptors. The rods and cones synapse with a layer of neurons, called *bipolar cells,* which in turn synapse with another layer of neurons, the *ganglion cells.* In addition to these clearly defined layers of neurons, there are other neurons, called *horizontal cells* and *amacrine cells,* that allow the horizontal transmission of information between the photoreceptor cells (rods and cones), bipolar cells, and ganglion cells. The axons of all of the ganglion cells come together to form the **optic nerve** that pierces the external layers of the eye as it leaves at the back of each eye (see Figures 15-2 and 15-3). Blood vessels that provide most of the eye's nourishment enter the eye in the area where the optic nerve leaves it. This area is called the **optic disc.** Because the optic disc contains no photoreceptors, it forms a "blind spot" on the retina.

The eye is divided into three chambers. The *anterior chamber* lies in front of the iris and lens, and the *posterior chamber,* between the iris and the ciliary body (see Figure 15–2). The anterior and posterior chambers are filled with a watery fluid, called *aqueous humor.* The aqueous humor is continuously synthesized by cells that form part of the ciliary body and drains into veins through a drainage canal, the *canal of Schlemm,* located in the wall of the anterior chamber. The aqueous humor provides nutrients for the cells of the lens and cornea, and the intracellular pressure it maintains gives the eyeball its shape. Behind the lens and the ciliary bodies is the third chamber, containing the *vitreous body.* The vitreous body is a transparent, gel-like substance that aids in the exchange of materials through the retina. In addition, because the vitreous body is a semisolid, it helps hold the lens in place and the inner layer of the retina tightly opposed to its outer, pigmented layer.

## Functioning of the Eye

Light entering the eye passes successively through the cornea, aqueous humor, the lens, and, finally, the vitreous body before impinging upon the inner layer of the retina. As light beams enter the cornea, they are bent (refracted), which helps focus them on the retina. Refraction of light rays by the cornea is, as it were, a coarse focusing adjustment. Fine focusing occurs as the light beams pass through the lens and is the result of changes in the shape of the lens, brought about by relaxation or contraction of the ciliary muscles.

Automatic adjustments of the eye for seeing at different distances are called *accommodations,* and they are carried out primarily by changes in the convexity of the lens. Because light rays from objects close to the eye enter the eye at wider angles than those from objects further from the eye, they have to be

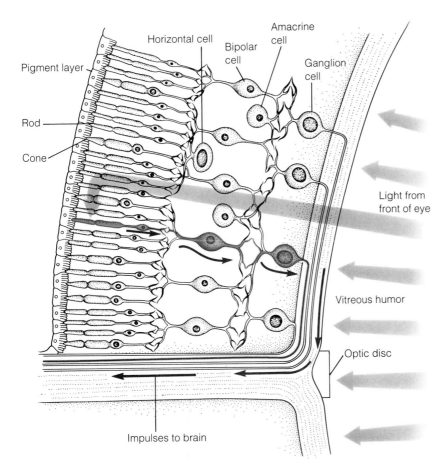

FIGURE 15–3 Diagram of the retina. Notice that light must pass through the neurons in the innermost layer of the retina in order to activate the rods and cones; nerve impulses travel in the opposite direction.

Pigment layer

Rod

Cone

Horizontal cell

Bipolar cell

Amacrine cell

Ganglion cell

Light from front of eye

Vitreous humor

Optic disc

Impulses to brain

# Headache

Headache is one of the more common forms of pain, and it has a variety of causes (see accompanying figure). Sometimes, it is a sign of a serious underlying disorder such as a brain tumor, a blood vessel aneurysm, or high blood pressure. It may result from a head injury, whiplash or, rarely, from eyestrain. It is a symptom of many infectious diseases, including meningitis, influenza, and sinusitis, to name just a few. And it can occur as a reaction to foods, food additives (such as monosodium glutamate, or MSG), and environmental chemicals.

Most headaches, however, are not caused by injury, disease, or sensitivities, but appear to be a result of tension brought on by fatigue, anxiety, or other stressful conditions. The vast majority of people have a simple tension headache at least several times a year, and many people have them frequently. Tension headaches occur when tension causes sustained contractions of skeletal muscles in the head and neck. These contractions may activate pain receptors in the skin and muscles that cover the skull, in the meninges that surround the brain, and in blood vessels associated with the brain, causing a throbbing pain in the forehead and temples that may feel as if the head were being squeezed by an evertightening metal band. Contrary to popular belief, headache pain does not emanate from the brain itself, since brain tissue does not contain

sensory nerve fibers. Nor does worry cause headaches directly; instead, it gives rise to the physical tension that leads to headache pain. Tension can also be a significant factor in "hangover" headaches caused by overindulgence in alcohol. Pain brought on by the toxic effect of metabolites of alcohol on blood vessels in the brain may cause muscle tension that adds to the

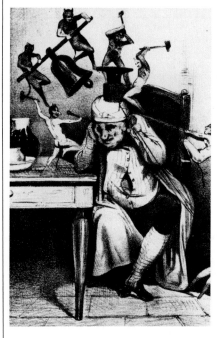

Lithograph from the nineteenth century showing the torments of headache pain.

severity of the headache. Tension headaches typically respond well to nonprescription pain relievers such as aspirin, acetaminophen (Tylenol, Datril, Anacin 3) and ibuprofen (Advil, Nuprin).

In addition to the common tension headache, some 40 million Americans suffer from severe, recurrent, disabling headaches. These are generally divided into three types: *migraine headaches, vascular headaches,* and *cluster headaches.*

Migraine headaches, experienced by about half of the people that have severe headaches, are characterized by a throbbing, incapacitating pain on one side of the head. The pain, which lasts from several hours to several days, is usually accompanied by nausea and vomiting as well as an extreme sensitivity to light and sound. In many people, the onset of pain is preceded by distinctive vision disturbances, including light flashes, blind spots, or zig-zag distortions, which can last for several minutes or hours and tend to disappear when the pain begins.

Migraine headaches generally begin in adolescence and stop occurring in the forties or fifties. They affect two and one-half times more women than men, and they tend to run in families. We do not understand why some people get migraine headaches or what triggers them. In some individuals, certain foods, such as chocolate, aged cheese, and red wine seem to bring on an attack. In other people, they

bent more in order to be focused on the retina. Thus, the lens must assume a more spherical shape to allow for greater bending when an object is close to the eye, as shown in Figure 15–4. Figure 15–4 also demonstrates that all images (other than a single point source) are focused on the retina both upside down and reversed left to right (mirror reversal); the brain must invert and reverse the image.

Light reaching the retina must pass through the nerve fibers and neurons in the inner layers of the retina before striking the photoreceptors—the rods and cones—in the outermost layer of the neural part of the retina. When the rods and cones are stimulated by light and generate a nervous impulse, the impulse

appear to be related to menstruation, to stress, and even to periods of relaxation after a stressful situation. In still others, they appear unrelated to any obvious precipitating factor. Whatever the trigger, a migraine headache occurs when the arterioles that supply the brain constrict and then dilate. The resulting disturbance in blood flow or the swollen arteries themselves are thought to cause the pain.

Vascular headaches, less severe than migraines, occur when the brain arterioles dilate (in contrast to migraines, in which constriction is followed by dilation). The frequency of vascular headaches and their severity varies from person to person. Attacks of vascular headaches may occur for several days in a row every month or so, during which a person typically wakes up in the morning with headache pain that may last for a number of hours each day. The pain ranges from mild to throbbing. Some researchers feel that vascular headaches may be caused by a virus that alternates periods of infectious activity, during which headache pain occurs, with quiescent periods.

Cluster headaches are characterized by severe pain centered in or around one of the eyes. The pain usually lasts several hours, and is often accompanied by tearing and redness in the painful eye and a stuffed nostril on the affected side. As their name suggests, cluster headaches occur in recurrent clusters. A person will typically have a headache every day for a period of weeks, followed by an interval of six months or so that is free of headache pain. Cluster headaches occur twenty times more often in men than in women. They are not hereditary and their cause is not known, although they are associated with heavy cigarette smoking.

Nonprescription medications are not effective in alleviating severe headache pains, and most headache specialists feel that narcotic painkillers like codeine and Demerol should not be used for chronic headaches because of the danger of addiction. Consequently, most treatment programs stress prevention. Beta blockers such as propanolol, prescribed for high blood pressure, have been used with moderately good success for preventing severe recurrent headaches of all types. Calcium blockers, used for treating angina and heart arrhythmias, also show promise in preventing cluster headaches and vascular headaches. Tricyclic antidepressant drugs are very effective in controlling vascular headaches. Acupuncture is reported to be extremely effective in China for treating and preventing recurrent headaches, but has had poor success in the United States.

Thermal biofeedback has helped many migraine sufferers. In this technique, a temperature-sensing electrode is attached to a finger, and the electrode transmits skin temperature readings to an instrument that beeps at a rate that is proportional to the temperature. Using the beeper as a guide, the patient learns to increase the temperature in the fingers, thus diverting blood from the dilated blood vessels in the brain to the hands, relieving the headache pain. Once having learned how to alter blood flow with the aid of the beeper, the patient can learn to do it without receiving feedback from the machine (see accompanying figure).

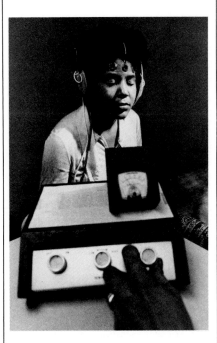

A person learning biofeedback techniques.

must then be conducted in the opposite direction to reach the fibers of the optic nerve (see Figure 15–3).

***Rods and Cones.*** Rods are sensitive to dim light and provide monochromatic vision, whereas cones require fairly bright light and are responsible for color vision. The fact that cones require bright light to function explains why we do not see color in dim light. The human retina has been estimated to contain about 7 million cones and 10 to 20 times as many rods. The light-sensitive portions of both rods and cones are oriented toward the outer, pigmented layer of the retina and contain the *visual pigments* that are responsive to light.

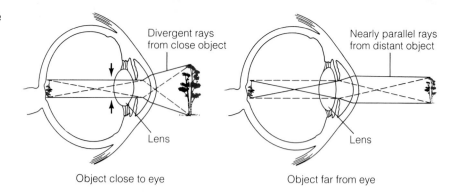

FIGURE 15-4 When objects close to the eye are viewed, the lens must assume a more spherical shape than with objects far from the eye. Also notice that all images are focused on the retina upside down and reversed left to right (mirror reversal).

Divergent rays from close object

Lens

Object close to eye

Nearly parallel rays from distant object

Lens

Object far from eye

The visual pigments in rods and cones, termed **rhodopsins,** consist of *retinal,* a vitamin A derivative, bound to a protein, called *opsin.* The absorption of light by rhodopsins splits retinal from the opsin. This molecular change causes an *increase* in polarization, or *hyperpolarization,* of the rod's or cone's plasma membrane by decreasing the membrane's permeability to $Na^+$ ions. (Hyperpolarization occurs because the cell accumulates more $Na^+$ ions than normal on its external surface).

When a rod or cone is in the dark—not stimulated by light—the cell is strongly *depolarized,* and a neurotransmitter is released steadily from the cell. The neurotransmitter, however, *inhibits* the firing of postsynaptic neurons in the retina. When a rod or cone is stimulated by light and becomes hyperpolarized, neurotransmitter release from it is inhibited, thus generating a nervous impulse in the postsynaptic neurons (Figure 15-5). Since the decrease in transmitter release is proportional to the brightness of light hitting the rod or cone, the response to light is graded. When no longer stimulated by light, the rhodopsins in rods and cones are resynthesized from retinal and opsin.

Rods have one type of rhodopsin that is stimulated by dim light. Cones, on the other hand, have three different types of rhodopsins, each having a different opsin subunit. One of these rhodopsins is sensitive to red light, one to green light, and one to blue light. Stimulation of various combinations of these rhodopsins leads to the perception of all the colors of the spectrum.

*Visual Acuity.* At the back of the eye, almost directly opposite the pupil, is a small depression in the retina, the central region of which is termed the **fovea centralis.** The photoreceptors in this region are all cones, and because they are thinner than in other regions of the retina, they are packed more densely. Moreover, the neurons in the retina diverge from the fovea centralis, with the result that the photoreceptors in the fovea centralis are more exposed to light than in other areas. Furthermore, there are no blood vessels in the fovea centralis. All three of these characteristics make the fovea centralis specialized for the greatest amount of visual acuity, and only those images formed in this area are clear and sharp. Thus, although we can see all objects in the field of vision, we can only see accurately and clearly in a small area in the center of the visual field. For example, as you read this sentence, you see clearly only a few letters at a time, and although you can see words on the entire page, you cannot see them clearly. The eye consequently must scan each line a few letters at a time. Indeed, our eyes must continually scan the environment in order for us to have an accurate image of it.

*Visual Pathways.* Impulses initiated in the photoreceptor cells of the retina leave the eye by the optic nerve. The optic nerves from each eye converge at the **optic chiasma,** where some fibers from each nerve pass over to the other side of the brain. The two fiber tracts leaving the optic chiasma, termed the **optic tracts,** pass to each side of the thalamus and from there to the primary

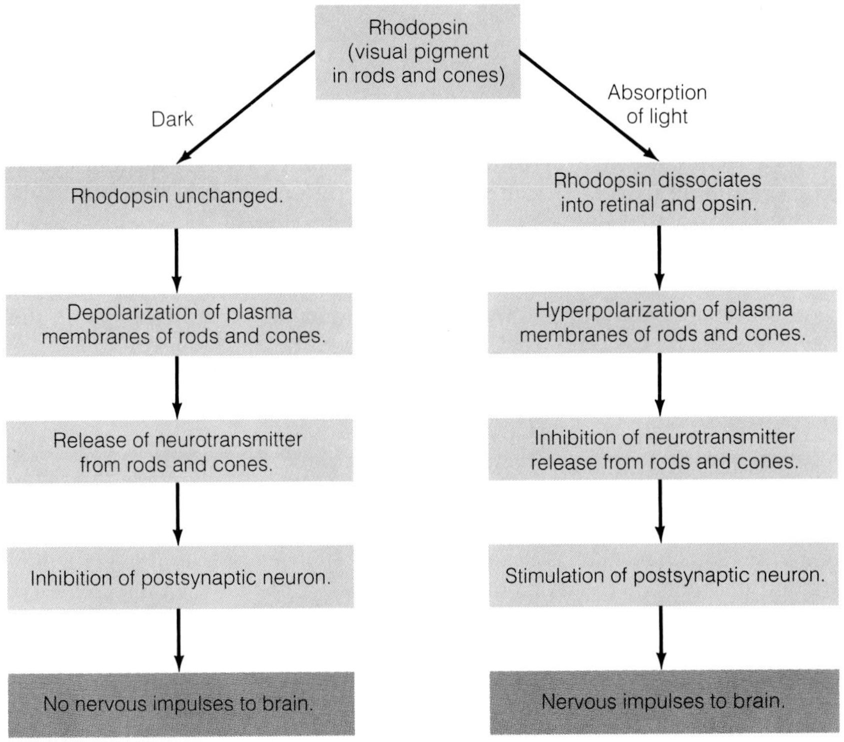

FIGURE 15–5 Summary of events that occur in rods and cones in the dark (left side) and as a result of the absorption of light (right side).

Rhodopsin
(visual pigment
in rods and cones)

Dark

Absorption
of light

| Rhodopsin unchanged. | Rhodopsin dissociates into retinal and opsin. |

| Depolarization of plasma membranes of rods and cones. | Hyperpolarization of plasma membranes of rods and cones. |

| Release of neurotransmitter from rods and cones. | Inhibition of neurotransmitter release from rods and cones. |

| Inhibition of postsynaptic neuron. | Stimulation of postsynaptic neuron. |

| No nervous impulses to brain. | Nervous impulses to brain. |

visual areas at the back of the occipital lobes of each hemisphere. Figure 15–6 shows how inputs from both eyes are connected to the visual cortex. The pathways are such that information from the right side of each eye is conveyed to the primary visual area in the right cerebral hemisphere, whereas information from the left side of each eye is conveyed to the primary visual area in the left cerebral hemisphere. Also notice, however, that images from the right side of an *object* that is viewed are processed in the primary visual area in the left cerebral hemisphere, and vice versa.

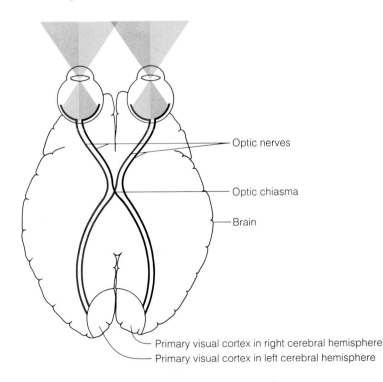

Optic nerves

Optic chiasma

Brain

Primary visual cortex in right cerebral hemisphere
Primary visual cortex in left cerebral hemisphere

FIGURE 15–6 Summary of connections of the eyes to the visual cortices. Notice that information from the right side of each eye converges on the visual cortex of the right cerebral hemisphere; information from the left side of each eye, on the visual cortex of the left cerebral hemisphere.

## Disorders of the Eye

***Problems of Accommodation.*** The most common disorders of the eye involve the inability to focus images on the retina (Figure 15–7). In *myopia,* or *nearsightedness,* the eyeball is too long for the accommodating power of the lens. Consequently, the image from a distant source focuses in front of the retina, resulting in poor distant vision but good near vision. Myopia runs in families and usually develops at about the age of 12, often worsening gradually until about age 20. It can be corrected by a concave lens fitted in front of the cornea. Near vision is not impaired by the presence of the lens since the eye is able to accommodate sufficiently to focus the image of a close object on the retina.

In *hyperopia,* or *farsightedness,* the eyeball is too short for the accommodating power of the lens. Distant objects are seen clearly, but close objects are focused behind the lens. The condition is usually present from birth and detected in early childhood. Like myopia, hyperopia has a strong hereditary component. Hyperopia may be corrected by convex lenses.

When the cornea or lens has an irregular curvature rather than a smooth one, *astigmatism* results. Astigmatism causes a blurring of vision, since light is focused unevenly on the retina. It is usually correctible with cylindrical lenses. Astigmatism is present from birth and does not grow worse with age.

FIGURE 15–7 Correction of accommodation problems by lenses.

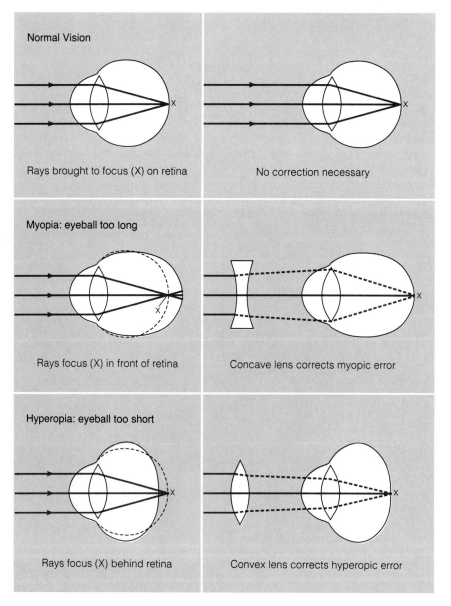

Normal Vision

Rays brought to focus (X) on retina

No correction necessary

Myopia: eyeball too long

Rays focus (X) in front of retina

Concave lens corrects myopic error

Hyperopia: eyeball too short

Rays focus (X) behind retina

Convex lens corrects hyperopic error

*Presbyopia* occurs as a result of hardening of the lens with age, a condition that reduces the ability of the lens to accommodate for seeing close objects. Presbyopia generally first becomes noticeable in the mid-forties and worsens gradually. As it worsens, printed material must be held further and further from the eye in order to focus it on the retina. Presbyopia is responsible for the observation by elderly people that their eyes are fine but their arms aren't long enough!

*Eye Infections.*   Infections of the eye and surrounding parts can cause troublesome symptoms and can sometimes be dangerous. Bacterial or viral infections of the conjunctiva cause *conjunctivitis,* characterized by reddening of the eye and inflammation of the eyelids. A *sty* is a bacterial infection of the follicle of an eyelash. *Ulcers,* or open sores, on the cornea can be caused by Herpes simplex virus, or they may result from a bacterial infection following injury to the eye. Such infections must be treated promptly to prevent permanent scarring of the cornea. If scarring of the cornea seriously affects vision, a *corneal transplant* may be performed. In this procedure, the damaged cornea is removed under general anesthesia, and a cornea taken from a cadaver is sewn in place. Because the cornea has no blood supply, rejection of the transplant rarely occurs since white blood cells and serum antibodies that mediate the rejection process do not reach the corneal tissue. About 10,000 corneal transplants are performed each year in the United States.

*Cataract.*   One fairly common disorder of vision, occurring most frequently in old age or as a complication of diabetes mellitus, is a clouding of the lens, called *cataract.* Cataracts usually occur in both eyes, and as the opacity of the lens gradually increases over a period of years, vision progressively deteriorates. Treatment of cataract involves surgical removal of the lens. This makes the eye markedly farsighted, and eyeglasses are then prescribed to correct this problem. (See Focus: *Monet's Cataracts.*)

*Glaucoma.*   **Glaucoma** is a severe eye disorder that occurs most commonly in people over 60. It is caused by a blockage of the canal of Schlemm that normally draws away excess aqueous humor. Pressure builds up in the anterior and posterior chambers of the eye, and pressure against the lens puts pressure, in turn, on the vitreous body. The increased pressure on the vitreous body collapses the blood vessels that nourish the retina, causing degeneration of retinal neurons.

Glaucoma may occur suddenly (acute glaucoma), but more commonly the canal of Schlemm becomes blocked over a period of years (chronic glaucoma). Untreated glaucoma may lead to severe vision impairment and blindness, but treatment with drugs or surgery can relieve intraocular pressure and prevent damage to the retina. Regular eye examinations that include a test for intraocular pressure can detect chronic glaucoma before damage to the eye occurs.

One to two percent of the population over 40 has chronic glaucoma. It is not known why the canal of Schlemm becomes obstructed in some individuals, but the disease has a strong hereditary component.

*Detachment of the Retina.*   Detachment of the retina is a rare condition in which the retina lifts away from the choroid layer. It may occur when a physical blow to the head or to an eye causes a small tear in the retina. Fluid accumulates between the retina and the choroid layer, gradually separating the two layers. Peripheral vision is usually lost first and the field of vision continues to narrow if the condition is not treated. The retina may be reattached to the choroid layer using a laser beam or a freezing probe.

*Colorblindness.*   Colorblindness is a hereditary disorder that is sex-linked (see Chapter 19) and is far more common in men than in women. It results

# Monet's Cataracts*

Claude Monet (1840–1926) is considered the most important French Impressionist landscape painter (see accompanying figure). Less concerned with portraying detail than with capturing the mood, or impression, of a scene, Monet used dabs of bright color to create shimmering, luminous patchworks that reflected nature's vibrancy. Throughout his long career, beginning in the 1860s and stretching some 60 years, Monet remained faithful to the impressionist view of nature.

When Monet was in his late sixties, he developed cataracts that began to cloud the lenses of both eyes and affect his vision. His paintings dating from that time show a blurring of forms somewhat more extreme than the deliberate imprecision of detail in his earlier work. As Monet's cataracts gradually worsened over the next 14 years, forms in his paintings became increasingly vague. He complained of difficulty in distinguishing colors, a problem that was particularly vexing to him, leading him in 1918 to lament:

*I no longer perceived colors with the same intensity, I no longer painted light with the same accuracy. Reds appeared muddy to me, pinks insipid, and the intermediate or lower tones escaped me. . . .*

*At first I tried to be stubborn. How many times, near the little bridge where we are now, have I stayed for hours under the harshest sun sitting on my camp-stool, in the shade of my parasol, forcing myself to resume my interrupted task and recapture the freshness that had disappeared from my palette! Wasted efforts. What I painted was more and more dark, more and more like an "old picture," and when the attempt was over and I compared it to my former works, I would be seized by a frantic rage and slash all my canvases with my penknife.*

By the summer of 1922, Monet's vision had deteriorated to the point that he could discern only light in his right eye and had 20/200 vision in his left eye, on which he relied to see and paint. The cataracts caused a yellow-brown opacity of the lenses, and he increasingly saw things in yellow tones. Several extremely yellow paintings date from this period.

In December, 1922, Monet decided to undergo cataract surgery on his right eye. Removal of the yellow-brown cataract in that eye allowed him to perceive colors, especially in the blue and violet range, that he had not seen for years. However, vision remaining in his left eye would still have had a yellow-brownish tint. Because Monet's cataract glasses were of poor quality, with high magnification in the right lens and none in the left, he could not

Claude Monet at age 86.

have seen with both eyes at once. He was forced to block off his left eye, which gave him vision that was overwhelmingly blue, perhaps because he was not used to seeing bluish tones. The color distortion frustrated him greatly.

*I see blue; I no longer see red or yellow. This annoys me terribly because I know that these colors exist, because I know that on my palette there is some red, some yellow, a special green and a certain violet. I no longer see them as I saw them previously, and I recall very well the colors that they gave me.*

Several of his paintings dating from this time are predominantly blue.

When in 1924 Monet was prescribed glasses with a yellow-green tint, the dominant blue tones gradually disappeared from his vision. From then until his death from lung disease in December, 1926, he was able to paint using his right eye without color distortion.

Monet's work is of special medical interest because of the effect of cataracts on his late style. As his vision and color perception changed because of his cataracts, so did the character of his paintings. This is not to belittle or dismiss the artistic merit of his later works, but to show that the way he interpreted the world around him was influenced by the way he saw it. It is ironic that Monet's late paintings, characterized by vague scenes and color distortions, are now considered to be a link to the abstract art of the twentieth century.

*This essay is based upon an article entitled "Monet's Cataracts" by James G. Ravin, M.D., appearing in the *Journal of the American Medical Association,* Vol 254, pp. 394–399, July 19, 1985.

from a defect in one or more of the three types of cones. The most common form of colorblindness, red-green colorblindness, involves an inability to distinguish red from green.

## The Eye Examination

In an eye examination, visual acuity is tested by determining an individual's ability to read letters of various sizes on a chart 20 feet away from the eye. Vision in each eye is designated as normal, or 20/20, if the person being tested is able to distinguish at 20 feet the smallest letters that a person with normal vision can read at 20 feet. Vision of 20/60 translates into an ability to read at 20 feet what a normal person can read at 60 feet. Since visual acuity can vary in each eye, it is measured in each eye separately.

In addition to determination of visual acuity, an eye exam may involve checking intraocular pressure and observing the condition of the retina and blood vessels of the eye. This latter test involves the use of an *ophthalmoscope,* an instrument that provides a strong light source by which a physician can look into the eye through the pupil (Figure 15–8). Such observation can detect anemia, high blood pressure, and deterioration or detachment of the retina.

## Taste Receptors

### Structure of Taste Receptors

Taste receptors are present in small onion-shaped organs called *taste buds* that are located primarily on the upper surface of the tongue. They are composed of several cell types, including elongated *taste receptor cells* and *supporting cells,* as well as *basal cells* which may give rise to one or both of the other cell types.

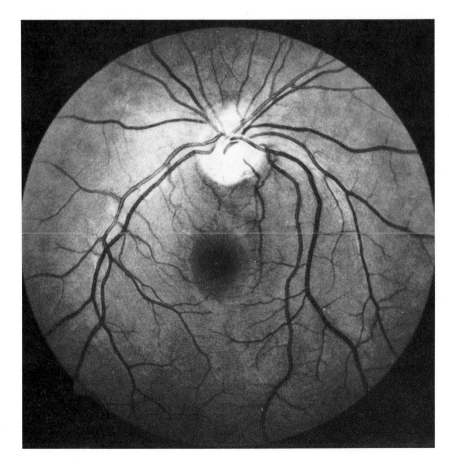

FIGURE 15–8 Appearance of a normal retina through an opthalmoscope, showing retinal blood vessels. The pale region is the optic disc; at the center of the dark, circular region is the fovea centralis.

Each taste bud opens onto the surface of the tongue through a small passageway called a *taste pore*. The ends of the taste receptor cells, which are folded into microvilli, project into the taste pore (Figure 15–9). The microvilli, also called *taste hairs,* are thought to be the sensitive part of the taste receptor cells.

In order for a substance to be tasted, it must dissolve in saliva or other fluid in the mouth and enter the taste pore, where it comes in contact with the microvilli (taste hairs) at the ends of the taste receptor cells. In some manner we do not fully understand, the taste receptor cells respond to a dissolved substance by becoming depolarized, generating a nervous impulse. A good deal of evidence suggests that the stimulation of taste receptor cells probably involves the attachment of the substance to specific receptor sites on the cell surface.

## Taste Sensation

Taste cells respond to four basic tastes: sweet, sour, bitter, and salty. We now know, contrary to what had been believed, that a single taste receptor cell is not rigidly specific for one basic taste, but may be sensitive, to varying degrees, to more than one basic taste. The wide variety of tastes we are able to perceive presumably results from the stimulation of taste receptor cells by various combinations of the four basic tastes. Furthermore, the flavor of food depends not only upon sensory information received from taste receptors but, as anyone who has ever had a stuffy nose can attest to, on information from olfactory receptors as well. Since taste and smell are interrelated, what appear to be unusual flavors may actually be smelled rather than tasted.

Receptors for the four basic tastes are present in high concentrations in particular areas of the tongue. As shown in Figure 15–10, receptors for sweet occur primarily at the tip of the tongue; receptors for sour, at the central edges of the tongue; receptors for salt, at the front and back edges of the tongue; and receptors for bitter, at the back of the tongue.

## Neural Pathways for Taste

The facial nerve (VII) carries impulses from taste receptors on the anterior two-thirds of the tongue, and the glossopharyngeal nerve (IX), from taste receptors on the posterior one-third of the tongue, to the medulla oblongata. From there, they are conveyed to the thalamus and then to the primary gustatory area in the cerebral cortex.

FIGURE 15–9 Diagram of a taste bud.

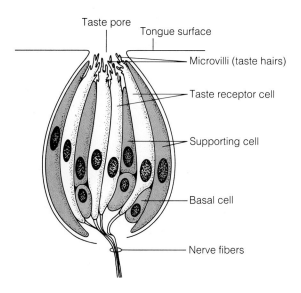

Taste pore
Tongue surface
Microvilli (taste hairs)
Taste receptor cell
Supporting cell
Basal cell
Nerve fibers

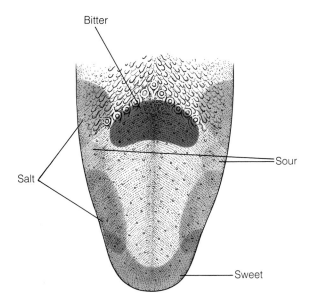

Bitter

Salt

Sour

Sweet

## Olfactory Receptors

### Structure of Olfactory Receptors

Olfactory receptors, responsible for the sense of smell, are present in the two **olfactory organs,** specialized regions of the mucous membrane that lines the roof of the two nasal cavities (Figure 15–11). Part of the mucous membrane of an olfactory organ is shown in cross-section in Figure 15–11c. It consists of three cell types: *basal cells* and *sustentacular* (supporting) *cells* interspersed with *olfactory cells.* The olfactory cells are neurons, the cell bodies of which are located in the mucous membrane that forms the olfactory organ. From the cell body of an olfactory cell, a dendrite runs to the surface of the mucous membrane and terminates in a bulblike expansion called the *olfactory knob,* from which long cilia project into the nasal cavity. The cilia are also referred to as *olfactory hairs.* Volatile substances in the air are believed to stimulate olfactory cells by attaching to specific receptor sites on the surface of the cilia and in some way triggering depolarization of the cell.

### Neural Pathways for Smell

Axons of the olfactory cells penetrate the lamina propria (supporting layer) of the mucous membrane and unite to form bundles of nerve fibers that are collectively referred to as the **olfactory nerves.** These pass through the cribriform plate, a sievelike patch of the ethmoid bone at the base of the cranium, and terminate in the **olfactory bulbs** of the brain, paired masses of gray matter lying underneath the frontal lobes of the cerebrum (see Figure 15–11b). The olfactory nerve fibers synapse with neurons in the olfactory bulbs, forming the **olfactory tracts,** which convey impulses to the primary olfactory areas of the cerebral cortex.

### Smell Sensation

Little is known about the mechanism by which different odors are perceived. To date, there is no evidence for the existence of different types of olfactory cells that are responsive only to specific odors.

Because the olfactory organs are located high up in the roof of the nasal cavities, only a small amount of inhaled air normally comes in contact with the olfactory organs. For this reason, many odors are perceived only weakly with

FIGURE 15–11 (a) Location of olfactory receptors at the roof of the nasal cavity. (b) An enlargement of (a), showing the olfactory nerves passing through the cribriform plate of the ethmoid bone and the olfactory bulb. (c) Details of the mucous membrane of the olfactory organ.

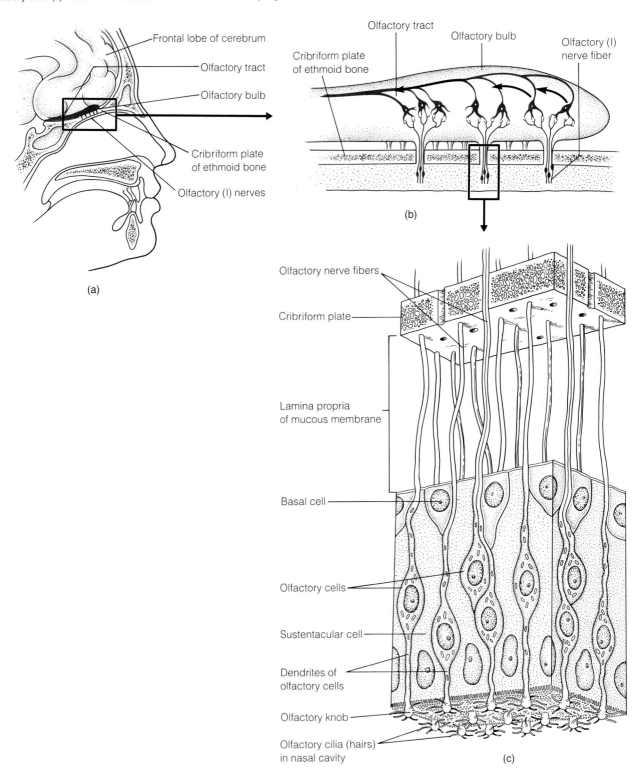

Frontal lobe of cerebrum

Olfactory tract

Olfactory bulb

Cribriform plate of ethmoid bone

Olfactory (I) nerves

(a)

Olfactory tract

Olfactory bulb

Cribriform plate of ethmoid bone

Olfactory (I) nerve fiber

(b)

Olfactory nerve fibers

Cribriform plate

Lamina propria of mucous membrane

Basal cell

Olfactory cells

Sustentacular cell

Dendrites of olfactory cells

Olfactory knob

Olfactory cilia (hairs) in nasal cavity

(c)

quiet breathing. However, if air is sniffed it is pulled into the upper regions of the nasal cavities, and the perception of odors is consequently greatly increased. When the lining of the nasal cavities becomes swollen and coated with a heavy layer of mucus during a cold or other respiratory infection, air does not easily reach the olfactory organs, and the sense of smell may be greatly reduced or eliminated. Olfactory receptors are normally sensitive to minute quantities of substances that reach them, but they adapt to stimulation completely in about one minute.

## The Ear: Hearing and Equilibrium

The ear is a sense organ with dual function: it is responsible for hearing and it is involved in the maintenance of balance and equilibrium. These two functions are performed by different parts of the ear.

### Hearing

The ear is divided into three separate compartments: the *outer ear,* the *middle ear* and the *inner ear* (Figure 15-12).

***The Outer Ear.*** The outer ear consists of the flattened flap of tissue (the **auricle** or **pinna**) that projects from the side of the head, and the **external auditory canal** that extends into the temporal bone of the skull. The skin that lines the external auditory canal contains ceruminous glands that secrete wax (cerumen), and there are hairs near the opening of the canal. The wax and hairs help prevent dust, insects, and other objects from entering the ear. At the internal end of the external auditory canal is a thin membrane, the **tympanic membrane** or **eardrum,** that separates the outer ear from the middle ear. Sound waves entering the external auditory canal cause the eardrum to vibrate; the pitch and loudness of the sound waves determine, respectively, the speed and amplitude of vibration of the eardrum. The higher the pitch of a sound, the more rapidly the eardrum will vibrate. In addition, the louder the sound, whatever its pitch, the greater will be the amplitude of the eardrum's vibration.

***The Middle Ear.*** The middle ear consists of a small chamber in the temporal bone, bounded on its outer side by the eardrum and on its inner side by the

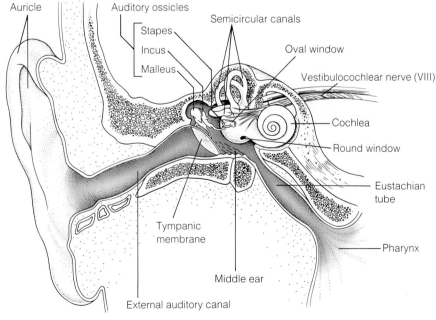

FIGURE 15-12 Structure of the ear.

bony wall of the inner ear. Within the middle ear are three tiny bones, the **auditory ossicles.** These are the **malleus** (= hammer), the **incus** (= anvil) and the **stapes** (= stirrup), named for their approximate shapes. The chamber of the middle ear is filled with air and is connected to the nasopharynx by means of a small channel, the **Eustachian tube.** Because the Eustachian tube is normally open to the atmosphere, the air in the middle ear is maintained at atmospheric pressure.

The auditory ossicles are connected to one another and suspended in the middle ear by ligaments and muscles. They form a "chain" of bones, connected at joints, that transmits the vibrations of the eardrum to the **oval window,** a small membrane-covered opening in the bone that forms the common wall of the middle and inner ears. Vibrations of the eardrum are transmitted, in sequence, from the malleus, which is attached firmly to the inner side of the eardrum, to the incus and then to the stapes. The footplate of the stapes, fitted against the oval window, transfers its vibrations to the oval window by pushing against it, somewhat like a piston in a cylinder. Because the middle ear is filled with air, and the inner ear is filled with fluid, vibrations in the air in the middle ear are thus converted into pressure variations in the fluids of the inner ear.

The ossicles do far more than simply transfer vibrations across the middle ear. Because vibrations in air are not readily transferred to a liquid (as occurs across the oval window), the force of vibration of the eardrum must be amplified if it is to create noticeable pressure changes in the fluid of the inner ear. This amplification occurs in two ways. First, the ossicle chain acts as a system of levers that increases the force of vibration by about 30 percent. More importantly, because the surface area of the eardrum is much larger than that of the oval window, there is a 17-fold increase in force when it is transferred from the eardrum to the oval window (force per unit area is increased).

In addition to its role in the transmission and amplification of vibrations from the outer to the inner ear, the middle ear serves to protect the delicate receptor apparatus in the inner ear from intense, sustained sounds. When sounds are loud, contraction of muscles in the inner ear increases tension on the eardrum, reducing its ability to vibrate, and alters the positions of the malleus and stapes, reducing the force by which the stapes pushes on the oval window.

Proper functioning of the middle ear depends upon maintaining atmospheric pressure on both sides of the eardrum so that it can vibrate freely. The Eustachian tube is crucial in this respect, for it allows air to pass into and out of the middle ear. However, the movement of air is somewhat inhibited by valvelike flaps that close off the Eustachian tube at its end in the nasopharynx. During fast altitude changes, as may occur in an airplane, the equalization of air pressure may be hastened by yawning, chewing, and swallowing, which help open the flaps.

When the Eustachian tube is blocked during a cold or other respiratory infection, some temporary hearing loss may occur. The effect on hearing of a blocked Eustachian tube may be particularly noticeable during rapid air pressure changes. In this situation, the inability to equalize air pressure on both sides of the eardrum may cause pain in the middle ear and temporary hearing loss. Although the function of the Eustachian tube is thus of great importance, it unfortunately provides an easy route by which upper respiratory infections are spread to the middle ear (see Chapter 10), causing a condition known as *otitis media.*

*The Inner Ear.* The inner ear consists of (1) a system of cavities in the temporal bone of the skull and (2) the contents of these cavities. The cavities in the bone form what is termed the *bony labyrinth.* There are two small openings, both covered with membranes, in the bony wall that separates the middle ear from the inner ear. The upper one is the oval window, mentioned above, and the lower one, the **round window.**

Suspended within the bony labyrinth is a series of interconnecting membranous sacs and tubes that form the *membranous labyrinth.* The

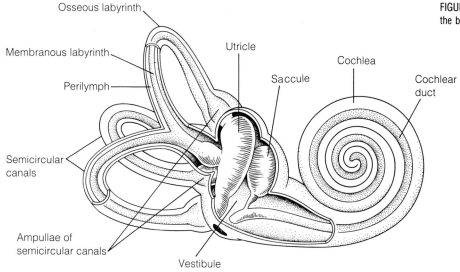

FIGURE 15–13 The membranous labyrinth within the bony labyrinth.

Osseous labyrinth

Membranous labyrinth

Perilymph

Semicircular canals

Ampullae of semicircular canals

Vestibule

Utricle

Saccule

Cochlea

Cochlear duct

membranous labyrinth is of similar pattern to the bony labyrinth but somewhat smaller (Figure 15–13). The space between the membranous labyrinth and the wall of the bony labyrinth is filled with a fluid called *perilymph;* the space within the membranous labyrinth is filled with a similar fluid called *endolymph.*

The part of the bony labyrinth that is crucial for hearing is the **cochlea,** a bony tube that is wound into a spiral of two and one-half turns. (*Cochlea* is the Latin word for snail's shell, which the cochlea resembles.) Suspended within the cochlea is a portion of the membranous labyrinth known as the *cochlear duct,* which is roughly triangular in cross-section and is filled with endolymph (Figure 15–14a). The floor of the cochlear duct is the *basilar membrane* and its roof is termed *Reissner's membrane.* Because the cochlear duct is attached to both sides of the cochlea, it separates the cochlea into two passageways called *scala* (= stairway). These are the *scala tympani* that lie below the basilar membrane of the cochlear duct and the *scala vestibuli,* above Reissner's membrane. The scala vestibuli and scala tympani are filled with perilymph and are connected by a small channel only at the tip of the cochlea; otherwise, the perilymph in both scalae are separated from one another along the entire course of the cochlea (Figure 15–14c).

Because the perilymph in the scala vestibuli is continuous with the perilymph that bathes the inner side of the oval window, the piston action of the stapes against the oval window creates pressure waves in the perilymph that are transferred into the scala vestibuli. The pressure waves in the scala vestibuli are transmitted across the cochlear duct and to the basilar membrane, which is deflected into the scala tympani; increased pressure in the scala tympani is then transmitted to the round window, which pushes outward (Figure 15–15).

The deflection of the basilar membrane activates receptor cells, called *hair cells,* that are part of the sensory structure for hearing called the **organ of Corti.** The organ of Corti, which lies above the basilar membrane along the entire length of the cochlear duct, is shown in cross-section in Figure 15–14b. Fine hairs that project from the hair cells are in contact with the overlying *tectorial membrane.* When the basilar membrane vibrates, the hair cells are displaced relative to the tectorial membrane; this displacement causes the hairs to be bent as they rub against the tectorial membrane. By some unknown mechanism, bending of the hair causes the hair cells to generate an action potential that is transmitted as a nerve impulse along afferent nerve fibers that run from the bases of the hair cells. The fibers join together as the cochlear

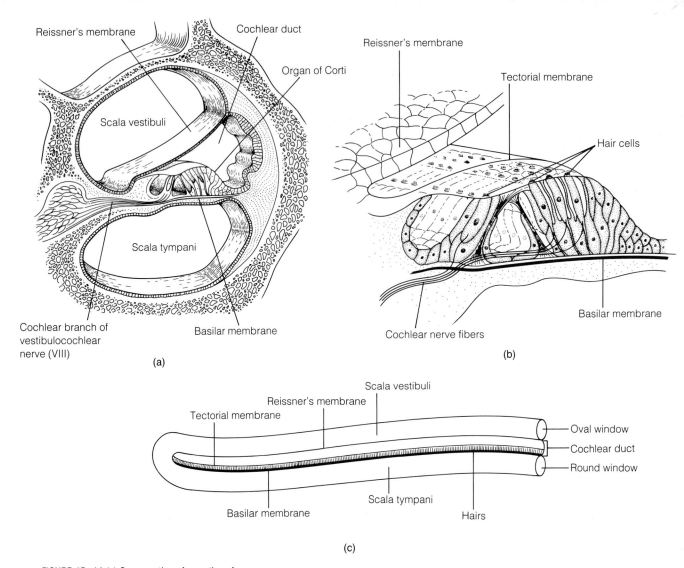

Reissner's membrane

Cochlear duct

Organ of Corti

Scala vestibuli

Scala tympani

Cochlear branch of
vestibulocochlear
nerve (VIII)

Basilar membrane

(a)

Reissner's membrane

Tectorial membrane

Hair cells

Basilar membrane

Cochlear nerve fibers

(b)

Scala vestibuli

Reissner's membrane

Tectorial membrane

Oval window

Cochlear duct

Round window

Basilar membrane

Scala tympani

Hairs

(c)

FIGURE 15–14 (a) Cross-section of a portion of
the cochlea. (b) Enlargement of the organ of
Corti. (c) A highly simplified diagram of the
cochlea as it would appear if stretched out.

portion of the *vestibulocochlear nerve* from each ear (cranial nerve VIII).
Impulses pass to the medulla oblongata, then to the thalamus, and finally to the
primary auditory areas of the cerebral cortex.

Since sound waves of various pitch cause the vibration of different regions
of the organ of Corti, a particular pitch is perceived as the stimulation of a
subgroup of hair cells. The highest pitched tones vibrate the basilar membrane
maximally near the base of the cochlea while the lowest pitched tones vibrate it
maximally near the apex. Sound intensity is perceived by the extent of
deflection of the basilar membrane; the louder the sound, the greater the
deflection.

***Disorders of Hearing.*** Hearing loss is of two basic types: *conductive* and
*sensorineural.* In conductive hearing loss, sound waves are not conducted
properly to the inner ear. This may be due to such conditions as the
accumulation of large amounts of wax in the ear canal, a ruptured eardrum,
fusion of the ossicles so that they do not vibrate properly, or damage to the
eardrum or ossicles as a result of a chronic middle ear infection. Conductive
hearing loss may also be due to *otosclerosis,* in which spongy bone grows at the
base of the stapes, fusing it to the oval window and preventing its vibration.
Otosclerosis can be treated surgically by removing the stapes and replacing it
with a miniature metal substitute.

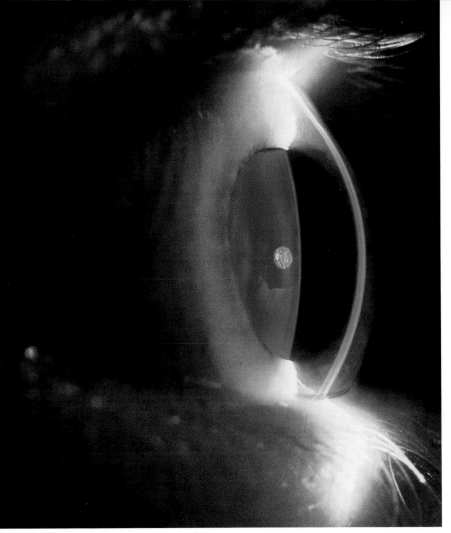

# THE HUMAN EYE

Poets see the eyes as mirrors of our inner nature; biologists view them as the organs of sight. This photograph of an eye, as seen from the side, seems to support both views.

Like a fine crystal watchglass, the living window that is the cornea (the thick, light blue line) covers and protects the front of the eye. It also acts as a crude focusing device that bends the rays of incoming light so that they are directed through the pupil and onto the lens (the purplish-gray, oval body). The space between the cornea and the lens is filled with a watery fluid, the aqueous humor, that puts pressure on the cornea, causing it to bow outward. Covering the edges of the lens is the iris, a circular, muscular diaphragm that gives the eye its color. When the iris widens in an automatic response to bright light, the space at its center—the pupil—becomes smaller. In dim light, the iris narrows, causing the pupil to dilate so that increased amounts of light enter the eye.

Whereas the cornea provides coarse focus, the lens permits fine focus. The lens is suspended from a ring of ligaments (suspensory ligaments), which are, in turn, attached to the ciliary muscles. When these contract, the tension in the suspensory ligaments is reduced, and the lens bulges. This response shortens the focal length so that near objects can be seen more clearly. Relaxation of the ciliary muscles has the opposite effect of thinning out the lens, thus allowing clear focus of distant objects.

After leaving the lens, light passes through the vitreous body, a clear, gelatinous substance that gives the eye its firmness. The light rays then impinge upon the retina at the back of the eyeball. The retina contains the rods and cones, specialized neurons that convert light stimuli into electrical impulses. These impulses are conveyed to the visual areas of the brain, which translate them into visual sensations.

Light from a laser beam penetrates an eye from right to left in order to weld the retina (left).

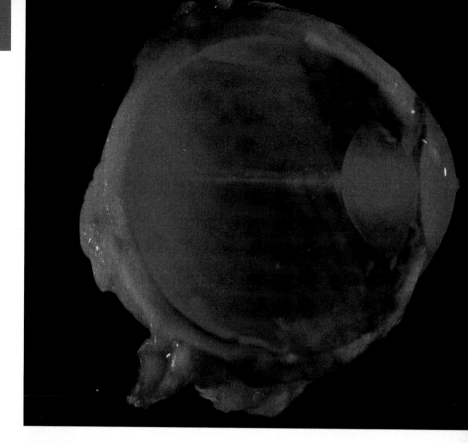

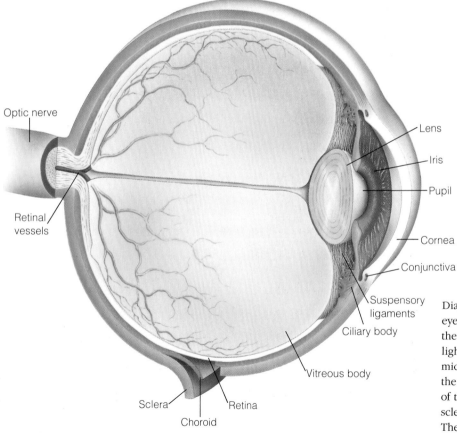

Optic nerve

Lens

Iris

Pupil

Retinal vessels

Cornea

Conjunctiva

Suspensory ligaments

Ciliary body

Vitreous body

Sclera

Retina

Choroid

Diagram of a cross-section through the eye. The eyeball has three main coats: the inner retina, which contains the light-sensitive rods and cones; the middle, darkly pigmented choroid; and the external sclera, which forms the right of the eye. At the front of the eye, the sclera becomes the transparent cornea. The conjunctiva is a fine membrane that lines the eyelids and covers the exposed part of the eyeball.

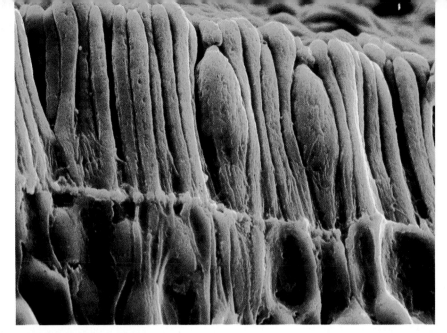

A scanning electron micrograph of a cross-section through part of the retina shows elongated rods and shorter cones. The rods, which are far more numerous than the cones, allow us to see shades of gray in dim light; the cones operate in bright light to give color vision.

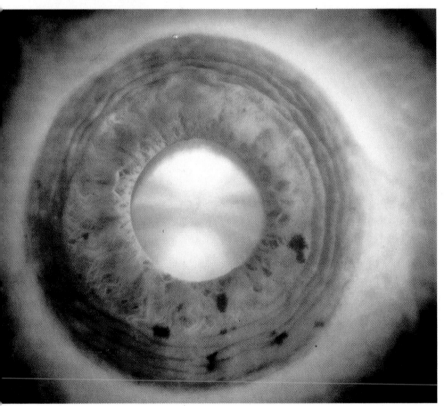

Photograph of the surface of a light brown iris. You can see radiating fibers and vessels, which are largely obscured by pigmented stromal cells. Dark spots are freckles.

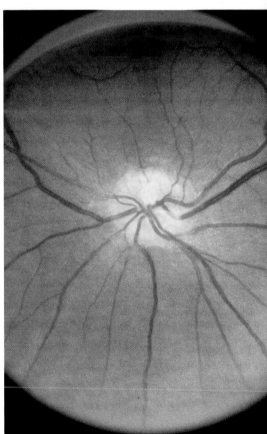

The back of a healthy eye, photographed through the pupil, shows retinal blood vessels arising from the optic disc, a light-colored circular area at the back of the retina. The optic disc is the region where optic nerve fibers and blood vessels enter the eyeball.

The Human Eye

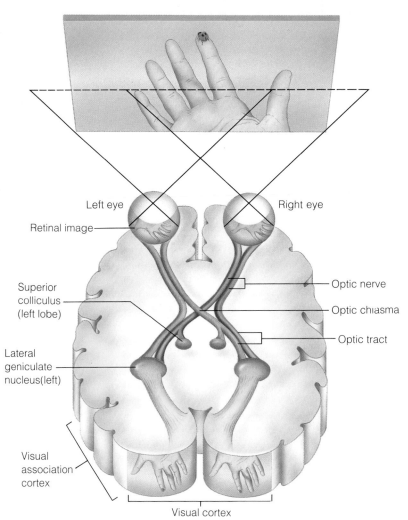

The diagram is a cross-section of the brain, viewed from above, showing the pathways by which images that reach the retina travel as electrical impulses to the visual areas of the brain. Impulses originating in the retinal rods and cones leave the eye by way of the optic nerve. The optic nerves from each eye come together at the optic chiasma, where some fibers from each nerve pass over to the other side of the brain. The two nerve tracts that leave the optic chiasma, termed the optic tracts, pass to the lateral geniculate nucleus in each cerebral hemisphere. From there, impulses are conducted to visual areas of the cerebral cortex. The pathways are such that information from the right side of each eye is conveyed to the primary visual area in the right cerebral cortex; information from the left side of each eye is conveyed to the visual area in the left cerebral cortex.

Sensorineural hearing loss results from damage to the cochlea or the auditory nerve. This occurs in some individuals with old age, but it can also be caused at any age by prolonged exposure to loud noises.

Sensorineural hearing loss is usually permanent and nontreatable. Mild conductive hearing loss can often be helped by a hearing aid that amplifies sounds. Other hearing aids, used in cases of severe conductive hearing loss, convert sounds into vibrations that are transferred by a vibrating pad through the bones of the skull to the inner ear.

## Equilibrium

While the cochlea is responsible for hearing, the posterior portion of the bony labyrinth (and the part of the membranous labyrinth it contains) is involved in the detection of movement and position of the head—information that is essential for maintaining equilibrium. This section of the inner ear consists of three membranous *semicircular canals* and two membranous sacs, the *saccule* and *utricle,* all situated within the bony labyrinth (see Figure 15–13). The utricle, which lies at the bases of the three semicircular canals, is connected to the saccule by a fine membranous channel, the *endolymphatic duct.* Endolymph fills these membranous structures, and they are suspended in perilymph that fills all of the space between these structures and the bony wall.

*Function of the Semicircular Canals.* The semicircular canals provide information on angular acceleration of the head, brought about by turning the head in any direction. Because the three semicircular canals are oriented at right angles to one another, a change in direction in any plane will affect at least one of them. An expanded end of each of the semicircular canals, called the *ampulla* (= little vase), contains a ridge of connective tissue, the *crista*. The

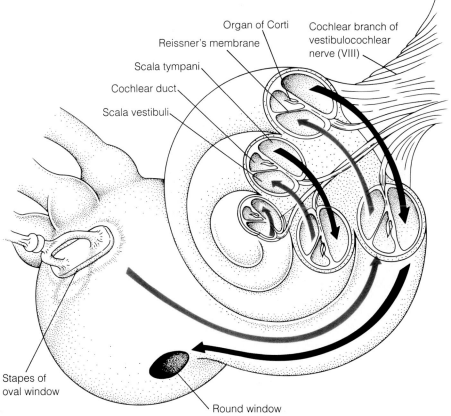

FIGURE 15–15 A section through the cochlea, showing the transfer of pressure waves from the oval window into the scala vestibuli, then to the scala tympani and to the round window.

Organ of Corti

Reissner's membrane

Scala tympani

Cochlear duct

Scala vestibuli

Cochlear branch of vestibulocochlear nerve (VIII)

Stapes of oval window

Round window

crista supports hair cells that act as receptors for angular acceleration, as follows. The hairs of the hair cells are covered by a gelatinous material. Turning the head moves the endolymph in one or more of the semicircular canals, causing the endolymph to push against the gelatinous mass. This action bends the hairs, stimulating depolarization of the hair cells and generating nerve impulses that travel along afferent fibers leading from the hair cells.

***Function of the Utricle and Saccule.*** Sensory organs in the utricle and saccule signal linear acceleration of the head, as occurs during acceleration in an automobile, and position of the head relative to gravity. These sensory organs consist of two small areas, termed *maculae* (sing: *macula*), one in the wall of the utricle and one in the wall of the saccule, containing hair cells. The hairs of the hair cells are embedded in a gelatinous material that contains a surface layer of small stones of calcium carbonate, called *otoliths* (*oto* = ear; *liths* = stones), which make the gelatinous material heavier. Linear acceleration of the head or a change of its position relative to gravity moves the dense, gelatinous otolith-containing material relative to the delicate hairs. When the hairs are bent, the hair cells are depolarized, generating a nerve impulse (Figure 15–16).

***Neural Pathways for Equilibrium.*** Impulses from the semicircular canals and from the utricle and saccule are conveyed over the vestibular branch of the vestibulocochlear nerve (VIII) to motor areas in the medulla and cerebellum. The cerebellum monitors this information and sends instructions to the motor cortex to bring about the contraction of those skeletal muscles needed to maintain equilibrium.

***Disorders of Equilibrium.*** Repeated changes in angular and linear acceleration can give rise to *motion sickness* by continually stimulating receptors concerned with the maintenance of equilibrium. Motion sickness results in nausea and vomiting, which can be severe. Drugs that prevent motion sickness are available, but are most effective if administered to susceptible individuals prior to beginning an activity that may result in motion sickness.

FIGURE 15–16 Portion of otoliths and hairs of the macula (a) when the head is in an upright position and (b) when the head is bent forward or undergoing linear acceleration.

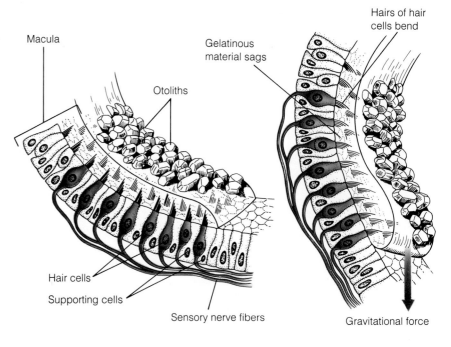

Macula

Otoliths

Gelatinous material sags

Hairs of hair cells bend

Hair cells

Supporting cells

Sensory nerve fibers

Gravitational force

(a) Head upright

(b) Head bent forward

## Study Questions

1. Name the five types of sensory receptors and indicate the function of each.
2. Indicate which types of sensory receptors are scattered throughout the body and which are located in sense organs.
3. a) Describe the structure of the retina.
   b) What is the optic disc?
   c) Why does the optic disc form a blind spot on the retina?
4. a) Explain how the eye accommodates for seeing at different distances.
   b) Describe the common eye problems that result from an inability of the eye to accommodate normally.
5. a) What is the fovea centralis?
   b) What is its importance to vision?
6. Describe how taste buds function.
7. a) Where are the olfactory organs located?
   b) How does their location influence the perception of odors?
8. Trace the path of sound waves as they pass through the ear from the ear canal to the organ of Corti.
9. Describe the function of the Eustachian tube.
10. Distinguish between sensorineural and conductive hearing loss.
11. Explain how (1) the semicircular canals and (2) the utricle and saccule function in the maintenance of equilibrium.

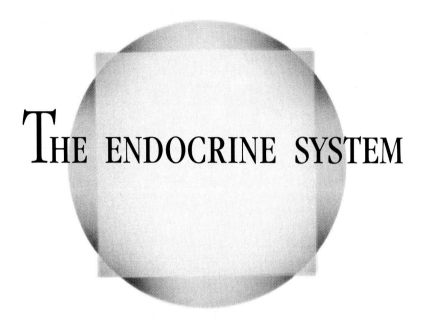

# THE ENDOCRINE SYSTEM

Coordination and regulation of all bodily activities is the shared responsibility of the nervous and endocrine systems; in fact, most activities are under the dual control of both systems. While the nervous system is involved primarily in rapid and more or less short-term adjustments of bodily activities, regulation over a longer time frame is performed by the endocrine system. Processes controlled by the endocrine system include reproduction, embryonic development, and growth. In addition, the endocrine system is largely responsible for the maintenance of metabolic homeostasis, which guarantees the relative stability of the body's internal chemical environment.

Not only do the functions of the nervous and endocrine systems overlap, but the two systems are integrated in complex ways. Nowhere is this integration more apparent than in the control of reproductive processes. The reproductive hormones of the endocrine system regulate most aspects of reproductive physiology. The secretion of the reproductive hormones is under the direct control of the hypothalamus of the brain, but the activity of the hypothalamus is, in turn, influenced not only by higher brain centers, but also by the circulating levels of the reproductive hormones themselves. Furthermore, some of the reproductive hormones influence sexual behavior by affecting specific areas of the brain. This type of complex interrelationship between the nervous and endocrine systems is the rule rather than the exception, as will become increasingly apparent in the discussion that follows.

## ENDOCRINE GLANDS

The endocrine system consists of eight different secretory organs, some of which are paired, called **endocrine glands.** The endocrine glands include the *pituitary, thyroid, parathyroids, adrenals* (including the adrenal medulla and the adrenal cortex), *islet cells of the pancreas, gonads, pineal,* and *thymus.* The location of the endocrine glands is shown in Figure 16–1. In addition to the endocrine glands, endocrine tissue is scattered in a wide range of locations throughout other tissues of the body.

### Hormones

All endocrine glands secrete regulatory chemicals called **hormones.** However, unlike exocrine glands such as sweat glands in which the secretory products are conveyed to their sites of action through ducts, the endocrine glands are ductless (see Figure 4–4). The hormones of endocrine glands are secreted into the tissue fluid that surrounds the endocrine cells and pass into blood vessels that pervade the glands. Transported through the circulatory system, hormones reach specific *target tissues,* whose activities they regulate. Indeed, it is the very fact that hormones travel through the circulatory system to influence target tissues that makes their action slower than that of nerves but allows for a more sustained effect.

Hormones belong to three groups of chemical substances. One group is derived from either tyrosine or tryptophan, both amino acids. Tyrosine is the source of the thyroid hormones as well as epinephrine and norepinephrine, secreted by the adrenal medulla; tryptophan gives rise to melatonin, secreted by the pineal gland. A second group, the steroid hormones, are chemical derivatives of cholesterol, and include hormones secreted by the testes, the ovaries, and the adrenal cortex. The third and largest group of hormones are peptides (short chains of amino acids), proteins, or glycoproteins (proteins with attached carbohydrates). These include hormones of the pituitary, parathyroids, and pancreatic islet. The hormones secreted by the major endocrine glands and the function of each hormone are summarized in Table 16–1. These hormones are discussed in detail in this chapter.

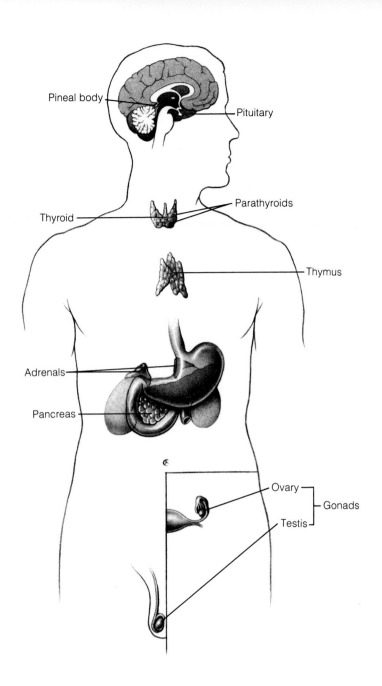

FIGURE 16–1 Locations of the major endocrine glands.

Pineal body

Pituitary

Parathyroids

Thyroid

Thymus

Adrenals

Pancreas

Ovary

Gonads

Testis

## Feedback Loops: The Regulation of Hormone Secretion

Small amounts of hormones in the blood have profound effects on the activity of target organs. Consequently, the endocrine system operates with elaborate feedback controls that precisely regulate hormone levels. In these so-called *feedback loops,* hormone secretion by an endocrine gland is regulated by the output of its target organ.

Feedback loops may be of two general types. The most common are **negative feedback loops,** in which an increase in the output, or function, of a target organ causes a reduction in the synthesis and release of the hormone that controls the organ (Figure 16–2). In **positive feedback loops,** an increase in target organ output increases the synthesis and release of the controlling hormone. Examples of negative feedback loops are the effects of insulin on blood sugar reduction and of glucagon on the elevation of blood sugar (see Chapter 11). In this system, a reduction in blood sugar levels inhibits insulin

TABLE 16-1 Hormones of the Major Endocrine Glands

| Gland | Hormone | Major Functions |
|---|---|---|
| **1. Pituitary** | | |
| Anterior lobe | Growth Hormone (GH) | Growth of tissue |
| | Prolactin (PRL) | Milk secretion by mammary glands; many general effects similar to growth hormone. |
| | Thyroid stimulating hormone (TSH) | Stimulates thyroid gland to synthesize and release thyroid hormone. |
| | Adrenocorticotropic hormone (ACTH) | Stimulates adrenal cortex to secrete glucocorticoids, important in responses to long-term stress. |
| | Follicle stimulating hormone (FSH) | In females, promotes maturation of ovarian follicles and estrogen secretion; in males, promotes sperm production by seminiferous tubules of testes. |
| | Luteinizing hormone (LH) | In females, triggers ovulation, maintains corpus luteum, stimulates secretion of estrogens and progesterone. In males, stimulates interstitial cells to secrete testosterone. |
| Intermediate lobe | Melanocyte stimulating hormone (MSH) | Role unclear in humans; may be involved in skin darkening in response to ultraviolet light. |
| Neural lobe | Antidiuretic hormone (ADH), vasopressin | Increases reabsorption of water by kidney tubules; elevates blood pressure. |
| | Oxytocin | Contraction of uterine smooth muscle during labor; milk ejection during lactation. |
| **2. Thyroid** | Thyroid hormones ($T_3$ and $T_4$) | Essential for normal growth and development; increase metabolic rate. |
| | Calcitonin | Lowers blood calcium and phosphate levels. |
| **3. Parathyroids** | Parathyroid hormone (PTH) | Raises blood calcium levels; lowers blood phosphate levels. |
| **4. Adrenals** | | |
| Medulla | Epinephrine (adrenalin) | Increases blood sugar levels; increases heart rate. |
| | Norepinephrine | Elevates blood pressure by constricting arterioles. |
| Cortex | Mineralocorticoids | Increase reabsorption of sodium and water from glomerular filtrate. |
| | Glucocorticoids | Increase carbohydrate stores; suppress inflammatory response. |
| | Gonadocorticoids | Similar to androgens (male sex hormones) and estrogens (female sex hormones). |
| **5. Pancreas** | Insulin | Increases transport of glucose into cells; lowers blood glucose levels. |
| | Glucagon | Raises blood glucose levels. |
| | Somatostatin | Retards movement of glucose and other nutrients from gastrointestinal tract to bloodstream. |
| **6. Gonads** | | |
| Testes | Testosterone | Determines male sexual characteristics. |
| Ovary | Estrogen | Determines female sexual characteristics. |
| | Progesterone | Maintains uterine endometrium and pregnancy; inhibits ovulation. |
| **7. Pineal** | Melatonin | Regulates biological clocks. |
| **8. Thymus** | Thymosins | Involved in proliferation and maturation of T-lymphocytes of the immune system. |

secretion by the islet cells of the pancreas while an increase in blood sugar levels inhibits glucagon secretion by pancreatic islet cells. The net effect of the action of insulin and glucagon is to maintain blood sugar at relatively constant levels. An example of positive feedback is in the control of events leading to ovulation, where ever higher levels of estrogen secretion by the ovary act to stimulate the secretion of LH by the pituitary, causing a spike in LH levels that triggers ovulation, as discussed in Chapter 17.

## PITUITARY GLAND

The **pituitary gland,** or **hypophysis,** is often called the "master gland" of the body. This designation, implying that the pituitary regulates all endocrine function, is inaccurate, but it is true that pituitary secretions influence a wide range of physiological processes. Whereas some pituitary hormones affect the

FIGURE 16–2 Operation of (a) negative and (b) positive feedback loops. In negative feedback, the product of the target cell *inhibits* further secretion by the endocrine gland; in positive feedback, the product of the target cell *stimulates* further endocrine secretion, which in turn triggers the production of additional target cell product and so on.

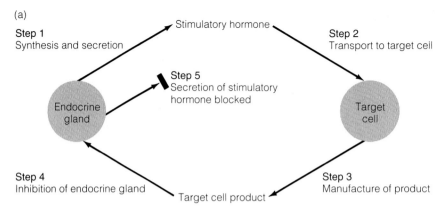

(a)

Step 1
Synthesis and secretion

Stimulatory hormone

Step 2
Transport to target cell

Step 5
Secretion of stimulatory hormone blocked

Endocrine gland

Target cell

Step 4
Inhibition of endocrine gland

Target cell product

Step 3
Manufacture of product

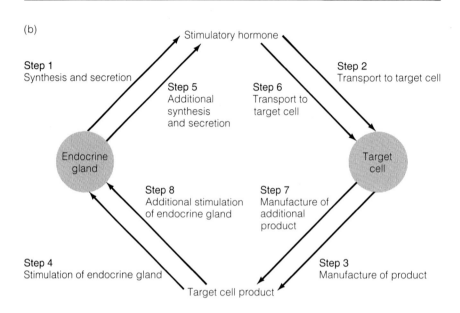

(b)

Stimulatory hormone

Step 1
Synthesis and secretion

Step 2
Transport to target cell

Step 5
Additional synthesis and secretion

Step 6
Transport to target cell

Endocrine gland

Target cell

Step 8
Additional stimulation of endocrine gland

Step 7
Manufacture of additional product

Step 4
Stimulation of endocrine gland

Target cell product

Step 3
Manufacture of product

activity of some of the other endocrine glands, other pituitary hormones act on nonendocrine tissues of the body.

The anatomy of the pituitary gland is shown in Figure 16–3. In adult humans, the pituitary consists of an *anterior lobe* and a *neural,* or *posterior lobe;* a third lobe, the *intermediate lobe,* is present during the fetal stage but disappears as a distinct structure in adults, its tissues becoming scattered in the other lobes. The pituitary gland is connected anatomically to the hypothalamus of the brain by means of the *neural stalk.* Indeed, the tissues of the neural stalk and of the neural lobe, as their names suggest, are derived from the nervous system and are essentially extensions of the hypothalamus. Specialized neurons that have their cell bodies in the hypothalamus send axons that run down the neural stalk and terminate in the neural lobe. These neurons, called **neurosecretory cells,** are specialized for the production of hormones. The axons of the neurosecretory cells form the *hypothalamo-hypophysreal tract* in the neural stalk.

The anterior lobe of the pituitary is composed of secretory tissue derived from tissue of the roof of the mouth and not from neural tissue. There is, however, a direct vascular connection between the hypothalamus and the anterior lobe via the *hypophyseal portal vessels* that convey blood from capillary beds in the lower portion of the hypothalamus to blood sinuses in the anterior lobe (see Figure 16–3).

## The Anterior Lobe

The anterior lobe synthesizes and secretes six hormones: *growth hormone (GH), prolactin (PRL), adrenocorticotropic hormone (ACTH), thyroid stimulating hormone (TSH), follicle stimulating hormone (FSH),* and *luteinizing hormone (LH).* Growth hormone and prolactin have effects in promoting the growth of tissues, whereas ACTH, TSH, FSH, and LH are trophic hormones (*trophic* = nourishing) that regulate the activity of other endocrine glands of the body.

### Growth Hormone

**Growth hormone (GH),** also known as *somatotropin,* is a protein hormone that promotes growth, particularly of bones and skeletal muscles, but also of other tissues, in children and adolescents. Its growth-promoting effects occur primarily by stimulating the entry of amino acids into cells and the synthesis of body proteins from these amino acids. GH also affects lipid and carbohydrate metabolism. It decreases the utilization of carbohydrates as energy sources and increases the breakdown and utilization of fats. This function is particularly noticeable following infancy, when growth is accompanied by the loss of "baby fat."

GH is required for normal growth. Infants who lack the ability to synthesize and secrete growth hormone eventually develop into miniature adults, sometimes attaining heights no more than two and one-half feet or so, but they are otherwise normal. Most children who lack GH respond well to treatment with the hormone during their preadolescent years. The recent commercial production of human GH by genetic engineering has provided a much needed source of the hormone for such therapy (see Chapter 1, *Focus: Benefits and Problems from Genetic Engineering*). Although GH is normally secreted during adult life, no serious consequences appear to result from absences of GH that begin in later life.

The production and secretion of excess GH during childhood, often as a result of a pituitary tumor, results in *gigantism.* Pituitary giants can be very large; one such individual is reported to have attained a height of 8 feet, 11 inches. Oversecretion of GH in adults, after growth has stopped, results in a

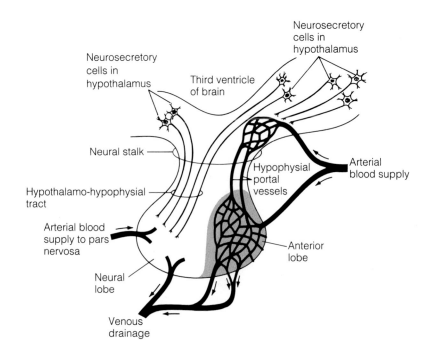

FIGURE 16–3 Diagram of the pituitary gland, showing the arrangement of blood vessels and hypothalamic neurosecretory cells.

condition known as *acromegaly*, which is characterized by a noticeable thickening of bones, particularly in the skull, fingers, arms, and legs.

### Prolactin

**Prolactin (PRL)** is a protein hormone that has a chemical structure very similar to that of growth hormone, and hence it is not surprising that it also has an important effect on growth, albeit a more specific effect than GH. PRL acts together with estrogen, progesterone, GH, glucocorticoids (hormones secreted by the adrenal cortex), placental lactogen (a prolactinlike hormone secreted by the placenta), and perhaps other hormones, to bring about full development of the mammary glands in the pregnant female prior to milk secretion. The secretion (manufacture) of milk by the mammary glands (as opposed to the ejection, or release, of milk from the mammary glands, a function of the hormone oxytocin) is also stimulated by PRL after the mammary glands have attained suitable growth. PRL levels rise throughout pregnancy, fall after delivery, and rise again in response to suckling. PRL secretion is maintained as long as breast feeding continues.

### Adrenocorticotropic Hormone

**Adrenocorticotropic hormone (ACTH)** is a protein consisting of a chain of 39 amino acids. Its primary role is to stimulate the synthesis and secretion of hormones called glucocorticoids from the adrenal cortex. ACTH also acts, as does GH, to stimulate the breakdown and utilization of fat deposits as energy sources.

### Thyroid Stimulating Hormone

**Thyroid stimulating hormone (TSH)** belongs to a family of glycoprotein hormones that includes follicle stimulating hormone and luteinizing hormone. TSH stimulates the synthesis and secretion of thyroid hormones from the thyroid gland.

### Follicle Stimulating Hormone

In females, **follicle stimulating hormone (FSH)** acts upon the ovaries to stimulate the development and maturation of an ovarian follicle each month; the follicle produces an egg cell, or ovum, that is released at ovulation. FSH also stimulates the ovarian follicle to secrete estrogen. In males, FSH promotes sperm production by the seminiferous tubules of the testes. The functions of FSH are discussed further in Chapter 17.

### Luteinizing Hormone

Like FSH, **luteinizing hormone (LH)** helps regulate reproductive processes in both females and males, as detailed in Chapter 17. In females, LH acts in conjunction with FSH to stimulate estrogen secretion by the ovarian follicle. LH triggers ovulation and converts the ruptured follicle into an endocrine organ, the **corpus luteum**, which secretes estrogen and progesterone. In males, LH stimulates *interstitial cells,* specialized endocrine cells packed between the seminiferous tubules of the testes, to secrete testosterone.

### Control of the Secretion of Anterior Lobe Hormones

Although the six anterior lobe hormones are synthesized in and secreted from the anterior lobe, their release from the anterior pituitary into the general circulation is controlled by chemical substances, termed *releasing factors.* The releasing factors are produced in the cell bodies of neurosecretory cells of the hypothalamus, transported along the axons of these cells, and discharged from axon terminals that are located near the hypothalamic capillary beds. These capillary beds empty into the hypophyseal portal vessels, which carry the releasing factors to the anterior lobe, where they have their effects.

Releasing factors have been isolated for all of the six anterior pituitary hormones (Table 16–2). In fact, some hormones have both stimulatory and inhibitory releasing factors. The secretion of releasing factors by the hypothalamus is controlled by complex feedback systems that are sensitive to neural input as well as to the blood concentration of hormones secreted by target organs. The synthesis of releasing factors and neural lobe hormones by neurosecretory cells of the hypothalamus is a good example of the tight functional integration of the nervous and endocrine systems.

## The Intermediate Lobe

The scattered intermediate lobe cells in human adults produce a protein hormone, **melanocyte stimulating hormone (MSH).** The function of MSH in humans is unclear. It may be related to the skin darkening ("tanning") that occurs following exposure to ultraviolet light in the sun or from other sources.

## The Neural Lobe

Neurosecretory cells in the hypothalamus synthesize two hormones found in the neural lobe: **antidiuretic hormone (ADH),** also known as **vasopressin,** and **oxytocin.** Both are small peptides composed of nine amino acids. Following their synthesis in the cell bodies of neurosecretory cells, ADH and oxytocin are packaged into vesicles that are transported along the cells' axons and stored in the axon terminals. Upon stimulation of the neurosecretory cells, ADH and oxytocin are discharged from the axon terminals and enter the blood vessels of the neural lobe, from which they are carried by the circulation to their target organs. Thus, the neural lobe does not synthesize ADH and oxytocin, but instead acts as a site for subsequent release of these hormones.

### Antidiuretic Hormone, or Vasopressin

As discussed in Chapter 12, ADH stimulates water resorption in the kidney by increasing the permeability of the cells of the distal convoluted tubules and collecting tubules of the nephrons. The net effect of ADH secretion is to withdraw water from the urine, leading to the conservation of water and the production of a more concentrated urine.

ADH deficiency, usually caused by disease or injury of the hypothalamus, leads to *diabetes insipidus.* This condition is characterized by excessive urination and unabated thirst. It responds well to treatment with ADH.

TABLE 16–2 Hypothalamic Factors Controlling the Release of Pituitary Hormones

| Hypothalamic Factor | Function |
| --- | --- |
| Thyrotropin releasing factor (TRF) | Release of TSH |
| Corticotropin releasing factor (CRF) | Release of ACTH |
| Gonadotropin releasing factor (GnRF) | Release of LH and FSH |
| Growth hormone releasing factor (GHRF) | Release of growth hormone |
| Somatostatin (SS) | Inhibition of growth hormone release |
| Prolactin releasing factor (PRF) | Release of prolactin |
| Prolactin release-inhibiting factor (PRIF) | Inhibition of prolactin release |
| MSH releasing factor (MRF) | Release of MSH |
| MSH release-inhibiting factor (MRIF) | Inhibition of MSH release |

## Oxytocin

Oxytocin stimulates the contraction of smooth muscle cells in the pregnant uterus during labor. It is also released into the circulation as a result of the stimulus of suckling (nursing a baby). Oxytocin brings about the release, or ejection, of milk from the mammary glands by causing the contraction of specialized cells that surround the mammary ducts where milk is stored. The action of oxytocin is discussed further in Chapter 18.

## THYROID GLAND

The **thyroid gland,** somewhat butterfly-shaped, consists of two lobes of tissue that lie on either side of the upper trachea; the two lobes are connected at the anterior of the trachea by a small mass of tissue called an *isthmus* (Figure 16–4). The gland receives blood from vessels arising from the common carotid arteries. These break up into a capillary network that pervades the gland and that drains into venules that lead into the internal jugular and brachiocephalic veins.

The thyroid gland is covered by a thin capsule of connective tissue that enters the substance of the gland and divides it into irregular masses of tissue. Within these masses, most of the endocrine tissue forms structures called *follicles* (Figure 16–5). Each follicle consists of a single, spherical layer of epithelial cells, called *follicle cells,* that surround a central cavity. Enclosing each follicle is a delicate layer of highly vascularized connective tissue. Interspersed between the follicles and the connective tissue that surround them are other endocrine cells called *C-cells,* or *parafollicular cells* (*para* = alongside of).

The follicle cells manufacture large glycoprotein molecules, called *thyroglobulin,* which are stored in the cavity of the follicle. Like all proteins, thyroglobulin is composed of amino acids, but unlike other proteins, it contains large amounts of iodinated tyrosine (the amino acid tyrosine to which iodine atoms are covalently bonded). While still a part of the thyroglobulin protein, iodinated tyrosines form covalently bonded pairs that are immediate

FIGURE 16–4 The thyroid gland and its blood supply.

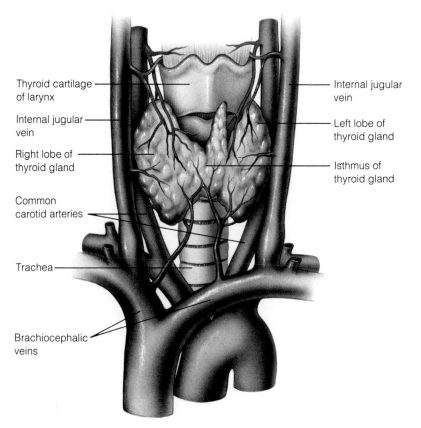

Thyroid cartilage of larynx

Internal jugular vein

Right lobe of thyroid gland

Common carotid arteries

Trachea

Brachiocephalic veins

Internal jugular vein

Left lobe of thyroid gland

Isthmus of thyroid gland

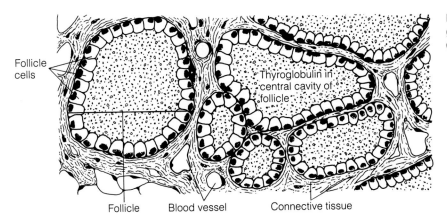

Follicle cells

Thyroglobulin in central cavity of follicle

Follicle    Blood vessel    Connective tissue

FIGURE 16–5 Secretion through the thyroid gland, showing thyroid follicles and surrounding connective tissue.

precursors to thyroid hormones. When the thyroid gland is stimulated to secrete thyroid hormone into the bloodstream, the follicle cells withdraw thyroglobulin from the central cavity and break it down, releasing the previously formed thyroid hormones. The two main thyroid hormones secreted in this way are **triiodotyrosine (T₃)** and **thyroxin (T₄).** The process of thyroid hormone synthesis and secretion is summarized in Figure 16–6.

FIGURE 16–6 Summary of steps in the synthesis and secretion of thyroid hormones by thyroid follicles.

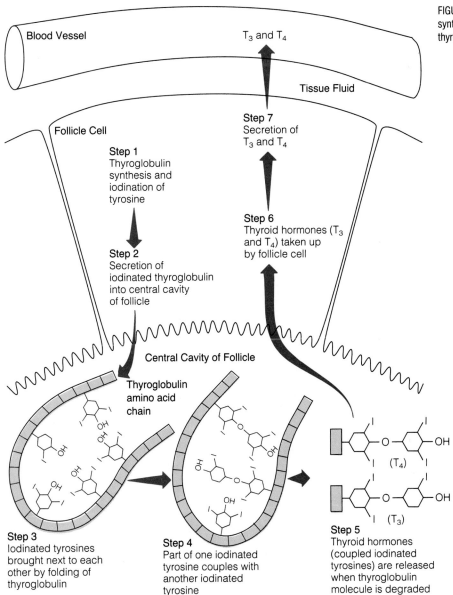

Blood Vessel

$T_3$ and $T_4$

Tissue Fluid

Follicle Cell

**Step 7**
Secretion of
$T_3$ and $T_4$

**Step 1**
Thyroglobulin
synthesis and
iodination of
tyrosine

**Step 6**
Thyroid hormones ($T_3$
and $T_4$) taken up
by follicle cell

**Step 2**
Secretion of
iodinated thyroglobulin
into central cavity
of follicle

Central Cavity of Follicle

Thyroglobulin
amino acid
chain

($T_4$)

($T_3$)

**Step 3**
Iodinated tyrosines
brought next to each
other by folding of
thyroglobulin

**Step 4**
Part of one iodinated
tyrosine couples with
another iodinated
tyrosine

**Step 5**
Thyroid hormones
(coupled iodinated
tyrosines) are released
when thyroglobulin
molecule is degraded

Thyroid Gland    333

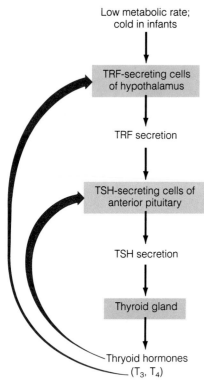

FIGURE 16-7 Control of thyroid hormone secretion. Heavy arrows represent negative feedback loops.

Thyroid hormones influence many functions. They are required, in conjunction with GH, for normal tissue growth and development. Thyroid hormones also have a major effect on metabolism. They increase metabolic rate by stimulating the breakdown of carbohydrates and lipids, and they raise body temperature.

The secretion of thyroid hormones is regulated by several feedback systems (Figure 16–7). The immediate stimulus to the secretion of thyroid hormones is TSH from the anterior pituitary; TSH, in turn, is secreted in response to **thyrotropin releasing factor (TRF)** secreted from the hypothalamus. The level of thyroid hormones in the blood is detected both by the hypothalamus and the anterior pituitary. Low blood levels of thyroid hormones stimulate, and high blood levels inhibit, TRF and TSH secretion. In addition, a lowering of the metabolic rate stimulates secretion of thyroid hormones by enhancing TRF secretion, whereas an increase in metabolic rate decreases secretion of thyroid hormones by depressing TRF secretion. Thyroid hormone release in response to cold occurs in infants, but apparently not in adults to any significant degree, if at all.

Deficiencies of thyroid hormones cause major abnormalities in development, growth, reproduction, and metabolism. A severe deficiency of thyroid hormones during childhood, usually as a result of pituitary or hypothalamic malfunction, leads to *cretinism,* characterized by dwarfism, physical deformity, and severe retardation due to failures in growth and development. Low body temperature and lethargy also result. In adults, hypothyroidism may cause physical and mental lethargy, a condition known as *myxedema.* Since iodine is required for the synthesis of thyroid hormones, dietary iodine deficiency can result in hypothyroidism. Although iodine deficiency is now prevented by the widespread use of iodized table salt, it was once a severe problem in Switzerland, the American Midwest, and other landlocked areas where seafood, a rich source of iodine, was not generally available. The importance of the thyroid hormones, and hence iodine, for brain development and optimal mental activity undoubtedly gave rise to the idea that fish is "brain food."

Hyperthyroidism, resulting from an excess of thyroid hormones, causes an abnormally high metabolic rate and corresponding hyperactivity ("nervous" energy). Enlargement of the thyroid gland, termed *goiter,* is a frequent symptom of hyperthyroidism, but may also occur in other thyroid disorders. For example, iodine deficiency may cause goiter because of excessive stimulation of the thyroid gland by continuously high levels of TSH, secreted because of a lack of thyroid hormones.

The parafollicular cells of the thyroid gland produce the protein hormone **calcitonin.** Calcitonin decreases calcium and phosphate levels in the blood by increasing the mineralization of bone, inhibiting the activity of osteoclasts (bone cells that break down, or resorb, bone), and increasing the excretion of phosphate. Calcitonin secretion is stimulated by high blood levels of calcium and is inhibited by low blood levels of calcium.

## PARATHYROID GLANDS

Embedded on the posterior sides of each lateral lobe of the thyroid gland are two small masses of tissue that form the **parathyroid glands** (Figure 16–8). The parathyroid glands contain two cell types. The most numerous, called *chief cells,* secrete **parathyroid hormone (PTH).**

PTH is a protein that is the principal agent in the regulation of blood calcium homeostasis. PTH secretion is stimulated by low levels of blood calcium. It enhances bone resorption by osteoclasts, thus transferring calcium and phosphate from bone tissue to the blood. It also acts on the kidneys to reduce the excretion of calcium in the urine and increase the excretion of phosphate. Since greater amounts of phosphate are excreted in the urine than are released from the bone, the net effect of PTH action is to increase blood levels of calcium and decrease blood levels of phosphate.

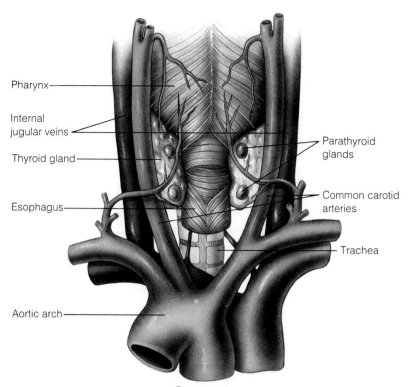

Pharynx

Internal jugular veins

Thyroid gland

Esophagus

Aortic arch

Parathyroid glands

Common carotid arteries

Trachea

Posterior view

PTH also facilitates the conversion of vitamin D into its active form. An inactive form of vitamin D is synthesized in the skin in response to exposure to ultraviolet light, then transported to the liver where it is converted into another inactive form, and finally transported to the kidney. The kidney, with the aid of PTH, converts the hepatic form of vitamin D into its active form. Vitamin D increases the absorption of calcium and phosphate by cells of the intestinal mucosa as well as the reabsorption of calcium and phosphate by the kidney tubules. The net effect of vitamin D is to decrease the loss of calcium and phosphate from the body. Thus PTH and vitamin D have similar actions with regard to calcium homeostasis (they both increase blood calcium levels), but have opposite effects on blood phosphate levels.

The effects of PTH and vitamin D on blood calcium levels are opposite to that of calcitonin. It is generally thought that PTH, in conjunction with vitamin D, is much more important than calcitonin in the regulation of blood calcium levels, since excessively high blood calcium levels do not normally occur in healthy people.

## ADRENAL GLANDS

The paired **adrenal glands** are attached to the superior end of both kidneys (Figure 16-9). The right adrenal gland is shaped like a pyramid; the left is semilunar. Both glands are covered by a thick layer of fatty connective tissue that is surrounded by an external fibrous capsule. Arteries that supply the adrenal glands are derived from the aorta and renal arteries. In cross-section, each adrenal gland is seen to consist of an outer *cortex* that forms most of the gland and an inner *medulla* (Figure 16-10a).

### Adrenal Medulla

The tissue of the **adrenal medulla** is composed of cells, called *chromaffin cells,* that are arranged in stringlike cords separated by blood sinuses (Figure 16-10b). The chromaffin cells secrete **epinephrine** and **norepinephrine,**

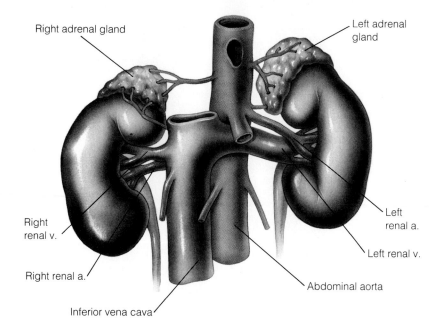

FIGURE 16-9 Location and blood supply of the paired adrenal glands.

Right adrenal gland

Left adrenal gland

Right renal v.

Left renal a.

Left renal v.

Right renal a.

Abdominal aorta

Inferior vena cava

two substances also synthesized by neurons for use as neurotransmitters. Unlike other endocrine glands in which secretion is not controlled by nerves, secretion of epinephrine and norepinephrine by the adrenal medulla occurs primarily as a result of the stimulation of preganglionic sympathetic fibers that innervate it. The adrenal medulla is of nervous system origin, and the chromaffin cells act essentially as postganglionic neurons, releasing their hormones in response to preganglionic stimulation.

The adrenal medulla is involved with the control of physiological reactions to short-term, immediate stress. Epinephrine and norepinephrine, secreted by the adrenal medulla in response to short-term stress, augment and prolong typical actions of the sympathetic division of the autonomic nervous system

FIGURE 16-10 (a) Cross-section through the adrenal gland. (b) Magnification of (a), showing the cell layers of the adrenal cortex and the cells of the adrenal medulla.

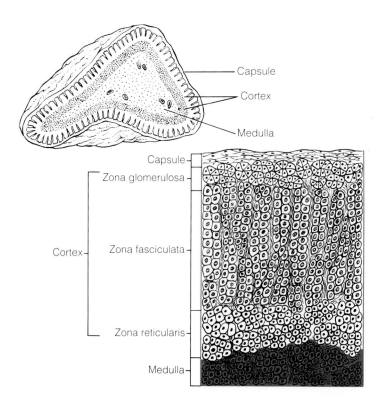

Capsule

Cortex

Medulla

Capsule

Zona glomerulosa

Cortex

Zona fasciculata

Zona reticularis

Medulla

that occur in "fight-or-flight" situations. These actions contribute to increased efficiency of operation for short periods to meet emergency demands of the body, as discussed in Chapter 13. About three-quarters of the total secretion of the adrenal medulla is epinephrine; the remainder is norepinephrine. These two hormones influence most systems of the body. Table 16–3 lists some of the most prominent effects of both hormones.

## Adrenal Cortex

The **adrenal cortex,** unlike the medulla, is devoid of nervous innervation. It develops from embryonic tissue similar to that from which the gonads arise, and like them, it secretes steroid hormones. The adrenal cortex is divided into three different layers: an outer *zona glomerulosa,* a middle *zona fasciculata,* and an innermost *zona reticularis* (Figure 16–10b). It is thought that each of these layers synthesizes a different type of steroid hormone. *Mineralocorticoids* are synthesized in the zona glomerulosa, *glucocorticoids* in the zona fasciculata, and *gonadocorticoids* in the zona reticularis.

## Mineralocorticoids

The **mineralocorticoids** are important in the maintenance of mineral homeostasis, particularly of sodium. *Aldosterone* is the principal mineralocorticoid in humans. It acts on the distal convoluted tubules and collecting tubules of the kidney to increase reabsorption of sodium into the blood. The reabsorption of sodium creates an osmotic pressure that pulls water out of the kidney tubules and into the blood. In addition, the positive charge on the sodium ions ($Na^+$) attracts negatively charged chloride ($Cl^-$) and bicarbonate ($HCO_3^-$) ions into the blood. Aldosterone also acts to increase the tubular secretion (elimination) of potassium and hydrogen ions.

The control of aldosterone secretion was discussed extensively in Chapter 12 (see Figure 12–7) as part of the renin-angiotensin system that regulates blood pressure. In that system, juxtaglomerular cells of the kidney, in response to low blood pressure, secrete renin, an enzyme that catalyzes the cleavage of the plasma protein angiotensinogen to angiotensin I. Angiotensin I is cleaved further to angiotensin II by a plasma enzyme. Angiotensin II elevates blood

TABLE 16–3 Major Actions of Epinephrine and Norepinephrine

| System | Major Actions | |
| --- | --- | --- |
| | Epinephrine | Norepinephrine |
| Circulatory | Increases cardiac output. Decreases peripheral resistance. Dilates coronary vessels. | No change in cardiac output. Increases peripheral resistance. Dilates coronary vessels. |
| Respiratory | Dilates bronchi. Stimulates rate and depth of breathing. | Dilates bronchi. Stimulates rate and depth of breathing. |
| Muscle | Prolongs contraction of skeletal muscle. Increases glycogen breakdown in skeletal muscle. | No effect on contraction of skeletal muscle. No effect on glycogen breakdown in skeletal muscle. |
| Digestive | Inhibits contractions of small and large intestine. Increases blood flow to liver. | Inhibits contraction of small and large intestine. No effect on blood flow to liver. |
| Urinary | Decreases blood flow to kidneys. | Decreases blood flow to kidneys. |
| Nervous | Increases anxiety. Increases blood flow to brain. | No effect on anxiety. Slightly decreases blood flow to brain. |
| Metabolic | Increases blood sugar. Increases rate of metabolism. | Increases blood sugar. Increases rate of metabolism. |

pressure by constricting arterioles and by stimulating the adrenal cortex to secrete aldosterone, which by increasing sodium and water reabsorption in the kidney, increases blood volume. Another stimulus to aldosterone secretion is an increase in blood potassium levels.

## Glucocorticoids

Whereas physiological reactions to short-term stress are mediated by the adrenal medulla in concert with the sympathetic nervous system, **glucocorticoids** from the adrenal cortex are involved in physiological responses to long-term stress, as occurs during starvation, severe injury, prolonged anxiety, and other such conditions. The glucocorticoids are a large group of steroid hormones that have wide effects on metabolism, and on the immune system and associated inflammatory responses to disease or injury.

The major human glucocorticoid is *cortisol,* also known as *hydrocortisone.* Cortisol and other glucocorticoids have two primary metabolic effects that make the body more resistant to stress. First, they function to elevate blood sugar, making more glucose available to meet the increased energy demands associated with stressful situations. Second, they increase the body's stores of glycogen, ensuring adequate energy reserves. The glucocorticoids produce these effects by stimulating the breakdown of storage fats and the release of fatty acids, as well as the breakdown of proteins to amino acids in muscle cells. The fatty acids and amino acids are transported to the liver where they are used for the manufacture of glucose, a process known as **gluconeogenesis.** The glucose that is formed may be released as blood sugar or used in the liver to form glycogen.

In addition to its effects on metabolism, cortisol and other glucocorticoids are *anti-inflammatory.* Inflammation, the body's normal response to infection or injury, involves the secretion of histamines and similar substances. These act to increase blood supply to the affected tissue and to stimulate the activity of white blood cells that combat infectious microbes and digest damaged tissue. (These processes are discussed in Chapter 21.) Glucocorticoids suppress these inflammatory responses, but in so doing, retard the body's ability to counteract damage caused by infection or injury. The anti-inflammatory action of glucocorticoids would appear to be beneficial in stressful situations where swelling and loss of function normally associated with inflammation could hamper a physical response to stress, such as running away from danger.

At doses considerably above physiological levels, glucocorticoids are *immunosuppressive;* that is, they suppress the activity of the immune system. They decrease the production of lymphocytes (white blood cells that secrete antibodies and attack infectious microbes), and they suppress the activity of lymphatic tissues, including the lymph nodes, spleen, thymus, and tonsils. Although glucocorticoids can be helpful in the treatment of autoimmune disease such as rheumatoid arthritis (see Chapter 4, *Focus: Connective Tissue Diseases*), they reduce resistance to infection.

Adrenal glucocorticoid release is activated in response to long-term stress. Stressful situations lead to the secretion of corticotropin releasing factor (CRF) by the hypothalamus. CRF, in turn, stimulates the release of ACTH from cells of the anterior pituitary. ACTH is carried by the bloodstream to the adrenal cortex, where it acts to bring about the release of glucocorticoids. High levels of glucocorticoids in the blood feed back on the hypothalamus and anterior pituitary to inhibit further glucocorticoid secretion; low levels of blood glucocorticoids are a stimulus to glucocorticoid secretion independent of a stressful situation. The control of glucocorticoid secretion is summarized in Figure 16–11.

## Gonadocorticoids

The zona reticularis of the adrenal cortex is capable of secreting small amounts of androgens (male sex hormones) and estrogens (female sex hormones).

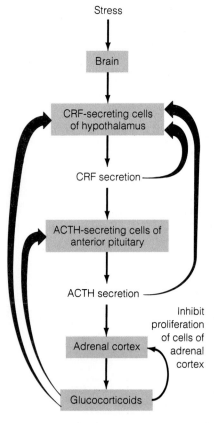

FIGURE 16–11 Control of glucocorticoid secretion by the adrenal cortex. Heavy arrows represent negative feedback loops.

These hormones are similar to those synthesized by the gonads, and hence are termed **gonadocorticoids.** Under normal conditions, the level of gonadocorticoid secretion is too low to have a noticeable physiological effect.

## Disorders of Adrenal Gland Function

The adrenal cortex is required for life, but the adrenal medulla is not. Consequently, undersecretion of the adrenal medulla is not generally associated with serious disorder. However, tumors of the adrenal medulla that result in hypersecretion of epinephrine and norepinephrine can cause symptoms of a prolonged "fight-or-flight" response, including nervousness, increased metabolic rate, high blood pressure, elevated blood sugar and excessive sweating. The body ultimately suffers fatigue and weakness from the extended mobilization of resources.

*Addison's disease* results from undersecretion of mineralocorticoids and glucocorticoids from the adrenal cortex. Symptoms of Addison's disease include weakness, weight loss, low blood pressure, and low blood sugar levels. Because of low levels of glucocorticoids, ACTH secretion is excessive and sustained, leading to increased pigmentation of the skin. This latter effect of ACTH is due to the fact that part of the chemical structure of ACTH is similar to that of MSH, which can stimulate skin pigment synthesis.

In *Cushing's syndrome,* there is excessive and prolonged secretion of glucocorticoids. This is often caused by a tumor of the glucocorticoid-secreting cells of the adrenal cortex or by a pituitary tumor that causes high levels of ACTH secretion. Cushing's syndrome is characterized by hyperglycemia (high blood glucose), high blood pressure, osteoporosis, increased thirst, increased urination, and distinctive patterns of fat accumulation in the face, abdomen, and back. The prolonged hyperglycemia of Cushing's syndrome may lead to the destruction of the pancreatic $\beta$-cells that secrete insulin, causing diabetes mellitus.

## ENDOCRINE PANCREAS

The **pancreas** functions as both an exocrine and endocrine gland. Its exocrine function involves the synthesis of pancreatic juice in saclike structures called acini, from which it is secreted through ducts into the duodenum. The structure of pancreatic acini, and the synthesis and secretion of pancreatic juice, are discussed in Chapter 11.

Scattered throughout the exocrine tissue of the pancreas are clusters of endocrine cells, called islets of Langerhans (Figure 16–12). Three types of

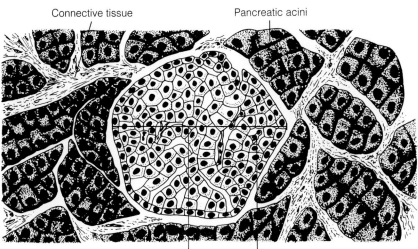

Connective tissue      Pancreatic acini

Islet of Langerhans     Blood vessel

FIGURE 16–12 Section through the pancreas, showing an islet of Langerhans and surrounding acini (exocrine tissue).

cells can be distinguished in the islets: *alpha cells,* which secrete *glucagon; beta cells,* which secrete *insulin;* and *delta cells,* which secrete *somatostatin.* All three of these hormones are involved in the regulation of blood glucose levels.

### Glucagon

**Glucagon,** secreted by the alpha cells, is a protein hormone that causes an increase in blood glucose levels. It acts by enhancing the conversion of liver glycogen into glucose, which then enters the blood. In addition, glucagon stimulates gluconeogenesis in the liver from noncarbohydrate sources such as amino acids and lactic acid; this newly synthesized glucose also enters the bloodstream.

The secretion of glucagon is regulated by the level of glucose in the blood. Low blood sugar stimulates the alpha cells to secrete glucagon, and high blood sugar inhibits glucagon secretion. Elevated fatty acid levels also decrease the secretion of glucagon.

### Insulin

**Insulin** is a protein hormone secreted by the beta cells of the islets of Langerhans. Its functions are essentially opposite to that of glucagon. It decreases blood sugar levels by increasing the transport of glucose from the blood into the body's cells, especially skeletal muscle cells, fat cells, and liver cells, and by accelerating the synthesis of glycogen from glucose in the liver. Insulin also increases the transport of amino acids into cells, stimulating protein synthesis, and stimulates the synthesis of lipids from glucose.

Insulin secretion is stimulated by high blood concentrations of glucose, and to a lesser extent by high blood levels of amino acids and fatty acids. Other hormones that produce hyperglycemia, such as GH and glucocorticoids, also stimulate insulin secretion.

### Somatostatin

**Somatostatin,** secreted by delta cells, is another protein hormone that is a component of the blood glucose regulation system of the pancreatic islets. It

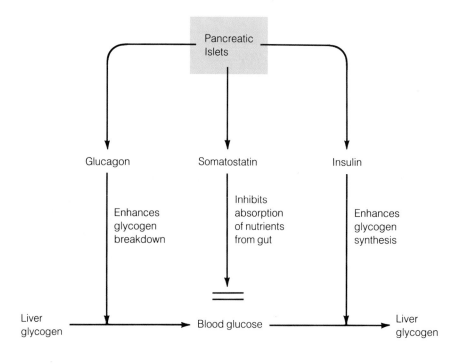

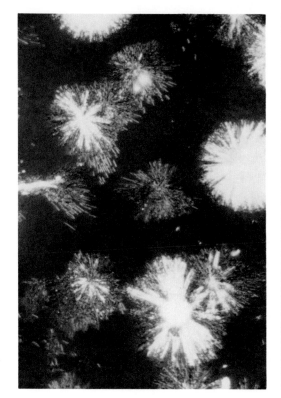
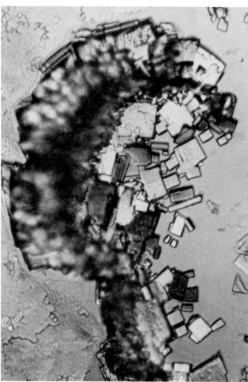

Shown here are crystals of estriol and estradiol; both are female sex hormones.

retards the movement of glucose, amino acids, and other nutrients from the gastrointestinal tract into the bloodstream by slowing the digestion of food and by reducing the rate of absorption of digested nutrients. Somatostatin also inhibits both glucagon and insulin secretion. The function of pancreatic somatostatin is unclear, although it has been suggested that reducing the rate of entry of nutrients into the circulation may allow time for the other islet hormones to coordinate a more efficient distribution of nutrients to the various cells of the body. Interestingly, pancreatic somatostatin is identical to hypothalamic somatostatin, which functions as growth hormone release-inhibiting factor.

### Disease of the Pancreatic Islets

The major disease of the pancreatic islets is *diabetes mellitus,* which results in an inability to use glucose as a source of energy. See *Focus: Diabetes Mellitus.*

## GONADS

The reproductive structures that secrete hormones are the *ovaries* in the female and the *testes* in the male. The ovaries are paired organs located on both sides of the pelvic cavity. They secrete *estrogen* and *progesterone,* steroid hormones that determine female sexual characteristics and, together with FSH and LH from the anterior pituitary, regulate reproductive cycles and maintain pregnancy.

The testes of the male are paired oval structures located in the scrotum. They produce *testosterone,* another steroid hormone that determines male sexual characteristics. Another hormone, *inhibin,* is secreted by the testes and functions to inhibit FSH secretion by the anterior pituitary. The gonadal hormones are discussed extensively in Chapter 17.

# Diabetes Mellitus

Insulin deficiency results in the disease *diabetes mellitus,* which occurs in two forms: *juvenile-onset diabetes (Type I diabetes),* the more serious form, and *maturity-onset diabetes (Type II diabetes).* About one million Americans have juvenile-onset diabetes which, as its name suggests, usually begins in childhood. It results from the destruction of the pancreatic beta cells that secrete insulin. Although the ultimate cause of beta cell destruction is not known, current ideas suggest that it results from an autoimmune reaction, perhaps triggered by a viral infection. In order to compensate for their lack of insulin, individuals with juvenile-onset diabetes must receive daily injections of insulin. Maturity-onset diabetes, appearing in adults often in association with obesity, affects some ten million Americans. It is usually due to an insensitivity to insulin; that is, the body does not respond optimally to the insulin it produces. The symptoms of maturity-onset diabetes are milder than those of juvenile-onset diabetes and often can be treated by diet and weight loss without the need for injections of insulin.

The symptoms of untreated juvenile-onset diabetes derive from the fact that, without insulin, glucose in the circulation cannot enter most cells. Consequently, glucose is not utilized by cells as a source of energy, nor can it be converted to glycogen in the liver for storage. This results in excessively high blood sugar levels and the excretion of glucose in the urine. Without glucose, cells must degrade fat, and eventually protein, as energy sources; excessive fat degradation leads to the accumulation in the blood of acetone and related by-products of fatty acid metabolism, called *ketone bodies.* These are excreted in the urine and by the lungs, giving an odor of acetone to the breath of diabetics who have not received enough insulin. The elimination of ketone bodies in the urine is accompanied by the osmotic loss of water, resulting in dehydration and accompanying thirst. Buffer systems in the blood are also overwhelmed, and the blood becomes acidic. The combination of dehydration and acidosis (blood that is too acidic) may result in diabetic coma and death.

Although insulin injection ameliorates many of the more severe effects of diabetes, wide fluctuations in blood glucose levels occur, resulting in alternating episodes of hyperglycemia and hypoglycemia. It is this instability in blood glucose levels that is thought to give rise to the serious and often fatal complications of long-term diabetes, including atherosclerosis, kidney damage, damage to the retina

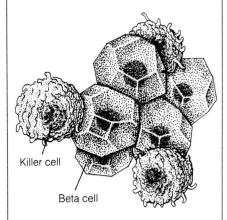

Killer cell

Beta cell

In juvenile-onset diabetes, killer cells of the immune system attack and destroy beta cells, which they incorrectly recognize as foreign.

of the eye, cataracts, and various neurological malfunctions, among others. In order to duplicate more closely the functioning of the normal pancreas and thus prevent large fluctuations in blood glucose levels, various artificial treatment devices are being developed. One of these is a mechanical device that continually monitors blood glucose levels and automatically administers appropriate amounts of insulin.

Another approach to treating juvenile-onset diabetes, still in an experimental stage, is to transplant a portion of the pancreas from a healthy donor onto the outer surface of the small intestine of an individual with diabetes. In a successful transplant, islet cells from the donor tissue will secrete insulin and regulate blood sugar at normal levels. However, as with any tissue transplant procedure, a major complication is the destruction of the transplant by the recipient's immune system, a process referred to as *tissue rejection.* In order to prevent rejection, most transplant recipients must take immunosuppressive drugs that, by suppressing the action of the immune system, have the unfortunate side effects of lowering the body's resistance to microbial infection and increasing the risks of cancer.

To avoid problems with tissue rejection, other researchers are experimenting with the transplantation of islet cells encased in a protective capsule. The capsule allows relatively small molecules such as glucose and insulin to pass through it, but prevents the entry of antibodies and tissue-destroying cells (lymphocytes) of the immune system. Experiments using encapsulated islet cells have been successful in mice and rats, and the technique may be available for human trials within several years.

## PINEAL GLAND

The **pineal gland** is a small, neuroendocrine structure that arises embryonically from the roof of the third ventricle of the brain, to which it remains connected by a thin stalk (Figure 16–13). It is shaped like a pine cone, from

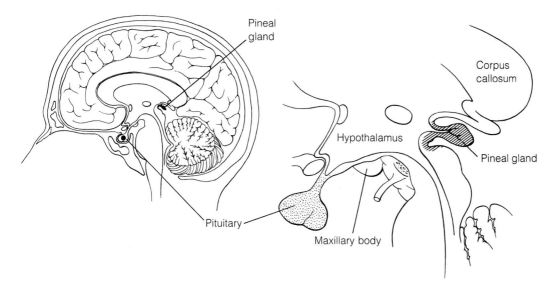

FIGURE 16–13 Location in the brain of the pineal gland.

which its name derives. The pineal gland is innervated by postganglionic fibers from the superior cervical ganglion of the sympathetic chain and acts essentially as an effector organ of the autonomic nervous system. The secretory cells of the pineal gland are called *pinealocytes,* which are present along with modified glial cells.

The function of the pineal gland in humans, long a mystery, has been somewhat clarified by recent work. It had been known for some time that the destruction of the pineal gland during childhood, as a consequence of injury or disease, results in precocious sexual maturity. The pinealocytes of the pineal gland are now known to secrete **melatonin,** a hormone derived from the amino acid tryptophan. Melatonin acts as an antigonadal hormone, preventing the maturation of the gonads until the time of puberty. It has also been shown that the secretion of melatonin in adults occurs in darkness but is inhibited by light. Experiments in animals have suggested a mechanism for the light-sensitivity of melatonin synthesis: light inhibits the release of norepinephrine at the postganglionic neurons that innervate the pineal gland, thus preventing melatonin secretion, whereas darkness promotes norepinephrine release in these same neurons and stimulates melatonin secretion. However, the relevance of daily light-dark cycles and of melatonin secretion to human reproduction has not been demonstrated.

Whatever the effect of melatonin on reproductive processes, there are indications that daily rhythms of melatonin secretion by the pineal gland may be important in the regulation of biological clocks, intrinsic rhythms in physiological activities that occur over a period of about a day. A good deal of evidence in experimental animals indicates that the patterns of melatonin secretion may provide the body with a physiological way of measuring the length of dark and light periods. Such measurements are crucial to the functioning of biological clocks, which are "set" with reference to daily light-dark cycles.

Data accumulated from studies of female athletes have suggested yet another possible effect of melatonin. It has been shown that blood levels of melatonin increase during exercise but decline to pre-exercise levels by about thirty minutes after completion of the exercise. There are indications that melatonin may inhibit the secretion of stress-related hormones such as ACTH, an effect that might explain why exercise is so effective in reducing stress.

## THYMUS

The **thymus gland** is a lymphatic organ with endocrine functions. It is a bilobed structure located beneath the sternum, anterior to the trachea (Figure 16–14). It is relatively large in children, attains its maximum size at puberty,

FIGURE 16-14 Location of the thymus gland.

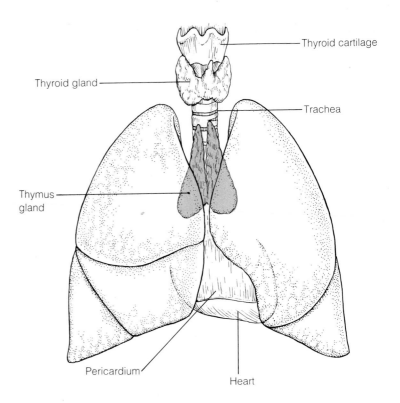

and then atrophies in adults, much of its substance being replaced by fat and connective tissue. Each lobe is surrounded by a fibrous capsule of connective tissue that penetrates the lobe and divides it into lobules, each of which contains an outer cortex, dense with cells, and an inner, less dense medulla. Both the cortex and medulla consist of lymphocytes interspersed among a branched framework of epithelial cells. There are many more lymphocytes in the cortex than in the medulla, a fact that accounts for the higher cell density in the cortex.

The thymus gland produces a number of related hormones referred to collectively as **thymosins.** Thymosins act to promote the functional maturation of a type of immature lymphocyte, called T-cells, which act to destroy invading microbes and abnormal host cells. The role of T-cells in the immune system is dealt with in Chapter 21.

## OTHER HORMONES

In addition to the clearly defined endocrine organs, there are a large number of endocrine cells scattered throughout other tissues of the body. These secrete regulatory substances that are now generally classified as hormones.

### Gastrointestinal Hormones

Endocrine cells dispersed throughout the mucosal lining of the gastrointestinal tract secrete a large number of hormones that are important in the control of digestive processes. The principal gastrointestinal hormones, discussed in Chapter 11, are *gastrin,* secreted by endocrine cells in the pyloric antrum region of the stomach, and responsible for the production of hydrochloric acid and gastric juice; *cholecystokinin,* secreted by the duodenal mucosa, and responsible for gallbladder contraction and the secretion of pancreatic juice rich in digestive enzymes; and *secretin,* also secreted by the duodenum, and the stimulus for the secretion of bicarbonate-containing pancreatic juice.

## Prostaglandins, Prostacyclins, and Thromboxane

**Prostaglandins, prostacyclins,** and **thromboxane** form a group of related hormones derived from long-chain polyunsaturated fatty acids. Prostaglandins were first discovered in low concentrations in the prostate gland, from which their name is derived, but they have since been detected in virtually all types of mammalian tissue. The two main prostaglandins are prostaglandin $E_2$ ($PGE_2$) and prostaglandin $F_{2\alpha}$ ($PGF_{2\alpha}$). These two hormones are involved in the control of smooth muscle contraction in a wide range of tissues. They play a major role in the distribution of blood between various tissues of the body by affecting smooth muscle in arterioles. $PGE_2$ stimulates local relaxation of arteriolar smooth muscle, thus increasing blood flow; $PGF_{2\alpha}$ stimulates its contraction, decreasing blood flow. $PGE_2$ and $PGF_{2\alpha}$ are secreted by tissues in response to various stimuli related to the flow of blood, including the lack of oxygen and blood loss.

There is growing evidence that exceptionally painful menstruation that includes severe cramps, nausea, vomiting, diarrhea, fatigue, and headache may be caused by the overproduction of uterine prostaglandins. These are thought to bring about the contraction of smooth muscle in the uterus, causing cramps; in the stomach, causing nausea and vomiting; and in the large intestine, causing diarrhea. Drugs that block the action of prostaglandins are effective in preventing these symptoms.

In addition to their effects on smooth muscle, prostaglandins influence a wide range of physiological processes. For example, they help mediate inflammation. Prostaglandins also act to stimulate steroid synthesis, increase the effectiveness of endogenous pain-causing chemicals, and effect norepinephrine release.

Prostacyclin $I_2$ ($PCI_2$), the major prostacyclin, is produced by the endothelial cells that form the walls of blood vessels. It prevents the disruption of blood flow by inhibiting the aggregation of blood platelets and by acting locally to dilate arterioles.

Thromboxane $A_2$ ($TXA_2$) is produced by blood platelets in response to chemical stimuli that are known to trigger blood clotting, such as the secretion of epinephrine and the presence of certain clotting factors. $TXA_2$ is thought to induce changes in the shape of platelets, causing their aggregation.

## Placental Hormones

During pregnancy (see Chapter 18), the placenta secretes *human chorionic gonadotropin (HCG),* which stimulates the ovaries to secrete estrogens and progesterone during the early months of pregnancy. Thereafter, the uterus itself secretes estrogens and progesterone, as well as *placental lactogen,* which has actions similar to that of prolactin.

## Atrial Natriuretic Factor

It has recently been shown that cardiac muscle cells of the atrium secrete several related peptide hormones, collectively referred to as **atrial natriuretic factor (ANF),** that induce diuresis (loss of water) and natriuresis (loss of sodium). ANF produces these effects by inhibiting aldosterone secretion from the adrenal cortex and ADH secretion from the posterior pituitary. It also causes dilation of arterioles. The net result of these actions is to reduce blood pressure by decreasing blood volume and increasing the total capacity of the blood vessels.

## Erythropoietin

**Erythropoietin** is a protein hormone that is secreted by the kidneys when they lack sufficient oxygen. It stimulates the production of red blood cells.

### Endogenous Opioid Peptides

The **endogenous opioid peptides** are hormones that bind to specific opiate receptors in the brain—receptors that also bind the opiate drugs such as morphine and heroin. The endogenous opioid peptides, including *enkephalins* and *endorphins,* are synthesized in the brain and anterior pituitary. They are secreted in response to painful stimuli, and play an important role in blocking the sensation of pain. They also have important effects on mood by producing a euphoria similar to that of the opiate drugs.

## MECHANISMS OF HORMONE ACTION

The effect of hormones is to alter the synthesis and/or activity of enzymes present in target cells and thus to change the metabolic capabilities of these cells. However, it is important to realize that hormones do not create any new potential in the target cell, but merely allow it to express a potential that has already been determined.

At the level of the target cell, hormones act in several ways. The effects of many protein and peptide hormones are mediated by an intracellular chemical, termed a *second messenger,* whereas steroid hormones and thyroid hormones influence cells by directly affecting the expression of hereditary information (DNA).

### Second Messenger Mechanism

Hormones as diverse as TSH, ACTH, LH, glucagon, parathyroid hormone, vasopressin, and epinephrine produce their effects by the **second messenger mechanism,** outlined in Figure 16–15. In the first step of the process, the hormone attaches to specific receptors present on the outside surface of the

FIGURE 16–15 The second messenger mechanism of hormone action.

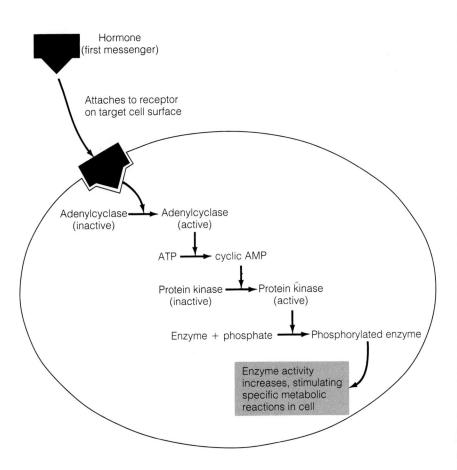

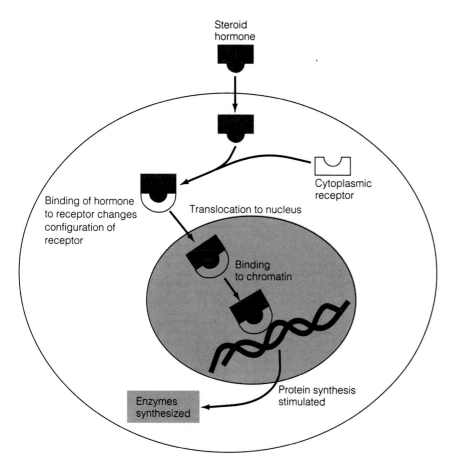

Steroid hormone

Cytoplasmic receptor

Binding of hormone to receptor changes configuration of receptor

Translocation to nucleus

Binding to chromatin

Protein synthesis stimulated

Enzymes synthesized

target cell's plasma membrane. However, the hormone does not enter the cell in order to act. Instead, the mere attachment of the hormone to its receptor activates an enzyme, termed *adenyl cyclase,* within the target cell. Adenyl cyclase catalyzes the removal of two phosphate groups from adenosine triphosphate (ATP), forming **cyclic AMP,** a special cyclic form of adenosine monophosphate. The increase in cyclic AMP levels within the cell activates other enzymes, the *protein kinases.* Protein kinases alter the activities of still other enzymes in the cell by attaching phosphate groups to some of the amino acids composing the enzymes. Because the hormone initiates the effect, it is referred to as the *first messenger* and cyclic AMP is the *second messenger.*

A significant fact about the second messenger mechanism is that different hormones affect their target cells in the same way: that is, by increasing cyclic AMP levels. The specificity of a particular hormone's effect must therefore lie in the selective responsiveness of its target cell.

## DNA Activation Mechanism

Steroid hormones (including testosterone, estrogen, progesterone, and the hormones of the adrenal cortex) and thyroid hormones do not act by the second messenger mechanism. Their mode of action, termed the *DNA activation mechanism,* is summarized in Figure 16–16. These hormones pass through the target cell's plasma membrane and combine with specific hormone receptors in the cytoplasm. Binding of the hormone with its receptor causes a conformational change in the receptor—a change that allows the hormone-receptor complex to enter the nucleus and bind to specific regions on the chromatin, triggering the activation of areas of DNA that code for the synthesis of specific enzymes. These enzymes allow the cell to perform specific metabolic functions. Because protein synthesis is required for the effects of steroid hormones, their actions are considerably slower than those of hormones that operate by the second messenger mechanism.

## Study Questions

1. How does the function of the endocrine system differ from that of the nervous system?
2. Give two specific examples of the integration between endocrine and nervous function.
3. Describe the ways that endocrine glands differ from exocrine glands.
4. Distinguish between negative and positive feedback loops.
5. **a)** List the three lobes of the pituitary gland.
   **b)** List the hormones secreted by each lobe and their functions.
6. **a)** What is the source of releasing factors?
   **b)** What is their function?
   **c)** In what ways are releasing factors similar to posterior pituitary hormones?
7. **a)** Describe the function of the thyroid gland.
   **b)** Summarize the process by which thyroid hormones are synthesized in, and secreted from, the follicles.
   **c)** Explain why iodine deficiency results in thyroid malfunction.
8. Explain the role of calcitonin, parathyroid hormone, and vitamin D in blood calcium homeostasis.
9. Describe the role of the adrenal glands in the physiological responses to short-term and long-term stress.
10. Describe how hormones of the pancreatic islets control blood sugar levels.
11. Indicate one function of the pineal gland in humans.
12. **a)** Describe the second messenger mechanism of hormone action.
    **b)** List several hormones that act by the second messenger mechanism.
13. Describe the DNA activation mechanism of steroid and thyroid hormones.

Section Three

# A NEW LIFE

# THE REPRODUCTIVE SYSTEMS

Reproduction is the process by which offspring are produced. Like other major physiological activities, reproduction involves specialized organs that function in a complex and highly integrated manner, and it is controlled or conditioned by a variety of internal and external factors. The reproductive system also differs in significant ways from other physiological systems. First, the reproductive system does not function throughout life. The reproductive organs are small and nonfunctional until puberty, the time at which sexual reproduction is possible, and reproductive function ends with menopause in females and declines gradually in aging males. In addition, success in reproduction requires functional coordination between the reproductive systems of *both* parents. Moreover, reproduction is neither required for the survival of the individual, nor is it obviously beneficial; in fact, in females, it may sometimes be physiologically detrimental. Nevertheless, reproduction is necessary for survival of the species, and as such it is an indispensable attribute of life.

In human reproduction, sperm produced in the testes of the male is deposited in the female's reproductive tract. Union of a single sperm with an egg cell produces a fertilized egg—a one-celled embryo—that implants in the lining of the uterus. Nourished in the mother's womb, the embryo grows by cell division and undergoes a gradual change in form. After approximately nine months, embryonic development is complete, and labor is initiated, resulting in the expulsion of the fully developed embryo from the uterus.

The cellular and physiological events culminating in fertilization involve two interrelated aspects: (1) the actual formation of germ cells (sperm and egg cells) and (2) the processes that lead to the fusion of sperm and egg. Understanding the first of these requires a knowledge of the cellular process of **meiosis**—division in cells that give rise to germ cells.

## MEIOSIS

### Comparison of Cell Division in Somatic Cells and Germ Cells

You will recall from Chapter 3 that the hereditary material of a cell is in the form of a number of DNA molecules located in the cell's nucleus. The information encoded in this DNA is identical in all somatic cells of the body (all cells except germ cells), and it represents a blueprint necessary for determining not only the structural and functional pattern of the cell itself, but also the general characteristics of the organism of which the cell is a part.

Each DNA molecule in a cell is complexed with protein to form a chromatin strand that condenses and coils during cell division to form a chromosome. Human somatic cells contain 46 chromosomes that are present as 23 pairs; such cells are termed **diploid** (*diplo* = double) because they possess two of each chromosome type. One chromosome of each pair is derived from the germ cell of the mother, and the other member of each pair is derived from the germ cell of the father. The two chromosomes of a pair are called **homologous chromosomes** because they have similar morphology (form) and because they carry **genes**—functional units of hereditary material—that determine similar traits, such as blood type and eye color.

Somatic cells divide by *mitosis,* a process outlined in Chapter 3 and reviewed in Figure 17–1. In mitosis, each chromatin strand (and the DNA it contains) is precisely replicated; the doubled chromatin strands, still attached to one another, condense into chromosomes; each chromosome is divided so that one of each of the doubled chromatin strands, called *sister chromatids,* is distributed to each daughter cell. (You may need to review the details of mitosis in Chapter 3 in order to recollect pertinent terminology.) The net result of mitosis, summarized in Figure 17–2, is that each new daughter cell comes to

FIGURE 17-1 The stages of mitosis.

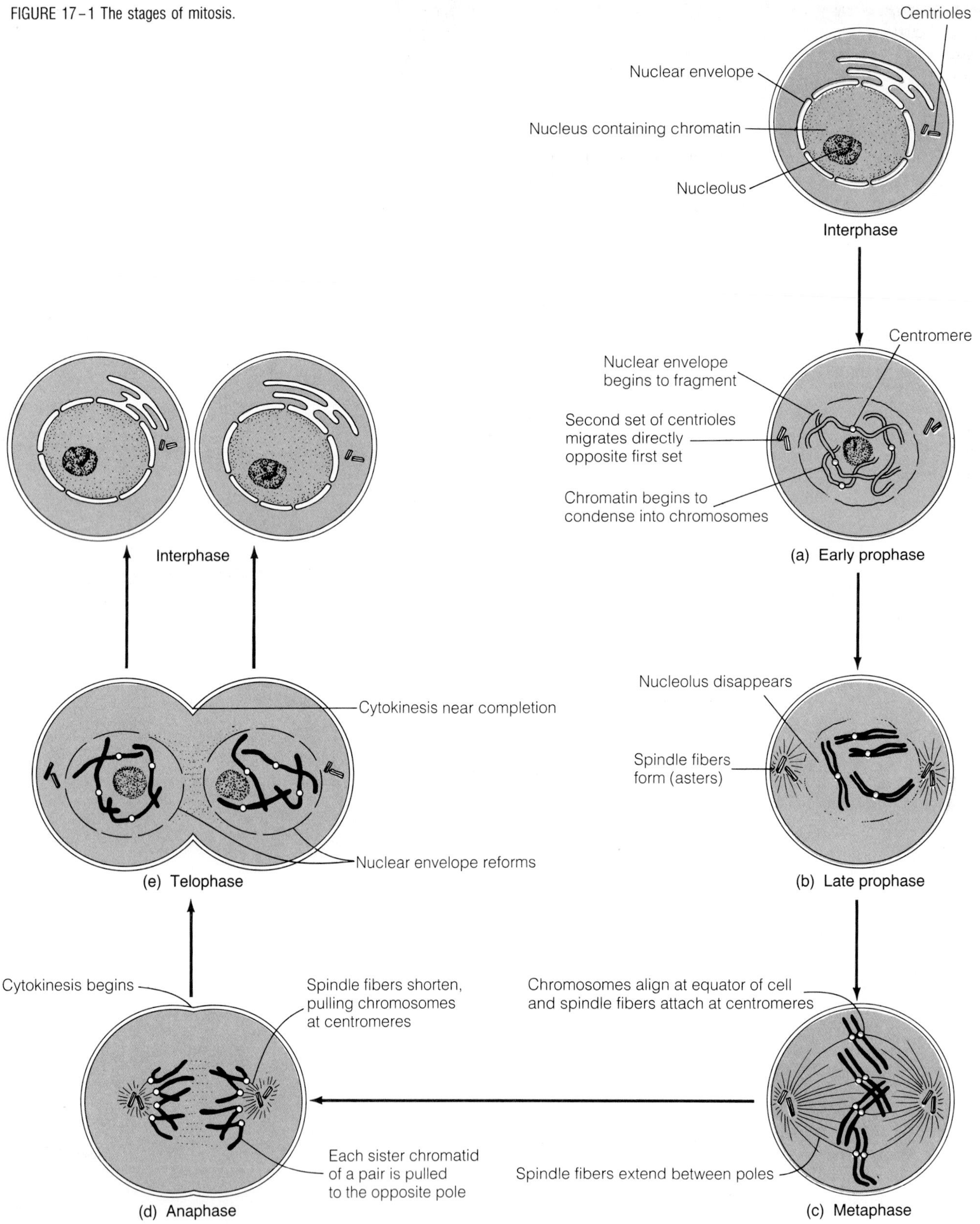

Centrioles

Nuclear envelope

Nucleus containing chromatin

Nucleolus

Interphase

Centromere

Nuclear envelope begins to fragment

Second set of centrioles migrates directly opposite first set

Chromatin begins to condense into chromosomes

(a) Early prophase

Nucleolus disappears

Spindle fibers form (asters)

(b) Late prophase

Interphase

Cytokinesis near completion

Nuclear envelope reforms

(e) Telophase

Cytokinesis begins

Spindle fibers shorten, pulling chromosomes at centromeres

Each sister chromatid of a pair is pulled to the opposite pole

(d) Anaphase

Chromosomes align at equator of cell and spindle fibers attach at centromeres

Spindle fibers extend between poles

(c) Metaphase

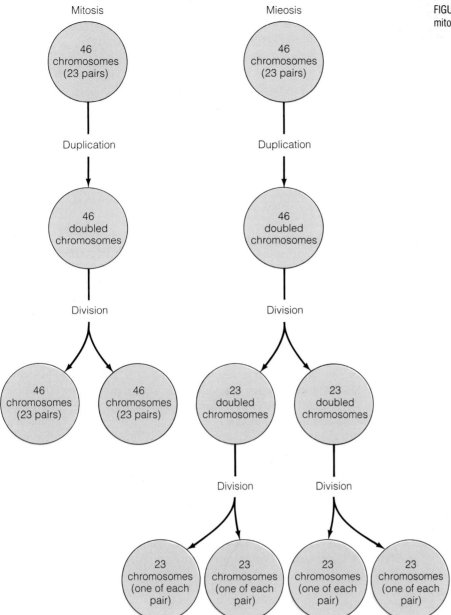

FIGURE 17-2 A schematic comparison of mitosis and meiosis.

possess 23 pairs of chromosomes (or chromatin strands) that are identical to those in the parent cell.

Since all somatic cells in the human body contain 23 pairs of chromosomes, as do all cells in an embryo, it is intuitively obvious that either the number of chromosomes in germ cells must be halved prior to the fusion of sperm and egg at fertilization, or half of the chromosomes donated by the sperm and egg must be eliminated shortly after fertilization. Otherwise, the number of chromosomes in human somatic cells would double at each generation. In fact, we know that germ cells contain only 23 chromosomes, corresponding to one member of each homologous pair. Because they contain only half of the somatic number of chromosomes, germ cells are termed **haploid** (*haplo* = single). Thus at fertilization, the normal chromosome number in somatic cells—23 pairs of chromosomes—is reconstituted.

Germ cells are produced from diploid precursor cells by a type of cell division called *meiosis*. Whereas mitosis involves duplication of all 23 pairs of chromosomes followed by a single nuclear division in which the duplicated chromosomes are equally distributed between two daughter cells, meiosis involves duplication of all 23 pairs of chromosomes followed by *two* consecutive nuclear divisions. Meiosis thus results in the production of four cells, each of which is haploid (see Figure 17–2).

## Stages of Meiosis

The various stages of meiosis are summarized in Figure 17–3. These stages are grouped into those associated with the first nuclear division *(meiosis I)* and those associated with the second nuclear division *(meiosis II).*

### Meiosis I

In meiosis I, the chromatin strands duplicate and a single nuclear division occurs. The following stages are evident.

***Interphase I.*** Prior to undergoing meiosis, a cell is said to be in *interphase I.* At some point in interphase I, the cell becomes committed to divide. Each of the chromatin strands is duplicated, but the two strands remain attached to one another at the centromere.

***Prophase I.*** In *prophase I,* the chromatin strands condense into chromosomes, and homologous chromosomes come to lie adjacent to one another and begin to pair along their lengths. This pairing of homologous chromosomes is called **synapsis.** Because each chromosome is composed of two chromatin strands (sister chromatids) as a result of doubling during interphase I, the pairing of homologous chromosomes creates a four-stranded structure known as a **tetrad** (Figure 17–4a). During synapsis, material from one chromosome may be reciprocally exchanged for similar material on the homologous chromosome, a process termed *crossing over* (Figure 17–4b and c). The significance of crossing over in the formation of germ cells is discussed below. In late prophase, the nuclear envelope disintegrates and spindle fibers begin to form.

***Metaphase I.*** The tetrads come to lie at the center of the cell, perpendicular to the spindle apparatus. Spindle fibers attach to each centromere in such a way that both sister chromatids of a single chromosome are attached to one pole and both sister chromatids of the homologous chromosome are attached to the other pole.

***Anaphase I and Telophase I.*** During *anaphase I,* one entire chromosome of each homologous pair is pulled to opposite poles. It is important to note that sister chromatids are not separated from one another in this division.

*Telophase I* is characterized by the formation of nuclear envelopes around each set of chromosomes. Division of the cytoplasm, having begun in anaphase I, is completed, producing two cells.

### Meiosis II

At the end of meiosis I, each cell has 23 chromosomes, each composed of two identical sister chromatids. Meiosis II results in the separation of sister chromatids from one another, forming haploid cells.

***Prophase II.*** The chromosomes remain coiled from telophase I, and the cell enters *prophase II* without an intervening interphase. During late prophase, the nuclear envelope degenerates, and a spindle apparatus begins to form.

***Metaphase II.*** The chromosomes line up in the center of the cell. Spindle fibers attach to each chromosome, connecting sister chromatids to opposite poles.

FIGURE 17–3 The stages of meiosis. Compare this figure with Figure 17–1, which shows mitosis. Black chromosomes are maternal and white are paternal. Notice that the chromosomal constitution of germ cells will differ depending upon (1) the extent of crossing-over during prophase I and (2) how the maternal and paternal chromosomes happen to align with respect to the two poles in metaphase I.

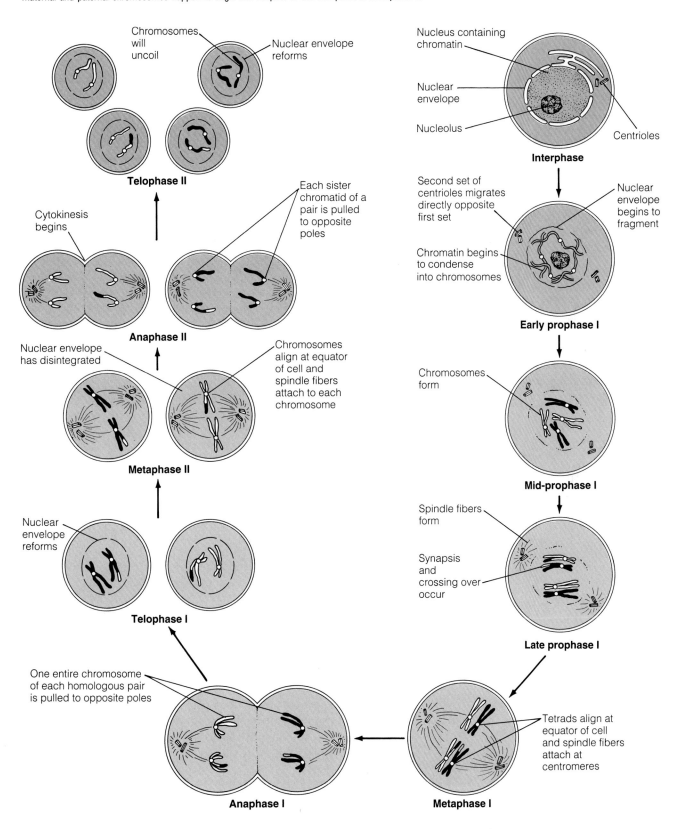

Chromosomes will uncoil

Nuclear envelope reforms

Telophase II

Cytokinesis begins

Each sister chromatid of a pair is pulled to opposite poles

Anaphase II

Nuclear envelope has disintegrated

Chromosomes align at equator of cell and spindle fibers attach to each chromosome

Metaphase II

Nuclear envelope reforms

Telophase I

One entire chromosome of each homologous pair is pulled to opposite poles

Anaphase I

Nucleus containing chromatin

Nuclear envelope

Nucleolus

Centrioles

Interphase

Second set of centrioles migrates directly opposite first set

Nuclear envelope begins to fragment

Chromatin begins to condense into chromosomes

Early prophase I

Chromosomes form

Mid-prophase I

Spindle fibers form

Synapsis and crossing over occur

Late prophase I

Tetrads align at equator of cell and spindle fibers attach at centromeres

Metaphase I

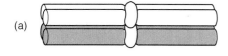

(a)

(b)

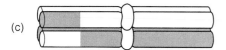

(c)

FIGURE 17–4 (a) Tetrad formation; (b) the reciprocal exchange of chromosomal material between homologous chromosomes by crossing-over; and (c) recombined chromosomes. Crossing-over occurs by breakage and reunion of the chromatin strands.

***Anaphase II and Telophase II.*** Sister chromatids are separated from one another and move to opposite poles. By the end of *anaphase II,* one complete haploid set of chromosomes has aggregated at each pole. *Telophase II* is characterized by the formation of nuclear envelopes around each set of chromosomes, the uncoiling of the chromosomes, and final division of the cytoplasm.

## Germ Cell Variability

Meiosis functions to reduce the chromosome number in germ cells prior to fertilization. However, reduction in chromosome number is not the only consequence of meiosis. Events that occur during meiosis also result in enormous variability in the hereditary makeup of germ cells produced by a single individual. This germ cell variability assures hereditary variability in offspring and makes it virtually impossible on a statistical basis for two siblings — or for any two human beings — to be genetically identical (except, of course, for identical twins, which arise from a single fertilized egg).

Two events that occur in meiosis I contribute to germ cell variability. In order to understand the significance of these events, we must recall the chromosomal constitution of the cells that undergo meiosis. Since each individual is a product of the fusion of two haploid germ cells, every diploid cell in an individual, including those that give rise to germ cells, possesses one chromosome of each homologous pair that is inherited from the mother *(maternal chromosome)* and one that is inherited from the father *(paternal chromosome).* When tetrads, consisting of both maternal and paternal chromosomes of a homologous pair, line up in the center of a dividing cell at metaphase I, it is purely a matter of chance as to whether the paternal or the maternal chromosome of any tetrad faces one or the other pole of the cell. Consequently, when homologous chromosomes are separated from one another in anaphase I, each pole will usually receive some mixture of maternal and paternal chromosomes; it will receive all paternal or all maternal chromosomes only rarely (study Figure 17–3).

Because the 23 maternal chromosomes and the 23 paternal chromosomes assort independently of one another during meiosis I, the number of different combinations of maternal and paternal chromosomes that may be produced during meiosis is $2^{23}$, or more than 8 million. Since a single male and a single female can each produce over 8 million different types of germ cells, two parents have the potential of producing, *on this basis alone,* over 64,000,000,000,000 genetically different offspring (8 million times 8 million).

Germ cell variability arising as a consequence of the independent assortment of maternal and paternal chromosomes is vastly increased by crossing over that occurs in prophase I. By interchanging sections of hereditary material between homologous chromosomes, crossing over has the effect of shuffling maternal and paternal genes prior to the assortment of chromosomes during meiosis I.

## DESIGN OF THE REPRODUCTIVE SYSTEMS

Sperm and egg cells, also referred to respectively as **spermatozoa** (sing: *spermatozoan*) and **ova** (sing: *ovum*), are produced by the *primary reproductive organs.* These are the *testes* in the male and the *ovaries* in the female. In addition to the production of germ cells, the primary reproductive organs synthesize and secrete *sex hormones.* These include *androgens,* which are produced mainly in males (and to a lesser degree in females), and *progesterones* and *estrogens,* which are produced primarily in females (and to a lesser degree in males). The sex hormones are necessary for normal development, growth, and function of not only the primary reproductive organs, but the so-called *accessory reproductive organs,* as well. The accessory reproductive organs form the *reproductive tract,* which includes the hollow structures through which sperm and eggs pass as well as the associated glands that release their secretions into these structures. Often included as accessory reproductive organs are the breasts of the female, which are necessary for nourishing the young.

Sex hormones are also responsible for the development of the *secondary sexual characteristics* that distinguish males from females but that are not essential for reproduction. These include such things as differences in the structure of the pelvic girdle, the size of the larynx, the distribution of body hair, the development of bones and muscles, and the deposition of fat in the breasts, hips, and other structures. The secondary sexual characteristics develop as the primary and accessory reproductive organs become functional at *puberty,* the time when reproduction becomes possible.

## MALE REPRODUCTIVE SYSTEM

The male functions in reproduction to produce sperm cells and to deposit them into the reproductive tract of the female. The anatomy of the male reproductive system is shown in Figure 17–5. It consists of (1) two testes, which produce sperm and that synthesize and secrete androgens; (2) a system of ducts, including two *epididymes* (sing: *epididymis*), two *vasa deferentia* (sing: *vas deferens*), two *ejaculatory ducts,* and the urethra, that lead from the testes to the exterior; (3) accessory glands, including two seminal vesicles, prostate, and two bulbourethral glands, that secrete fluid in which sperm are suspended; and (4) the *penis,* a copulatory organ, by means of which sperm are deposited in the female's vagina. Notice that the urethra is shared by the urinary and reproductive systems.

### Testes and Scrotum

The **testes** are suspended in the **scrotum,** a pouch consisting of a thin layer of skin that covers, and is closely united to, an underlying smooth muscle layer. Internally, the smooth muscle layer forms a partition that divides the scrotum into two cavities, one for each testis. Contraction of the smooth muscle in the

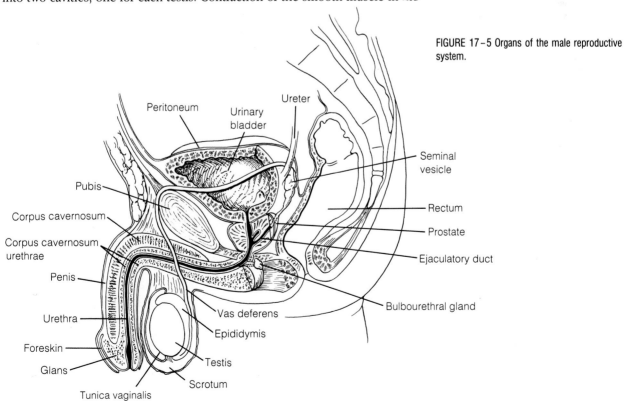

FIGURE 17–5 Organs of the male reproductive system.

wall of the scrotum in response to cold, fright, and various other stimuli causes the scrotum to become less pendulous and firmer, with a wrinkled wall.

## Development and Descent of the Testes

During early embryonic life, the testes arise in depressions on either side of the posterior abdominal wall, near the kidneys and behind the peritoneum. By the end of the eighth month of development, the testes have usually descended into the scrotum by passing through the **inguinal canals,** tunnels in either side of the lower abdominal wall. As each testis descends, it pulls with it the vas deferens as well as blood vessels, lymphatic vessels, and nerves that supply it. These structures, enclosed in a connective tissue sheath, form the *spermatic cord.* Also during descent, each testis becomes covered with a double membrane, the *tunica vaginalis,* derived from the peritoneum (see Figures 17–5 and 17–6).

*Cryptorchidism (Undescended Testes).*    Occasionally, the testes do not descend into the scrotum during fetal life, leading to a condition known as *cryptorchidism* or *undescended testes.* They may descend during the first few years of life, or they may fail to descend at all. Because sperm production requires a temperature somewhat lower than internal body temperature, descent of the testes into the scrotum is necessary for fertility. Consequently, undescended testes are usually lowered into the scrotum surgically if they have not descended by about the age of five. If left in the abdominal cavity after that age, the sperm-producing capacity of the testes may be permanently impaired. For some reason not clearly understood, males born with undescended testes have a much higher incidence of testicular cancer as adults than other males.

*Inguinal Hernia.*    The inguinal canals generally close before or shortly after birth. However, the site of the obliterated canal remains a weak point in the abdominal wall that may herniate. In one type of *inguinal hernia,* a loop of intestine may protrude into the inguinal canal.

## Structure of the Testes

Beneath the tunica vaginalis, each testis is covered by a fibrous connective tissue capsule, called the *tunica albuginea,* from which partitions push inward

FIGURE 17–6 Internal structure of the testis and epididymis. Sperm cells, sloughed from the walls of the seminiferous tubules, enter the lumen of the tubules and pass through a converging network of channels into the epididymis.

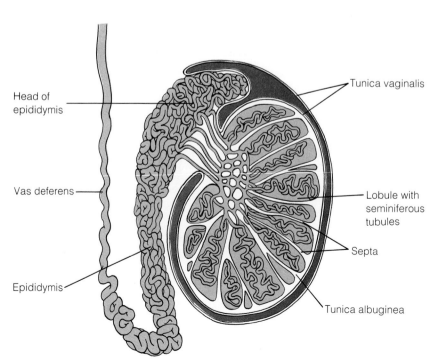

Head of epididymis

Vas deferens

Epididymis

Tunica vaginalis

Lobule with seminiferous tubules

Septa

Tunica albuginea

to divide the testes into several hundred lobules. Each lobule contains one or several highly coiled **seminiferous tubules,** where sperm are produced. The seminiferous tubules are supported by connective tissue containing groups of *interstitial cells* that are the site of androgen synthesis.

## Duct System

As they leave the lobules, the seminiferous tubules join together in the testis, forming a network of passageways. These drain into 15 to 20 tubules that leave the testis and pass to the head of the epididymis, where they unite to form the single *duct of the epididymis.* The duct pursues a tortuous course through the epididymis, emerging from the epididymis as the **vas deferens.** The vas deferens from each testis passes through the inguinal canal as part of the spermatic cord, enters the pelvic portion of the abdominal cavity, and then passes over and behind the urinary bladder. Just before it enters the prostate, each vas deferens is joined by a duct from a seminal vesicle, forming a common duct, the *ejaculatory duct.* Both ejaculatory ducts enter the prostate at its midline and then join the urethra. The urethra leaves the prostate, and just as it enters the penis it is joined by ducts from the paired *bulbourethral glands* (*Cowper's glands*).

## Accessory Glands

The seminal vesicles and prostate secrete fluids that, mixed with sperm cells, form the **semen** that is released upon ejaculation. The **seminal vesicles** are 5 to 7 cm long, and are composed of highly folded convoluted tubes, lined with secretory epithelium. They secrete a viscous, yellowish fluid that is stored and released upon ejaculation. In addition to serving as a vehicle for the transport of sperm, the seminal secretions contain, among other substances, fructose and prostaglandins. Fructose serves as a nutrient source for the sperm cells, and prostaglandins, by stimulating the contraction of smooth muscle in the female reproductive tract, may aid in the transport of sperm.

The **prostate,** about the size and shape of a horse chestnut, produces a thin, whitish fluid containing high concentrations of enzymes that liquefy semen. This secretion is stored in the prostate and is released through several dozen ducts into the urethra at the time of ejaculation.

In older men, the prostate commonly enlarges and becomes stiff and inflexible, obstructing the flow of urine through the urethra. Urination may become difficult and painful, and repeated failure to empty the bladder completely may lead to *cystitis* or *pyelonephritis* (see Chapter 12). Surgical removal of the prostate may be necessary to alleviate these problems.

On occasion, men with symptoms of an enlarged prostate are found to have cancer of the prostate. This cancer is rare in young men, but becomes increasingly common with age. Unlike most other cancers, prostate cancer grows slowly, seldom spreads to other organs of the body, and rarely causes serious health problems. It can usually be contained for long periods of time by treatment with estrogens.

The **bulbourethral glands,** also known as Cowper's glands, secrete a thick, clear, slippery fluid that is released following erection and that helps lubricate the vagina during sexual intercourse.

## Penis

The **penis** is composed mainly of three cylinders of erectile tissue (see Figure 17–5). Two of these, known as the *corpora cavernosa* (sing: *corpus cavernosum*), lie side by side; the third, carrying the urethra and hence called the *corpus cavernosum urethrae,* lies medially below the other two. The cylinders of erectile tissue are bound together by elastic connective tissue that is overlayed by the skin of the penis. The erectile tissue contains connective tissue partitions surrounding blood sinuses that become engorged with blood during erection of the penis.

The corpus cavernosum urethrae is expanded at the end of the penis to form the **glans.** The glans, possessing many sensory nerve endings, is covered by a cuff of skin, the *foreskin* or *prepuce.* The foreskin is often removed surgically in a procedure termed *circumcision.*

## Spermatogenesis

**Spermatogenesis,** the production of sperm, occurs in the seminiferous tubules of the testes. If a seminiferous tubule is examined in cross-section under the microscope, the wall of the tubule is seen to contain a large number of cells in various stages of cell division (Figure 17–7). These cells are the developing sperm cells that, as they mature, enter the lumen of the seminiferous tubule.

### Structure of Seminiferous Tubules

The external border of a seminiferous tubule is bounded by a typical basement membrane. Inside of this supporting membrane, and in contact with it, are epithelial cells of the so-called **germinal epithelium.** These cells are rapidly dividing, diploid germ cells called *spermatogonia* that may divide in one of two

FIGURE 17–7 (a) Cross-section through a human testis, showing seminiferous tubules and interstitial cells. (b) Diagram of a section of the wall of a seminiferous tubule. The relationship of developing sperm cells to Sertoli cells is shown.

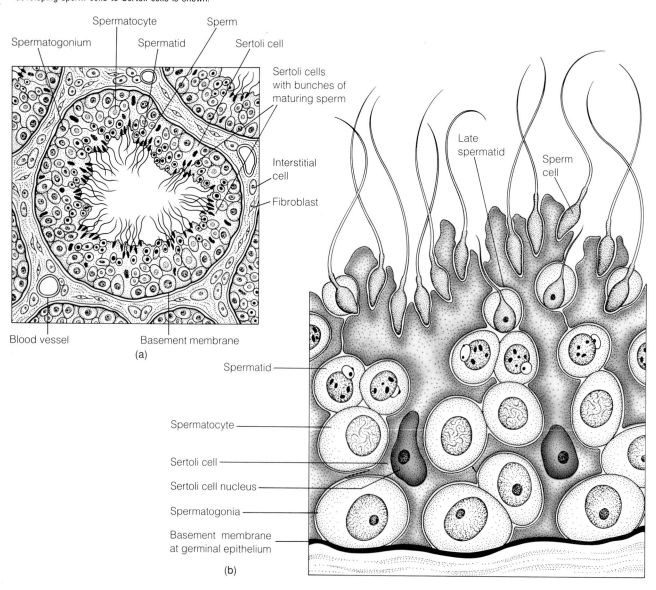

FIGURE 17-8 Summary of spermatogenesis.

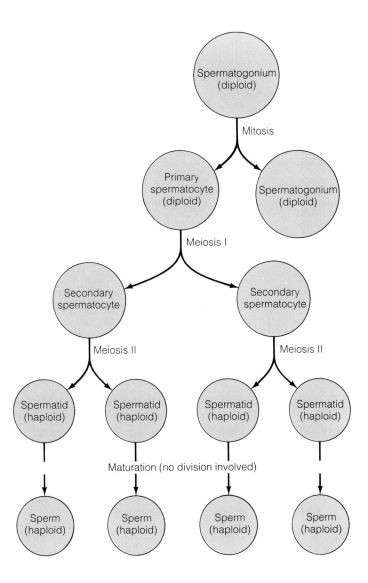

ways. Some undergo meiosis within the wall of the tubule, giving rise to sperm cells (spermatozoa), whereas others undergo mitosis to produce more identical spermatogonia, thus assuring that the supply of spermatogonia is not depleted. Thus, spermatogonia are not only self-replacing, but provide a continuous supply of sperm cells, amounting to hundreds of millions each day.

## Stages of Spermatogenesis

Spermatogenesis is summarized in Figure 17-8. Spermatogonia begin their transformation into sperm cells by increasing markedly in size, forming cells called *primary spermatocytes*. Each primary spermatocyte undergoes meiosis I, giving rise to two *secondary spermatocytes*. These are smaller than primary spermatocytes and almost immediately after their formation, each secondary spermatocyte undergoes meiosis II to form two haploid spermatids. As the transformation from spermatogonia to spermatids occurs, cells in each subsequent stage are pushed progressively away from the basement membrane toward the lumen of the tubule (see Figure 17-7b). In the innermost region of the tubule's wall, spermatids are transformed into sperm cells by the loss of cytoplasm and the development of structures typical of sperm cells (Figure 17-9). Once spermatids have developed into sperm cells, the latter are released into the lumen of the seminiferous tubules. The sperm cells move

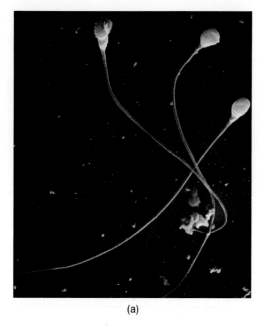

(a)

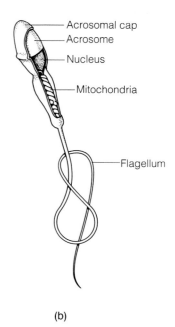

— Acrosomal cap
— Acrosome

— Nucleus

— Mitochondria

— Flagellum

(b)

FIGURE 17–9 (a) Scanning electron micrograph of human sperm cells, (b) Details of sperm structure.

through the seminiferous tubules and collecting ducts in the testes to the epididymis, where they are stored for several weeks while undergoing further maturation, and attaining full motility and capacity for fertilization. The transformation from spermatogonium to functional sperm cells takes about 74 days in humans.

In addition to germ cells, the walls of seminiferous tubules contain very large cells, called **Sertoli cells,** that extend across the wall of the tubule from the basement membrane to the lumen (see Figure 17–7). The Sertoli cells are intimately associated with the developing germ cells, which are embedded in depressions in the plasma membrane of Sertoli cells. The Sertoli cells form the blood-testis barrier.

### The Blood-Testis Barrier

The walls of seminiferous tubules have no direct blood supply. Nevertheless, spermatogonia and Sertoli cells, both in contact with the basement membrane of the seminiferous tubules, are able to receive nutrients and oxygen from tissue fluid that diffuses through the basement membrane from the connective tissue between adjacent sections of seminiferous tubule; this tissue fluid is, of course, derived from capillaries that supply the connective tissue. However, within the wall of the seminiferous tubule just inside the area occupied by spermatogonia, the plasma membranes of adjacent Sertoli cells are fused tightly together, forming an impermeable ring. This cellular ring constitutes the *blood-testis barrier,* since it prevents further diffusion of tissue fluid into the inner regions of the wall. As spermatogonia begin their transformation into sperm cells, the primary spermatocytes squeeze through temporary openings they create at the junctions between Sertoli cells. Once inside the impermeable ring created by the Sertoli cells, the primary spermatocytes and their meiotic products can only obtain nutrients and oxygen that diffuse *through* the cytoplasm of the Sertoli cells. It is believed that this arrangement may be important in preventing harmful substances in the blood from reaching the developing germ cells, since everything reaching those cells must pass through—and is presumably screened by—the Sertoli cells. Because of the nutritive role they play in spermatogenesis, Sertoli cells are often termed *nurse cells.*

## Delivery of Sperm to the Female Tract

The deposition of sperm by the male into the vagina of the female involves two processes: *erection* of the penis and *ejaculation* of semen into the vagina.

### Erection

**Erection** of the penis is initiated by the stimulation of parasympathetic nerves that cause dilation of the arterioles supplying the blood sinuses of the penis; simultaneously, sympathetic nerves that bring about arteriole constriction are inhibited. As blood engorges the sinuses in the three corpora cavernosa, the major veins of the penis are compressed, minimizing outflow and causing erection.

The stimulation of parasympathetic nerves that initiate erection can occur by a spinal reflex pathway activated by the stimulation of touch receptors in the glans of the penis. In addition, cerebral pathways involving sight, sound, smell, and other stimuli associated with sexual excitement may override the sensory pathway of the spinal reflex, producing erection without the stimulation of touch receptors in the penis. Alternatively, cerebral input may prevent erection. In fact, failure to attain or sustain an erection, a condition known as **impotence,** is most often due to psychological factors that result in the stimulation of sympathetic nerves to the arterioles in the penis and the inhibition of parasympathetic ones. However, impotence may also be caused by certain drugs such as alcohol and tranquilizers that interfere with higher brain function, or it may be a result of various chronic illnesses, including hypertension, diabetes, and others.

### Ejaculation

Repeated stimulation of touch receptors in the head of the erect penis eventually triggers **ejaculation,** another spinal reflex. In the first stage of ejaculation, referred to as *emission,* semen is emptied into the urethra by contraction of the accessory glands and genital ducts. This is followed by a series of rhythmic contractions of smooth muscle in the urethra, which expel the semen.

Ejaculation occurs at the climax of sexual excitement, termed **orgasm.** Orgasm is associated with intense pleasure, and involves a variety of physiological changes, including increases in heart rate, blood pressure, respiration rate, and blood flow to the skin, as well as generalized skeletal muscle contraction. After ejaculation, the lack of parasympathetic stimulation triggers constriction of the arterioles supplying the penis. As more blood leaves the blood sinuses than enters them, the penis once again becomes flaccid. Gradually, other physiological parameters return to normal. Ejaculation is usually followed by a refractory period during which another erection is not possible. The refractory period is several minutes in the young, sexually active male and up to several hours in a man in his sixties.

Particularly in teenagers, but also in older males who abstain from sexual intercourse, spontaneous ejaculation may occur occasionally during sleep, an event referred to as a nocturnal emission or "wet dream." Folklore to the contrary, there is no evidence that ejaculation leads to physical weakness; this idea apparently arose as propaganda against masturbation.

### Fertility

**Fertility** refers to the ability of sperm to fertilize an egg. Fertility may be distinguished from **potency,** which refers to the ability to attain an erection and to sustain it long enough to ejaculate during intercourse. Normal fertile males generally have about 100 million sperm per milliliter of semen, and males with less than 20 million per milliliter are generally sterile. It is not understood why such a large number of sperm cells is required for fertilization, since only one sperm fertilizes an egg. However, recent evidence has suggested that other factors such as the motility and morphology of the sperm may be as important in determining fertility as sheer numbers.

## Control of Male Reproductive Function

Male reproductive function is regulated in a complex way by hormones secreted from the hypothalamus of the brain, the anterior pituitary gland, and the testes. The maturation of the male reproductive system begins in the teenage years with increased production and release of *gonadotropin releasing hormone* (GnRH) from the hypothalamus. GnRH is carried in the bloodstream to the anterior pituitary gland, where it stimulates the secretion of *follicle stimulating hormone* (FSH) and *luteinizing hormone* (LH). (FSH and LH are referred to as **gonadotropins** or **gonadotropic hormones** (*tropic* = acting upon), because they stimulate the activity of the gonads. Both hormones are secreted in males and females, but they are named for their roles in the female. For some time it was thought that another hormone, referred to as *interstitial cell stimulating hormone* (ICSH), substituted for LH in the male. However, LH and ICSH are now known to be identical, and consequently reference to ICSH has largely been dropped.

FSH and LH each act upon the testes in specific ways. FSH acts on the seminiferous tubules, probably by binding to the Sertoli cells, to stimulate spermatogenesis. However, normal spermatogenesis also requires the presence of androgens, primarily the hormone **testosterone.** Testosterone is secreted by the interstitial cells in response to LH, acting together with *prolactin,* another anterior pituitary hormone. Thus, testosterone is required not only for the development and maintenance of the testes, the accessory reproductive organs, and the secondary sex characteristics, but also for normal function of the reproductive organs in the sexually mature male. Testosterone production declines gradually beginning at about the age of 40, but reproductive function is still maintained at advanced years. Indeed, some men in their eighties have been known to father children.

Blood levels of GnRH, FSH, LH, and androgens are regulated by a complicated series of feedback controls. These are summarized in Figure 17–10. If, in response to GnRH, the blood levels of FSH and LH become too high, these latter two hormones feed back on the hypothalamus to inhibit the further secretion of GnRH; this, in turn, reduces the secretion of FSH and LH. In addition to this coordinate control of LH and FSH, the level of each hormone can be regulated independently of the other. Thus, high levels of testosterone act on the pituitary to inhibit further LH secretion, whereas high levels of FSH are thought to stimulate the Sertoli cells to secrete a hormone, called *inhibin,* which prevents further FSH release by the anterior pituitary.

These regulatory feedback mechanisms in the male serve to maintain the secretion of FSH, LH, and testosterone at more or less steady, continuous levels. Consequently, spermatogenesis occurs steadily and continuously. This situation is in direct contrast with that in the female, where large monthly swings in LH and FSH levels produce a cyclic maturation and release of egg cells.

## FEMALE REPRODUCTIVE SYSTEM

The female reproductive system consists of (1) two *ovaries,* that produce egg cells (ova) and that synthesize the hormones estrogen and progesterone; (2) two *oviducts* or *Fallopian tubes* through which ova, released from the ovaries, are transported to the uterus and where fertilization normally occurs; (3) a *uterus,* in which the fertilized egg is nourished and protected during embryonic development; (4) a *vagina,* which receives the erect penis during intercourse and into which semen is ejaculated; (5) external genitalia; and (6) *mammary glands,* used for nourishing the newborn baby. These structures

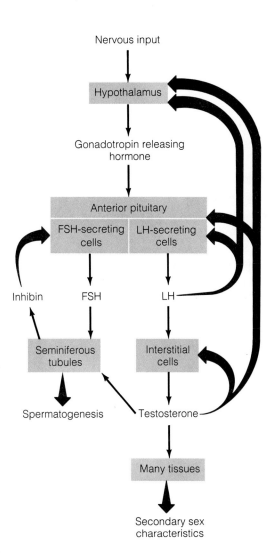

Nervous input

Hypothalamus

Gonadotropin releasing
hormone

Anterior pituitary

| FSH-secreting cells | LH-secreting cells |

Inhibin    FSH         LH

Seminiferous
tubules

Interstitial
cells

Spermatogenesis    Testosterone

Many tissues

Secondary sex
characteristics

FIGURE 17–10 Endocrine interrelationships in the male reproductive system. Thin arrows indicate hormone secretion, heavy dark arrows indicate final effects, and heavy light arrows signify negative feedback loops.

(with the exception of the mammary glands) are shown in Figure 17–11. Note that the reproductive and urinary systems of the female, as opposed to those of the male, are entirely separate.

## Ovaries

The **ovaries** lie in shallow depressions on either side of the lateral wall of the pelvic cavity. They are somewhat flattened, ovoid bodies, averaging about 4 cm long, 2 cm wide, and 1 cm thick. Each ovary is covered by a layer of epithelium derived from the parietal peritoneum; two layers of typical parietal peritoneum are fused together at the top and back of the ovary to form the *mesovarium* that connects the ovary to the body wall and that carries blood vessels, lymphatics, and nerves to and from the ovary. Because the ovaries project into the peritoneal cavity (between the parietal and visceral layers of the peritoneum), an ovum released from the surface of an ovary at the time of ovulation enters the peritoneal cavity, from where it is swept into a Fallopian tube, as mentioned below.

One of the most common conditions that affect female reproductive organs is the development within the ovary of benign, fluid-filled tumors, called *ovarian cysts*. Frequently ovarian cysts produce no symptoms, but they may cause irregular vaginal bleeding and some pain, particularly if they are large. Their presence may disrupt the cyclic hormone production by the ovary that is necessary for normal ovulation. Ovarian cysts generally require no treatment, but they may be removed surgically if they are troublesome.

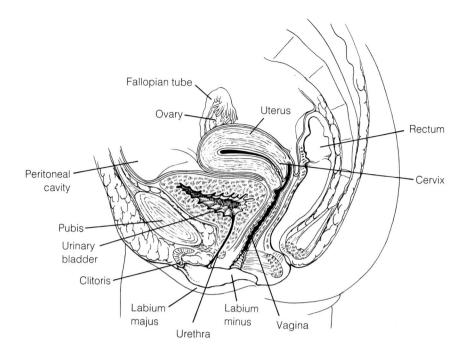

FIGURE 17-11 Pelvic organs of the female reproductive system.

Fallopian tube

Ovary

Uterus

Rectum

Peritoneal cavity

Cervix

Pubis

Urinary bladder

Clitoris

Labium majus

Labium minus

Vagina

Urethra

## Fallopian Tubes

Lying near each ovary in the peritoneal cavity is the trumpet-shaped opening to an **oviduct,** or **Fallopian tube** (Figure 17–12). The border of the opening possesses fingerlike projections that sweep the ovulated ovum into the Fallopian tube.

Each Fallopian tube is about 4 inches long, and its wall consists of two layers of smooth muscle lined with ciliated mucous membrane that is thrown into extensive longitudinal folds (Figure 17–13). Movement of the cilia and contraction of the smooth muscle in the wall of the Fallopian tubes helps propel the ovum toward the uterus, into which the Fallopian tubes lead.

FIGURE 17-12 Partially cut away, dorsal view of the female reproductive organs. An egg cell, released from the surface of the ovary at ovulation, enters a Fallopian tube and is transported to the uterus (arrow).

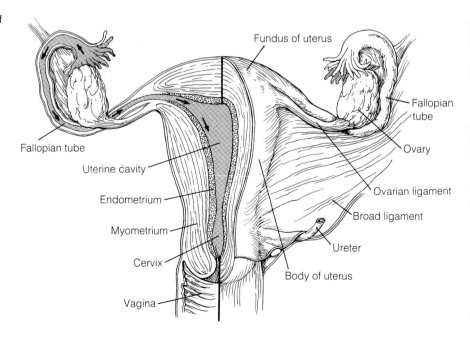

Fundus of uterus

Fallopian tube

Ovary

Fallopian tube

Uterine cavity

Endometrium

Ovarian ligament

Broad ligament

Myometrium

Ureter

Cervix

Body of uterus

Vagina

## Uterus

### Structure

The Fallopian tubes empty into either side of the upper uterus (see Figure 17–12). The **uterus** is a hollow, thick-walled muscular organ, about the size and shape of a pear. It is divided into an upper part, the *body,* and a lower part, the *cervix.* The part of the body above the openings of the Fallopian tubes is called the *fundus.*

The wall of the uterus consists of three layers. From the outside in, these are (1) a layer of visceral peritoneum that is continuous at each side of the uterus with the *broad ligament,* a sheet of connective tissue covered with peritoneum that connects both sides of the uterus to the body wall; (2) a thick coat of smooth muscle, the *myometrium;* and (3) a mucous membrane, or **endometrium,** that lines the body and the fundus of the uterus. The structure of the endometrium varies with the phase of the menstrual cycle, as discussed below.

### Diseases or Abnormal Conditions of the Uterus

One out of five women over 30 develop benign tumors of the uterus, termed *fibroids.* These may grow very slowly over a long period of time or may reach the size of a tennis ball or larger within several years. Small fibroids usually produce no symptoms, but larger ones may cause heavy menstrual bleeding. Fibroids may be removed surgically.

*Cancer of the uterus,* including *endometrial cancer* and *cancer of the cervix,* is the second most common cancer of the reproductive organs in women, behind breast cancer. Both endometrial and cervical cancer occur most frequently in women over the age of 50. Symptoms of both diseases are similar. Menstruating women may experience bleeding between periods, abnormal menstrual flow, or unusual vaginal discharge. In postmenopausal women, the most frequent symptom is vaginal bleeding. If left untreated, both cancers will spread deep into the uterus and to adjoining structures.

Cancer of the cervix, arising in its mucous membrane lining, may be detected before symptoms appear by a test known as a *Pap smear,* named for its originator, George Papanicolaou. In the Pap smear, a few cells are scraped from the tip of the cervix and examined for abnormalities under a microscope. If cancerous cells are detected, the uterus is removed surgically, a procedure known as *hysterectomy.* Because a Pap smear can detect cervical cancer in its

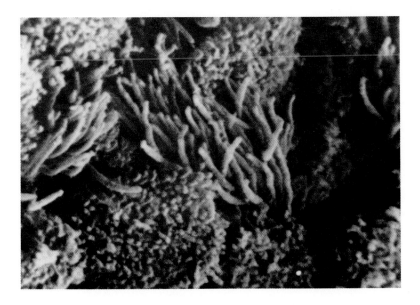

FIGURE 17–13 Scanning electron micrograph of the ciliated epithelium of the Fallopian tubes.

early stages while it is 100 percent curable, a yearly Pap smear is recommended by most physicians although others feel that a somewhat longer interval between smears is appropriate.

A tentative diagnosis of endometrial cancer is confirmed by a procedure known as *dilatation and curettage* or *"D and C."* In a "D and C," the cervical opening is dilated, and the endometrium of the uterus is scraped with a curette. The scrapings are then examined under a microscope. Endometrial cancer grows and spreads slowly, and it is curable by hysterectomy in four out of five women if it is detected in its early stages. Consequently, consultation with a physician at the first appearance of symptoms is imperative.

### Vagina

The **cervix** opens into the vagina, a 3- to 4-inch long tube composed of an internal lining of mucous membrane and an external smooth muscle coat surrounded by fibrous connective tissue. The mucous membrane contains glands that lubricate the inner surface of the vagina during sexual intercourse. The walls of the vagina are normally collapsed upon themselves.

Because it communicates with the exterior, the vagina not infrequently becomes infected with various microorganisms. Vaginal "yeast infections," caused by the fungus *Candida albicans,* often occur when prolonged antibiotic therapy kills acid-producing bacteria that normally reside in the vagina and keep the growth of the fungus in check. The main symptoms of yeast infections are itching and irritation of the vagina and a thick whitish discharge. Yeast infections are usually cured with an antifungal cream. Another common vaginal infection, referred to colloquially as "trich," is caused by the protozoan *Trichomonas.* Symptoms of *Trichomonas* infection are similar to those of yeast infections, but the discharge is usually greenish-yellow and has an unpleasant odor. Because the infection may be transmitted at sexual intercourse by asymptomatic males, sexual partners must both be treated with antiprotozoal drugs.

### External Genitalia

The external genitalia of the female are collectively termed the **vulva.** The vulva consists of five structures (Figure 17–14). The **mons pubis,** a rounded bulge in front of the pubic symphysis, becomes covered with hair at the time of

FIGURE 17–14 External genitalia of the female.

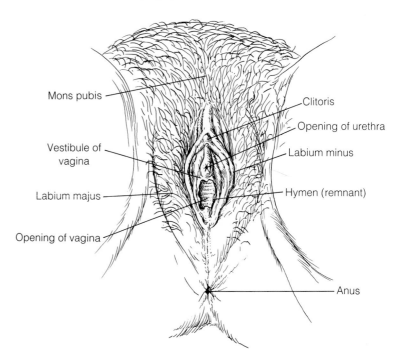

Mons pubis

Vestibule of vagina

Labium majus

Opening of vagina

Clitoris

Opening of urethra

Labium minus

Hymen (remnant)

Anus

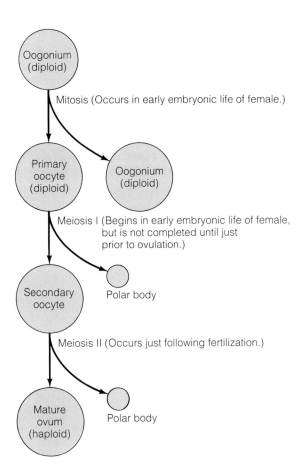

FIGURE 17–15 Summary of oogenesis. Unlike spermatogenesis, in which each primary spermatocyte gives rise to four mature sperm (see Figure 17–8), each primary oocyte gives rise to only a single mature ovum; other chromosomes are eliminated in polar bodies that degenerate.

Oogonium
(diploid)

Mitosis (Occurs in early embryonic life of female.)

Primary
oocyte
(diploid)

Oogonium
(diploid)

Meiosis I (Begins in early embryonic life of female,
but is not completed until just
prior to ovulation.)

Secondary
oocyte

Polar body

Meiosis II (Occurs just following fertilization.)

Mature
ovum
(haploid)

Polar body

puberty. It is formed under the influence of estrogen by the deposition of a mound of fat under the skin. The **labia majora** (sing: *labium majus*) are a set of two skin folds covered with hair on their outside surface and smooth on their inside. They correspond embryologically to the scrotum of the male. Internal to the labia majora is an additional set of smaller skin folds, the **labia minora** (sing: *labium minus*). Between the anterior ends of the labia minora is the **clitoris,** an erectile structure that is embryologically similar to the penis and that distends during sexual excitement.

Between the labia minora and posterior to the clitoris is a cleft, termed the *vestibule of the vagina.* The vestibule contains the urethral and vaginal openings, and, prior to initial sexual intercourse, a fold of mucous membrane, the **hymen,** that partly covers the opening to the vagina. The inner edges of the hymen almost contact each other, and the opening to the vagina appears as a space between them. Mucus-secreting glands open into the vestibule between the hymen and the labia minora. The slippery secretion of these glands aids the penetration of the penis into the vagina.

### Breasts

The structure of the breasts and their function in lactation will be discussed in the next chapter.

### Oogenesis

### Differences Between Oogenesis and Spermatogenesis

Oogenesis, the process by which ova are produced in the ovary, is summarized in Figure 17–15. Unlike the situation in males, no new germ cells are formed in females after birth; instead, all potential ova are formed while the female is still an embryo.

During early embryological development, female germ cells, called **oogonia,** proliferate in the ovary by mitosis until a million or so are formed. Shortly thereafter, near the end of the third month of gestation, the oogonia develop into larger cells termed **primary oocytes.** Soon after primary oocytes are formed, they begin meiosis by entering the prophase stage of meiosis I; however, after their chromatin has replicated, they go into a prolonged state of meiotic arrest. In fact, meiosis I is completed *only* in those cells that are about to be ovulated. When it does occur, the division of the cytoplasm is very unequal. One of the daughter cells, the **secondary oocyte,** retains most of the cytoplasm, while the other daughter cell, termed a **polar body,** is very small and has almost no cytoplasm.

Meiosis II begins in the secondary oocyte only after it has been penetrated by a sperm cell (fertilized), an event that normally occurs in the Fallopian tubes. This division also occurs unevenly, giving rise to a mature ovum and a second polar body. Although both polar bodies degenerate shortly after they are produced, their formation provides a means by which chromosomes may be eliminated without the loss of much cytoplasm, which is needed by the fertilized egg during early embryological development.

It is perhaps worth emphasizing that a mature ovum is formed only after fertilization; it is a primary oocyte that is ovulated and a secondary oocyte that is fertilized.

Only about 400 primary oocytes are ovulated during a woman's reproductive years, and the rest degenerate. Because all primary oocytes are present at birth and no new ones are formed later, egg cells ovulated in a woman's later reproductive years are much older than those ovulated just after puberty. It has been suggested that aging of egg cells may be responsible for the observation that the frequency of certain congenital defects such as Down's syndrome (Chapter 18) increases with the age of the mother.

### Changes Occurring in the Ovary During Oogenesis

If an ovary is observed in cross-section under the microscope, numerous primary oocytes are seen surrounded by a single layer of epithelial cells, termed *follicular cells* or *granulosa cells.* A primary oocyte together with its follicular cells is termed a **primary follicle** (Figure 17–16a).

In response to hormonal stimulation in the reproductively mature, non-pregnant female, several primary follicles begin to develop every 28 days. Development of the follicle is characterized by the proliferation of additional follicular cells and an increase in size of the primary oocyte. When the primary oocyte is about twice its original size, it becomes surrounded by a thick envelope, the **zona pellucida,** that is apparently secreted by both the oocyte and the follicular cells. Cytoplasmic processes from the inner layer of follicular cells penetrate the zona pellucida and make contact with microvilli on the surface of the oocyte, allowing the oocyte to receive necessary nutrients. Follicular cells thus serve a sustaining role, much as do the Sertoli cells in the seminiferous tubules (Figure 17–16b).

As the follicle develops, new follicular cells continue to form, and a connective tissue *theca* (= box) develops around the follicle. The theca becomes organized into two layers: an inner layer, the *theca interna,* made up of more or less rounded cells, and the *theca externa,* of spindle-shaped cells. The theca externa is more fibrous and contains fewer capillaries than the theca interna.

Fluid begins to accumulate between the follicular cells as the follicle approaches maturity. Eventually, a fluid-filled space, the *antrum,* comprises much of the follicle (Figure 17–16c and d). The primary oocyte, surrounded by the zona pellucida and several layers of follicular cells, projects into the antrum at the tip of a little peninsula of follicular cells. At this stage, the follicular cells that line the antrum are called the *membrana granulosa.* Just prior to ovulation, the fully developed follicle has come to balloon out on the surface of the ovary, and the outer wall of the ovary at the site of the bulge is very thin. At

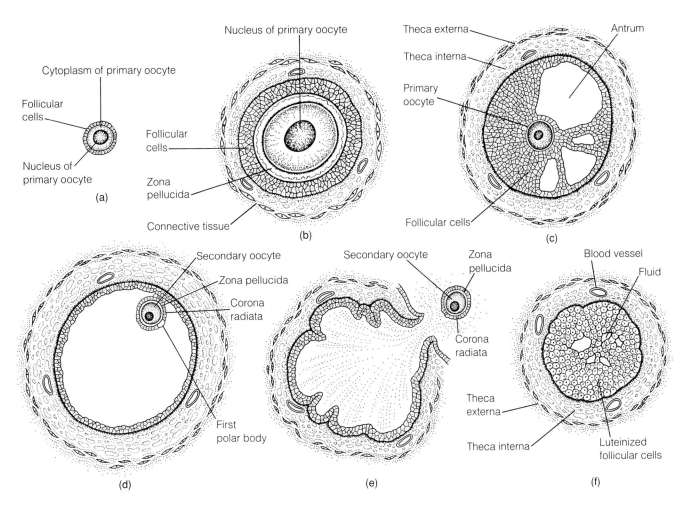

Cytoplasm of primary oocyte

Follicular cells

Nucleus of primary oocyte

(a)

Nucleus of primary oocyte

Follicular cells

Zona pellucida

Connective tissue

(b)

Theca externa

Theca interna

Primary oocyte

Antrum

Follicular cells

(c)

Secondary oocyte

Zona pellucida

Corona radiata

First polar body

(d)

Secondary oocyte

Zona pellucida

Corona radiata

Theca externa

Theca interna

(e)

Blood vessel

Fluid

Luteinized follicular cells

(f)

FIGURE 17-16 Cellular events occurring in the ovary prior to and following ovulation. (a) Primary follicle; (b) through (d), maturation of follicle; (e) rupture of follicle at ovulation; and (f) corpus luteum. (a) and (b) are not drawn to the same scale as are (c) through (f).

ovulation the ovarian wall ruptures (Figure 17–16e), and the secondary oocyte, still surrounded by follicular cells that are referred to collectively as the **corona radiata,** is liberated into the abdominal cavity close to the opening of the nearby Fallopian tube.

Rupture of the follicle may sometimes be associated with some abdominal pain, termed *mittelschmerz* (literally, middle pain; that is, pain in the middle of the menstrual cycle). Usually only one follicle reaches maturity and ovulates every 28 days, but occasionally more than one will ovulate. If two or more eggs are fertilized at more or less the same time, multiple births of nonidentical individuals (e.g., fraternal twins) will result.

After the follicle ruptures at ovulation it collapses, allowing the edges of the wound to heal together. Cells of the membrana granulosa and theca interna enlarge greatly, and they accumulate lipid and yellow pigment. Capillaries grow in among these cells, and fibroblasts invade the area, forming connective tissue. The resulting structure is an endocrine gland, the **corpus luteum** (*corpus* = body; *luteum* = yellow), shown in Figure 17–16f.

The corpus luteum develops for 10 to 12 days following ovulation, attaining a diameter of about 2.0 cm. If fertilization has occurred and pregnancy results, the corpus luteum is maintained until the end of pregnancy; if it has not occurred, the corpus luteum begins to degenerate after the 10- to 12-day period. Degeneration of the corpus luteum is followed by a new round of follicular maturation.

## The Menstrual Cycle

Cyclic changes within the ovary are accompanied by precisely coordinated cyclic changes in the state of the uterus; collectively, all of these changes constitute the **menstrual cycle.** While follicles are maturing in preparation for ovulation, the uterus is attaining a functional state that will permit implantation of the fertilized egg.

Although the length of the menstrual cycle can vary greatly both among different women and within the same woman, its average length is 28 days. Because the beginning of menstrual bleeding (**menstruation** or *"the period"*) is easily pinpointed, it has become established convention to designate this event as Day 1 of the cycle. On this time scale ovulation occurs, on average, on Day 14.

## Hormonal Regulation of the Menstrual Cycle

Ovarian and uterine events that occur during the menstrual cycle are regulated and coordinated by the same hormones from the hypothalamus and from the anterior pituitary that control the male reproductive system. These hormones are the hypothalamic hormone GnRH and the pituitary hormones FSH and LH. However, in the female, two hormones produced in the ovaries—*estrogen* and *progesterone*—replace the androgens of the male. (It should be noted parenthetically that the term *estrogen* is used as a collective noun and includes a number of chemically related hormones with similar physiological effects.) **Estrogen** is necessary for the development and maintenance of the ovaries, accessory reproductive organs, and female secondary sex characteristics; the major effect of **progesterone** is to bring about functional changes in the uterus.

In discussing the hormonal regulation of the menstrual cycle, it is difficult to consider simultaneously the control of both ovarian and uterine events. Consequently, we will first discuss how changes in the ovary are regulated and then consider uterine changes.

## Regulation of Ovarian Changes

Changes in the ovary during the menstrual cycle are divided into two phases (Figure 17–17). The **follicular phase** lasts for the first 14 days of the menstrual cycle and ends at ovulation. It is characterized by the development of the ovarian follicle. The 14 days following ovulation, during which the corpus luteum functions as an endocrine gland and then begins to degenerate, is called the **luteal phase.**

The hormonal regulation of the follicular and luteal phases occurs as follows:

1. At the beginning of the follicular stage (Day 1 of the menstrual cycle), GnRH is released from the hypothalamus and stimulates the anterior pituitary gland to secrete FSH and LH into the bloodstream. FSH and LH act together to stimulate estrogen secretion by cells of the follicle. Estrogen, in conjunction with FSH and LH, is required for the continued growth and development of the follicle. GnRH release, which initiates the cycle, is regulated by various feedback mechanisms (discussed shortly) and by a cyclic rhythm that is an inherent property of the hypothalamus of the female.

2. As the follicle enlarges during the follicular stage, estrogen secretion increases progressively. Low blood concentrations of estrogen, as occur during the first 7 to 10 days of the cycle, feed back on the pituitary to cause a modest decline in FSH secretion.

3. For the last few days of the follicular stage, blood concentrations of estrogen reach very high levels. These levels stimulate the hypothalamus to secrete GnRH and also directly influence the pituitary to increase LH secretion. As a result LH production surges, and there is an accompanying smaller increase in FSH production.

4. One day after a sharp LH peak, ovulation occurs.

**Endometrium**

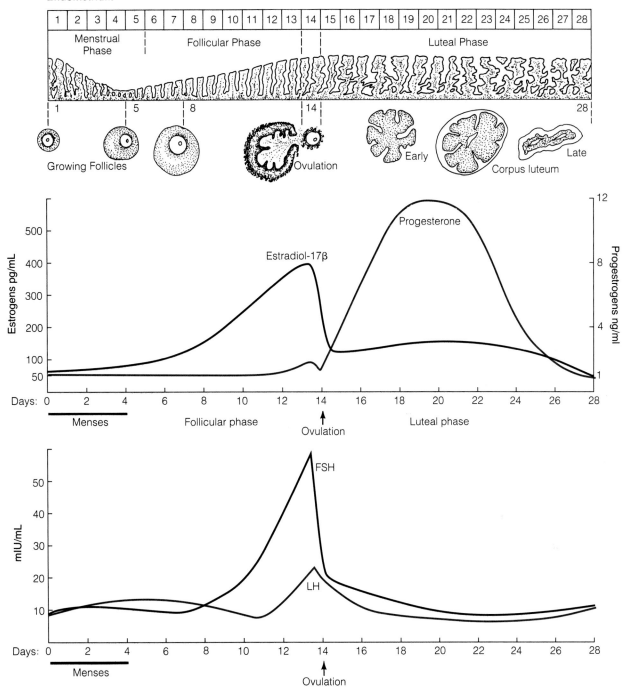

FIGURE 17-17 Hormonal regulation of changes in the ovary and uterus during the menstrual cycle. See text for details.

5. Following ovulation, the ruptured follicle is converted into a corpus luteum, which secretes large quantities of estrogen and progesterone.

6. High blood levels of estrogen *and* progesterone feed back on the hypothalamus, reducing GnRH release, and on the pituitary itself to diminish the release of LH and FSH. Lowered concentrations of LH and FSH prevent the development of a new follicle (and oocyte) during the luteal phase. Small amounts of LH continue to be secreted to maintain the corpus luteum.

7. If fertilization does not occur, the corpus luteum degenerates during the last days of the cycle. It is not clear why the corpus luteum degenerates, but it has been proposed that the corpus luteum may have a built-in life span.

Degeneration of the corpus luteum results in lowered concentrations of estrogen and progesterone, triggering the release of GnRH and the initiation of a new cycle.

If pregnancy does occur, a new hormone secreted by the placenta, termed *human chorionic gonadotropin* (HCG), maintains the corpus luteum. Continued high levels of estrogen and progesterone secretion by the corpus luteum, and subsequently by the placenta, prevent the development of a new follicle until pregnancy ends.

## Female Infertility

The two most frequent causes of female infertility are the failure to ovulate and blockage of the Fallopian tubes. New techniques designed to overcome these problems are discussed in *Focus: In Vitro Fertilization — New Treatment for Infertility* in Chapter 18.

## Uterine Changes During the Menstrual Cycle

The estrogen and progesterone produced by the ovary has profound effects on the uterus. Estrogen stimulates the growth of the myometrium and the endometrium, whereas progesterone, acting after estrogen-induced growth of the uterus has begun, stimulates changes in the newly proliferated endometrium that make it able to support a fertilized egg.

The uterus goes through three distinct stages during the menstrual cycle (see Figure 17–17). Beginning at Day 1 of the cycle, these are the *menstrual phase,* the *proliferative phase,* and the *progestational phase.*

***The Menstrual Phase.***    At the beginning of menstruation, the endometrium is 4 to 6 mm thick, and it consists of a thin lining of epithelial cells that cover the *endometrial stroma,* a layer of connective tissue. Tubular glands extend from the deepest layer of the stroma to the epithelial surface, where they empty their secretions. Most of the endometrium, the so-called *functional layer,* is sloughed at menstruation. However, the *basilar layer* of the stroma — a thin layer lying next to the myometrium — does not degenerate during menstruation. After menstruation, the functional layer regenerates from the basilar layer. The remnants of the tubular glands in the basilar layer are the source of the epithelial cells that spread out at the surface of the stroma to form the new epithelial lining.

The menstrual phase is initiated by the degeneration of the corpus luteum, begun during the last few days of the previous cycle. As the secretory function of the corpus luteum declines, there is a sharp fall in estrogen and progesterone levels (see Figure 17–17). Lack of these hormones causes constriction of the highly coiled arterioles (spiral arteries) that supply the functional layer of the endometrium; the basilar layer has its own arterial supply, and hence is not affected by constriction of the spiral arteries. Because of the loss of its blood supply, the functional layer begins to deteriorate. Concomitantly, the blood vessels in the functional layer expand, and as their walls deteriorate, bleeding into the endometrium occurs. This blood, along with necrotic tissue from the endometrium, is eliminated through the vagina and constitutes the menstrual flow. The menstrual phase lasts an average of four days during which some 100 mL of blood is lost, although this latter figure can be extremely variable even from one menstruation to another in the same woman.

***The Proliferative Phase.***    At the time that menstruation is beginning, a new follicle starts to develop in the ovary. The secretion of estrogen by the new follicle begins to rise by the end of the menstrual stage. Estrogen begins to stimulate the proliferation of a new functional layer from the 1 mm-thick basilar layer. At ovulation, which marks the end of the *proliferative phase,* the endometrium is 2 to 3 mm thick. New tubular glands have also regenerated.

*The Progestational Phase.*    The progestational phase lasts from the time that the corpus luteum is formed after ovulation until the beginning of the menstrual phase. Events that occur during the progestational phase depend upon the simultaneous synthesis of progesterone and estrogen by the corpus luteum. Acting together, progesterone and estrogen bring about a further thickening of the endometrium and an increase in its vascularization. The uterine glands accumulate increased amounts of glycogen-containing secretions that are released onto the surface epithelium. Presumably, glycogen is used as an energy source by the implanting embryo.

During the last several days of the progestational phase, the corpus luteum degenerates. This brings about constriction of the spiral arteries and other events that foreshadow the beginning of menstruation.

## Menopause

At the average age of 47 (usually between the ages of 40 and 55) menstrual cycles start to become irregular as the function of the ovary becomes sporadic. After months or several years, menstrual cycles cease altogether, an event called **menopause.** (The term *menopause* is frequently and incorrectly used to designate the period of time that precedes and follows the complete cessation of menstrual cycles.) After menopause, follicles no longer mature, and consequently ovulation does not occur and pregnancy is not possible.

We do not understand the events that trigger menopause. Menopausal women have high blood levels of pituitary gonadotrophins, but their ovaries have lost responsiveness to these hormones. Estrogen levels decline during the time preceding menopause as ovarian function diminishes and new follicles—the primary source of estrogen—no longer develop or do so only irregularly. After menopause, some estrogen synthesis continues, but it gradually diminishes, resulting in the eventual atrophy of the reproductive organs. Lack of estrogen following menopause is also often associated with osteoporosis (see *Focus: Osteoporosis* in Chapter 6).

During the time before and after menopause when estrogen levels are falling, many women experience "hot flashes" that produce an uncomfortable feeling of warmth associated with heavy sweating. Hot flashes are caused by inappropriate dilation of blood vessels in the skin. Some women experience headaches, heart palpitations, and bouts of irritability, anxiety, and depression. Usually, adaptation to the lowered levels of sex hormones occurs after several years, and unpleasant symptoms associated with menopause disappear. Occasionally, hot flashes and other symptoms are so severe that a physician may prescribe estrogen replacement therapy. However, the use of estrogen to treat menopausal symptoms has recently declined as evidence has accumulated that estrogen treatment can cause uterine (endometrial) cancer in postmenopausal women.

## Problems Related to the Menstrual Cycle

In most girls, **menarche,** or the beginning of menstrual periods, takes place between the ages of 11 and 14. At first, ovulation occurs irregularly and so do menstrual periods, but after a year or two they generally assume a more or less regular pattern.

*Amenorrhea.*    The temporary or permanent absence of menstrual periods is termed **amenorrhea.** *Primary amenorrhea* refers to the failure of menstrual periods to develop. If it is not due simply to a late menarche, it is usually the result of some abnormality in the reproductive system or in its hormonal control. *Secondary amenorrhea* refers to the situation where menstrual periods cease in a woman who has had regular periods. Pregnancy or menopause are the most common, natural causes of secondary amenorrhea, and they should always be ruled out before other alternatives are investigated.

# Herpes Simplex Viruses: Oral and Genital Herpes

Among the best known of the viruses that cause recurrent infections in humans are two closely related strains of herpes viruses, herpes simplex I and II. Herpes simplex I usually causes oral herpes, which is characterized by the formation of "cold sores" or "fever blisters" on the lips or on adjacent mucous membranes of the mouth. Genital herpes infections, resulting in similar blisters on the genitals, buttocks, and thighs, are most frequently caused by herpes simplex II. However, it has recently been discovered that each of the two viruses may also infect the other site.

The pattern of infection by both types of herpes simplex is similar. The virus is transmitted to uninfected tissue from blisters on infected tissue by direct cell-to-cell contact. After the virus has entered cells of a previously uninfected individual, it may remain dormant for an extended period of time, producing no obvious symptoms. However, when infected cells divide, the virus is replicated and distributed to the daughter cells. Eventually, the virus enters an active stage, during which it multiplies and destroys infected cells, causing blisters that contain fluid and large numbers of new virus particles. These infect neighboring cells, which in turn are destroyed as a result of viral replication.

Symptoms of the initial infection typically last several weeks before subsiding as the immune system brings the infection under control. The virus then enters a latent stage within nearby nerve cells. In the case of oral herpes, the virus migrates to the trigeminal ganglia located near the brain; genital herpes viruses invade ganglia near the spinal cord. During their latent stage, herpes simplex viruses are not infectious.

The virus remains in the latent stage for an indeterminate period of time. It may never leave the latent stage, but more typically it reinvades the initial site of infection and resumes active replication after a number of months or years. Attacks subsequent to the initial one generally are milder and show symptoms for only about a week. We do not understand the molecular mechanisms that determine latency of the virus, but we do know that recurrences of active infection can be brought on by stress, other diseases, exposure to ultraviolet light, and hormonal changes that occur during the menstrual cycle or as a result of the use of birth control pills.

On occasion, herpes simplex infections may lead to other, more serious conditions. Viruses in the trigeminal ganglia may infect the brain, causing a severe form of encephalitis, or the eye, causing infections that can damage vision. Genital herpes in pregnant women is particularly serious if it recurs at the time of birth, since newborns may become infected with the virus during delivery. Herpes infections are lethal for more than 50 percent of newborns, and those that survive such infections have a 50 percent chance of blindness or brain damage. To prevent herpes infection of newborns, caesarean sections are recommended for mothers who have active herpes lesions at the time of birth.

There is currently no adequate treatment for the 20 million Americans that suffer from genital herpes. The drug acyclovir (tradename: Zovirax), the only treatment approved by the FDA, is available as an ointment or as oral capsules. The ointment, spread on the blisters, only somewhat reduces the severity of symptoms. The oral capsules are considerably more effective, reducing the severity or frequency of attacks, or both, when taken daily. However, these effects last only as long as the treatment is continued, and as of yet the oral capsules have not been approved for daily use in excess of six months. Several herpes vaccines are currently under development, but these are aimed at preventing infection in uninfected individuals and are not expected to be particularly effective for those who already have the disease.

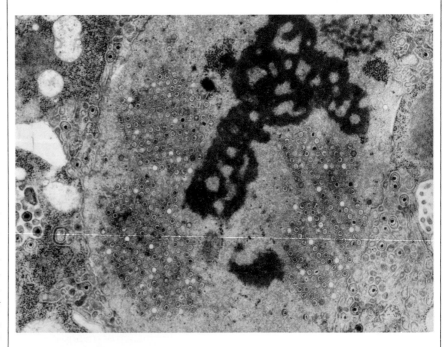

A herpes-infected cell.

In young women who are not pregnant, secondary amenorrhea may be caused by a number of factors or conditions that disrupt normal ovarian function. These include such things as severe emotional distress, a large weight loss, severe illness, intense athletic training, and abnormalities or tumors of the hypothalamus, pituitary gland, or ovaries. In addition, women who stop using oral contraceptives may have a temporary secondary amenorrhea.

In itself, amenorrhea is no risk to health and requires no treatment. However, it must always be kept in mind that amenorrhea may be due to some underlying disease that should be treated. Moreover, women with amenorrhea usually are not able to become pregnant, although treatment with fertility drugs can be attempted if pregnancy is desired.

*Variations in Frequency and Severity of Menstruation.*    Other disorders of the menstrual cycle include infrequent periods, common in women approaching menopause; exceptionally painful periods, sometimes caused by fibroids or pelvic infection; and unusually heavy or prolonged periods that may occur at menopause or be caused by fibroids or pelvic infection. Some or all of these conditions are often experienced by otherwise normal women at some time in their reproductive lives.

*Premenstrual Syndrome.*    From a week or more to several days before menstruation begins, many women experience symptoms that constitute *premenstrual syndrome*. Premenstrual syndrome may include such physical symptoms as increased fluid retention, pain in the pelvic region, and enlarged breasts that may be tender to the touch. In addition, some women report mood changes such as increased irritability and depression. Variability in the severity of premenstrual syndrome can be great. Once dismissed as mere hypochondria, complaints about premenstrual symptoms are now taken seriously. They are now thought to be due to hormonal variations that occur during the menstrual cycle. Limiting salt intake has been shown to reduce premenstrual fluid retention, and pain killers may be taken to relieve discomfort. If symptoms are severe, oral contraceptives may be prescribed. These reduce the severity of premenstrual syndrome, presumably by preventing large swings in hormonal levels.

*Toxic Shock Syndrome.*    In recent years, an extremely rare life-threatening infection caused by the bacterium *Staphylococcus aureus* has been identified in some individuals. Called *toxic shock syndrome,* it occurs when the bacterium produces a powerful toxin under suitable growth conditions. Symptoms include a high fever, a sunburnlike rash, and very low blood pressure. Toxic shock syndrome is treated with antibiotics, and measures commonly employed for shock, including massive fluid replacement. It was originally identified in menstruating women using tampons and thought to occur exclusively in that group of individuals. It has since been identified in other menstruating women as well as in men and nonmenstruating women.

The role of tampons in toxic shock syndrome is not clearly understood. Some evidence suggests that certain components present in some highly absorbent tampons may encourage the production of toxins by *Staphylococcus.*

## Sexual Response in the Female

Sexual intercourse is associated in the female, as in the male, with specific physiological changes. During sexual excitement, the nipples and the clitoris become erect, and the vaginal epithelium and the glands that empty into the vestibule of the vagina produce mucousy secretions. At the height of sexual excitement during intercourse, orgasm may occur. This is associated with rhythmic contractions of the smooth muscle in the vagina and uterus, and with other psychological and physiological changes similar to those in the male. However, no event analogous to ejaculation occurs.

# SEXUALLY TRANSMITTED DISEASES

*Sexually transmitted diseases,* also known as *venereal diseases,* are most commonly spread by some form of sexual contact. The most important sexually transmitted diseases are chlamydia, gonorrhea, syphilis, genital herpes, and AIDS. Chlamydia, gonorrhea, and syphilis, discussed shortly, are caused by bacteria; genital herpes and AIDS are caused by viruses. Genital herpes is discussed extensively in *Focus: Herpes Simplex Viruses: Oral and Genital Herpes* in this chapter and AIDS is discussed in *Focus: Acquired Immune Deficiency Syndrome (AIDS)* in Chapter 21.

## Chlamydia

The most prevalent of all sexually transmitted diseases is *chlamydia,* which affects three million or more Americans each year and up to 10 percent of all college students. It is caused by the bacterium *Chlamydia trachomatis,* which is easily transmitted between sexual partners.

When properly diagnosed, the disease is cured by a ten-day course of erythromycin or tetracycline. Unfortunately, the symptoms of chlamydia—a viscous discharge from the penis or vagina and a painful or burning sensation experienced by both men and women during urination—mimic the symptoms of gonorrhea in both sexes or of urinary tract infections in women. Since gonorrhea and urinary tract infections are not ordinarily treated with tetracycline or erythromycin, misdiagnosed chlamydial infections are not cured. Chlamydia can now be identified by several simple diagnostic tests that have been developed recently.

If left untreated, chlamydial infections usually do no permanent damage in men, but in women they may cause *pelvic inflammatory disease,* an infection and inflammation of the uterus, Fallopian tubes, ovaries, and surrounding tissues. In this condition, which may be caused by other sexually transmitted infections as well, the reproductive organs may become scarred, interfering with normal reproductive processes. Scarring of the Fallopian tubes increases the chances that a fertilized egg may implant there, causing a dangerous ectopic pregnancy (see Chapter 18); alternatively, scarring may be so severe as to block the Fallopian tubes so that fertilization cannot occur.

Pregnant women with chlamydia may pass the infection to their babies during delivery, resulting in a severe conjunctivitis that may cause blindness if left untreated.

## Gonorrhea

The causative organism of gonorrhea is the bacterium, *Neisseria gonorrheae.* In males, gonorrhea may be asymptomatic, but more commonly it is manifested by active infection of the urethra, making urination painful and causing a cloudy, purulent discharge from the penis. In severe cases, the urethra may be damaged to such an extent that its lumen is narrowed, and urination becomes extremely difficult. Gonorrhea in the female usually begins as an infection of the cervix. In about 30 percent of infected women it shows no symptoms at all; in others, symptoms may be limited to a slight vaginal discharge that often goes unnoticed. The insidious nature of gonorrhea in females and to some extent in males greatly increases transmission of the disease among sexually active individuals.

If gonorrhea is untreated, it may spread to the prostate gland, vasa deferentia, epididymes, and testes in the male or cause pelvic inflammatory disease in females. In both males and females, the bacteria can enter the bloodstream, frequently spreading to the joints and causing a form of arthritis.

Although more than a million new cases of gonorrhea are reported by physicians each year, the true incidence is undoubtedly much greater than this since many individuals do not seek treatment. Treatment involves the use of large doses of antibiotics, usually penicillin or ampicillin. In recent years, there has been an alarming increase in strains of *Neisseria gonorrheae* that are resistant to particular antibiotics, as mentioned in Chapter 21.

## Syphilis

Syphilis is caused by the bacterial spirochete *Treponema pallidum,* and it is usually spread by sexual contact. The bacterium commonly gains entry into an uninfected individual by penetrating mucous membranes of the vagina, the urethra of the penis, and the mouth.

Untreated syphilis occurs in three sequential stages, termed *primary syphilis, secondary syphilis,* and *tertiary syphilis.* Primary syphilis, occurring in the first two to four weeks after initial infection, is characterized by the appearance of a firm, red, painless skin ulcer at the site of entry, usually a genital organ. This ulcer is termed a *chancre* (pronounced shanker). At this stage, the disease is readily transmissible by contact with the chancre. The chancre typically heals spontaneously by several weeks after appearance (Figure 17–18).

Secondary syphilis occurs two to six weeks after the chancre has healed. It is characterized by a variety of skin rashes and lesions, and by fever, headache, general malaise, and swollen lymph nodes in the neck, armpit, and groin. Again, the disease is highly contagious by contact with skin lesions. At this stage bacteria also may cross the placenta and enter the fetal circulation, causing *congenital syphilis* that often results in malformations as well as serious permanent damage to the child as a result of contracting the disease. The symptoms of secondary syphilis usually disappear within a month to a year after their initial appearance. About one-third of infected individuals recover completely at this stage, while in others the disease enters an asymptomatic latent, or dormant, stage that may lead to tertiary syphilis.

After many years of latency, the symptoms of tertiary syphilis may become evident. Tissue lesions may develop in the central nervous system, circulatory system, or skeletal muscles, and may lead to paralysis, insanity, aortic aneurysms, blindness, and other disabling conditions. Once the lesions of tertiary syphilis appear, treatment is largely ineffective.

Syphilis is usually completely cured by penicillin, tetracycline, or erythromycin if treatment is initiated before the symptoms of tertiary syphilis appear. Because many individuals have not received treatment during the primary and secondary stages and consider themselves cured while the disease is, in fact, in its latent state, blood tests for syphilis are indispensable for diagnosis and treatment. One such blood test is required before marriage certificates are issued in the United States and other countries.

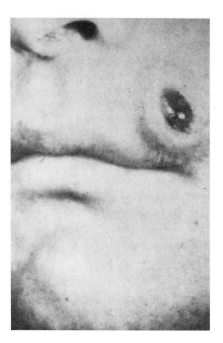

FIGURE 17–18 A chancre typical of primary syphilis.

## Study Questions

1. Compare the stages of meiosis and mitosis.
2. a) What is the purpose of meiosis?
   b) What are some of the consequences of meiosis?
3. List the primary reproductive organs, the accessory reproductive organs, and the secondary sexual characteristics of males and females.
4. What is the importance of the scrotum to spermatogenesis?
5. Compare spermatogenesis and oogenesis.
6. Describe the components and the operation of the blood-testis barrier.
7. Outline the endocrine control of male reproduction.
8. Describe the structure of the uterus.
9. a) What is the purpose of the Pap smear?
   b) How is it performed?
10. Describe the events that occur in the ovary and uterus during the menstrual cycle and indicate how hormones control these events.

# PREGNANCY AND EMBRYONIC DEVELOPMENT

The formation of a new individual begins at fertilization, when a single sperm cell penetrates and unites with a recently ovulated oocyte to produce a one-celled embryo. After fertilization, which normally occurs within a Fallopian tube, the embryo begins the process of embryonic development.

Embryonic development involves a series of precisely timed and coordinated events by which the embryo increases in mass and changes in form and functional capacity as its cells divide and specialize. In the initial stages following fertilization, the developing embryo migrates along the Fallopian tube toward the uterus, a journey that takes several days. Upon reaching the uterus, the embryo remains loose in the uterine cavity for a few days more before attaching to and invading the endometrium of the uterus, a process termed **implantation.**

Although it initially obtains nutrients from digestion of the endometrium, the embryo soon becomes surrounded by a new organ, the placenta, specialized for providing nourishment to the embryo. Developing partly from embryonic cells and partly from endometrial cells, the highly vascularized placenta provides a site for the exchange of materials between the circulatory systems of the mother and the developing embryo. At the placenta, water, nutrients, $O_2$, and other life-sustaining substances are transferred from the circulatory system of the mother to that of the embryo, whereas $CO_2$ and other waste products resulting from the embryo's metabolism are transferred in the opposite direction. The placenta thus serves as the life support system for the developing embryo—the embryo's respiratory, digestive, and excretory systems all wrapped up in one. Sustained by the placenta, covered by protective membranes, suspended in cushioning fluids, and carried within the mother's uterus, the embryo is provided with an ideal environment while undergoing development.

While the mother, by way of the placenta, sustains the life of the embryo, the process of embryonic development itself is largely independent of the mother. It is determined and regulated by the developmental program encoded in the embryo's DNA, its hereditary material. Every event that occurs during normal embryonic development is ultimately a consequence of the unfolding of this developmental program. The embryo's DNA, half of which is inherited from the mother and half from the father, thus contains the entire set of instructions—the blueprint as it were—for the formation of a new individual.

After about nine months of embryonic development, the developmental program has run its course, and a new individual has been formed: the microscopic one-celled embryo has been transformed into some seven pounds of mature fetus. Birth, the process by which the new individual emerges from its mother's uterus, is initiated by the onset of the strong contractions of the uterine smooth muscle that characterize labor.

Following birth, the baby becomes physiologically independent of its mother. Yet it is nevertheless entirely helpless, completely dependent on its parents to supply it with food and other necessities, keep it warm, and otherwise protect it. Before commercially prepared sterile formulas were available, breast feeding by the natural mother or a surrogate one was virtually essential for survival. Not until the youngster is many years old is it capable of surviving independently. Indeed, providing for the young after birth is every bit as important in successful reproduction as is the actual production of a new individual during pregnancy.

Although reproduction is necessary for the survival of the human species, overpopulation is a growing concern in many areas of the world where food, water, and other necessities of life are in short supply. Moreover, some individuals feel that their personal goals and aspirations as well as the opportunities available to their children are limited by a large family. These concerns have led to the development of birth control, which will be discussed at the end of this chapter.

# FERTILIZATION

You will recall from the previous chapter that at ovulation the oocyte, released into the peritoneal cavity, is swept into the end of the Fallopian tube by undulating movements of the elongated projections that extend from the border of the tube's opening (see Figure 17–12). Once in contact with the end of the Fallopian tube, the oocyte is moved along it by the beating of the cilia that line the inner wall of the tube. The oocyte moves along the Fallopian tube very slowly, taking about three days to reach the uterus. Because the human oocyte is probably fertile for only 10 to 15 hours after ovulation, it is clear that fertilization must normally occur within the Fallopian tube.

## Sperm Transport

Once sperm are ejaculated into the vagina, sperm are transported through the uterus and into the Fallopian tube within an hour or so. Since sperm cells on their own can only move at a rate of about 3 mm per hour, other factors must be responsible for their rapid transport through the female genital tract. It is thought that sperm may be transported as a result of smooth muscle contractions in the uterus and Fallopian tubes, perhaps initiated by prostaglandins in the semen.

Of the hundreds of millions of sperm cells present in the ejaculate, only a few hundred, or at most a few thousand, ever reach the Fallopian tube. This extremely low survival rate may account, at least in part, for the observation that men with low sperm counts have reduced fertility.

Although sperm cells undergo maturation while in the epididymes, newly ejaculated sperm are not capable of fertilizing an oocyte unless they have resided in the female reproductive tract for a number of hours. This acquisition by sperm of the capacity to fertilize an egg is termed *capacitation*. Capacitation is not well understood, but it may somehow facilitate the sperm's ability to penetrate the cellular layers (follicle cells) surrounding the egg and the zona pellucida, as discussed below. In any case, capacitated sperm within the Fallopian tubes are able to fertilize an oocyte for up to 48 hours.

## Penetration of the Oocyte by a Sperm Cell

Once a sperm cell is in the vicinity of an oocyte, movement of the sperm's flagella is required for final approach and penetration of the oocyte. Penetration of the oocyte involves several steps. After the anterior end of the sperm makes contact with the layers of cells that surround the oocyte, the sperm's plasma membrane and the underlying acrosomal vesicle rupture (see Figure 17–9 for a review of sperm structure). Released from the acrosomal vesicle are enzymes capable of dispersing the cellular layers surrounding the oocyte by digesting the intercellular material that connects the cells to one another. The sperm must then bind to specific attachment sites on the zona pellucida, an event that is followed by enzymatic digestion of a path through the zona pellucida. When it contacts the plasma membrane of the oocyte, the sperm cell is engulfed by the oocyte and internalized by an endocytotic process (Figure 18–1). As the sperm is taken into the oocyte, membrane-bound organelles that lie directly under the egg's plasma membrane, called *cortical granules,* fuse with the plasma membrane, releasing their contents at the surface of the oocyte beneath the zona pellucida. The substances released from the cortical granules prevent the entry of additional sperm, apparently both by blocking sperm binding sites on the zona pellucida and by modifying the structure of the zona pellucida, making it more difficult to be penetrated.

Recent research has shown that rupture of the cortical granules is triggered by the release of calcium ions from a bound state within the egg cytoplasm, a response presumably initiated by sperm contact. In addition to this effect, the higher levels of free calcium ions may initiate the developmental sequence by increasing the activity of certain enzymes, promoting protein synthesis and

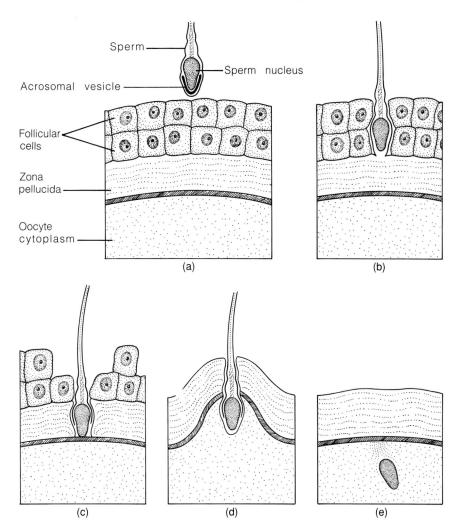

Sperm

Sperm nucleus

Acrosomal vesicle

Follicular cells

Zona pellucida

Oocyte cytoplasm

(a)

(b)

(c)

(d)

(e)

FIGURE 18-1 Penetration of an oocyte by a sperm cell. (a) Sperm approaches surface of egg. (b) Acrosomal vesicle ruptures, releasing enzymes that disperse nearby follicular cells; sperm binds to surface of zona pellucida. (c) After acrosomal enzymes digest a path through the zona pellucida, sperm contacts oocyte plasma membrane. (d) Oocyte plasma membrane begins to engulf sperm cell. (e) Sperm nucleus entirely within oocyte cytoplasm.

affecting the cell's metabolism in other ways. This triggering of embryonic development is called **activation.**

Penetration of sperm into the ovum also initiates the second meiotic division, which transforms the secondary oocyte into a mature ovum. The haploid nuclei of the sperm and the egg (referred to as **pronuclei** after the sperm has penetrated the egg) fuse, producing a single diploid cell, or **zygote.** Fertilization is thus complete.

In recent years, it has been possible to fertilize mature human egg cells in the laboratory and to obtain viable zygotes that will implant and develop to term in the mother's uterus or in the uterus of a surrogate mother. This technique, called *in vitro fertilization,* has so far allowed hundreds of infertile couples to have children (see *Focus: In Vitro Fertilization: New Treatment for Infertility*).

## EMBRYONIC DEVELOPMENT

### Development Before Implantation

The fertilized egg is in the oviduct for about three days, during which it undergoes a number of cell divisions (Figure 18–2). These early divisions are called *cleavages* because they result in a progressively larger number of smaller cells without an increase in the total mass of the embryo. By the time the embryo reaches the uterus, it has undergone four or five cleavages and consists

# In Vitro Fertilization: New Treatment for Infertility

Infertility, defined as the inability to conceive within a year or more of trying, is a concern of a large number of Americans. Some two and a half million American couples—about one in seven—are infertile. The infertility rate, although difficult to measure because a large proportion of the population is using contraception, is probably modestly higher than it once was, primarily as a result of an increase in the incidence of sexually-transmitted infections, which may lead to infertility.

Male infertility accounts for about half of the infertility cases. Most commonly, this is due to low sperm counts, the presence of morphologically abnormal or immotile sperm, and blocked sperm ducts. Among women, infertility may be caused by hormonal imbalances that interfere with the production and release of mature egg cells or with the implantation of a fertilized egg. Uterine problems such as an abnormally shaped uterus, the presence of fibrous cysts, or the scarring of the uterus as a result of a venereal infection can prevent fertilization or implantation. However, the single most common reason for infertility in females, affecting some 500,000 women, is absent, blocked, or scarred Fallopian tubes.

Many infertility problems can be successfully treated by providing advice on the optimal timing and frequency of sexual intercourse, by administering hormones and other drugs, and by surgery to repair defects in reproductive organs. However, for those women whose Fallopian tube defects cannot be repaired by surgery, *in vitro* fertilization (*in vitro* = in glass; outside the body) may be the only way to conceive.

The procedure used for *in vitro* fertilization is not complicated. First, the woman is given a fertility drug such as Clomid (an FSH-like drug) to stimulate the maturation of a number of ovarian follicles. As egg cells in the follicles approach maturity, the woman is given a single injection of human chorionic gonadotrophin (HCG), which stimulates processes leading to the

Elizabeth Carr, America's first test tube baby, was born on December 28, 1981.

final release of mature eggs from the follicles. In about 33 hours, just before the mature eggs would have been released, they are removed surgically from the ovary. This is done by making a small incision in the abdomen and inserting a *laparoscope,* a long hollow tube with a light on the end, through which the ovary can be seen. Mature follicles, visible as bulges on the surface of the ovary, are aspirated with a long syringe needle that is inserted through a second incision. Eggs that are recovered are washed, placed in culture medium in separate plastic petri dishes, and stored in an incubator at body temperature. The husband then produces a sperm sample that is washed and diluted to simulate conditions in the upper Fallopian tubes where fertilization occurs.

After the eggs have incubated for about six hours, several drops of diluted sperm are added to each of the dishes in the incubator. Fertilization occurs within the next 24 hours. When the fertilized eggs

have developed to between two and eight cells, one or more of the eggs is gently inserted into the uterus, using a long plastic cannula placed through the vagina. Progesterone is often administered to help make the uterus receptive to implantation.

*In vitro* fertilization is performed in some 200 clinics around the world. It is an expensive procedure, costing more than $5000 for each attempt. Costs can quickly escalate, however, since the highest success rates are about 20 percent if one embryo is inserted into the uterus, and about twice that figure if three are inserted. Although multiple insertions increase the success rate, they may also lead to multiple births.

The first baby to be born as a result of *in vitro* fertilization was Louise Brown, born in England on July 25, 1978. In the United States, *in vitro* fertilization has resulted in the birth of more than a thousand babies.

*In vitro* fertilization can also be a solution for male infertility caused by low sperm counts, since only a small number of sperm, something on the order of 50,000, are needed for the procedure. Most clinics are also willing to use sperm donated from a sperm bank if the husband is sterile. There are, of course, other possibilities. One might be the use of eggs from a female donor, sperm from the husband, and implantation of the embryo into the sterile wife. However, the legal and ethical ramifications of such alternative procedures remain unclear.

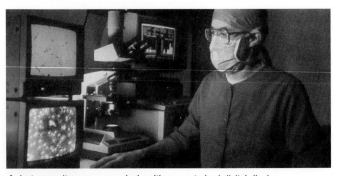

A doctor monitors sperm analysis with computerized digital display.

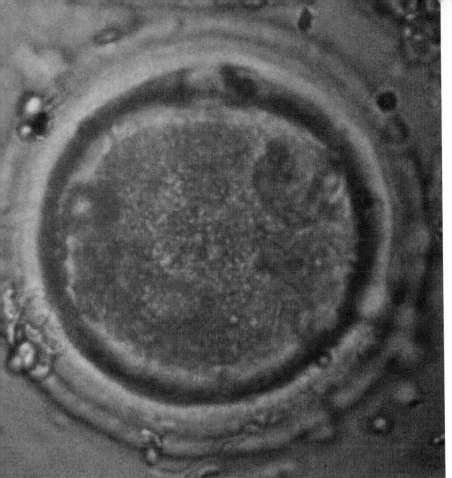

# HUMAN DEVELOPMENT

The development of a human embryo, from its first beginnings as a fertilized egg to the birth of a baby, has been variously described as a mystery and a miracle. While these are words generally inappropriate to a scientific mode of thought, they are not entirely inaccurate from a biological standpoint. Human development is a mystery in the sense that we still do not understand how it occurs; that is, we do not know how information that is encoded in the DNA of the fertilized egg is decoded as a precisely timed developmental program. The biological miracle of development is that this phenomenally complex process works correctly most of the time, fine-tuned to near perfection by millions of years of evolution.

Although we still do not fully understand the remarkable transformation that occurs during human development, we have observed and described many of its stages. The series of photographs on the accompanying pages help us appreciate the intricate changes that occur during development, while at the same time emphasizing the mystery and miracle of such a complex biological process.

Our photo essay begins at the second week of development. Prior to that time, the fertilized egg has migrated down the Fallopian tube while undergoing a series of rapid cell divisions that transform it, by three days after fertilization, into a solid clump of about 16 cells, termed a **morula**. The morula floats free in the uterine cavity for the next three days. During this time, cell division continues, and the morula is rearranged into a hollow ball of cells that surround a central, fluid-filled cavity. The layer of cells forming the outer wall will become the fetal part of the placenta, while a small cluster of cells, attached to the inner side of the wall, will form the body of the embryo and some of its surrounding membranes. At Day 6 or 7, the developing embryo begins to implant, a stage represented in the first photograph.

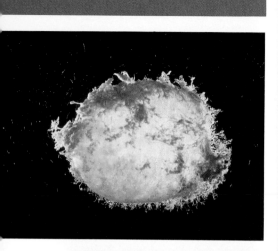

At the beginning of the second week, the embryo has developed extensive membranes that lie in close contact with the mother's tissues (top left). The delicate projections of the chorion have penetrated the mother's tissues in all directions as they take nutrients and oxygen from her blood and deposit their own metabolic wastes for her system to carry away.

Well into the third week (bottom left) the chorionic membrane has continued to penetrate the mother's endometrium. The chorion, shown radiating outward, carries blood vessels that lie closely intertwined with those of the mother, but the two do not join. The balloon-like structure is the yolk sac. The human embryo need carry only enough food to last through its first few weeks since food will later be derived from the mother's blood.

The embryo at the fourth week (right). It lies protected in its amniotic sac. The dark eye is prominent and the enormous brain lies tucked against the embryonic heart. The embryo by this time has already developed primordial cells that will form its own gametes. By now the tubular heart has begun its first timorous beats.

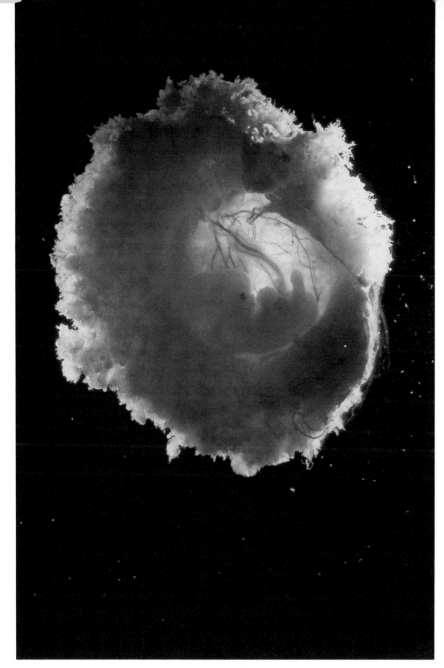

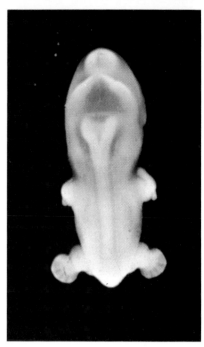

The human embryo is here shown at forty-two days with the surrounding membranes removed (above). It is about half an inch (or sixteen millimeters) long. This is a dorsal view showing the enormous head (which helps direct the growth of the rest of the body) and the spinal cord extending to the rump. Notice the paddlelike appendages. Fingers and toes are already apparent as the tissue between them dies as a result of a mysterious chromosomal timing mechanism.

At about six weeks (left) the extensive vascular system is clearly visible leading from the projections of the embryonic chorion to the embryo itself. The organ below the eye is the now-looped heart. The embryo is still so water-laden that its tissues are virtually transparent.

At six weeks (right), with the amnion removed, the fingers are apparent and the bulbous brain still dominates the embryo. Notice the "tail" tucked under the abdomen. (It is destined to disappear.) The tiny pit above the arm will become the ear.

At about this time the embryo is extremely vulnerable to all sorts of chemical agents. An increasing number of drugs have been found to produce congenital abnormalities, for example, in the growth of limbs. Even X rays, at this time, may endanger the development of the embryo. Certain diseases are also particularly dangerous. For example, if a mother contracts German measles during the fourth to twelfth week of her pregnancy, the result may be deformities in the offspring's eyes, heart, and brain.

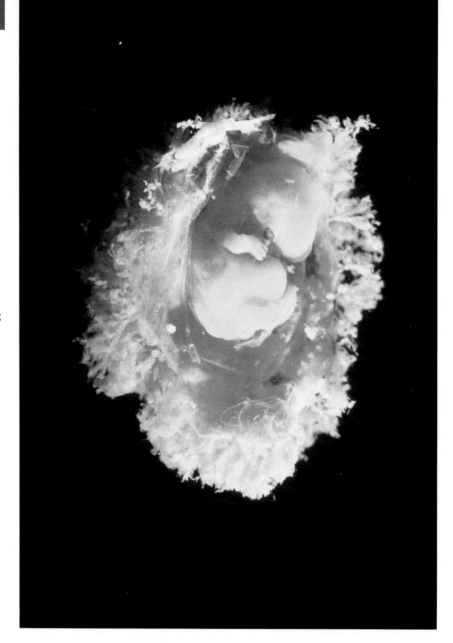

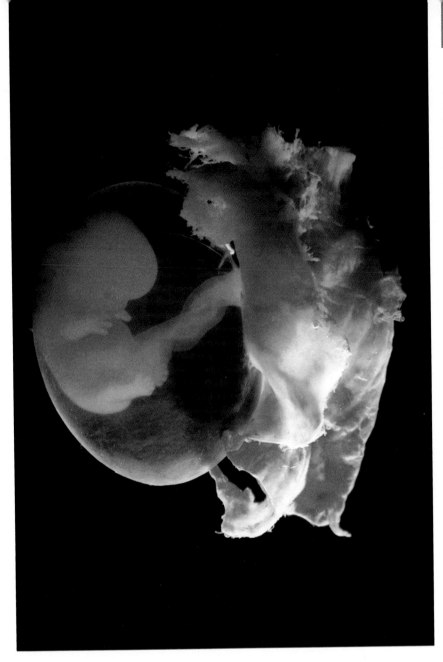

At about seven weeks (left), the embryo, afloat in its amniotic fluid, is clearly anchored to its placenta by the twisted umbilicus through which great blood vessels pass. The abdomen is swollen due to the rapid growth of the liver, the main blood-forming organ at this time.

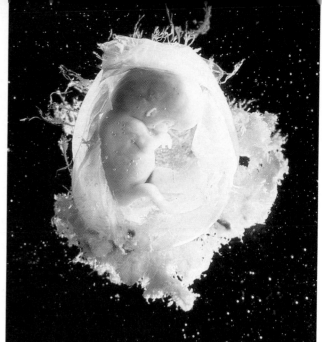

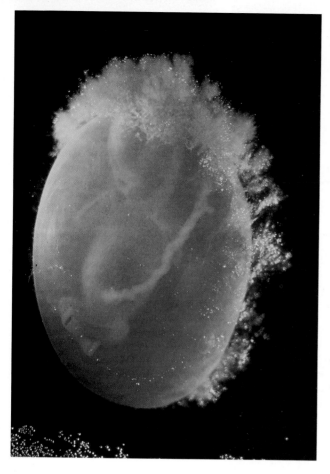

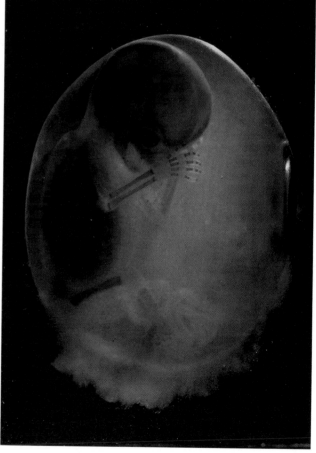

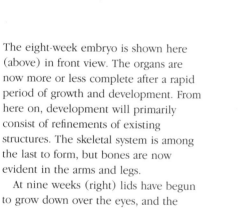

The eight-week embryo is shown here (above) in front view. The organs are now more or less complete after a rapid period of growth and development. From here on, development will primarily consist of refinements of existing structures. The skeletal system is among the last to form, but bones are now evident in the arms and legs.

At nine weeks (right) lids have begun to grow down over the eyes, and the outer ear begins to form. Because the plates of the skull have not fused, the head is rather flexible. During this third month, the fetus may begin to move, wave its arms and legs, and may even suck its thumb. It is now beginning to fill its amniotic space and will soon assume the typical, upside-down fetal posture.

At ten weeks (lower right), the skeleton is well along in its development. The long bones begin developing independently, growing from areas near their ends. They will join,

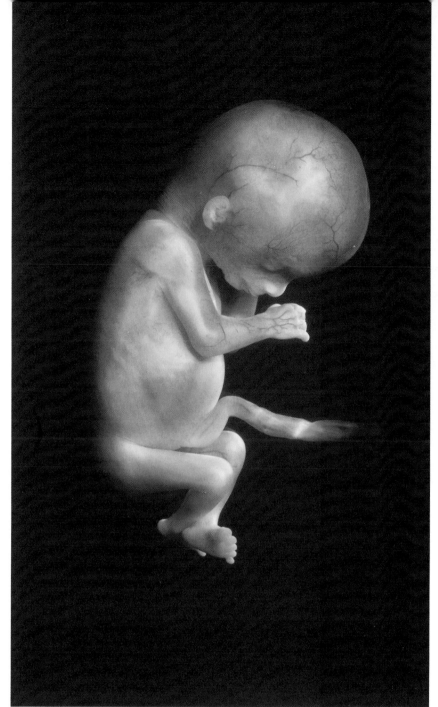

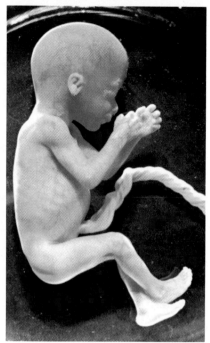

forming joints, later. In fact, the joints may not be firmly abutted by the time the baby is born. The head is still disproportionately large and will remain so, to a decreasing degree, through childhood. Notice the coiled umbilicus lying near the cuboid ankle bones showing as dark spots. The wrist bones have not yet begun to form and the jaw structure is weak indeed.

At fourteen weeks (above) the fetus is fist-sized. Ribs and blood vessels are visible through the translucent skin. The vigorous movements of the fetus can now be felt by the mother. The delicate skin is actually covered with a cheesy protective coating. Refinements such as fingerprints and fingernails have not yet developed.

By the end of five months (above right) the fetus is covered with fine, downy hair and its head may have already started to grow its own crop. It has already started the lifetime process of discarding old cells and replacing them with new ones. The heart is beating now at a rate of 120 to 160 times per minute.

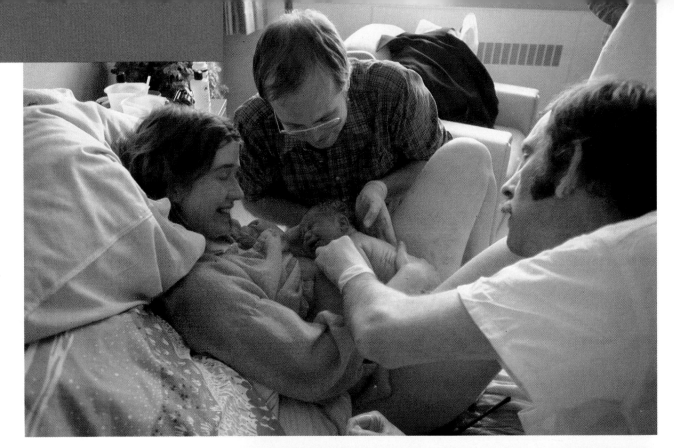

Delivery is a much too clinical term for what the mother is experiencing both physically and emotionally. It is a culmination of long months of changes. The infant is still not completely developed, but it is able to survive apart from the mother.

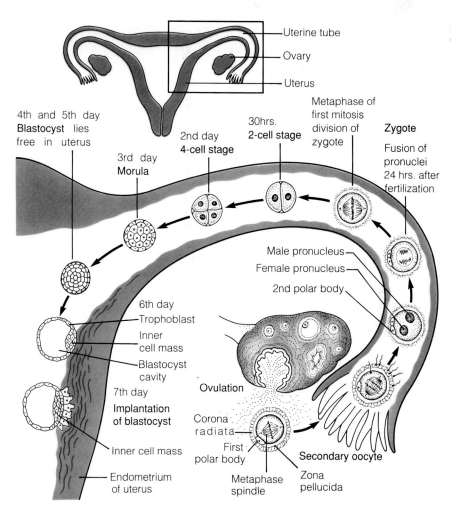

of a solid clump of cells, referred to as a **morula** (*morus* = mulberry). The morula stage is illustrated by the three-day embryo in Figure 18–2. The embryo floats free in the uterine cavity for an additional two or three days, nourished by secretions of the uterine glands. During this time, fluid accumulates between some of the cells, forming an embryo referred to as a **blastocyst** (see Figure 18–2). The blastocyst consists of an outer layer of cells, the **trophoblast** (*tropho* = nourishment) that will be involved in the formation of the placenta, and an *inner cell mass* attached eccentrically to the inside of the embryo proper as well as to certain embryonic membranes. As the blastocyst expands in size with the accumulation of fluid, it bursts out of the zona pellucida, exposing microvilli on the trophoblast cells that apparently aid in implantation; indeed, the zona pellucida must be lost before implantation can occur.

## Implantation and Development of the Placenta

### Normal Sequence of Events

The events occurring during implantation and development of the placenta are summarized in Figure 18–3. Implantation begins six or seven days after fertilization.* Trophoblast cells adjacent to the inner cell mass adhere to the epithelial surface of the endometrium, most commonly high up in the uterus, toward the fundus. As the trophoblast cells proliferate, first in the area of initial

---

*In our discussion, we will consider fertilization to be Day 1 of embryonic development. This differs from the convention used in medical practice, where pregnancy is dated from the onset of the last menstrual period.

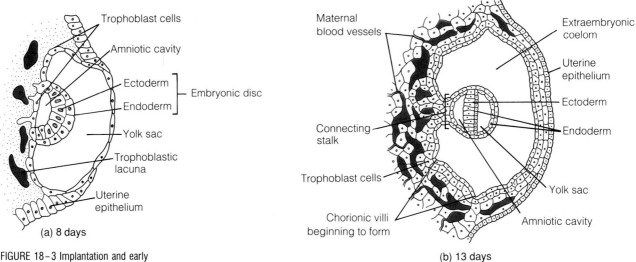

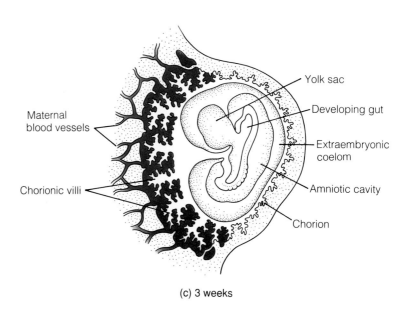

FIGURE 18–3 Implantation and early development of the placenta and embryo. (a) About eight days following fertilization; (b) about 13 days, and (c) about three weeks. Diagrams are not drawn to scale.

contact with the endometrium and later on the entire surface of the blastocyst, fingerlike processes of trophoblast cells invade and erode the endometrial epithelium and underlying stroma (Figure 18–3a). As mentioned, the enzymatic digestion of endometrial tissue that occurs during implantation provides nutrients for the developing embryo. Eventually, by about the ninth or tenth day following fertilization, the blastocyst becomes completely buried within the endometrial stroma, and the epithelial surface grows over the embedded blastocyst. During implantation, the uterus is in its progestational stage, and in reaction to the implanting blastocyst, the uterus proliferates extensively and becomes increasingly vascularized, especially in the area surrounding the implanted blastocyst.

As implantation is proceeding and the trophoblast expands, it ceases to be a compact mass, instead becoming permeated by a system of cavities called *lacunae*. Continued erosion of endometrial tissue by the trophoblast causes the deterioration of blood capillaries in the endometrium, and these bleed into the lacunae. In this way, nutritive substances in the maternal blood become available to the developing embryo. By about two weeks after fertilization, veins and arteries that lie deeper in the endometrial stroma are eroded by the growing trophoblast, establishing the circulation of maternal blood through the lacunae; that is, maternal arteries now supply blood to the lacunae, and maternal veins drain it away (Figure 18–3b).

During the third week of development, trophoblast tissue that remains between the lacunae contributes to the formation of structures called **chorionic villi** (Figure 18–3c). The villi eventually form treelike systems that spread into the blood-filled lacunae. Blood vessels derived from embryonic tissue develop within the villi such that each villus contains an extensive capillary network supplied by an arteriole and drained by a venule. The capillaries are lined with endothelium and surrounded by loose connective tissue that forms the core of the villus. The blood vessels supplying the villi soon become connected to the developing circulatory system of the embryo proper by running through a peninsula of tissue called the *connecting stalk* that links the trophoblast with the developing embryo. The connecting stalk later becomes the umbilical cord that carries the embryo's blood to and from the placenta in the umbilical arteries and vein. The chorionic villi and their associated structures form a membrane system known as the **chorion.** The chorion covers all structures that arise from the fertilized egg and it forms a membranous barrier between the circulatory systems of the embryo and the mother.

At the beginning of the fourth week of development, the heart of the embryo begins to beat, and a few days later blood starts to circulate through the villi. As this occurs, materials may be exchanged between the embryo's blood circulating through the villi and the mother's blood in which the villi are bathed.

As mentioned above, the implanted blastocyst initially disappears within the endometrium, but as the embryo and trophoblast grow, they form a swelling that bulges into the cavity of the uterus. During the early stages of the development of the placenta, villi form over the entire surface of the blastocyst. By about 12 weeks, as the developing embryo continues to increase in size, villi begin to disappear on the surface of the chorion that lies between the embryo and the cavity of the uterus. In this region of the endometrium, the maternal blood supply is limited. However, on the side of the chorion facing the myometrium, the maternal tissues have a good blood supply, and it is here that the chorionic villi and the uterine wall continue to develop in an interlocked fashion, forming the disc-shaped placenta by about 16 weeks.

As the villi elongate during pregnancy, the thickness of the placenta increases. At birth, the placenta is 15 to 20 cm in diameter, about 3 cm thick, and weighs about one and one-quarter pounds (Figure 18–4). It occupies about 30 percent of the internal surface of the expanded uterus. In the fully developed placenta, the chorionic villi have a combined surface area of approximately 50 square feet.

## Physiology of the Placenta

There is one important point about the physiology of the placenta that is so often misunderstood that it needs to be stressed. Within the placenta, *there is no mixing of the mother's and the embryo's blood.* Instead, the chorionic villi are bathed in pools of maternal blood, and nutrients, water, minerals, $O_2$, $CO_2$, hormones, and other materials exchanged (in either direction) between this blood and the blood of the embryo must pass through the **placental barrier**—the tissue layers of the chorion and the endothelial cells that line the embryonic capillaries within the villi. Most blood cells cannot pass through the placental barrier nor can most large molecules. One exception to this latter limitation are maternal serum antibodies. Produced in the mother in response to the presence of bacteria, viruses, and other foreign substances, antibodies in the mother's serum can be transferred to the developing embryo and protect it from diseases to which the mother is immune.

It was once thought that the placental barrier was so selective that the developing embryo was effectively isolated from any potentially harmful substances that were present in the mother's circulation. However, we now know that most drugs taken by the mother, or environmental chemicals that

FIGURE 18–4 Photograph of a full-term placenta.

dissolve in her blood, are transferred across the placenta and enter the circulatory system of the developing embryo. A number of viruses that infect the mother are also able to cross the placenta and infect the tissues of the embryo. Such drugs, chemicals, and viruses can have detrimental effects on the development of the embryo (as discussed in a later section of this chapter entitled "Abnormal Development: Birth Defects"). In addition, not all maternal antibodies that cross the placenta are beneficial to the embryo; some are responsible for Rh and other blood type incompatibilities, as discussed in Chapter 21.

## Ectopic Pregnancy

An **ectopic pregnancy** (*ectopic* = out of place) occurs when the fertilized egg implants in a tissue other than the endometrium of the uterus. About 1 in every 200 pregnancies is ectopic, and about 95 percent of these occur by implantation of the embryo in the wall of the Fallopian tube. The embryo can obtain sufficient nourishment in the Fallopian tube to sustain it during early stages of development, but as the embryo grows in size it can tear the wall of the Fallopian tube, causing severe and sometimes fatal hemorrhage. Ectopic pregnancies are usually discovered during the first month or two of pregnancy with the onset of abdominal pain, often accompanied by vaginal bleeding. Once the diagnosis is confirmed, ectopic pregnancies must be terminated surgically as soon as possible. Ectopic pregnancies in the Fallopian tube are more likely if the tubes have been scarred by infection or if they are congenitally abnormal.

## Hormones of Pregnancy

The placenta synthesizes many hormones, at least three of which are essential for normal pregnancy. Shortly after implantation the trophoblast, followed later by cells of the chorionic villi that are derived from the trophoblast, secrete human chorionic gonadotrophin (HCG). HCG is similar in chemical structure and in most of its actions to luteinizing hormone (LH) from the pituitary gland. Beginning about two weeks after fertilization, blood levels of HCG begin to increase rapidly, reaching a peak at six to eight weeks, and then declining, equally rapidly, until about a month later. During the rest of pregnancy, relatively constant low levels of HCG are maintained until labor begins.

HCG is important in maintaining the corpus luteum past the time that it normally would have degenerated if pregnancy had not occurred. Continued secretion of estrogen and progesterone by the corpus luteum during early pregnancy inhibits FSH and LH release by the pituitary and consequently prevents menstruation and further ovulation. The drop in HCG levels after the first six to eight weeks of development coincides with the secretion by the placenta of increased levels of estrogen and progesterone. Indeed, placental estrogen and progesterone levels continue to increase throughout pregnancy. Although the corpus luteum remains functional during pregnancy, its contribution to the secretion of estrogen and progesterone is negligible by about ten weeks after fertilization. In fact, the ovaries (containing the corpus luteum) may be removed after this time without affecting the pregnancy.

Progesterone is synthesized entirely within the placenta from cholesterol extracted from the mother's blood. Estrogen, on the other hand, is synthesized from intermediate compounds, some of which are produced in the placenta and some, in the adrenal glands of the developing embryo.

## Pregnancy Tests

The increased HCG levels that occur during early pregnancy have long been used as a basis for pregnancy tests. Older pregnancy tests relied upon the fact that HCG in the urine of pregnant women stimulates ovulation in mice, rabbits, and toads, and the release of sperm in male frogs and toads.

Newer pregnancy tests detect HCG levels directly in urine by the use of antibodies that specifically bind to HCG. One widely used test is diagrammed and explained in Figure 18–5. Antibodies to HCG, needed for performing the test, are obtained from the serum of rabbits or some other animal by injecting these animals with HCG. In step 1 of the test, a urine sample, obtained from a woman who is being tested, is mixed with a solution containing antibodies to HCG. In step 2, this mixture is added to a suspension of latex particles on which HCG molecules have been absorbed. If the urine contains detectable amounts of HCG, as it would during early pregnancy, the HCG in the urine is bound in step 1 to the HCG antibody, forming an HCG-HCG antibody complex. Consequently, when the HCG-coated latex particles are added to the mixture in step 2, the HCG on the latex particles cannot form a complex with the HCG antibodies, because these antibodies have been tied up in a complex with HCG in step 1. Accordingly, the latex particles remain as a fine suspension in the mixture (Figure 18–5a). On the other hand, if no HCG is present in the urine, then the HCG antibodies remain free after the completion of step 1 of the test. These antibodies, when added to the HCG-coated latex particles in step 2, bind to the HCG on the surface of the particles, causing a visible clumping of them (Figure 18–5b).

(a) Urine of pregnant woman

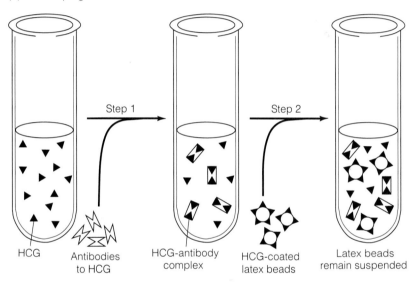

FIGURE 18–5 Pregnancy testing, outlined in this diagram, is designed to detect human chorionic gonadotropin (HCG), a hormone secreted only during pregnancy.

(b) Urine of nonpregnant woman

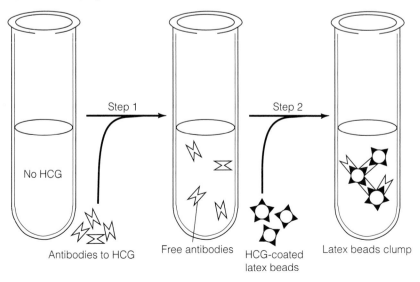

This test is most reliable when used later than four weeks after fertilization (two weeks after the first missed menstrual period). Kits for pregnancy testing, now available over the counter, use methodologies similar to the one described above. More sensitive tests are available that can detect HCG in blood as early as one week following fertilization.

## Stages of Embryonic Development

The entire developmental sequence, from fertilization to birth, is a continuum. Nevertheless, it is often helpful for purposes of discussion to divide embryonic development into three arbitrary stages. The first stage, the **period of germ layer formation,** begins at fertilization and encompasses about the next two and one-half weeks. During this stage implantation occurs, and the trophoblast begins forming the placenta. Simultaneously, the cells of the inner cell mass that make up the embryo proper are becoming differentiated into three kinds of cells, forming the so-called primary germ layers (ectoderm, endoderm, and mesoderm, as discussed below).

The second stage of development is the **embryonic period,** extending from the end of the previous period through the eighth week. It is characterized by the appearance of most of the organs of the body, at least in rudimentary form. The final stage, lasting from the ninth week of development until birth, is termed the **fetal period.** Although the fetal period involves the development of some new structures, it is primarily a period of refinement, growth, and maturation of structures whose development was initiated in the embryonic period. During the fetal period the developing embryo is commonly referred to as a **fetus,** although the term embryo is entirely appropriate throughout the entire developmental period from fertilization to birth.

## Germ Layer Formation

Near the beginning of the second week of development, about halfway through the implantation sequence, the inner cell mass differentiates into two layers (see Figure 18–3a). Cells next to the cavity of the blastocyst form an inner layer of small, cuboidal cells called **endoderm** (*endo* = inner; *derm* = skin), while cells closer to the trophoblast layer form the **ectoderm** (*ecto* = outer) of tall, columnar cells. Together, these two layers comprise the **embryonic disc.**

As the embryonic disc is forming, a cavity appears between cells of the ectoderm and the trophoblast. This cavity, filled with fluid, is called the **amniotic cavity,** and the cells that make up its wall form a membrane called the **amnion.** The amnion and the cavity it encloses will later form a fluid-filled, protective sac that surrounds the embryo. While the amniotic cavity is forming, endodermal cells begin to spread out from the edges of the embryonic disc, coming to line the entire cavity of the blastocyst. The migration of endodermal cells forms the **primary yolk sac,** part of which will become the lumen of the gastrointestinal tract. Later, toward the end of the second week, a third cavity, the **extraembryonic coelom,** will develop between the cells that line the yolk sac and the trophoblast. The extraembryonic coelom soon expands to become the largest cavity inside the trophoblast. As a result of its expansion, the embryonic disc with its associated amniotic cavity and yolk sac becomes suspended by a group of cells, the connecting stalk, that will become the umbilical cord with its associated vessels (as mentioned earlier). This stage is illustrated in Figure 18–3b.

Toward the beginning of the third week, on about the sixteenth day, the embryonic disc, now more oval than disc-shaped, develops a groove along its long axis (Figure 18–6a). This groove is called the **primitive streak.** Ectodermal cells migrate toward this groove, into it, and then spread sideways between the ectodermal and endodermal layers, forming the third primary germ layer, the **mesoderm** (Figure 18–6b).

The formation of the three germ layers is an important milestone in development, since it represents the beginnings of the process of cell

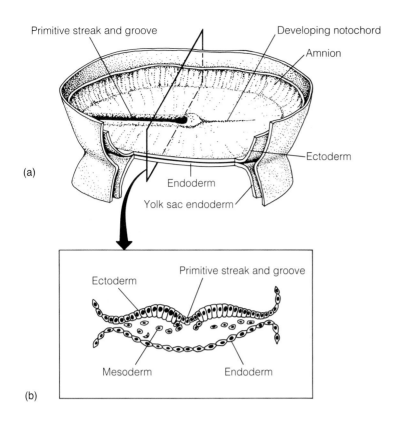

Primitive streak and groove  Developing notochord

Amnion

(a)

Ectoderm

Endoderm

Yolk sac endoderm

**(b)**

Ectoderm  Primitive streak and groove

Mesoderm  Endoderm

FIGURE 18-6 Formation of the primitive streak and the mesodermal germ layer. (a) The embryonic disc at about 16 days, showing the primitive streak. (b) Cross-section through the embryonic disc and primitive streak in the area designated in (a).

specialization (differentiation). Each germ layer will subsequently give rise to specific structures, as outlined in Table 18–1.

## The Embryonic Period

During the embryonic period there is rapid growth, and the basic features of the major organ systems and of external shape appear. The events that occur during this period are extremely complicated and cannot be described in great detail. This discussion is instead designed to provide a general summary of how the embryo comes to assume its basic form. Figure 18–7 summarizes development during the embryonic period.

The formation of the three primary germ layers by Day 16 or 17 is immediately followed by events that lead to the development of the nervous system. Cells within the mesodermal layer that lies between the ectoderm and endoderm start to aggregate along the midline of the embryonic disc to form a rod, the **notochord,** which later contributes to the formation of the bones of the vertebral column (see Figure 18–6). The formation of the notochord is an

TABLE 18-1 Derivatives of the Three Embryonic Germ Layers

| Ectoderm | Mesoderm | Endoderm |
|---|---|---|
| Nervous system | Muscles | Digestive system |
| Epidermis of skin and | Skeletal system | Respiratory system |
| structures derived from | Connective tissues | Thyroid gland |
| epidermis such as hair, | Circulatory and | Parathyroid gland (origin unclear; |
| sweat glands, mammary glands | lymphatic systems | secretory cells |
| Pituitary gland | Urinary system | may be ectodermal) |
| Adrenal medulla | Reproductive system | Pancreatic islets (origin |
| | Adrenal cortex | unclear; may be |
| | | ectodermal) |

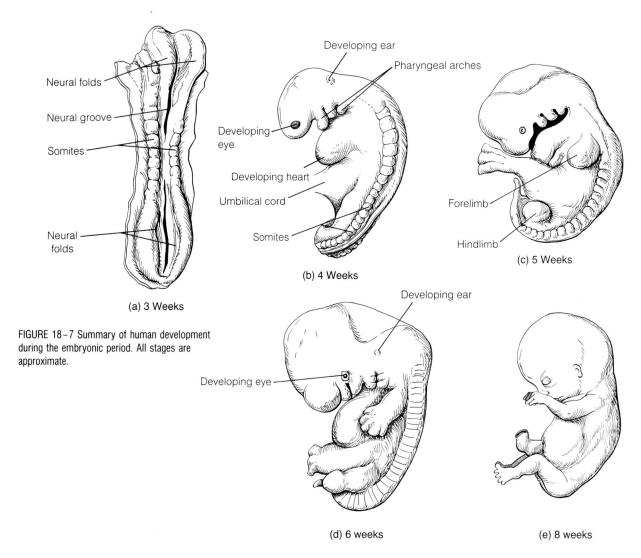

Neural folds

Neural groove

Somites

Neural folds

(a) 3 Weeks

Developing ear

Pharyngeal arches

Developing eye

Developing heart

Umbilical cord

Somites

(b) 4 Weeks

Forelimb

Hindlimb

(c) 5 Weeks

Developing ear

Developing eye

(d) 6 weeks

(e) 8 weeks

FIGURE 18–7 Summary of human development during the embryonic period. All stages are approximate.

important developmental event, for not only does it establish the midline of the embryo, but it also causes, or *induces,* the overlying ectoderm to begin differentiating into embryonic nervous tissue.

The induction of nervous tissue by the notochord first becomes apparent as ectodermal cells lying above the notochord form, along the dorsal length of the embryo, a shallow plate with thickened edges. This plate is called the **neural plate** and its edges, the **neural folds.** The neural folds grow together and fuse, forming a hollow tube, the **neural tube,** that will eventually become the brain and the spinal cord; other cells along the ridges of the neural folds will migrate throughout the embryo to give rise to the neurons of the peripheral nervous system. At about three weeks, the neural tube has formed in the middorsal region of the embryo, but the folds have not yet fused at the front and back ends (Figure 18–7a). The entire neural tube is complete by about Day 28. By this time, the major parts of the brain have begun to develop, as have the eyes and the ears. Unlike the other systems of the body, the nervous system continues to develop, increasing in complexity throughout most of the fetal period.

As the notochord forms, columns of mesoderm on either side of the notochord become thickened, and these soon become separated into paired segments of mesoderm, called **somites** (Figure 18–7a and b). Somites form from the anterior end of the embryo toward the posterior end until about Day 30 when 42 to 44 pairs are present. The somites contribute to the formation of the dermis of the skin, the vertebrae and ribs, and skeletal muscle.

Several days after the embryonic mesoderm forms, it begins to give rise to blood vessels in the embryo and in the developing placenta. The heart is formed from one of these vessels. Although the heart begins to beat as early as about Day 21, it initially consists of only a single chamber and does not acquire its definitive structure until the end of the sixth week. As blood vessels develop, mesodermal cells trapped within them give rise to blood cells.

At the beginning of the fourth week, about the time that the heart begins to beat, the primitive gut starts to develop. This occurs as the lateral edges of the embryonic disc fold down, fusing on the ventral side of the embryo and, as they do so, pinching off a portion of the yolk sac that becomes the tube of the gut (Figure 18–3c). By the end of the fifth week, the gut tube has differentiated into the pharynx, esophagus, stomach and intestines, and the liver, pancreas, and gallbladder have also developed as outpocketings of the gut tube. Also by this time, other outpocketings of the gut have formed the trachea and the lung buds. The lungs develop from these latter structures by the end of the seventh week, but they do not become functional until birth.

As the lateral edges of the embryonic disc fold down to form the gut, the anterior and posterior portions of the disc also begin to fold under. At the anterior end, this fold gives rise to the top of the head and upper part of the face. Ridges of tissue, extending laterally and ventrally around the pharynx, develop in the fourth week; these are the **gill arches,** or **pharyngeal arches,** that give rise to the jaws (lower part of the face), larynx, portions of the ear, and nearby structures (Figure 18–7b). By the eighth week, the face has formed in outline (Figure 18–7e). (The development of the face is discussed in Chapter 6 in connection with the occurrence of cleft palate.)

The kidneys begin to develop in the fourth week, reaching their definitive form in the seventh week. Early in the fifth week, the forelimb and hindlimb buds first appear as small swellings; by the end of the seventh week, the forelimbs have fingers that are fairly well developed. The lower limbs lag somewhat behind the forelimbs in development.

Last of the major systems to form is the reproductive system. The gonads begin to differentiate in the sixth week and are completed in the eighth week in the male; in the female, the gonads do not develop until the twelfth week. The external genitalia have not attained their final form by the end of the eighth week, but they are distinct enough so that the two sexes can be differentiated.

The amount of development that has occurred by the end of the embryonic period is truly incredible. In a scant eight weeks, the embryo has progressed from a single cell to a multicellular organism with all major structures and organ systems present.

## The Fetal Period

The thirty weeks of the fetal period are characterized by growth and continued refinement of structural details. The eight-week embryo is about 40 mm in length (crown to rump) and weighs about 5 grams. At 12 weeks, the fetus is 87 mm in length and weighs 45 grams. It is quite human in appearance, but the head is too large.

By the end of the sixteenth week, the fetus is somewhat better proportioned, and it has increased during the previous four weeks about one and one-half times in length (to 140 mm) and about four and one-half times in weight, to 200 grams (Figure 18–8). From 20 to 38 weeks, the growth rate slows and only limited differentiation continues to occur.

## How Does Embryonic Development Occur?

During embryonic development, cells become differentiated: the fertilized egg, an unspecialized cell, develops into a multicellular organism composed of more than a hundred different cell types, each specialized in structure and function to play particular roles within the organism. Although we do not understand precisely how differentiation occurs in a developing embryo, we do understand some of the factors that are involved. We have known for a long time that the DNA present in the fertilized egg and in all cells of the body (except haploid germ cells) is identical, and that it contains the information necessary to define not only the structural and functional pattern of the cell

FIGURE 18–8 Human fetus of 16 weeks.

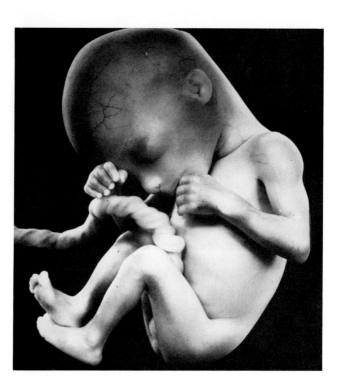

itself, but also the general characteristics of the entire organism. The developmental program is, of course, part of this information.

If the DNA in all cells is identical and determines the characteristics of each cell type, then it follows that cellular differentiation must result from *differential expression* of the information in DNA; that is, certain information is used, or expressed, in one cell type that is not expressed in others. Thus, for example, red blood cells—and only red blood cells—express that part of the hereditary information that allows the synthesis of hemoglobin, whereas liver cells express information that provides them with the unique ability to produce bile and to do other things that only liver cells do.

Although differentiation results from the differential expression of hereditary information, we still do not understand how a specific cell type comes to express the particular information that it does. The interactions that occur between cells within the developing embryo and between the embryo and its external environment (as perceived via the placenta) must somehow determine the course of cellular differentiation. As development proceeds and cell division occurs, tissues become increasingly differentiated. The first step in differentiation is the appearance of the three primary germ layers. Interactions of the three germ layers with one another bring about, or induce, further differentiation, and so on during development. These interactions probably involve the production by one cell of a molecule that acts upon other cells to induce a different pattern of gene expression. Such an effect of one tissue on another is well illustrated by the influence of the notochord in inducing the differentiation of overlying ectoderm into nervous tissue; if the notochord does not develop, or develops abnormally, the differentiation of nervous tissue will not occur, or will be abnormal.

In addition to differentiation, other processes occur during embryonic development. The final shape and functional capacity of the embryo also depends upon such factors as the directed migration of cells within the embryo, differential growth rate of various cells and tissues, the programmed death of various cells, and so forth. We currently understand very little about these processes.

## Abnormal Development: Birth Defects

A number of agents may disrupt normal development, causing death of the embryo or birth defects of varying severity, some of which may be fatal. Such agents are called **teratogens** (*teratos* = monster). Birth defects include obvious structural abnormalities that may or may not affect function, as well as visually less apparent abnormalities that hamper function and lead to such conditions as mental retardation and learning disabilities.

Two to five percent of individuals have significant birth defects, the exact percentage depending upon how the term "significant" is defined. Considering the complexity of the process of embryonic development, it is surprising that most individuals are born completely normal.

## Susceptibility to Teratogens

It is essential to understand that an embryo is not equally susceptible to a teratogen at all stages of its development. Indeed, the timing of exposure to a potential teratogen determines whether it will cause a birth defect at all and, if it does, the type of birth defect that occurs. If exposed to a teratogen during approximately the first two weeks of development, an embryo is likely either to recover completely, meaning that it is born without a birth defect, or to abort spontaneously at an early stage. A teratogen is most likely to cause a severe birth defect if exposure occurs during the embryonic period when most of the organ systems are developing. In general, the earlier the exposure during the embryonic period, the more severe will be the defect. The chances of a teratogen causing a birth defect decrease steadily during the fetal period. After about 18 weeks of development, exposure to a potential teratogen usually has little, if any, apparent detrimental effect.

A woman often does not even suspect she is pregnant, particularly if she has irregular periods, until several weeks after the first missed menstrual period—a time corresponding to the end of the fourth week of development. During this time interval, the embryo has a particularly high risk of being damaged by teratogens. Consequently, a woman who is attempting to become pregnant or has a good chance of becoming pregnant should take care to limit exposure to potential teratogens, such as those described in the following section.

## Causes of Birth Defects

Some birth defects are hereditary; that is, they are caused by mutations, or permanent changes, in the DNA of the germ cells of one or both parents. However, evidence now suggests that most birth defects are due to environmental agents such as chemicals (including drugs), viruses, and radiation.

*Chemicals.* A large number of chemicals and drugs have been shown to cause birth defects in experimental animals, and a few of these are known to cause birth defects in humans. Two particularly well-documented teratogens in humans are *thalidomide,* the effects of which were discussed in Chapter 1, and *ethyl alcohol.*

Numerous studies have demonstrated the teratogenicity of ethyl alcohol, the intoxicating substance in alcoholic beverages. It has been estimated that as many as a third of the offspring of chronic alcoholic women display a group of symptoms, known collectively as *fetal alcohol syndrome,* which include mild to moderate mental retardation, retarded postnatal growth, and characteristic facial abnormalities (Figure 18–9). About one infant in 750 births has fetal alcohol syndrome. Recent evidence has suggested that even moderate or light drinking during pregnancy may be associated with a low birth weight and an increased risk of miscarriage. Because it is not known whether there is some low amount of alcohol that can be taken safely during pregnancy, the Surgeon General has warned pregnant women to avoid all alcohol.

Many chemicals and drugs are suspected human teratogens, based upon their effect in experimental animals. However, unequivocal proof that a substance is harmless or harmful in humans is difficult to obtain (see Chapter 1). Consequently, prudence dictates that pregnant women should avoid all drugs unless recommended by a physician and should attempt to minimize

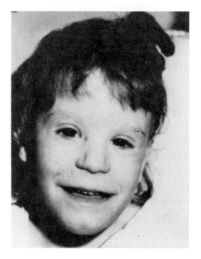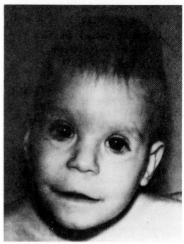

FIGURE 18-9 These mentally retarded children have facial abnormalities characteristic of fetal alcohol syndrome. These include a narrow forehead, a low nasal bridge, a short nose, short eye slits, folds of skin at the inner corners of the eyes (epicanthic folds), an absence of ridges running between the nose and the mouth, a thin upper lip, and a small midface.

contact with cosmetics, cleaning fluids, paints, and other substances containing chemicals that may be inhaled or absorbed through the skin.

*Viruses.*   *Rubella,* the German measles virus, has long been known to cause birth defects in humans. German measles is usually a mild disease, lasting for a few days and characterized by a slight fever and a rash of flat, reddish spots. However, if German measles is contracted by a pregnant woman, the fetus can develop a variety of birth defects, including mental retardation and abnormalities of the heart, eyes, and ears. During the first three weeks of development, infection by rubella is likely to cause the embryo to abort, but if it does not, birth defects are likely. If rubella is contracted during the first month of development, 70 percent of babies have one or more severe birth defects. In the following weeks, the risk decreases somewhat erratically, with no increased risk after the first six months of gestation (Figure 18-10).

Because of the danger of rubella to pregnant women, preschool children are now routinely immunized against the virus. In addition, a blood test can be performed on women of childbearing age to check their immunity to rubella. If they are not immune, they should be immunized at least three months before becoming pregnant.

*Radiation.*   X rays and any other penetrating radiation, such as radiation from some radioactive elements, cause injury to DNA, including mutations and chromosome aberrations. Consequently, germ cells as well as the embryo itself may be damaged. As with any teratogen, the nature of the injury depends on the time of exposure and the dose. In the embryonic period, irradiation can cause a wide variety of birth defects, and high doses during the fetal period can interfere with brain development, causing serious mental defects.

It is crucial that all pregnant women avoid exposure to X rays and other penetrating radiation unless such exposure is absolutely necessary. Furthermore, exposure to penetrating radiation at any age prior to reproduction should be minimized, as it can cause permanent mutations and chromosome aberrations in germ cells. It should also be kept in mind that exposure to radiation can cause cancer (Chapter 22).

## Prenatal Diagnosis

In recent years, a number of techniques have been developed for the prenatal diagnosis of certain genetic disorders and birth defects. This information can be used by parents for determining whether or not they wish to have a pregnancy terminated.

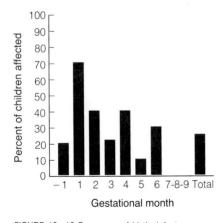

FIGURE 18-10 Frequency of birth defects as a function of the gestational month of maternal rubella infection. A gestational month of −1 refers to the month prior to fertilization. Because infected infants have been born to women who had recovered from rubella several weeks before their last menstrual period (approximately two weeks prior to fertilization), it seems that the virus can be harbored for many days after the mother's illness.

## Amniocentesis

*Amniocentesis* is a commonly used method for examining the chromosomal and genetic makeup of a fetus. Typically performed between the thirteenth and sixteenth weeks of development, the procedure involves the insertion of a hypodermic needle through the abdominal and uterine walls into the amniotic cavity and the withdrawal of a small amount of amniotic fluid. Skin cells and other cells sloughed from the fetus are present in the amniotic fluid, and these may be cultured in the laboratory for biochemical and chromosomal analysis (Figure 18–11). Some 200 disorders can be detected by amniocentesis, including such serious conditions as Down's syndrome (characterized by severe mental retardation), Tay-Sachs disease (characterized by progressive neurological deterioration leading to death by the age of three), and neural tube defect (in which the neural tube does not fuse properly, resulting in severe brain or spinal cord defects). Many women who are at risk for bearing a child with a genetic disorder are now choosing to have amniocentesis performed.

A drawback to amniocentesis is that it cannot be performed before the thirteenth week of development when the amount of amniotic fluid is sufficient and there are enough living cells of fetal origin. Moreover, a complete chromosomal analysis takes an additional three weeks or so. Consequently, results are not available until about the sixteenth week of development or later. The risk of complications in terminating a pregnancy at such a late date is substantially higher than if the termination is performed earlier.

## Chorionic Villi Sampling

A newly developed, still experimental technique called *chorionic villi sampling* overcomes some of the problems of amniocentesis by permitting prenatal diagnosis at an early stage in pregnancy. Done between the sixth and eighth weeks of development, chorionic villi sampling involves removing a small piece of embryonic tissue from a chorionic villus and analyzing it for biochemical and chromosomal abnormalities.

Other methods of prenatal diagnosis involve the visualization of the fetus by ultrasound waves (Figure 18–12), sampling of fetal blood to detect

FIGURE 18–11 Summary of the technique of amniocentesis.

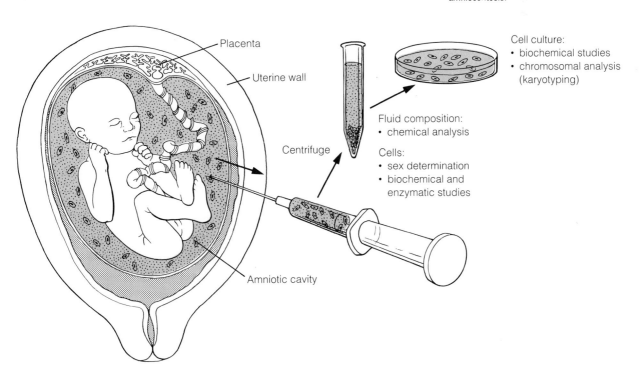

Placenta

Uterine wall

Centrifuge

Amniotic cavity

Cell culture:
- biochemical studies
- chromosomal analysis (karyotyping)

Fluid composition:
- chemical analysis

Cells:
- sex determination
- biochemical and enzymatic studies

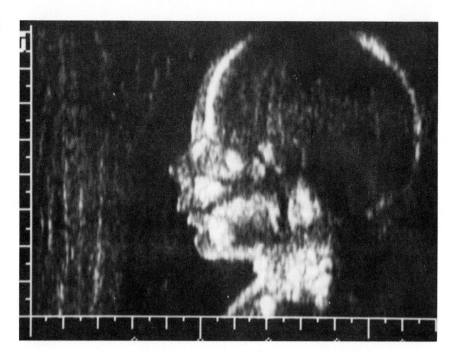

FIGURE 18–12 An ultrasound image of the head of a 20-week-old fetus.

abnormalities of hemoglobin synthesis, and analyzing maternal blood to detect the presence of fetal metabolites and proteins that may be indicative of certain fetal abnormalities.

## Problems of Pregnancy

### Miscarriage and Stillbirth

A miscarriage is defined as the spontaneous termination of pregnancy at any time before the eighteenth week of development. However, most miscarriages occur during the first 12 weeks. Miscarriages result when the placenta separates from the endometrium. Although the cause of most miscarriages is not known, in some cases they may be due to improper attachment of the placenta or abnormal development of the embryo.

If a baby is delivered alive after the eighteenth week of development but substantially before term (38 weeks), the delivery is termed *premature;* if the baby is born dead, it is called a *stillbirth.*

### Pre-eclampsia and Eclampsia

*Pre-eclampsia* is a disorder of late pregnancy characterized by abnormal fluid retention and high blood pressure in the mother. If severe, pre-eclampsia may cause headaches, blurred vision, nausea, and vomiting. Untreated pre-eclampsia may lead to *eclampsia,* in which the mother's blood pressure reaches very high levels, causing convulsions and unconsciousness. Eclampsia is a serious condition that can be fatal to the mother and the fetus. Pre-eclampsia and eclampsia are also called *toxemia of pregnancy.* The cause of these conditions is unknown.

### Birth (Parturition)

In recent years there has been a trend toward *natural childbirth,* in which medical intervention during childbirth is kept at a minimum. This trend has stimulated the interest of expectant parents in the process of birth, and consequently a number of surprisingly thorough books have been written on the subject. Therefore, we will provide here merely an outline of the birth process.

On the average, 38 weeks elapse from the time of fertilization to birth. However, this interval is somewhat variable, and it is not uncommonly several weeks shorter or longer than this. Inaccuracies in determining the time of birth ("due date") also derive from the fact that it is difficult to estimate the time when ovulation, and hence fertilization, occurred, particularly in women with irregular periods.

Throughout pregnancy the uterus undergoes weak contractions. These are not noticeable until the last several weeks of pregnancy when they increase in frequency and strength, culminating in the process called **labor,** by which the fetus and the placenta are expelled from the uterus. Several weeks before the onset of labor, the position of the fetus and its surrounding membranes shifts toward the cervix. In nine out of ten births, the head of the fetus is pressed against the cervix and is the first part of the fetus to pass out of the uterus at the time of birth.

The onset of labor is signalled by increased uterine contractions that may occur at more or less regular intervals, 20 or so minutes apart. Another sign that labor is beginning, or soon to begin, is the expulsion of the cervical plug of mucus that has prevented the entry of bacteria and other potentially harmful agents into the uterus during pregnancy. Tinged with blood, the expelled plug is called a "show." Also at the beginning of labor, or during it, the chorion and amnion burst, releasing the amniotic fluid through the vagina, an event termed "breaking water."

## Stages of Labor

Labor is said to occur in three stages (Figure 18–13). In the first stage, contractions of the uterus help widen the canal of the cervix. This occurs as the wall of the cervix thins and merges with the wall of the uterus, an event termed *effacement.* In addition, the cervix opens up, or *dilates.* Both effacement and dilatation may begin several weeks before birth. When the cervix has dilated to about four inches (10 cm), the first stage of labor has ended. The first stage takes 12 hours on the average for the initial birth and four to eight hours for subsequent births. However, there is enormous variability in these time estimates.

The second stage of labor, which ends with the delivery of the baby, is accompanied by strong and frequent contractions that sweep down the uterus, shrinking its size and thus expelling the fetus. These contractions are often accompanied by an urge on the part of the mother to "push" the baby out of the uterus by increasing abdominal pressure. The second stage usually lasts up to an hour for first births and 30 minutes for subsequent ones. After delivery of the baby, the umbilical cord is clamped and cut. Usually the baby begins to breathe immediately, an event that allows the baby to exist independent of the placenta.

In the third stage of labor, contractions of the uterus separate the placenta from the uterine wall, and the placenta, now referred to as the **afterbirth,** is expelled, usually within a few minutes after delivery of the baby. Delivery of the placenta is accompanied by some bleeding from the ruptured blood vessels of the uterus, but contractions of the uterus serve to collapse and seal these vessels.

Labor proceeds without difficulty in most cases. Occasionally it requires some medication for pain, or perhaps an *episiotomy,* in which the skin at the open end of the vagina is cut to prevent a more serious and painful tear; this incision is sewn up after birth. In a minority of cases, some condition may interfere with normal delivery. For example, the mother's pelvis may be too narrow to allow the baby's head to pass through it. Another relatively common problem is *malpresentation,* in which the fetus is oriented other than head down in the uterus. In this case, some structure other than the head may emerge first. Because the head has the largest diameter of any body part, it may

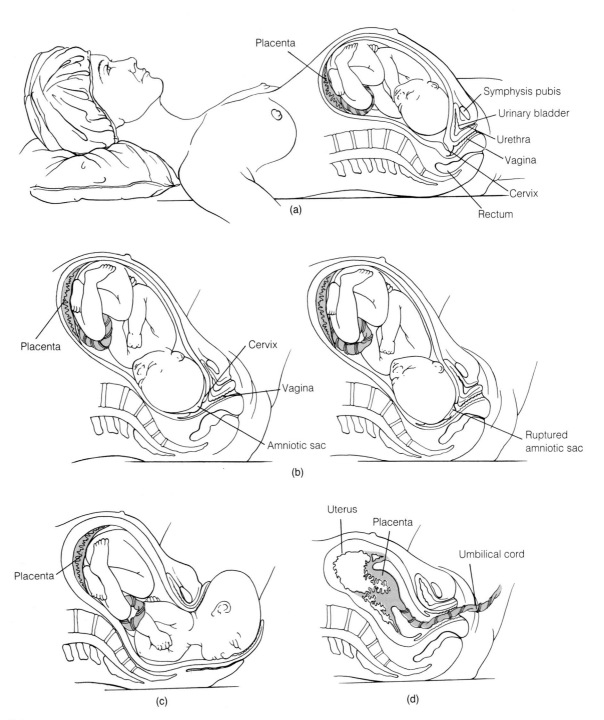

Placenta
Symphysis pubis
Urinary bladder
Urethra
Vagina
Cervix
Rectum

(a)

Placenta
Cervix
Vagina
Amniotic sac

Ruptured amniotic sac

(b)

Placenta

(c)

Uterus
Placenta
Umbilical cord

(d)

FIGURE 18-13 Stages of labor. (a) Position of the fetus before labor begins. (b) Dilation of the cervix and rupture of the amniotic sac. (c) Passage of the fetus through the birth canal. (d) Delivery of the placenta.

then have difficulty passing through the birth canal. Another danger with malpresentation is that the umbilical cord may be compressed between the fetus and the wall of the birth canal, disrupting the flow of $O_2$ to the fetus. In this event, the baby must be delivered by emergency caesarean section, as described shortly.

### Premature Labor

Labor may occur before term, resulting in the birth of a *premature infant* that has developed less than 35 weeks. About a third of all premature labors are caused by pre-eclampsia and eclampsia. Multiple births also are frequently premature. Abnormal attachment of the placenta, high blood pressure, diabetes, cigarette smoking, poor nutrition, and excessive consumption of alcohol are other causes of premature labor.

The longer a premature infant has developed in the uterus, the better its chances for survival. Lack of surfactant production in the lungs of premature newborns (respiratory distress syndrome; see Chapter 10) is a common cause of death in premature infants. With extensive medical care, babies as small as 12 ounces have been known to survive, although this is rare.

## How is Labor Initiated?

The factors that initiate labor are not well understood. Recent research has focused on the roles of the hormones oxytocin and prostaglandins in the onset of labor. Oxytocin, secreted by the posterior pituitary gland, is known to stimulate contraction of uterine muscle, as do some of the prostaglandins produced by uterine tissue in pregnant women. One model for the onset of labor suggests that, under the influence of rising levels of estrogen, the number of cell surface receptors capable of binding oxytocin increases dramatically in uterine tissue during the final stages of pregnancy; when a certain threshold level of receptors is reached, the uterine smooth muscle becomes receptive to the stimulating effects of oxytocin, which is secreted throughout pregnancy. Simultaneously, the binding of oxytocin to its receptors in the maternal portion of the placenta stimulates the synthesis of prostaglandins, which diffuse into the adjacent myometrium and enhance oxytocin-induced contractions. This model, which stresses the importance of oxytocin receptors in the initiation of labor, is supported by the fact that the uterus of a pregnant woman at term is some 50-fold more sensitive to oxytocin than is the uterus of a nonpregnant woman.

In certain cases in which labor does not occur spontaneously or the contractions are too weak to expel the fetus, oxytocin may be administered. Where labor is excessively prolonged or some other condition threatens the life or health of the baby or the mother, a *caesarean section* may be performed. In this procedure, the baby is removed from the uterus through an incision in the abdominal and uterine walls.

## Multiple Births

Twin infants are born in about 1 out of every 90 pregnancies in the United States; triplets in about 1 out of 8000. Seventy percent of twins are **fraternal twins,** resulting from the simultaneous fertilization of two eggs, each by a different sperm. **Identical twins,** on the other hand, arise when a single zygote splits into equal halves during early development. This splitting usually occurs during the early blastocyst stage and involves division of the inner cell mass. Because fraternal twins derive from different eggs, each with its own trophoblast, each fraternal twin has its own placenta; identical twins usually share one placenta between them, since the inner cell mass divides after the trophoblast has formed (Figure 18–14).

## LACTATION

The process by which milk is supplied to nourish a baby is called **lactation.** Lactation is the function of the breasts and is regulated by complex interactions between the endocrine and nervous systems.

## Breasts

The breasts consist of mammary glands that are surrounded by fatty and fibrous connective tissue. The breasts are supplied by nerves, blood vessels, and lymphatic vessels, and are covered with skin. A central nipple protrudes from the anterior surface of each breast and is surrounded by a roughened and pigmented area of skin called the *areola.*

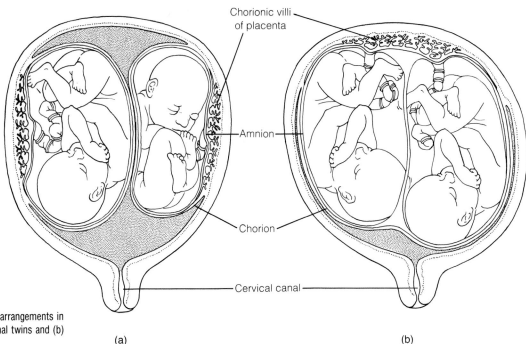

Chorionic villi
of placenta

Amnion

Chorion

Cervical canal

FIGURE 18–14 Placental arrangements in multiple births. (a) Fraternal twins and (b) identical twins.

(a)                                                                    (b)

## Mammary Glands

The function of the mammary glands is to secrete milk for nourishing the young. Like sweat glands, the mammary glands are exocrine glands derived from ingrowths of the surface epithelium. The structure of mammary glands varies enormously with their functional state. In their fully developed state during lactation, the mammary glands in each breast form 15 to 20 lobes (Figure 18–15). Each lobe consists of extensive clusters of saclike structures, the *alveoli,* which secrete milk into small ducts. These ducts converge into one large *lactiferous duct.* As it approaches the nipple, the lactiferous duct of each lobe dilates just beneath the areola to form a *lactiferous sinus* that serves to store milk in lactating women. From the sinus, each lactiferous duct again narrows before opening separately at the tip of the nipple. Surrounding the alveoli and much of the duct system are contractile cells called *myoepithelial cells.* Although these cells, unlike smooth muscle cells, are derived from epithelial tissue, they contain contractile filaments similar to those of smooth muscle. The contraction of myoepithelial cells serves to squeeze out milk during nursing.

Before puberty, the mammary glands consist only of a few short, blind-ended ducts that radiate from the nipple. When estrogen secretion increases dramatically at puberty, the breasts become enlarged as fat is deposited within them. Some growth and branching of the mammary duct system also occurs, but alveoli are not formed. During pregnancy, the mammary glands become extensively differentiated with the development of most of the ducts and alveoli complete by the end of the sixth month of pregnancy. Expansion of the breasts during the latter months of pregnancy is due almost entirely to the accumulation of a secreted fluid called *colostrum,* which is rich in proteins, including antibodies.

## Endocrine Control of Breast Development

The influence of the endocrine system on breast development is too extensive to consider in detail. Structural changes in the mammary glands that occur at the time of puberty are promoted by estrogen, growth hormone, and hormones from the adrenal cortex. During pregnancy, these hormones as well as progesterone, prolactin from the anterior pituitary, a prolactinlike hormone secreted by the placenta, and perhaps other hormones act together to bring about breast development.

## Milk Secretion

During the first few days after birth, the mammary glands secrete colostrum. They then begin to secrete milk under the influence of several hormones, the most important of which is prolactin secreted in large amounts after birth.

Milk is a complex liquid that provides all of the nourishment required by the baby. It contains proteins, fats, the sugar lactose, vitamins, salts, and water. Some of the proteins are maternal antibodies that are absorbed intact through the baby's intestinal epithelium, as mentioned in Chapter 11.

## Milk Ejection

In a lactating woman, milk is secreted more or less continuously, and it accumulates in the alveoli and ducts. The stimulus of nursing, or sometimes even the anticipation of nursing, triggers the release of oxytocin from the posterior pituitary. Oxytocin brings about contraction of the myoepithelial cells, squeezing milk into the lactiferous sinuses and making it available for the nursing baby. This process is termed *milk let-down.*

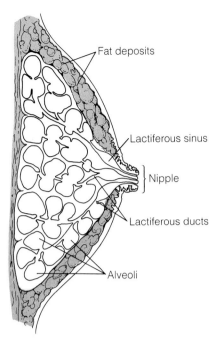

FIGURE 18–15 Mammary glands during lactation. After weaning, the alveoli regress.

## Maintenance of Lactation

The secretion of prolactin, necessary to maintain lactation, is triggered by nervous stimuli that are generated by nursing and act on the hypothalamus to prevent secretion of prolactin-inhibiting releasing hormone (see Chapter 16). In the control of milk production, demand determines supply. Thus, continual emptying of the mammary glands increases milk production. When nursing is discontinued and the breasts are no longer emptied, the structure of the mammary glands gradually returns to the prepregnancy state.

During lactation, the high levels of prolactin prevent hormonal cycling and ovulation in about 50 percent of nursing women. Although lactation is thus an effective birth control method *in some women,* it definitely is not so in others, contrary to popular belief.

## Diseases of the Breast

*Breast cancer* is the second most common form of cancer in women, with about 1 in 11 women developing the disease during their lifetimes. It usually begins as a growth in one of the mammary ducts, eventually forming a lump beneath the skin that most often is discovered upon self-examination. Untreated, breast cancer eventually spreads to the lymph nodes in the adjacent armpit and from there, to other organs. Women whose female relatives have had breast cancer, who have not breast-fed a baby, or who have a particularly late menopause are somewhat more prone to developing breast cancer than women without these risk factors. Moreover, the risk of breast cancer increases in all women with age. For women at risk and for women over 40, periodic X-ray examinations of the breast *(mammography)* are advised since these exams can detect tumors too small to be felt.

Treatment for breast cancer is now somewhat controversial. The standard treatment is *mastectomy* (removal of the breast), sometimes including removal of lymph nodes in the adjacent armpit *(modified mastectomy)* or removal both of lymph nodes and of underlying chest muscles *(radical mastectomy).* Which surgical option is chosen depends upon the size of the tumor and the degree to which it has spread. Radiation treatment or chemotherapy is often used following surgery, with the aim of killing any remaining cancerous cells. If the cancer has not spread to adjoining lymph nodes, 70 percent of patients are alive 10 years after such treatment.

More recently, studies in Italy and the United States have shown that with tumors less than three-quarters of an inch in diameter, a procedure involving surgical removal of only the tumorous lump *(lumpectomy)* followed by radiation therapy is just as effective in the treatment of these tumors as mastectomy. Consequently, some physicians are now recommending this procedure for small tumors, but most physicians feel that caution still justifies a mastectomy in all cases. If further studies confirm the success of more limited surgery and encourage adoption of the procedure, it is thought that some women who are liable to procrastinate in seeking treatment for fear of losing a breast will consult a physician earlier, substantially improving the results of treatment.

All lumps in the breast are not cancerous; in fact, 80 percent are not cancerous. Other lumps include benign tumors of various types, cysts (fluid filled, saclike tumors), and breast abcesses (infected areas in the mammary glands of lactating women), among others.

## BIRTH CONTROL

**Birth control** refers to any manipulation of the reproductive process that lowers the frequency of childbirth (Table 18–2).

### Preventing Union of Sperm and Egg

Until the development of oral contraceptives, most birth control methods aimed at preventing the union of sperm and egg without otherwise interfering with the reproductive process. These methods include the following:

1. *Vaginal diaphragm,* a latex cup that is inserted in the vagina and covers the tip of the cervix, thus preventing the entry of sperm into the uterus.
2. *Condom,* a latex sheath that covers the penis and sequesters the ejaculate.
3. *Withdrawal* of the penis from the vagina prior to ejaculation.
4. The *rhythm method,* which involves abstaining from sexual intercourse between Days 10 and 17 of the menstrual cycle when ovulation is statistically most likely to occur. The frequent failure of the rhythm method is due to the fact that the time of ovulation is often unpredictable.

TABLE 18–2 Effectiveness of Various Methods of Birth Control

| Method | Pregnancies/100 women/year | |
|---|---|---|
| | Theoretical effectiveness* | Use effectiveness** |
| Oral contraceptive (combined pill) | 0.34 | 4–10 |
| Condom plus foam | <1 | 5 |
| Intrauterine device (IUD) | 1–3 | 5 |
| Condom | 3 | 10 |
| Diaphragm with jelly or cream | 3 | 2–20 |
| Spermicidal foam or cream | 3 | 2–30 |
| Withdrawal | 9 | 20–25 |
| Rhythm | 13 | 21 |
| None | 90 | 90 |

Source: Hatcher, R.A., et al (1984) *Contraceptive Technology 1984–85,* 12th edition, Irvington Publ. Inc., New York.

*Maximum effectiveness when used correctly.

**Effectiveness in actual use conditions.

5. *Jellies, foams,* and *creams* designed to kill sperm. Introduced into the vagina, such spermicidal products have a high failure rate when used alone. However, when used in conjunction with a diaphragm, condom, or the rhythm method, spermicides increase the effectiveness of these methods.

6. *Postcoital douche,* intended to wash sperm from the vagina. This method is virtually ineffective, mainly because success requires that it be done within a minute or so of ejaculation.

Another way of preventing sperm from reaching the egg is *surgical sterilization.* This is a permanent method of birth control that prevents the transport of eggs from the ovary or sperm from the seminiferous tubules. In males, the vas deferens is cut and ligated in a local operation called a *vasectomy* that takes 15 minutes or so in a physician's office. After a vasectomy, seminal fluids are still produced and ejaculation occurs normally, but the ejaculate contains no sperm. Surgical sterilization of women is most simply done by sealing off the Fallopian tubes with electrical cauterization in a local operation that requires only overnight hospitalization.

It is important to note that surgical sterilization in no way disrupts normal hormone production or reproductive function; it merely physically obstructs the transport of sperm and eggs. For this reason, surgical sterilization is seen as a desirable option by some individuals. However, it should be used only by those who are sure they do not want children, since it has proven difficult or impossible to regain fertility by surgically reuniting the altered ducts.

## Oral Contraceptives

*Oral contraceptives,* known collectively as *"the pill,"* prevent ovulation by interfering with the normal release of FSH and LH. The most frequently used oral contraceptive is the *combination pill* containing small amounts of both synthetic estrogen and progesterone. The pills are taken from the fifth to the twenty-fourth day of the cycle and then discontinued for seven days, during which a few days of light menstrual bleeding will occur.

Oral contraceptives are virtually 100 percent effective if used properly. Some evidence suggests that women over 35 who take oral contraceptives are somewhat more likely to die of heart attacks and strokes due to intravascular clotting than other women of the same age. Serious side effects in women under 30 are rare. Women starting on "the pill" may experience nausea, vomiting, fluid retention, tender breasts, and headache, but these problems generally subside after a short time. Only time will tell whether the disruption of normal hormone patterns brought about by oral contraceptives has serious consequences as yet unnoticed.

## Prevention of Implantation

Several birth control methods are designed to prevent implantation of the fertilized egg. For example, a "morning after" pill containing synthetic estrogens can be taken within 72 hours of sexual intercourse and continued for five days. This treatment reduces progesterone production, making the uterus inhospitable for implantation. However, the "morning after" pill is only appropriate for emergency situations such as rape, since the estrogens cause nausea and vomiting and are potentially carcinogenic.

*Intrauterine devices,* or IUDs, also interfere with implantation. In this method of birth control, a small metal or plastic device is inserted by a physician into the uterus. The presence of the device in the uterus causes changes in the endometrium that discourage implantation. IUDs cause increased menstrual bleeding, pain, and bleeding between periods in some women. In addition, they have been associated with severe pelvic infections (pelvic inflammatory disease) that may cause sterility after IUD removal and an increased rate of ectopic pregnancies. Because of these problems, the use of IUDs has been discouraged in the United States in recent years.

## Abortion

*Abortion* is the artificial termination of pregnancy. It is not considered a preferred method of birth control, but it is legally available to those who choose it. In 1973, the United States Supreme Court affirmed the right of a woman to obtain an abortion up to the twenty-fourth week of pregnancy (22 weeks of development) and later in cases where the mother's life is in jeopardy as a result of the pregnancy.

The simplest type of abortion is called *suction curettage,* and it can be performed up to the tenth week of development (twelfth week after the last menstrual period). In this technique, a tube connected to a vacuum line is inserted into the uterus to remove the embryo and associated tissue. Between the tenth and eighteenth week of development, the preferred method of abortion is a D and E (dilatation and evacuation). It is a combination of the conventional D and C (dilatation and curettage, described in Chapter 17) and the suction method. When abortions are done after the thirteenth week of development, the risks of complications rise.

## Research in Birth Control

A great deal of research activity centers on attempts to devise new methods of birth control. Suppression of sperm production by drugs and the implantation of pellets that release steroid hormones and other contraceptive chemicals over long periods of time are but two of many possible birth control methods that are now being investigated.

## Study Questions

1. Describe the mother's role in the development of an embryo.
2. Outline the sequence of cellular events that occurs when a sperm cell penetrates and fuses with an oocyte.
3. Summarize the process by which the placenta develops.
4. a) What is the placental barrier?
   b) How effective is it?
5. Describe how antibodies to HCG are used in pregnancy tests.
6. Indicate the three stages of embryonic development and summarize the overall processes that occur in each stage.
7. Why are (1) the formation of the three germ layers and (2) the formation of the notochord particularly important events in embryonic development?
8. a) What is differentiation?
   b) How does it occur?
9. Explain why the developing embryo is more susceptible to damage by teratogens at some developmental stages than at others.
10. List the main causes of birth defects and give a specific example of each.
11. a) Describe how amniocentesis is performed.
    b) Other than ethical considerations, what serious drawback is there to amniocentesis?
    c) List three methods other than amniocentesis that can be used in prenatal diagnosis.
12. Describe the events that occur during the three stages of labor.
13. a) Distinguish between breasts and mammary glands.
    b) Are mammary glands exocrine or endocrine in nature?
14. Describe the controversy that has recently developed concerning treatment for breast cancer.
15. Describe the various methods of birth control and indicate the relative effectiveness of each.
16. What physiological variable accounts for the frequent failure of the rhythm method of birth control?

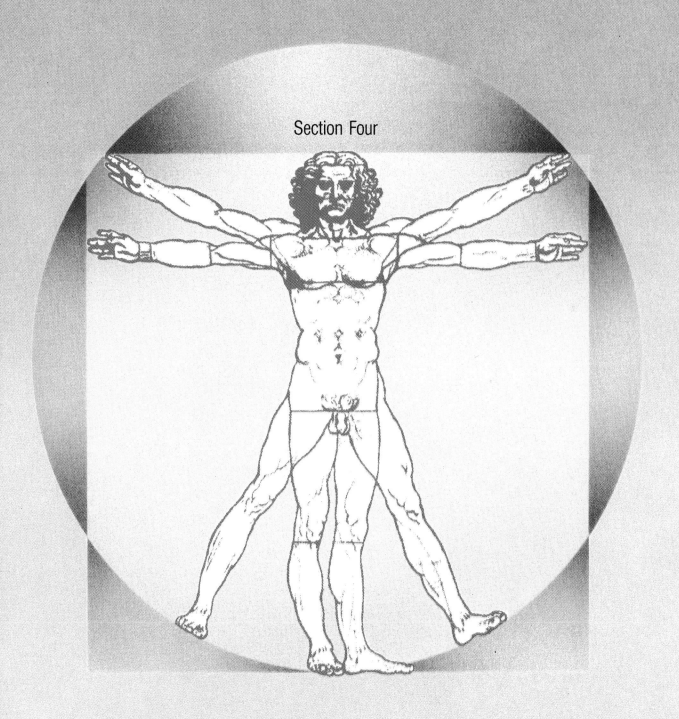

Section Four

# THE BLUEPRINTS OF LIFE

# Basic principles of heredity

Perhaps no area of human biology is so subject to misconceptions as is heredity. Because the rules that govern inheritance are not intuitively obvious, explanations for observed patterns of inheritance have been fabricated in the myths and superstitions of folklore. Consequently, most individuals do not understand how traits are inherited. Research in basic biology and medicine, however, has unraveled many of the basic mechanisms of heredity. Understanding these mechanisms provides us not only with the ability to understand how traits are inherited, but it allows us to predict the probability of appearance of particular traits. Moreover, it permits us to understand the basis of genetic diseases and suggests means for their treatment and prevention.

## THE CONCEPT OF THE GENE

### DNA: The Hereditary Material

In the middle 1940s, Oswald Avery, Colin MacLeod, and Maclyn McCarty performed experiments with the bacterium *Diplococcus pneumoniae* (also known as *pneumococcus*) which proved unequivocally that DNA is the cell's hereditary material. In these experiments, they used strains of pneumococcus that differed with respect to pathogenicity and cellular morphology. Pathogenic strains of pneumococcus cause bacterial pneumonia in mice and other organisms. Furthermore, the cells of pathogenic pneumococci are covered by a surface capsule of polysaccharide; the presence of this polysaccharide capsule is easily detected by the fact that these cells, when grown on the surface of a solid agar medium, form colonies with a smooth border. Because cells of all pathogenic strains of pneumococcus display a capsule and show similar growth characteristics, they are referred to as *S cells,* or *smooth cells.* Cells of nonpathogenic strains, on the other hand, are not encapsulated, form colonies with rough edges when grown on agar, and are termed *R cells,* or *rough cells.* The presence or absence of a capsule is a hereditary characteristic. Interestingly enough, the presence of a capsule determines pathogenicity, since phagocytic white blood cells cannot attach to the capsular material and hence cannot destroy the bacterial cells.

Avery, MacLeod, and McCarty extracted DNA from cultures of S cells and added the DNA to growing cultures of R cells. This treatment changed, or *transformed,* the R cells into S cells; that is, the previously unencapsulated, nonpathogenic R cells acquired the ability to form a capsule, becoming pathogenic (Figure 19–1). In related experiments, Avery, MacLeod, and McCarty were able to show that the added DNA, and not some protein or other macromolecular contaminant, was responsible for the transformation. We now know that the S cell DNA, a portion of which determines the synthesis of capsular material, is taken up by the growing R cells and permanently incorporated into their hereditary material.

### DNA, Genes, and Chromosomes

DNA molecules are composed of two strands of nucleotides, wound together in the form of a double helix. (The structure of DNA is discussed in detail in Chapter 20). The DNA of human cells is broken up into a number of separate molecules, each complexed with proteins to form *chromatin strands* that cannot be clearly distinguished in the nondividing interphase nucleus. The chromatin strands twist and coil during cell division to form **chromosomes,** each with characteristic morphology. Human somatic cells contain 23 pairs of chromosomes; the two members of a chromosome pair are termed *homologous chromosomes.* Of the 23 pairs of human chromosomes, there are 22 pairs, called **autosomes,** and two **sex chromosomes** that determine an individual's sex.

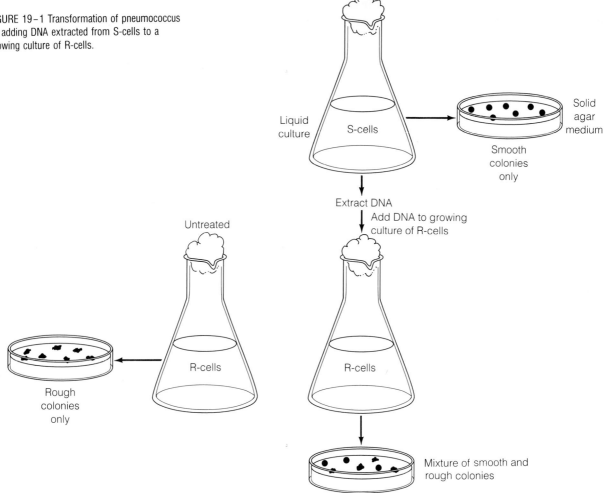

FIGURE 19–1 Transformation of pneumococcus by adding DNA extracted from S-cells to a growing culture of R-cells.

Liquid culture

S-cells

Solid agar medium

Smooth colonies only

Extract DNA

Add DNA to growing culture of R-cells

Untreated

Rough colonies only

R-cells

R-cells

Mixture of smooth and rough colonies

In females, the two sex chromosomes are identical in morphology and are referred to as **X chromosomes;** males have one X chromosome and a much smaller **Y chromosome** (Figure 19–2).

Linear sections of a DNA molecule form functional units of heredity called **genes.** Each gene determines the synthesis of a particular protein that endows the organism with specific structural and/or metabolic characteristics. Because genes are sections of DNA, genes are said to exist on chromosomes.

## HEREDITARY MECHANISMS

The first scientific work to establish the mechanism of heredity was published in 1866 by Gregor Mendel, an Augustinian monk in the town of Brno in what is now Czechoslovakia. On the basis of extensive experiments on the transmission of hereditary traits in garden peas, Mendel deduced a set of genetic principles that laid the groundwork for all of genetic theory. Subsequent work has shown that Mendel's principles, forming the basis of a genetic system that is now referred to as **Mendelian genetics,** are generally applicable to all organisms that reproduce sexually.

### Mendel's Principles

### Genes in Pairs

Mendel's experiments led him to the conclusion that units of heredity, (now called *genes)* exist in pairs. For the vast majority of traits (with rare exceptions in the male, noted a little later), an individual has two genes, one inherited from each parent. Although Mendel knew nothing about the relationship between genes and chromosomes, we now know that a particular gene is a section of DNA located at a specific region, or **locus,** on a particular chromosome, and that the gene's pair exists on the homologous chromosome at the same location. Since at fertilization, an individual receives one member of each pair of chromosomes from each parent, the individual will inherit two genes for each trait.

## Dominance and Recessiveness

Another of Mendel's principles states that a gene can exist in alternative forms. We now call these alternative forms **alleles.** In some cases, but not in all, one allele of a gene pair is expressed, masking the effect of the other allele. In this case, the former allele is said to be **dominant,** while the latter is referred to as **recessive.** In other cases, neither allele masks the effect of the other, a condition termed **incomplete dominance.**

For a single pair of genes, the two alleles may be identical or they may be different. If they are identical, the individual is said to be **homozygous** for that particular gene. On the other hand, if the two alleles are different, the individual is **heterozygous** for the gene pair.

## Segregation of Alleles

As a third principle, Mendel proposed that the members of a gene pair segregate, or separate, randomly from one another during the formation of germ cells. Thus, any particular germ cell will receive any one member of a gene pair, but not both.

This principle is easily understood by reference to what we know about the formation of germ cells. During meiosis, each germ cell receives any one member of each pair of homologous chromosomes. Furthermore, the particular chromosome of a pair that a germ cell receives is randomly determined, based merely on how the maternal and paternal chromosomes happen to orient during metaphase I (see Figure 17–3). Since each member of a homologous pair of chromosomes contains one member of a gene pair, the two genes of a pair will segregate from one another during meiosis.

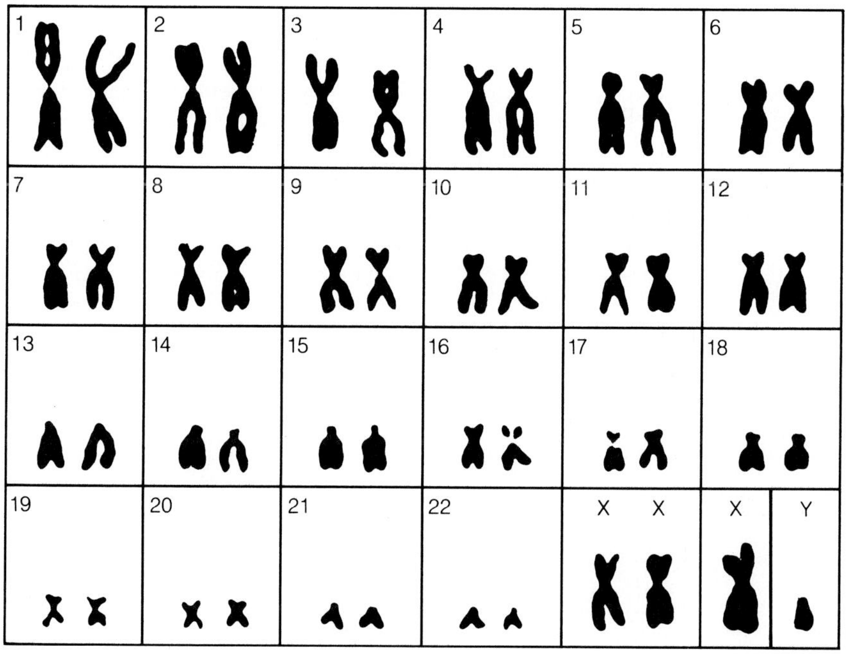

FIGURE 19–2 The chromosomes of human somatic cells. Females and males both contain pairs 1 through 22; in addition, females contain two X chromosomes, whereas males contain one X and one Y chromosome.

A number of hereditary traits are determined by single gene pairs that are located on one of the homologous pairs of 22 autosomes. These traits, termed *simple autosomal Mendelian traits,* may be determined by alleles that are recessive, dominant, or incompletely dominant.

## Albinism: An Autosomal Recessive Trait

Let us illustrate the inheritance of a simple autosomal trait by examining the transmission of **albinism,** a trait appearing in about 1 in 38,000 whites and 1 in 22,000 blacks in the United States. An individual with albinism, referred to as an *albino,* synthesizes little or no melanin, the pigment in skin, hair, and eyes. As a result, albinos have very pale skin, white hair, and eyes that are pink or the palest of blue. Their pigmentless skin is highly susceptible to sunburn and prone to develop skin tumors; without protective pigment within the eyes, the retina is easily damaged by strong light, leading to vision problems.

Of several forms of albinism, the most common results from an absence of a functioning tyrosinase enzyme that is needed in the pathway of melanin synthesis (see Figure 20–9). Individuals with this form of albinism are homozygous for a recessive allele of the normal gene that determines the synthesis of a functional tyrosinase enzyme. Indicating the alleles for normal pigmentation and albinism in standard genetic notation, we can designate the normal dominant allele by a capital letter, such as *A,* and its recessive allele by a lower case of the same letter, such as *a.* With regard to this one trait, albinos thus have a genetic constitution, or **genotype,** of *aa;* they have inherited a recessive allele from both parents.

In matings between an albino individual and a normal individual with no family history of albinism, all children have normal pigmentation. The reason is as follows: the albino parent has a genotype of *aa;* when germ cells are formed by this individual during meiosis and the chromosome number is halved, all germ cells will contain a single *a* allele. The normal parent, on the other hand, possessing a genotype of *AA,* will produce germ cells containing the *A* allele. Children of these two individuals, formed by the union of sperm and egg cells (containing paternal and maternal chromosomes, respectively) will have the genotype *Aa.* Because the *A* allele is dominant over the *a* allele, the effects of the *a* allele will be masked, and all children will have normal pigmentation. However, the genotype of the children (*Aa*) differs from that of the normal parent (*AA*), and consequently the children are *carriers* of the recessive allele. A **carrier,** by definition, possesses a recessive allele that is not expressed in that individual, but can be passed on to progeny.

The appearance of an individual with regard to a genetic trait is referred to as the individual's **phenotype** (*pheno* = show). In the situation discussed above, the normal parent and all children thus have a normal phenotype (although their genotypes differ), while the affected parent has an albino phenotype. Figure 19–3 summarizes the genotypes and phenotypes in the mating between an albino and a normal individual. Notice that the progeny would have the same genotypes and phenotypes, no matter which parent is the albino.

Let us now examine the consequences of matings between two carriers or between a carrier and an albino. In the first instance, both carriers have the

FIGURE 19–3 Summary of the results of matings between a normal and albino individual. $F_1$ refers to the first filial generation.

|  |  |  |
|---|---|---|
| AA × aa | | Parental genotypes |
| Normal   Albino | | Parental phenotypes |
| pigmentation ↓ | | |
| All Aa | | $F_1$ Genotype |
| All Normal Pigmentation | | $F_1$ Phenotype |

genotype *Aa*. When germ cells, sperm or eggs, are made by either individual, the two alleles will segregate randomly from one another. Because either allele has an equal probability of entering a germ cell, about half of the germ cells of both individuals should possess the *A* allele, while the other half should possess the *a* allele. Since it can be shown that progeny result from random encounters between sperm and egg cells, then in matings between heterozygotes (carriers), about 25 percent of the progeny will have *AA* genotypes, 50 percent will have *Aa* genotypes, and 25 percent with have *aa* genotypes. Phenotypically, 75 percent of the progeny will have normal pigmentation and 25 percent will be albino. These results are diagrammed in Figure 19–4. A genotypic ratio of 1:2:1 and a phenotypic ratio of 3:1 is characteristic of the inheritance seen in matings between two carriers of a recessive allele.

Mating results such as those shown in Figure 19–4 can be predicted in a simpler, alternative way by constructing what is termed a *Punnett square*. By writing possible genotypes of the sperm on the top of a square table and those of egg cells on one side, proportions of progeny can be determined by diagramming the various combinations of sperm and egg cells (Figure 19–5). Results are of course identical to those shown in Figure 19–4.

A Punnett square for matings between a carrier and an albino is shown in Figure 19–6. In this case, the genotypic and phenotypic ratios are identical, both 1:1.

In all matings just diagrammed, reciprocal crosses (in which the genotype of the male and female parent are switched) yield identical results. Furthermore, the various genotypes and phenotypes occur with equal frequency among male and female progeny. Both circumstances are characteristic of genes located on autosomal chromosomes.

## Other Autosomal Recessive Traits

In addition to albinism, there are almost 500 other known autosomal recessive traits in humans that are determined by single gene pairs. Several of the more well known of these are Tay-Sachs disease (see Chapter 2), phenylketonuria (see Chapter 20), and cystic fibrosis, a disabling disease characterized by the production of a thick mucus that clogs respiratory passages and the ducts of digestive glands, hindering respiration and digestion. Another trait determined by an autosomal recessive gene is blue eyes, which occur when melanin pigment in the front layer of the iris is lacking.

Although most deleterious autosomal recessive traits are rare in the overall population, they are more frequent in marriages between relatives (consan-

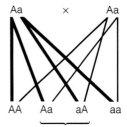

FIGURE 19–4 Genotypic and phenotypic ratios among progeny of two individuals, both heterozygous for albinism.

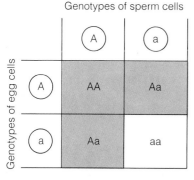

Genotypic ratio is 1AA:2Aa:1aa
Phenotypic ratio is 3 normal:1 albino

FIGURE 19–5 A Punnett square showing predicted genotypes in matings between two carriers of the recessive allele for albinism. Shaded squares indicate phenotypically normal individuals.

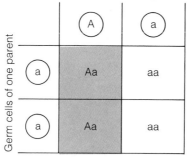

Genotypic ratio is 1Aa:1aa
Phenotypic ratio is 1 normal:1 albino

FIGURE 19–6 Punnett square of a cross between an albino individual and a carrier of the recessive allele for albinism. Shaded squares indicate individuals with normal phenotypes.

guineous marriages). This is because descendants of a common ancestor that possessed a particular recessive trait are more likely to be carriers of the trait than are other individuals in the population. Consequently, consanguineous marriages tend to bring together the recessive alleles.

### Autosomal Dominant Traits

About 450 other human traits are known to be caused by dominant autosomal genes. Some harmless traits caused by dominant autosomal genes are wooly hair (hair that is tightly kinked and very brittle), dimpled cheeks, clockwise rotation of the hair whorl at the top back of the head, the ability to roll the tongue into a U-shape without using the lips, a white forelock, and a protruding lower lip, called the Hapsburg lip (Figure 19–7). *Brachydactyly*, or short-fingeredness, was the first dominant genetic trait to be recognized in humans. One severe disease caused by a dominant gene is Huntington's chorea, characterized by mental and motor deterioration (see Chapter 14). Because the onset of Huntington's chorea is delayed, with symptoms first appearing between the ages of 30 and 45, many people carrying the gene for Huntington's chorea have already reproduced by the time they are aware they have the disease.

One characteristic of many dominant genes is that they vary widely in the degree to which they are expressed in different individuals. For example,

FIGURE 19–7 Inheritance of a protruding lower lip in members of the Hapsburg Dynasty. (a) Emperor Maximilian I (1459–1519); (b) Maximilian's grandson, Emperor Charles V (1500–1558); (c) Archduke Charles of Teschen (1771–1847); and (d) Charles of Teschen's son, Archduke Albrecht (1817–1895).

(a)

(b)

(c)

(d)

Abraham Lincoln is thought to have had a mild form of Marfan's syndrome, a hereditary connective tissue disorder caused by a dominant gene; the same disorder was fatal at an early age in one of his brothers (see *Focus: Connective Tissue Diseases* in Chapter 4).

## Incomplete Dominance

Incomplete dominance, in which neither allele is dominant over the other, is characteristic of two of the alleles that determine the ABO blood group (the genetics of other blood groups will be discussed in Chapter 21). Three alleles, usually denoted as $I^A$, $I^B$, and $I^O$, are involved in the inheritance of the ABO blood group. These three alleles are located at the same locus on a particular autosome, and every individual, of course, has a pair of ABO alleles. The $I^A$ and $I^B$ alleles determine the presence, respectively, of A and B proteins on the surface of red blood cells; the $I^O$ allele is associated with the absence of A and B surface proteins.

$I^A$ and $I^B$ are dominant over $I^O$. An individual with $I^A I^A$ or $I^A I^O$ genotype has A type blood; $I^B I^B$ and $I^B I^O$ individuals belong to blood type B. $I^O I^O$ individuals are of blood type O. However, in an individual with the genotype $I^A I^B$, both genes are expressed, resulting in the production of both proteins A and B on the red blood cell surface. Thus, the $I^A$ and $I^B$ alleles show incomplete dominance toward one another, and their presence together determines the AB blood type.

Blood types are often used as evidence in paternity suits. If paternity is disputed between two men, it is sometimes (but not always) possible to rule out one or the other as the father. Suppose, for example, that a woman with type A blood gives birth to a child with type AB blood. Since the mother has type A blood, her genotype must be either $I^A I^A$ or $I^A I^O$. Her child with AB blood must have inherited the $I^A$ allele from her and the $I^B$ allele from its father. Consequently, the father must have had blood type B ($I^B I^B$ or $I^B I^O$) or AB ($I^A I^B$) and could not have had blood type A or O ($I^O I^O$).

## Sex-Linked Inheritance

## Sex Determination

In addition to the 22 pairs of autosomes, each human cell has two sex chromosomes. In the male, these are an X chromosome and a much smaller Y chromosome. Females, on the other hand, have two X chromosomes and no Y chromosome.

In the formation of gametes during meiosis, the two sex chromosomes segregate from one another to enter daughter cells. Because there are 2 X chromosomes in female cells, all egg cells receive a single X chromosome. Sperm cells, however, receive either an X or a Y chromosome, and they do so in approximately equal numbers. At fertilization, it is the chromosomal constitution of the sperm that determines the sex of the newly formed individual: if a Y-containing sperm fertilizes an egg, a male results, whereas fertilization by an X-containing sperm results in a female. An interesting historical note on the genetics of sex determination is that Henry the Eighth should have blamed himself instead of his wives for his lack of male offspring. The prediction of sex ratios in families is discussed in *Focus: Predicting Sex Ratios in Families*.

## Sex Differentiation

The gonads are first recognizable during the fifth week of development as bilateral thickenings of epithelium at the edges of the primitive kidneys. In the male, these early gonadal structures are sexually indifferent until the seventh week of development when they begin to differentiate into testes, and soon the rest of the male accessory reproductive organs begin to form. The situation is somewhat different in the female, in which the ovaries differentiate from the

# Predicting Sex Ratios in Families

Calculating the probability of the occurrence of an event is a confounding exercise, for the outcome often seems counterintuitive. The prediction of sex ratios in families is no exception. Most people, for example, are under the impression that if a couple already has four girls, they have an exceedingly good chance of having a boy on the next try. Their reasoning is based upon the assumption that a family of five girls and no boys would be a very unlikely occurrence. Although the latter assumption is correct, it has no bearing on the former situation, and in fact the popular impression is wrong.

Confusion results from the difference between calculating probabilities as you go (on each child) or in advance. When a couple has its first child, there are two possibilities: either it will be a girl or it will be a boy. If we assume that equal numbers of girls and boys are born, an assumption that is roughly true, then the chance that the first child will be a girl is one out of two (1/2). Each time the couple has a child, the chance of its being a girl is 1/2. The odds of having a girl at each birth remain the same, since each birth is an independent event; that is, the sex of previous children has no effect on the sex of subsequent ones. It is just like tossing a coin: at each toss there are two possibilities (heads or tails), and the result of each toss is independent of other tosses. Thus, if a couple has had four girls in a row, the chance that the fifth child will be a boy is 1/2, precisely the same as is the chance that it will be a girl.

Calculating the probabilities of particular sex ratios in an *anticipated* family is, however, an entirely different matter. Mathematically, we can show that the probability of the occurrence of two independent events is the product of the probabilities of the separate events. Thus, in an anticipated family of five children, the chance that it will consist of five girls is one out of thirty-two (1/32), calculated as follows. The chance that the first child will be a girl is 1/2; the chance that the second child and each subsequent child will be a girl is also 1/2. Consequently, the chance of having five girls is the product of the separate probabilities, or

$$1/2 \times 1/2 \times 1/2 \times 1/2 \times 1/2 = 1/32.$$

Thus, if you are anticipating a family of five children, the probability that you will have five girls is rather low. Once you have four girls though, the chance of having a fifth is comparatively good.

What is the probability of having this family of all boys?

undifferentiated gonads at about the eleventh or twelfth week of development and after the formation of the female accessory reproductive organs.

The development of male reproductive organs is controlled by a specific gene on the Y chromosome, the so-called *HY gene.* The HY gene produces a protein that triggers the development of the testes from the indifferent gonads. Once the testes develop, the secretion of testosterone and another hormonal substance brings about the differentiation of the male accessory organs. In the absence of the HY gene and the protein it produces, the indifferent gonads develop into ovaries. Thus, female development occurs at the appropriate time in the absence of stimulation by the HY protein.

In the United States, the sex ratio at birth is not equal to one. On average for American whites, 105.6 males are born for every 100 females; for American blacks, the ratio of males to females is somewhat less, 103.3 to 100. We do not know what accounts for skewing of the sex ratio. There is no evidence that meiosis yields other than equal numbers of X- and Y-bearing sperm. One hypothesis suggests that Y-bearing sperm may be more efficient at fertilizing eggs, perhaps because they are somewhat lighter than sperm containing the larger X-chromosome and hence can travel at a faster rate.

## Sex-Linked Traits

There are a number of genes located on the sex chromosomes, primarily on the X-chromosome, that have nothing to do with the determination of sex. The traits determined by these genes are called *sex-linked traits* because their inheritance is linked to the inheritance of sex. Unlike the case with pairs of autosomes, the X and Y chromosomes do not have many homologous regions, and consequently genes present on the X-chromosome do not generally have alleles on the Y chromosome, nor do genes on the Y chromosome have alleles on the X chromosome. Because of the general lack of paired alleles on the X and Y chromosomes, genes present on the X chromosome are termed *X-linked genes* and those on the Y chromosome are termed *Y-linked genes.* Y-linked genes of a father are transmitted to all of his sons, but to none of his daughters, simply because the father's single Y chromosome is always inherited only by sons. X-linked genes of the father are inherited by daughters but never by sons, whereas X-linked genes of the mother are transmitted to offspring of both sexes (Figure 19–8).

***X-linked Inheritance.*** Genes on the X-chromosomes can be either dominant or recessive. In females, recessive X-linked genes are only expressed in their homozygous state, but males express all X-linked genes, whether recessive or not, since the male has only a single X chromosome.

About one hundred abnormal traits are determined by genes on X chromosomes. Most of these genes are recessive. One well known X-linked trait is *hemophilia.* In the most common type of hemophilia, called *classic hemophilia,* blood clotting time is greatly slowed due to the lack of a functional form of a plasma protein known as *antihemophilic globulin* or *Factor VIII.* Individuals with hemophilia bleed profusely even from minor injuries. In addition to uncontrolled bleeding from open cuts, bruises of the skin or ruptures of capillaries in internal organs may result in massive internal bleeding, particularly into joints. Not long ago, hemophiliacs had a poor survival rate, but they now can be treated with injections of the missing clotting factor.

Although hemophilia is a rare disease, it was prevalent in a number of the royal families of Europe, transmitted by descendants of Queen Victoria, who was a carrier of the disease (Figure 19–9). Victoria had one affected son and two daughters who were carriers. Because there was no history of hemophilia in Victoria's ancestors, it is thought that the gene probably arose as a mutation in an X chromosome in cells of the germ line of one of her parents or in one of her X chromosomes during her embryonic life.

The transmission of rare recessive X-linked traits like hemophilia shows a characteristic pattern. It is expressed in a man, it is not expressed in any of his children, and it then reappears in some of his daughter's sons but not in any of his son's children. This somewhat complicated pattern of inheritance is explained in Figure 19–10. As can be seen, sons are either normal (if they have the normal gene) or hemophiliacs (if they have the recessive gene), whereas daughters are carriers. Daughters can only be affected if their mothers are carriers and their fathers are hemophiliacs, a very unlikely event. See *Focus: A Boy's Disease that Changed the Course of History.*

Another recessive X-linked trait that is relatively common is red-green colorblindness, an inability to distinguish red from green. In the United States, about 1 in 12 males and 1 in 250 females have some degree of colorblindness.

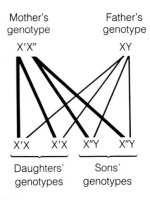

FIGURE 19–8 Inheritance of X and Y chromosomes by sons and daughters. The mother's two X chromosomes are designated X′ and X′′ to distinguish them from one another and from the X chromosome of the father.

# A Boy's Disease that Changed the Course of History

**A**ugust 12, 1904, was a day of celebration throughout Russia. Churchbells peeled and cannons boomed, announcing the birth of Alexis, the fifth child and only son of Tsar Nicholas II and his wife, Alexandra. Alexis was the first male heir to be born to a reigning Russian monarch for more than two hundred years. His birth seemed to portend a bright future for Imperial Russia and for the royal family.

The joy of his parents was soon to be tempered. When Alexis was six weeks old, he began to hemorrhage from his navel. The bleeding lasted for a full day and continued intermittently for two more days before it finally stopped. The suspicion that he had inherited hemophilia from his mother, a granddaughter of Queen Victoria of England, was confirmed in the months that followed. As Alexis began to crawl and then to walk, the slightest bruise or bump would turn into an ugly dark blue swelling filled with blood that had failed to clot as it flowed from tiny breaks in superficial blood vessels within the skin. Often, his joints would become swollen with blood that seeped into them. This caused excruciating pain, destroying bone, cartilage, and other joint tissues, and seriously weakening his limbs. Internal bleeding often did not stop until the pressure of the blood, confined internally, became great enough to prevent further bleeding from the broken blood vessels. The loss of blood, even from a minor bruise, was often so severe that Alexis was frequently at the brink of death.

When Alexis was three and near death from a bleeding episode that would not stop, his parents, in desperation, summoned Gregory Rasputin, a reputed holy man well known in St. Petersburg society. When Rasputin arrived, he began to talk to Alexis. "The child . . . began to bubble with laughter. Rasputin laughed too. He laid his hand on the boy's leg and the bleeding stopped at once. 'There's a good boy,' Rasputin said. 'You'll be all right.'"

During the next ten years, Rasputin was called upon frequently to heal Alexis during times of crisis, and his appearance often brought about dramatic improvement in the boy's health. It is not clear what Rasputin did. It has been proposed that by using hypnosis or suggestion, Rasputin was able to calm Alexis, which would tend to reduce bleeding.

Rasputin was a thoroughly disreputable character, a drunk and a womanizer, a holy man only in name. But because he could heal their son when no one else could, Nicholas and Alexandra thought of him as an extraordinary individual. "Rasputin took the empire by stopping the bleeding of the Tsarevich [Alexis]," wrote J. B. S. Haldane, the noted British geneticist. Rasputin came to have great influence on the royal family and particularly on Alexandra, who tolerated his meddling and heeded his advice on matters far removed from her son's health. He encouraged Alexandra to convince her husband to resist further demands for governmental reform which had begun in 1905 when absolute power had been wrested from the Tsar with the formation of the Dumas, the parliament which was formed to advise the emperor on legislation. By denying all calls for reform of the monarchy, Nicholas paved the way for the Russian revolution.

The Russian revolutionary, Alexander Kerensky, once said, "If there had been no Rasputin, there would have been no Lenin." And as historian Robert Massie adds: "If this is true, it is also true that if there had been no hemophilia, there would have been no Rasputin."

Rasputin.

Nicholas II with Alexandra and their children in 1905.

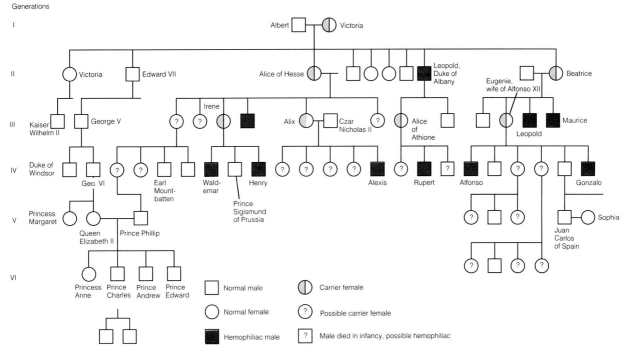

FIGURE 19–9 The pedigree of Queen Victoria, tracing the transmittance of hemophilia among members of the royal families of Europe.

***Y-linked Inheritance.*** In addition to the HY gene that determines the development of the male gonads, only one other Y-linked gene has been unambiguously identified. This gene determines the presence of heavy, stiff hairs on the rims of the ears (hairy ear rims), a trait that has no apparent physiological consequence.

## Sex-Influenced Traits

Some traits that are determined by autosomal genes show variable expression depending upon the sex of the individual. Such traits are termed *sex-influenced. Pattern baldness,* the loss of hair on the top of the head but not on the sides, is usually described as a sex-influenced trait.

The inheritance and expression of pattern baldness, as it is usually explained, is summarized in Table 19–1. If baldness is regulated by a single gene pair—a fact that inheritance patterns support—we can designate the gene for baldness as B′ and its normal allele as B. Homozygous BB individuals

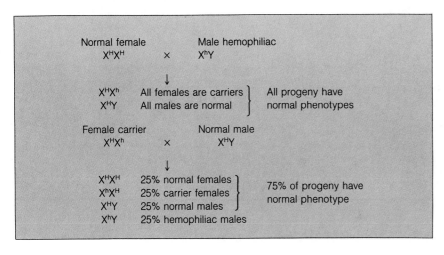

FIGURE 19–10 Inheritance of hemophilia, an X-linked trait. By convention, the superscripts represent genes present on the X chromosome. The normal gene for blood clotting is designated as H and its recessive allele for hemophilia as h. All X-linked recessive traits show a similar pattern of inheritance.

TABLE 19–1 Inheritance of Pattern Baldness, a Sex-Influenced Trait

| Genotype* | Sex | Phenotype |
|-----------|-----|-----------|
| BB | male | normal |
| BB | female | normal |
| B′B | male | bald |
| B′B | female | normal |
| B′B′ | male | bald |
| B′B′ | female | bald |

*B is the normal gene for hair growth, and B′ is its allele for baldness.

of both sexes have normal hair retention, whereas homozygous B′B′ individuals of both sexes develop baldness. However, heterozygous individuals (BB′) show different expression in the two sexes. In males, the heterozygote develops baldness, while heterozygous females have a normal phenotype. Presumably, the presence of testosterone or some other sex hormone or hormones in the male can interact with the genotype to produce baldness in heterozygous individuals.

It is important to note that there is not unanimous agreement among geneticists on how pattern baldness is expressed. Because pattern baldness appears to occur less frequently in women than would be predicted by the foregoing explanation, some geneticists feel that pattern baldness is not expressed in the female, whatever her genotype, except under abnormal physiological conditions in which testosterone levels are elevated. In these cases, baldness would presumably be expressed in females with both BB′ and B′B′ genotypes.

## Multifactor Inheritance

A number of characteristics are not inherited as single genes. Instead, a number of genes interact to produce a final phenotype. Inheritance of this type is called **multifactor,** or **polygenic, inheritance.** Multifactor inheritance determines traits such as height and skin pigmentation. For example, matings between individuals of medium height can produce progeny varying in height from very tall to very short. However, if a large number of progeny from such matings were examined, their heights would form a bell-shaped distribution curve, with most progeny of medium height and the fewest at the two extremes (Figure 19–11). Such a distribution pattern merely indicates that it is more likely for parents of medium height to have progeny of medium height, although other heights are possible.

FIGURE 19–11 The height distribution of progeny of individuals of medium height.

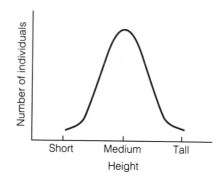

# CHROMOSOMAL ABNORMALITIES

Alterations in the number and structure of chromosomes in germ cells may result from spontaneous errors in meiosis and from exposure to certain environmental agents, such as ionizing radiation and some drugs.

## Alterations in Chromosome Number

If homologous chromosomes do not separate in anaphase of meiosis I or II (an event termed *nondisjunction*), the unseparated pair will enter one daughter cell whereas the other daughter cell will receive none of that chromosome type (Figure 19–12). The combination of either type of these abnormal germ cells with a normal gamete will result in a new individual with various phenotypic abnormalities due to an improper number of chromosomes in the zygote.

In the situation in which a gamete containing two chromosomes of one type and one chromosome of all other types unites with a normal gamete, the new individual will have three chromosomes of one type and a pair of all other types. This condition is termed **trisomy,** and particular types of trisomy are named for the chromosome that is tripled. Hence, individuals with trisomy-21 have three copies of chromosome 21 but a normal number (two) of all other

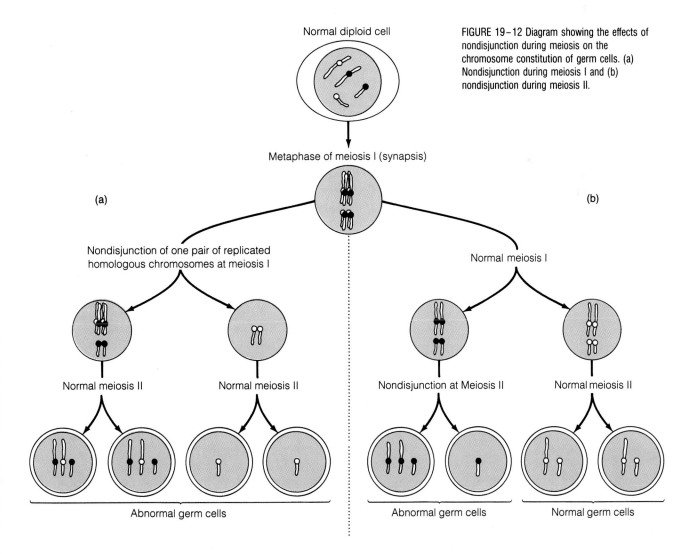

FIGURE 19–12 Diagram showing the effects of nondisjunction during meiosis on the chromosome constitution of germ cells. (a) Nondisjunction during meiosis I and (b) nondisjunction during meiosis II.

FIGURE 19–13 Karyotype of an individual with Down's syndrome. Notice the three copies of chromosome 21.

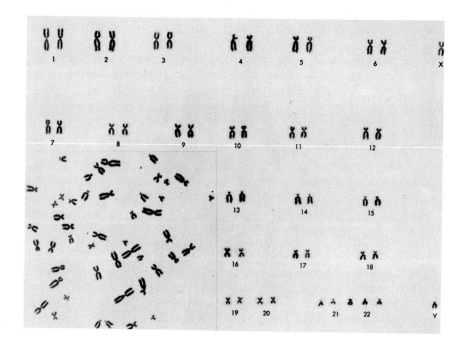

FIGURE 19–13 Karyotype of an individual with Down's syndrome. Notice the three copies of chromosome 21.

chromosomes (Figure 19–13). On the other hand, if an abnormal gamete lacking one kind of chromosome but possessing one member of all other pairs unites with a normal gamete, **monosomy** results, with particular types of monosomy in this instance named for the chromosome of which there is only one copy.

Trisomies and monosomies of autosomal chromosomes have severe effects on the phenotype, and most of these individuals are aborted spontaneously during early development, although some do survive. Trisomies and monosomies of the X chromosome and various other alterations in chromosome number involving the sex chromosomes are, in general, better tolerated, and their effects, while often profound, may not significantly affect survival.

### Abnormal Numbers of Autosomes

*Down's Syndrome.*   The most common type of individuals with abnormalities in the number of autosomes are those with *trisomy-21,* also referred to as *Down's syndrome* (see Figure 19–13). Individuals with Down's syndrome have a characteristic appearance, including a broad face with eyes slanting upwards at their outer corners, broad hands with short fingers and unique palm creases, and a short stature. They are invariably severely mentally retarded with an IQ that usually does not exceed 70. Many have circulatory and other physical abnormalities as well, and they are prone to leukemia and respiratory infections. With modern medical treatment, their life span averages about 40 years.

About 1 individual in 700 births has Down's syndrome, although the incidence of the condition is actually much higher, since as many as 1 in about 40 embryos that abort before 20 weeks are affected with Down's syndrome. However, the risk of having a child with Down's syndrome is not spread evenly among members of the population but increases dramatically with maternal age, particularly after about age 35 (Figure 19–14). Because of the strong correlation of Down's syndrome with maternal age, amniocentesis or chorionic villi sampling may be recommended for older women who are willing to have an abortion. Once thought to be independent of paternal age, evidence now suggests that men over 55 also have an increased risk of fathering a child with Down's syndrome.

Whether the defect is carried in the sperm or the egg, the cause of Down's syndrome appears to be nondisjunction of chromosome 21. It is not clear why nondisjunction is more likely to occur in older individuals than younger ones. Some investigators have proposed that older individuals on average have had greater exposure to environmental agents such as radiation and chemicals that increase the rate of nondisjunction.

*Trisomies 18 and 13.* Other trisomies, considerably rarer than Down's syndrome, have been reported. Two of these, trisomies 18 and 13, are both associated with severe mental retardation and a variety of physical defects.

## Abnormal Numbers of Sex Chromosomes

Nondisjunction of sex chromosomes during meiosis can result in individuals with other than the normal number of sex chromosomes. Nondisjunction of X chromosomes during anaphase of meiosis I or II results in the formation of some eggs with two X chromosomes and others with no X chromosome. If an XX egg is fertilized with a Y-bearing sperm, *Klinefelter's syndrome* results. Because they have a Y chromosome, individuals with Klinefelter's syndrome are males, but they do not develop fully male characteristics at puberty. Their sex organs are reduced in size, their body build is somewhat like that of females, and they may have some breast development. Spermatogenesis is abnormal, resulting in sterility. They also have an increased frequency of mental retardation. About 1 in 400 live male births have Klinefelter's syndrome.

Union of an XX egg with an X-bearing sperm yields a fertilized egg with *trisomy X*. Such individuals are females, many of whom are normal in appearance, fully fertile, and unaware of their trisomic condition. Others, however, have menstrual disturbances and may be sterile. Trisomy X occurs in roughly 1 in 1200 live female births.

If an egg with no X chromosome is fertilized with a Y-bearing sperm, a zygote with only a single Y sex chromosome results. The embryo aborts at an early stage, since genes on the X chromosome are needed for survival.

A no-X egg fertilized by an X-bearing sperm produces a female with *monosomy X,* or *Turner's syndrome.* Turner's syndrome occurs in 1 out of 2500 live births. At birth, individuals with Turner's syndrome have a thick fold of skin on either side of the neck, but they otherwise appear relatively normal. However, the female sex organs and secondary sexual characteristics do not

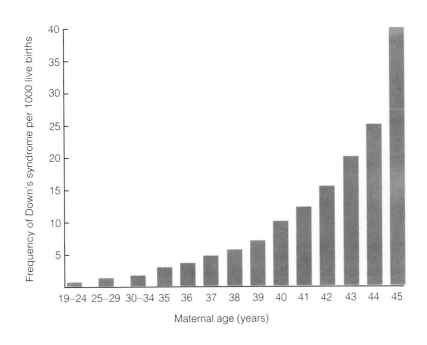

FIGURE 19–14 Risk of Down's syndrome in live births as a function of maternal age.

FIGURE 19–15 The results of nondisjunction of the sex chromosomes in males. In (a), nondisjunction occurs at meiosis I; in (b), nondisjunction occurs at meiosis II. Notice that synapsis of the X and Y chromosomes occurs only at their ends.

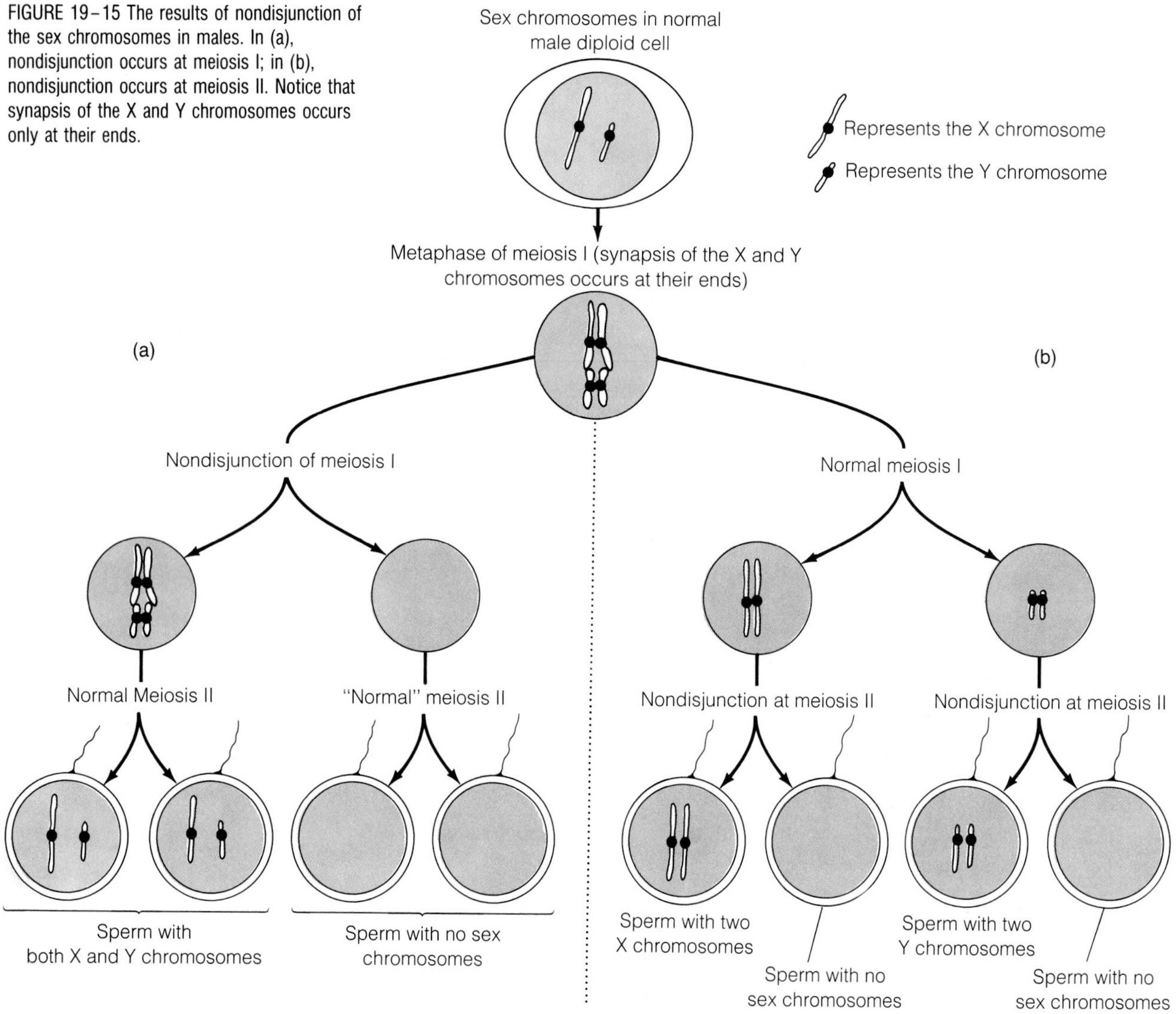

Sex chromosomes in normal male diploid cell

Represents the X chromosome

Represents the Y chromosome

Metaphase of meiosis I (synapsis of the X and Y chromosomes occurs at their ends)

(a)

(b)

Nondisjunction of meiosis I

Normal meiosis I

Normal Meiosis II

"Normal" meiosis II

Nondisjunction at meiosis II

Nondisjunction at meiosis II

Sperm with both X and Y chromosomes

Sperm with no sex chromosomes

Sperm with two X chromosomes

Sperm with no sex chromosomes

Sperm with two Y chromosomes

Sperm with no sex chromosomes

develop at puberty, remaining juvenilized. Consequently, such individuals are completely sterile.

Nondisjunction occurring in the male leads to development of sperm with an abnormal number of sex chromosomes (Figure 19–15). If the X and Y chromosomes do not separate at the first meiotic division of spermatogenesis, some of the resulting sperm will have both X and Y chromosomes and others will have no sex chromosomes. Fertilization of a normal egg with a sperm bearing both X and Y chromosomes produces an individual with Klinefelter's syndrome (XXY); fertilization of a normal egg with a sperm containing no sex chromosomes yields Turner's syndrome.

Nondisjunction occurring at meiosis II of spermatogenesis can give rise to some sperm with two Y chromosomes, some with two X chromosomes, and others with no sex chromosome (see Figure 19–15). If a YY sperm fertilizes a normal egg, an individual with *XYY syndrome* results. XYY individuals are males with below average IQ but above average stature. A great deal of controversy has surrounded studies which show that the prevalence of XYY individuals in institutions for the criminally insane is considerably higher than their prevalence in the general population. Consequently, it has been suggested that XYY individuals may be prone to violent criminal behavior, but this suggestion has not been adequately verified. We do know that the great majority of XYY males lead normal lives.

## Variations in Chromosome Structure

During crossing over in meiosis, chromosomes undergoing synapsis regularly break and recombine, an event that leads to a shuffling of the genes on maternal and paternal chromosomes (see Chapter 17). Errors in this normal event can fragment chromosomes in abnormal ways, significantly altering the structure of chromosomes in germ cells. Furthermore, ionizing radiation and drugs such as LSD are also known to fragment chromosomes (Figure 19–16).

Fragmentation of a chromosome may result in the *deletion* of a portion of the chromosome. Large deletions, in which many genes are lost, are likely to have severe, even fatal, effects for a developing embryo. Small deletions may cause minor defects or no obvious ones, but they do allow the expression of recessive genes present on the homologous chromosome in the same region as the deletion.

Chromosome structure can also be altered by a **translocation,** in which a section of a chromosome is transferred to another region on the same chromosome or to an entirely different chromosome. The movement of genes from their normal locations on chromosomes to other locations may significantly alter the expression of the translocated genes in a developing embryo. For example, it is now known that one form of chronic myelocytic leukemia (see Chapter 8) in children is associated with a translocation of a section of the long arm of chromosome 22 to the long arm of chromosome 9. Presumably, the translocation somehow disturbs normal controls on cell growth and differentiation, resulting in cancer.

## EXPRESSION OF THE PHENOTYPE

Although genes are the essential determinants of biological form and function, it is incorrect to think of the phenotype of an individual as a direct and invariable consequence of its genotype. The genotype represents a biological potential, but the final expression of that potential depends upon a complicated interaction between the genotype and the organism's environment, both internal and external. For example, an individual with a genotype that would otherwise determine large stature may attain only slight stature if nutritionally deprived during the growing period. In many cases, however, the relationship between genotype and the environment is less apparent. For instance, we have mentioned previously that a particular trait such as Marfan's syndrome may be fully expressed in some individuals but only partially in others, even though both individuals have the same genotype with regard to that trait. Indeed, there are many examples of this phenomenon.

Such variability in gene expression may be due to a variety of factors. We know that interactions between genes may have profound effects on gene expression. There are genes that function specifically to activate, suppress, or modify the expression of other genes. Furthermore, the position of a gene on a chromosome may influence its expression, as noted previously. Sometimes, the expression of one gene is required for the expression of another gene or a group of genes. Such sequential gene expression is crucial during embryonic development, and it is well illustrated in the case of the development of the eye, where expression of genes determining lens development is an absolute requirement for continued normal eye development. In this regard, a host of environmental agents including viruses, chemicals, and radiation may modify or interfere with normal gene expression.

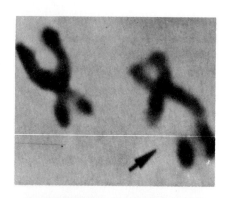

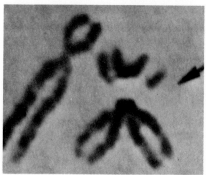

FIGURE 19-16 Chromosomal aberrations (arrows) in the leukocytes of a long-time user of LSD. More than 20 percent of this man's leukocytes had chromosomal aberrations.

A related aspect of phenotypic expression is the question of the relative importance of the genotype and the external environment on the development of complex human characteristics such as intelligence, personality, and behavior. This so-called nature-nurture controversy (*nature* = genotype; *nurture* = environment) has been at the center of much debate for the past 50 years or so. In humans, one of the only ways of obtaining experimental data on this question is to study the expression of traits in fraternal and identical twins raised together and separately. Since fraternal twins are genetically dissimilar and identical twins are genetically identical, it is often possible to show whether a trait has a genetic component. However, it has proven impossible to demonstrate the relative importance of heredity and environment in most instances. The current view of the nature-nurture controversy is that most complex human traits are influenced by both the genotype and the environment.

## GENETIC COUNSELING

Information about the symptoms and transmission of genetic diseases can be obtained from human geneticists and physicians who specialize in *genetic counseling*. Couples may be referred to genetic counselors if they are concerned about the possibility of having a child affected by a genetic disease. Frequently, such concern is generated by the fact that one or both of the partners have a particular genetic disease or a family history of the disease.

Understanding the mode of transmission of any genetic condition is a necessity for useful genetic counseling. With some diseases, a genetic counselor may be able to provide definitive information without even assembling a family history. For example, if a trait is known to be determined by an autosomal dominant gene, then one of the partners must be affected by the trait in order to pass it on to progeny; a history of the disease, even in the families of both parents is irrelevant. If, on the other hand, one parent is affected and one is not, then statistically half of their children will inherit the disease and half will not. This means that each child would have a 50 percent chance of having the disease.

In the case of a disease caused by an autosomal recessive gene, analysis of the situation may be more complex, and family pedigrees for the disease may be required. If one partner is affected (and hence is homozygous recessive) and the other partner has no family history of the disease, then it is usually safe to conclude that the normal partner is homozygous dominant. Hence, all of the couple's children would be carriers (heterozygotes), but none would be affected. If, however, both partners have a family history of the disease but neither is affected, both may be carriers. Identifying carrier individuals is important, since two carriers will have a 1 in 4 chance of having an affected child (see Figure 19–4). Sometimes, the issue of heterozygosity can be unequivocally resolved from family histories. For instance, an individual with an affected parent must be a carrier. In other instances, pedigrees will not supply the answer.

Genetic screening for heterozygotes is now possible for a variety of conditions, including such serious ones as Tay-Sachs disease, cystic fibrosis, and sickle cell anemia. With the results of these tests in hand, two heterozygous individuals may decide that the probability of having an affected child is too great for them to tolerate and will not choose to have children. Others may opt for amniocentesis or chorionic villi sampling (see Chapter 18) to screen for the disease, with the idea of aborting an affected fetus. Still others may feel that preventing conception or intervening with the pregnancy is not desirable, appropriate, or in conformity with their moral or religious values. However, even these latter individuals may find comfort in the knowledge provided by genetic counselors.

1. Describe how Avery, MacLeod, and McCarty proved that DNA is the hereditary material of the cell.
2. a) How do chromatin and chromosomes differ?
   b) What are homologous chromosomes?
3. Describe Mendel's principles.
4. Define a) genotype, b) phenotype, and c) carrier individual.
5. A couple has two phenotypically normal children. The couple's third child is an albino. What is the probability that their fourth child will be phenotypically normal? (Hint: see *Focus: Predicting Sex Ratios in Families*).
6. A woman with blood type AB has a child with blood type A. What are the possible genotypes of the father?
7. a) What is the HY gene?
   b) How is it important to sex differentiation?
8. a) From whom does a male hemophiliac inherit the gene for hemophilia? Explain.
   b) From whom does a female carrier of hemophilia inherit the gene for hemophilia? Explain.
9. Explain why there is no hemophilia in the present royal family of England.
10. a) What is nondisjunction?
    b) What is the significance of nondisjunction in germ cell formation?
11. a) Describe the chromosomal abnormality associated with Down's syndrome.
    b) What factors are associated with an increased risk of having a child with Down's syndrome?
12. Why do you suppose that abnormalities in the number of sex chromosomes generally produce less serious effects than abnormalities in the number of autosomes?
13. Two individuals plan to have a child. One has a history of Huntington's chorea in the family; the other does not. If you were a genetic counselor, what could you tell them about their chances of having a child with Huntington's chorea?

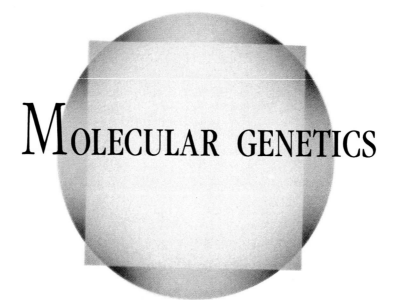

# MOLECULAR GENETICS

In many of the previous chapters, we have mentioned the central role of DNA as the cell's hereditary (genetic) material. We have noted that DNA fulfills this role by containing, in coded form, the information needed for the synthesis of all the proteins an organism possesses. These proteins, the linchpins of life's machinery, perform crucial functions as enzymes, structural components, regulatory molecules, transport molecules, and antibodies, among others.

DNA need only carry the information to synthesize proteins, for it is these versatile molecules that determine the metabolic capabilities and structural characteristics of the cells themselves and of the organism that the cells compose. Not contained in DNA are instructions for synthesizing other types of macromolecules or for performing countless other metabolic tasks. Such functions are determined by the presence of specific enzymes that catalyze these reactions; the enzymes, of course, are synthesized by information that *is* contained in DNA.

## EXPRESSION OF THE GENETIC MATERIAL

Although the role of DNA as genetic material was established in the middle 1940s, it was not until some twenty years later that we understood how information is coded in DNA and how it is decoded by the cell for use in synthesizing proteins. The "cracking" of the genetic code may be the single most significant breakthrough in our understanding of cellular function. Certainly its ramifications have been immense. Not only has it led to an understanding of fundamental mechanisms of heredity, but it has opened up an entire new area of research activity, termed *genetic engineering,* by which it has proved possible to transfer DNA experimentally from one organism to another.

Genetic engineering has only been done for about a decade, but in that time it has substantially increased our basic knowledge of gene structure and of the regulation of gene activity. In addition, it has led to a number of important practical developments, including the commercial production of medically important human proteins by bacterial cells, and it promises eventually to yield significant advances in the treatment of hereditary disease by the replacement of defective genes with normal ones. In order to discuss these important developments in biology and medicine, it is necessary first to understand how hereditary information is coded in DNA and decoded in specific patterns of protein synthesis. These and related topics that explain heredity in molecular terms form the field of *molecular genetics.*

### The Chemical Structure of DNA

You will recall from Chapter 2 that, as determined by James Watson and Francis Crick, the DNA molecule is composed of two polynucleotide strands, wound together in the form of a double helix and stabilized by hydrogen bonds between the two chains (see Figure 2–8). A polynucleotide chain consists of a linear sequence of nucleotides; each nucleotide, in turn, consists of deoxyribose, phosphate and a nitrogenous base (see Figure 2–13). Because the nucleotides in DNA have either adenine (A), guanine (G), cytosine (C), or thymine (T) as their nitrogenous bases, there are four different types of DNA nucleotides. Each polynucleotide strand is constructed in such a way that the deoxyribose and phosphate portions of the nucleotides lie on the outside of the helix, forming the backbone of the polynucleotide chain; the nitrogenous bases of each nucleotide lie on the inside of the helix (Figure 20–1).

The hydrogen bonds that hold the strands together in a helical configuration occur between nitrogenous bases that lie opposite one another on separate strands, and these hydrogen bonds always form in specific ways. Adenine on one strand always forms a hydrogen bond with thymine on the

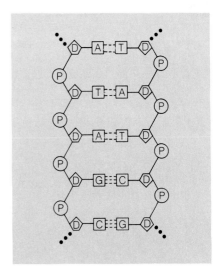

FIGURE 20–1 Diagram of the chemical structure of a portion of a DNA molecule. D, deoxyribose sugar; P, phosphate; A, T, G, and C are the nitrogenous bases adenine, thymine, guanine, and cytosine, respectively. Hydrogen bonds (dashed lines) between nitrogenous bases on opposite strands hold the two strands together.

other strand, and guanine on one strand always forms a hydrogen bond with cytosine (see Figure 20–1). As a consequence of these specific base-pairing rules, the sequence of nitrogenous bases (or nucleotides) on one polynucleotide strand specifies the sequence of nitrogenous bases on the other strand. Because of this arrangement, the two polynucleotide strands are said to be *complements* of one another, and the paired bases are termed *complementary bases.*

## DNA Replication

When Watson and Crick first published the molecular structure of DNA in 1952, they noted that the specific base pairing between nitrogenous bases could provide a mechanism for the replication of DNA. They proposed that each of the two strands in a DNA molecule, if unwound from the other, could serve as a pattern, or *template,* for the synthesis of a new strand, since each nucleotide along the strand would have a natural affinity, because of possible hydrogen bond formation, with a nucleotide containing the complementary base. Thus, for example, if a nucleotide containing cytosine were present on one of the original strands, it would attract the nucleotide containing guanine, and so on. The covalent linkage of the attracted nucleotides to one another, end to end, would then result in the production of a new strand. In this type of replication, a newly replicated DNA molecule would consist of one of the original strands (a parental strand) and a newly synthesized strand. Since half of the original parent molecule is conserved upon replication, this type of replication is termed *semiconservative.*

Semiconservative replication of DNA was confirmed with the discovery of *DNA polymerase enzymes* that are capable of catalyzing the synthesis of new polynucleotide strands, using another strand as a template. They do this by linking nucleotides together, one at a time and sequentially down the strand, once the proper nucleotides are attracted to the DNA template. During replication, the entire parental DNA molecule does not unwind before synthesis begins; instead, as small areas of the helix are unwound progressively down the molecule, the new DNA strand is synthesized (Figure 20–2).

## The Genetic Code and Protein Synthesis

Protein synthesis occurs in two stages. In the first stage, termed **transcription,** the genetic code—the sequence of nucleotides in a polynucleotide strand of DNA—is transferred to an RNA molecule, termed **messenger RNA (mRNA).** RNA, like DNA, is also composed of a linear sequence of nucleotides that differ in minor ways from DNA nucleotides, as described shortly. Because the nucleotide sequence of the mRNA is determined by the sequence of nucleotides in DNA, the genetic code becomes encoded as a nucleotide sequence in RNA. Once the code has been transferred to mRNA, DNA is no longer involved in the process of protein synthesis.

In the second stage of protein synthesis, mRNA leaves the nucleus and migrates to the cytoplasm, where the code it contains is decoded as specific sequences of amino acids in proteins. The actual manufacture of proteins, using the coded information in mRNA, is called **translation,** and it occurs on ribosomes in the cytoplasm.

## Transcription

Because it was known that the DNA of eukaryotic cells is confined to the nucleus and proteins are made in the cytoplasm, it was not immediately obvious how DNA was involved in protein synthesis. As a solution to this problem, the French biologists Francois Jacob and Jacques Monod suggested in 1961 that there must be a chemical "messenger" that transports the information coded in DNA to the cytoplasm. Shortly thereafter, messenger RNA was isolated.

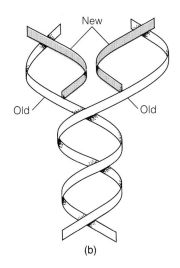

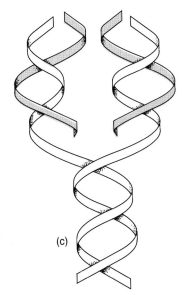

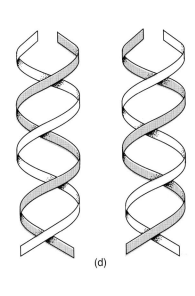

FIGURE 20-2 Replication of DNA. As the original molecule unwinds, two new strands are synthesized, using each of the original strands as templates.

(a)

(b)

(c)

(d)

New

Old

Old

Messenger RNA, like most other RNAs, is a single polynucleotide strand, composed of four different types of nucleotides arranged in linear sequence. The nucleotides contain ribose, phosphate, and one of the four nitrogenous bases: adenine (A), guanine (G), cytosine (C) or uracil (U). See Figure 2–13 for a review of RNA structure.

The synthesis of mRNA during transcription is catalyzed by *RNA polymerase,* an enzyme that functions much as DNA polymerase in the replication of DNA. RNA polymerase binds to one of the strands in a DNA helix at specific *promoter regions* that serve as initiation sites for transcription. The binding of RNA polymerase causes a localized unwinding of the DNA helix, and the RNA polymerase begins to travel along the nucleotide strand. As it does so, it constructs a mRNA strand by linking together nucleotides containing nitrogenous bases that are attracted by hydrogen bond formation to opposite bases on the DNA strand. In the base pairing that occurs, A, G, T, and C on the DNA strand form hydrogen bonds, respectively, with U, C, A, and G on RNA nucleotides. During mRNA synthesis, the DNA helix is unwound only at the site of transcription, and the helix reforms after RNA polymerase passes. Transcription continues until the RNA polymerase encounters a specific nucleotide sequence on the DNA strand that signals termination of synthesis. Apparently, only one of the two DNA strands contains useful information that is transcribed in this way.

The final result of transcription, summarized in Figure 20–3, is the synthesis of a mRNA strand possessing a base sequence that is complementary to that of the DNA strand from which the RNA is transcribed. Hence, the information contained in the nucleotide sequence of DNA is preserved in its mRNA transcript.

## Translation

Once mRNA becomes attached to a ribosome in the cytoplasm, the code in mRNA is translated to produce proteins with specific sequences of amino acids. Clearly, if the sequence of nucleotides in a mRNA molecule determines the sequence of amino acids in a protein, then specific nucleotides, or groups of nucleotides, must somehow specify the incorporation of particular amino acids into proteins during protein synthesis, just as a sequence of three dots in Morse code specifies the incorporation of the letter S into a word.

## Codons

A great deal of research has demonstrated that a sequence of three nucleotide bases constitutes a code word, or **codon,** which specifies one amino acid.

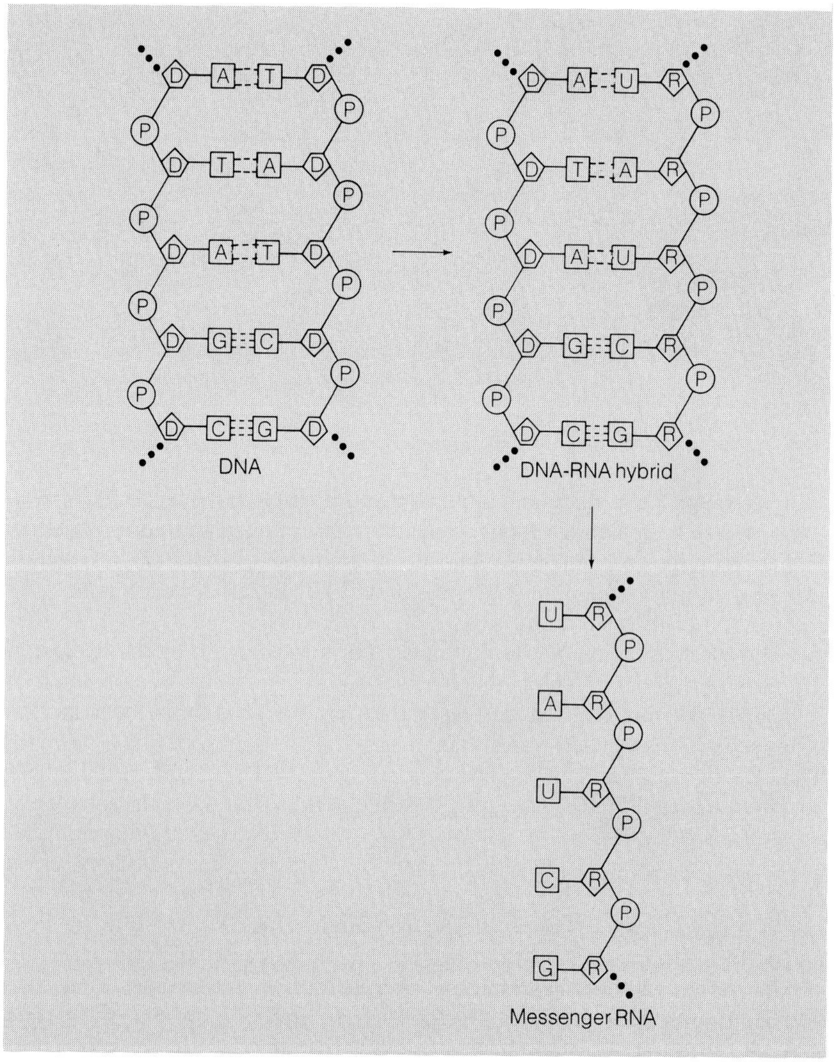

FIGURE 20–3 Stages of transcription, the process by which messenger RNA is synthesized. In actuality, mRNA is synthesized progressively down the DNA chain at regions where the DNA molecule is transiently unwound. D, deoxyribose; R, ribose; P, phosphate; A, T, U, G, and C are the nitrogenous bases adenine, thymine, uracil, guanine, and cytosine, respectively.

Thus, for example, the sequence AAA in mRNA specifies the amino acid lysine, and the sequence UUU specifies phenylalanine. Since a codon consists of three bases, each of which may be either A, G, C, or U, 64 possible codons exist ($4 \times 4 \times 4 = 64$). Of these, 61 code for amino acids, and three signal termination of protein synthesis (as discussed shortly). Because there are only 20 amino acids in naturally occurring proteins, many amino acids are designated by more than one codon; hence, the genetic code is said to be *degenerate*. The code is also *universal*; that is, it is the same in all organisms.

## Transfer RNA and Anticodons

At the ribosome, the mRNA codons are read sequentially, beginning with an *initiator codon* near one end of the mRNA. The amino acids specified by each codon are brought to the ribosome and coupled with peptide bonds. However, particular codons do not directly attract their corresponding amino acids to the ribosome. Instead, a special class of RNA molecules, termed **transfer RNAs (tRNAs),** transport amino acids to the ribosome for assembly into proteins. Each of the 20 amino acids in proteins is transported by a different transfer RNA molecule that is named for the amino acid it carries. For instance, alanyl-tRNA carries alanine, histidyl-tRNA carries histidine, and so forth.

Although the different types of tRNAs have unique nucleotide sequences, they have many structural similarities. All consist of a short polynucleotide strand that bends back upon itself at several points, allowing the formation of hydrogen bonds between some of the base pairs. The hydrogen bonding endows tRNA molecules with a three-dimensional structure containing three nucleotide loops; consequently, the entire structure has been likened to a cloverleaf (Figure 20–4). The crucial structural difference between different tRNA molecules is the nucleotide sequence in one of the looped areas, called the *anticodon loop.* This loop contains a sequence of three bases, forming the **anticodon,** which is specifically involved in the recognition of mRNA codons.

The important point is that each different tRNA possesses a unique anticodon loop and transports a specific amino acid. Thus, there is a correspondence between the amino acid carried by a tRNA and the anticodon that the tRNA possesses. For example, the tRNA that carriers lysine (lysyl-tRNA) has the anticodon UUU; histidyl-tRNA has the anticodon GUG.

Amino acids become attached to their respective tRNAs in an energy-requiring process that is catalyzed by specific enzymes in the cytoplasm. A tRNA molecule that is carrying an amino acid is said to be *charged,* whereas a tRNA without an attached amino acid is *uncharged.*

The three bases forming the anticodon of a tRNA molecule recognize mRNA codons by base pairing with them. For example, the anticodon UUU of a charged lysyl-tRNA molecule base pairs with a AAA codon in mRNA. Such specific base pairing between anticodons of tRNAs and codons in mRNA, occurring at the ribosome, serves as the mechanism by which specific amino acids are brought to the ribosome to be linked into protein.

## Initiation, Chain Elongation, and Termination

A ribosome contains two binding sites, called the *P site* and the *A site,* that can each receive a charged tRNA molecule. In the first step of translation (Figure 20–5), the mRNA orients itself so that the initiator codon (AUG) is positioned at the P site and the second codon in the sequence is positioned at the A site. Charged tRNAs are inserted at the P and A sites; these tRNAs must, of course, have anticodons that can base pair with the mRNA codons positioned at the two sites.

In the next step, a peptide bond is formed between the two amino acids by the transfer of the amino acid on the tRNA at the P site to the amino acid that is still attached to the tRNA at the A site. The tRNA at the P site, having lost its amino acid, leaves the ribosome. The tRNA at the A site, now carrying the first two amino acids of the protein that is being synthesized, moves to the empty P

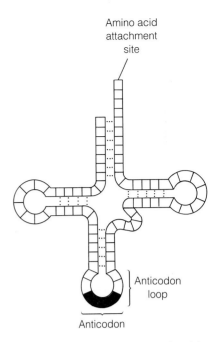

FIGURE 20–4 Diagrammatic representation of the structure of a transfer RNA molecule, showing the position of the anticodon (colored boxes) in the anticodon loop and the site where a specific amino acid attaches. Each box represents one nucleotide; dashed lines are hydrogen bonds between complementary nitrogenous bases.

FIGURE 20–5 Summary of the steps involved in the translation of mRNA at the ribosome to form protein.

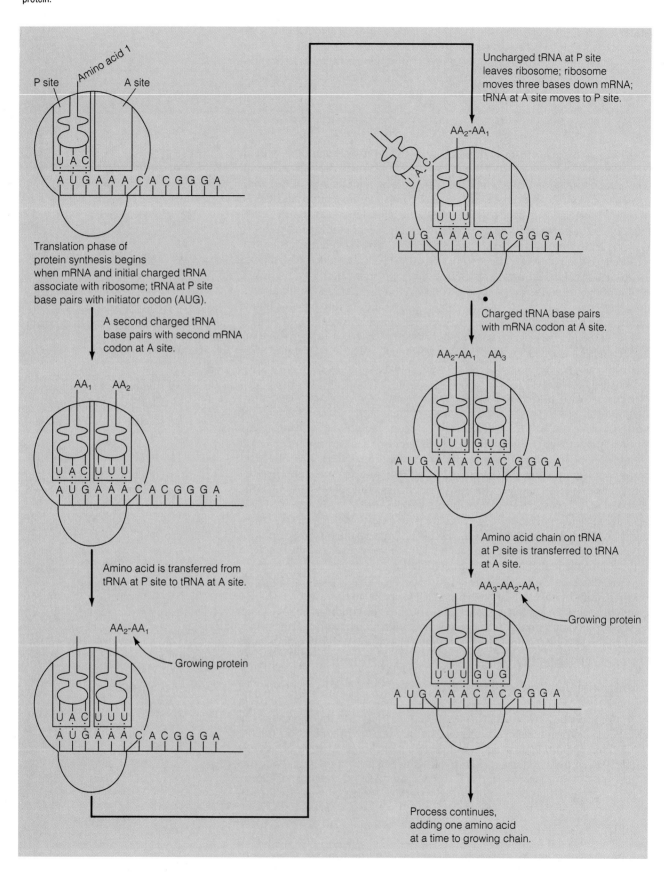

P site   Amino acid 1   A site

U A C

A U G A A A C A C G G G A

Translation phase of
protein synthesis begins
when mRNA and initial charged tRNA
associate with ribosome; tRNA at P site
base pairs with initiator codon (AUG).

A second charged tRNA
base pairs with second mRNA
codon at A site.

AA₁   AA₂

U A C U U U

A U G A A A C A C G G G A

Amino acid is transferred from
tRNA at P site to tRNA at A site.

AA₂-AA₁

Growing protein

U A C U U U

A U G A A A C A C G G G A

Uncharged tRNA at P site
leaves ribosome; ribosome
moves three bases down mRNA;
tRNA at A site moves to P site.

AA₂-AA₁

U A C

U U U

A U G A A A C A C G G G A

Charged tRNA base pairs
with mRNA codon at A site.

AA₂-AA₁   AA₃

U U U G U G

A U G A A A C A C G G G A

Amino acid chain on tRNA
at P site is transferred to tRNA
at A site.

AA₃-AA₂-AA₁

Growing protein

U U U G U G

A U G A A A C A C G G G A

Process continues,
adding one amino acid
at a time to growing chain.

site, and the ribosome moves three nucleotides down the mRNA. This movement positions a new mRNA codon (the third codon) at the A site, to which the appropriate charged tRNA can bind. At the next step, the growing amino acid chain, attached to the tRNA at the P site, is transferred intact to the amino acid carried by the tRNA at the A site, forming a peptide bond.

A continuation of this process adds amino acids, one by one, to the growing chain. Note particularly that during protein synthesis, the growing chain remains bound to a tRNA molecule.

Termination of synthesis occurs when a *termination codon* (UAA, UGA, or UAG) is encountered by the ribosome. Since there are no tRNAs that recognize those codons, no new amino acids can be added to the chain. Instead, the termination codons are recognized by protein *release factors*. The release factors activate an enzyme that splits the chemical bond between the last amino acid of the completed protein and the tRNA to which it is attached.

## THE MOLECULAR BASIS OF GENETIC MUTATION

Because of the crucial role DNA plays in normal function, survival depends upon accurate maintenance of the information stored in DNA. Since the information in DNA derives from its nucleotide sequence, this sequence must be reliably conserved in order to prevent serious malfunction. Although cells have a number of mechanisms to assure that DNA is replicated faithfully and that damage to DNA is repaired, on occasion, the nucleotide sequence of DNA becomes altered. Such alterations are termed **mutations** (= changes).

### Types of Mutations

Changes in the base sequence of DNA are likely to change the information encoded in DNA. Such changes may be of several types (Figure 20–6). The *substitution* of one base for another, also termed a *point mutation,* may cause one amino acid to be substituted for another; on the other hand, a point mutation that has the effect of changing one codon to another that codes for the same amino acid will not change the encoded information. Other mutations

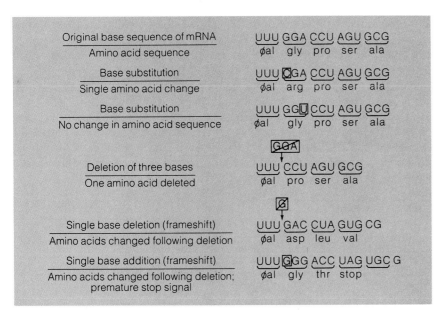

FIGURE 20–6 Illustration of several mechanisms by which mutations in DNA, by changing the bases in the mRNA transcript, can alter the sequence of amino acids in the protein product. Amino acid abbreviations are as follows: φal, phenylalanine; gly, glycine; pro, proline; ser, serine; ala, alanine; asp, aspartic acid; leu, leucine; val, valine; arg, arginine; thr, threonine.

involve the *deletion* or *addition* of one or more bases. If three bases, or some multiple of three bases, is deleted from or added to the DNA, then whole codons will be deleted or added, and the end result will be the deletion or addition of specific amino acids in the protein product. However, the deletion or addition of a number of bases other than three or a multiple thereof will shift the reading frame of the mRNA transcript. Such *frameshift mutations* result in a change in all amino acids past the site of mutation. In another type of mutation, called a *translocation,* a section of DNA is moved from one region to another on the same or a different DNA molecule. Such position changes may break genes in two or may alter the expression of intact, translocated genes.

Deletions and additions, and particularly frameshift mutations, are likely to result in the production of a protein that is dysfunctional (see Figure 20–6). A single point mutation may severely affect the biological function of a protein product, but it more often has a less pronounced effect, if any at all. Whatever the mutation, its effect is much more likely to be deleterious than advantageous, since mutations result from *random* changes in DNA.

Mutations in somatic cells of the body have substantially different consequences than mutations in germ cells. Although somatic mutations are inherited by descendants of the mutated somatic cell, they do not accumulate in the population since they are not passed on to the next generation. However, somatic mutations may have deleterious effects on the individuals in which they occur. Among other effects, somatic mutations have been implicated in the development of cancer, and they may be involved in the initiation of aging processes. Germ cell mutations, on the other hand, are transmitted to progeny and consequently can affect the future survival of both our progeny and the human species.

## Causes of Mutations

Some mutations are a result of intrinsic processes that occur normally within cells, independent of external agents. Mutations of this type are termed *spontaneous mutations.* Spontaneous mutations occur in human germ cells at a rate that varies on average from 1 to 10 per million gametes formed.

Several causes of spontaneous mutations have been identified. Random molecular collisions between water molecules and nitrogenous bases in DNA can disrupt covalent bonds linking adenine and guanine to the backbone of DNA chains, resulting in the removal of an estimated 5000 of these bases per day from the DNA of each human cell. Other chemical reactions involving reactive compounds that are formed as a result of normal metabolism can convert the nitrogenous base cytosine to uracil. Faulty base pairing during replication may result in the substitution of one base for another. Such errors may be caused by defective DNA polymerases or the tendency of some nitrogenous bases to occur on rare occasions in alternative molecular forms that do not observe the normal base-pairing rules. In this latter case, a rare form of adenine can form a hydrogen bond with cytosine, and a rare form of thymine can form a hydrogen bond with guanine (Figure 20–7).

Most mutations are not spontaneous ones. Instead, they are caused by external chemicals or other external agents called **mutagens;** these mutations are termed *induced mutations.* Many different mutagens have been identified. Although we do not understand how most mutagens cause mutations, we do know a good deal about the action of some of them.

## Chemical Mutagens

A number of chemicals react with nitrogenous bases in DNA and modify them chemically. For example, nitrous acid causes point mutations by converting cytosine to uracil. Another group of well-studied chemical mutagens acts by slipping in between adjacent bases in the DNA molecule and distorting its geometry. When the DNA replicates, deletions and insertions may result.

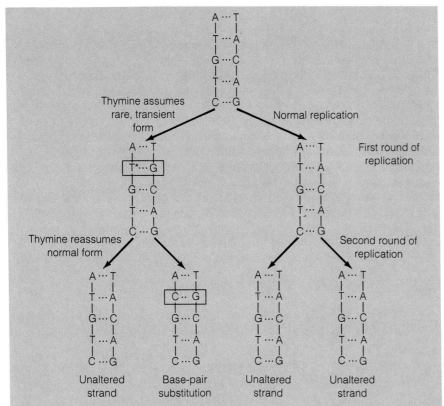

FIGURE 20–7 A mutation can occur during DNA replication when thymine assumes a rare, transient molecular form (T*) that permits it to pair, during synthesis of a new strand, with guanine, as shown in the first round of replication above. During the second round of replication, thymine will most probably assume its normal form, again pairing with adenine. These events convert an AT pair to a GC pair in one of the replicated molecules.

## Ultraviolet Light

Another well-studied mutagen is ultraviolet light, present in sunlight and in artificial sources such as the ultraviolet lights in tanning booths. When two thymine bases lying next to each other on the same polynucleotide strand absorb ultraviolet light, they may become covalently linked, forming what are termed thymine dimers. By reducing the distance between adjacent nucleotides, the thymine dimers may cause additions or deletions during replication.

## Ionizing Radiation

X rays, cosmic rays, gamma rays, alpha- and beta-particles, and other types of high energy radiation are able to penetrate tissues. When they do so, they may remove electrons from neutral molecules with which they collide, forming ions or highly reactive fragments, known as *free radicals,* that contain unpaired electrons. Ions and free radicals may react with and change the nitrogenous bases in DNA, causing point mutations, and they may break the DNA molecule, causing additions, deletions, and translocations. It has been shown recently that certain substances including vitamins A, C, and E, and salts of selenium and zinc may protect cells against the effects of ionizing radiation by preventing the formation of free radicals.

In recent years, there has been considerable debate concerning the genetic risks of exposure to high and low levels of ionizing radiation. Although some evidence suggests that all levels of ionizing radiation — even the lowest — can cause mutations in DNA, other data suggest that there is a threshold level below which genetic damage does not occur. Until the debate is resolved, most biologists contend that it is prudent to avoid all unnecessary radiation, no matter how low the dose.

## Genetic Repair Mechanisms

The foregoing discussion has indicated that damage to the DNA of human cells is extensive. Including both spontaneous and induced events, thousands of changes occur in the DNA of each cell every day. If all of these changes resulted in permanent mutations, the result would soon be disastrous. However, we now know that the vast majority of changes in DNA are repaired before they become permanently established at the time of replication. This function is performed by naturally occurring *repair systems;* these are so efficient that only a few permanent mutations accumulate in each cell every year.

Repair systems are able to repair a wide range of damage to DNA. For example, thymine dimers produced by ultraviolet light are removed from DNA in several ways. One repair enzyme, which requires visible light for its activation, can cleave the covalent bonds between thymine dimers. In another repair system, an enzyme recognizes and excises the thymine dimers along with a few neighboring nucleotides. The gap in the nucleotide strand is then filled in by DNA polymerase, which uses the undamaged complementary strand

FIGURE 20–8 Summary of the excision-repair of DNA damaged by the formation of thymine dimers.

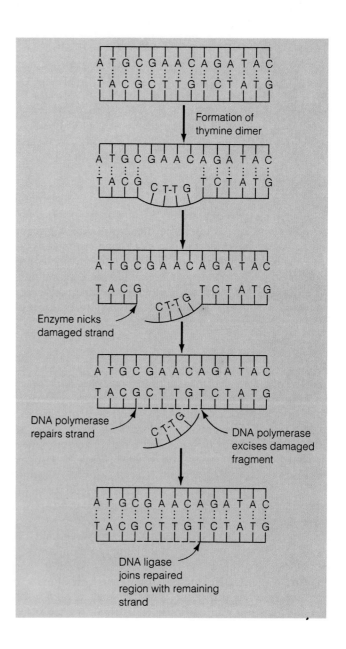

as a template, and by another enzyme, *DNA ligase*, which repairs the break in the polynucleotide strand (Figure 20-8).

The importance of genetic repair systems is illustrated by the human disease *xeroderma pigmentosum*. In this rare condition, caused by autosomal recessive genes, the skin is extraordinarily sensitive to ultraviolet light from the sun or from artificial sources. When individuals with xeroderma pigmentosum are exposed to the sun, they develop multiple cancerous skin tumors; the disorder is so serious that most affected individuals do not survive past their teens, although there has been considerable success recently in the suppression of tumor development by the meticulous use of protective clothing and sunscreen. A number of studies have shown that individuals with xeroderma pigmentosum lack one or more of the enzyme systems needed to repair ultraviolet damage to DNA.

## GENETIC DISEASE

Mutations in germ cell tissue give rise to a variety of inherited disorders *(genetic diseases)*. For many of these disorders, we now understand the end result of the mutation; that is, we have been able to identify an altered protein that is functionally defective and thus deprives the organism of some normal function. The discovery that many diseases are intrinsic to the organism and are not caused by external agents such as microorganisms was one of the most significant breakthroughs in medical science. Although many genetic diseases have been studied, we will consider as examples two particularly well understood groups: those involving abnormal metabolism of the amino acids tyrosine and phenylalanine, and those concerned with the synthesis of abnormal hemoglobins.

### Abnormal Metabolism of Tyrosine and Phenylalanine

Archibald Garrod, a physician in England at the beginning of this century, was the first to correlate a specific metabolic defect with a specific genetic condition. Garrod's classic book *Inborn Errors of Metabolism,* published in 1909, detailed his studies on the abnormal metabolism of the amino acids tyrosine and phenylalanine. One disease in this group that Garrod studied extensively was *alkaptonuria*. In this rare, autosomal recessive disorder, affected individuals are not able to break down homogentisic acid, also called alkapton, derived from the normal metabolism of phenylalanine and tyrosine; normal individuals, on the other hand, break down homogentisic acid to other products that are excreted (Figure 20-9). Individuals with alkaptonuria excrete homogentisic acid in their urine, and when the urine is exposed to air, it turns a black color due to the formation of oxidation products of homogentisic acid. The oxidation products also accumulate in cartilaginous tissues, causing a symptomless darkening of the ears, nose, and whites of the eye.

By supplementing the diet of affected individuals with protein or with the amino acids phenylalanine or tyrosine, Garrod was able to show that the concentration of homogentisic acid increased in the urine. These and other experiments led Garrod to the conclusion that the enzyme required for the breakdown of homogentisic acid was lacking. We now know that alkaptonuria is due to a mutation in the gene that codes for the enzyme homogentisic acid oxidase.

Two related disorders of tyrosine and phenylalanine metabolism are *albinism,* also studied by Garrod, and *phenylketonuria* (PKU), first described in 1934. As shown in Figure 20-9, albinism results from a defective enzyme in one of the steps by which DOPA is used to synthesize melanin pigments.

In phenylketonuria, phenylalanine cannot be converted to tyrosine, a reaction normally catalyzed by the liver enzyme phenylalanine hydroxylase,

FIGURE 20-9 Normal pathways (greatly simplified) involved in the metabolism of the amino acids phenylalanine and tyrosine. Mutations affecting enzymes that catalyze specific reactions in the pathways may block a pathway, causing various disease states including alkaptonuria, phenylketonuria (PKU), and albinism, as indicated here.

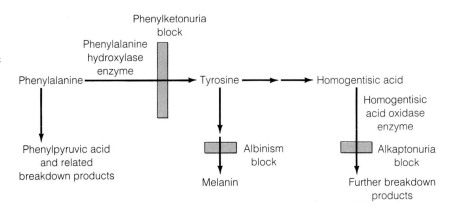

which is absent or deficient in affected individuals. The result is an increase in phenylalanine levels in the blood that may reach fifty times normal levels. In addition, accumulated phenylalanine may be degraded to phenylpyruvic acid and related compounds, as shown in Figure 20–9. In some manner that is not understood, high levels of phenylalanine or of its degradation products, if present during childhood, will adversely affect brain development, causing severe mental retardation. In order to prevent mental retardation, PKU must be discovered and treated very soon after birth. To this end, a simple blood test for PKU is required for newborns in all states of the United States. If an infant is found to have PKU, an occurrence in about 1 of 10,000 births, mental retardation can be averted by placing the infant on a diet low in phenylalanine. The result of the enzyme deficiency in PKU is thus overcome.

## Hemoglobin Disorders

Genetic mutations that result in abnormal hemoglobin production have been widely studied. One such mutation results in *sickle cell anemia,* discussed extensively in *Focus: Sickle Cell Anemia and the Importance of Cell Shape;* see Chapter 3. Individuals with sickle cell anemia have erythrocytes that become sickle-shaped at low oxygen levels. The sickle-shaped cells may clog capillaries and prevent normal oxygen flow to the tissues, causing severe tissue damage and pain. They are also more fragile than normal cells and tend to disrupt easily, resulting in severe anemia. Untreated sickle cell anemia is usually fatal in childhood, but with modern medical care many affected individuals survive beyond the age of 50.

Sickle cell anemia is inherited as a Mendelian trait in which the normal allele and the mutant, sickle cell allele show incomplete dominance toward one another (Table 20–1). Thus, individuals with two sickle cell alleles have sickle cell anemia, whereas those with one normal and one sickle cell allele have a mild form of the disease, called *sickle cell trait,* which rarely causes problems. The sickle cell allele is found primarily in people of African descent,

TABLE 20-1 Genotypes and Phenotypes of Individuals with Normal and Sickle Hemoglobin

| Genotype | Phenotype | |
| --- | --- | --- |
| | Hemoglobin | Disease State |
| AA | Normal | None |
| AS | Mixture of normal and sickle | Sickle cell trait |
| SS | Sickle only | Sickle cell anemia |

although it is also prevalent in some regions of Saudi Arabia, India, Greece, and in southern Italy. About 8 percent of American blacks carry one sickle cell allele and consequently have sickle cell trait; 1 in about 600 black infants in the United States are homozygous for the sickle cell allele and are therefore afflicted with sickle cell anemia.

The molecular basis of sickle cell anemia was revealed in work done by Linus Pauling and Vernon Ingram. (Remember that hemoglobin consists of two identical alpha chains and two identical beta chains, each with attached heme groups; these components are oriented in a specific three-dimensional configuration.) In 1949, Pauling demonstrated that the hemoglobin of individuals with sickle cell anemia differed in charge from that of normal individuals. Ingram, working in the middle 1950s, expanded Pauling's observation by showing that the charge difference between the two hemoglobins is a result of the substitution of one amino acid in the beta chain for another. The amino acid substitution has since been shown to be due to a point mutation in the DNA of affected individuals. Although the mutation results in the alteration of only one amino acid out of a total of 287 in the alpha and beta chains, the function of sickle cell hemoglobin is severely impaired. At low oxygen levels, sickle cell hemoglobin forms fibrous aggregates that lead to a collapse in the normal shape of the red blood cells.

There is an interesting sidelight to the story of sickle cell anemia. For a long time, biologists had questioned why a mutant gene with such severe effects was not lost from the population. They reasoned that without medical care, all individuals with sickle cell anemia would die at an early age without reproducing and that individuals with sickle cell trait, although only mildly affected, would be at a reproductive disadvantage to normal individuals. This latter assumption was proven incorrect when it was shown that in regions where malaria is a frequent cause of death, most individuals in the population have sickle cell trait. We now know that individuals with sickle cell trait have a high resistance to malaria, apparently because red blood cells containing a mixture of normal and sickle hemoglobin, as is the case in sickle cell trait, are not easily attacked by the malarial parasite. Thus, the potentially harmful sickle cell gene has survived in the population because it confers resistance to malaria when present in the heterozygous state.

In addition to sickle cell hemoglobin, over a hundred other hemoglobin variants have been detected in humans. Most of these involve a single amino acid change and are due to point mutations. Because most of these mutations are exceedingly rare, they are only seen in the heterozygous state and cause few, if any, symptoms.

## GENETIC ENGINEERING

The unraveling of the molecular structure of DNA and of the mechanism of protein synthesis led, in the middle 1970s, to the development of genetic engineering techniques, by which DNA from any organism could be introduced into bacterial cells. More recently, DNA has been transferred between various eukaryotic species.

### Techniques of the Genetic Engineer

*Recombinant DNA techniques* used in the transfer of foreign DNA to bacteria are merely test tube extensions of the natural processes by which DNA molecules break and recombine during mating in bacteria and during crossing-over in meiosis in eukaryotic organisms. By using naturally occurring bacterial enzymes that break DNA, it is possible to splice foreign DNA into that of bacteria.

Recombinant DNA techniques use strains of the bacterium *Escherichia coli* *(E. coli)* that possess in addition to the large, circular DNA molecule of all bacteria, small pieces of circular DNA called **plasmids.** Plasmids are on the order of a hundred times smaller than the main DNA molecule, and they replicate autonomously from it. They contain genes that, while not essential for normal survival and growth, provide the bacterium with the ability to live in a broader range of environments than would otherwise be possible. For example, plasmids contain genes that confer resistance to specific antibiotics and the ability to metabolize certain alternative nutrient sources.

In addition to a plasmid-containing *E. coli* strain, recombinant DNA techniques require the use of specific enzymes, termed *restriction enzymes.* Normally present in many bacteria, restriction enzymes function to cleave DNA, but they do not do so indiscriminately. Rather, each different restriction enzyme cleaves double-stranded DNA only at a specific nucleotide sequence that it recognizes. These *restriction sequences* consist of some four to eight nucleotide pairs that have a palindromic configuration; that is, the sequence of nucleotides in one DNA strand is the same as the sequence in the complementary strand when read in the opposite direction (Figure 20–10a). Because most restriction enzymes cut one DNA strand at one end of the restriction sequence and the other DNA strand at the other end, the cleaved DNA has short lengths of unpaired nucleotides at both ends (Figure 20–10b and c). In addition, the fact that restriction sequences are quite short means that any sequence is likely to occur, merely by chance, a good number of times in a large piece of DNA. Consequently, the DNA will be chopped by a particular restriction enzyme into a number of segments, all possessing unpaired, single-stranded extensions.

Restriction enzymes, isolated from *E. coli,* are used to incorporate foreign DNA into a plasmid, as summarized in Figure 20–11. First, a large number of identical plasmids containing a single specific restriction sequence are isolated and incubated in a test tube with the restriction enzyme specific for that sequence; the enzyme will open up the plasmid circle. Second, large pieces of foreign DNA containing many copies of the identical restriction sequence are also incubated with the same enzyme. This treatment will cut the foreign DNA into a number of pieces. Third, the opened plasmids are mixed in a test tube with the chopped up foreign DNA. Because the unpaired ends of the opened plasmids and of the pieces of foreign DNA are complementary to one another, the two ends of a piece of foreign DNA may become aligned, by hydrogen bonding, with the two ends of an opened plasmid. Such hybrid molecules are

FIGURE 20–10 The cleavage of a restriction sequence (shown in color) by a restriction enzyme. Restriction sequences are palindromic in the sense that the nucleotide sequence on one DNA strand (here, GGATCC) is the same as that on the complementary strand, when read in the opposite direction.

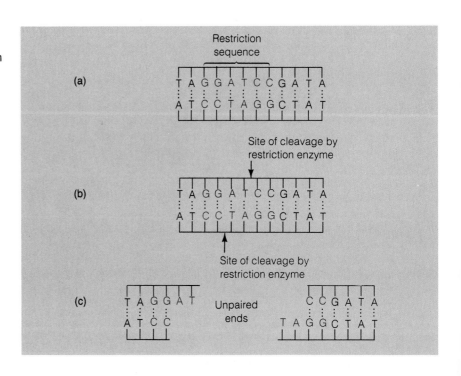

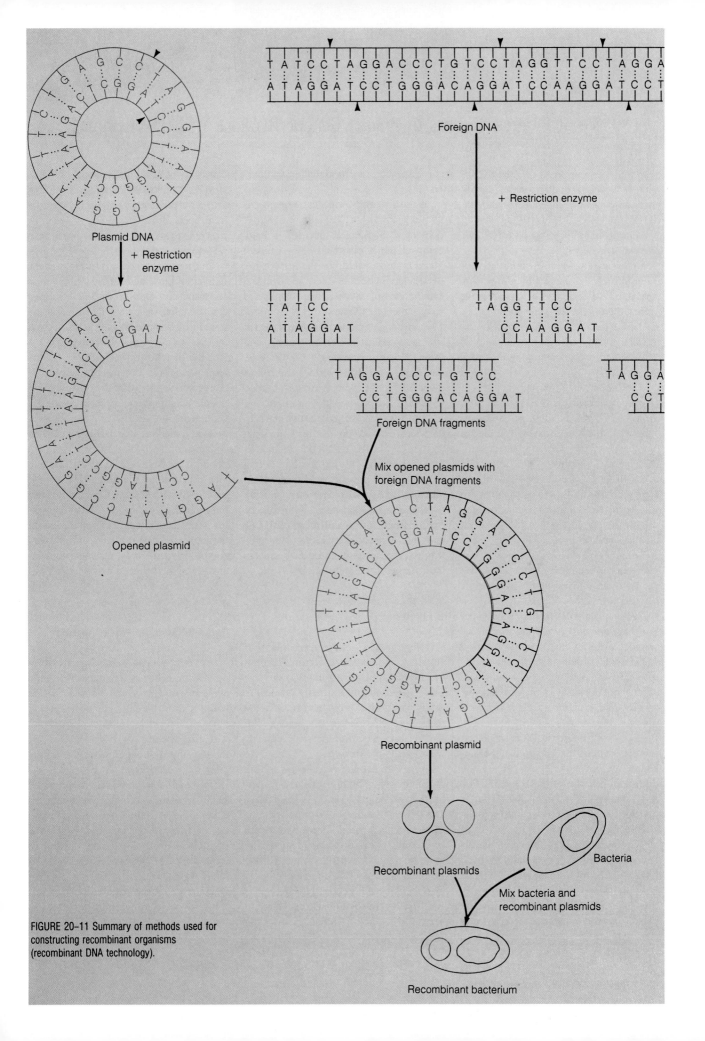

FIGURE 20–11 Summary of methods used for constructing recombinant organisms (recombinant DNA technology).

# Price of Progress: Efforts to Predict Genetic Ills Pose Medical Dilemmas

**W**ith impressive regularity, science is finding ways to identify apparently healthy individuals who are likely to develop a deadly or debilitating hereditary disease or to pass it on to their offspring.

Unfortunately, science's ability to spot these diseases is outstripping by many years its ability to do very much to prevent or treat them. This lag is creating complex medical, legal and moral dilemmas involving family relationships, confidentiality of medical data, proper counseling and much more.

The rapidly multiplying number of people caught in this time gap between diagnosis and possible therapy requires an "interim ethic," says LeRoy Walters, a bioethicist at the Kennedy Institute of Ethics at Georgetown University. He would like a set of principles that "allows large social institutions to avoid unreasonable harm, while at the same time protecting the liberty and welfare rights of individuals and families."

Congress's Office of Technology Assessment has been studying the subject for many months, and it plans to issue a report late this year [1987].

Recombinant DNA technology is largely responsible for identifying a score of disorders with a genetic cause or important genetic component, including diseases as disparate as Huntington's and manic depression. For many of these, predictive tests are in use; for others, the search for a valid test is continuing.

For a few diseases, such as sickle-cell anemia, scientists have located and isolated the defective gene that causes the disorder. Once the gene has been identified, researchers can begin figuring out its function and then start searching for possible therapy.

With many diseases, though, scientists have only identified a so-called genetic marker—an easily located bit of DNA that is so consistently inherited with the defective gene as to leave no doubt that the person has the gene itself. Huntington's

and cystic fibrosis are identified by markers.

In still other instances, research is pointing toward a "genetic predisposition" to such common disorders as hardening of the arteries and some cancers. These findings tend to be the least definitive so far, since they seem to involve a combination of several genetic factors and environmental "triggers."

Occasionally, particularly in the case of genetic predispositions, diagnosis and testing can dictate changes in life style, diet or drugs that can reduce the risk of disease. But for many serious inherited diseases, little or no treatment exists.

Until recently, genetic testing has had its widest impact on the birth process. Testing of worried would-be parents before conception or testing of a fetus during pregnancy can provide welcome reassurance for some at-risk couples, but a dreadful choice—to abort or not—for others. Now, though, more and more testing—and the quandaries that go with it—involves the health of individuals already born. Some of the questions:

## Who Wants to Know If Early Death Is Likely?

Symptoms of Huntington's disease usually don't appear until a person's late 30s or early 40s, but then lead to progressive nerve deterioration and early death. Recently developed pre-symptomatic tests, using a genetic marker, can predict with 96% accuracy whether a person eventually will develop the disease.

Surveys of people with a family history of Huntington's find high percentages saying they would like to know their chances. The test can remove stressful uncertainty, either freeing potential victims of worry or, if positive, allowing them to decide more intelligently whether to marry, have children, change jobs—in general, better plan their remaining healthy years.

"Huntington's is a diagnosis I hate to make," says Gilbert Omenn, dean of the

University of Washington School of Public Health. "But so many families thank me. They say, 'You have to understand what we have been through with the uncertainty. Now at least we can plan to deal with what we know is really there.'"

But testing programs under way at the Johns Hopkins Medical Institutions and Massachusetts General Hospital suggest that the percentage actually willing to be tested may be far smaller than the surveys indicate. Of 349 at-risk adults told by Johns Hopkins that the test was available, only 65 have asked for it. Because nothing can be done to prevent or delay the disease, many people apparently prefer to live with uncertainty than to learn they are doomed.

Some medical specialists worry that certain people who find out they're sure to develop a late-onset disease won't be able to handle the news. A number of people diagnosed with Huntington's try to commit suicide, and some succeed. "Uncovering genes that regulate human vulnerability to grave illnesses such as Huntington's chorea or Alzheimer's disease could increase the incidence of suicide as well as selective abortion," says University of Illinois bioethicist Marc Lappe.

## Can a Relative Be Forced to Help?

Genetic probes based on markers rather than defective genes require a "linkage" analysis involving blood or tissue samples from parents, grandparents and other close relatives, including at least one with the disease. An individual seeking the test must persuade these other family members to cooperate.

But what if a family member refuses? Will the person wanting the test go to court or try to compel the reluctant relative, or simply hate him or her for life? In the Johns Hopkins program, one man wanted the test, but his twin didn't. For a long while, the father of a young man wouldn't cooperate, arguing that his son wasn't up to coping with a positive test report.

"Suppose the person has cystic fibrosis, and you want to find out whether the siblings are at risk," says John Graham, a geneticist at Dartmouth Medical School. "You stir up a lot of guilt within the family as soon as you ask parents or grandparents for a blood sample. There's a lot of reluctance to cooperate."

Occasionally a sense of guilt or shame or some other factor makes a person testing positive reluctant to tell other relatives who might themselves be at risk and need to be tested. Philip Reilly, medical director of the Shriver Center for Mental Retardation, says the person being tested may have not only a moral but "a legal obligation to warn the appropriate relatives." He suggests a court might "permit a woman to sue her sister for failing to warn her about a pertinent reproductive risk, the knowledge of which would have given her an opportunity to avoid the birth of a severely impaired child."

What is a doctor's responsibility to these relatives if the patient refuses to tell them? "I leave it in the hands of the person who is consulting me," says Dr. Graham. "I think I am good enough to get the point across that the results are going to come out sooner or later, and that if the information isn't shared now, the eventual rift will be all the greater."

And if the doctor can't persuade the patient? In 1983, a presidential commission suggested that while confidentiality should be a general rule, doctors and other health-care providers should alert relatives at serious risk, and a number of medical officials agree. But Lori Andrews, an American Bar Foundation specialist, declares: "Many physicians want to inform relatives, but I caution them they're taking a big chance" of being sued for invading a patient's privacy.

### How Much Should Patients Be Told?

As with most medical procedures, a person considering genetic testing must give "informed consent" — not just agreeing to take the test but understanding just what the test shows and how reliably, the stresses of coping with a positive result, possible treatments, and the stresses that might result from not taking the test.

Explaining all this is demanding. The Johns Hopkins Huntington's program, for example, gives people extensive counseling before even asking for a decision on whether they want to take the test.

"Informed consent is an enormous problem," says Joan Marks, who oversees a Sarah Lawrence College program training graduate students to be genetic counselors. "If a patient doesn't understand what he is being told, how can he give informed consent? And many physicians are just not that great at communicating."

### What About Doctors Who Are Ill-Informed?

With proper counseling so critical and genetic diagnosis developing so fast, many specialists worry that the average doctor won't stay up-to-date and will fail to inform and guide patients adequately. Some may refer patients to genetic counselors,

| Decade of Discovery | Inherited Diseases | |
|---|---|---|

During the past 10 years, a genetic cause or important genetic ingredient has been discovered in a score of disorders. Some breakthroughs:

1978 Sickle-cell anemia; Beta thalassemia
1979 Alpha thalassemia
1981 Alpha-1-antitrypsin deficiency; growth hormone deficiency
1983 Huntington's disease; Duchenne muscular dystrophy; phenylketonuria (PKU)
1984 Hemophilia
1985 Cystic fibrosis; adult polycystic kidney disease; retinoblastoma
1986 Chronic granulomatosis
1987 Alzheimer's disease; manic depression

Rough estimates of the number of Americans afflicted by some of the diseases with a genetic cause or an important genetic component:

| Disease | Number Affected |
|---|---|
| Adult polycystic kidney disease (congenital kidney defect) | 300,000 to 400,000 |
| Sickle-cell anemia | 50,000 |
| Cystic fibrosis | 30,000 |
| Huntington's disease | 25,000 |
| Duchenne muscular dystrophy | 20,000 to 30,000 |
| Hemophilia | 20,000 |
| Phenylketonuria | Below 10,000 |
| Alzheimer's disease | 2–4 million* |
| Manic depression | 1–2 million* |

*Heredity may account for only a portion of these cases

Source: Integrated Genetics Inc.

Sources: Office of Technology Assessment and various disease foundations

# Price of Progress (cont'd)

but as testing expands in coming years, the number of qualified counselors will almost certainly fall far short of needs. Last year [1986], Ms. Marks says, she had 25 graduates for 70 jobs available.

So far, most genetic testing is conducted by university and other specialized labs. But several biotechnology firms are hard at work on diagnostic kits for use in doctor's offices and nonspecialized labs. Thus, far more future testing will be done by doctors and lab technicians who may not have much experience in this field, and many geneticists worry that reliability will inevitably suffer.

But others say these fears are exaggerated. "To practice modern medicine you have to keep up with many things—new drugs, new equipment, new operations," Dr. Omenn says. "Genetic information isn't all that difficult to keep up with."

Some see a strong role for government in this area, to make sure that biotechnology firms don't oversell their tests' abilities and to ensure a high standard of laboratory work.

## Can Third Parties Insist on Tests?

As more and better genetic probes emerge, insurance companies and employers surely will push to use them. Employers won't want to hire or promote high-risk people likely to have late-onset disorders that will cut productivity and push up the company's health-insurance and workmen's compensation costs. Insurers will seek to limit their risks by insisting on genetic testing as a condition of life-insurance coverage and possibly health-insurance coverage also, and then charging higher premiums or denying coverage to those likely to develop costly or fatal illnesses. A debate is inevitable over whether such third parties should have the right to use genetic-test results and the extent to which doctors can breach a patient's privacy by giving results to insurers or employers without the patient's consent.

Many argue that physicians must keep such information confidential. "The right to privacy includes the expectation that genetic information will not be obtained or disseminated without the patient's consent," says Dr. Reilly of the Shriver Center. To protect patients, he urges legislation that would limit the right of third parties to request genetic tests or to get access to the results of such tests.

But others believe that employers and insurers are entitled to genetic-test data and that in some cases the interests of society should override individual rights to confidentiality. For example, shouldn't airlines be free to require testing to screen out pilots with inherited predisposition to early heart attack? "If you're a frequent flyer, you might see your safety as more important than the pilot's right to privacy," says Jason Brandt, a neuropsychologist in charge of the Huntington's program at Johns Hopkins.

## Will Everyone Eventually Be Tested?

For the immediate future, testing will usually be done for specific individuals with family history of a disease or other reason to suspect a genetic disorder. Most current testing uses genetic markers, and since these tests involve family members, they are too unwieldly for screening of the general population. Moreover, present tests are still too expensive and unreliable for general screening programs.

But as testing reliability improves and more effective treatments emerge, there are bound to be proposals to test entire segments of the population—all teenagers, for example.

Already, there are pointers in this direction. All states require routine testing of newborns for phenylketonuria, because early detection and treatment of the enzyme deficiency can reduce brain damage. A National Institutes of Health panel recently recommended that all newborns be screened for sickle-cell anemia; early detection and treatment can limit fatal infections.

"There is more and more activity in state legislatures to require prenatal or newborn screening for more and more disorders, on a cost-benefit argument" that early detection and treatment can limit future costs, says Ms. Andrews.

But many people question whether saving money is valid reason to invade a person's privacy, and contend that benefit-cost ratios must be very large to justify mandatory screening without consent. Others argue that general-population testing is likely to produce too many false results.

Obviously, many of the issues to be resolved in connection with genetic testing—widespread screening, confidentiality, adequate counseling and others—are similar to those now being debated on testing for acquired immune deficiency syndrome. Some think policies now being developed on AIDS will set a pattern for genetic testing, despite substantial differences between the two areas.

"AIDS is blazing this giant pathway about how much society really needs to know about all these risks," says Dartmouth's Dr. Graham.

But Thomas Murray, director of the Center for Medical Ethics at Case Western Reserve University, is more cautious. "All the right issues are being raised in the AIDS debate, and the right arguments are being made," he says. "But I think the results are going to be a pretty mixed bag. I wouldn't want them carried over" to genetic testing.

By Alan L. Otten, *Staff Reporter of* The Wall Street Journal.

then stabilized by the use of DNA ligase that catalyzes the covalent linkage of the polynucleotide strands of the plasmid and the foreign DNA end to end.

After foreign DNA is permanently incorporated into one or more plasmids, the recombinant plasmids must be returned to an *E. coli* cell. This is done by incubating the recombinant plasmids with specially treated *E. coli* cells, some of which will occasionally take up a plasmid from the medium. Once inside the cell, a plasmid will replicate along with the main DNA when the bacterium divides. Consequently, large numbers of recombinant plasmids can be isolated from growing cultures of the bacteria.

Because the foreign DNA is cut by a particular restriction enzyme into many different pieces, various recombinant plasmids will contain different sequences of foreign DNA. Consequently, methods must be used to select, among bacteria that have picked up recombinant plasmids, those bacteria containing foreign DNA sequences that are desired.

These recombinant DNA techniques can also be used with synthetic DNA. For example, DNA that codes for a particular protein can be synthesized chemically or by using the enzyme *reverse transcriptase* that catalyzes the synthesis of double-stranded DNA using mRNA as a template. Such an approach permits the incorporation of hand-tailored DNA sequences into bacteria. In another technique, it is possible to insert DNA from one eukaryotic species into the DNA of another by using recombined viruses rather than plasmids as vectors for the transfer.

## Benefits of Genetic Engineering

The significance of recombinant DNA technology depends on the fact that the structure of DNA and its mode of replication as well as the genetic code and the mechanism of protein synthesis are fundamentally similar in all organisms. Consequently, bacteria containing foreign DNA can be used as factories for obtaining multiple copies of this DNA or for synthesizing proteins for which the DNA codes.

In many recombinant DNA experiments in which foreign DNA is transferred to bacteria, the sole purpose is to obtain large quantities of easily purified foreign DNA. The preparation of multiple copies of purified genes has allowed us to determine their nucleotide sequences; this is impossible with only small amounts of DNA. Sequence determination has revealed a great deal of new information about the structure of genes and the way proteins are synthesized. Moreover, it has allowed us to study how genes are expressed—information that is crucial to an understanding of such processes as embryonic development and cancer. Sequence analysis is also useful in studying the genetic basis of hereditary diseases, and it may lead to new methods for the detection of mutated genes in prospective parents or in embryos (see *Focus: Price of Progress—Efforts to Predict Genetic Ills Pose Medical Dilemmas*).

Genetic engineering has already proven to be a commercially useful method for obtaining large quantities of proteins that are needed in medical treatment. The importance of this approach lies in the fact that there is no readily available source of human proteins, other than those few that can be extracted from blood. In addition, most proteins are too large to be synthesized chemically in the laboratory by attaching amino acids sequentially to one another; even if methods became available for such synthesis, it would be necessary to know the structure of the protein before it could be synthesized. These problems can be overcome by the transfer into bacteria of human genes that code for particular proteins. In their bacterial hosts, these genes can be expressed, yielding the desired proteins. Some human proteins that have been synthesized to date by recombinant DNA methods are insulin, growth hormone, and several interferons (antiviral proteins). The synthesis of many more proteins, both human and of other species, is currently being undertaken by commercial biotechnology companies (Figure 20–12).

By transferring genes from one species to another, it is possible to construct entirely new organisms. For example, it may be feasible to transfer

FIGURE 20–12 Equipment used in a commercial biotechnology company for large-scale extraction and purification of proteins synthesized by recombinant organisms.

bacterial genes that code for nitrogen fixation (the ability to capture essential nitrogen from nitrogen gas in the air) to agriculturally important organisms such as corn. Such a transfer might eliminate the need for nitrogen-containing fertilizers.

Ultimately, it may be possible to use recombinant DNA technology to replace mutant genes in humans by normal ones. This might be done in somatic cells, allowing some amelioration of the symptoms of genetic disease, or even in the fertilized egg or in germ cells, thus preventing progeny from inheriting defective genes. Genetic engineering has already resulted in advances in our understanding of molecular genetics and in biomedical research beyond the expectations of its proponents.

## Dangers of Genetic Engineering

Shortly after the development of recombinant DNA techniques, there was much discussion in the scientific community about the potential dangers of such research. Some scientists felt that the possibility of obtaining undesirable or harmful genetic combinations greatly outweighed any benefits that might arrive from the use of this technology. They wondered, for example, what might be the consequences if mammalian tumor virus genes were accidently transferred to *E. coli,* a bacterial resident of the human intestine. Could these genes be picked up by human cells and be spread indiscriminately throughout the population? Was there not the potential for some unforeseen disaster?

These and other questions resulted in the adoption of strict rules by the National Institutes of Health to govern the conduct of recombinant DNA research, particularly with regard to the containment of recombinant organisms. Furthermore, special mutant strains of *E. coli* were isolated for use as DNA recipients. These strains cannot manufacture many of the cellular constituents they need, and hence they cannot survive unless provided with a complex mix of specific organic nutrients—a mix of nutrients that would not normally be available in nature. They also are killed if exposed to ultraviolet light in sunlight. The use of these strains, unable to survive for any length of time in a natural environment, further reduces the risk of unintentional contamination of the environment by recombinant *E. coli.*

In addition to the extra measure of safety provided by these precautions, it is argued that recombinant DNA research involving *E. coli* as host organisms is probably not intrinsically dangerous. Because restriction enzymes occur naturally, and *E. coli* in the gut is exposed to DNA from many sources, it seems likely that recombinant events must occur frequently in nature without adverse effects. Presumably, naturally occurring recombinant *E. coli* do not compete effectively with other *E. coli* strains and are therefore eliminated.

Even if genetic engineering using *E. coli* as the recipient organism is not dangerous, some biologists are still concerned that genetic engineering involving animal viruses and higher eukaryotic organisms could lead to potentially hazardous situations. These biologists think that genetic engineering should be severely restricted or banned entirely. Because recombinant DNA techniques have only been used for about fifteen years, we do not have a great deal of experience to guide us as we attempt to resolve the controversy over the safety of these techniques. However, there have to date been no reported health hazards involving recombinant DNA.

## Study Questions

1. Describe the structure of DNA.
2. a) What is the consequence of the specific base pairing in DNA?
   b) How is it important in the replication of DNA?
3. Summarize the events that occur during a) transcription and b) translation.
4. Why is the genetic code said to be universal and degenerate?
5. a) What are mutations?
   b) What is the molecular basis of mutations?
   c) Describe three general types of mutations.
6. Why is a frameshift mutation likely to result in the synthesis of a dysfunctional protein?
7. List several causes of a) spontaneous and b) induced mutations.
8. a) What are genetic repair systems?
   b) Describe how one type of repair system operates.
   c) What is the importance of repair systems to human disease?
9. What is meant by the term "inborn errors of metabolism?"
10. a) What is the molecular basis of phenylketonuria?
    b) How does phenylketonuria lead to mental retardation?
    c) How can the effects of phenylketonuria be prevented?
11. Write a short essay describing the inheritance, causes, and symptoms of sickle cell anemia.

12. Explain why the sickle cell mutation has survived and spread in populations even though it can lead to severely deleterious physiological effects.
13. a) What is genetic engineering?
    b) Describe how it is performed.
14. a) What are several benefits of genetic engineering?
    b) What are its potential dangers?
    c) Why do most biologists contend that its benefits outweigh its potential dangers?

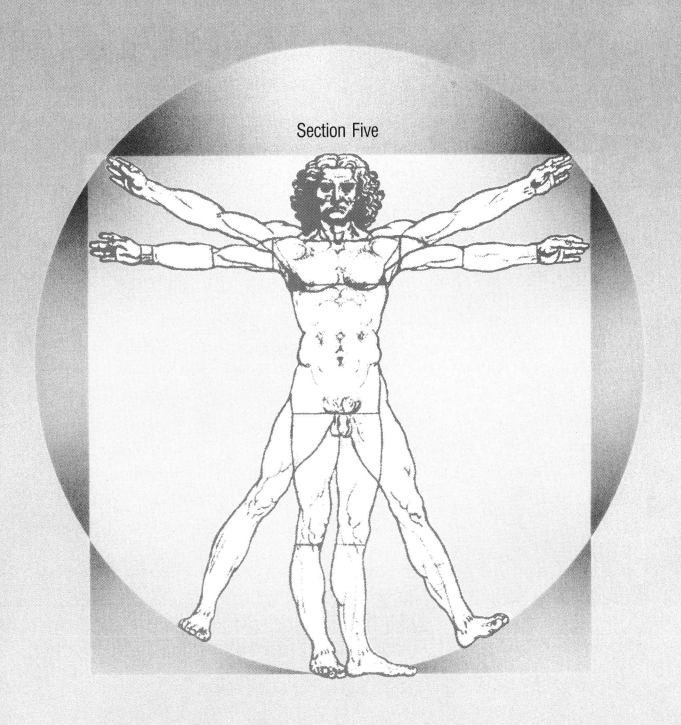

Section Five

# LIFE THREATENED

# DISEASE AND THE IMMUNE SYSTEM

The body is constantly assaulted by a variety of harmful agents. Much of the food and water we ingest is contaminated with toxic chemicals capable of injuring or killing the body's cells. The air we breathe may also contain harmful chemicals in addition to dust and other particulate matter that irritates our respiratory systems. Radiation from natural and man-made sources strikes our tissues, damaging cells and their macromolecule constituents, notably DNA.

Our bodies have specialized ways of destroying or counteracting the effects of many of these noxious agents. We have seen, for example, that the liver detoxifies many toxic chemicals, while others are eliminated in the urine. The airways of the respiratory system filter out much of the particulate matter from air. Cells themselves contain chemicals and elaborate enzyme systems that inactivate the potentially damaging compounds formed by radiation and that repair radiation-induced damage to DNA.

In addition to these and similar mechanisms that protect the body from a large number of injurious physical and chemical agents, the body has other mechanisms that function primarily to protect it from the major and ever-present threat of invasion by a host of microorganisms. The nutrient-rich fluids of the body, maintained at a suitable pH and temperature, provide an optimal environment for the growth of microorganisms. These damage tissue directly or secrete metabolic products that have injurious effects. This chapter will discuss the role of microorganisms in disease and the protective mechanisms used by the body in defense against microorganisms and certain other foreign substances that invade the body.

## DISEASE-PRODUCING MICROORGANISMS

History has recorded epidemics of infectious disease that have ravaged mankind (Figure 21–1). The bubonic plague of the fourteenth century reduced Europe's population by a third; smallpox killed some sixty million Europeans during the course of the eighteenth century. Cholera, typhus, diphtheria, influenza, pneumonia, and other infectious diseases exacted terrible tolls. As late as 1900, infectious disease still accounted for nearly 50 percent of all the deaths in the United States (Figure 21–2).

During the first half of the twentieth century, the threat of infectious disease was dramatically reduced by the widespread adoption of sewage treatment; filtration and chlorination of public water supplies; pasteurization of dairy products; improved methods for storing, handling, and preparing food; antiseptic procedures in hospitals; and other such modern health practices. The development of a wide range of safe and effective vaccines and of antimicrobial drugs has added important weapons to the fight against microbes. Advances in the prevention and treatment of infectious disease have been so remarkable that now only about 5 percent of the deaths in the United States are attributed to infectious disease.

While mortality from infectious disease has been greatly reduced, the incidence of infectious disease is still extremely high. Venereal disease has reached near epidemic proportions in the United States, with millions of new cases of gonorrhea, syphilis, chlamydia, and genital herpes reported annually. Few of us escape a single year without an acute respiratory infection caused by influenza virus, streptococcus, or other microorganisms. Clearly, infectious disease even today poses a serious health problem.

### Pathogenic Cellular Microbes

Microorganisms that cause disease in humans include bacteria, fungi, protozoa, and viruses. Bacteria, fungi, and protozoa are cellular organisms, all of which cause infection by interacting with their human host in similar ways. Viruses,

FIGURE 21–1 Devastating epidemics of bubonic plague have been recorded since biblical times. This print depicts scenes in London during the Great Plague of 1665.

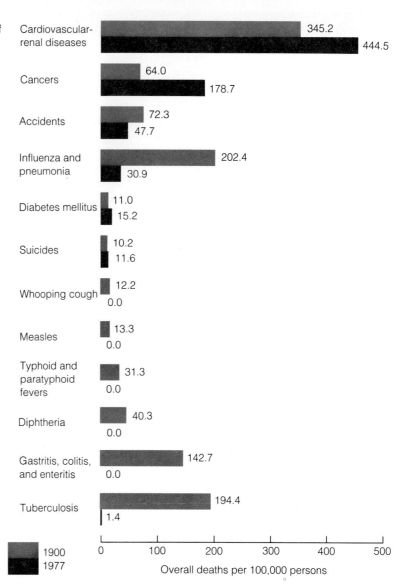

FIGURE 21–2 Comparison of the major causes of death in the United States in 1900 and 1977.

on the other hand, are not composed of cells and cause infection by mechanisms that are fundamentally different from those used by the other three types of pathogenic microbes. Consequently, we will consider viruses separately in a later section.

## Our Microbial Residents

A variety of cellular microorganisms are commonly associated with the human body. These microbes, termed the *resident,* or *indigenous, flora,* are normally present on the surface of the skin and in parts of the gastrointestinal, respiratory, urinary, and genital tracts.

Ordinarily, the resident flora live in harmless association with the human body and are considered to be nonpathogens. However, it is important to point out that all of these organisms are potentially pathogenic. The unrestrained growth of any single microbial type is normally prevented by competition among the different types of resident microbes for available nutrients and by the natural defense mechanisms of the body.

Conditions that disrupt the balance between various microbial types or that reduce the effectiveness of host defense mechanisms may encourage the proliferation of usually harmless microbes, resulting in infection. For example, therapy with antimicrobial drugs, which kill or suppress the growth of susceptible microbes, may allow the uncontrolled proliferation of other microbes that are resistant to the action of the drug, thus producing a *secondary infection*. Similarly, host defense mechanisms may be compromised by malnutrition, illness, trauma, radiation, and suppression of the immune system by the use of immunosuppressive drugs, increasing host susceptibility to infection by otherwise harmless resident microorganisms.

One example of an indigenous microbe that can become pathogenic is the yeastlike fungus *Candida albicans,* commonly found on the skin and mucous membranes in various parts of the body. *Candida* infections of the mucous membranes of the tongue and mouth may occur in infants during the first few days after birth before the normal bacterial flora have been established or in older individuals after treatment with antibiotics, causing a condition known as *oral thrush,* characterized by raised, creamy-yellow patches of fungal growth. A similar candidal infection of the vagina, commonly referred to as "yeast infection," produces itching, irritation, and vaginal discharge. It often occurs during extended antibacterial therapy that kills acid-producing bacteria that normally reside in the vagina and maintain an acid environment that prevents fungal growth. Hormonal changes that occur during pregnancy or as a consequence of the use of oral contraceptives can also suppress the growth of these acid-producing bacteria, disposing women to candidal yeast infections. In addition to localized infections, systemic infections of *Candida* can develop in most internal organs. The mortality rate from such systemic infections is quite high, but the incidence is very low, occurring only in severely ill individuals or after prolonged treatment with antibacterial or immunosuppressive agents.

Human beings also encounter a variety of nonindigenous microorganisms that are present in food, water, air, and soil, or are carried by animals. Most of these are incapable of producing disease in humans, but some do. For example, the soil bacterium *Clostridium tetani* can cause tetanus if it contaminates a wound. A related bacterium, *Clostridium botulinum,* may grow in improperly preserved food, secreting a deadly toxin that causes *botulism.* A protozoan, *Entamoeba histolytica,* may contaminate water supplies, causing amebic dysentery, a form of severe diarrhea that may result in serious illness or death. Some nonindigenous fungi can grow on the skin, causing diseases such as ringworm (Figure 21–3), of which athlete's foot is one type. Other nonindigenous fungi can cause severe systemic infections that may prove fatal if not brought promptly under control.

Many nonindigenous microorganisms capable of causing disease are harbored by humans and may be transmitted from one infected individual to another, particularly under crowded conditions or during winter when people are confined indoors for extended periods of time. In some cases, the transmission of microbes may require direct contact with infected individuals, but in other instances, transmission can occur from contact with towels, utensils, and clothing, or through contaminated food and water.

Some microbes are characteristically harbored by a few individuals, termed *disease carriers,* who show no symptoms of the disease but may infect others. Typhoid and diphtheria are two diseases that are commonly transmitted by symptomless carriers. One famous carrier of typhoid was Mary Mallon, better known as "Typhoid Mary," a cook in the early 1900s. Before she was quarantined by imprisonment in Riverside Hospital, North Brother Island, New York, she infected numerous people who ate contaminated food she had prepared.

Although many nonindigenous microbes can infect vigorous, healthy individuals, serious infection in hospitals today is caused most often by indigenous microorganisms such as the bacteria *Escherichia coli, Staphylococcus aureus,* and *Proteus mirabilis.* Ironically, infection by indigenous microbes is encouraged by the extensive use of new medical technologies and drugs. Intravenous catheters, widely used for delivering drugs and life-supporting fluids, can spread indigenous microbes from the skin to the circulatory system. Irradiation for cancer, the use of immunosuppressive drugs in transplant patients, and treatment of inflammatory diseases such as rheumatoid arthritis

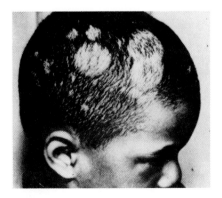

FIGURE 21–3 Ringworm, a fungal infection of the skin.

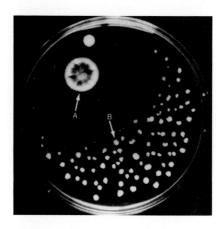

FIGURE 21-4 The antibiotic penicillin was discovered in 1928 by Alexander Fleming when one of his petri plates on which bacteria were growing became contaminated with *Penicillium* mold. As in the petri plates shown above, Fleming noted that bacterial growth (B) was inhibited in the region surrounding the *Penicillium* colony (A). This inhibition is due to the secretion, by the mold, of penicillin into the semi-solid growth medium.

with corticosteroids and other anti-inflammatory agents, all decrease host defense mechanisms to infection. In addition, the extensive use of antimicrobial drugs can lead to severe secondary infections, as discussed earlier.

## Identification of Pathogenic Microorganisms

More than a hundred years ago, the German physician and microbiologist Robert Koch established a scientific procedure for proving that a particular microorganism causes a certain disease. Koch's procedure is still valid today and is considered the ultimate proof of pathogenicity. It involves the following steps: (1) A particular microorganism must be found in individuals with a particular disease, but not in healthy ones. (2) The microbe, isolated from diseased individuals, must be grown in the laboratory. (3) Pure, cultured microorganisms must produce the particular disease when used to inoculate a new, susceptible animal. (4) It must be possible to re-isolate the microorganism from this diseased animal and grow it in the laboratory in pure culture.

## Antimicrobial Therapy

Antimicrobial drugs are widely prescribed to treat infections by bacteria, fungi, and protozoa. *Antibiotics* kill or suppress the growth of bacteria; *antifungal agents* and *antiprotozoal agents* act against fungi and protozoa, respectively. Within each of these three classes of antimicrobial agents are individual drugs with differing specificities. For example, some antibiotics are effective only against a narrow range of bacteria or even a single bacterial type, while others, known as *broad-spectrum antibiotics,* will inhibit the growth of a wide range of bacterial species. Examples of this latter class of antibiotics are penicillin, tetracycline, and erythromycin (Figure 21-4).

The availability of a wide range of antimicrobial agents has led to a general misunderstanding about the usefulness of these drugs. No matter how judiciously or promptly antimicrobial therapy is initiated, the complete elimination of infecting microbes must be performed by the host. Antimicrobial drugs generally only reduce the proliferation of infecting microbes but do not eradicate them. Therapy with antimicrobial drugs "buys time," preventing large numbers of pathogens from overwhelming the host during the time that host defense mechanisms, particularly those involving the immune system, are developing their full potential to eliminate the infecting microbes. The importance of host defense mechanisms to the eradication of infection is illustrated by our experience with transplant patients and patients undergoing some types of cancer therapy. Because their immune systems are suppressed by drugs or radiation, these individuals may develop serious infections that often prove fatal even with extensive antimicrobial therapy. A similar situation is recognized in children who are born with defective immune systems and in individuals with AIDS, whose immune systems are damaged by disease.

It has been recognized for years that many microbes display an alarming potential to develop resistance to antimicrobial agents. For example, bacteria that cause gonorrhea, and certain others that cause meningitis were once extremely susceptible to sulfa drugs and penicillin, but now many strains of both bacteria show resistance to these antibiotics. Indeed, it has been noted that many types of bacteria slowly develop resistance to most antibiotics. The unwarranted use of antibiotics for trivial infections exacerbates this problem by exposing bacteria to antibiotics to which the bacteria may then begin to develop resistance.

## Bacteria

Bacteria form a group of organisms called **prokaryotes.** Most of the prokaryotes are single-celled, but some are colonial, meaning that a number of cells exist in loosely organized colonies. The cellular structure of prokaryotes differs from that of all other organisms, the **eukaryotes.** Prokaryotic (bacterial) cells are much smaller than eukaryotic ones, ranging in size from about 0.2 to 3.0 $\mu$m. The smallest of these is just barely detectable in the light microscope. Eukaryotic cells, on the other hand, may be very large (e.g., a chicken's egg), but most eukaryotic tissue cells range in size between roughly 10 and 40 $\mu$m.

In addition to their smaller size, bacterial cells are simpler in structure than eukaryotic cells. As shown in Figure 21–5, bacterial cells do not contain a membrane-bound nucleus, nor is their DNA organized into chromosomes. Instead, "naked" DNA is present in a region termed the nucleoid that is not strictly delineated from the cytoplasm. Most other intracellular structure is also absent in bacterial cells. They do not contain membrane-bound organelles such as mitochondria and lysosomes, nor do they possess Golgi complexes, endoplasmic reticulum, and other complex membrane systems. Some bacteria may have simple internal membranes such as the mesosomes that form from infoldings of the plasma membrane (see Figure 21–5). The plasma membrane of bacterial cells is surrounded by a rigid cell wall, composed primarily of carbohydrates and amino acids. (Green plants and fungi possess a cell wall composed of cellulose, and animal cells have no cell walls.) The integrity of the bacterial cell wall is usually required for survival.

The differences in structure and, to a lesser extent, in function between bacterial and eukaryotic cells provide the basis for the action of many antibiotics. For example, penicillin specifically inhibits bacterial cell wall synthesis in susceptible strains. Other antibiotics may inhibit enzymes or metabolic processes specific to microbial function.

In the Western Hemisphere, bacterial infection is by far the most common form of infectious disease. Some of the more important of these diseases are listed in Table 21–1.

## Fungi

The fungi are single-celled or multicellular organisms that include the slime molds, molds, yeasts, rusts, and mushrooms. Most fungi, of which there are more than 100,000 species, are nonpathogenic and live as **saprophytes,** obtaining organic nutrients from dead or decaying organic matter. However, others infect a wide variety of plants and animals. A number of infectious diseases of humans caused by fungi have been mentioned, including candidal infections, and ringworm.

Although some fungi are true pathogens, causing active infections that can be transmitted from one individual to another, most fungi cause disease only

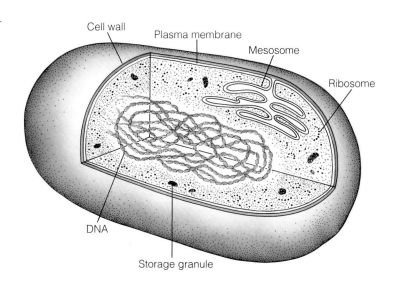

FIGURE 21–5 Cellular structure of a "typical" bacterial prokaryote. Compare with Figure 3–1, which shows a typical eukaryotic cell.

TABLE 21-1 Infectious Diseases Caused by Bacteria

| Bacterium | Disease |
|---|---|
| *Staphylococcus aureus* | Can infect any part of the body. Common cause of boils, other skin infections; wound infections; sinusitis, pneumonia; endocarditis; septicemia; toxic shock syndrome. |
| *Streptococcus* (variety of species) | Sore throat, scarlet fever, endocarditis, impetigo. Most common pneumonia. |
| *Neisseria meningitidis* | Bacterial meningitis |
| *Neisseria gonorrheae* | Gonorrhea |
| *Clostridium tetani* | Tetanus (lockjaw) |
| *Escherichia coli* | Both are present in intestinal flora; |
| *Proteus vulgaris* | common cause of peritonitis, cystitis |
| *Salmonella typhi* | Typhoid fever |
| *Shigella dysenteriae* | Shigellosis (bacterial dysentery) |
| *Vibrio cholerae* | Cholera |
| *Bordetella pertussis* | Whooping cough |
| *Yersinia pestis* | Bubonic plague |
| *Haemophilus influenza* | Meningitis, pneumonia, urinary tract infections |
| *Mycobacterium tuberculosis* | Tuberculosis |
| *Mycobacterium leprae* | Leprosy |
| *Corynebacterium diphtheriae* | Diphtheria |
| *Legionella pneumophila* | Legionnaire's disease |
| *Chlamydia trachomatis* | Conjunctivitis |
| *Trepenoma pallidum* | Syphilis |

when large numbers of spores invade a host or when normal resident fungi, such as *Candida albicans,* grow out of control. Indeed, the extensive use of antibiotics and immunosuppressive drugs has led to a sharp increase in fungal infections caused by resident fungal flora.

The products of fungi can also be dangerous to humans. The inhalation of fungal spores may cause allergic symptoms similar to that of pollen allergies ("hay fever"). Furthermore, a number of fungi produce deadly poisons or hallucinogenic chemicals.

## Protozoa

The protozoa form a large group of single-celled animals occupying moist environments such as lakes, pools, and wet soils. Although thousands of species of protozoa have been identified, only about 35 cause disease in humans. Pathogenic protozoa are present throughout the world, but the most troublesome ones are usually indigenous to tropical or subtropical regions. Some protozoal diseases have been carried by travelers to temperate regions including the United States.

Protozoa cause diseases of the blood, central nervous system, gastrointestinal tract, urogenital tract, and other organs. Perhaps the most common infectious disease in humans is malaria, caused by a protozoan that infects the red blood cells and other tissues, and is carried by the Anopheles mosquito (Figure 21-6).

## Viruses

Viruses are known to cause disease in many types of organisms, and in fact may well infect all organisms. Important viral diseases in humans include influenza, viral hepatitis, oral and genital herpes, measles, mumps, chicken pox, shingles, polio, rabies, mononucleosis, and the common cold. Widespread immunization programs have greatly reduced the incidence of some viral diseases, particularly the childhood ones. Smallpox, a viral disease that was once a scourge, has been eradicated worldwide as a result of an extensive vaccination effort (see *Focus: The Eradication of Smallpox*).

Although the incidence of some dangerous viral diseases has been reduced, a large number of viral diseases still afflict humans. Some of these, including the common cold and influenza, affect virtually all people from time to time.

## Structure of Viruses

Viral particles do not have a cellular structure, nor do they have the metabolic machinery characteristic of cells. Their only function is replication, and they are able to replicate only within a suitable host cell. Outside of a living cell, they are essentially inert, dormant particles—macromolecular complexes

FIGURE 21-6 A simplified diagram of the life cycle of *Plasmodium*, the causative organism of malaria. When an *Anopheles* mosquito bites an individual with malaria, infectious *Plasmodium* gametes are ingested by the mosquito. Fused gametes enter the mosquito's salivary gland, producing spores that may be transmitted by a mosquito bite to a new human host. The spores develop into various cellular forms that can infect tissue cells or red blood cells. Symptoms of the disease, including headache, fever, and chills, occur when infected red blood cells are disrupted, releasing *Plasmodium* cells into the plasma.

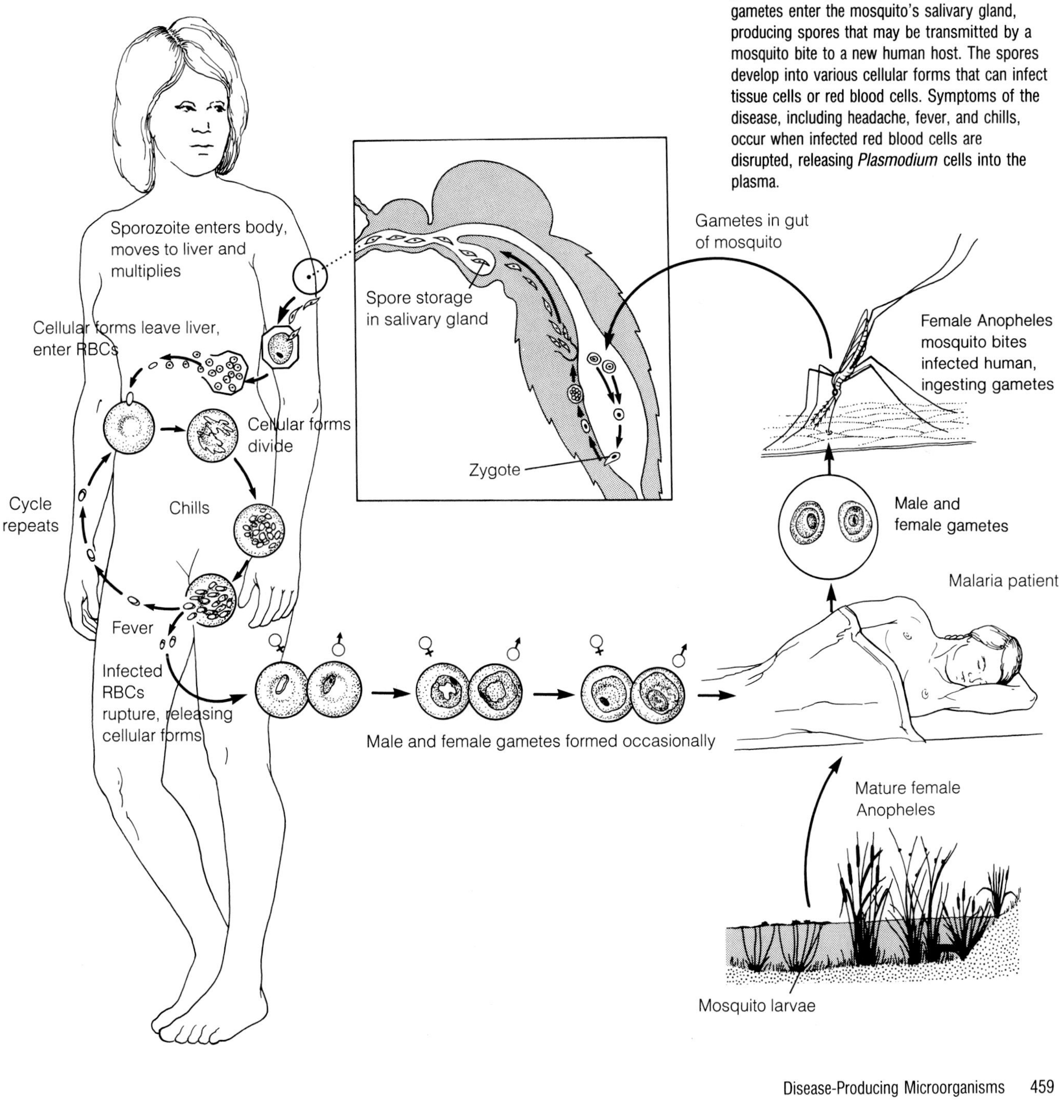

Sporozoite enters body, moves to liver and multiplies

Spore storage in salivary gland

Gametes in gut of mosquito

Female Anopheles mosquito bites infected human, ingesting gametes

Cellular forms leave liver, enter RBCs

Cellular forms divide

Zygote

Cycle repeats

Chills

Male and female gametes

Fever

Malaria patient

Infected RBCs rupture, releasing cellular forms

Male and female gametes formed occasionally

Mature female Anopheles

Mosquito larvae

# The Eradication of Smallpox

Smallpox, a disease that has devastated human populations since ancient times, remained a serious threat well into the middle of the twentieth century. Although universal vaccination programs eliminated it from North America and Europe in the 1940s, it was still endemic as late as 1967 in some 33 countries where there were an estimated 10 to 15 million cases (see map).

Smallpox is caused by a virus that is transmitted directly in saliva, nasal discharges, and exhaled droplets of infected individuals, or indirectly on contaminated towels, linens, and clothing. The characteristic symptoms are a pustular rash accompanying a high fever. Once smallpox is contracted, there is no cure, and the disease must run its course. The more severe form of the disease, referred to as variola major, kills 25 to 40 percent of its victims, while the milder form, variola minor, is fatal to only about 1 percent. The only real weapon against smallpox is vaccination, which protects individuals from infection and thus prevents spread of the virus.

In 1967, the World Health Organization began a campaign to eradicate smallpox. Although this goal seemed ambitious, there were a number of reasons why it appeared attainable. The smallpox virus infects only human beings, there are no asymptomatic human carriers, and no other living organism is known that can harbor the virus and thus serve as a reservoir of infection. In addition, the symptoms of the disease are easily recognizable, and the period of infectivity is well defined, lasting only three to four weeks. Finally, a highly effective, easily administered vaccine is available.

At first, the eradication program concentrated on widespread immunization in countries with the highest incidence. As outbreaks of smallpox decreased in frequency and became more localized, the strategy of the campaign shifted to containment of the disease by identifying smallpox victims and vaccinating potential contacts. This approach allowed the elimination of smallpox from regions where less than 10 percent of the population was vaccinated.

By late 1976, smallpox had been eradicated in all areas of the world except mountainous regions of Somalia. On October 26, 1977, the world's last known patient with endemic smallpox developed a rash in Merka, Somalia. When intensive surveillance for two years failed to identify any additional cases of naturally transmitted smallpox, smallpox was officially certified as eradicated—the only disease ever eliminated through human efforts.

There is a tragic postscript to the story. In the summer of 1978, a worker in a London research laboratory that still possessed smallpox virus contracted smallpox and died shortly thereafter. Although quarantine of the smallpox victim and vaccination of her contacts prevented spread of the virus, the incident illustrates the potential dangers of the existing laboratory stocks of smallpox virus. The possibility of a serious smallpox epidemic caused by a laboratory accident is evermore threatening with the recent abandonment of routine smallpox vaccination in all countries.

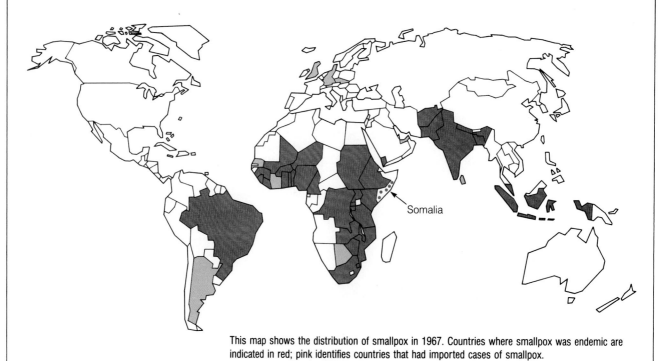

Somalia

This map shows the distribution of smallpox in 1967. Countries where smallpox was endemic are indicated in red; pink identifies countries that had imported cases of smallpox.

FIGURE 21-7 External morphology of some representative viruses.

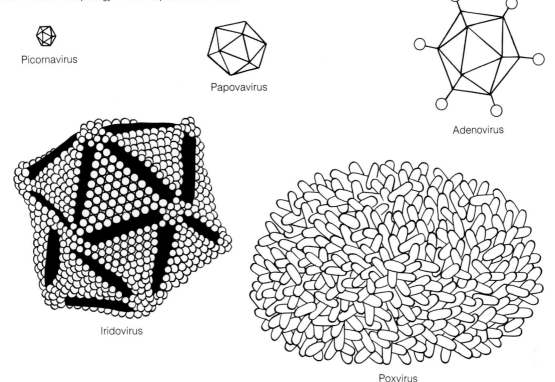

Picornavirus

Papovavirus

Adenovirus

Iridovirus

Poxvirus

displaying none of the properties normally associated with living cells. Most viruses can exist for very long periods of time, often years, in the dormant state; they can be isolated, purified, and stored, usually without damage.

Different types of viral particles vary widely in size and morphology (Figure 21-7). All are extremely small and have a relatively simple structure. Animal viruses, for example, range in size from about 20 to 300 nm, and are thus on the order of a hundred to a thousand times smaller than a typical animal cell. In fact, the small size of viruses made their identification extremely difficult. They were originally characterized as chemical agents capable of passing through porcelain filters that excluded all cellular material. Indeed, the term *virus* means poisonous fluid. It wasn't until the development of the electron microscope some three decades ago that viral particles were observed.

All animal viruses are composed of one or more nucleic acid molecules surrounded by a protein coat. Depending upon the particular type of virus, the nucleic acid may be either RNA or DNA, and it may contain information for synthesizing as few as four or five proteins or as many as several hundred. The protein coat, also called the **capsid,** is formed of protein subunits, the **capsomeres.** Some types of viruses have, in addition, a lipoprotein membrane called the **envelope** external to the protein coat (Figure 21-8). The envelope

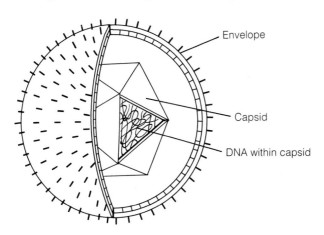

Envelope

Capsid

DNA within capsid

FIGURE 21-8 Structure of herpesvirus. The capsid, which encloses the viral DNA, is covered by an external envelope. The envelope possesses spikelike structures, composed of protein, that are necessary for the initial attachment of the virus to the cell it infects.

is usually derived from fragments of the host plasma membrane or nuclear envelope that become associated with the virus at the time of its release from the host cell. Present in the external coat of some viral particles are enzymes (proteins) that are involved in penetration of the host cell or in the synthesis of viral nucleic acid.

## Virus Types

Viruses fall into two broad groups: the *cytocidal viruses* and the *oncogenic viruses*. Cytocidal viruses, described below, usually kill the cells they infect. The oncogenic viruses, discussed in Chapter 22, do not kill cells, but instead convert them into tumor cells.

## Infection by Cytocidal Viruses

Viral infection occurs in five stages: (1) adsorption, (2) penetration, (3) synthesis of viral macromolecules, (4) assembly of viral particles, and (5) release of new viral particles.

Adsorption involves the attachment of a dormant virus to the surface of a host cell. This attachment requires the presence of specific binding sites, called **receptor sites,** on the cell surface of the host. Attachment of the virus occurs as a result of specific chemical bonding between surface components of the virus and the receptor site, much as a key fits a lock. In the absence of appropriate receptor sites for a particular virus, virus attachment cannot occur, and the cell is resistant to infection by that virus. Indeed, most viruses can infect only a small range of cell types because other cells do not have the necessary receptor sites. Infection by poliovirus, the causative agent of poliomyelitis, is a good example of the importance of receptor sites to the infective process. Because poliovirus can attach only to cells of primate origin, all other groups of animals are resistant to the virus.

Following adsorption, the virus must penetrate the host cell. We do not know how this occurs for all types of viruses, but for some, the virus is taken into the cell by endocytosis. Once inside the cell, the nucleic acid is released from the viral particle by removal of the protein coat, a process referred to as *uncoating* the virus. The free nucleic acid either remains in the cytoplasm or migrates to the nucleus, depending upon the virus type.

After invasion of the host cell has occurred, many viruses immediately begin to replicate. During replication, the viral nucleic acid directs the synthesis of enzymes and macromolecular components needed to form additional viral particles. Because the virus has no metabolic machinery of its own, it can synthesize these substances only by using raw materials, energy sources, and protein-synthesizing machinery present in the host cell. It thus diverts the host's metabolism for its own purposes. In the process, the virus suppresses the synthesis of proteins and other macromolecules of the host, breaks down host macromolecules to provide additional sources of energy and raw materials, and otherwise deranges the host cell's metabolism. Depending upon the virus type, the host cell is damaged to varying degrees, often irreversibly.

After many copies of viral nucleic acid and coat proteins are synthesized within the host cell, they begin to assemble spontaneously to form large numbers—often hundreds or even thousands—of new viral particles. Once assembled, the new viral particles are released from the host cell. Many viruses actually rupture *(lyse)* the host's plasma membrane, resulting in immediate destruction and death of the cell. Released viral particles then reinfect surrounding cells, repeating the replication cycle (Figure 21–9). In this manner, infection of a single cell can lead to the destruction of increasing numbers of cells.

Although most disease-causing viruses actually lyse their host cells during release of new particles, some do not. Instead, they may be extruded through the plasma membrane of the host cell by exocytosis, often acquiring, in the

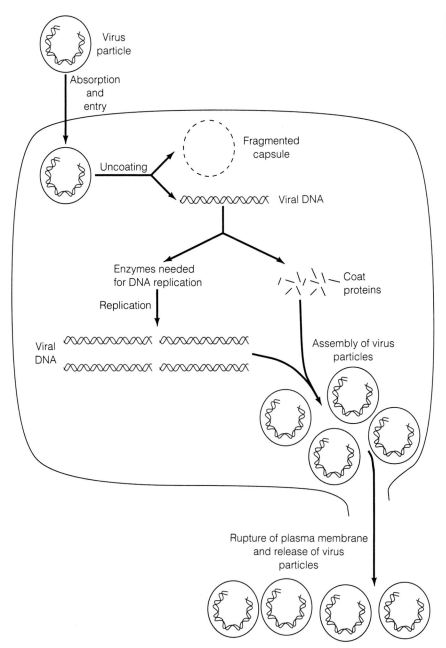

FIGURE 21-9 Life cycle of a typical DNA-containing virus.

Virus particle

Absorption and entry

Uncoating

Fragmented capsule

Viral DNA

Enzymes needed for DNA replication

Coat proteins

Replication

Viral DNA

Assembly of virus particles

Rupture of plasma membrane and release of virus particles

extrusion process, an outer envelope that is derived from the cell's plasma membrane (Figure 21–10). In some instances, viruses that are released by exocytosis are produced in the host cell on a slow, continuous basis, and the host cell is not immediately killed, but may survive for long periods of time until it has suffered severe metabolic damage. Viruses of this latter type cause *persistent infections,* one of the best known examples of which is serum hepatitis, caused by hepatitis B virus.

There are other variations on the basic theme of viral replication. For example, some viruses do not immediately begin to reproduce following penetration of a host cell. Instead, they remain dormant for long periods of time, causing no apparent damage to the infected cell. After a period of weeks, months, even years, these viruses may be activated. They leave the dormant state and begin replication—a process that culminates in host cell death and brings about recognizable symptoms. Following this acute stage of infection, the virus reinfects surviving cells and once again enters the dormant state, to be reactivated at some later time. Viruses with life cycles characterized by alternate phases of dormancy and replication cause what are termed *latent infections.*

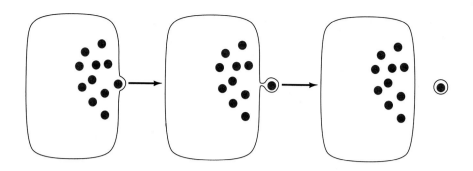

FIGURE 21–10 Extrusion of a virus particle through the plasma membrane of an infected cell.

Unlike persistent infections in which infectious viral particles are produced continuously at a slow rate, latent infections yield communicable viral particles only during the active, replicative stages. Viruses that cause latent infections include the chicken pox virus, measles virus and herpes simplex viruses (see *Focus: Herpes Simplex Viruses: Oral and Genital Herpes,* in Chapter 17).

Slow viruses are another type of cytocidal virus that cause diseases characterized by extremely long incubation periods, often 20 or 30 years or more—hence the designation "slow." They result in degeneration of the central nervous system that inevitably proves fatal. Some of these diseases are caused by slow viruses that resemble conventional viruses. Others, however, are caused by transmissible agents, presumably viruslike, that have not been identified. Because they are resistant to inactivation by heat and many chemicals that inactivate conventional virus particles, these viral-like agents have been termed *unconventional viruses.* One slow viral disease caused by an unconventional virus is *kuru,* identified in the cannabalistic Fore tribe of New Guinea. In kuru, the infectious particles are transmitted by the ingestion of brain tissue during cannabalistic rituals.

### Treatment of Viral Infections

Attempts to treat viral infections with drugs have been largely unsuccessful. Although some antiviral agents are available, they are generally useful only with very few viruses, and even then, success is often sporadic. The inability to find effective antiviral agents is presumably a result of the close association between viruses and their host cells. Successful antimicrobial agents exploit metabolic differences between microbe and host, inhibiting microbial function while leaving host function unaffected. However, because virus replication depends upon host metabolism, it has proven impossible to find effective antiviral compounds that are not also extremely toxic to the host.

One substance that may prove to be useful for treating a wide range of viral infections is *interferon.* Interferon is a protein synthesized by host cells in response to viral infection. The presence of viral nucleic acid in a host cell serves to stimulate interferon synthesis, which occurs within several hours of infection. Once manufactured by the host cell, interferon is secreted and binds to receptor sites on the surface of uninfected cells. This binding triggers the synthesis of new antiviral proteins that, upon subsequent infection of the cell, inhibit the manufacture of viral macromolecular components (Figure 21–11). The cell that first synthesizes interferon may also be stimulated to make antiviral proteins, perhaps leading to lowered production of viruses during infection.

Interferon synthesis is not virus-specific. It occurs in response to infection by a wide range of viruses, and the antiviral proteins synthesized as a result of interferon stimulation are also effective against many virus types. The interferon of each animal species is distinct; thus, human interferon protects human cells but has little effect on cells of other animals. Until recently, it has been impossible to obtain large amounts of human interferon, since it had to be extracted from human tissue. However, human interferon is now being manufactured using recombinant DNA technology (see Chapter 20) by splicing human interferon genes into bacteria, and it is currently being tested in clinical situations.

## NONSPECIFIC DEFENSE MECHANISMS AGAINST DISEASE

The body has a number of general defense mechanisms that help protect it against invasion by microorganisms. Because these defense mechanisms are not specific to any one type of microbe, they are termed *nonspecific*. **Nonspecific defense mechanisms** can, in turn, be divided into *external defense mechanisms* that discourage invasion of the body by microbes and *internal defense mechanisms* that operate once microbes have penetrated the body.

### External Defense Mechanisms

Surfaces exposed to the external environment provide the body's first line of defense against infection. The skin, with its external layer of keratinized epithelium, serves as a barrier to the penetration of microorganisms. This barrier is generally so effective that infection of the skin, or of the body by microbes entering through the skin, does not usually occur unless the surface of the skin is broken by a wound or by certain other conditions such as excessive dryness. In some cases, if the skin becomes moist and soft, certain pathogens, especially fungi, can cause skin infections, as occurs for example in athlete's foot.

In addition to providing a mechanical barrier, the skin has other properties that discourage the survival of pathogenic organisms that may come in contact with the skin's surface. As noted above, the indigenous microbial flora of the

FIGURE 21–11 Cellular events associated with the synthesis and action in virus-infected cells.

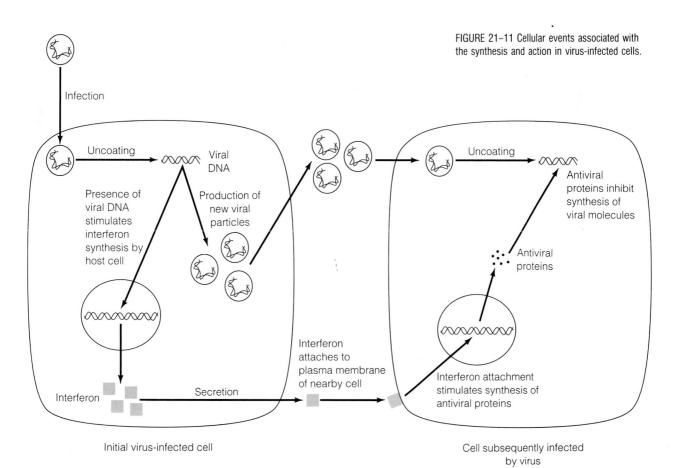

skin, usually nonpathogenic, discourage the growth of pathogens by helping to maintain an acid environment on the skin's surface, by secreting antimicrobial substances and by competing in a general way with pathogenic microbes. In addition, the secretions of the sebaceous glands (see Chapter 5) contain long-chain fatty acids that kill many bacterial strains.

The penetration of mucous membranes by microbes is substantially easier than the penetration of the skin, for mucous membranes are not protected by a keratinized layer of cells. Nevertheless, a number of external defense mechanisms prevent penetration of mucous membranes. The indigenous microbial flora of the upper respiratory tract and lower parts of the genital, urinary, and gastrointestinal tracts are particularly important in helping to discourage infection by pathogenic microbes. In addition, we have noted the role of the ciliated lining of the respiratory tract in the removal of inhaled particles, and the fact that stomach acid kills most microorganisms that are swallowed.

The eye provides another external surface in contact with microbes in the environment. Infection of the eye is largely prevented by the secretion of tears, which constantly flush bacteria into the nasopharynx through the tear duct. Tears also contain large amounts of lysozyme, an enzyme that lyses many bacterial types.

It is now recognized that the penetration of epithelial surfaces, particularly mucous membranes, must be preceded by the attachment of microbes to specific attachment sites on the epithelial cells. (The attachment of viruses to receptors on the surfaces of host cells, mentioned earlier, is merely one example of this phenomenon.) It recently has been suggested that nonpathogenic flora may reduce penetration by pathogens by covering up attachment sites needed for infection by pathogenic microbes.

## Internal Defense Mechanisms

Although external barriers to the penetration of microbes are extremely effective, some microbes do penetrate into or through epithelial surfaces. When this occurs, the microbes stimulate an **inflammatory response** by the body. The classic symptoms of inflammation are *redness, heat, swelling, pain,* and *impairment of function.* These symptoms may be dramatic and clear cut and constitute what is known as *acute inflammation; chronic inflammation* involves a more subdued, long-term reaction.

### Inflammation and the Role of Neutrophils and Monocytes

Let us now consider the sequence of events that gives rise to the symptoms of inflammation in tissue infected with pathogenic microorganisms. Suppose a finger is pierced with a splinter that passes through the protective epithelial layer of the skin and penetrates the connective tissue below. Bacteria on the splinter, as well as those on the surface of the skin, are pushed into the connective tissue. Here, the nutrient-rich tissue fluid provides an ideal environment for the growth of the bacteria, which begin to multiply and damage surrounding cells.

Inflammation is initiated by the release of *histamine* from damaged tissue and disrupted **mast cells.** Mast cells, present in loose connective tissue, are thought to be basophils that leave the circulation and reside in the connective tissue (see Chapter 8 for a review of white blood cell types). One effect of histamine is to cause the local dilation of blood vessels. Dilation increases the blood supply to the injured tissues, causing the area to become red and hot. Histamine also brings about an increase in the permeability of the venules and capillaries near the injury, allowing increased amounts of plasma to pass into the tissue fluid. This process, referred to as **exudation,** causes swelling and resultant pain as nerve endings are irritated and stretched.

Shortly after exudation has occurred, **neutrophils** can be seen sticking to the lining of the affected venules and capillaries. They then begin to move like amoebas through the blood vessel walls by extending a fingerlike projection,

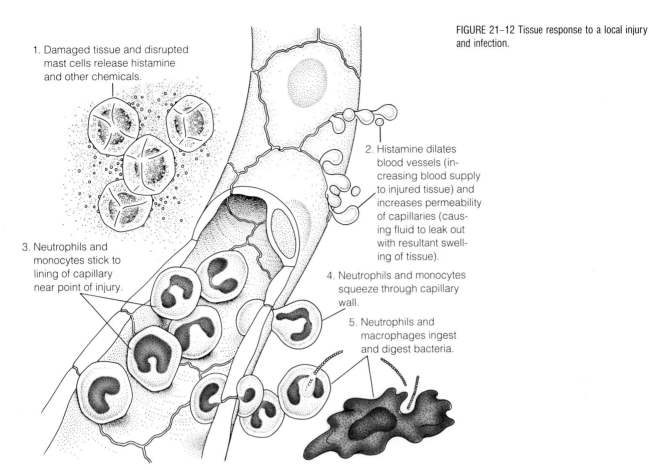

FIGURE 21-12 Tissue response to a local injury and infection.

1. Damaged tissue and disrupted mast cells release histamine and other chemicals.

2. Histamine dilates blood vessels (increasing blood supply to injured tissue) and increases permeability of capillaries (causing fluid to leak out with resultant swelling of tissue).

3. Neutrophils and monocytes stick to lining of capillary near point of injury.

4. Neutrophils and monocytes squeeze through capillary wall.

5. Neutrophils and macrophages ingest and digest bacteria.

called a pseudopodium, between plasma membranes of adjacent cells and then squeezing through (Figure 21–12). Neutrophils leave the blood because they are attracted to the bacteria by chemicals secreted by the bacteria themselves and by neutrophils already in the area. In addition, neutrophils are attracted to the damaged area by *kinins;* these are small proteins, produced by many cell types and present in blood plasma, which pass into the tissue fluid when exudation occurs.

Upon contacting bacteria, neutrophils ingest them by endocytosis (see Chapter 3), usually referred to as **phagocytosis** (*phago* = eat) when large particles are ingested. The engulfed bacteria come to lie within a membranous sac in the cytoplasm, referred to as a *phagosome.* Cytoplasmic lysosomes containing digestive enzymes then fuse with the phagosome, resulting in the digestion of the engulfed bacteria (Figure 21–13).

Accompanying or soon following the neutrophils out of the blood vessels are **monocytes.** Once in the tissues, monocytes increase in size to form large phagocytic cells called **macrophages.** These join the pool of macrophages that normally reside in all tissues. Macrophages can reproduce and are specialized for engulfing microorganisms as well as exhausted neutrophils, bits of dead tissue, and other cellular debris (Figure 21–14). There is evidence that macrophages are preferentially attracted to some microorganisms while neutrophils are attracted to others.

Although bacteria are usually eliminated quickly by neutrophils and macrophages, sometimes the neutrophils and macrophages destroyed by bacterial toxins. The dead leukocytes, along with dead bacteria and cellular debris, form a yellowish viscous substance called *pus.* Pus drains away from wounds open to the surface, but in those not open to the surface it accumulates as an abscess and must be drained surgically. Bacteria that are not ultimately eliminated by leukocytes will continue to multiply and may enter the blood-

FIGURE 21-13 Ingestion of a bacterium by a neutrophil. (a) A neutrophil contacts a bacterium. (b) The bacterium is engulfed by the neutrophil. (c) The engulfed bacterium comes to lie within a sac, called the phagosome, that is lined by a fragment of the neutrophil's plasma membrane. (d) A cytoplasmic lysosome fuses with the phagosome. (e) The lysosome releases digestive enzymes that degrade the bacterium.

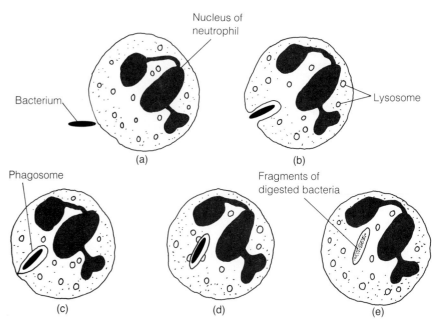

stream, causing a condition known as *septicemia* (literally bacteria in the blood) that is fatal if left untreated.

In serious infections, certain substances called *pyrogens* are formed from bacterial toxins and from the breakdown products of neutrophils. Pyrogens are absorbed into blood and carried to the temperature control center in the brain, causing temperature elevation or *fever*. Several observations have suggested that fever is an important component of the body's response to severe infection, since elevated temperatures reduce the reproduction rate and viability of infecting microbes. Substances identical or similar to pyrogens also stimulate the release of mature and immature neutrophils and monocytes from bone marrow. This causes *leukocytosis*, an increase in the number of white blood cells in the circulation.

Another important internal defense mechanism involves the production of interferon by cells infected with viruses. The action of interferon was outlined earlier.

### The Complement System

A group of plasma proteins, constituting the *complement system,* are important mediators of the nonspecific inflammatory response. These proteins are normally present in the blood in an inactive state, and they may be activated in a sequential series of reactions to produce complement molecules that mediate virtually every step of the inflammatory response, including dilation of blood

FIGURE 21-14 Scanning electron micrographs of a macrophage engulfing a colony of bacteria (a-c).

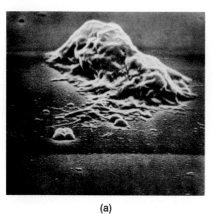

(a)

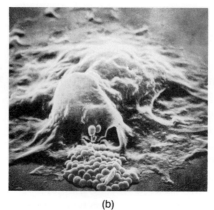

(b)

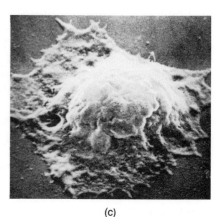

(c)

vessels, attraction of white blood cells to damaged areas, enhancement of phagocytosis, and so forth. Activation occurs when a circulating plasma protein that is not a complement protein binds to carbohydrate molecules present on the surface of invading microbes. This binding changes the conformation of the plasma protein so as to confer upon it the capacity to initiate one of the early complement steps.

Although the activation of the complement system is an important aspect of the nonspecific inflammatory response, the complement system is also important in the immune response, as discussed shortly.

## SPECIFIC DEFENSE MECHANISMS: THE IMMUNE SYSTEM

The nonspecific defense mechanisms we have just discussed help prevent the entry of infecting microbes into the body and, by means of the inflammatory response, attempt to eliminate any microbes that do manage to enter. The presence of microbes in the body also triggers an **immune response,** by which the body specifically recognizes the microbes as foreign and initiates events that lead to their destruction or inactivation. The immune response, which thus plays a crucial role in combatting disease, constitutes a **specific defense mechanism** because it protects the host against specific harmful agents. These agents include not only microorganisms, but toxins of microorganisms, other foreign macromolecules, tissues or organ transplants, and even abnormal cell types of the body, such as damaged, mutated, or tumorous cells.

Proper operation of the immune system, then, utterly depends upon its ability to recognize foreign substances or abnormal body cell types as different from normal body cells—to distinguish, as it were, "self" from "nonself." This function is performed by **lymphocytes,** a class of white blood cells. Cells that are precursors to lymphocytes arise in the bone marrow from the same stem cell that gives rise to other blood cell types; the precursor cells then migrate to peripheral lymphatic tissues and organs, where they proliferate and mature. These tissues and organs, whose detailed structure and function will be discussed in a later section, include the thymus, lymph nodes, spleen, tonsils, and smaller, scattered patches of lymphatic tissue.

The immune response consists of two very different, but interrelated components: **humoral immunity** and **cell-mediated immunity.** In humoral immunity, certain types of lymphocytes produce antibodies directed against specific harmful agents, particularly bacteria and their toxins, some viruses, and foreign macromolecules. Humoral immunity is so named because the antibodies are secreted into and carried in the circulation. Cell-mediated immunity, on the other hand, involves the destruction of certain harmful cell types by the direct action of another class of lymphocytes. Cell-mediated immune responses destroy fungal cells, host cells that are infected with intracellular microbes, transplanted or grafted tissues, and abnormal body cell types. Although we will discuss each type of immunity separately, it must be kept in mind that an effective immune system depends upon the interaction of both components.

### Humoral Immunity

The lymphocytes that are responsible for humoral immunity are termed **B-lymphocytes.** They are derived from stem cells that begin to differentiate into lymphocytes within the bone marrow, and then migrate to the various lymphatic organs and tissues (except the thymus) where they mature into B-lymphocytes. The presence of specific foreign substances within the body triggers the enlargement, proliferation, and differentiation of certain B-lymphocytes to form groups of identical antibody-secreting **B-cells,** or **plasma cells** (Figure 21–15). The antibodies, which are carried in the blood as plasma proteins, are capable of binding specifically with the particular foreign substance, or **antigen,** that elicited their production; the binding of antibody to

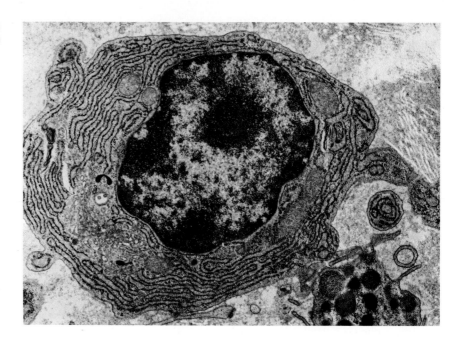

FIGURE 21–15 Electron micrograph of a plasma cell, or activated B-cell. Note that the cytoplasm is filled with an extensive network of rough endoplasmic reticulum, the site of antibody synthesis.

antigen leads to the inactivation or eventual destruction of the antigen by mechanisms discussed shortly.

## Antibody Structure

Antibodies are a large group of proteins, called *immunoglobulins* (Ig). Five classes of Ig molecules, each consisting of thousands of unique antibody molecules, have been identified; these classes are IgG (immunoglobulin G), IgM, IgA, IgD, and IgE, where the letters correspond to the Greek letters gamma, mu, alpha, delta, and epsilon, respectively. The most abundant antibodies in plasma are IgG. IgM molecules represent the first class of antibodies appearing in the serum in response to an antigen, and they provide most of the specific immunity to bacteria and extracellular viruses. IgA molecules are present in secretions such as saliva, tears, and mucus of the respiratory, digestive, and urogenital tracts. They provide the first line of immunological defense against bacteria and viruses. The benefit of the other two classes of immunoglobulin molecules, IgD and IgE, is not known, but IgE molecules are known to be involved in certain allergic responses.

The chemical structure of IgG antibodies has been extensively studied. Each IgG antibody molecule is composed of four polypeptide chains, consisting of two identical *short chains,* also termed *light (L) chains,* and two identical *long,* or *heavy (H), chains.* Both the light and heavy chains in an IgG molecule have two regions that can be identified: a *constant region,* whose amino acid sequence does not vary from one IgG molecule to another, and a region of variable amino acid sequence *(variable region).* The four chains composing an IgG molecule are held together in a Y configuration, such that the variable part of each light chain is associated with the variable part of a heavy chain (Figure 21–16). The variable regions of the four chains form two identical **antigen binding sites** at the ends of each of the arms of the Y. The binding of antigen to antibody occurs as a result of a specific chemical interaction between the antigen and the variable region.

## Function of Humoral Antibodies

How do antibodies function to protect the body against invading pathogens? The specific attachment of antibody to an antigen on the surface of a microbe leads to a number of responses that result in the destruction or inactivation of the microbe or its toxins.

***Activation of the Complement System.*** By far the most important result of the formation of an antigen-antibody complex is the enhancement of the inflammatory response, particularly by the activation of the complement system. Antigen-bound antibody possesses a binding site in the stem of the Y to which the first complement molecule in the complement sequence can bind. Binding of the first complement molecule to this site initiates the entire series of complement reactions. Once formed, complement components facilitate phagocytosis of the microbes or kill the microbes directly.

***Facilitation of Phagocytosis.*** The attachment of antibodies to surface antigens on the microbe also directly stimulates the activity of phagocytes, independently of the complement system. The net result is increased phagocytosis of the microbes.

***Neutralization.*** Another important aspect of antibody function is the direct neutralization of bacterial toxins and viruses. In order to harm a cell, toxins must combine with membrane receptors on the host cell, as must a virus in order to invade a host cell. By combining with toxins and viruses, antibodies may neutralize their effects by preventing their attachment to membrane receptors.

***Formation of Antigen-Antibody Lattices.*** The interaction of antibodies with microbes, toxins, and other foreign materials may also lead to extensive, interlocking lattices of antigen-bound antibody molecules (Figure 21–17). When these lattices become so large that they precipitate out of solution in the body fluids, they are readily phagocytized.

## Active and Passive Immunity

As mentioned, exposure to a particular antigen results in the proliferation of certain B-lymphocytes to form large numbers of cells that differentiate into antibody-secreting plasma cells. However, some of the B-lymphocytes that

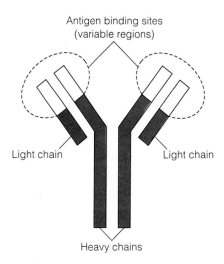

FIGURE 21–16 Schematic diagram of the structure of IgG antibodies. The colored areas of the light and heavy chains are the constant regions.

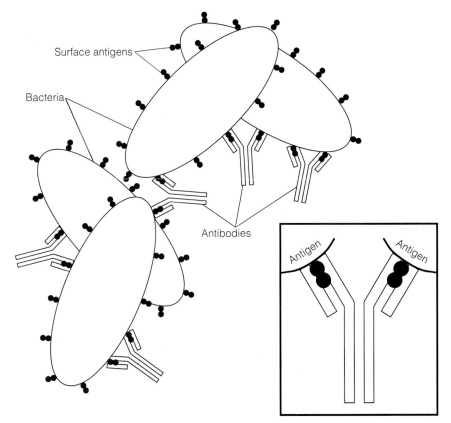

FIGURE 21–17 Schematic drawing of the formation of extensive antigen-antibody complexes. Inset: Greatly simplified representation of specific "lock and key" type interaction between antigen and antibody.

proliferate during this activation process do not differentiate fully into plasma cells. Instead, they give rise to "memory" cells that are stored in lymphoid tissues. "Memory" cells are responsible for immunological memory, by which the body "remembers" antigens to which it has been previously exposed. Upon reexposure to the same antigen, "memory" cells differentiate rapidly into antibody-secreting plasma cells, resulting in immediate production and secretion of large amounts of antibody that quickly react with the antigen before it has time to harm the host. This rapid immunological response upon reexposure to an antigen compares with the two or three days or more that are required for the optimal synthesis of antibody upon first exposure to the antigen.

The rapid response time of immunological memory provides the body with a high level of resistance to repeated infection by most microbes. Resistance of this type, acquired by previous exposure to an antigen, is called **active immunity.** It is because of active immunity that we contract certain infectious diseases such as measles and chicken pox only once. Unfortunately, we are repeatedly susceptible to certain infectious agents (for example, the influenza virus) since they continually evolve new surface antigens to which the body has not previously been exposed.

Immunological memory also provides the basis for **immunization,** also called **vaccination.** In this procedure, a dead or weakened pathogenic microbe, a harmless microbe with surface antigens similar to those of a pathogenic one, or antigens derived from a pathogenic microbe or its toxin are injected or given orally. These substances stimulate the production of plasma cells and "memory" cells. The latter provide immunological resistance upon exposure to the live, pathogenic microbe. The widespread use of immunization has greatly reduced the incidence of many dangerous infectious diseases.

If antibodies from one individual are transferred to another, the recipient acquires the immunity of the donor. Such immunity is termed **passive immunity.** Newborn infants have passive immunity acquired from the mother because maternal antibodies are able to cross the placenta during development. In addition, breast-fed infants acquire passive immunity by receiving antibodies in their mother's milk. In both of these cases, the maternal antibodies help prevent infection during the first several months after birth, when the baby's immune system has not yet developed full effectiveness. Sometimes, an individual who has been exposed to a dangerous disease or who is receiving drugs that suppress the activity of the immune system may receive injections of antibodies to help ward off the threat of infection. While passive immunity can be helpful in short-term situations, injected antibodies deteriorate within several weeks. In addition, the injected antibodies may themselves elicit an immune response that can be dangerous.

## Cell-Mediated Immunity

Cell-mediated immunity requires the activity of **T-lymphocytes.** T-lymphocytes arise from stem cells in the bone marrow that migrate to the thymus gland. There, the stem cells begin to differentiate into lymphocytes; they then migrate to the peripheral lymphoid tissues where they multiply and complete their differentiation into T-lymphocytes. When stimulated by contact with an antigen on the surface of a foreign or abnormal cell type, some of these T-lymphocytes enlarge and proliferate as **activated T-cells.** Instead of secreting antibodies, activated T-cells actually combine with the antigen on the cell surface in order to have their effects.

## T-cell Types

Three types of T-cells, each with a specific function, have been identified. These are **killer T-cells, helper T-cells,** and **suppressor T-cells.** Killer T-cells, upon combination with an appropriate antigen, release a group of chemicals called *lymphokines.* Some lymphokines are toxins that are able to kill cells directly. Other lymphokines act to enhance the inflammatory response by attracting neutrophils and monocytes to the area and by stimulating the phagocytic activity of macrophages. Thus, the effect of lymphokines is much like the activation of the complement system by antigen-antibody complexes. In addition to secreting lymphokines, killer T-cells may also secrete large amounts of interferon. Killer T-cells are primarily responsible for the destruction of fungi, parasites, cells infected with intracellular microbes, and organ transplants; they may be important in the destruction of tumors, as mentioned shortly (Figure 21–18).

Although some antigens stimulate the differentiation of B-lymphocytes into antibody-secreting plasma cells, other antigens do not. The activation of B-lymphocytes by these latter antigens requires the presence of helper T-cells. It is not clear how helper T-cells have this effect, but it has been proposed that contact between helper T-cells and antigen may cause the T-cells to secrete substances that bring about differentiation of B-lymphocytes into plasma cells.

Suppressor T-cells inhibit, or suppress, the response of plasma cells to the presence of antigen. Consequently, antibody production is reduced. The action of suppressor T-cells is thus important in preventing the overproduction of antibodies.

As with humoral immunity, each type of activated T-cell has memory cells that speed up the immune response upon subsequent exposure to the antigen. Thus, active immunity also exists for the cell-mediated response. Passive immunity can be conferred by administering activated T-cells isolated from an immune individual.

## Graft Rejection

In addition to protecting the body against certain types of infectious disease, cell-mediated immunity is responsible for **graft rejection,** the recognition and destruction of tissue or organ transplants. This function is, of course, detrimental to the host in cases where surgical transplantation of tissues or organs from one individual to another is performed to remedy certain pathological conditions in the recipient. However, it must be remembered that such transplants do not occur in nature, and the body has no way of differentiating between useful and harmful foreign cells.

Graft rejection occurs because of differences in surface antigens on the cells of the donor and on those of the recipient. Particularly important in graft rejection are surface glycoproteins called *histocompatibility antigens*. Histocompatibility antigens, which are genetically determined, are similar on all

FIGURE 21–18 (a) Scanning electron micrograph of a killer T-cell attacking two cancer cells. (b) Destruction of the cancer cell begins with the appearance of blebs (bubblelike structures) on its plasma membrane. (c) A cancer cell in the late stages of destruction.

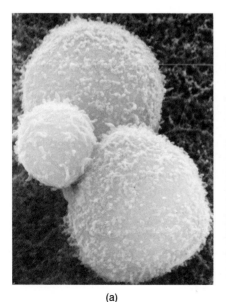

(a)

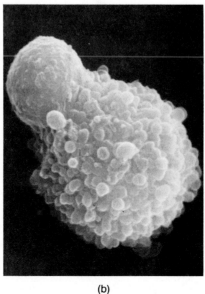

(b)

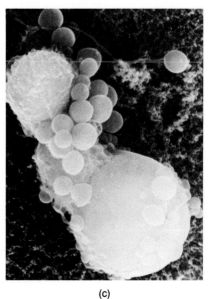

(c)

nucleated cells of a single individual, but differ between all individuals, except identical twins. The histocompatibility antigens provide, as it were, a "monogram" that distinguishes grafted cells from normal body cells. When killer T-cells of the body recognize foreign monograms on grafted cells, they attack and destroy the transplant.

In order to reduce the severity of graft rejection, histocompatibility antigens of the donor and the recipeint may be matched so as to minimize differences in those antigens that are most important in graft rejection. The identification of histocompatibility antigens is usually performed on an individual's white blood cells and is called *tissue typing*. Close matching of histocompatibility antigens reduces the severity of graft rejection, but does not eliminate it. Other methods of minimizing graph rejection include radiation of the lymphoid tissues and use of immunosuppressive drugs, both of which decrease T-cell numbers. However, these treatments also reduce the number of B-cells in the body and thus increase susceptibility to infection by diminishing the humoral immune response.

## Immune Surveillance

A process similar to graft rejection may also be involved in the detection and destruction of tumor cells by killer T-cells. It has been known for some time that the conversion of normal cells to cancerous ones is accompanied by the appearance of new cell surface antigens. There is evidence to suggest that these new antigens make the cells vulnerable to attack by T-cells. It has consequently been proposed that killer T-cells may be involved in a process known as **immune surveillance,** whereby they constantly examine cell surfaces for the appearance of new surface antigens and destroy any such cells that they detect. One researcher has suggested that everyone may get cancer once a day, but that cancer cells are normally destroyed through immune surveillance. Presumably, cancer would result when a cancer cell fails to be recognized as abnormal by the immune system. The role of T-cells in immune surveillance has not been conclusively demonstrated, and some recent research has identified a class of lymphoid cells termed *natural killer cells,* that are distinct from T-cells and that may perform immune surveillance.

## Role of Macrophages in the Immune System

Macrophages have an important, yet poorly understood, role in the response of lymphocytes to antigens. Antigenic material is apparently taken up by macrophages, which then somehow process it and "present" it to both B- and T-lymphocytes in such a way as to stimulate an immunological response from these cells. In addition, macrophages secrete a variety of substances that affect lymphocyte function.

## Activation of Lymphocytes by Antigens

In the previous discussion, we have skirted an important question regarding lymphocyte function: How are lymphocytes activated by specific antigens? The response of lymphocytes to antigens is explained by the **clonal selection theory,** proposed by Macfarlane Burnet in 1957 and now widely accepted (Figure 21–19).

According to the clonal selection theory, the body contains a large number of lymphocytes, each of which is capable of responding to a single, specific antigen. In the absence of the antigen to which it can respond, a lymphocyte circulates in the blood in a relatively inactive state. However, when the antigen enters the body for the first time, the antigen binds to a specific antigen receptor that is present on the cell surface of the lymphocyte. The binding of antigen to the receptor stimulates the proliferation and differentiation of the lymphocyte into a large number of activated lymphocytes, all of which respond to the same antigen. Such a group of genetically identical cells is called a **clone.** The

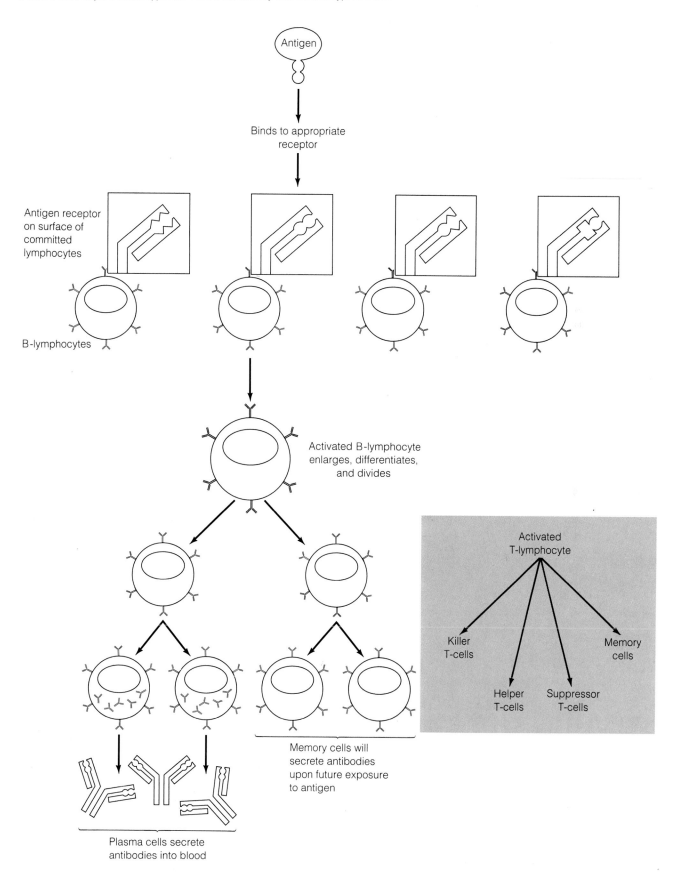

FIGURE 21-19 Summary of the clonal selection theory as it applies to the activation of B-lymphocytes by antigen. Inset: Activation of T-lymphocytes by antigen is similar to that of B-lymphocytes, although the identity of the surface receptor on T-lymphocytes is not known. Activated T-lymphocytes give rise to killer T-cells, helper T-cells, suppressor T-cells, and memory cells for each type of T-cell.

Antigen

Binds to appropriate receptor

Antigen receptor on surface of committed lymphocytes

B-lymphocytes

Activated B-lymphocyte enlarges, differentiates, and divides

Memory cells will secrete antibodies upon future exposure to antigen

Plasma cells secrete antibodies into blood

Activated T-lymphocyte

Killer T-cells

Helper T-cells

Suppressor T-cells

Memory cells

activation of a B-lymphocyte by the antigen will result in a clone of plasma cells capable of secreting antibodies that interact with the specific antigen; if a T-lymphocyte is activated by the antigen, a clone of activated T-cells will result, all of which are also capable of interacting with the particular antigen. In each case, memory cells are also produced as a result of the activation process. When the antigen has been destroyed or neutralized, the lymphocytes return to their inactive state, but the presence of memory cells will allow a fast immunological response upon subsequent contact with the antigen.

The important point of the clonal selection theory is that each lymphocyte is committed to respond to a particular antigen *before* it has encountered that antigen. Because there are millions of different lymphocytes, each capable of responding to a different antigen, the immune system provides a broad defensive screen.

## Diseases and Disorders of the Immune System

An effective immune system requires both humoral and cell-mediated responses. In combination, these two immune responses can provide a successful defense against most invaders. However, there are certain conditions, called *immune deficiency diseases,* that are characterized by the absence of a normal immune response.

### Immune Deficiency Diseases

In *congenital agammaglobulinemia,* a rare genetic disease, there are no functioning B-cells and consequently antibodies are not formed, but cell-mediated immunity is normal. Children with this disease are unable to mount an effective immune response against many types of bacteria, but they are able to reject transplants and to destroy some antigens that are attacked by T-cells. On the other hand, *congenital thymus deficiencies,* also genetic and rare, result in impaired cell-mediated immunity but relatively high levels of circulating antibodies. Some individuals have deficiencies of both the humoral and cell-mediated immune systems. All of these immune deficiency diseases are usually fatal early in life, although a few children have been raised in protective plastic bubbles that provide a sterile environment and hence prevent infection.

Another disease of the immune system, *acquired immune deficiency syndrome (AIDS),* has been identified in recent years. Individuals with AIDS succumb to various infections and a rare type of cancer [see *Focus: Acquired Immune Deficiency Syndrome (AIDS)*].

### Allergies

**Allergies,** or *allergic responses,* represent maladjustments of the immune system, in which a violent response is mounted in the presence of largely harmless antigens. Numerous substances trigger allergic responses, from pollen, animal dandruff, and dust, to drugs and food.

Upon initial exposure to one of these antigens, the body responds by the production of plasma cells that synthesize immunoglobulins of the IgE class and by the production of memory cells. The IgE antibodies circulate in the bloodstream, eventually attaching to the surface of mast cells; this attachment occurs in a way that does not interfere with the antigen-binding site of the antibody. When the same antigen is encountered again, the antigen combines with the IgE antibodies attached to mast cells, causing the mast cells to release histamine and related chemicals. Furthermore, the presence of antigen activates the memory cells, leading to the rapid synthesis of large amounts of additional IgE antibodies. These antibodies attach to other mast cells that will release histamine during another cycle of exposure to antigen.

The chemicals released by mast cells bring about a typical local inflammatory response, characterized by dilation and increased permeability of blood vessels with resultant swelling, redness, and warmth. The symptoms of the allergy depend upon the location of the inflammatory response. For example, airborne antigens such as pollen, animal dandruff, or dust combine with IgE-attached mast cells in the respiratory passages; the chemicals released from the mast cells bring about swelling of the nasal membranes, production of large amounts of tissue fluid, and contraction of the smooth muscle lining the air passageways. These physiological events cause the nasal congestion, sneezing, runny nose, and difficulty in breathing that are characteristic of allergies to airborne substances.

Sometimes, the allergic response is so extreme that chemicals released from mast cells enter the circulation and are carried throughout the body. This may result in a life-threatening reaction called *anaphylactic shock,* characterized by extremely low blood pressure and constriction of bronchioles. Treatment for anaphylactic shock requires the immediate initiation of procedures that will elevate blood pressure to normal levels and permit the free flow of air through respiratory passages.

Allergies may be treated with *"allergy shots,"* the injection of small quantities of antigen on a regular basis. This procedure is called *desensitization.* Steady exposure to the antigen stimulates the formation of high levels of plasma IgG antibodies that compete with IgE antibodies for antigen. When the individual comes in contact with the antigen, the IgG antibodies combine with the antigen before it is able to bind with IgE antibodies on mast cells.

## Autoimmunity

In certain diseases, the body may produce antibodies or activated T-cells that are directed against the body's own tissues. This condition is called **autoimmunity,** and results from a breakdown of the mechanisms by which the body distinguishes "self" from "nonself." Tissue damage due to autoimmunity can be severe. It is now recognized that many diseases are caused by autoimmunity, including rheumatoid arthritis, lupus, and several thyroid diseases.

## Transfusion Reactions and Blood Type Incompatibility

### ABO System

You will recall from Chapter 19 that there are four major blood types—A, B, AB, and O—that are genetically determined and that constitute the so-called ABO system. These blood types are based upon the presence or absence of certain antigens on the surface of an individual's red blood cells. As summarized in Table 21–2, Type A blood possesses antigen A; Type B possesses antigen B; Type AB possesses both antigens A and B; and Type O possesses neither A nor B antigens.

In the typical immune response, an individual would synthesize antibodies against a particular antigen only upon exposure to the antigen. However,

TABLE 21–2 Properties of ABO Blood Type System

| Blood Type | Surface Antigen | Plasma Antibody | Compatible Transfusions (type that can be accepted) |
|---|---|---|---|
| A | A | Anti-B | A,O |
| B | B | Anti-A | B,O |
| AB | A and B | Neither anti-A nor anti-B | A,B,AB,O |
| O | Neither A nor B | Anti-A and Anti-B | O |

# Acquired Immune Deficiency Syndrome (AIDS)

Toward the middle of 1978, physicians in several cities in the United States became aware of unusual disease symptoms in a small number of young homosexual men. Many of these men possessed a rare form of cancer, called Kaposi's sarcoma. In addition, they were susceptible to serious, often fatal, microbial infections. One particularly devastating infection, common in the group of afflicted individuals but rare in the general population, was a severe pneumonia caused by the protozoan *Pneumocystis carinii*.

As more individuals with these symptoms were identified, it became clear that the disease had one underlying manifestation: the immune system of victims was severely depressed. This conclusion was supported by several lines of epidemiological evidence. First, Kaposi's sarcoma had

been a very rare disease in the United States. When it had occurred, it primarily affected older men of Mediterranean origin, in whom it responded well to therapy, or individuals whose immune systems had been depressed by cancer chemotherapy or use of immunosuppressive drugs. In this latter group of individuals, as in the young men afflicted with the new disease, Kaposi's sarcoma responded poorly to chemotherapy, often causing death within two years. Second, the severe microbial infections seen in the disease victims not only were caused by microbes that did not usually produce serious illness in otherwise healthy individuals, but followed a progressively relentless course. Even with extensive antibiotic therapy, these infections often proved fatal. Because symptoms of the newly described disease suggested an acquired suppression of the

immune system—a suppression that did not have a hereditary basis—the disease was termed acquired immune deficiency syndrome or AIDS.

As more cases of AIDS were identified in the early 1980s, it became clear that certain groups of individuals were at high risk for contracting AIDS. These included homosexual and bisexual males with large numbers of sex partners (the largest single group, accounting for about 70 percent of all cases), male and female intravenous drug users, Haitian immigrants, and hemophiliacs. The early epidemiological data suggested that AIDS was caused by an infectious agent, presumably a virus, that is transmitted by sexual contact, in blood (on the dirty needles shared by intravenous drug users), and in blood products (such as the blood clotting factor that hemophiliacs must take). This mode of

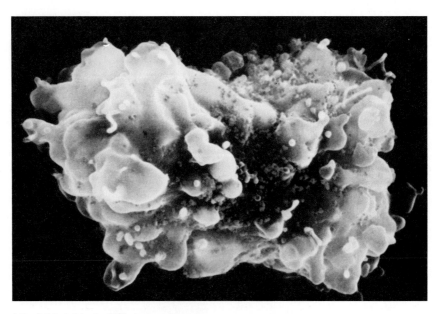

A T-cell infected by the AIDS virus.

transmission has been confirmed with the more recent identification of AIDS in heterosexual men and women who have had sexual contact with AIDS victims and in children of women with AIDS.

In 1984, a virus, now called human immunodeficiency virus, type 1 (HIV-1), was isolated from the blood of AIDS victims and was firmly implicated as the cause of AIDS. Although HIV-1 can infect many cell types, it preferentially infects and destroys helper T-cells that assist in activating B-lymphocytes to produce antibodies; the loss of helper T-cells, relative to suppressor T-cells, accounts for the substantial suppression of the immune system seen in AIDS. HIV-1 has now been isolated from blood, semen, saliva, and tears of AIDS patients. However, all epidemiological data indicate that the virus is transmitted in blood or blood products, or by sexual intercourse. There is no evidence to indicate that the virus is transmitted by casual contact, as, for example, in saliva released in sneezes and coughs.

Initial infection by HIV-1 may be followed by a short illness similar to acute mononucleosis, characterized by fever, swollen glands, sore throat, diarrhea, and pain in the muscles and joints. When these symptoms disappear after a few weeks, the person is left with antibodies to HIV-1 but appears otherwise healthy. It is not known whether this transient illness appears in all individuals when they are first infected with AIDS. Whatever the case, the virus enters a latent stage after infection is established. With the evidence we now have, it appears that the latent stage may continue indefinitely in many infected individuals, who remain healthy. In some individuals, however, the virus may become active and lead either to AIDS or to a milder, chronic infection, called AIDS-related complex (ARC), characterized by fever, swollen lymph glands, and a feeling of exhaustion. ARC may continue indefinitely, disappear, or turn into AIDS.

After someone has been infected by HIV, the virus may be transmitted to others, whether or not there are symptoms of disease. Blood tests for AIDS detect antibodies to HIV, and not the virus itself. Consequently, there is a period of several weeks following infection, but before antibodies appear in the blood, during which an infected person can transmit the virus to others but will test negative for the disease.

Epidemiological data show that the risk of developing AIDS following infection by HIV increases with time. From evidence accumulated so far, it has been predicted that about 15 percent of infected individuals will develop AIDS within five years of infection, 24 percent will develop AIDS within six years, and 36 percent within six and one-half years. It is difficult to make predictions for longer time periods because of the relative newness of the disease.

As of April, 1988, some 81,000 cases of AIDS have been reported in 137 countries. Somewhat more than 50,000 individuals have contracted AIDS in the United States alone; of these, about half have died. The World Health Organization has estimated that as many as 10 million people worldwide are infected by HIV, and the Center for Disease Control estimates that about 1,500,000 Americans harbor the virus.

Worldwide, epidemiologists have noted two general patterns of infection. In the first pattern, seen in Western industrialized countries and in Australia and New Zealand, most individuals with AIDS are homosexual and bisexual men, intravenous drug users, and the sexual partners of these individuals. In the second pattern, evident in Central Africa, Haiti, and parts of South America, AIDS afflicts primarily heterosexual men and women in approximately equal numbers, and it is transmitted mainly by heterosexual intercourse. About five percent of the people in urban populations in Central Africa are AIDS victims. Furthermore, in central West Africa, much of the AIDS is caused by a second AIDS virus, designated HIV-2. There have been very few cases of AIDS to date in the Middle East, Asia, and the Soviet Union, and little is known about transmission patterns in these areas.

Currently, there is no cure for AIDS, and the disease is considered universally fatal. AZT (*a*zidodeoxy*t*hymidine) is the only drug that has been approved by the FDA for AIDS treatment. It prolongs the life of AIDS patients with pneumocystis pneumonia by bringing about a partial, but temporary, restoration of immune system function. A number of other antiviral drugs are now in the testing stages. Attempts to develop a vaccine for AIDS have so far been frustrated by the fact that the virus can change its structure rapidly, and hence a vaccine that is effective against one substrain of the AIDS virus may not be effective against another.

Because the AIDS virus is transmitted only through sexual intercourse and in blood and blood products, it is relatively easy to protect against infection. Condoms will prevent the transmission of HIV during sexual intercourse. Intravenous drug users must avoid sharing hypodermic needles with others. Transmission of the virus in transfused blood and in other blood products (such as blood clotting factors used by hemophiliacs) has been reduced to a very low level now that these substances are routinely screened (see *Focus: Blood Transfusions and Artificial Blood* in Chapter 8).

# Acquired Immune Deficiency System (AIDS), cont'd.

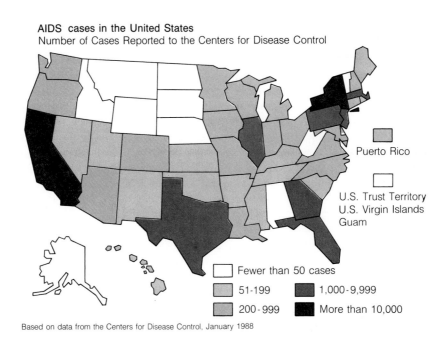

**AIDS cases in the United States**
Number of Cases Reported to the Centers for Disease Control

Puerto Rico

U.S. Trust Territory
U.S. Virgin Islands
Guam

- Fewer than 50 cases
- 51-199
- 200-999
- 1,000-9,999
- More than 10,000

Based on data from the Centers for Disease Control, January 1988

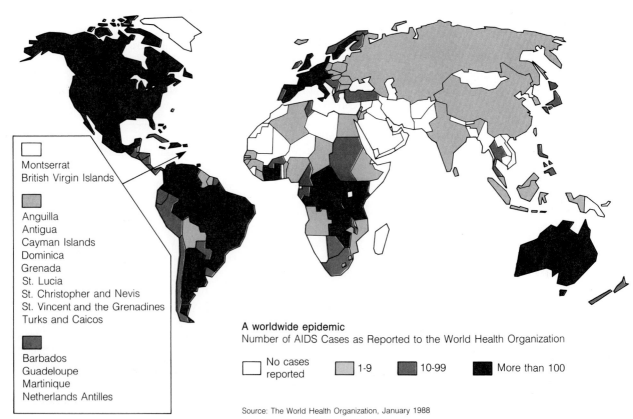

Montserrat
British Virgin Islands

Anguilla
Antigua
Cayman Islands
Dominica
Grenada
St. Lucia
St. Christopher and Nevis
St. Vincent and the Grenadines
Turks and Caicos

Barbados
Guadeloupe
Martinique
Netherlands Antilles

**A worldwide epidemic**
Number of AIDS Cases as Reported to the World Health Organization

- No cases reported
- 1-9
- 10-99
- More than 100

Source: The World Health Organization, January 1988

the ABO system is an exception to this rule. *Even without previous exposure to antigen,* all individuals possess plasma antibodies directed against any of the antigens in the ABO group that their red blood cells do not carry. Thus, for example, an individual with Type A blood possesses a high concentration of plasma antibodies directed against the B antigen (anti-B antibodies). In similar manner, Type B blood contains anti-A antibodies, type O blood has both anti-A and anti-B antibodies, and type AB blood has neither antibody type (see Table 21–2).

Because these antibodies are naturally present in blood of certain types, the transfusion of blood from one individual to another may lead to *immune transfusion reactions,* in which antibodies in the recipient's blood cause the agglutination, or clumping, of the donor's red blood cells. Agglutination is followed by hemolysis (rupture) of the red blood cells and the release of hemoglobin into the plasma. This type of reaction causes chills, fever, and shock. Furthermore, the presence of hemoglobin in the plasma may block the kidney tubules, impairing kidney function.

To avoid transfusion reactions, the *red blood cells of the donor* must not possess antigens that can react with *antibodies present in the recipient's blood.* Transfusions that will not cause immune reactions are summarized in Table 21–2. Because type AB blood possesses neither the A nor B antibody, individuals with type AB blood are called "universal recipients" because, in an emergency, they can receive blood of any type. Conversely, because type O individuals have red blood cells without the A or B antigens, they are "universal donors."

Transfusions may, of course, result in another incompatibility that we have not discussed: the presence of antibodies in the donor's blood plasma that are specific to antigens on the recipient's red blood cells. If only small amounts of blood are transfused, such an incompatibility is usually insignificant, since antibodies in the donor's blood are greatly diluted in the recipient's plasma. However, this type of incompatibility may cause a minor immune reaction.

ABO incompatibilities between mother and fetus may cause *hemolytic disease of the unborn.* If the blood of the fetus possesses red blood cell antigens that the mother's red blood cells do not possess, maternal plasma antibodies directed against these antigens may pass through the placenta into the fetal circulation. These antibodies destroy fetal red cells, and the infant is delivered with anemia which, however, is much milder than in the case of Rh incompatibility (discussed shortly). Soon after birth, the anemic infant may develop jaundice, a yellowing of the skin. Jaundice occurs because hemoglobin that is released into the plasma upon hemolysis is converted into bilirubin, a yellow pigment. Before the baby is born, bilirubin in the infant's blood passes across the placenta and is converted into bile in the mother's liver. At birth, however, the infant's liver cannot process all of the bilirubin that is produced, and it accumulates in the plasma, tinting the skin. At high levels, bilirubin may cause permanent brain damage. Newborn infants with jaundice are exposed to ultraviolet light, which induces the synthesis of enzymes in the liver that convert bilirubin into bile.

ABO incompatibility occurs most frequently among type O women carrying fetuses with type A or type B blood. Some studies have suggested that ABO incompatibilities may sometimes cause spontaneous abortions during early pregnancy.

## Rh System and Incompatibilities

Another group of red blood cell surface antigens forms the Rh (Rhesus) system, named for the Rhesus monkey in which similar surface antigens were discovered. Although three different antigens are involved, usually only the D-antigen is important. *Rh-positive* ($Rh^+$) individuals possess the D-antigen on their red blood cells, while *Rh-negative* ($Rh^-$) individuals do not. Rh incompatibilities can, under certain circumstances, cause serious immune reactions,

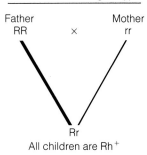

Rh⁺ father with RR genotype

Father        Mother
RR      ×       rr

Rr
All children are Rh⁺

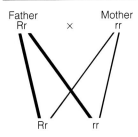

Rh⁺ father with Rr genotype

Father        Mother
Rr      ×       rr

Rr              rr

50% of children are Rh⁺
50% of children are Rh⁻

FIGURE 21–20 Inheritance of Rh blood types.

resulting in hemolytic disease. However, immune reactions involving the Rh system are different than those of the ABO system, since blood of an Rh⁻ individual does not have naturally occurring anti-D antibodies. Instead, these antibodies are synthesized only in response to exposure to the D-antigen.

Inheritance of the Rh system follows simple Mendelian patterns. The gene that determines the synthesis of the D-antigen is designated R and is dominant to its recessive allele (r). Consequently, individuals with RR and Rr genotypes are Rh⁺, while rr individuals are Rh⁻. Potential victims of hemolytic disease are Rh⁺ children born to Rh⁻ women. Such a situation can only occur if the father is Rh⁺ (Figure 21–20).

An Rh⁺ fetus carried by an Rh⁻ mother represents a potentially incompatible situation, since the red blood cells of the fetus possess the D-antigen while those of the mother do not. If the mother were to possess anti-D antibodies in her plasma, then these antibodies could cross the placenta and enter the blood stream of the fetus, initiating hemolytic disease. However, an Rh⁻ mother produces anti-D antibodies only upon exposure to Rh⁺ blood cells. During pregnancy, the maternal and fetal blood systems are completely separate, and the Rh⁺ red blood cells of the fetus are too large to cross the placenta. But when the placenta pulls away from the uterus at birth, placental vessels rupture, and fetal red blood cells may enter the maternal circulation in substantial number. The presence of Rh⁺ red blood cells in the mother's circulation stimulates a typical immune response in the mother, resulting in the production of anti-D antibodies. These may be synthesized by the mother for years.

Since the mother normally does not possess anti-D antibodies in her plasma prior to the delivery of her first Rh⁺ child, the first Rh⁺ child is usually not afflicted with hemolytic disease. In rare situations, the first Rh⁺ child *can* have hemolytic disease if the mother has previously developed anti-D antibodies as a result of having an earlier pregnancy involving an Rh⁺ fetus that was aborted or having a prior transfusion of Rh⁺ blood.

Because the mother possesses anti-D antibodies following the birth of her first Rh⁺ child, subsequent Rh⁺ children are usually afflicted with hemolytic disease (Figure 21–21). Although most fetuses with hemolytic disease survive until birth, they are born with severe anemia and are likely to develop neonatal jaundice. Because the fetus responds to the loss of its red blood cells by releasing into its circulation large numbers of immature red blood cells termed *erythroblasts,* this hemolytic condition is referred to as *erythroblastosis fetalis.* It is also called *hemolytic disease of the newborn.*

Babies born with erythroblastosis fetalis are treated by *exchange transfusion,* in which much of the baby's blood is replaced by Rh⁻ blood at birth. The Rh⁻ red blood cells in the transfused blood will not react with the maternal anti-D antibodies in the baby's circulation, and the Rh⁻ blood will tide the baby over the time required for the degeneration of the anti-D antibodies. Gradually, the normal production of red blood cells by the baby will replace the Rh⁻ cells with Rh⁺ ones.

A preventive treatment for Rh incompatibility is now used extensively. Immediately following the birth of her first Rh⁺ child or following an abortion where Rh incompatibility may exist, the mother is injected with anti-D antibodies. (The antibody preparation is referred to as *RhoGAM;* Rho· is an alternative designation for the D antigen, and GAM refers to immunoglobulin G.) The injected anti-D antibodies destroy the fetal Rh⁺ red blood cells before the mother's immune system can respond to them.

An interesting effect has been noticed in situations where the mother and fetus have both ABO and Rh incompatibilities: the relatively mild ABO incompatibility provides protection against the more serious Rh incompatibility. This apparently occurs because maternal antibodies to the A- or B-antigens carried by the infant's red blood cells destroy these cells, which are also Rh⁺, when they enter the mother's circulation at the time of birth. Thus the ABO incompatibility has the same effect as does the injection of RhoGAM.

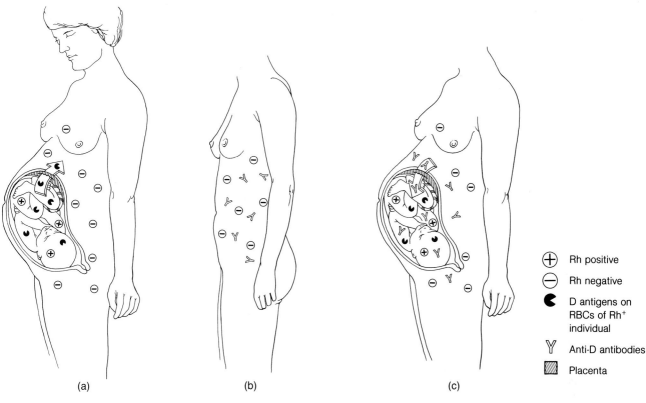

(a)　　　　　　　　(b)　　　　　　　　(c)

Rh positive

Rh negative

D antigens on RBCs of Rh⁺ individual

Anti-D antibodies

Placenta

FIGURE 21–21 Summary of events that occur during Rh incompatibility. (a) If a woman who is Rh⁻ carries an Rh⁺ child for the first time, placental rupture during childbirth may allow fetal red blood cells to leak into the mother's circulation. (b) The mother's immune system produces antibodies directed against the D antigens (anti-D antibodies) on the fetal red blood cells. (c) Upon a subsequent pregnancy with an Rh⁺ child, these antibodies enter the fetal circulation and destroy fetal red blood cells.

## Monoclonal Antibodies

In 1975, immunologists Cesar Milstein and Georges Köhler devised a technique for obtaining large amounts of a single antibody by fusing an activated B-cell with a myeloma cell (a cell from a bone marrow cancer). The hybrid cell that results, termed a **hybridoma,** retains properties of both of the cells that formed it; that is, it synthesizes a single type of antibody, and it shows the vigorous, unrestrained growth and division of all cancer cells. Antibody synthesized by a hybridoma is called **monoclonal antibody** because it is the product of a clone of cells that originates from one B-cell. Until the development of the hybridoma technique, it was impossible to obtain large amounts of a single antibody because activated B-cells will not thrive in tissue culture for any significant length of time.

Monoclonal antibodies are currently revolutionizing diagnostic testing because they allow the rapid and accurate detection of extremely small amounts of any antigen for which they are specific. Used in tests to detect minute amounts of specialized products of certain cancers, monoclonal antibodies can facilitate diagnosis of these cancers at an early stage. Monoclonal antibodies have also greatly simplified the process of tissue typing, since their selective binding can be used to detect specific antigens with ease.

Monoclonal antibodies are currently being used experimentally in the treatment of cancer and other diseases, and they show great promise. For example, recent experiments have demonstrated that the injection of monoclonal antibodies developed against leukemic cells can, in some cases, effectively control certain types of leukemia. Other experiments have used monoclonal antibodies to search out cancer cells. In one such experiment, a particular type of tumor was implanted in an experimental animal. When monoclonal antibodies, developed against the same tumor cells, were injected into these animals, the monoclonal antibodies specifically migrated and attached to the implanted tumor. If the same experiment is done with a monoclonal antibody that has a radioactive chemical attached to it, the position of the tumor in the body can be determined by scanning the body with a machine that detects radioactivity.

The ability of monoclonal antibodies to home in on specific antigens also promises to be useful in cancer chemotherapy. Many effective chemotherapeutic agents are extremely toxic to normal cells as well as to cancerous ones, a fact that often limits their usefulness. However, if a chemotherapeutic agent is attached to monoclonal antibodies that are specific to a certain cancer cell type, the monoclonal antibodies could be used to deliver the toxic agent specifically to the cancer cells. Such an approach is termed *"targeted drug therapy."*

## LYMPHATIC TISSUES

Several organs and tissues of the body are involved in the production and storage of lymphocytes and play a central role in the body's immunological response to foreign substances. These organs and tissues are part of the lymphatic system, which was discussed earlier as a part of the circulatory system (see Chapter 9). The lymphatic organs include the *lymph nodes, spleen,* and *thymus,* all of which are encapsulated structures. In addition, there are a number of nonencapsulated nodules (small masses) of lymphatic tissue that are primarily located in association with the loose connective tissue of mucous membranes lining the respiratory, digestive, and urogenital tracts.

### Lymph Nodes

Lymph nodes are the only lymphatic organs that have lymphatic vessels emptying into them. Because of this feature, lymph nodes are exposed to antigens present in lymph, and they play an important part in the body's immune reaction to these antigens. Lymph nodes are sometimes called *lymph glands,* but this is inaccurate, since they are not involved in secretion.

Lymph nodes vary in size from several millimeters to several centimeters in diameter, and are oval or bean-shaped. Some lymph nodes are located singly along medium-sized lymphatic vessels, but most are arranged in clusters along the course of the main lymphatic vessels that flow into the thoracic and right lymphatic ducts (see Chapter 9). These clusters of lymph nodes form six major groups: the cervical lymph nodes, axillary lymph nodes, thoracic lymph nodes, abdominal lymph nodes, pelvic lymph nodes, and inguinal lymph nodes (Figure 21–22).

The structure of a lymph node is shown in Figure 21–23. It consists of a mass of lymphatic tissue enclosed in a fibrous capsule. Fibrous partitions, or *trabeculae,* extend inward from the capsule, providing support and dividing the node into compartments. Lymph enters the node through a number of *afferent lymphatic vessels* that penetrate the capsule; flows through microscopic spaces, the *lymphatic sinuses,* that permeate the lymphatic tissue; and leaves the node through one or two *efferent lymphatic* vessels. As shown in the diagram, the lymphatic vessels are equipped with flaplike valves that prevent the backflow of lymph.

The tissue within the compartments of the node consists of networks of reticular fibers in which dense masses of cells, mostly lymphocytes, are held loosely in place. These masses are called *nodules.* The nodules are areas where, upon exposure to antigens in the lymph, B-lymphocytes and T-lymphocytes proliferate and differentiate into plasma cells and activated T-cells, respectively. These various types of lymphocytes enter the lymphatic sinuses where they are added to the lymph.

In addition to the production of lymphocytes, lymph nodes serve to filter the lymph as it passes through the lymphatic sinuses. The lymphatic sinuses are crisscrossed by reticular fibers to which macrophages cling. Bacteria and other

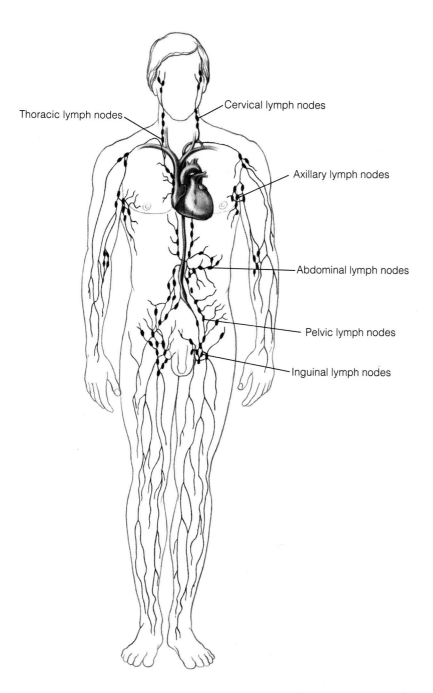

FIGURE 21-22 Location of the six major groups of lymph nodes.

Thoracic lymph nodes

Cervical lymph nodes

Axillary lymph nodes

Abdominal lymph nodes

Pelvic lymph nodes

Inguinal lymph nodes

microbes, cellular debris, and foreign particles such as dust are strained from the lymph when they become trapped in the fibers and are phagocytized.

The filtration function of lymph nodes is important in preventing the spread of localized infection throughout the body. However, the action of macrophages may cause inflammation, enlargement and tenderness of lymph nodes located on lymphatic vessels that drain the infected area. Indeed, this is an important symptom of infection. Infection may also result in the inflammation of lymphatic vessels themselves, causing red streaks on the skin where infected vessels course. This condition, known as *lymphangitis,* is a sign that infection is not being contained by the lymph nodes, and it signals the possibility of *septicemia* (blood poisoning) if the invading pathogen is carried in the lymph to the blood stream. Malignant cells that break away from cancerous tumors may also be filtered from lymph by lymph nodes. If they are not destroyed in the lymph nodes, these cells can multiply in the nodes to form new cancerous tumors. Because of this possibility, surgeons routinely remove surrounding lymph nodes with cancerous tissue.

FIGURE 21–23 Lymph node structure.

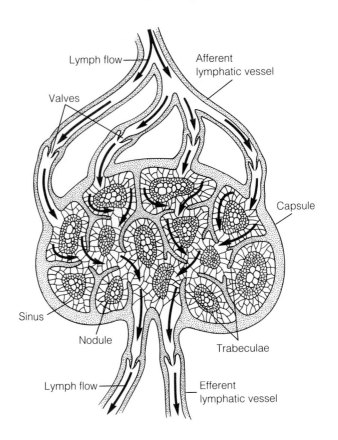

Lymph flow

Afferent lymphatic vessel

Valves

Capsule

Sinus

Nodule

Trabeculae

Lymph flow

Efferent lymphatic vessel

## Spleen

The spleen is located posterior to the left side of the stomach, just inferior to the diaphragm. It is usually about the size and shape of a clenched fist. It is soft and, because it contains much blood, dark purple in color. Branches of the splenic artery enter the spleen at several points along the spleen's *hilus,* a long fissure in its medial side. No lymphatic vessels enter the spleen, but lymphatic vessels and veins leave the spleen in association with the arteries that enter it. The veins eventually join to become the splenic vein; the lymphatic vessels drain tissue fluid from the spleen.

The surface of the spleen is covered by a fibrous and elastic connective tissue capsule, from which trabeculae, also formed of connective tissue, extend into the tissue of the spleen. The trabeculae divide the spleen into lobes and serve as supporting structures for vessels and nerves that enter and leave the spleen. The interior of the spleen is filled with a highly vascular tissue called the splenic pulp (Figure 21–24). There are two types of pulp and both are clearly visible with the naked eye: *red pulp* and *white pulp.*

Most of the splenic pulp is red pulp. It consists of a meshwork of reticular fibers (connective tissue) that supports vast numbers of red blood cells and macrophages as well as reticular cells (which secrete the reticular fibers), lymphocytes, granular leukocytes, and platelets. These aggregations of cells, forming the more or less solid tissue of the red pulp, are referred to as the *splenic cords.* Permeating the cords are thin-walled blood vessels, termed the *splenic sinusoids,* that empty into veins draining the spleen. Thus, the red pulp has a structure much like a mushy Swiss cheese—the sinusoids corresponding to the holes in the cheese, and the splenic cords to the cheese itself.

Within the red pulp are small areas of white pulp. The white pulp, which is the actual lymphatic tissue of the spleen, may be distributed as tiny islands in the red pulp or along the sheaths of arteries that leave the trabeculae and enter the red pulp. The white pulp contains T-lymphocytes and B-lymphocytes.

Although the spleen is not essential to life, it normally plays an important role in the filtration of blood. Blood leaves arteries that terminate in *marginal zones* where red and white pulp converge, percolates through this splenic pulp, and finally passes through the walls of the sinusoids, returning to the circulation. During this process damaged or aging red blood cells are phagocytized by macrophages in the red pulp. The macrophages break down the hemoglobin, returning most of the iron to the circulation and converting the heme to bilirubin, which is transported to the liver to become a constituent of bile. At the same time, antigens in the blood are exposed to lymphocytes in the white pulp, with the resultant production of plasma cells and activated T-cells. In addition, lymphocytes that are produced in the spleen enter outgoing lymphatic vessels.

The spleen has several additional minor functions. It stores about 200 mL of blood that can be expelled into the circulation in the event of severe hemorrhaging. Furthermore, the spleen serves as a hemopoietic (blood forming) tissue in fetal life, and retains the potential for blood cell formation in adults under certain pathological conditions.

## Thymus

The thymus is a pinkish-gray, bilobed, roughly triangular organ that is situated immediately posterior to the sternum (see Figure 16–14). Its internal structure was described in detail in Chapter 16. Lobules of tissue, formed by fine partitions of connective tissue that extend inward from the capsule surrounding

FIGURE 21–24 Location and structure of the spleen. (a) Position of the spleen in relation to other abdominal organs. (b) Enlargement of the spleen, showing its external structure. (c) Cross-section through a portion of the spleen.

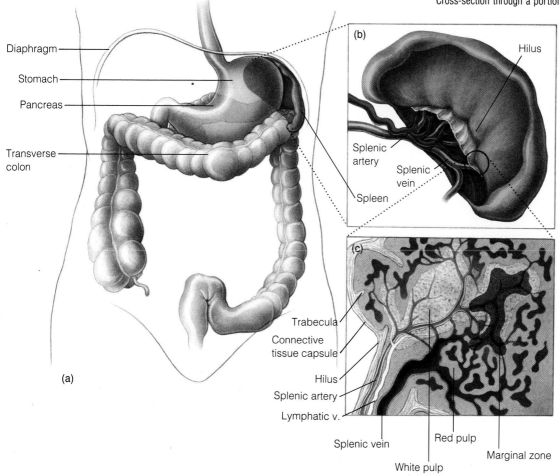

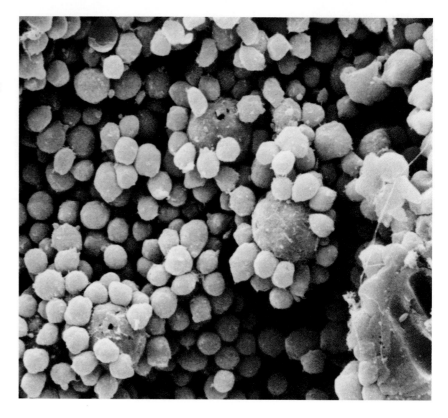

FIGURE 21–25 Scanning electron micrograph of the cortex of the thymus. Large numbers of lymphocytes (Ly) surround a few macrophages (Ma). From TISSUES AND ORGANS: A text-atlas of scanning electron microscopy, by Richard G. Kessel and Randy H. Kardon. W. H. Freeman and Company, Copyright © 1979.

each lobe, consist of an outer cortex and an inner medulla. The cortex, containing densely packed lymphocytes, is the active site of T-lymphocyte production (Figure 21–25). The newly formed T-lymphocytes migrate to the medulla; from there, they enter the blood circulation, where they are carried to the lymph nodes, spleen, and other lymphatic tissues. There are no afferent lymphatic vessels that convey lymph to the thymus, but efferent vessels drain tissue fluid from the thymus.

The thymus secretes a group of hormones, known collectively as *thymosins*. These hormones are believed to play an important role in the differentiation of T-lymphocytes and their maintenance in other lymphatic tissues.

## Nonencapsulated Lymphatic Tissue

While the lymph nodes and spleen are exposed to antigens present in the lymph and blood, respectively, the nonencapsulated nodules of lymphatic tissue serve primarily to protect the body against invaders that may penetrate various mucous membranes of the body. The largest groups of such tissues are the *palatine, pharyngeal,* and *lingual tonsils* that surround the opening to the respiratory and digestive tracts. The tonsils are described in Chapters 10 and 11. Other small nodules of lymphatic tissue are scattered in the loose connective tissue of the mucous membranes lining the respiratory, digestive, and urogenital tracts, and are located in a few other sites as well.

The nonencapsulated lymphatic nodules contain both B- and T-lymphocytes. In addition, there are regions within the nodules where these lymphocytes proliferate and differentiate.

## Study Questions

1. **a)** What are resident flora?
   **b)** Under what conditions can they become pathogenic?
2. Describe Koch's procedure for identifying a pathogenic microbe.

3. What significant health problems have been created by the overuse of antibiotics?
4. How does the cellular structure of prokaryotes differ from that of eukaryotes?
5. What characteristics of the smallpox virus permitted its eradication?
6. Describe the structure of a "typical" animal virus.
7. List the five stages of infection by a cytocidal virus and summarize the events that occur at each stage.
8. Distinguish between latent and slow viruses.
9. Explain how interferon inhibits viral infections.
10. List several external defense mechanisms of the body.
11. List the five classic symptoms of inflammation.
12. Describe how neutrophils and monocytes combat infection by microbes.
13. Distinguish between humoral and cell-mediated immunity.
14. **a)** Describe the structure of an IgG antibody.
    **b)** What is the function of the variable regions?
15. **a)** What is the molecular basis of active immunity?
    **b)** Distinguish active from passive immunity.
16. Indicate the functions of each of the three T-cell types.
17. How are T-cells important in graft rejection?
18. Write a short essay describing Macfarlane Burnet's clonal selection theory.
19. What is the rationale behind allergy shots?
20. Explain why $Rh^+$ children born to $Rh^-$ women are at risk for hemolytic disease, whereas $Rh^-$ children born to $Rh^+$ women are not.
21. **a)** What is a monoclonal antibody?
    **b)** How is it produced?
    **c)** How are monoclonal antibodies used to treat disease?
22. Explain the function of lymph nodes.
23. Distinguish between the red and white pulp of the spleen.
24. Describe the role of the thymus.

# CANCER

Directly or indirectly, cancer touches the lives of us all. One quarter of all Americans will develop cancer during their lifetimes, and about one in six will die of it. Behind these grim statistics, though, is some encouraging news. In the last several years, we have made impressive strides in understanding the cause of cancer. This information, coupled with knowledge we have acquired about the operation of the body's immune system, shows promise of leading to new treatments and methods of prevention.

## WHAT IS CANCER?

Popular beliefs to the contrary, cancer is not a modern disease. Indeed, it may well be as old as the earliest multicellular plants and animals. The petrified bones of dinosaurs that lived more than 60 million years ago show cancerous growths. The Java man (*Homo erectus*), a representative of an early species of humans, suffered cancer of the thigh bone, as shown in its 300,000-year-old remains. Evidence of cancer is also present in numerous mummies and skeletons of ancient periods (Figure 22–1).

The first descriptions of cancer appear in Egyptian papyri, dating as far back as 2500 B.C. The Ebers papyrus (1500 B.C.) suggested treating cancer by using a wooden instrument that was applied to the tumor and spun rapidly in order to burn the tumor out. Greek and Roman physicians also recognized cancer and prescribed a variety of treatments. Cancer derived its name during the dominance of Greek and Roman medicine. "Cancer" is the Latin word for crab and apparently derives from the observation of Hippocrates that breast cancer in its advanced stages consists of a central lump of tissue with processes, or arms, extending from it, resembling the form of a crab.

Although cancer is clearly not a modern disease, it is considerably more prevalent today than it was in the past. Since 1900, the death rate from cancer in the United States has more than tripled. Cancer is now second only to heart disease as the most common cause of death; in 1900, cancer was the sixth most common cause of death in the United States (Figure 22–2).

The rise in mortality from cancer during the years between 1900 and the present is directly attributable to a decrease in mortality from infectious diseases. More than 50 percent of the deaths in 1900 in the United States were due to infectious diseases. With widespread acceptance of the germ theory of disease in the early twentieth century came vast improvements in public health practices. This led to a dramatic reduction in the incidence and severity of infectious diseases, with the result that deaths from such diseases now account for only about one in twenty.

The reduction in mortality from infectious disease has led to an increase in the average life span of Americans from about 50 years in 1900 to over 70 years today. In addition, the age structure of the American population has changed dramatically during this period. The number of people 65 years of age or older has more than tripled from 4 percent in 1900. As shown in Figure 20–2, mortality from cancer rises dramatically with age. Put another way, an older individual is more likely to die of cancer than a younger one. Thus, as the average age of the population has risen, so has the death rate from cancer.

### Benign and Malignant Tumors

Cancer is a disease characterized by the development of malignant tumors. A **tumor,** or **neoplasm,** is any abnormal growth of tissue that lacks functional coordination with the rest of the body. For example, a tumor of adipose tissue will not release its fat even in a starving individual; a tumor of an endocrine gland may or may not produce hormones, but if it does, the level of hormone production is not regulated to meet the metabolic demands of the body.

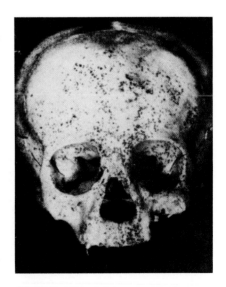

FIGURE 22–1 This skull from a pre-Columbian Inca Indian shows dark scars produced by the spread of melanoma, a type of skin cancer, throughout the body.

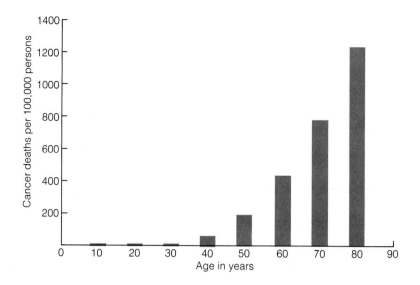

FIGURE 22–2 Number of cancer deaths per 100,000 individuals of each age during 1979 in the United States.

Two general types of tumors can be distinguished, based upon the degree to which they interfere with normal physiological functioning. **Benign tumors** grow without actively destroying neighboring tissue or disrupting the general metabolism of the body. They are generally surrounded by a fibrous capsule and remain localized at the site of origin. Although benign tumors can grow to very large sizes, they are usually not life-threatening unless they are inoperable and occur in an area where their growth destroys normal tissue by physically compressing it. For example, in a benign brain tumor, death may occur as a result of disabling pressure on the brain as the tumor grows.

**Malignant tumors,** also called **cancerous tumors** or **cancers,** grow in an unregulated fashion, infiltrating and destroying normal tissues and organs. Cancerous tumors, unlike benign ones, do not usually remain localized, especially during the latter stages of their development. Instead, they have the capacity to become widely disseminated in the body by a process known as **metastasis.** In fact, it is their ability to metastasize, or spread, that makes cancerous tumors so dangerous.

It is important to point out that the distinction between benign and malignant tumors is not always a clear one, since there are tumors with intermediate properties that cannot be unambiguously classified even by an experienced pathologist. Actually, tumors fall into a spectrum of types from the highly malignant to the extremely benign. Even among clearly malignant tumors, the degree of malignancy may vary strikingly. For example, basal cell tumors, a type of skin cancer, grow slowly and rarely metastasize, while melanoma, a skin cancer of melanin-producing cells, grows rapidly and has often metastasized by the time it is recognized.

## Types of Malignant Tumors

Malignant tumors are classified by the type of tissue in which they originate. Four major classes have been defined: *carcinomas, sarcomas, leukemias,* and *lymphomas.*

**Carcinomas,** which constitute about 85 percent of all cancers, are derived from the epithelial tissues (the embryonic ectoderm and endoderm) that cover free surfaces, line tubes or cavities, and form parts of various organs. This includes tissue that makes up the skin; lines ducts and vessels as well as organs of the digestive tract, the urinary tract, and genital tract; forms glands and nerves; and is present in such organs as the lungs, breasts, kidneys, liver, and brain (Figure 22–3).

**Sarcomas** are cancers that are derived from connective tissue, excluding blood-forming tissues and lymphatic tissues. Sarcomas include cancers of bone, cartilage, tendons, and muscle, and they represent about 2 percent of all cancers.

Cancers that originate from blood cell-forming tissue in the bone marrow and lymphatic system are called **leukemias.** Leukemias, accounting for roughly 4 percent of all cancers, are characterized by the presence of large numbers of abnormal white blood cells in the circulation and in the blood-forming tissues, where their proliferation crowds out normal cell types.

**Lymphomas** are solid, malignant tumors that arise in the lymphatic tissues, usually in lymph nodes. They are formed by the proliferation of abnormal lymphocytes that remain localized in lymphatic tissue and, at least initially, do not spill out into the blood. About 5 percent of cancers are lymphomas.

## Properties of Cancer Cells

As compared to normal cells, cancer cells have four distinguishing properties. They are permanently changed into a cancerous state, their rate of growth and cell division is not properly regulated, they do not differentiate correctly, and they tend to invade and destroy other tissues of the body.

### Transformation

A cancer cell arises from a change in a normal cell. This change is referred to as **transformation.** Whatever the cause of transformation, it is passed on from one cell to another during cell division. Hence, descendants of cancer cells are cancerous.

### Growth and Division

Cancer cells grow and divide with less restraint than cells of the tissues from which they are derived. This does not mean that all types of cancer cells divide at a similar, maximal rate. Indeed, many normal cell types grow and divide more rapidly than many cancer cells. It is not the rate of division *per se* that distinguishes cancer cells from normal ones, but the fact that cell growth and division are not coordinated with the requirements of the body. In almost all normal tissues of the adult organism, there is a gradual *turnover* of cells; that is, as cells that form the tissue reach the end of their normal life span, they must be replaced by new cells. (Recall, however, that nerve cells, cardiac muscle cells,

FIGURE 22–3 Invasion of normal breast tissue by cells of a breast carcinoma.

and skeletal muscle cells are not capable of division in the adult, and hence cells lost by disease, injury, or normal degenerative processes cannot be replaced.)

Cell turnover is well illustrated by the example of red blood cells. An average adult has some 25 trillion red blood cells that have a life span of about 120 days. Thus, on the average, 25 trillion red cells must be replaced every 120 days, or about 2.5 million every second. If additional red blood cells are lost as a result of hemorrhage or some pathological process, then more red blood cells must be replaced, such that the total number of red blood cells in the circulation remains constant. The important point is that the rate of red blood cell production, which occurs by the division and maturation of precursor stem cells in the bone marrow (see Chapter 8), is precisely regulated to equal the rate of red blood cell loss. In any normal tissue in which cell turnover occurs, the rate of cell division is regulated closely, but flexibly. By some means that we do not understand, cancer cells have lost these regulatory mechanisms.

### Differentiation

Cancer cells do not differentiate properly and consequently do not perform the specialized functions or attain the particular properties of the cells from which they are derived. In the normal situation, the replacement of cells lost from a tissue usually occurs by the division of stem cells, as we noted above in the case of red blood cell production. Stem cells are relatively undifferentiated cells, similar to the undifferentiated cells that give rise during embryonic development to the 100 or so different cell types of the body. When a stem cell divides, it gives rise, on the average, to one stem cell and to another cell that begins to differentiate along a genetically determined pathway. This pathway involves a number of cell divisions and a successive restriction in function and structure, ultimately to form a cell that stops dividing and that performs specialized functions. (Refer to Chapter 18 for a more detailed discussion of the process of cellular differentiation.)

Cancer cells arise as a result of events that disrupt normal differentiation. They do not complete all steps of the normal pathway of differentiation, but instead remain blocked at some early stage. This occurrence gives rise to a poorly differentiated cell that remains poorly differentiated throughout its lifetime and passes this defect on to its descendants. Because they are not completely differentiated, cancer cells retain the capacity to divide that is lost by most fully differentiated, mature cells. In addition, cancer cells that are blocked at an early stage of differentiation tend to divide faster and to be more malignant (metastasize more readily) than cells that are blocked at a later stage.

### Metastasis

Another characteristic of cancer cells is their ability to metastasize, as mentioned briefly above. Metastasis occurs in several steps (Figure 22–4). It begins when cells from the original tumor invade nearby tissues, eventually penetrating the walls of blood or lymphatic vessels. Individual cells or small clumps of cells are carried through these vessels to distant sites, where they become lodged in capillaries. There, they pass through the walls of the capillaries and invade normal tissues, where they establish secondary tumors. The growth of secondary tumors is supported by the development of new blood vessels that penetrate the tumor, providing essential nutrients and oxygen, and removing wastes.

When metastasizing cells invade a lymphatic vessel, they are often trapped in a nearby lymph node where they proliferate. For example, when a breast cancer metastasizes, tumor cells are found in axillary lymph nodes that are on drainage routes from the original tumor. Although lymph nodes temporarily restrict metastasis, some cells will eventually escape from the nodes and spread throughout the body.

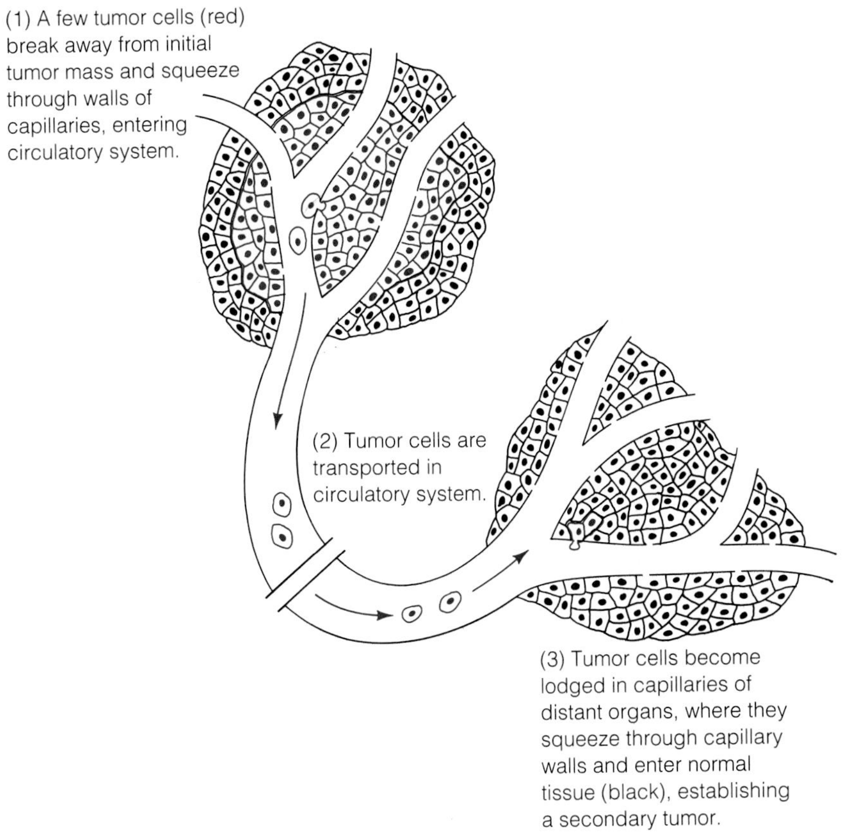

(1) A few tumor cells (red) break away from initial tumor mass and squeeze through walls of capillaries, entering circulatory system.

(2) Tumor cells are transported in circulatory system.

(3) Tumor cells become lodged in capillaries of distant organs, where they squeeze through capillary walls and enter normal tissue (black), establishing a secondary tumor.

FIGURE 22–4 Events occurring during metastasis, the process by which malignant tumor cells at the original site of a tumor spread to other areas of the body.

Two characteristics of cancer cells encourage metastasis. First, cancer cells are usually more motile than normal cells. Second, cancer cells, unlike normal cells, do not have a strong tendency to adhere to one another or to other cells of the body. Presumably, both of these characteristics are a consequence of cell surface changes that occur as a result of transformation.

Metastasis is the most devastating aspect of cancer. Most cancer patients die not from the effect of the original tumor, but from metastasis. Metastasis is very difficult to treat for a number of reasons. In many cancers, metastasis has already occurred by the time the disease is first diagnosed. In addition, secondary tumors that result from metastasis are often difficult to locate or are present in organs that cannot tolerate effective concentrations of chemotherapeutic agents (anticancer drugs). Most importantly, recent research has shown that cells of a single tumor are biologically heterogenous, meaning that they contain subgroups of cells with different biological properties. The cells of different subgroups can vary in their response to drug treatment, in the ease with which they metastasize, and in their ability to evade the body's immune system. Secondary tumors arising from different subgroups of cells may have very different properties, and thus a single chemotherapeutic agent or treatment regimen is not liable to eliminate all of the secondary tumors. In fact, recent evidence suggests that treatment for cancer may favor the emergence of new subgroups of cells with enhanced metastatic potential.

For reasons we do not understand, metastasis occurs in particular patterns. Lung cancer, for example, often metastasizes to the brain, whereas breast cancer most frequently metastasizes to the lungs, bones, and liver.

## Molecular Events That Lead to Malignancy

Because transformation to the malignant state is a permanent change in a cell, it presumably results either directly or indirectly from mutations in the cell's DNA or from the altered expression of information coded in DNA. This conclusion has considerable support. As we shall see, most chemical agents that cause cancer are also mutagens. In addition, recent research on some cancer-causing viruses has demonstrated that they may cause cancer by bringing about the inappropriate expression of dormant genetic information that is involved in the control of cell division and differentiation.

## CAUSES OF CANCER

Cancer was once thought to be an inescapable disease, an inevitable consequence of aging or heredity. Although it is certainly true that some rare cancers are genetically determined and that susceptibility to cancer may be inherited, hereditary factors alone account for only a few percent of all cancers. We now recognize that the vast majority of cancers are caused by environmental agents that we encounter during our lives.

Evidence to support this conclusion comes from studies that compare the incidence of cancer in different geographical areas of the world. For example, the incidence of stomach cancer in Japan is four times higher than in the United States. On the other hand, breast and colon cancer are both about five times more frequent in the United States than in Japan. Differences in the incidence of specific types of cancer in different countries had long been explained as a result of differences in the hereditary susceptibility to cancer of various population groups. However, recent studies of people who migrate from one country to another show that migrants tend to acquire the pattern of cancer incidence characteristic of their adopted countries. For instance, Japanese who migrate to the United States acquire lower rates of stomach cancer and show an increase in the incidence of colon and breast cancer. Sons and daughters of these emigrants show a cancer pattern even closer to that of Americans (Table 22–1). The change in cancer incidence among Japanese living in the United States coincides with the abandonment of their traditional diet in favor of an American diet. These epidemiological studies indicate that, for the most part, cancer is related to long-standing features of life-style including such things as diet, social habits, individual behavioral characteristics, technological practices, average life expectancy, and endemic infectious diseases.

The link between cancer and exposure to certain environmental agents also explains why the incidence of cancer increases with age. In general, the older a person is, the greater is his or her cumulative exposure to environmental agents that cause cancer.

### Carcinogenic Agents

Environmental agents that cause cancer are called **carcinogenic agents** or **carcinogens.** These are commonly grouped into three general categories:

TABLE 22–1 Cancer Mortality in Japanese Immigrants to the United States[a]

| Cancer type | Native Japanese | Children of Japanese immigrants | General U.S. population |
|---|---|---|---|
| Stomach | 100 | 38 | 17 |
| Colon | 100 | 288 | 489 |
| Breast | 100 | 136 | 591 |

[a] From Haenszel, W. & M. Kurchara (1968). *J. Nat'l Cancer Institute 40*, 43. Data for stomach and colon cancer are for men only; breast cancer, for women. Values are standardized relative to mortality in native Japanese, which has been assigned an arbitrary value of 100.

*chemicals, radiation,* and *viruses.* Before discussing each of these groups of carcinogenic agents, we will consider how carcinogens are detected.

## Testing for Carcinogens

Most of the proven human carcinogens were first implicated by epidemiological (population) studies that trace the incidence of cancer in large groups of individuals exposed to specific environmental agents. One of the first such studies was published in 1775 by the British physician Percivall Pott. Pott reported a high incidence of scrotal cancer among men who had been chimney sweeps as youngsters. Common practice of the day was for young boys to clean chimneys by wriggling naked through them. In the process, creosote and soot lodged in folds of the scrotal skin, dissolving in natural skin oils and causing cancer.

As useful as epidemiological studies have been in detecting carcinogens, their value is limited because of the great difficulty in obtaining supporting data. Epidemiological studies require large numbers of individuals and a pronounced trend, and they can be confounded by differences in life-style, heredity, and personal histories. In addition, they do not detect carcinogens until many people have been harmed.

To overcome problems with epidemiological studies, a variety of other tests have been developed to detect carcinogens. They all depend upon the assumption that those things that are carcinogenic in experimental organisms are carcinogenic in man—an assumption that has proven to be generally valid (see Chapter 1).

*Animal Testing.* In these tests, experimental animals are treated with a suspected carcinogen, and an untreated group serves as a control. Mice are typically used and are observed for the development of tumors throughout their life span of about three years. In order to obtain reliable data, a minimum of about 400 animals must be tested. Testing a single substance in this way may cost in excess of $400,000 and may occupy the time of several professionals for the duration of the experiment.

*Short-Term Tests.* Because of the expense and length of time required for animal testing, several short-term tests that take only a few days or weeks are commonly used as initial screening devices. These employ mammalian cells or microorganisms growing in artificial medium.

One of the most useful of the short-term tests was devised by molecular biologist Bruce Ames. The **Ames test** takes several days and costs only a few hundred dollars. It uses bacteria as test organisms and assumes that substances that are carcinogenic in animals are also mutagens in all organisms, including bacteria. In fact, of 300 substances with known carcinogenic potential in humans, 90 percent are mutagens in bacteria; conversely, all mutagens in bacteria are carcinogens in animals.

In the Ames test, hundreds of millions of cells of a mutant strain of *Salmonella,* which is genetically incapable of synthesizing the amino acid histidine, are treated with the suspected carcinogen. If the substance is a mutagen, it will cause many mutations, one type of which is a mutation that will once again allow the bacteria to synthesize histidine (Figure 22–5). The percentage of bacteria capable of synthesizing histidine following treatment with a given dose of test substance is an indicator of the mutagenicity of the chemical.

In recent years, it has been shown that many chemicals become carcinogenic only when metabolized to an active form; that is, the body converts them into carcinogens. This occurs as a result of the inadvertent action of enzymes in the liver that normally are involved in the detoxification of harmful substances. Chemicals that are carcinogenic only when converted to an active form are called **precarcinogens.** Because of the importance of liver enzymes in the

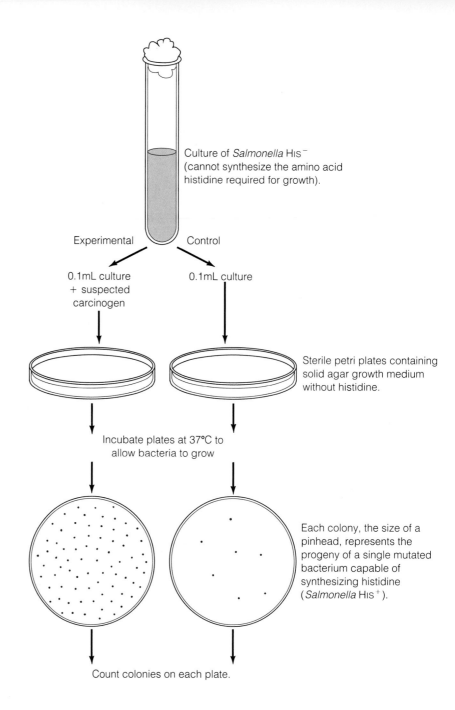

FIGURE 22–5 Summary of the Ames test for suspected carcinogens.

Culture of *Salmonella* His⁻ (cannot synthesize the amino acid histidine required for growth).

Experimental    Control

0.1mL culture + suspected carcinogen

0.1mL culture

Sterile petri plates containing solid agar growth medium without histidine.

Incubate plates at 37°C to allow bacteria to grow

Each colony, the size of a pinhead, represents the progeny of a single mutated bacterium capable of synthesizing histidine (*Salmonella* His⁺).

Count colonies on each plate.

activation of precarcinogens, suspensions of mammalian liver cells are now commonly added to the bacterial test system in the modified Ames Test.

## Carcinogenic Chemicals

The effectiveness of a chemical carcinogen in causing cancer depends upon such things as dose, length of exposure, the tendency of the substance to concentrate in certain organs, the ability of the body to detoxify or to activate the substance, and so forth. An important characteristic of chemical carcinogens is that there is usually a long delay between initial contact with the carcinogen and the appearance of cancer. In humans, the delay may be a few years or as many as 40 years. It is this long delay that makes the identification of chemical carcinogens difficult.

Although a large number of chemicals with carcinogenic potential have been identified, only a few of these are of practical concern to the general populace. These troublesome carcinogens are ones that are widely distributed and present in effective concentrations in the environment. We will emphasize these important carcinogens in the following discussion.

***Tobacco Smoke.*** The relationship between tobacco smoke and cancer is now firmly established. In fact, more cancer deaths are attributed to tobacco smoke than to any other single carcinogen. Tobacco smoke is associated directly with cancer of the lung, mouth, larynx, esophagus, pancreas, bladder, and kidney. (Table 22–2). Overall, smoking is responsible for approximately 41 percent of all male cancer deaths and for 14 percent of all female cancer deaths in the United States. These figures mean that between 25 and 30 percent of *all* cancer deaths in the United States are caused by smoking!

Lung cancer is by far the most frequent cancer caused by tobacco smoke. Some 32 percent of all cancer deaths in males and about 12 percent of those in women are from lung cancer directly attributable to smoking. Cancer of the lung in women has recently overtaken breast cancer as the most frequent cause of cancer death.

The increase in lung cancer mortality rate for smokers as compared to nonsmokers is variable and depends, as with exposure to any carcinogen, on a variety of factors including length and degree of exposure (length of time smoking has continued and amount smoked). For example, several studies of long-term, heavy smokers have shown lung cancer mortality rate increases of some 20-fold over nonsmokers. These rates must also be looked at in the context of general air pollution. Lung cancer among all individuals occurs more frequently in areas where air pollution is heavy, but tobacco smokers living in polluted areas have a much higher incidence of lung cancer than nonsmokers. Figure 22–6 shows the correlation between cigarette smoking and deaths from lung cancer in England between 1900 and 1970. These curves demonstrate the fact that roughly 20 years elapse between the time a group of individuals begins to smoke and the time a measurable increase in lung cancer deaths is evident.

Tobacco smoke contains several thousand chemicals, at least 30 of which are carcinogenic in test systems. Other chemicals in cigarette smoke appear to act as tumor promoters (described shortly). In addition to these effects, a great deal of evidence suggests that tobacco smoke impairs the function of the immune system and thus may weaken the body's defenses against cancer.

***Carcinogens in Food.*** It has been estimated that 30 to 40 percent of human cancers are related to diet. Although this figure derives from theoretical calculations and is largely unsubstantiated, there is no doubt that our food contains a variety of carcinogens or substances that contribute to the formation of carcinogens.

TABLE 22–2 Cancer Deaths Attributed to Smoking

| | Males | | Females | |
|---|---|---|---|---|
| Site of cancer | Estimated deaths in 1981 | % attributed to smoking[a] | Estimated deaths in 1981 | % attributed to smoking |
| lung | 77,000 | 97 | 28,000 | 74 |
| mouth | 6,300 | 78 | 2,850 | 46 |
| esophagus | 5,800 | 83 | 2,300 | 50 |
| pancreas | 11,500 | 28 | 10,500 | 22 |
| larynx | 3,100 | 99 | 600 | 57 |
| bladder | 7,300 | 28 | 3,300 | 22 |
| kidney | 4,900 | 28 | 3,200 | 22 |
| all tumors | 227,500 | 41 | 192,500 | 14 |

Sources: Reif 1979; American Cancer Society 1980

[a]For males, the figures include fatalities attributed to smoking cigars and pipes, as well as cigarettes.

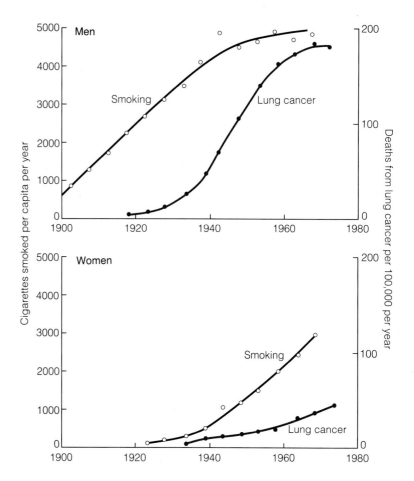

FIGURE 22–6 The relationship between smoking and lung cancer deaths in England. Increased rates of lung cancer mortality, although clearly correlated with cigarette smoking, are not evident until members of a population have been smoking for about 20 years.

1. *Natural Carcinogens.* Some foods are known to contain natural substances with proven carcinogenic potential in test systems. Tea, for example, contains the carcinogen tannin, mushrooms contain hydrazine, and sassafras tea contains safrole. Either alcohol is carcinogenic itself or alcoholic beverages contain one or more substances that are carcinogenic. This conclusion is supported by studies which show that moderately heavy drinkers have a two- to three-fold increased risk of cancer of the mouth, larynx, and esophagus as compared to nondrinkers. Drinking alcoholic beverages in conjunction with smoking greatly increases the risk of cancer at these sites. Recent studies have also shown that women who consume as few as three drinks a week have a 40 to 50 percent increased risk of breast cancer.

2. *Additives and Preservatives.* Some substances that are added to food as preservatives or to enhance taste or appearance may be carcinogens or may be converted to carcinogens under various circumstances. For example, nitrite is used to preserve fish and meats that cannot be heated without destroying their flavor or natural color. Although nitrites are not themselves carcinogenic, they may react with amino acids in the digestive tract to produce *nitrosamines,* a class of substances that is strongly carcinogenic. A particularly high concentration of nitrosamines is found in fish preserved with nitrite and stored in vinegar. It has been suggested that ingestion of large amounts of pickled fish, which is a relatively common food in Japan, may be at least partially responsible for the comparatively high incidence of stomach cancer in that country.

3. *Carcinogens Produced by Cooking.* Normal cooking can bring about the formation of carcinogens. Frying of bacon produces nitrosamines. Smoked or charcoal-broiled meats and fish contain some of the same carcinogenic hydrocarbons that are present in cigarette smoke.

4. *Contaminants.* Pesticides and herbicides, some of which are carcinogenic in animals, may be present in food as contaminants. Until several years ago, the hormone diethylstilbestrol (DES) was added to livestock feed to increase weight gain, and some animals had traces of DES in their meat. DES has been linked to vaginal cancer in daughters of women who took DES during pregnancy to prevent miscarriage—a use, ironically, for which DES was not effective.

Peanuts, rice, and other grains stored in hot, humid climates can become contaminated with the fungus *Aspergillus flavus* which produces a toxin, called *aflatoxin,* that is highly carcinogenic in all animals including humans. In parts of the world where aflatoxin concentrations in food are not closely regulated as they are in the United States, the incidence of liver cancer is extremely high.

5. *Dietary Fats and Cancer.* A diet high in fats, typical of the diet of most Americans, has been linked to an increased incidence of cancers of the breast, uterus, colon, and rectum. The reasons for this relationship are not clear.

***Carcinogens in the Air.*** The types and concentrations of carcinogens present in outside air vary greatly from one locale to another. Urban areas generally have a high concentration of pollutants from motor vehicle exhaust, and many of these are carcinogens. In industrialized areas, there may be carcinogens in the air from industrial gases, smoke, and dust.

Air within houses of smokers often has a considerably higher concentration of pollutants than outside air. Recent research has shown that tobacco smoke in the air significantly increases the risk of lung cancer among nonsmoking spouses of smokers.

***Carcinogens in Drinking Water.*** A variety of chemicals that are carcinogenic in test systems have been isolated, usually in small amounts, from drinking water supplies. Most of these result from the contamination of ground water by industrial wastes or by agricultural chemicals such as herbicides or pesticides.

***Carcinogens in the Workplace.*** About 44,000 chemical substances are used commercially in the United States. Some 7000 of these have been tested, and 1500 show some carcinogenic activity.

A number of industrial chemicals have been directly linked to cancer in industrial workers. About 17 percent of long-term asbestos workers can be expected to die of lung cancer; however, the risk of lung cancer among asbestos workers is 10-fold greater for those who smoke than for those who don't. Other cancers for which there is an increased risk include bladder cancer in aniline dye workers, liver cancer in workers in the plastics industry who are exposed to vinyl chloride, and lung cancer in metal workers involved in chromium and nickel refining.

***Carcinogenic Drugs.*** A number of drugs have carcinogenic activity. These include estrogens used to treat postmenopausal symptoms and certain endocrine problems (Chapter 17), and DES, now used as a postcoital contraceptive, or "morning after pill."

***Tumor Promoters.*** Some substances are not carcinogenic themselves but promote the activity of carcinogens. These substances are called **tumor promoters.** Tumor promoters were discovered when it was shown that cancer can be caused by combining a single exposure to a known carcinogen, but in an amount so small that cancer would not normally result, with prolonged exposure to small quantities of another substance, the tumor promoter. Thus, although tumor promoters do not cause cancer, they amplify the effects of low doses of carcinogens.

The existence of tumor promoters coupled with the fact that there is normally a long lag time between exposure to a carcinogen and the appearance of cancer has led to the conclusion that carcinogenesis is a multistep process. This process is thought to occur in two stages: *initiation,* which results in an irreversible cellular change after a limited exposure to a carcinogen, and *promotion,* which is reversible and must occur for an extended period in order to bring about the changes that allow cells to multiply in an unrestrained fashion. It is important to note that carcinogens, by definition, can bring about both initiation and promotion. Thus, a cell that has been initiated by exposure to a small dose of carcinogen can be converted into a cancer cell by additional exposure to the carcinogen or by exposure to a tumor promoter.

Tumor promoters include phenobarbital, which promotes liver cancers, and the artificial sweetener saccharin, which promotes bladder tumors. Cigarette smoke contains not only carcinogens, but also tumor promoters. In fact, many researchers think that the concentrations of carcinogens in cigarette smoke are not great enough to account for the high incidence of cancer associated with smoking. The observation that exsmokers experience a progressively decreasing risk of developing cancer as time passes, as shown in Figure 22–7, is consistent with an important role for tumor promoters in the development of cancer. Presumably, lengthy exposure to tumor promoters in cigarette smoke is needed for cancer to develop.

The cellular mechanisms of initiation and promotion have been studied extensively. Evidence suggests that initiation may occur as a result of a mutation in the cell's DNA. Significantly, recent research on some promoters has shown that they affect the activity of a cellular enzyme that is involved in the control of cell growth and differentiation.

***Detoxification of Chemical Carcinogens.*** A great deal of debate has centered on a controversy questioning the role of very small amounts of chemical carcinogens in causing cancer in humans. Because the incidence of cancer is extremely low at low doses of carcinogen, it is impossible to obtain reliable data that demonstrate a causal link between low amounts of carcinogen and cancer. Consequently, one group of researchers concludes that, in the absence of specific data, the safest course of action is to assume that trace amounts of carcinogen do lead to cancer, albeit at low rates. Other scientists argue that the human body has ways of detoxifying small amounts of carcinogens and thus that carcinogens in low doses are inconsequential. Although detoxification apparently does occur with a few carcinogens such as vinyl chloride and

FIGURE 22–7 The rate of lung cancer deaths among exsmokers drops dramatically within the first few years after smoking is stopped, and it continues to decline thereafter at a slower but steady rate.

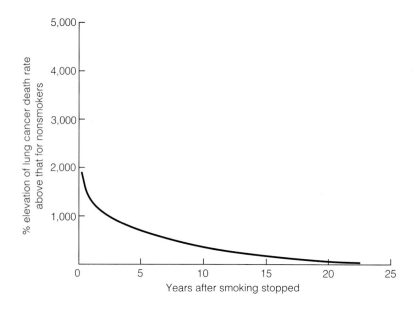

TABLE 22–3 Sources of Exposure to Ionizing Radiation in the United States

| Source of radiation | Average dose (millirem/year) |
|---|---|
| I. Natural sources of radiation<br>   Cosmic rays<br>   Terrestrial radiation<br>   Internal radiation | 102 |
| II. Radioactive contamination due to nuclear explosions | 4 |
| III. Radioactive contamination due to nuclear power plants | 0.003 |
| IV. Radiation from occupational exposure | 0.8 |
| V. Medical irradiation | 73 |
| VI. Miscellaneous sources of radiation | 2 |

chloroform, it has not been demonstrated with a large number of carcinogens. At present, the tendency is to favor the former view, even if it may involve overreaction.

## Radiation and Cancer

Radiation that is capable of ionizing atoms or breaking molecules into highly reactive fragments can cause cancer. The types of ionizing radiation to which a human may be exposed are grouped into two categories. *Natural radiation* includes ultraviolet light from the sun, emissions from radioactive elements in the soil or water, and cosmic rays. The other category of radiation, *man-made radiation,* consists of medical X rays and radiation produced by industrial and military sources. Natural radiation accounts for about 60 percent of the radiation to which an average person is exposed during a lifetime, and man-made radiation accounts for the remainder (Table 22–3).

In the induction of cancer by radiation, a number of factors are important. These include, among others, total dose, the rate at which the dose is received, the type of radiation, and certain host-specific factors such as age, sex, and the ability of the body to repair radiation damage.

***Ultraviolet Light.*** Ultraviolet light from the sun or from tanning lamps is the main cause of skin cancer. Hence it should not be surprising that the Western ideal of sun-tanned beauty, espoused since World War II, has resulted in a rapid rise of skin cancer in the United States and Europe during the past forty years. For example, skin cancer rates in Denmark, where detailed cancer records have been kept, have almost doubled since 1945. It is important to note, however, that these statistics apply primarily to light-skinned individuals. People with dark skin (Blacks, Orientals, and Indians) rarely get skin cancer because the pigment in the epidermal layer of their skin shields underlying layers from ultraviolet light. Although the tanning of light skin in response to ultraviolet light is due to an increase in epidermal pigment, the tanning response does not occur rapidly enough, and fades too quickly, to protect the skin from ongoing injury that may result in cancer (see *Focus: Skin Cancer*).

***Radioactive Elements.*** One of the most tragic examples of the role of radioactive elements in causing cancer occurred in the watchmaking industry. From 1913 to 1929, about a thousand young women worked as painters of luminous dials on wristwatches and clocks. In order to make the paint luminous, the radioactive element radium was added to it. The painters routinely ingested the radium when they moistened the tips of their paint-brushes with their tongues in order to keep fine points on the brushes. As a consequence, many of these women subsequently died of bone cancer.

Atomic bomb explosions can also produce high levels of radioactive

# Skin Cancer

**S**kin cancer is by far the most prevalent type of cancer in the United States, with more than 400,000 new cases reported each year. About 80 percent of these are *basal cell carcinomas,* which occur in the basal cells that give rise to the epithelial layer of skin. Basal cell carcinoma usually begins as raised, transluscent, pearly tumors. These may eventually become ulcerated, forming a nodule with a raw, moist center that may bleed but does not heal. Basal cell carcinomas are the least dangerous form of cancer. They grow slowly and rarely metastasize (spread). If they are promptly removed using local surgery, the cure rate approaches 100 percent.

The second most common form of skin cancer, representing about 15 percent of the total, are *squamous cell carcinomas.* These tumors form from cells in the upper layers of the epidermis. Squamous cell carcinomas generally begin as raised, reddish, scaly patches that may ulcerate at the center. They are somewhat more malignant than basal cell carcinomas, but they can usually be cured by surgery.

Most basal cell and squamous cell carcinomas are caused by long-term exposure to ultraviolet light. They occur most frequently in sailors, farmers, outdoor construction workers, and others who spend long hours in the sun. They are also more common among people who live in particularly sunny regions or in tropical latitudes where the sunlight is intense than among those living in more temperate, less sunny areas. Ninety-five percent of these cancers occur on areas of the skin that are usually exposed to the sun, such as the face, back of the neck, and back of the hands. They are rarely found in areas usually covered by clothing. Most of the increase in basal cell and squamous cell carcinomas since World War II is undoubtedly a result of our penchant for sunbathing. However, other carcinogenic chemicals in the air may also play a role. In addition, recent depletion of the atmospheric ozone layer, which absorbs ultraviolet light in the sun, may be an important factor.

The most dangerous form of skin cancer is *malignant melanoma.* It accounts for only about 5 percent of skin cancer in the United States, but is responsible for the majority of skin cancer deaths because it metastasizes so readily. Malignant melanoma occurs when melanocytes, the pigment-producing cells of the skin, become cancerous. It usually develops as a change in a pre-existing mole (moles arise from melanocytes). Normal moles are usually less than one-quarter inch in diameter, have smooth regular borders, are roughly round in shape, and possess a uniform brown coloration. Early warning signs for malignant melanoma in moles include growth to a size larger than one-quarter inch, an uneven border, an irregular shape, and a mix of colors, especially dark brown and black, sometimes intermingled with whitish, reddish, or bluish hues.

About 26,000 cases of malignant melanoma were diagnosed in 1987. This represents an alarming increase over the past sixty years. In 1930, an American with light skin had a lifetime risk of developing melanoma of about 1 in 1500; today, the risk is about 1 in 150, and rising at an overall rate of about 2 percent per year.

Although the development of non-melanoma skin cancer is related to total accumulated exposure to sunlight (or

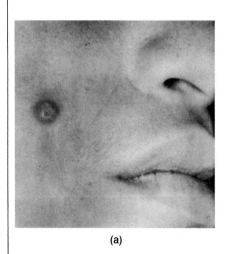 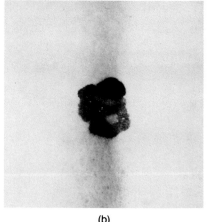

(a) Benign mole and (b) malignant melanoma. The melanoma has an irregular border, uneven shape, large size, and mixture of colors.

substances. Not surprisingly, survivors of the bombs that devastated Hiroshima and Nagasaki showed, in subsequent years, increased rates of a variety of tumors, especially leukemia (Figure 22–8).

*X rays.* Diagnostic X rays are the major source of man-made radiation to which most individuals are exposed. X rays have long been linked with an increased risk of cancer of a variety of types. Since risk is proportional to dose, it behooves everyone to minimize *unnecessary* exposure to medical X rays.

other ultraviolet light sources), this does not appear to be the case with melanoma. Instead, most melanomas appear to arise as a result of intermittent exposure to particularly intense sunlight. A recent Harvard study showed that even one case of a severe, blistering sunburn before the age of 20 can double the risk of melanoma. Several other studies have shown an increased incidence of melanoma among indoor workers who get sunburned on weekends or during winter vacations, but no increased risk, or even a decreased risk, of melanoma in outdoor workers who are exposed to the sun more or less continuously. The importance of intermittent intensive sunlight exposure to the development of melanoma is further supported by the observation that most melanomas arise on the back, not on the areas of the body that usually receive the most sun exposure (the face, back of the neck, and back of the hands).

Another important risk factor for melanoma is the presence of an unusual type of mole—large, irregularly shaped, and variably pigmented—which is a precursor to malignant melanoma. These moles, called dysplastic nevi (*dysplastic* = abnormally developed; *nevi* = moles), are present in 1 percent of white adults. Melanoma presumably arises as a consequence of the action of ultraviolet light on dysplastic nevi. Heredity can be an important factor in the development of dysplastic nevi and melanoma. Studies on the genetics of melanoma have suggested that the disease can be inherited as an autosomal dominant trait. It has been shown that 90 percent of individuals who inherit the gene develop dysplastic nevi.

Although melanoma is a serious cancer, it can be cured if it is detected and removed at an early stage before metastasis has occurred. The *average* survival rate for melanoma victims is now about 75 percent, up from roughly 40 percent or less 40 years ago. The improvement in survival rate almost certainly results from earlier diagnosis, since surgical treatment has not changed, nor has other therapy become available.

To prevent melanoma as well as other types of skin cancer, dermatologists recommend using sunscreen lotions conscientiously and avoiding exposure to the sun between the hours of 10 AM and 3 PM, when sunlight is most intensive. People with skin that burns and does not tan, as well as those with dysplastic nevi, should avoid sunbathing entirely and use sunscreens with a blocking factor of 15. (The number 15 indicates that it will take fifteen times longer to burn when the product is used.) Protecting the skin of children is particularly important, since most damage is done early, and most of a lifetime's sun exposure occurs before the age of 30.

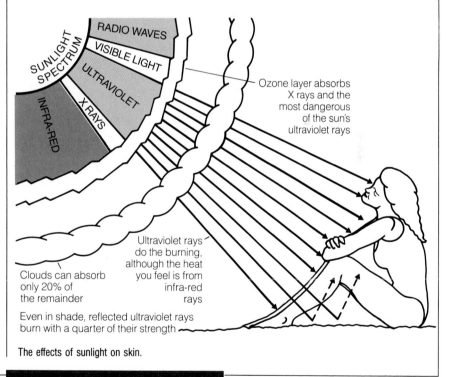

The effects of sunlight on skin.

Particularly damaging are diagnostic X rays of fetuses, which cause an increased rate of leukemia and other cancers during early childhood.

***Mechanism of Cancer Induction by Radiation.*** It has been known for some time that low doses of ionizing radiation cause mutations in DNA. High doses also cause mutations but, in addition, result in visible chromosome damage. The result of mutation and chromosome damage is presumably to bring about alterations in the normal processes that control cell division and differentiation.

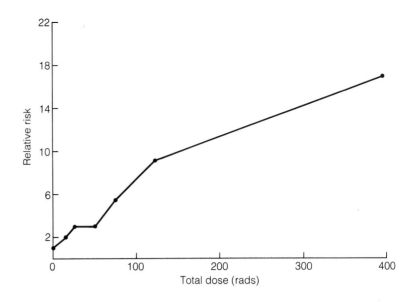

FIGURE 22–8 Medical records of survivors of the Hiroshima atomic bomb blast show that the relative risk of leukemia rises dramatically with the total dose of radiation received.

## Viruses and Cancer

There are two general types of animal viruses: *cytocidal viruses* and *oncogenic viruses.* **Cytocidal viruses,** described in Chapter 21, cause such common human diseases as measles, mumps, chicken pox, and influenza. Infection of a cell by a cytocidal virus results in the production and release of new virus particles and in eventual cell death. **Oncogenic** (*onco* = tumor; *genic* = causes) **viruses,** or tumor-producing viruses, do not always kill the cells they infect. Instead, the virus changes host cells into tumorous ones by altering normal controls over reproduction and differentiation. The transformation of a normal cell into a tumorous one by an oncogenic virus may, but does not always, lead to the production and release of new virus particles.

It had long been known that oncogenic viruses cause some types of *benign* tumors in all vertebrates that have been studied, including humans; in addition, they had been shown to cause *cancer* in all of these vertebrates *except* humans. Only recently has an oncogenic virus (HTLV-I; see discussion that follows) been firmly established as the cause of one type of human cancer (leukemia). Furthermore, there is now strong presumptive evidence linking several other oncogenic viruses to human cancer.

Oncogenic viruses are similar in structure to cytocidal viruses, consisting of nucleic acid surrounded by one or more external coats composed of protein and sometimes other macromolecules. The nucleic acid in oncogenic viruses may be DNA or RNA, a property that is commonly used to classify oncogenic viruses into two groups: **oncogenic DNA viruses** and **oncogenic RNA viruses.**

*Oncogenic DNA Viruses.*    Two types of oncogenic DNA viruses important to humans are *papovaviruses* and *herpesviruses.*

1. *Papovaviruses.* One type of papovavirus, the *papilloma virus,* has been known for a long time to cause genital warts as well as other types of benign warts of the skin and mucous membranes. Recently, however, the papilloma virus has been strongly implicated as a cause of cervical cancer. Studies in the United States and Europe have identified papilloma virus in about 90 percent of cervical cancers that were examined. Moreover, papilloma virus is also present in 90 percent of cervical dysplasias— abnormal cell types that represent a stage in the progression to cancer. The association between cervical cancer and an infectious agent is further supported by epidemiological data. Not only are women with multiple sex partners at highest risk for cervical cancer, but women who marry men

whose previous wives had cervical cancer are three to four times as likely to contract cervical cancer as other women. A link has also been shown, although less firmly, between the presence of genital warts (and hence infection by the papilloma virus) and cancer of the vagina, vulva, and penis.

2. *Herpesviruses.* A large number of viruses all belonging to the herpes group have been identified. Various herpesviruses cause cancer in frogs, chickens, rodents, monkeys, and pigs. As mentioned in Chapter 21, five major kinds of herpesviruses infect humans. One of these, the Epstein-Barr virus, has been suspected for some time as the causative agent for certain human cancers.

The Epstein-Barr virus infects some 80 percent of the people of the world. Infected individuals develop antibodies to the virus, but usually there is no evidence of sickness. However, in the United States and other countries, a small proportion of those infected with Epstein-Barr virus contract infectious mononucleosis, a disease lasting several weeks that is characterized by the production of large numbers of abnormally formed nongranular leukocytes. In parts of central West Africa, the Epstein-Barr virus is strongly implicated in causing Burkitt's lymphoma, a type of lymphoma that is primarily confined to children and young adults. It has been demonstrated that cancer cells in more than 99 percent of Africans with Burkitt's lymphoma carry antigens of the Epstein-Barr virus. However, only those individuals with malaria or some other chronic infectious disease contract Burkitt's lymphoma. It has been postulated that chronic infection weakens the immune system, allowing the Epstein-Barr virus to express its oncogenicity.

The Epstein-Barr virus is also associated with cancer of the nasopharynx in parts of Africa, Southeast Asia, and China. As with Burkitt's lymphoma in West Africa, all individuals who develop nasopharyngeal cancer are infected with Epstein-Barr virus, but only a small proportion of infected individuals contract the cancer. There is some evidence to suggest that there may be certain dietary factors common to those contracting nasopharyngeal cancer.

It thus appears that the result of infection by Epstein-Barr virus varies, depending upon conditions in the host organism. Under some conditions, yet unidentified, it may lead to infectious mononucleosis, but in conjunction with chronic infection or certain dietary carcinogens, Burkitt's lymphoma or nasopharyngeal cancer may result.

***Oncogenic RNA Viruses (Retroviruses).*** Another type of virus that has been linked to cancer in humans is a member of a group of oncogenic RNA viruses (retroviruses) called C-type viruses. C-type viruses, of which many types have been isolated, have been shown to cause leukemias, lymphomas, and sarcomas in virtually all laboratory birds and mammals, and in many other animals including nonhuman primates. For example, leukemia in cats is caused by a feline C-type virus that may be transmitted to an uninfected cat through the saliva of an infected one.

A great deal of recent research has demonstrated that one type of human leukemia, a T-cell malignancy called adult T-cell leukemia, is caused by a human C-type retrovirus. Interestingly, the virus is HTLV-I (human lymphotropic virus, type 1; see Figure 22–9), a close relative of HIV-1 and HIV-2 that cause AIDS (see *Focus: Acquired Immune Deficiency Syndrome (AIDS)* in Chapter 21).

Adult T-cell leukemia is a rare disease first identified in a relatively isolated population in southern Japan. It has since been recognized in other countries, including the United States. A study in 1988 estimated that one in 4000 blood donors in the United States may carry the HTLV-I virus. The American Red Cross has since announced that it will start screening donated blood for HTLV-I as soon as a screening test is approved by the FDA. Like AIDS, adult T-cell leukemia is transmitted through blood and by sexual intercourse.

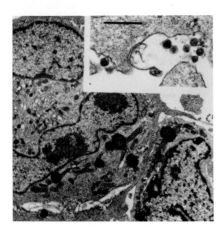

FIGURE 22–9 Human T-cell infected with HTLV-I. Inset is at higher magnification, showing HTLV-I particles.

***Transformation by Oncogenic DNA Virus.*** Studies on the infection and transformation of animal cells by oncogenic DNA viruses have provided many details of the cellular and molecular events associated with these processes. The following general pattern emerges. Once an oncogenic DNA virus has entered a host cell, the DNA is released from the virus particle and becomes physically integrated into one of the host cell's chromosomes. Part of the viral DNA is then transcribed into messenger RNA and this is translated into proteins (see Chapter 20 for details of protein synthesis). One or more of these proteins, called *transforming proteins,* brings about transformation by disrupting normal controls on cell division and differentiation. Infection by an oncogenic DNA virus may, but need not, result in the production of new viral particles that escape from the cell.

By being integrated into the chromosomal DNA of the host, the viral DNA becomes a permanent part of the genetic material of the host cell. During cell division, the viral DNA is replicated and distributed to daughter cells along with the cell's regular chromosomal DNA. This fact accounts for the observation that transformation to the cancerous state represents a permanent change.

***Transformation by Retroviruses.*** Oncogenic RNA viruses (retroviruses) transform normal cells by a process similar to that of oncogenic DNA viruses. However, there is one important difference. Since the genetic material of an oncogenic RNA virus is single-stranded RNA and not DNA, the RNA must be converted into a DNA copy before it can be integrated into a host chromosome. The conversion of viral RNA into DNA occurs in a series of steps summarized in Figure 22–10. It requires an enzyme called *reverse transcriptase* that is contained within the virus particle and synthesized according to information present in viral RNA. Integration into a host cell chromosome of the DNA copy of the viral RNA is followed by the synthesis of transforming protein, as in the case of oncogenic DNA viruses.

***Viral Oncogenes.*** Studies of retroviruses have recently yielded important discoveries concerning the molecular basis of cancer. A group of retroviruses, called *highly transforming viruses,* have been identified that are capable of transforming cells with high efficiency. Some 20 highly transforming viruses have been isolated from a wide variety of organisms, including chickens, mice, rats, cats, and monkeys. They are each known to contain a single gene, which encodes a protein that is responsible for the transformation of normal cells into cancerous ones. This gene is called an **oncogene.**

The highly transforming viruses are particularly efficient at transforming cells because their oncogenes are expressed at high levels; that is, a large amount of transforming protein is made. We now know that the high level of oncogene expression occurs because the oncogene lies next to another viral gene, a **viral promoter gene,** that promotes, or activates, the expression of the oncogene. (Do not confuse viral promoter genes with tumor promoters.)

***Cellular Oncogenes.*** Cells can be infected by another type of retrovirus that has a viral promoter gene but *no* oncogene. When infection occurs, the cell may be transformed into a cancerous one! This observation, coupled with results from other experiments, has led to the conclusion that normal, uninfected animal cells have their own oncogenes that can cause cancer when activated by a viral promoter gene. In fact, oncogenes of animal cells and of cancer viruses form a related group of genes that are found throughout all higher animal species, including humans.

Why are oncogenes present in animal cells? It seems unlikely that their normal function is to cause cancer, for that would cause the death of the organism. It has been postulated that oncogenes function during very early embryonic development to cause rapid proliferation of unspecialized cells. In other words, cells of the early embryo have properties much like cancer cells, and their cancerlike growth is determined by the expression of oncogenes. The

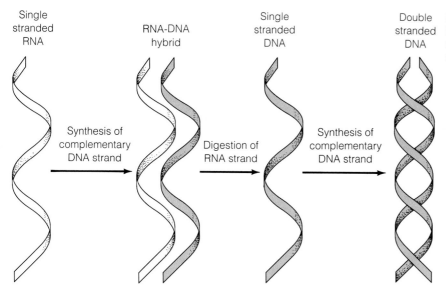

FIGURE 22–10 The enzyme, reverse transcriptase, catalyzes all three of the above reactions, resulting in the synthesis of double-stranded DNA from single-stranded RNA.

oncogenes are presumably activated by cellular promoter genes—which we know exist—that are similar in function to viral promoter genes and that come to lie near oncogenes by moving from one region of a chromosome to another (jumping genes). In later stages of development when cell division slows and differentiation begins, oncogenes are presumed to be inactivated by movement of the cellular promoter gene to a position away from the oncogene. This theory goes on to propose that cancer results when oncogenes are inadvertently reactivated either by a viral promoter gene or a cellular promoter gene when one or the other comes to lie near the oncogene (Figure 22–11). It has been suggested that chemical carcinogens and radiation may, in some cases, cause cancer by acting as a stimulus to the movement of cellular promoter genes.

## Cancer Prevention and Treatment

At the present time, prevention is the most effective way of dealing with cancer. Since most cancer is a direct result of life-style, alteration of life-style to avoid unnecessary exposure to environmental carcinogens should do much to

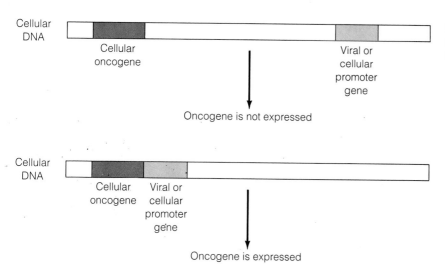

FIGURE 22–11 Diagram summarizing the idea that the position of promoter genes may be crucial to the expression of oncogenes.

## Main Nutritional Recommendations of the National Research Council Intended to Reduce the Risk of Cancer

1. Lower intake of fats from 40 percent of total calories to 30 percent. Evidence links excessive fat consumption to cancers of the colon, breast, and prostate.
2. Increase consumption of foods containing fiber, including whole grain cereals, fruits, and vegetables. Evidence links low fiber diets to colon cancer.
3. Consume fruits and vegetables high in vitamin C and in beta-carotene, a chemical that is converted by the body into vitamin A. Citrus fruits and peppers are good sources of vitamin C; beta-carotene is present in dark-green leafy vegetables such as spinach and in deep-yellow vegetables and fruits, including carrots and apricots. Vitamin C, vitamin A, and beta-carotene act as antioxidants, which may block the formation of free radicals that can damage DNA and act as tumor promoters.
4. Consume more vegetables of the cabbage family, including cabbage, broccoli, cauliflower, and brussel sprouts. Unidentified chemicals in these vegetables reduce the general incidence of cancer.
5. Eat less salt-cured and smoked fish and meat, including sausages, smoked fish and ham, bacon, and hot dogs. Evidence links consumption of these foods to cancers of the stomach and esophagus.
6. Reduce the consumption of alcoholic beverages. Consumption of alcohol, particularly in combination with tobacco smoking, is linked to cancers of the upper gastrointestinal tract.

prevent the development of most cancers. The National Research Council has proposed various dietary recommendations that are designed to reduce the risk of cancer. These are summarized in *Focus: Main Nutritional Recommendations of the National Research Council Intended to Reduce the Risk of Cancer.*

The best treatment of cancer involves early detection. If a cancer is detected early when it is small and before it has metastasized, chances of a complete cure are excellent. To increase the chances of early detection and treatment, the American Cancer Society has published seven warning signals that suggest cancer, as listed in Table 22–4.

The three means of treating cancer are *surgery, chemotherapy,* and *radiation,* alone or in combination. Surgery involves removal of the diseased tissue and of any lymph nodes or other tissues to which the cancer may have spread. Chemotherapy and radiation depend upon the premise that these treatments will kill cancer cells in preference to normal cells. One particularly promising form of treatment is the use of monoclonal antibodies to search out and destroy cancerous cells (see Chapter 21).

TABLE 22–4 Cancer's Seven Warning Signals

1. Change in bowel or bladder habits.
2. A sore that does not heal.
3. Unusual bleeding or discharge.
4. Thickening or lump in breast or elsewhere.
5. Indigestion or difficulty in swallowing.
6. Obvious change in wart or mole.
7. Nagging cough or hoarseness.

## Study Questions

1. How did the misconception arise that cancer is a modern disease?
2. Distinguish between a) benign and b) malignant tumors.
3. List the four major classes of malignant tumors and give several examples of tumors of each class.
4. List the four major properties of malignant cells and briefly describe each.
5. List the three major causes of cancer.
6. a) What is the Ames test?
   b) How is it performed?
   c) What are the advantages of the Ames test over other testing methods?
   d) What are its disadvantages?
7. Why are chemical carcinogens difficult to identify by using epidemiological data?
8. a) What is the single most common cause of lung cancer in the United States?
   b) Indicate evidence that supports this conclusion.
9. a) What is a tumor promoter?
   b) How does it differ from a carcinogen?
10. a) What is meant by "detoxification" of carcinogens?
    b) How does "detoxification" sometimes increase the carcinogenicity of chemicals?
11. a) What is the main cause of skin cancer?
    b) Indicate evidence that supports this conclusion.
12. Distinguish between cytocidal and oncogenic viruses.
13. Describe the evidence which suggests that the Epstein-Barr virus may cause cancer in humans.
14. Describe the molecular mechanism by which a) oncogenic DNA viruses and b) retroviruses transform cells.
15. a) What disease does HTLV-I cause?
    b) To what other virus is HTLV-I closely related?
16. a) What are cellular oncogenes?
    b) What are cellular promoter genes?
    c) How might oncogenes and promoter genes cause cancer?

# AGING AND DEATH

At about the time that growth stops and an individual reaches maturity, the inexorable process of aging begins. The manifestations of aging, including changes in appearance, impairment of the senses, and a general loss of strength and stamina, are readily apparent and clearly distinguish an older individual from a younger one. Although we can recognize characteristics associated with aging, we understand very little about the aging process. We do not know how it starts or why it occurs, and we consequently do not know if or how the effects of aging can be controlled. We are beginning to discover some of the molecular, cellular, and organismal events that occur during aging, but we have yet to determine whether these events are causes or consequences of aging.

## THE PROCESS OF AGING

**Gerontology,** the study of aging, has only attracted the serious attention of biologists and medical scientists within the last 30 years or so. Prior to that time, the nature of aging was not, by and large, the subject of genuine scientific inquiry. Most of those purporting to study aging were charlatans and quacks, promising elixirs of life and other impossible cures for the aging process. Consequently, the few scientists who ventured into the field were not considered reputable, and their research was largely ignored by the greater scientific community.

The long neglect of research on aging has had unfortunate consequences. At a time when our population is aging dramatically, and when diseases associated with aging have come to dominate clinical practice in the West, we find ourselves grossly unprepared to deal with the medical problems of the aged (Figure 23–1). Clearly, research on the nature of aging has immense importance, now more than ever.

The practical purpose of research on aging does not lie primarily in attempting to prolong life or even to counteract the aging process, as significant as these goals may be. Rather, by understanding the aging process, we will be better equipped to treat its associated diseases and to ameliorate its effects. Our ultimate aim is to keep the aged alive in a state of physical health and mental competence as long as possible.

### Definition of Aging

Aging has been defined as the sum total of all changes that occur after maturity and that lead to functional impairment and an increased probability of death. Functional impairment is manifested as increasing lack of resistance with age to stress, injury, and disease. Hence, the net effect of aging is a decrease in life expectancy with increasing age.

### Life Expectancy and Rate of Aging

Medical science has greatly increased the life expectancy of humans. This fact is illustrated in Figure 23–2, which shows human survival curves for the United States at various times in the twentieth century. Survival curves are constructed by determining the proportion of individuals of any age group that survive in the population. The age at which 50 percent of the individuals in a population are still alive is termed the **median life span,** which is equivalent to the *median life expectancy at birth* (see Figure 23–1). Put another way, the average individual dies at an age equivalent to the median life span. This age is currently about 75 in Western countries.

It is important to understand that the dramatic increase in life expectancy that has occurred since the beginning of this century is due almost exclusively to a reduction in mortality from malnutrition and infectious disease, as we

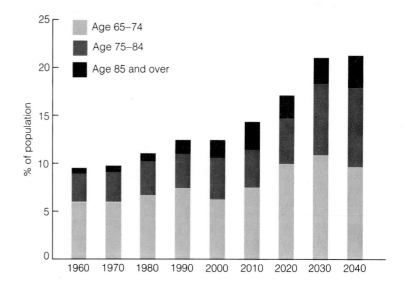

FIGURE 23–1 Historical and projected percentages of elderly people in the United States.

noted in Chapter 16. However, there is no evidence to suggest that the maximum life span of humans—referring to the age achieved by the most long-lived members of the population—has increased over the centuries. The longest-lived humans age at a rate that is similar to that in all previous historical periods, and they still die at about 100 years of age; the highest documented age is 120 years. Hence, there appears to be a biological limit on the human life span that cannot be overcome, whether or not specific diseases are cured or prevented. It would consequently seem that attempts to reduce the *rate* of aging would be a more fruitful avenue of research activity than attempts to increase maximum life span.

### Effects of Aging on Individuals

There are a variety of changes that occur in aging individuals, but these do not occur in a consistent way, often appearing at different times in different people. For example, we all know people who do not look their age and others who look older than their age. Even within the same individual, different characteristics associated with aging may appear at different times. Some individuals in their twenties have gray hair but no other obvious signs of aging, while other much older individuals may be physically frail "without a gray hair on their heads."

FIGURE 23–2 Human survival curves for the United States, comparing the year 1901, when infectious disease was a major cause of death, with later years after improvements in public health and medicine. The median life expectancy at birth of individuals born in 1901, 1948, and 1975 is equivalent to the age at which one-half of the population still survives.

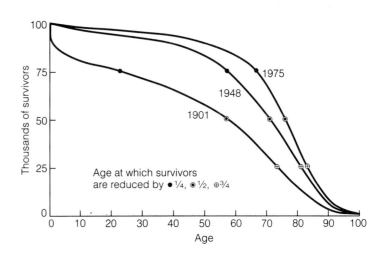

## Physiological Changes

The most obvious changes that occur during aging are in physical appearance, including the wrinkling of skin, the graying and thinning of hair, and a stooped posture. Besides physical changes, many other characteristics are associated with aging. Notable among these are impairment of the senses. Both hearing and vision deteriorate with age. For example, visual acuity diminishes steadily and reasonably constantly from age 25 (Figure 23–3a). Hearing declines steadily from adolescence until about 50 years of age, after which the decline proceeds more slowly.

Another manifestation of aging is loss of muscular strength. Muscular strength peaks between the ages of 20 and 30, and declines continuously thereafter (Figure 23–3b). Age-related decreases in muscular strength are due to loss of skeletal muscle cells and a reduction in the number of myofibrils per cell. These changes are accompanied by alterations in the neuromuscular junctions, reduced density of the long bones and vertebrae, and deterioration of cartilages and articulating surfaces.

Decline in the circulatory and respiratory systems is also revealed during aging (Figure 23–3c and d). There is a gradual loss of cardiac output, which is particularly apparent during exercise. Whereas a young adult may increase cardiac output during exercise from about 5 liters per minute to as much as seven times this value, cardiac output of the elderly cannot respond adequately to the oxygen demands of contracting muscles—a situation that puts severe restraints on the amount of physical work an elderly person can perform. The maximum amount of air that can be moved into and out of the lungs (the vital capacity of the lungs) declines 40 percent between the ages of 20 and 80, while the amount of oxygen absorbed at the lungs decreases more than 60 percent during the same interval.

The kidneys are another site of age-related changes (Figure 23–3e). These include a gradual reduction in the number of functioning nephrons and

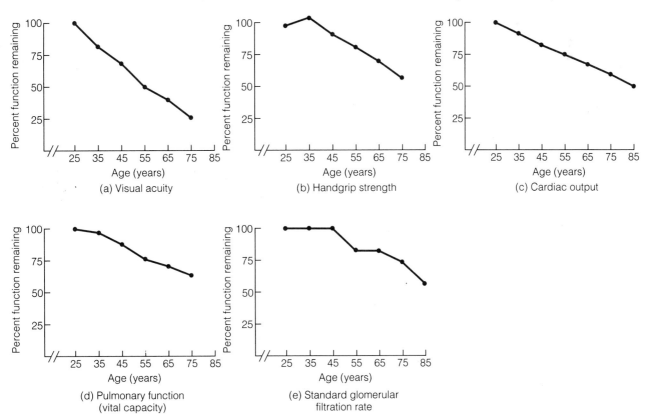

FIGURE 23–3 Graphs illustrating decline in physiological function with age. Values obtained at age 25 are taken as 100 percent.

(a) Visual acuity

(b) Handgrip strength

(c) Cardiac output

(d) Pulmonary function (vital capacity)

(e) Standard glomerular filtration rate

Aging effects on skin. Facial skin changes in noticeable stages. With age, skin loses its elasticity and becomes coarser and thinner.

declines in the rates of renal blood flow, glomerular filtration, tubular secretion and tubular reabsorption. These physiological changes lead to a progressive impairment of kidney function in aging individuals.

The function of the immune system also deteriorates progressively with age. There is a marked decline not only in the immune response to newly encountered antigens, but also in immunological memory. Furthermore, autoimmune reactions increase with age, reflecting the inefficiency of the immune system. Increases in autoimmunity are evidenced by the fact that normal healthy people as they age have elevated blood levels of antibodies to "self" *(autoantibodies)*. Indeed, many characteristic diseases of old age, such as rheumatoid arthritis, are associated with autoimmune reactions.

The function of the central nervous system, as determined by physiological and behavioral testing, appears to become increasingly inefficient with age. Mental reaction time, memory, and manipulative ability show age-related declines. However, current evidence cannot attribute the declining function to neuronal aging.

In addition to changes in various organ systems, there are also marked differences between the young and old in the ability to maintain physiological homeostasis, particularly under situations of physiological stress. Older individuals are not able to regulate acid-base levels, ion concentrations, temperature, and blood sugar, and other metabolite levels as precisely as younger individuals.

### Aging and Disease

Independent of physiological changes that occur during aging, there is an increased incidence of many types of disease, including arthritis, osteoporosis, atherosclerosis, and cancer (Figure 23–4). The fact that the maximum human life span has not increased from previous historical periods suggests that these diseases are not the cause but a consequence of aging; if they caused aging, then we would expect advances in the treatment of these diseases to prolong the maximum human life span, which they do not. Although disease and injury might increase the rate of aging to some small degree, their effect is not substantial. It consequently seems reasonable to view the physiological changes that occur in aging individuals as changes that would happen even in the absence of disease.

### Aging Changes in Tissues

In attempting to account for the physiological changes that occur in aging individuals, a number of researchers have examined connective tissues and cells from a variety of other tissue types for age-related changes. A discussion of these changes follows.

***Changes in Connective Tissue.***    As you recall, connective tissue is found in all parts of the body where it functions to bind together, support, and protect other tissues and organs. It is composed of cells surrounded by a nonliving intercellular matrix containing proteinaceous fibers such as collagen and elastin that are secreted by fibroblasts.

Although a variety of age-related changes have been noted in connective tissue, especially in the composition of the intercellular material, a great deal of research has focused on collagen. Collagen is a major component of skin, bone, cartilage, tendons, and ligaments, and it accounts for more than 25 percent of total body protein. The basic structural unit of collagen is *tropocollagen,* a rod-shaped molecule composed of three polypeptide chains wound around each other in a triple-stranded helix (Figure 23–5a). Collagen fibrils are formed of parallel bundles of tropocollagen molecules, arranged in a staggered array (Figure 23–5b and c).

Newly formed collagen fibrils can be readily extracted from connective tissue, but as collagen ages it becomes quite insoluble. The reason for this insolubility is the formation of covalent cross-linkages within and between tropocollagen molecules that form the collagen fibrils (Figure 23–5d). The amount of cross-linking in collagen varies with the physiological function and age of the connective tissue from which it is derived. As an individual ages, collagen accumulates cross-linkages. In fact, the extent of cross-linking in collagen from a particular tissue is a good indicator of age in higher animals. As collagen becomes increasingly cross-linked, the connective tissue becomes less flexible. This inflexibility of connective tissue increases the rigidity of other tissues that the connective tissue permeates and surrounds. It has been proposed that the loss of flexibility of various tissues of the body may account for many age-related changes in physiological function. This idea is discussed in the next section.

A related change in connective tissue that has been extensively studied is the breaking strain of tendons; that is, the force required to break tendons. It has been found that the breaking strain of tendons decreases steadily during adult life. This manifestation of aging is so regular that it can be used as an extremely reliable indicator of age.

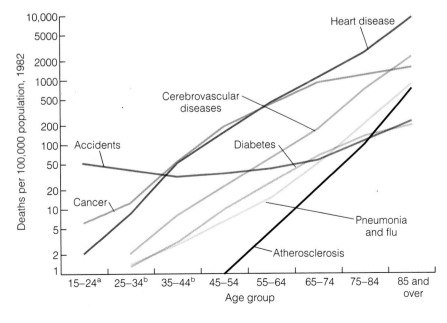

FIGURE 23–4 Deaths in the United States from accidents and disease as a function of age.

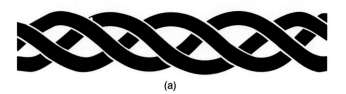

(a)

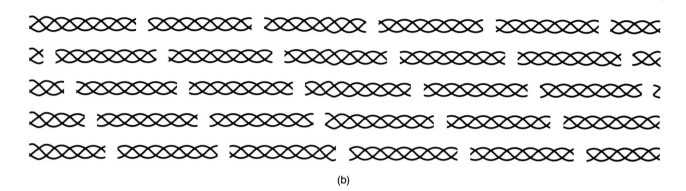

(b)

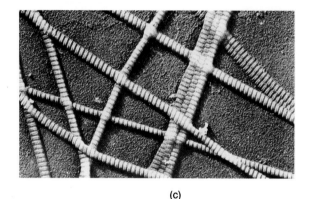

(c)

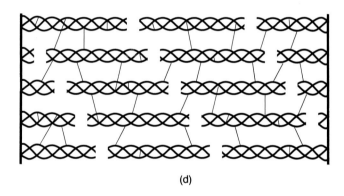

(d)

FIGURE 23–5 Structure of collagen. (a) A molecule of tropocollagen. (b) An array of tropocollagen molecules forming a collagen fibril. Spaces between individual tropocollagen molecules correspond to cross-striations seen in collagen fibrils (c) as viewed in the scanning electron microscope. (d) Crosslinking within and between tropocollagen molecules.

***Changes in Cells of Other Tissues.*** Nonrenewing tissues, in which cells are not replaced in the adult, show different aging characteristics than renewable tissues. In nonrenewing tissues, such as brain, skeletal muscle, and cardiac muscle, there is a loss of cells with aging. In addition, cells accumulate brownish granules, called *lipofuscin granules,* that increase in concentration with the age of the cell. Lipofuscin granules contain many different types of lipids and a significant amount of protein, including a number of degradative enzymes. The origin of lipofuscin granules is not known, but it has been postulated that they may represent depots of degraded materials. Whatever their origin and function, they apparently accumulate substances that cannot be removed from the cell, and hence they increase in concentration as the cell ages.

Cells in tissues that are renewable, such as epidermis, intestinal epithelium, and blood cell-forming tissue, show age-related changes that are less well-defined than those in nonrenewable tissues. There seems to be some decrease in the ability to produce new cells in older individuals as compared to younger ones. Cells of renewable tissues do not accumulate lipofuscin granules.

### Aging of Cells in Tissue Culture

A large amount of data collected by biologist Leonard Hayflick over the past 20 years has led to the conclusion that human diploid cells have only a limited life span when grown in tissue culture. Using fibroblasts from human embryonic lung tissue, Hayflick demonstrated that cell division ceased and cellular deterioration began after the cultured cells had undergone approximately 50 doublings (Figure 23–6). Control experiments indicated that the death of the cells was not due to deficiencies of the culture medium.

In other experiments, Hayflick compared the growth of cultures of fetal lung fibroblasts with fibroblasts from adult lung tissue. He was able to show that the latter cell types stopped proliferating and began to deteriorate after about 20 doublings, compared with the 50 doublings seen for fetal fibroblasts (see Figure 23–6). This result would be predicted if cells are capable of only a limited number of cell divisions, since cells from adults would already have divided a number of times prior to being cultured. In addition, some researchers have demonstrated an inverse correlation between the age of the donor and the doubling potential of skin fibroblasts. Interestingly enough, skin fibroblasts taken from individuals with Werner's syndrome, a form of inherited premature aging, also have a greatly reduced doubling potential when grown in tissue culture. Hence there appears to be a relationship between the life span of cells in culture and the age of the organism from which the cells are derived. This suggests that there may be changes in aging cells that restrict their regenerative potential.

## Healthy Aging

Much of the research on aging has focused on *average* age-related losses in physiological function. This emphasis has given rise to the inaccurate stereotype that aging people must suffer through a long period of physical and mental deterioration, which prevents them from leading a happy and meaningful existence. Gerontologists have come to realize that the effects of age-related deficits have been exaggerated. In the first place, many aging individuals do not show anything like the average functional losses characteristic of the entire group. Furthermore, the physiological deficits that result from normal aging do not, except at their extreme, substantially compromise physical and mental capabilities. Healthy aging—aging that minimizes functional losses—requires those things that are necessary for health during earlier stages of life. These include good nutrition, regular exercise, healthy personal habits, competent medical care, and high levels of emotional and social involvement.

## THEORIES OF AGING

The foregoing discussion illustrates a variety of age-related changes that occur in the molecules, cells, tissues, and organs of the human body. Nevertheless, it has not been possible to identify which, if any, of these changes are causes of

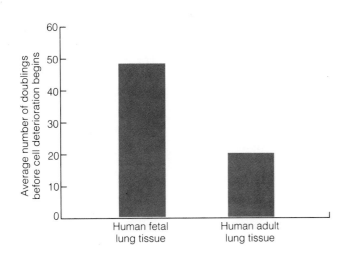

FIGURE 23–6 Growth of fetal and adult lung fibroblasts in tissue culture.

aging. Virtually all researchers agree that many of the documented changes in aging individuals are consequences of aging, not causes. Gray hair, for instance, is not a cause of old age nor is wrinkled skin, but no consensus has developed among researchers as to the most likely cause of aging. Instead, a large number of different theories have been proposed to account for the aging process. Each theory has some experimental support, but none can explain all of the known facts about aging. Indeed, aging may have a number of causes, and more than one aging theory may be correct—or none of them may be. As speculative as these theories are, they are useful in organizing our thinking and suggesting experimental approaches.

## Programmed Aging Theories

The *programmed aging theories,* of which there are many somewhat different versions, suggest that the life span and rate of aging of cells, and consequently of the organisms they compose, are genetically determined. These theories view aging as a continuation of the developmental process that becomes activated at the time of fertilization and that proceeds through a series of genetically programmed events.

Two lines of evidence provide the primary support for programmed aging theories. First is the observation that maximum life span and rate of aging vary enormously among different species. The fact that the maximum life span of a mouse is about four years and that of a human is about 100 years implies that genetic differences between species are at least in part responsible for aging. Second, there is a precedent for genetically programmed cell death. It occurs in embryonic development during the shaping of the embryo. An event such as the opening of the eyelids involves the preprogrammed death of cells in the region where the slit between the two lids will appear.

One of the more plausible versions of programmed aging theories suggests that aging is an unfortunate result of normal processes of development and differentiation. For example, if proper functioning of connective tissue requires the formation of cross-links between collagen molecules in order to strengthen the tissue, then cross-linking must be a programmed genetic event that occurs during differentiation. Presumably, a cross-linking enzyme would be synthesized early in the developmental program. However, if the activity of this enzyme is not inhibited at maturity, collagen fibrils will eventually become excessively cross-linked, causing rigidity of the connective tissue. According to this view, cellular components that are essential for optimum function are ultimately damaged as an unavoidable consequence of the expression of the genetically determined developmental program.

Other investigators have focused the general idea of programmed aging to propose that aging is a result of the failure of genetically programmed neuroendocrine control systems. This idea is supported by data showing changes in endocrine function in older individuals. For example, menopause, which occurs in aging women, is initiated by age-related changes in the functioning of the pituitary gland.

## Error and Damage Theories of Aging

A large number of other theories, grouped here under the heading of *error and damage theories,* propose that aging results not from a preprogrammed genetic process, but from the random accumulation of damage to important molecular constituents of living organisms. This damage would eventually lead to disruption of normal metabolic function. A variety of mechanisms have been proposed to account for random molecular damage.

### Attack by Free Radicals

Some researchers claim that *free radicals,* very reactive chemicals that are produced as transient intermediates during normal metabolism or as a result of radiation or exposure to certain environmental chemicals, may lead to the formation of cross-linkages in DNA and proteins, which would then interfere with the functioning of these molecules. Furthermore, some experiments show that attack by free radicals may cause oxidative damage to cellular membranes and lead to gradual loss of function. Support for the concept of free radical damage to membranes comes from the observation that vitamin E, vitamin A, vitamin C, and artificial antioxidant compounds, which protect membranes from damage, enhance longevity to some degree in experimental animals.

### Mutation in DNA

Other researchers have theorized that cellular DNA may be damaged by the gradual accumulation of mutations caused by environmental mutagens. As an individual ages and mutations become more numerous, greater numbers of abnormal proteins would be synthesized, eventually interfering with proper functioning.

### Error Catastrophe Theory

Another error and damage theory, first imposed by Leslie Orgel in the 1960s, has had wide influence. Called the *error catastrophe theory,* it suggests that errors in proteins that are essential to the process of protein synthesis may be particularly important in causing age-related deterioration of cells, because they would lead to "error catastrophes" in which many different types of defective proteins would be produced (Figure 23–7).

### Cross-Linking in Collagen

Another version of error and damage theories holds that aging is due to the accumulation of cross-linkages in collagen. Proponents of this idea theorize that the cross-linking of collagen causes aging by reducing the flexibility and mobility of many tissues and organs. For example, a less flexible heart might contract less efficiently, accounting for age-related decreases in cardiac output. Similarly, the ability of the lungs to expand might be reduced, causing a decrease in vital capacity. Extensively cross-linked connective tissue in arterial walls might make them more rigid and increase the resistance to blood flow in the arteries, creating hypertension and reducing blood supply to organs. A more rigid connective tissue might also account for the brittleness of articular cartilage that is associated with degenerative joint disease in the aged.

Perhaps most important, excessively cross-linked collagen might impede the flow of such substances as gases, nutrients, metabolites, hormones, and antibodies to and from cells through the extracellular tissue fluid. Since the movement of substances between cells and the circulatory system is essential for all physiological activity, impairment of this process might explain the age-related physiological decline in many organs and systems. A trivial effect of the cross-linking of collagen might also be the wrinkling of skin, caused by the loss of flexibility of connective tissue of the dermis.

### Autoimmune Theory of Aging

We noted previously that the immune system undergoes characteristic age-related changes, including a loss of responsiveness to antigens and an increase in autoimmunity. The *autoimmune theory of aging* proposes that autoimmune reactions may be responsible for degeneration of tissue in aging individuals.

### Multiple Causes for Aging

Increasingly, it is felt that aging may have a variety of causes. Some scientists have consequently proposed mechanisms for aging that combine several aging theories. In one view, aging may be caused by the accumulation of random

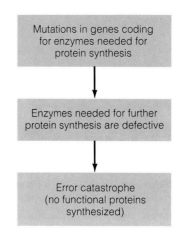

FIGURE 23–7 Summary of Orgel's error catastrophe theory of aging.

# Euthanasia

A 90-year-old man, suffering from Alzheimer's disease, falls and breaks his hip in a struggle with nurses who are trying to bathe him. Bedridden, he develops pneumonia. Should he be treated aggressively with antibiotics in an attempt to save his life?

An 85-year-old woman, mentally alert but crippled with arthritis, bedridden, and in severe and constant pain, wishes to die. Should her physician, who agrees that she has no prospect for recovery, be allowed to administer a lethal injection to her? Should she be treated with antibiotics if she develops pneumonia?

A 75-year-old woman has taken care of her 82-year-old husband who has been in a coma for four years. Exhausted, depressed, and informed by his physician that her husband will most probably never recover, she shoots him. Should she be sent to jail for murder?

These are three fictional episodes, but events similar to these have occurred and will be encountered more frequently in the future as our population ages. The questions they raise have stimulated a great deal of discussion about euthanasia.

The word "euthanasia" is derived from the Greek and literally means "easy death." Ethicists commonly distinguish between passive and active euthanasia. Passive euthanasia involves a deliberate decision to withhold medical treatment that might prolong life. An example of passive euthanasia would be a decision not to administer antibiotics to a person with pneumonia. Active euthanasia involves doing something with the aim of shortening a person's life. Lethal injection is an example of active euthanasia.

As with any ethical dilemma, there are no clear answers and opinions vary. Some people feel that life should be sustained if at all possible while others take a more limited view. Typical of the latter approaches are the conclusions of a group of lawyers, ethicists, and medical personnel who met in 1987 at the Hastings Center, an

institution for bioethics. They decided that an informed, mentally competent person who is terminally ill, critically ill, or gravely impaired should have the right to refuse medical treatment or have treatment withdrawn. While agreeing that the removal of life-sustaining treatment is permissible, participants in the panel opposed the use of assisted suicide or active euthanasia, even when the motive is compassion. Most ethicists draw a similar line of permissible

action between passive euthanasia, on the one hand, and active euthanasia or assisted suicide on the other.

The legal, medical, and ethical issues of euthanasia will not easily be resolved, and they are sure to be discussed with increasing urgency. Some individuals who wish to choose death under certain circumstances have decided to execute a "living will" to express their wishes (see accompanying figure).

---

## LIVING WILL DECLARATION

### To My Family, Physician, and Medical Facility

I, _____, being of sound mind, voluntarily make known my desire that my dying shall not be artificially prolonged under the following circumstances:

If I should have an injury, disease, or illness regarded by my physician as incurable and terminal, and if my physician determines that the application of life-sustaining procedures would serve only to prolong artificially the dying process, I direct that such procedures be withheld or withdrawn and that I be permitted to die. I want treatment limited to those measures that will provide me with maximum comfort and freedom from pain. Should I become unable to participate in decisions with respect to my medical treatment, it is my intention that these directions be honored by my family and physician(s) as a final expression of my legal right to refuse medical treatment, and I accept the consequences of this refusal.

Signed _____ Date _____

Witness _____ Witness _____

### DESIGNATION CLAUSE (optional*)

Should I become comatose, incompetent, or otherwise mentally or physically incapable of communication, I authorize _____

presently residing at _____ to make treatment decisions on my behalf in accordance with my Living Will Declaration. I have discussed my wishes concerning terminal care with this person, and I trust his/her judgment on my behalf.

Signed _____ Date _____

Witness _____ Witness _____

*If I have not designated a proxy as provided above, I understand that my Living Will Declaration shall nevertheless be given effect should the appropriate circumstances arise.

molecular damage, both in cells and within connective tissue. However, it is envisioned that the rate at which damage accumulates may be dependent both upon genetic factors, such as the susceptibility of cellular molecules to damage and the effectiveness of molecular repair mechanisms, and upon environmental factors, such as diet and exposure to radiation and chemicals.

## DEATH

Whatever the precise cause or causes of aging, the deterioration of function that occurs as a consequence of aging leads ultimately to death of the individual, even in the absence of specific disease.

Death is difficult to define. Implicit in all definitions is the idea that cessation of vital function is irreversible. Beyond this qualification, defining death involves subjective criteria, as does defining life (see Chapter 1). We must attempt to determine which physiological events constitute irreversible loss of vital function.

Agreeing on a definition of death is of particular importance in cases where life support systems are being used to sustain life (see *Focus: Euthanasia*). Life support systems for critically ill patients may involve such simple measures as the intravenous administration of fluids, nutrients, and drugs, or the use of more complicated technology such as dialysis machines to replace kidney function or respirators to supplement or replace respiration (Figure 23–8).

As the borderline between life and death becomes increasingly blurred with the use of life support systems, defining death becomes more difficult, yet more necessary than ever. The inability to agree on the criteria that constitute death is reflected in the laws of the 50 states of the United States. Twenty-five states have laws that define death as the cessation of heartbeat and respiration, while the remaining 25 equate death with an irreversible cessation of total brain function.

FIGURE 23–8 Modern life-support systems.

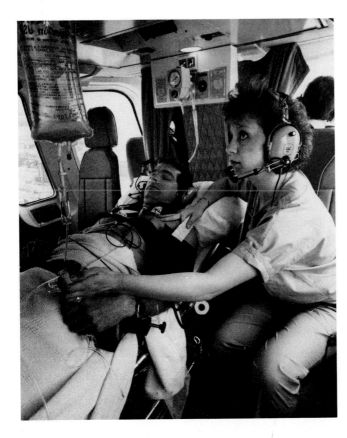

Even when we can decide upon a definition of death, applying it is not always straightforward. Individuals suffering barbiturate poisoning, for example, may show cessation of brain function as determined by an electroencephalogram for as long as a day, but may then recover fully. It is thus imperative that clinical criteria for death be clearly established.

## Study Questions

1. a) What is the cause of the increase since 1900 in the median life expectancy at birth?
   b) Has the maximum life span increased during the same period? Explain your answer.
2. a) Describe five physiological changes that are associated with aging.
   b) Do these changes appear to be the cause or result of aging?
   c) Explain your answer to b.
3. Describe the results of Leonard Hayflick's experiments on the growth of cells in tissue culture and explain their relevance to aging studies.
4. Write a short essay describing in detail two possible theories of aging.
5. Why is death difficult to define?

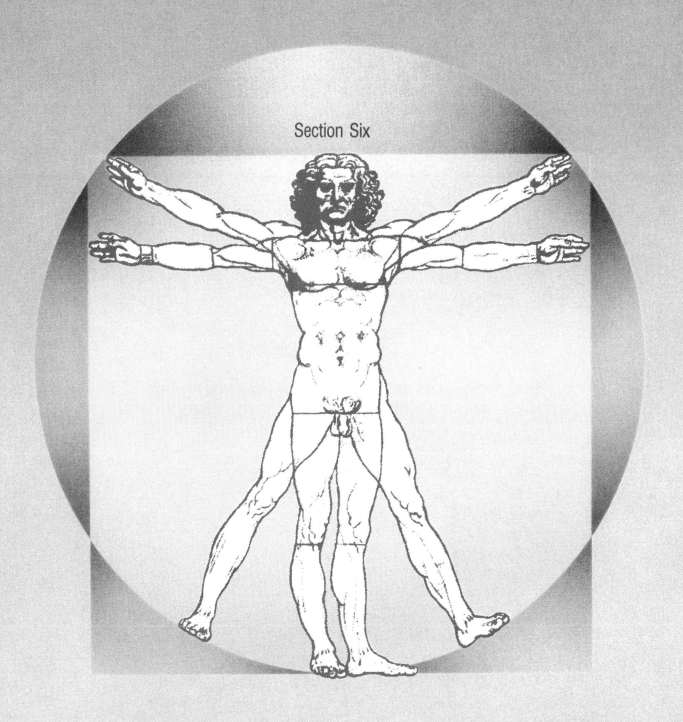

Section Six

# THE PAST, THE PRESENT, THE FUTURE

# HUMAN EVOLUTION

In Chapter 1, we emphasized the fundamental similarities between all living organisms. This unity in the biological world underlies a more obvious diversity. Even the most casual observer is struck by the vast array of living types. Living organisms range in size from the minuscule to the enormous; they come in virtually every color and in a great variety of shapes. As a group, living organisms can flourish in practically every environment on earth. There are land dwellers and water dwellers; there are organisms that can live in scalding hot springs, on the frozen arctic tundra, or in the high salinity of the Great Salt Lake. The list of types, each with its own particular characteristics, is enormous. Indeed, almost two million types of living organisms have been described in the biological literature, and there are many other millions of types, now extinct, that have lived in times past. The evolutionary biologist G. G. Simpson has estimated that 500 million to 4 billion types of organisms have existed since life originated on earth.

Diversity and unity are the dual themes of the living world. Why, we may ask, is this so? The answer to our question may be found in the theory of evolution, by far the most significant unifying principle of biology.

## EVOLUTION: THE UNIFYING CONCEPT OF BIOLOGY

The theory of biological evolution states that living organisms change continually and gradually in character with the passage of time and thus become increasingly different from their early ancestors. In Charles Darwin's words, evolution is "descent with modification" (see *Focus: Darwin's Theory of Natural Selection*). By contending that diverse types of organisms have originated by the gradual modification of common ancestors, evolutionary theory implies that all living organisms are related to one another, and thus provides a unifying concept in biology.

Evolution, then, is a historical process. It describes the historical development of groups of organisms—a process referred to as **phylogeny.** According to evolutionary theory, the unity of the living world is derived from the fact that all living organisms are related to one another, and hence share a basic similarity in design. Diversity, on the other hand, is a result of phylogenetic change and is achieved by *superficial* modifications of the fundamental design. Unity is thus the stable superstructure on which diversity is built.

So much evidence has accumulated to support the theory of evolution— evidence from the fossil record and from comparative studies of the anatomy, physiology, and biochemistry of various types of living organisms—that it is now considered axiomatic in biology that evolution is a natural process that has occurred in the past and is continuing even today (see *Focus: How Embryonic Development Reveals Evolution's Imprint*).

## THE ORIGIN AND EVOLUTION OF LIFE

### Geological Time Scale

Tracing the history of life requires reference to a time scale. Although a number of different systems have been used to describe geological time, most geologists now divide the history of the earth into four major *eras.* The eras are separated from one another by times of major geological upheavals. These resulted in the formation of mountain ranges, radical changes in sea level, and drastic alterations in climate. Beginning with the formation of the earth approximately 4.6 billion years ago, the major eras are the *Precambrian* (*pre* = before; *cambrian* = Cambrian Period), *Paleozoic* (*paleo* = ancient; *zoic* =

TABLE 24-1 The Geological Time Scale and Major Events in the History of Life

| Era | Period | Epoch | Time Boundaries (millions of years) | Event |
|---|---|---|---|---|
| Precambrian | | Early | — 5000 — | |
| | | | — 4600 — | Formation of earth and planets of solar system. |
| | | | | Origin of life; oldest fossils (3.1–3.2 billion years old) resembling modern bacteria. |
| | | | — 2500 — | |
| | | Middle | — 1700 — | Many diverse groups of bacteria. |
| | | Late | | Fossils of first organisms with complex cell structure (eukaryotes); toward end, increasing numbers of protists and multicellular marine invertebrates, including sponges, worms. |
| | | | — 570 — | |
| Paleozoic | Cambrian | | | Appearance of most invertebrate types; first marine algae. |
| | | | — 500 — | |
| | Ordovician | | | First jawless fishes; diversification of marine algae. |
| | | | — 430 — | |
| | Silurian | | | First jawed fishes; first land plants; first wingless insects. |
| | | | — 395 — | |
| | Devonian | | | First amphibians; first conifers (nonflowering plants). |
| | | | — 345 — | |
| | Mississippian | | | First reptiles; insects common; widespread forests of seed ferns and conifers. |
| | | | — 320 — | |
| | Pennsylvanian | | | |
| | | | — 280 — | |
| | Permian | | | Mammal-like reptiles appear. |
| | | | — 225 — | |
| Mesozoic | Triassic | | | First dinosaurs; conifers dominant. |
| | | | — 190 — | |
| | Jurassic | | | First mammals and birds; first flowering plants. |
| | | | — 136 — | |
| | Cretaceous | | | First modern birds; vast oak and maple forests. |
| | | | — 65 — | |
| Cenozoic | Tertiary | Paleocene | | Higher mammals arise and diversify; flowering plants reach full development; forests dominant, then decline toward end with spread of grasslands. |
| | | | — 54 — | |
| | | Eocene | | |
| | | | — 38 — | |
| | | Oligocene | | |
| | | | — 26 — | |
| | | Miocene | | |
| | | | — 21 — | |
| | | Pliocene | | |
| | | | — 2 — | |
| | Quaternary | Pleistocene | | |
| | | | — .01 — | |
| | | Holocene | | Extinction of great mammals; age of humans. |

life), *Mesozoic* (*meso* = middle), and *Cenozoic* (*ceno* = modern). Within each era were times of less extensive geological changes that separate *periods;* within these periods, *epochs* are separated by more localized modifications (Table 24–1).

## The Origin of Life

It is probably fair to say that the vast majority of modern biologists accept the so-called *materialistic theory* of life's origin. First proposed in 1924 by the Russian biologist A. I. Oparin and independently several years later by the British biologist J. B. S. Haldane, the materialistic theory suggests that the first living organisms originated from nonliving matter by a gradual process of *chemical evolution.*

## Chemical Evolution

This evolutionary process is thought to have begun with the formation of the earth approximately 4.6 billion years ago. A great deal of evidence from astronomy has suggested that the atmosphere of the primitive earth was probably a reducing one; that is, it contained no free oxygen gas ($O_2$), but consisted primarily of methane ($CH_4$), ammonia ($NH_3$), hydrogen gas ($H_2$), and water vapor ($H_2O$), along with smaller amounts of other gases, such as hydrogen sulfide ($H_2S$), nitrogen ($N_2$), and carbon monoxide ($CO$). It is assumed that simple organic compounds were formed from chemical reactions between these gases, using energy sources available in the environment. These included ultraviolet light from the sun, electric discharge from thunderstorms (lightning), and heat from volcanic activity.

As simple organic compounds formed and became concentrated in the primitive oceans, organic chemical substances of ever increasing complexity were formed. These substances gradually achieved a loose structural organization and a capacity for self-reproduction. Eventually, the first cellular entity evolved, signaling the beginning of biological evolution. Once living organisms had evolved, this process could have been repeated only in isolated environments, because any newly developed form would quickly be eliminated by more efficient competitors. A summary of the origin of life as proposed by the materialistic theory is shown in Figure 24–1.

## Simulation Experiments

Although we do not know—and cannot know—the precise details of how life evolved, there is good circumstantial evidence to suggest that the overall scheme of the materialistic theory is a reasonable one. Its wide acceptance among biologists is based in large part on the ease with which organic compounds characteristic of contemporary living organisms can be produced in laboratory experiments, using starting materials and experimental conditions that simulate those thought to have been present during the early history of the earth. These so-called *simulation experiments,* performed by a large number of investigators, have produced not only many of the simple organic building blocks of present-day organisms, but also larger organic molecules and polymers similar or identical to the complex macromolecules found in biological systems. In addition, it has even been possible, again by simulating primitive earth conditions, to form aggregates of organic compounds that display a few of the physical and chemical characteristics common to living cells.

*Synthesis of Small Organic Molecules.* The first simulation experiment was conducted in 1953 by Stanley Miller, then a graduate student of Harold Urey at the University of Chicago. To perform his experiment, Miller constructed a

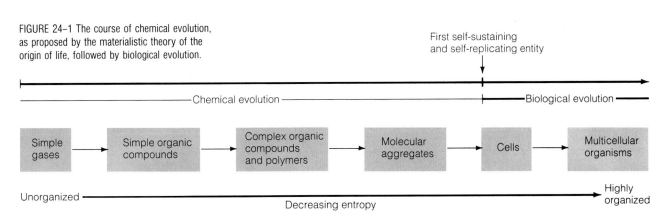

FIGURE 24–1 The course of chemical evolution, as proposed by the materialistic theory of the origin of life, followed by biological evolution.

First self-sustaining and self-replicating entity

———— Chemical evolution ————   ——Biological evolution——

Simple gases → Simple organic compounds → Complex organic compounds and polymers → Molecular aggregates → Cells → Multicellular organisms

Unorganized ———— Decreasing entropy ————→ Highly organized

# Darwin's Theory of Natural Selection

**C**harles Darwin must be included among a handful of truly great scientists, and the impact of his work has been immense. Darwin's evolutionary theory—the theory of natural selection—not only provides a unifying principle for all of biology, but it has changed our very thinking about the nature of humans themselves.

The theory of natural selection is most succinctly summarized by Darwin in the "Introduction" to his book *On the Origin of Species by Means of Natural Selection,* published in 1859. The passage is worth careful reading.

*As many more individuals of each species are born than can possibly survive; and as, consequently there is a frequently recurring struggle for existence, it follows that any being, if it vary in any manner profitable to itself, will have a better chance of surviving and thus be* naturally selected. *From the strong principle of inheritance, any selected variety will tend to propagate its new and modified form.*

The logic of Darwin's argument may be summarized as follows:

1. The reproductive potential of living organisms is extremely high, but the size of a population (defined as a group of organisms of a single type, or species) is limited by available resources. Hence there must be a "struggle for existence"—a competition among members of a species for available resources necessary for survival.

2. Variation in type is characteristic of individuals of a species. Such variation may include morphological, physiological, behavioral, and other parameters.

3. Those individuals of a species that possess a hereditary variation that in some way—no matter how minor—makes them better equipped than other members of the species to compete for available resources will tend to survive and *successfully reproduce their kind* in greater pro-

Charles Darwin.

portion than do other forms. This process, which favors the survival and successful reproduction of those individuals best adapted to the environment, was termed *natural selection* by Darwin.

To support his concept of natural selection, Darwin emphasized its similarity to the artificial selection used by animal breeders in developing breeds with specific characteristics. He pointed out, for example, that animal breeders had developed a number of distinct breeds of pigeons from the ancestral rock pigeon by selecting at each generation those individuals as breeding stock that possessed the most extreme variation in the desired characteristics.

Darwin recognized that variation can potentially provide the difference that gives an organism an edge in competition to survive and successfully reproduce its kind. To take a simple example, a variation in color that might make an individual animal less noticeable to predators than are other members of its species would

tend to increase the chance of that individual's survival and reproduction. Variation is thus sorted out by natural selection to bring about the long-term changes in living organisms that we call evolution.

Natural selection will tend not only to favor the survival and reproduction of those individuals with advantageous variations, but to eliminate those with undesirable variations. The consequence of natural selection is that a species will be modified in the direction of the most advantageous variants. The rate at which such evolutionary change occurs depends upon the intensity of the selection pressure. Forms that are highly unsuited to their environment are eliminated faster than those that are less so. Such a process of evolutionary change, occurring over a long enough time, may eventually give rise to the formation of new species.

Although natural selection may act upon any variation in the population, only *heritable* variations have any significance from an evolutionary standpoint; that is, in order to bring about evolutionary change, a variation must be passed on to future generations. Nonheritable variation, such as variation that is a result of characteristics acquired during an individual's lifetime (for example, loss of a leg by amputation), may be crucial to the survival of the individual, but it cannot lead to evolutionary change.

Darwin himself recognized that success in leaving offspring that reach reproductive age is a more important component of natural selection than survival of the individual. Indeed, we now realize that natural selection operates through a differential rate of reproduction among members of a population. In the often misunderstood phrase "survival of the fittest," the fittest are those that are the most *reproductively fit.*

It is important to realize that natural selection is not some mystical force, changing species in a predetermined manner. It is the environment that determines the intensity and direction of natural selection. Obviously environments change with time, and a trait that is selectively advantageous in one environment is not necessarily so in another, as illustrated in our earlier discussion of sickle cell anemia (see Chapter 20).

Darwin's theory of natural selection was incomplete in one very important respect: Darwin gave no satisfactory explanation of how variation comes about. In the middle 1800s when Darwin formulated his theory, virtually nothing was known about the mechanism of heredity. Gregor Mendel's experiments on breeding in peas, which provided the theoretical foundation for modern genetics, were published in 1866, but his work was ignored until it was rediscovered at the turn of the century. By the 1930s, the principles of modern genetics had been firmly established, and geneticists realized that the ultimate source of hereditary variation is random change, or mutation, in the hereditary material of germ cells. Darwin's theory of natural selection, aligned with modern genetics so as to attribute variation to random change in the hereditary material, is known as *neo-Darwinism*.

### Neo-Darwinism: A Summary

Evolution involves the *chance* production of hereditary variation among the members of a species, followed by the natural selection of those individuals who are reproductively more fit. These individuals will thus, on the average, contribute to the next generation a greater proportion of individuals that reach reproductive age than do forms that are less adapted to the environment (see accompanying figure).

Differential reproduction arising from randomly occurring hereditary variation is, then, the key to evolutionary change as proposed by neo-Darwinian theory. The point cannot be overemphasized, for it is a failure to understand this central concept that has resulted in countless misinterpretations of the evolutionary process.

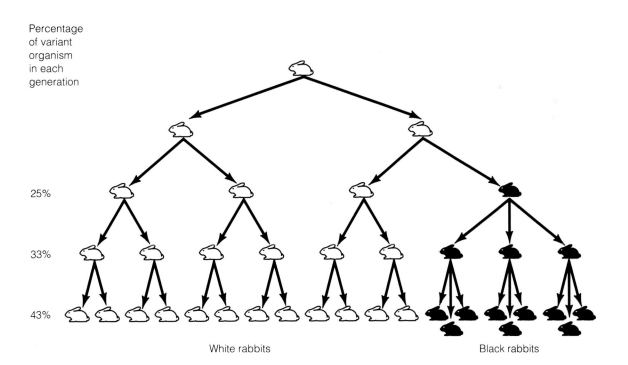

Percentage of variant organism in each generation

25%

33%

43%

White rabbits

Black rabbits

Illustration of evolutionary change by differential reproduction as proposed by neo-Darwinian theory. If an organism possessing a chance hereditary variation (black symbols) leaves, on average, more offspring that reach reproductive age than do other individuals in the population (white symbols), the variant type will form, in succeeding generations, an increasingly large proportion of the population. Competition for resources between the two types may lead to the extinction of the nonvariant type.

glass discharge apparatus, shown in Figure 24–2. Water was added to the flask (a), and the entire apparatus was then evacuated of all air. Hydrogen, methane, and ammonia were added through the stopcock (d) without allowing any air to enter the apparatus. The water in the small flask was then heated to boiling, producing steam that circulated clockwise through the apparatus. The spherical sparking chamber (e) contained two tungsten electrodes (c), separated by a gap of about 1 cm. These were attached to a high frequency sparking coil, and a continuous spark discharge was produced between the electrodes. The convection currents, caused by the circulation of the steam through the apparatus, moved the gases past the sparking electrodes. Steam was condensed below the sparking chamber in the vicinity of the water-cooled condenser (b). In this way, the nonvolatile products formed in the sparking chamber were washed through the trap and into the small flask, where they remained. Any volatile products would continue to recirculate, along with the steam and original gases, through the apparatus and past the spark.

Miller's experiment was designed to mimic as closely as possible conditions believed to have existed during early stages of chemical evolution. The spark served as a source of energy for chemical synthesis, and the mixture of reduced gases, containing no free oxygen, was designed to simulate the composition of the primitive atmosphere. The flask of boiling water, representing early stages in the history of the ocean, served as a means of producing steam and circulating the gases in the apparatus. With some imagination, the apparatus might be regarded as representing the synthesis of organic compounds by the action of lightning in the upper atmosphere, followed by washing of the products into the primitive ocean by rain.

FIGURE 24–2 The spark discharge apparatus used by Stanley Miller in an experiment designed to simulate primitive earth conditions.

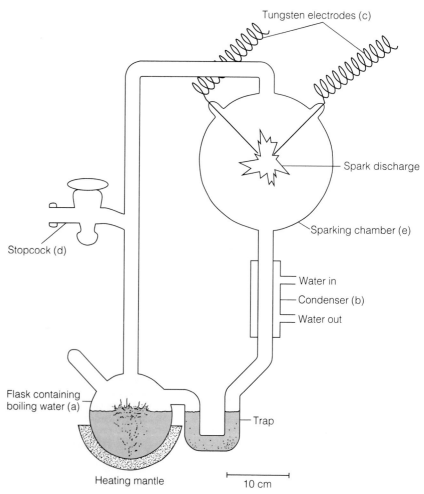

Tungsten electrodes (c)

Spark discharge

Sparking chamber (e)

Stopcock (d)

Water in
Condenser (b)
Water out

Flask containing boiling water (a)

Trap

Heating mantle

10 cm

The experiment was allowed to run for about a week with continuous sparking, after which the accumulated products in the small flask were subjected to a detailed chemical analysis. A number of organic compounds of biological interest were identified, including the amino acids glycine, alanine, and aspartic acid, and a number of other simple organic compounds, which included urea, lactic acid, and succinic acid, important metabolic compounds in living organisms (see Chapter 2).

In the decades since the initial report of Miller's research, numerous simulation experiments have been performed in an attempt to produce other organic compounds of importance to living systems. In addition to amino acids, many other compounds including sugars, nitrogenous bases, fatty acids, and porphyrin ring structures (as in the heme of hemoglobin) have been produced using a variety of simple gaseous mixtures and a number of different energy sources. The significant point is that conditions leading to the production of biologically important organic compounds need not be carefully defined. We thus do not have to invoke some highly improbable set of circumstances in order to find experimental support for the materialistic theory.

*Synthesis of Macromolecules.* In addition to the production of small organic compounds required for the evolution of simple living systems, more complex organic molecules have been formed from these simple ones, again using conditions that may have existed on the primitive earth. For example, heating mixtures of dry amino acids has yielded amino acid polymers called *proteinoids,* similar in many ways to biologically synthesized proteins. Polymers of nucleotides and of glucose have also been obtained in simulation experiments.

*Formation of Molecular Aggregates.* In yet other simulation experiments, aggregates of complex organic molecules have been obtained; these constitute localized systems partially isolated from their surroundings. One type of stable molecular aggregate can be formed from proteinoids dissolved in a boiling salt solution. When the solution is allowed to cool, vast numbers of small spherical structures called *microspheres* separate out of solution. Microspheres possess a number of properties similar to those of living cells. They are separated from their environment by a surface membrane, are similar in size and shape to some forms of bacteria, undergo division by budding off, and display osmotic properties, shrinking in solutions that are hyperosmotic to the solution in which they were formed and swelling in hypo-osmotic solutions. It is thought that other types of molecular aggregates could have been formed in the primitive ocean as a result of the adsorption (adhesion) of organic compounds to clays and other minerals, and by *coacervation,* a process by which concentrated droplets of organic materials can form in an aqueous medium.

In whatever way molecular aggregates might have formed, they are envisioned as providing a matrix, or framework, for the incorporation of organic substances from the environment. As the aggregates increased the complexity of their organization, they eventually gave rise to self-replicating and self-sustaining living systems—the ancestral living organisms.

Although the results of simulation experiments may support the materialistic theory, they do not constitute proof that events on the primitive earth necessarily followed the postulated course. The transition from molecules to cells is particularly difficult to envision, and although many theories have been proposed, further research is necessary before we can say that we understand how cellular entities evolved. For the moment, the basic outline of the materialistic theory seems to be a reasonable hypothesis for the origin of life on earth.

## The History of Life

Most of what we know about the history of life comes from the study of fossils—remains, impressions, or traces of organisms that have been preserved in the earth's crust (Figure 24–3, p. 536).

## How Embryonic Development Reveals Evolution's Imprint

As in the case of humans, the life of most multicellular animals begins with fertilization—the fusion of an egg and sperm cell—forming a single-celled embryo that does not resemble its parents. The embryo then undergoes a complex process involving cell division, growth, and reorganization by which it is transformed into a shape more or less characteristic of the adult. The sequence of stages involved in the transformation of the fertilized egg to an adult individual is known as an organism's **ontogeny,** or **ontogenetic development.**

In the eighteenth and early nineteenth centuries, studies by embryologists had shown that temporarily distinct stages of ontogenetic development could be described and studied. A leader in this field was the German embryologist Karl Ernst von Baer who undertook a comparative study of the ontogenetic stages in a number of different vertebrate types. Based upon his studies, von Baer formulated in 1828 some principles of comparative physiology that have since become known as the "laws of von Baer." These are:

1. During ontogenetic development, the more generalized traits of an organism appear before the more specialized traits.

2. The more specialized traits of an organism always develop from the more generalized traits.

3. As ontogenetic development proceeds, each animal becomes progressively more unlike other animals.

4. Early embryos of a species resemble the early embryos of animals of lower taxonomic status.

The first three laws of von Baer state that the basic structural organization (generalized traits) of an organism is established during the earliest stages of ontogenetic development. This basic organization is then modified to form specialized traits that are unique to a particular type of organism. The consequence of this developmental pattern is that an organism becomes increasingly more unlike other organisms as ontogenetic development proceeds. For instance, in the ontogeny of a human, it is initially recognizable only as a vertebrate (a more generalized character); at a much later stage in development, more specialized traits appear that identify the organism as a mammal, then a primate; finally, the species is identifiable.

The fourth law of von Baer states that the early ontogenetic stages of a developing embryo are similar to the early ontogenetic stages of lower forms. For example, a higher vertebrate such as a human being goes through early stages in its ontogenetic development that are closely comparable to early stages in the ontogeny of fish, amphibians, and reptiles (lower vertebrates) as seen in the accompanying figure. Von Baer's fourth law can be summarized by the following statement: *embryos recapitulate (repeat) the early ontogeny of their ancestors to varying degrees.*

There are a number of examples of von Baer's laws, but perhaps the best is the development of gills and related structures in vertebrates. Gills form the respiratory apparatus of water-breathing animals, including most of the fishes and many larval amphibians. Their function is to absorb oxygen gas from the surrounding water for use by the organism and to give up carbon dioxide and other wastes to the environment. As such, gills serve the same function as the lungs of terrestrial animals.

The gills develop from a series of pouches, or outpocketings, that bulge out from the walls on either side of the pharynx. Pharyngeal pouches develop in all vertebrate embryos (see accompanying figure). However, in terrestrial vertebrates (adult amphibians, reptiles, birds, and mammals) the pharyngeal pouches do not give rise to gills. Instead, they are rearranged during ontogeny to form other structures. In embryonic reptiles and birds, several of the five or six pairs of pharyngeal pouches typically open to the exterior, but the openings later close prior to birth. (As gills develop in fish, all pharyngeal pouches open to the exterior and remain permanently open.)

In embryonic mammals, one or two pairs of pouches may open briefly to the exterior, but in general they do not open at all. As a rare developmental abnormality in humans, the pharyngeal pouches may overdevelop and form clefts that persist as openings in the neck region. In mammals, the most anterior pair of pharyngeal pouches is retained as the Eustachian tube, which connects the pharynx to the middle ear. Remnants of the other pouches persist in terrestrial vertebrates as certain glandular structures, including the thyroid and parathyroid glands. The *pharyngeal arches,* bars of supporting tissue that develop in fish between the pharyngeal pouches, also form in terrestrial vertebrates, but they are later modified into jaw components and supporting structures of the neck.

---

The oldest rocks in which fossils have been discovered are sedimentary rocks of the Fig Tree and Onverwacht geological formations, located in the Transvaal of South Africa. These formations, which are about 3.1 to 3.2 billion years old, contain microscopic organic fossils that closely resemble present-day bacteria (Figure 24–4). Today's bacteria are considered the most primitive living organisms because they are primarily unicellular and possess a prokaryotic cellular structure that is much simpler than that of all other types of

Why should a terrestrial vertebrate that never develops gills pass through embryonic stages in which it forms pharyngeal pouches that even may open briefly to the exterior? Indeed, why do virtually all organ systems show similar patterns of development in all vertebrates? Such an ontogenetic pattern is logically consistent with evolution. If all vertebrates are related to one another through common descent, we would expect the early, and hence more fundamental, ontogenetic stages to be similar in all vertebrates. Mammals, for instance, have retained a basic organizational plan that they share in common with their fishlike ancestors; this plan is used during early development to form generalized structures that are subsequently modified to form specialized traits.

Embryos *do* repeat to varying degrees the embryonic stages of their ancestors, and this recapitulation is best explained by evolutionary theory. Von Baer's laws and their supporting evidence thus provide strong support for the occurrence of evolution.

Comparable stages in the ontogeny of a fish, salamander (amphibian), tortoise (reptile), rabbit (mammal), and human (mammal). Note that the early stages of higher forms such as the rabbit and human are very similar to the early stages of lower forms.

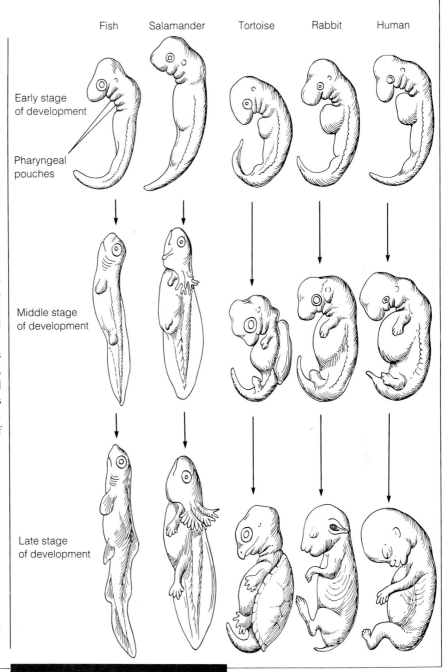

Fish  Salamander  Tortoise  Rabbit  Human

Early stage of development

Pharyngeal pouches

Middle stage of development

Late stage of development

organisms (the eukaryotes). It is noteworthy that fossils of more complex organisms have not been found in the Fig Tree and Onverwacht formations, suggesting that more advanced types of organisms were not present approximately 3 billion years ago.

As we examine progressively younger Precambrian rocks, we see an increased diversity and complexity of fossil organisms. Fossils discovered in the Late Precambrian Bitter Springs formation in Northern Territory, Australia

FIGURE 24–3 These footprints of two walking individuals, discovered in Northern Tanzania by Mary Leakey and dated to about 3.7 million years ago, represent the earliest fossil evidence of bipedalism. The big toes, instead of projecting sideways like those of apes, point straight ahead, like those of humans.

polishing scratch

FIGURE 24–4 A rod-shaped fossil about 3.1 billion years old from the Fig Tree geological formation of South Africa. It is similar in size and shape to modern rod-shaped bacterial species.

(about one billion years old), include 23 certain or probable species of bacteria. In addition, two genera of green algae and two possible species of fungi were reported, discoveries of particular interest because present-day green algae and fungi have a eukaryotic cell structure that is considerably more complex than that of bacteria. It is thus apparent not only that bacteria were highly diversified by this time, but that more complex organisms were represented.

The fossil record encompassing the last 400 million years of the Precambrian shows an increasing number of reasonably complex unicellular organisms such as amoebae and traces of simple multicellular invertebrates such as sponges and worms. In the Cambrian period (570 to 500 million years ago) most major groups of invertebrate animals are represented. The most primitive fishlike vertebrates appear in the Ordovician period (500–430 million years ago). In the Silurian (430–395 million years ago), the jawed fishes appear, as well as some extremely complex invertebrates such as insects. The first amphibians appear in the Devonian (395–345 million years ago); the first reptiles in the Pennsylvanian (300–280 million years ago); then mammals and birds appear by Jurassic times (190–135 million years ago). The diversification of plants follows a similar pattern, with simple unicellular organisms appearing first in the fossil record, followed by progressively more complex types. A summary of the major events in the history of life, as reconstructed from the fossil record, is shown in Table 24–1.

## HUMAN EVOLUTION

How did our own species, Homo sapiens, evolve? The question is a difficult one to answer, not because there is any dispute among biologists that *Homo sapiens* originated by an evolutionary process, but because evidence bearing on the question is so fragmentary. Fossils of human ancestors are rare, often consisting of only a portion of a skull, jaw, or other bony structure. Consequently, there is considerable controversy about the identity of many fossils and their relationship to other fossil finds. Muddling the issue even further are the results of new techniques, used to test biochemical similarities between existing species, that often do not yield information consistent with that inferred by the fossil record. The story of human evolution is thus far from clear, and any reconstruction of the events leading to the evolution of modern man must be considered speculative.

### Relationship Between Humans and Other Organisms

Studies of human structure, physiology, biochemistry, and embryological development have long clarified the general relationships between human beings and other living organisms. Our fundamental body plan, that of a *vertebrate,* is shared by all fishes, amphibians, reptiles, birds, and mammals. Within the vertebrate group, we are aligned most generally with the *placental mammals.* We belong to the *primate* order of mammals, and within that group, we are most similar to the modern anthropoid apes including gibbons, orangutans, gorillas, and chimpanzees. We are the only living members of the genus *Homo;* our species is *sapiens.* A complete taxonomy (classification according to natural relationships) of humans is shown in Table 24–2. Knowledge of this scheme will help in understanding the discussion that follows.

## Our Mammalian Ancestors

The fossil record tells us that mammals evolved from a reptilian ancestor at the end of the Triassic Period, about 200 million years ago (see Table 24–1). These first mammals, about the size of a mouse, were tree-dwelling (arboreal) and nocturnal. Like present-day mammals, they were characterized by the presence of hair, mammary glands in the female, and differentiated teeth. They existed in a terrestrial environment dominated by large numbers of diverse reptilian types, including the dinosaurs. For as long as the reptiles remained in ascendancy, mammals evolved slowly and uneventfully.

Toward the end of the Cretaceous Period some 65 million years ago, the dinosaurs and many other reptiles suddenly disappeared from the fossil record. These mass extinctions have been variously attributed to climatic changes, predation of reptile eggs by mammals, and a shower of comets hitting the earth and raising dust clouds that blocked out sunlight for years. Extinction of the reptiles was accompanied by a very rapid evolution of the mammals. By the end of the Cretaceous Period there were three main groups of mammals in existence: the **monotremes,** or egg-laying mammals, the **marsupials,** or pouched mammals, and the **placental mammals.**

TABLE 24–2 Classification of Modern Humans

| Taxonomic category | Latin designation | Characteristics and types |
|---|---|---|
| Kingdom | Animalia | Primarily motile, multicellular, heterotrophic nutrition. Includes all animals. |
| Phylum | Chordata | Bilateral symmetry, hollow dorsal nerve cord; notochord and pharyngeal pouches present sometime during life. Includes protochordates (e.g., sea squirt, amphioxus) and vertebrates. |
| Subphylum | Vertebrata | Segmented backbone and spinal cord; supporting internal skeleton. Includes fishes, amphibians, reptiles, birds, and mammals. |
| Class | Mammalia | Warm-blooded animals with hair and mammary glands. Includes monotremes (egg-laying mammals), marsupials (pouched mammals), and placental mammals. |
| Order | Primates | Highly developed brain; feet and hands with five digits; opposable first digit on feet or hands, or both; eyes directed forward giving binocular vision. Includes prosimians (e.g., lemurs, tarsiers, lorises) and anthropoid monkeys, apes, and humans. |
| Suborder | Anthropoidea | Flattened face (snout absent); specialized structures of nasal passages; increasing development toward prehensile hands and/or feet. Includes New World Monkeys, Old World Monkeys, anthropoid apes, and humans. |
| Superfamily | Hominoidea | Large size; barrel-shaped chest; tailless; distinctive molar pattern. Includes anthropoid apes and humans. |
| Family | Hominidae | Upright stance; reduced canines; foramen magnum under skull; eye ridges and skull crests reduced. Includes two genera: Homo and the fossil genus Australopithecus. |
| Genus | Homo | Large brain; increased development of chin and nose; well-developed head. Includes fossil and modern forms. |
| Species | sapiens | Largest brain relative to size; high forehead; jaw less pronounced. Includes modern humans and Neanderthal humans. |

FIGURE 24–5 The first primates are thought to have resembled the present-day tree shrew, shown in this photograph.

## Our Primate Ancestors

Among the placental mammals present by the late Cretaceous were the earliest primates. It is generally accepted that they evolved from a line of arboreal insectivores (the insectivores today are a group of nocturnal placental mammals that includes shrews, hedgehogs and moles).

### Prosimians

The first primates were small, squirrel-like mammals with bushy tails and somewhat enlarged brains for their size. They presumably showed some development toward prehensile fingers and toes and the opposability of the first digits (Figure 24–5). They have been classified in the suborder *Prosimii,* most clearly related to present-day prosimians—lemurs, tarsiers, and lorises.

The early prosimians spread quickly over large areas of the Old World (eastern hemisphere) and New World (western hemisphere), retaining their arboreal existence, large eyes, sharp snouts, and bushy tails. The prosimians of today have very restricted ranges compared to those of their earlier ancestors. For example, lemurs, once common in Europe and North America, are now found only on the island of Madagascar; tarsiers, widely distributed in northern continents, are now present only in the Philippines and East Indies.

### Anthropoidea

By the Oligocene Epoch, some 38 million years ago, the prosimians had given rise to two other lines, the first members of the suborder *Anthropoidea* (see Table 24–2). One of these lines consisted of ancestors to the New World monkeys and the other of ancestors to the Old World monkeys, the anthropoid apes and humans (Figure 24–6). The fossil record suggests that these two lines evolved independently of one another from different lines of prosimians. The New World monkeys, considered the most primitive of the two groups, are arboreal and have prehensile tails and nostrils that are far apart and flared outward. They are referred to as *platyrrhine* monkeys (*platy* = broad, flat; *rhine* = nose). New World monkeys are now found in the tropical forests of Central and South America, and are represented by such species as the capuchin and squirrel monkeys. The other anthropoid line—the Old World monkeys of Africa and Southern Asia, the anthropoid apes and humans—consists of organisms with nostrils that are set close together and open downward; they are called *catarrhines* (*cata* = down; *rhine* = nose). Their tails, if present, are short and never prehensile (adapted for grasping).

### The Hominoid Line

The catarrhine group of anthropoids eventually diverged into the Old World monkey line including both arboreal and ground-dwelling forms, and another line, the hominoids (superfamily *Hominoidea*) that led to the anthropoid apes and humans (see Figure 24–6). The fossil record is unclear about the timing of this divergence, but it has been variously suggested to have occurred sometime between 25 and 35 million years ago.

### Dryopithecus

The earliest known fossil hominoid, dating back some 20 million years ago, was discovered by Louis Leakey in the 1930s in an island in Lake Victoria. It has been classified in the genus *Dryopithecus* (*dryo* = tree; *pithecus* = ape). Other dryopithecine fossils, dated as late as 12 million years ago, have since been found in India and Europe. The dating of these fossil finds suggests that the dryopithecines evolved in Africa and then spread throughout Europe and Central Asia during the Miocene and Early Pliocene Epochs.

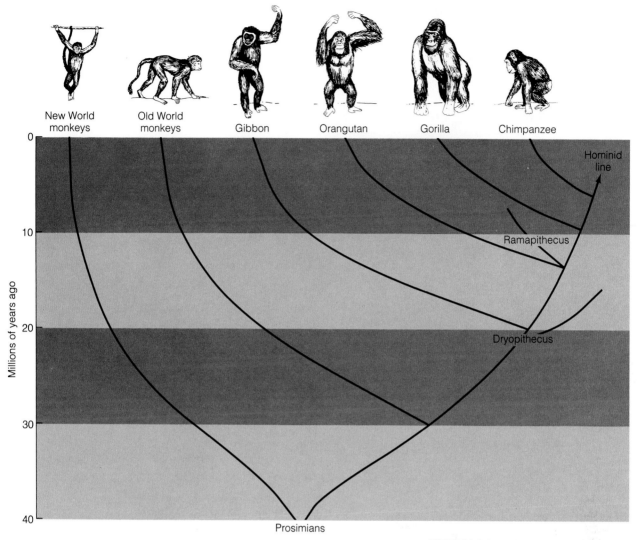

New World monkeys    Old World monkeys    Gibbon    Orangutan    Gorilla    Chimpanzee

Hominid line

Ramapithecus

Dryopithecus

Millions of years ago

Prosimians

FIGURE 24–6 A tentative scheme of anthropoid evolution.

The dryopithecines include several distinct forms, varying in size from a small to a large gorilla. They lacked most of the specialized features that distinguish the modern anthropoid apes from humans. Their skulls did not have the well-developed crests and heavy brow ridges that are characteristic of modern apes (Figure 24–7). Compared to the anthropoid apes, their canines were smaller and their arms shorter; the latter characteristic was an adaptation for running and climbing in their arboreal environment rather than swinging. Their thin tooth enamel, similar to that of modern gorillas and chimpanzees, indicates that they were fruit eaters. The characteristics of their skeletal remains suggest that the dryopithecines may have been common ancestors to the modern great apes and humans (see Figure 24–6).

### Ramapithecus

Another fossil hominoid genus, *Ramapithecus,* is represented by teeth and fragments of the jaw and face that have been discovered in India, Pakistan, Africa, and Turkey. These fossils cover a time span from about 15 to about 8 million years ago. Ramapithecus was primarily a ground-dwelling ape, some forms of which may have been bipedal. Its face was flatter with less of a muzzle than that of modern apes, its tooth enamel was thick, and its canines were small (see Figure 24–7). These characteristics, similar to those of later hominids and

FIGURE 24–7 Skulls of a gorilla (a), *Dryopithecus* (b), and *Ramapithecus* (c).

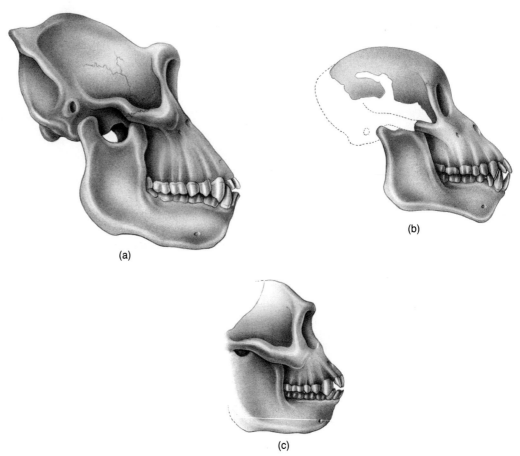

(a)

(b)

(c)

humans, once suggested to many anthropologists and biologists that ramapithecines were the first hominids—our oldest direct ancestors. However, several lines of recent evidence cast substantial doubt on this conclusion. First, *Ramapithecus* now appears too old to be a hominid, since it existed well before the time that the hominid line is presumed to have diverged from ancestors of our closest anthropoid ape relatives (hominoids). For example, divergence times in the hominoid lineage have been estimated by Charles Sibley and Jon Ahlquist of Yale University, using data obtained from a technique that compares similarities between DNA nucleotide sequences in present day anthropoid apes and humans. Their data suggest that the gibbon diverged from its hominoid ancestors 18 to 22 million years ago, followed by the orangutan (13 to 16 million years ago), then the gorilla (8 to 10 million years ago), and finally the chimpanzee (6 to 8 million years ago). These divergence times were used to construct the scheme of anthropoid evolution shown in Figure 24–6.

Recent analyses of hominoid tooth enamel also suggest that *Ramapithecus* was not a hominid. It appears from these studies that thick tooth enamel, rather than being a distinctly hominid characteristic as previously thought, is a primitive character of the common ancestor to orangutans, gorillas, chimpanzees, and humans, and consequently cannot be used to identify hominids. A third line of evidence tending to decertify *Ramapithecus* as a hominid is the fact that *Sivapithecus,* a fossil genus closely related to *Ramapithecus,* is now known to have facial features aligning it to the orangutan.

### The Hominid Line

The fossil record is virtually nonexistent between 8 and 4 million years ago. Since hominids had already appeared by about 4 million years ago, there are no fossils documenting the hominid-anthropoid ape divergence.

## Australopithecines

The earliest fossils indisputably identified as hominids date from about 4 million years ago in Africa. Almost half of the skeletal remains of an adult female who lived some 3.7 million years ago were discovered by Donald Johanson and his colleagues in the Afar region of Ethiopia. Nicknamed "Lucy," it was formally classified by Johanson as *Australopithecus afarensis* (*Australo* = southern; *pithecus* = ape; *afarensis* = from the Afar region). *Australopithecus afarensis* was about three feet tall with a body of human proportions. Its brain size of 500 cc was marginally larger than that of a similarly sized African ape. The structure of its pelvis, which is humanlike, and of the knee joint indicates that it walked fully erect. Its teeth were primitive with long canines and a rectangular arc similar to that of the modern apes. Its reconstructed skull looks quite similar to that of a modern gorilla (Figure 24–8).

Fossils of a somewhat younger age, all found in Africa, include at least three species of *Australopithecus: A. africanus, A. robustus,* and *A. boisei.* The structure of their legs and pelvises suggests that all three species were bipedal. They all apparently coexisted from about 3 million years ago to 1 million years ago, when they became extinct. *A. africanus,* thought to be the oldest of the three species, was about four feet tall, weighed 50 to 90 pounds, and possessed a brain size, on average, of about 450 cc, comparable to that of a modern ape of the same size. Its dental arcade is U-shaped like that of humans; its canines are smaller, its face flatter, and its cranium higher than that of apes. *A. robustus,* as its name suggests, was a more robust form (heavier skeleton) with a heavy jaw

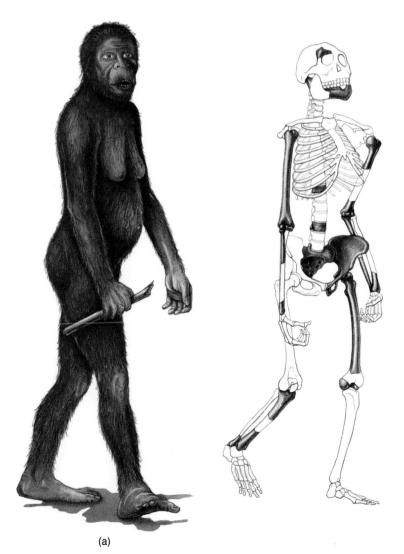

(a)

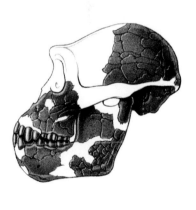

FIGURE 24–8 (a) A model of Lucy (*Australopithecus aferensis*) is shown. (b) Reconstructed skeleton and skull of *Australopithecus afarensis.* Compare the structure of this skull to that of a gorilla (Figure 24–7).

(b)

and strong teeth; it weighed about 150 pounds and stood some 5 feet tall. *A. boisei,* at 200 pounds and five and one-half feet tall, was even more robust. It is generally agreed that both robust species were evolutionary dead ends.

### Homo Habilis

In 1960, Louis and Mary Leakey discovered a 1.8 million-year-old hominid skull in Olduvai Gorge, Tanzania, with a brain size of about 700 cc, some 50 percent larger than the *A. africanus* brain. On the basis of its large brain, the Leakeys became convinced that the skull was that of the earliest known human, and they designated their fossil discovery *Homo habilis* (*Homo* = man; *habilis* = skillful). Other anthropologists argue that the skull is merely that of another form of *Australopithecus*. More recent discoveries have not resolved the issue. Whatever the case, the hominid lineages are far from clear. Some anthropologists contend that *A. afarensis* gave rise both to other australopithecine species and to *Homo* (Figure 24–9a). Others suggest that *A. afarensis* was ancestral to *A. africanus* and that *Homo* arose from *A. africanus* (Figure 24–9b). Still others, especially the Leakey family and its supporters, argue that *Homo* probably arose too early for any of the australopithecines to be its direct ancestor, and suggest that a yet undiscovered hominid gave rise both to *Australopithecus* and *Homo* (Figure 24–9c).

Despite disagreement on the exact hominid lineage, there is general agreement that the hominids were distinguishable from their more primitive ancestors in a number of ways. First, structural modifications allowed the

FIGURE 24-9 Several possible schemes detailing hominid evolution.

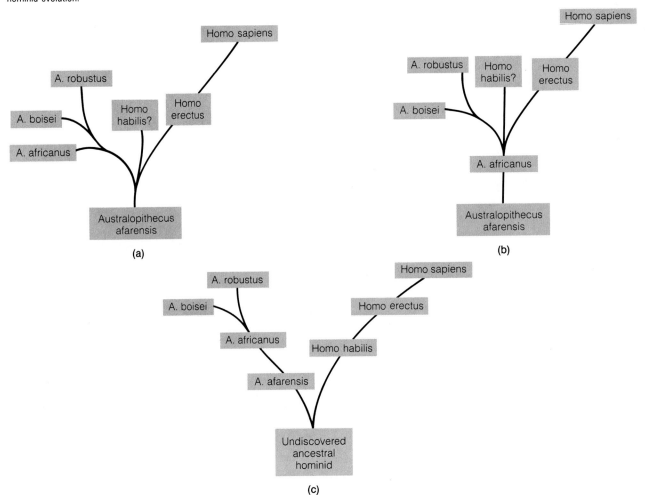

hominids to stand upright and move with a bipedal gait. Attainment of an upright, bipedal gait is considered an important event in the developmental history of hominids. It freed the upper limbs for carrying food, infants, and weapons, and it permitted travel for great distances while exploring the environment in search of water, food, and shelter. Second, the absence of specializations in the jaw and dentition suggests a varied, omnivorous diet. Third, precise measurements of cranial dimensions and of the inner contours of early hominid skulls shows an expansion of association areas on the top and side of the brain. This reorganization of the brain suggests increased abilities to analyze and respond to sensory information, and may have permitted language and a primitive cultural behavior in the earliest hominids. Supporting the evidence for increased intelligence in the australopithecine hominids, as compared to their earlier ancestors, is the discovery of tools made from stones that were chipped to make sharp edges; these sharpened stones could have been used for cutting and scraping.

## Undeniably Homo

### Homo Erectus

The first unequivocal example of the *Homo* genus, *Homo erectus,* appears in 1.6 million- to 300,000-year-old geological formations in many areas of the Old World. Indeed, there is general agreement among anthropologists that *Homo erectus* was the direct ancestor to more recent forms of human beings (see Figure 24–9).

*Homo erectus* was a little over five feet tall, stood fully erect, and walked bipedally. Its brain size averaged about 1000 cc, but later forms tended to possess larger brains than earlier forms (modern humans have an average brain size of about 1400 cc). Compared to modern humans, *Homo erectus* had a skull that was more massive, with a less protruding chin, a flatter, more receding forehead, and more prominent eyebrow ridges (Figure 24–10). The fossil evidence suggests that *Homo erectus* originated in Africa, migrating from there to Asia and other regions. They were hunters who constructed and used simple stone tools such as axes, scrapers, and choppers. Two well-known examples of *Homo erectus* are *Java man,* an early representative of the species, and *Peking man,* who lived between 700,000 and 400,000 years ago.

### Homo Sapiens

Early forms of *Homo sapiens* (archaic *Homo sapiens*) emerged about 400,000 years ago. Two fossil specimens of *Homo sapiens* dated 250,000 and 200,000 years old show, relative to *Homo erectus,* increased brain size, lightening of the skull bones, rounding of the cranium, and flattening of the plane of the face.

***Homo sapiens neanderthalensis: Neanderthals.***   There is a hominid fossil gap from about 200,000 to 80,000 years ago. After this time, the fossil record shows that for a period from 80,000 to 40,000 years ago, a distinctive type of *Homo sapiens* was widespread in Eurasia. They were called *Neanderthals,* a name derived from the Neander Valley in Germany where their fossil remains were first discovered. Although there was a good deal of physical variation between Neanderthals from different regions, they closely resembled modern humans in stature and brain size, but had a somewhat thickened skull, heavy brow, and low forehead (see Figure 24–10). Excavation of anthropological sites indicates that the Neanderthals had developed an extensive culture. They lived in social groups in caves or camps, made a variety of tools, and hunted and gathered food. They treated the sick, and had complex rituals for burial of the dead.

***Homo sapiens sapiens: Modern humans.***   About 40,000 years ago, distinctive Neanderthal fossils disappeared. The earliest fossils of modern humans,

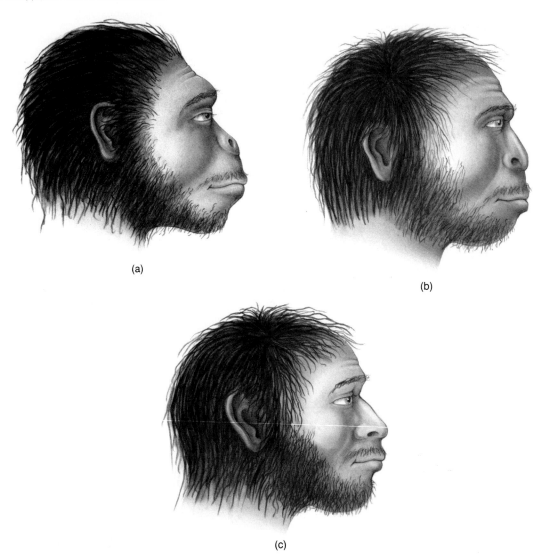

(a)

(b)

(c)

called *Cro-Magnons,* date back about 25,000 years. Cro-Magnons used sophisticated tools, had well-developed language, lived in cave dwellings decorated with art, and showed other evidence of existing in a complex societal structure.

There is no good evidence to explain the rapid disappearance of the Neanderthals. Some anthropologists suggest that fully modern humans evolved in other parts of the world and invaded Europe and northern Asia, wiping out the Neanderthals or interbreeding with them. Others contend that the Neanderthals evolved independently of other human groups into modern Europeans (Cro-Magnons).

## Evolution Now and In the Future

The past 40,000 years or so of human evolution have not been characterized by significant changes in physical form. Rather, we have undergone remarkable cultural evolution that has allowed us increasingly to adapt to environmental change without being subjected to the pressures of natural selection. To take one of the simplest examples, the use of clothing has insulated us not only literally from the cold, but also from environmental selection pressures that might otherwise lead, through evolution, to biological solutions for dealing with cold environments (such as the development of fur). Many hereditary diseases and defects that would once have been disadvantageous to survival and reproduction are no longer so because they are now correctible or treatable by modern medicine. Such conditions range from minor defects in vision and hearing to potentially fatal diseases such as diabetes, phenylketonuria, and hemophilia. As we move into the age of biotechnology and learn to correct genetic defects, we will insulate ourselves still further from the effects of natural selection. However, cultural evolution also has its down side. Nuclear weapons, pervasive pollution, and overpopulation threaten our very existence, as will be discussed in the next chapter.

## Study Questions

1. How can the diversity and similarities of living organisms be rationalized by evolutionary theory?
2. Outline the materialistic theory of life's origin.
3. a) Describe Miller's classic simulation experiment.
   b) Why are the results of Miller's experiment important evidence in support of the materialistic theory?
4. a) What are microspheres?
   b) Explain why their development might have been an important step in the origin of life.
5. a) What does the distribution of fossil organisms tell us about the origin of life on earth?
   b) What does it tell us about biological evolution?
6. What was the most significant event in the evolution of mammals?
7. Starting with the early prosimians, construct a tentative scheme outlining the major steps of anthropoid evolution.
8. What is the importance of *Australopithecus afarensis* to our understanding of hominid evolution?
9. a) Trace the lineage of the *Homo* genus from its first appearance to modern times.
   b) How reliable is your scheme?

# HUMAN ECOLOGY

So far in this book, we have discussed the human organism in a very limited way. We began with the smallest particles of which humans are built, the subatomic particles with which nuclear physicists concern themselves. We learned about the atoms and molecules of chemists and biochemists. Then, we examined the first structures we can see with an electron microscope, the organelles that form the functional parts of cells. Then came the cells themselves, the tissues in which they come together, and the organs built of tissues. From organs are built organ systems, which make up organisms such as individual human beings.

What we have not considered is how human beings and other living things are organized beyond the individual level. The organisms of any species, or type, are members of groups we call **populations.** All the populations of all the species in a particular area comprise a **community.** A community plus its inanimate environment is an **ecosystem.** All the living things on earth comprise our planet's **biosphere** (Figure 25–1).

In this final chapter, we will explore **human ecology,** the interactions of humans with their environment, in terms of the various levels of ecological complexity: population, community, ecosystem, and biosphere. In the process, we will find that certain aspects of human ecology are best understood at specific levels. Let us begin at the population level, with human population ecology.

## HUMAN POPULATION ECOLOGY

Biologists define a **species** as a group of organisms that resemble each other and are capable of interbreeding *in nature* to produce viable and fertile offspring. Thus, by definition, individuals constituting a species do not normally or successfully interbreed with individuals of other species.

A **population** is a group of organisms of a single species that *do* interbreed. That is, they live in the same locale and have the opportunity to interact and mate. Species are typically, but not always, divided into a number of populations. Individuals *do* mate within each population. Individuals belonging to different populations presumably are capable of mating, but do not because they never have the opportunity.

Human beings belong to a single species, for any two normal, healthy individuals of the opposite sex are capable of mating and having fertile offspring. Until fairly recently, in the historical sense, the human species was divided into many populations whose members never met. Today, most of the human species is a single population. Nevertheless, many people who study human population ecology still speak of smaller populations, such as those of Washington, DC, or of China, as well as of the world population.

### A History of Increase

Human population ecology, often called **human demography,** focuses primarily on the factors that determine the size of human populations, including birth rates, death rates, and age structures. By measuring these aspects of a population, demographers can pinpoint the causes of population change and predict the growth patterns of a population. Since the size of a population greatly affects its impact on the environment, this is a critical area of study.

The history of the human population is a history of growth. From the first beginnings of *Homo sapiens* some 400,000 years ago, the earth's human population has grown to more than five billion.

The reason for this immense growth in population is very simple in demographic terms. The growth rate of any population is the difference

FIGURE 25-1 Levels of biological organization.

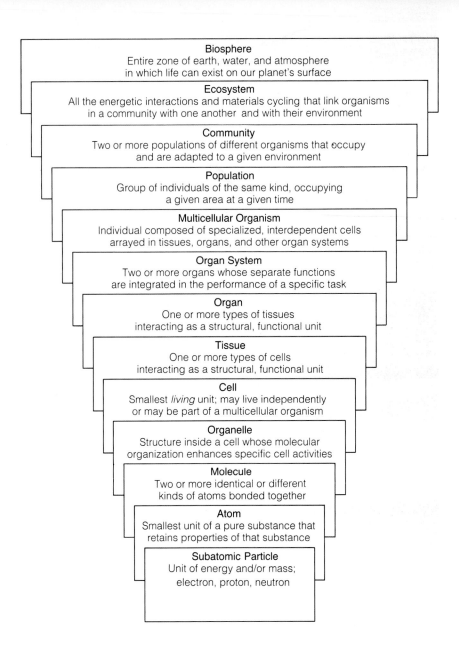

between how fast the population adds members and how fast it loses members. For the world as a whole, the growth rate is the difference between the birth rate and the death rate, as follows:

$$\text{Growth rate} = \text{Birth rate} - \text{Death rate}$$

For less-than-worldwide populations, the equation must also include the immigration and emigration rates.

*World* population growth always indicates that the **birth rate** (number of births per unit time) has exceeded the **death rate** (number of deaths per unit time), allowing for a net increase in the number of humans. To explain any growth of the world population, we need to look for factors that increase the birth rate and/or decrease the death rate.

The overall growth history of the world population is shown in Figure 25-2. There are clearly two phases: an early phase, up through the Middle Ages, when population growth was slow and gradual, and a later phase, from the end of the Middle Ages to the present, when growth was very rapid.

The gradual increase of the human population in the early phase was due to two main factors that stimulated modest growth. First, our ancestors developed tools and became skilled in their use. The first tools were probably carefully chosen rocks; gradually, they became more sophisticated. These tools increased the hunting skills of early humans and therefore improved the quality of their diet. The immediate effect of this improvement was increased survivorship (a decreased death rate) and increased fertility (an increased birth rate). The latter effect exists because in well-nourished populations, as compared to poorly nourished ones, women begin menstruation at younger ages and continue their cycle more regularly. They thus have more opportunities to have children. They also are more likely to have full-term pregnancies and to enter menopause later in life. It is thus not surprising that well-fed populations increase their size.

The second cause of early human population growth was the development of simple agriculture. Cultivating plants instead of relying on wild foods gave people a more reliable and ample food supply. This too meant improved survivorship and fertility and hence a larger population. However, by the time of the ancient Greeks and Romans, the human life expectancy was still only about 30 years, and less than half of all newborns survived to the age of one. The human population was still kept in check by disease, famine, and war.

## Limits to Growth

The human population entered the exponential growth phase several hundred years ago. The causes were major improvements in agriculture and in technology, the opening of new areas of the world (such as the Americas) to settlement by technologically advanced humans (Europeans), improvements in sanitation, and the development of modern medicine. These factors combined to reduce the human death rate and increase the human birth rate. At the same time, they increased the number of humans the world could hold; that is, they increased the *global carrying capacity.*

Unfortunately, the carrying capacity of the earth cannot be infinite. Current projections say that the human population will exceed the earth's carrying capacity (probably around 15 billion people) during the twenty-first century. At that time, the population must level off. First, however, it seems likely to rise past the global carrying capacity. It must then decline to below that level, perhaps far below it.

## The Basic Growth Pattern

The growth pattern just described is typical of all populations of organisms. First, there is a period of slow growth, the **lag phase.** Then the growth curve shoots upward, coming to resemble a "J." This is the *J-shaped growth curve,* or pattern of *exponential growth,* during which the population increases by a fixed percentage in each successive time interval. However, at some point the growth

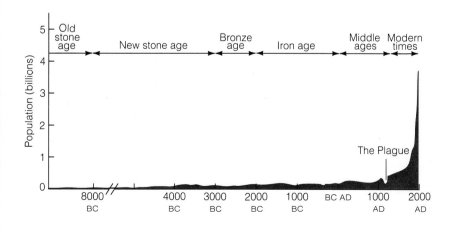

FIGURE 25–2 Human population growth, 8000 BC to present.

of a population reaches a ceiling because any environment, local or worldwide, can only support a certain number of individuals of any species in terms of food, water, and shelter sites. Populations of organisms, including humans, must eventually slow their growth. The J-shaped curve must level out, as shown in Figure 25–3. The size of the population when the curve levels out defines the carrying capacity of the environment, the population size the environment can support indefinitely. The curve may or may not overshoot and oscillate around the carrying capacity, as illustrated by the dotted line in Figure 25–3.

## The Need to Control Growth

The large size and rapid growth of the human population will continue to have a major impact on the environment. This is especially true in developing countries, like many in Africa and South America, where human population growth is concentrated. In these countries, the birth rate and the fertility rate (the number of children per woman per lifetime) are higher than in most developed countries. For example in Africa, the birth rate is 4.5 percent and the fertility rate is 6.3 percent; in Europe, the birth rate is 1.3 percent and the fertility rate is 1.8 percent. In both regions, the death rate is low (1.6 percent in Africa; 1.0 percent in Europe), largely because of the immense effectiveness of vaccines and sanitation in preventing childhood illnesses. The growth rate— the difference between the birth and death rates—is consequently much greater in Africa, and in most developing countries, than it is in most developed countries.

The world population as a whole is growing. We can analyze the population growth in different regions of the world by constructing age structure diagrams, which summarize the numbers of men and women in five-year age groups. The age structure of a country with a high population growth will yield a diagram similar to that in Figure 25–4a, in which, beginning at the youngest ages, there are decreasing numbers of individuals at each five-year interval. A country with a stable population will show age groups all about the same size (Figure 25–4b). If younger age groups are smaller than older ones, the population is shrinking.

Several developing countries have recognized the problem of excessive growth and have instituted strict legislation to limit their growth rates. For example, in India the minimum age of marriage has been increased from 18 to 21 years for males and from 15 to 18 years for females. Delaying marriage in this way reduces the number of years that a female can bear children. Unfortunately, India's attempt to curtail population growth has not been very successful.

In China, whose population exceeds one billion people, an annual growth rate of 2.5 percent (meaning a population doubling time of 28 years) has recently decreased to 1.4 percent (a doubling time of 48 years). This decrease was caused by a massive campaign to delay marriage until the mid- to late-twenties and to limit the number of births to one child per couple. The Chinese

FIGURE 25–3 All populations grow in the same pattern: after an initial slow lag phase comes a J-shaped exponential growth phase. When the population nears the environment's carrying capacity, the curve becomes S-shaped. It may or may not oscillate around the carrying capacity (dotted line).

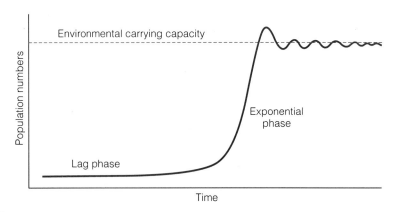

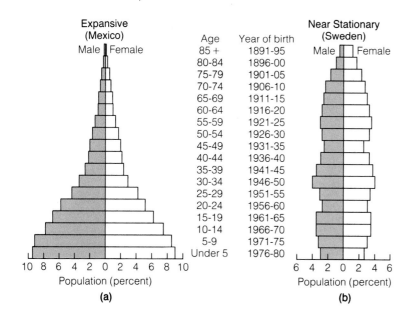

government encouraged its citizens to cooperate with these restrictions by attaching financial penalties to couples with more than one child and by making birth control methods readily available. China's campaign to reduce the rate of population growth has been fairly successful, but not popular, and it is now being deemphasized.

The world population of over five billion is currently estimated to be increasing at a rate of 1.7 percent per year. Although this rate seems small, it indicates that the world population will double in the next 41 years. As the world population increases, so will its impact on the environment. Although this impact has existed for as long as the human species, it has never been as pronounced as it is today. To understand this essential concept, we must now look at the next ecological level, the biological community of which humans are a part.

## HUMAN COMMUNITY ECOLOGY: HUMAN INTERACTIONS WITH OTHER SPECIES

Humans interact with other living organisms. They therefore belong to an **ecological community,** defined as the populations of plants, animals, and microorganisms living and interacting in a given locality. Among the most fundamental of these interactions is that of the eater and the eaten, which allows us to define the *food chain,* one of the most basic concepts in ecology.

A **food chain** is precisely what its name implies, a sequence of living organisms interrelated in their feeding habits. In the sequence, types of living organisms are followed by the types that eat it. A simple example of a food chain would have grass forming the lowest level of the sequence. The grass is eaten by grasshoppers, which are eaten by birds. The birds in turn are consumed by cats.

### Trophic Levels

We call the levels of a food chain **trophic** (feeding) **levels** (Figure 25-5). They represent stages in the transmission of energy and materials from organism to organism, and they permit us to organize and describe communities simply, in terms of energy flow. That is, we can classify the organisms in a community according to how they obtain their energy.

The first trophic level of any food chain is occupied by **autotrophs** such as plants, which manufacture the organic compounds they need by using energy

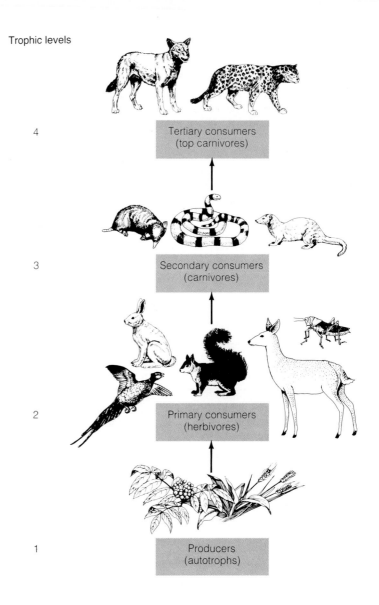

FIGURE 25-5 The levels of a food chain.

Trophic levels

4    Tertiary consumers
(top carnivores)

3    Secondary consumers
(carnivores)

2    Primary consumers
(herbivores)

1    Producers
(autotrophs)

directly from the sun. All higher levels in the food chain are occupied by **heterotrophs,** organisms that obtain their food (or, more precisely, the organic compounds they need) by eating other organisms.

Among the higher levels, the second level of a food chain belongs to the *primary consumers,* or **herbivores,** animals that eat plants. The third level is that of the *secondary consumers,* or **predators.** On the fourth level we find the *tertiary consumers,* and so on. Finally, in all food chains there are decomposers, primarily bacteria, fungi, and other microorganisms, that break down dead organic material.

Humans also belong to food chains. They can be primary consumers when they eat plants, or secondary consumers if they eat herbivorous animals such as cattle. They can also be tertiary consumers if they eat carnivores such as seals (a major component of the diet for some Eskimos). Most humans—like many other animals—occupy more than one trophic level, playing the dual role of primary consumer (eating plants) and secondary consumer (eating meat); they are called **omnivores.**

### Ecological Pyramids

All food chains share certain characteristics. Among the most obvious of these characteristics are their structural features, or trophic levels. But in a more fundamental way, they hold in common the laws of thermodynamics, which explain basic energy dynamics. The second law of thermodynamics, which states that some usable energy is lost whenever energy is utilized, is an especially strong influence on the structure of food chains. The second law dictates the relative amount of usable energy embodied in each level of a food chain. (See Chapter 2 for a more detailed discussion of the laws of thermodynamics.)

Because of the second law of thermodynamics, each trophic level in a food chain must contain less usable energy than the level that precedes it. Within a food chain much more energy is tied up in autotrophs than in herbivores, and more in herbivores than in carnivores. The difference in energy content of successive trophic levels is expressed in a rule of thumb: Of the solar energy captured by plants, only about 10 percent is converted into the organic material of primary consumers; in turn, only about 10 percent of the energy of primary consumers is converted into secondary consumers. The remaining 90 percent of the energy in any trophic level's food supply goes to running that level's metabolic activities and is ultimately lost as heat to the environment. Ten percent goes into growth and reproduction, into the creation of organic material that can serve in turn as food to the next level up the chain.

Clearly, this pattern of energy flow limits the length of a food chain. Most food chains have a maximum of five levels, because once the fourth or fifth level is reached, there is little energy left to sustain yet another trophic level.

It is thus possible to diagram a food chain as an *ecological pyramid.* In a *biomass pyramid,* the lowest level of the pyramid represents the amount of organic material present as the bodies of producers, or plants (1000 pounds in Figure 25–6). The second level represents the amount of material in the primary consumers (100 pounds). The third level represents the amount of material in the secondary consumers (10 pounds). The fourth level, if it is present, holds a single pound of tertiary consumers. If we represent the material in the bodies of the organisms in this food chain in terms of its energy content, we see a very similar second pyramid, an *energy pyramid.* An energy pyramid differs from a biomass pyramid only in that its numbers measure units of energy (calories), not mass. A third kind of ecological pyramid, a *numbers pyramid,* represents the numbers of producers, primary consumers, and so on.

## Biological Magnification

We see an additional characteristic of food chains when we realize that consumers fail to break down or excrete, some substances. For instance, lead and other heavy metals accumulate in bones. Many pesticides and other toxic chemicals, which are soluble in lipids but not in water, accumulate in fat. Clearly, how much of such substances an animal contains depends on how much is in its food. Knowing this, it is easy to see that the successive levels in a food chain must contain successively greater amounts of these substances, a phenomenon termed **biological magnification.**

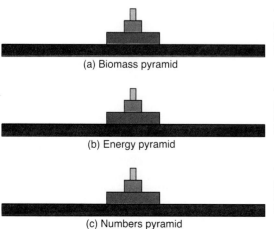

Tertiary consumers: 1 lb
Secondary consumers: 10 lbs
Primary consumers: 100 lbs
Producers: 1000 lbs

(a) Biomass pyramid

Tertiary consumer: 1 kcal
Secondary consumer: 10 kcal
Primary consumer: 100 kcal
Producer: 1000 kcal

(b) Energy pyramid

Tertiary consumers: 1
Secondary consumers: 10
Primary consumers: 100
Producers: 1000

(c) Numbers pyramid

FIGURE 25–6 Ecological pyramids. (a) Biomass pyramid; (b) energy pyramid; and (c) numbers pyramid.

# DDT: The Profile of a Pesticide

In 1939, a Swiss chemist, Paul Mueller, developed DDT (dichlorodiphenyltrichloroethane), the first major organic pesticide. DDT was heralded as a miracle. —it was inexpensive, effective at low doses, long-lasting, easy to apply, and lethal to a wide range of insect pests. Indeed, DDT, manufactured secretly during World War II by Great Britain and the United States, initially produced spectacular results. For example, in Naples in 1944 an outbreak of typhus (a disease spread by body lice) was successfully controlled with DDT. In India, the use of DDT to kill the *Anopheles* mosquito (which carries the malaria protozoan) reduced the incidence of malaria from 100,000 cases per year to 50,000. So it came as no surprise when Mueller was awarded the 1944 Nobel Peace Prize.

Unfortunately, the miracle was too good to be true. The first drawback to be detected was the development of strains of insect pests that were resistant to DDT. This occurred because even though DDT is lethal to insects, it rarely kills all the insects exposed to it. A few are genetically resistant to the poison (that is, they have particularly thick exoskeletons that prevent DDT absorption, or they are able to detoxify DDT). These survivors then interbreed and pass their resistance on to another generation of insect pests. In 1947, DDT-resistant house flies were discovered. Today, more than 400 species of insects are resistant to DDT.

The first response to the resistance phenomenon was to increase the dose. Unfortunately, this helped only temporarily, because eventually pests evolved that were resistant to even higher doses. Several cycles of such DDT-induced selection for resistance then produced pests that could not be killed by this pesticide at any economically practical dose. When people switched to other pesticides, the pests became resistant to those as well. Today, in many areas of the world, insect pests are immune to all of the common pesticides.

Biological magnification of DDT makes the problem worse because pests do not exist in isolation. They are prey to numerous predators, who, when their food contains DDT, build up much larger doses of poison. Consequently, a large proportion of the predator population dies, and as the resistant pest variety appears, it flourishes because the natural restraints on its population growth (predators) are missing. When predators such as ladybugs or lacewings disappear, we rarely notice them.

But we do notice when songbirds are no longer in our yards. It was their absence that provoked Rachel Carson to write her famous book, *Silent Spring*, in 1962. Later, researchers learned that though many birds died because the DDT was toxic to them, others vanished because the poison interfered with reproduction, often by making eggshells too thin to support the weight of a nesting bird.

Sadly, though Carson's book helped bring about a ban on DDT in the U.S. and taught us to take great caution with other pesticides, DDT is still with us. It is long-lasting, or persistent. Persistence may be an advantage to a farmer who wishes to spray a crop only once during a growing season, but it is a disadvantage when it allows the poison to accumulate in the soil, and be transferred to non-pest bird foods such as earthworms. Persistence is also a disadvantage when it permits the poison to drain from the soil into streams and lakes, thus entering aquatic food chains that end in fish-eating animals such as eagles and humans. DDT itself is relatively nontoxic to humans, but related chemicals have been shown to cause neurological disorders, birth defects, cancer, and even death

The most striking example of this phenomenon occurs when toxins (such as DDT or PCBs) that humans have introduced into the environment become concentrated in the third and fourth trophic levels. For example, DDT was once a widely used pesticide. By the 1960s, it had reached a concentration of 0.000003 ppm (parts per million) in the water, but had attained concentrations of 25 ppm and more in fish-eating birds such as eagles and ospreys (tertiary consumers). Such levels are enough to stop reproduction in these birds and to decimate their populations. Although wildlife biologists are successfully restoring the populations of many of these species, they can hope only to preserve, in zoos, a few specimens of such species as the California condor (see *Focus—DDT: The Profile of a Pesticide*).

A second example of environmental toxins that become concentrated in the upper trophic levels of food chains are PCBs (polychlorinated biphenyls). These compounds are soluble in fats and degrade slowly, two properties that make them ideal for biological magnification. PCBs have become concentrated in predatory fish such as the coho salmon of the Great Lakes. Researchers have

One response to the problems caused by DDT was to develop new types of pesticides. Two other major groups of pesticides are the organic phosphates (such as malathion and parathion) and the carbamates (such as carbaryl, commonly known as Sevin). Both of these have the advantage of being less persistent in the environment than DDT: however, both act as nerve toxins and have been shown to cause birth defects and genetic damage in humans. In addition, insects can also develop resistance to these compounds.

Increasingly, attempts have been made to control insect pests with minimal use of pesticides, an approach termed integrated pest management(IPM). IPM relies on a variety of techniques, many of them fitting under the label of "biological control," to reduce pest populations. Among the techniques useful in IPM are:

1. Releasing natural predators of the insect pest, including parasites and disease-causing organisms.

2. Using sex attractants (the chemicals the female pest itself uses to attract mates) to decoy the males into traps. Spraying a field with sex attractants may also make it difficult for the males to find the females.

3. Developing genetic varieties of plants that are resistant to the pest.

4. Altering the time of planting to avoid peak pest periods.

5. Planting low-value trap crops (as when alfalfa is planted next to cotton to attract lygus bugs and lower their impact on the cotton).

6. Increasing crop diversity so that the amount of food available to any one pest is reduced.

7. Applying pesticides only when predators and parasites are not present or when other methods fail.

Crop dusters are frequently used to spread pesticides over large areas of farmland.

recommended that these fish not be consumed by women in their child-bearing years since PCBs are thought to cause birth defects.

The phenomenon of biological magnification has important implications for feeding human populations. Plant foods are likely to contain the lowest levels of toxic pollutants such as pesticides, heavy metals, and PCBs. At the same time, a vegetarian diet can support more humans than can a carnivorous diet, since 90 percent of the energy in plants is lost in the transfer of energy from plants to the second trophic level. We should realize however, that if all human beings became vegetarians, the earth's carrying capacity would not increase ten-fold, because for most people, the bulk of their diet is already plant matter.

We should also realize that animals can make more energy available to humans than can plants, for animals can turn indigestible plant materials, such as cellulose, into meats. For this reason and others, people process animal food more efficiently than plant food. They can assimilate only 20–60 percent of the energy in plant foods, but 50–90 percent of the energy in meat, milk, eggs, and

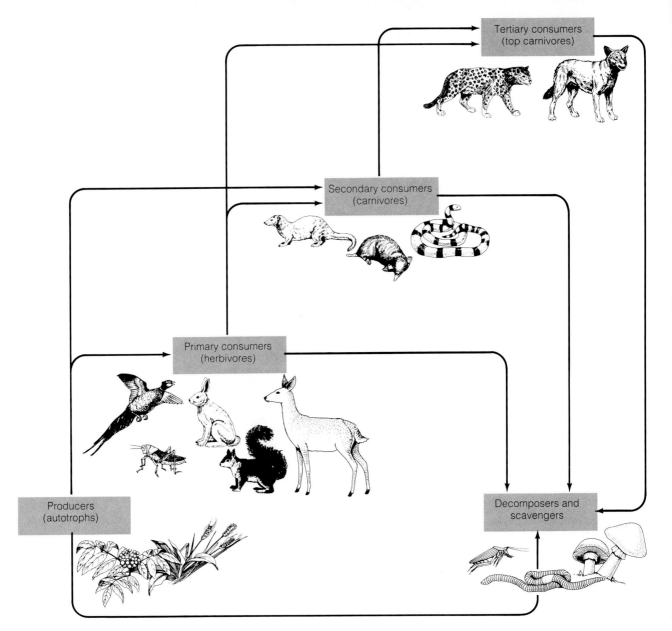

FIGURE 25-7 A food web.

## Food Webs

cheese. Thus, to get the same amount of energy from plant and animal foods, an individual would have to eat a greater amount of plant food than of animal food. Idealized food chains do not exist in nature. The truth is that herbivores eat some meat, as when a mouse, which is usually herbivorous, eats a grasshopper. Carnivores eat some plants, as when a dog consumes a little grass. Omnivores eat both primary and secondary consumers. To sketch the relationship between the eaters and the eaten accurately requires a diagram more like a spider's web than a simple chain (Figure 25-7).

Not surprisingly, ecologists use *food webs* to describe the true trophic dynamics of a whole community. Some communities have relatively simple food webs; others have more complex ones. Interestingly, as human society has become more and more industrialized, the food webs of which humans are a part have become simpler. Early humans, living in tropical areas, probably

*To pronounce the "!", click your tongue in the back of your throat.

relied on a wide variety of animal and plant foods and therefore were a part of a very complex food web. We have some evidence of this from studying current primitive cultures. For example, the !Kung* Bushmen of South Africa eat over 70 plant, 33 mammalian, seven reptile, one amphibian, and several bird and insect species. The diet of the Tasaday, an Australian Aborigine group, includes 11 frog species and the roots of 29 different plant species, in addition to numerous other plant and animal foods. In contrast, most modern societies gain most of their dietary diversity from three or four animal species and approximately 20 different plant species.

## Human Impact on Ecological Communities

Early humans lived in scattered small groups, obtaining food by hunting game, gathering wild plants, and collecting other resources they needed from the environment. Because of the small size of the human population and its subsistent level of existence, the impact of these early humans on the environment and other ecological communities was relatively small.

Humans began to have an impact on other living things millennia ago. By that time, hunters had learned to set fire to grasslands to chase herds of wild cattle and buffalo over cliffs. They had also developed weapons and the idea of teamwork, and they were depleting the local populations of animals.

As the numbers of humans and the level of their technology increased, their impact became more pronounced. It is widely believed that humans caused the mass extinction of many different species such as the wooly mammoth, the saber-toothed tiger, the cave bear, and the giant beaver during the Pleistocene. The same process apparently wiped out the horse and camel, once native to North America.

A number of alternative explanations for these extinctions have been proposed (such as climate change), but the overkill hypothesis is still widely accepted. It is supported by the fact that extinctions by overkill have continued into more recent times. A notable example is the passenger pigeon, which was once one of the most common birds in the forests of the Eastern United States and was extensively killed for food and sport; the last one died about 1914. The dodo of Mauritius fell victim to hungry sailors (as well as to the rats and dogs that accompanied human settlers) and died out in the seventeenth century. Steller's sea cow of the Bering Sea was such a rich source of oil that it was hunted into extinction late in the eighteenth century. In the mid-1900s, hunting drove the Carribean monk seal into extinction.

The rate at which humans are causing extinctions is increasing as the human population size increases. Although overkill or overexploitation remains a cause of extinction (and may soon kill off Africa's rhinoceroses and elephants, among species), extinction increasingly results from habitat destruction due to development, exploration, and farming. Humans are destroying suitable habitat for thousands of species and thereby causing their demise.

Currently, the extinction rate has intensified due to the large-scale destruction of the tropical rain forest. But habitat destruction is a worldwide problem. It occurs everywhere—even in the United States. The loss of wetlands in the Mississippi delta region of Louisiana is one example. These wetlands are highly productive breeding grounds for birds, shrimp, crabs, fish, and numerous other marine organisms. Losing the breeding grounds means lessening the productivity of the fisheries that provide food and employment to thousands of people and failing in our responsibility to preserve the resources of nature for our descendants. (see *Focus—The Human Effect on Ecological Communities: The Extinction of Species*).

## HUMAN ECOSYSTEM ECOLOGY

In our discussion of populations and communities, we have focused on living organisms in isolation. We have not considered their inanimate environment, the source of their nutrients, energy, water, and oxygen; the site of their

The passenger pigeon, once killed for food and sport, became extinct around 1914.

# The Human Effect on Ecological Communities: The Extinction of Species

We—the human species—share the earth with literally millions of other species. Yet we are not always good neighbors. We are famous for wiping out other species by hunting them to extinction, preempting the resources they need, and destroying their habitats.

Millennia ago, the human impact on other species was mainly that of the hunter, as discussed elsewhere in this chapter. More recently, we have begun to take resources other species need. An example arises from the way we manage tree farms and forests, harvesting wood for human use so efficiently that the supply of dead trees vanishes. Unfortunately, wood ducks, ivory-billed woodpeckers, and other creatures rely on holes in these dead trees for their nesting sites. Without nesting sites, they cannot reproduce, unless—as in the case of wood ducks—people supply them with artificial nesting sites.

Habitat destruction is a worldwide problem, though it is worst in the tropical rain forests. The rain forests comprise the world's most biologically diverse habitats. Where a typical forest in the northeastern United States may have 10 tree species in a hectare (2.47 acres), a hectare of rain forest may support over 100 tree species. This environment also supports thousands of species of other plants and animals not to mention fungi and bacteria. However,

In spite of great efforts by local citizens, seabirds that become covered with oil almost invariably die.

despite this immense diversity of living things, some 13,000 tropical plant species are listed in danger of extinction (only 4500 temperate plants are so listed).

Tropical rain forests occur in Africa, Central and South America, Malaysia, and Australia. In prehistoric times, they covered approximately 5 million square miles. Today, they have been reduced to 3.5 million square miles. The loss continues at a rate of 25,000 square miles per year (an area the size of the state of West Virginia). It is estimated that the tropical rain forest of Malaysia will be gone by the year 2000. With that rain forest will go untold species of plants and animals. In general, scientists estimate that each time an area is reduced to one-tenth of its original size, the diversity of species within it is halved.

The destruction of forest is called **deforestation.** In tropical regions, it happens for three main reasons. First is the conversion of forest to cropland for farming and to pasture for raising cattle. Second is the need for timber. Third is the need for vast quantities of wood pulp by

shelters; the matrix in which they must set their roots. When we expand our vision to encompass both organisms and their environment, we begin to see the level of ecological organization known as the **ecosystem.**

## Ecosystems

Ecosystems follow two basic principles. The first is that energy flows through them. It enters when plants use sunlight to produce organic matter by photosynthesis, passes from plant to herbivore to carnivore to decomposer, and finally is converted to heat, which is then lost. The second principle is that nutrients such as carbon, phosphorus, sulfur, and calcium are constantly recycled. Plants absorb such nutrients from the air, water, or soil, incorporate them into organic molecules, and can pass them on to animals. Through their wastes and decomposed bodies, animals return nutrients to the environment, from which plants can use them again. We will see how this recycling works in two cases—carbon and water—when we discuss the biosphere.

Ecologists generally think of ecosystems as fairly small, self-contained units, such as a pond, field, or forest. These units depend on their own internal

the paper industry. Deforestation is exacerbated by the need of less developed countries for exports—of exotic woods, wood pulp, and beef—and the money they can provide. Importing nations such as the United States are therefore partly to blame for tropical deforestation.

Why should we care about whether our activities destroy habitats and species? The answer is simple: We have always depended on the plants and animals with which we share the earth, and we always will. They provide us not only with food, shelter, and other resources, but also with the pleasures we gain from contemplating them or vacationing among them. In addition, they give us medicines: Currently, 45 percent of all medicines are derived entirely or in part from animals, plants, or fungi. New species yet to be discovered, and old ones awaiting more detailed study may provide important drugs for treating human illnesses. Many marine invertebrates and rain forest plants may be potential sources of anti-cancer drugs. Wild relatives of our cultivated plants may provide genetic information invaluable for devel-

oping crops that are resistant to bacteria or insects. For example, agriculturalists are currently raising wild strains of the cacao plant (the source of chocolate) in hope of finding a strain that is resistant to the pests

that commonly ravage this important crop. Sadly, once a species becomes extinct, the resources and benefits it may provide are lost to us forever.

Clear-cutting, a controversial lumbering practice, causes erosion, degrades water supplies, and depletes the soil of nutrients.

recycling mechanisms to maintain their supply of nutrients; only water and sunlight (and small amounts of other materials) enter from outside the ecosystem's boundaries. A self-contained ecosystem can be as simple as a terrarium, consisting of a layer of soil, a few plants and rocks, and perhaps a small lizard; or an aquarium, containing gravel, water, plants, and fish (Figure 25–8). Either type of ecosystem, if not overloaded with plants or animals, can persist indefinitely, with no input of animal food, plant fertilizer, or water. All it needs is light.

All ecosystems provide important services to human populations. Vegetation reflects heat to moderate desert climates. Trees evaporate water from their leaves to cool the air beneath them and any house sheltered by their shade. Swamps cleanse and filter water. Both swamps and forests store water against drought. Plants bring mineral nutrients from the soil into the human diet. Plant-eating animals keep the plants from overrunning the world, and carnivores and diseases keep the herbivores in check.

Ecosystems have two groups of components. Their **biotic,** or living, components are plants, animals, and other organisms. Their **abiotic,** or nonliving, components include precipitation, wind, and temperature (weather

FIGURE 25–8 A simple aquarium fits the definition of an ecosystem as a community of organisms plus their inanimate environment. The recycling of nutrients and water in such a model ecosystem means that it can persist indefinitely with no input other than light.

and climate), the amount of available sunlight or energy, the atmosphere with its content of oxygen and carbon dioxide, and the soil and water with their nutrients.

## Biomes

A **biome** is a group of ecosystems that share a similar climate and many of the same types of plants and animals. Most of the ecosystems in a biome therefore closely resemble each other. Some exceptions are the ponds and lakes embedded in a forest or mid desert oases. The ecosystems in a biome also tend to be near each other.

Because biomes depend on climate, they stripe Earth's continents with broad bands as shown in Figure 25–9. The tropical rain forest biome straddles the equator wherever there is plenty of rainfall. To the north and south lie deserts, and then savannah, chaparral, and temperate deciduous and coniferous forests. To the far north and south lie taiga and tundra. At the poles is the polar biome, marked by cold, ice, six-month nights, and sea-based food chains.

Biomes also stripe the flanks of mountains. The reason is that temperature—a major determinant of local climate—drops with increasing elevation, just as it does with increasing latitude (Figure 25–10). Still other biomes divide the oceans into the lighted, life-rich surface layers and the dark, deep layers, and divides the lakes and rivers into still and moving water.

## HUMAN IMPACT ON THE BIOSPHERE

The **biosphere** is the sum of all Earth's life. It extends from the tops of the highest mountains to miles beneath the surface of the sea, where communities of bacteria, worms, clams, and other creatures surround volcanic vents (Figure 25–11). Strictly speaking, the biosphere does not include the environments of life—it is a sum of communities. (The **ecosphere** is the sum of ecosystems or biomes.) Yet many people use the term biosphere to include the environment, and so will we.

Thinking of the biosphere as an all-encompassing ecosystem allows us to grasp the impact of humans on the living world. The abiotic components of any ecosystem include the atmosphere, the hydrosphere, and the lithosphere (the land part of the earth). Each of these three abiotic components of the world, as we will see in the following examples, has been in some way altered by human activities.

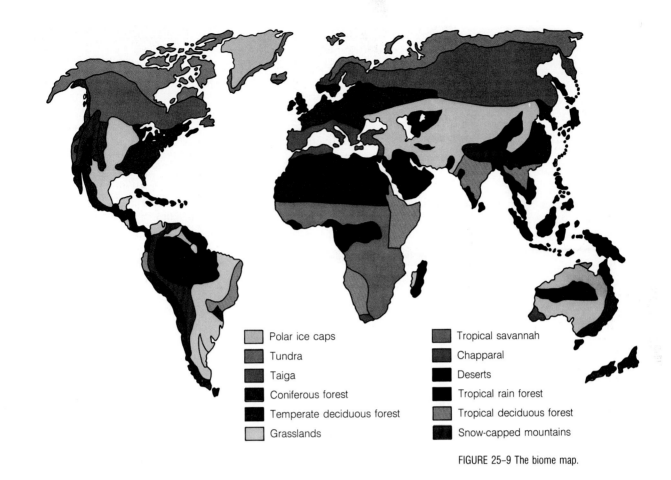

| | | | |
|---|---|---|---|
| ▢ | Polar ice caps | ▢ | Tropical savannah |
| ▢ | Tundra | ▢ | Chapparal |
| ▢ | Taiga | ▢ | Deserts |
| ▢ | Coniferous forest | ▢ | Tropical rain forest |
| ▢ | Temperate deciduous forest | ▢ | Tropical deciduous forest |
| ▢ | Grasslands | ▢ | Snow-capped mountains |

FIGURE 25–9 The biome map.

## The Atmosphere and the Greenhouse Effect

Earth's atmosphere contains, in addition to trace amounts of several gases, about 78 percent nitrogen gas, 21 percent oxygen gas, and .03 percent carbon dioxide, and it is the main abiotic reservoir—the *atmospheric pool*—for the elements nitrogen, oxygen, and carbon. These elements are continuously

FIGURE 25–10 The mountain biome.

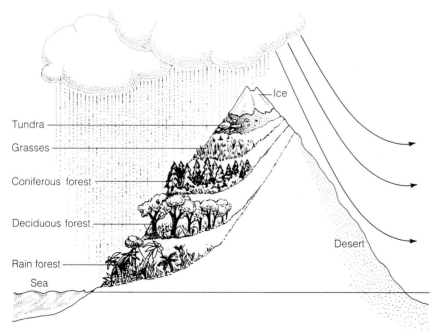

FIGURE 25-11 Deep-sea vent community.

being transferred from this atmospheric pool to living organisms and then back to the atmospheric pool.

The composition of the atmosphere is not fixed. It can be—and has been—altered by biological events. Four billion years ago, for instance, the atmosphere was dramatically different from what it is today, consisting mainly of nitrogen gas, ammonia, water vapor, carbon dioxide, and methane. It lacked oxygen entirely. Then organisms evolved that gave off oxygen gas as a by-product of photosynthesis, and oxygen was able to accumulate in the atmosphere.

Modifications of the atmosphere continue today. These modifications are caused either directly or indirectly by humans. An excellent example is the alteration of the carbon dioxide levels in the atmosphere. In recent times, the carbon dioxide levels have been steadily increasing (Figure 25-12). Even though carbon dioxide is only a small percentage of the total atmospheric composition, small changes can have large effects on world climate.

## Carbon Cycle

To understand the effect of atmospheric carbon dioxide on climate, we need first to examine the carbon cycle (Figure 25-13). Carbon is removed as carbon dioxide from the atmospheric pool as plants use it to manufacture sugars (and other organic compounds) during photosynthesis. From the plants, the carbon is transferred to higher levels of the food chain, returned to the atmospheric pool as carbon dioxide during the oxidation of glucose, or buried in sediments (as carbonate rocks, coal, oil, and natural gas).

## Atmospheric Carbon Dioxide

A number of human activities have significantly increased the amount of carbon dioxide in the atmosphere. First, the burning of fossil fuels, particularly in the last century, has released vast quantities of carbon dioxide into the atmosphere. The manufacture of cement from limestone has had a similar effect. The fossil fuels and limestone were laid down in the earth's sedimentary rocks over millions of years. By burning the fuels in automobiles, homes, and industrial plants, and processing the limestone, we release their carbon all at once and unbalance the biosphere's natural homeostatic mechanisms. It may take thousands or millions of years to remove the added carbon dioxide from the air.

Another factor causing an increase in the amount of carbon dioxide in the atmosphere is deforestation, particularly of tropical rain forests.This process reduces the number of trees that are drawing carbon dioxide from the atmospheric pool. In addition, deforestation also makes dead wood available to

FIGURE 25-12 Changing levels of carbon dioxide in the atmosphere from 1958–1981.

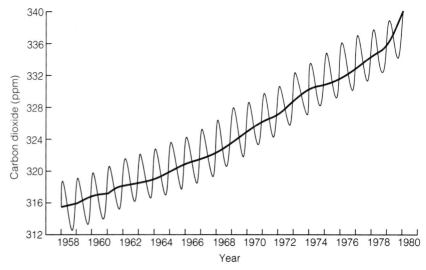

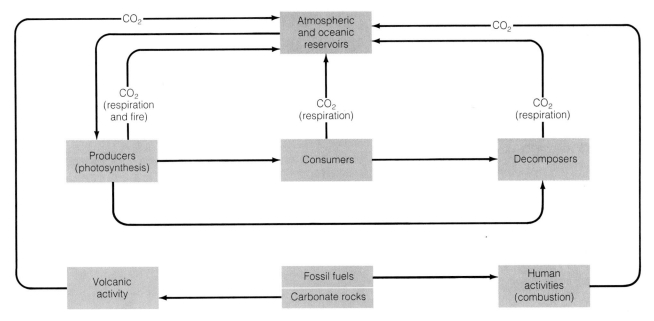

FIGURE 25-13 The carbon cycle.

termites and decomposers, who release all the carbon stored in the wood as carbon dioxide. Indeed, some scientists have estimated that termites release twice as much carbon dioxide as does the burning of fossil fuels.

### Greenhouse Effect

The main predicted effect of increased atmospheric carbon dioxide is an increase in the global temperature. This carbon dioxide-induced warming is called the *greenhouse effect* because carbon dioxide acts something like the glass in a greenhouse. As shown in Figure 25-14, solar radiation, passing through the earth's atmosphere, eventually strikes the surface of the earth. Some of the incoming solar radiation is reradiated as infrared (heat) rays that normally go out into space. However, if an infrared ray hits a molecule of carbon dioxide, it is absorbed. Later, the carbon dioxide molecule radiates the ray, but not necessarily toward space. About half the time, the ray travels toward the earth. The net effect is that the carbon dioxide keeps heat from escaping.

Although scientists have predicted a global warming trend due to the greenhouse effect, there is tremendous uncertainty in their predictions. The main problem is that it is difficult to measure accurately the different components of the carbon cycle. In particular, the ocean plays an important role in buffering the amount of carbon dioxide in the atmosphere. That is, when the level of carbon dioxide in the atmosphere increases, the ocean absorbs carbon dioxide; when the level decreases, the ocean releases carbon dioxide. We do not know how much carbon dioxide the ocean can ultimately absorb. Nor do we know how fast the ocean can respond to changes in the atmosphere. Other sources of uncertainty include the roles of other gases, such as methane, that also have "greenhouse" effects, and of particulate matter, added to the atmosphere by human industry (as a form of air pollution) and by volcanoes. This particulate matter, because it reflects sunlight from the atmosphere and reduces warming, may counteract the greenhouse effect. Unfortunately, no one can say just how great this counter-greenhouse effect may be.

The best estimates to date suggest that by the end of the next century, the greenhouse effect will warm the earth as a whole by up to eight degrees Fahrenheit (4.5 degrees Celsius), with perhaps three times that amount of warming near the poles. This may be enough to have dramatic effects. As glaciers and polar ice caps melt, the sea level will rise by between .7 meters (2.3

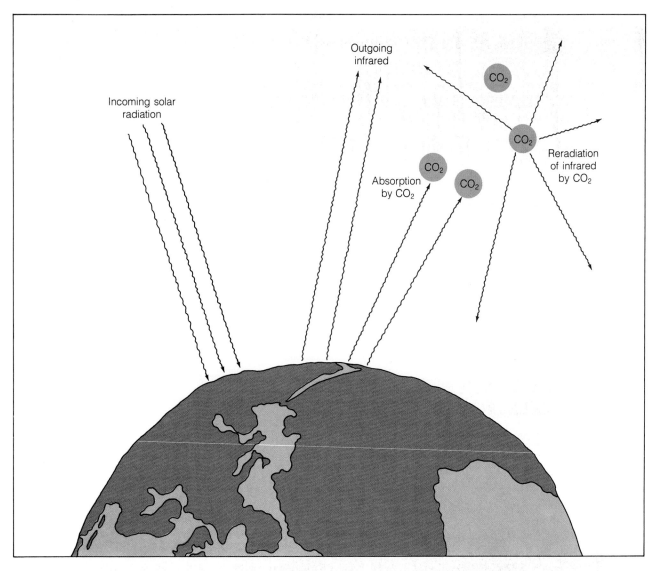

FIGURE 25-14 The greenhouse effect.

feet) and 2.4 meters (8 feet) by the year 2100. This will flood many low-lying coastal areas. At the same time, the warming may make more northerly regions (such as Canada and Siberia) more suitable for agriculture while drying out current "corn belts" such as the American Midwest.

### The Hydrosphere and Acid Rain

As with carbon and many other substances essential to life, water is endlessly recycled. In this case, the abiotic reservoir is the hydrosphere. The cycle passes from living things to the atmosphere and back to the hydrosphere. The hydrosphere includes water both on the ground, as the reservoirs of oceans, lakes, rivers, and streams, and in the air, as incoming precipitation. To understand how the hydrosphere has been altered by human activities, we first need to look at the basic water cycle (Figure 25–15).

### Water Cycle

The main pool of water is the world's oceans. Water evaporates from the ocean surface, as well as from smaller bodies of water, the surface of the soil, and the leaves of plants. As warm air carrying water vapor rises, it cools. This cooling causes the water to condense out of the air to form clouds. Eventually, the water returns to the ocean and land as precipitation. Water hitting the land can be absorbed by plant roots, can drain into bodies of water, or can percolate deep into the ground to replenish underground **aquifers,** zones of porous rock that serve as groundwater reservoirs.

Vegetation and soil are critical in regulating the water cycle by performing three important functions.

1. They regulate the flow of water. As large pulses of storm water strike the earth, the water percolates through the leaves of a forest and through the soil. Thus, the storm's water is released more gradually into streams, so that flooding is minimized.
2. They change the quality of the water. When rainwater falls through the atmosphere, it reacts with the carbon dioxide in the air to form a weak acid with a pH of 5.6. This is the natural acidity of rainwater. That is, water that is in equilibrium with the carbon dioxide in the atmosphere will have a pH of 5.6. This slightly acidic water is neutralized as it moves through a forested ecosystem. Chemical reactions that occur on the surfaces of leaves and branches of plants and in the soil itself neutralize the hydrogen ions in the incoming rainwater.
3. The evaporation of water from stomata (a process called **transpiration**) moderates the temperature by cooling the air.

Humans have altered the hydrosphere in a number of ways. Most obviously, we have polluted various bodies of water. One spectacular example is the Cuyahoga River, which once bore so many flammable pollutants that it actually caught fire. In addition, humans are depleting groundwater reserves (Figure 25–16). In Texas and its surrounding states, people are withdrawing water from the Ogallala aquifer much more rapidly than that water is being replenished, and scientists estimate that the aquifer could run dry within 40 years. The results of such depletion are lack of water for human use in agriculture, at home, and in industry. When an aquifer is near a coast, depleting its fresh water contents can allow sea water to flow into the aquifer.

## The Effects of Acid Rain

The final example of human interference with the water cycle is the phenomenon of *acid rain,* which is drawing increasing public concern. Acid rain is any rainwater that has a pH below 5.6, the pH of unpolluted rainwater. Rainwater

FIGURE 25–15 The water cycle.

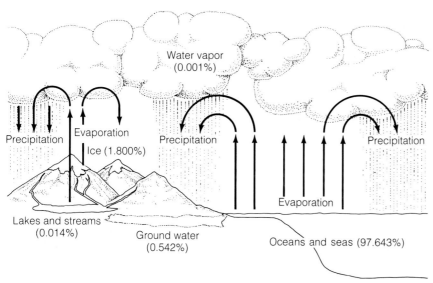

Water vapor
(0.001%)

Precipitation        Evaporation        Precipitation                    Precipitation
                     Ice (1.800%)

                                                   Evaporation

Lakes and streams
(0.014%)
            Ground water
            (0.542%)        Oceans and seas (97.643%)

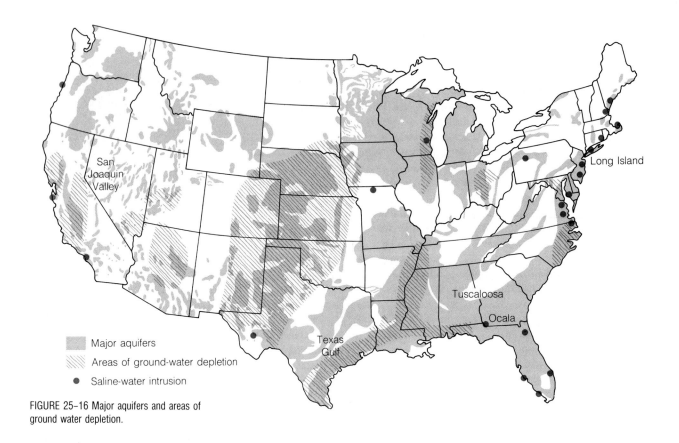

Major aquifers

Areas of ground-water depletion

Saline-water intrusion

San Joaquin Valley

Long Island

Tuscaloosa

Ocala

Texas Gulf

FIGURE 25–16 Major aquifers and areas of ground water depletion.

with a pH between 4 and 5 is now typical in New England. The most acidic rain ever recorded was at a site in Scotland; it had a pH of 2.4 (about the same as lemon juice). Even in remote wilderness areas on the western slopes of the Rockies, rainwater with a pH as low as 3.6 (roughly that of vinegar) has been recorded.

Acid rain is due mainly to the vast amounts of nitrogen and sulfur oxides that are spewed into the atmosphere by human activities. These oxides are by-products of the combustion of coal, gasoline, and other fossil fuels. They combine with oxygen and water in the air to form sulfuric and nitric acids. These acids are then washed out of the air by rain.

The amounts of sulfur and nitrogen oxides released into the atmosphere have increased dramatically. In 1940, 22 million tons of sulfur oxides and 6.7 million tons of nitrogen oxides were released into the atmosphere. By 1976, these amounts had increased to 32 million tons of sulfur oxides (a 45-percent increase) and 25 million tons of nitrogen oxides (a 273-percent increase). Sulfur oxides come mostly from electric power plants; the main source of nitrogen oxides is automobile exhaust.

Because changes in acidity can influence a wide range of chemical reactions, the effects of acid rain are complex and difficult to predict. However, we have already seen some of its consequences.

One of the earliest observed effects of acid rain was the loss of fish from small lakes and streams in the Adirondack Mountains, which are downwind from many large industrial areas. In 1940, 96 percent of the high-elevation lakes in this area had active fish populations. By 1970, only 30 percent of these lakes had fish. Although initially the decline in fish populations was blamed on beavers and the logging industry, recent studies have shown that acid rain is the main cause of the decline.

Acid rain reduces fish populations by killing fish eggs. In addition, acid rain percolating through soil and draining into lakes triggers the release of aluminum ions, which are very toxic to adult fish. The concentration of aluminum in the lakes of the Adirondacks, as well as in Canadian and

Scandinavian lakes, has been shown to be enough to cause the demise of many fish populations.

Several other effects of acid rain have been observed or predicted:

1. Scientists have shown that acid rain increases the leaching of nutrients from forest soils, and they predict that this will cause a decline in the growth rate and productivity of forests.
2. Acid rain has been shown to decrease the photosynthetic rate of crop plants and thus might have important effects on food production.
3. Acid rain has increased the weathering of classical architectural structures such as the Parthenon. Its effects can even be seen on statues, monuments, and tombstones well under a century old.

## The Lithosphere: Soil Pollution and Erosion

The land component of the biosphere, the lithosphere, includes the Earth's soil and rock. The lithosphere is the reservoir or pool for calcium, phosphorus, and other essential nutrients, all of which, like carbon and water, cycle through the biosphere (Figure 25–17).

Human activities affect the lithosphere mainly through changes in the soil. These changes include soil pollution (as by hazardous wastes) and soil loss or erosion. Both pollution and erosion reduce the supply of arable land. They therefore also reduce food production and food supply. In the United States alone, three billion tons of topsoil are lost each year to water erosion. This is the equivalent of about 500,000 hectares of arable land per year. These losses need to be counteracted by using soil conservation techniques, developing perennial crops that reduce the vegetation-free periods when the soil is most vulnerable to erosion, and developing crops that can grow on marginal lands.

Land also needs to be protected from other causes of loss. In the United States, every hour, 90 hectares of actual or potential farmland are converted to nonfarm uses, such as housing developments, shopping centers, parking lots, and highways. In a very real sense, such land is no longer part of any ecosystem, or of the biosphere.

The effects of acid rain are clearly evident on this weathered statue.

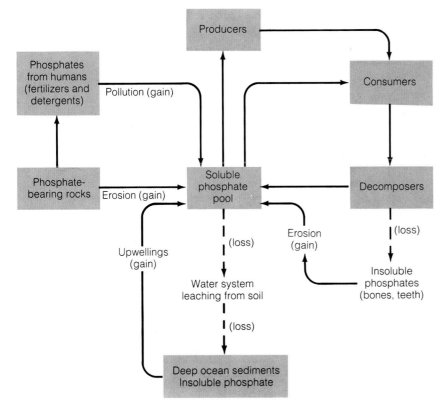

FIGURE 25–17 The phosphorus cycle.

## Our Collective Responsibility

If we fail to protect the earth's land, water, and atmosphere from the effects of human activities, then the hold of human life on the earth will become increasingly precarious. This is why it is important to understand human ecology at the levels of population, community, ecosystem, and biosphere. At each level, important ecological principles apply to all living things, including human beings. At the population level, these principles govern the growth of and the limits on populations, including the human population. At the community level, food chains, food webs, and the interactions of humans with other organisms dictate the place of humans in the vast pattern of nature. At the ecosystem level, biotic and abiotic factors define the independent subdivisions of the living world, and the natural processes that recycle basic nutrients define a need for human responsibility.

If we neglect that responsibility, if we forget our place in nature, if we ignore the built-in limits on population growth, then we will destroy all other living things, our world, and ourselves.

## Study Questions

1. Is the biosphere more like a community or an ecosystem? Why?
2. What technological developments united the many populations of the human species into a single population?
3. a) What are the main factors that were responsible for human population growth prior to the Middle Ages?
   b) What factors currently explain world population growth rates?
4. a) How does the demography of a developing nation differ from that of a typical developed nation?
   b) How do these differences in demography affect population growth rates?
5. Why are most food chains limited to four or five trophic levels?
6. Frances Moore Lappe argues that vegetarian diets can sustain a greater number of humans than diets that include meat. Use your understanding of food chain dynamics to support her argument.
7. a) Under what circumstances might the first level of a numbers pyramid be smaller than the second?
   b) When might the second level be smaller than the third level?
8. Define biological magnification and describe the properties a compound must have in order to be biologically magnified.
9. Describe several methods of pest control that can be used instead of chemical pesticides.
10. a) List as many as you can of the animal and plant species that are part of the American diet.
    b) Which ones do people eat the most of?
11. How have the causes of extinction changed between the nineteenth and twentieth centuries?
12. What will happen if you "overload" an aquarium with plants? With fish?
13. Define the following terms: ecosystem, natural ecosystem, and biosphere.
14. Diagram a basic nutrient cycle. Indicate on your diagram the parts of the cycle that are being altered by human activities.
15. Why is it so difficult to predict the consequences of increased levels of carbon dioxide in the atmosphere or the effects of acid rain?
16. a) Why would the climate changes caused by the greenhouse effect cause shifts in population distribution?
    b) In political power?
17. Describe two ecosystem services provided by a forested woodland area.
18. a) In what sense is it accurate to say that a parking lot is no longer part of the biosphere?
    b) In what sense is it inaccurate?

# GLOSSARY

**abiotic** the nonliving components of an ecosystem (25)

**acetylcholine** chemical released during the stimulation of a muscle cell by a motor neuron (7)

**acetylcholinesterase** enzyme present in the junctional folds of the sarcolemma (7)

**acid** substance that releases hydrogen ions when placed in water (2)

**action potential** an electrical impulse generated in a nerve or nerve cell (13)

**activated T-cells** T-lymphocytes that have enlarged and proliferated in response to contact with an antigen (21)

**activation** the triggering of embryonic development following penetration of the oocyte by a sperm cell (18)

**active immunity** immunity that has been acquired by previous exposure to an antigen (21)

**active transport** type of membrane transport that requires cellular energy to move substances across the plasma membrane against a concentration gradient (3)

**adipose tissue** specialized type of loose connective tissue that consists almost entirely of fat cells (4)

**adrenal cortex** the outer portion of the adrenal gland (16)

**adrenal gland** one of two endocrine glands attached to the superior end of both kidneys (16)

**adrenal medulla** the inner portion of the adrenal gland (16)

**adrenocorticotropic hormone (ACTH)** hormone that acts to stimulate the synthesis and secretion of glucocorticoids from the adrenal cortex and to stimulate the breakdown and utilization of fat deposits; produced in the anterior lobe of the pituitary gland (16)

**aerobe** organism that uses $O_2$ as the ultimate electron acceptor in the oxidation of energy sources (2)

**afferent neuron** neuron that carries impulses from other parts of the body to the central nervous system; also called a sensory neuron (13)

**afterbirth** a placenta that has been expelled from the uterus following delivery of a baby (18)

**air sinus** in the skull, one of four pairs of hollow chambers linked to the nasal cavity and lined with mucous membrane; lightens the weight of the skull and serves as a resonance chamber for the voice (6)

**albinism** inherited trait that produces an albino phenotype (19)

**albino** organism in which skin, hair, and eyes lack pigment; caused by an inherited inability to synthesize melanin (5)

**alleles** alternative forms of a gene (19)

**allergy** maladjustment of the immune system, in which a violent response is mounted to the presence of largely harmless antigens (21)

**all-or-none response** property of muscle cells by which they contract fully or not at all in response to a stimulus of a particular strength (7)

**amenorrhea** the temporary or permanent absence of menstrual periods (17)

**Ames test** one of the most useful short-term tests used to detect carcinogens (22)

**amnion** during the period of germ layer formation, membrane formed by the cells lining the amniotic cavity (18)

**amniotic cavity** during the period of germ layer formation, the fluid-filled cavity that appears between cells of the ectoderm and the trophoblast of the blastocyst (18)

**anabolic pathway** metabolic pathway that requires energy and is involved in the synthesis of more or less complex cellular components from simpler nutrient molecules; also called biosynthetic pathway (2)

**anaerobe** organism, mostly bacteria, that can only live in environments in which $O_2$ is absent or present in extremely low concentrations (2)

**aneurysm** a balloonlike swelling of a weakened artery wall (9)

**antagonistic pair** a pair of muscles in which one muscle contracts while the other muscle relaxes (7)

**anterior horn** one of two ventral extensions of the butterfly-shaped region of gray matter in the spinal cord; also called a ventral horn (13)

**antibody** plasma protein that provides immunity against disease; also called gammaglobulin (8)

**anticodon** in transfer RNA, sequence of three nucleotide bases involved in the recognition of mRNA codons (20)

**antidiuretic hormone (ADH)** in the kidneys, secretion that accelerates the rate of water reabsorption by increasing the permeability of the distal convoluted tubules and the collecting tubules to water; secreted from the posterior lobe of the pituitary gland when the body loses too much water; also called vasopressin (12, 16)

**antigen** foreign substance that elicits the production of antibodies (21)

**antigen binding site** site at which an antigen binds to an antibody (21)

**anus** the posterior opening of the digestive tube (11)

**aorta** artery that receives blood from the left ventricle of the heart; the body's largest blood vessel (9)

**appositional growth** type of cartilage growth in which chondrocytes synthesize intercellular substance that is deposited on the surface of existing cartilage (4)

**aquifer** underground zone of porous rock that serves as a groundwater reservoir (25)

**arachnoid** the middle of the three meninges protecting the spinal cord (13)

**arachnoid villus** (pl. *villi*) in the brain, projection of the arachnoid through which cerebrospinal fluid enters the dural sinuses (14)

**arrector pili muscle** small bundle of smooth muscle fibers that causes body hair to erect (5)

**arteriole** a medium-sized branch of an artery (8)

**artery** major blood vessel used to carry blood from the heart (8)

**articulation** the connection of two bones; also called a joint (6)

**ascending tract** in the white matter of the CNS, bundle of axons that transmits sensory information to the brain (13)

**asexual reproduction** reproduction achieved by a single organism (1)

**association neuron** a neuron that transmits impulses from one part of the CNS to another before synapsing with an efferent neuron; also called an interneuron (13)

**aster** during prophase, starlike spindle fibers seen radiating from each centriole pair (3)

**atom** smallest unit of a chemical element having all the properties of that element; atoms make up molecules (2)

**atomic orbital** space an electron occupies in orbiting the nucleus of an atom (2)

**atrial natriuretic factor (ANF)** group of related hormones secreted by the cardiac muscle cells of the atrium; reduces blood pressure by decreasing blood volume and increasing total capacity of the blood vessels (16)

**atrioventricular (AV) node** group of specialized cardiac muscle cells, at the base of the right atrium near the wall between the two atria, that conducts the impulse for contraction from the atria to the ventricles (9)

**atrium** (pl. *atria*) in the heart, one of two thin-walled chambers that receive blood from veins (9)

**atropine** drug that blocks the action of the parasympathetic division by competing with acetylcholine for receptors on the postsynaptic membranes (13)

**auditory ossicle** inside the middle ear, one of three tiny bones, the malleus, the incus, or the stapes (15)

**auricle** in the outer ear, the flattened flap of tissue that projects from the side of the head; also called the pinna (15)

**autoimmunity** immune response directed at the body's own tissues (21)

**autonomic nervous system** the division of the peripheral nervous system that controls involuntary functions (13)

**autosome** within a somatic cell, one chromosome of twenty-two pairs of chromosomes that contain hereditary information about all traits but sex (19)

**autotroph** organism that is able to synthesize all the organic compounds it needs from carbon dioxide, water, and other simple inorganic compounds; occupies the first trophic level in any food chain (2, 25)

**axial skeleton** the bones situated along the axis of the body; functions to support and protect organs of the head, neck, and trunk (6)

**axis** the second cervical vertebra; allows the head to turn from side to side (6)

**axon** nerve fiber that usually conducts an electrical impulse away from a neuron (13)

**axon collateral** side branch of an axon (13)

**ball-and-socket joint** joint in which the rounded head of one bone articulates with a cuplike cavity in another bone; allows angular movement in all directions (6)

**baroreceptor** specialized nerve cell, located in the walls of the aortic arch and of the carotid sinuses, that registers changes in blood pressure; also called pressoreceptor (9)

**base** substance that releases hydroxide ions when placed in water (2)

**basement membrane** fibrous membrane that connects sheets of epithelial tissue to underlying connective tissue (4)

**B-cells** B-lymphocytes that enlarge and secrete antibodies in response to the presence of foreign substances; also called plasma cells (21)

**benign tumor** tumor that grows without actively destroying normal tissue or disrupting the body's general metabolism (22)

**beta blocker** class of drug that blocks the action of beta-adrenergic receptors and thus inhibits sympathetic stimulation of the heart and blood vessels (13)

**biceps** large muscle that flexes the forearm; lies anterior to the humerus (7)

**bicuspid valve** heart valve that separates the left atrium from the left ventricle; also called the mitral valve (9)

**biological catalyst** substance that speeds up a specific chemical reaction within a cell; also called an enzyme (2)

**biome** a group of ecosystems that share a similar climate and many of the same types of plants and animals (25)

**biosphere** all the living organisms on Earth (25)

**biotic** the living components of an ecosystem (25)

**birth control** any manipulation of the reproductive process that lowers the frequency of childbirth (18)

**birth rate** number of births per unit of time (25)

**blastocyst** embryo formed about the time of implantation, when fluid accumulates between an outer layer of cells called the trophoblast and an inner cell mass (18)

**blood cell** in the circulatory system, a specialized cell suspended in an aqueous fluid; classified as a free connective tissue (4)

**B-lymphocytes** the lymphocytes that are responsible for humoral immunity (21)

**body system** a group of organs that performs a specific function (4)

**bone** a type of connective tissue made up of osteocytes, which secrete a collagenous intercellular substance that quickly becomes mineralized and extremely hard and rigid (4)

**Bowman's capsule** within a nephron, the part of the renal tubule that surrounds the glomerulus (12)

**bronchiole** a branch of a bronchus that has a diameter of about 1 millimeter or less (10)

**buffer** in living cells, dissolved substance or mixture of substances that regulates the pH of intracellular and extracellular fluids (2)

**bulbourethral gland** in the male reproductive system, one of two glands that secrete a thick, clear, slippery fluid that is released following erection and helps lubricate the vagina during sexual intercourse; also known as Cowper's glands (17)

**calcitonin** thyroid hormone that helps regulate the blood levels of calcium (16)

**callus** in the repair of a fractured bone, a collar of new tissue that bridges the area between the bone's broken ends (6)

**calorie** the amount of heat required to raise one gram of water one degree Centigrade; a measure of heat energy (2)

**canaliculus** (pl. *canaliculi*) in the bones, little channel extending between lacunae (6)

**cancellous bone** bone tissue in which a latticework of trabeculae forms a porous, coral-like structure (6)

**cancer** tumor that grows in an unregulated fashion, infiltrating and destroying normal tissues and organs; also called a malignant tumor or cancerous tumor (22)

**cancerous tumor** tumor that grows in an unregulated fashion, infiltrating and destroying normal tissues and organs; also called a malignant tumor or cancer (22)

**capillary** a small blood vessel that branches off an arteriole; capillaries form a vast network of blood vessels permeating body tissues (8)

**capsid** protein coat surrounding the nucleic acid molecules within a virus (21)

**capsomeres** protein subunits forming virus capsid (21)

**carcinogen** agent that causes cancer; also called carcinogenic agent (1, 22)

**carcinogenic agent** environmental agent that causes cancer; also called carcinogen (22)

**carcinoma** cancer derived from the epithelial tissues; represents about 85 percent of all cancers (22)

**cardiac control center** regulates cardiac output; one of the reflex centers in the medulla oblongata (14)

**cardiac muscle** muscle found in the walls of the heart and the pulmonary veins (4)

**carrier** person who carries a recessive allele that is not expressed in the individual but can be passed on to progeny (19)

**cartilage** a type of connective tissue made up of an extremely dense network of collagenous fibers embedded in a firm gelatinous material (4)

**catabolic pathway** metabolic pathway that breaks complex chemical substances into simpler ones, releasing energy that may be used to drive the cell's energy-requiring processes (2)

**celiac ganglion** a type of collateral or prevertebral ganglion (13)

**cell cycle** the sequence of events that occurs during somatic cell division; made up of interphase, mitosis, and cytokinesis (3)

**cell-mediated immunity** component of immune response in which a class of lymphocytes directly attacks and destroys certain harmful cell types (21)

**cell theory** theory that all living organisms are composed of cells and cell products (1)

**central chemoreceptor** in the brainstem, respiratory receptor that monitors $H^+$ levels, and thus indirectly monitors $CO_2$ levels, in the blood (10)

**central fissure** in the cerebrum, fissure dividing the frontal lobe from the parietal lobe; also called the fissure of Rolando (14)

**central nervous system (CNS)** the brain and the spinal cord (5, 13)

**centriole** in a cell, one of two cylindrical structures that lie at right angles to each other near the nucleus of a nondividing cell; made up of microtubules (3)

**centromere** during mitosis, the region at which the replicated chromosomes remain attached to one another (3)

**cerebellar hemisphere** one of two lobes of the cerebellum (14)

**cerebellum** located superior to and posterior to the medulla and the pons; functions at the subconscious level to monitor, coordinate, and control voluntary and involuntary movement (14)

**cerebral cortex** in the cerebrum, the thin, outer layer of gray matter (14)

**cerebral dominance** phenomenon in which one side of the brain may have significantly greater responsibility for one function than for another (14)

**cerebral peduncle** one of two fiber bundles in the ventral part of the midbrain (14)

**cerebrospinal fluid** modified tissue fluid that fills the space between the pia mater and the arachnoid, as well as the ventricle of the brain (13, 14)

**cerebrum** portion of the forebrain occupying most of the space in the cranial cavity (14)

**cerumen** wax found in the external canal of the ear (5)

**cervix** the part of uterus inferior to the body of the uterus (17)

**chemical bond** the attractive force that holds atoms together in a molecule (2)

**chemical reaction** transformation that occurs when chemical bonds are broken or new chemical bonds are formed (2)

**chondrocyte** cartilage cell (4)

**chorion** membrane system formed by the chorionic villi and their associated structures; covers all structures that arise from the fertilized egg and forms a membranous barrier between the circulatory systems of the embryo and the mother (18)

**chorionic villus** (pl. *villi*) treelike system of blood vessels formed from the trophoblast; part of the chorion (18)

**choroid** blood vessels and pigment covering the posterior two-thirds of the eye's vascular layer (15)

**choroid plexus** in the brain, membranous structure that produces cerebrospinal fluid and penetrates into the lumen of the four ventricles (14)

**chromosome** the folded and coiled chromatin strands containing most of the cell's hereditary information; found in the cell during cell division (3, 19)

**ciliary body** in the eye, the suspensory ligaments and ciliary muscles that support and change the shape of the lens; found in the anterior portion of the vascular layer (15)

**cilium** (pl. *cilia*) short, motile filamentous projection at the cell surface that serves to move substances past stationary cells (3)

**circulatory system** the heart and all vessels that circulate blood and lymph fluids through the entire body (4)

**cirrhosis** chronic deterioration of the liver; decline in liver function can lead to liver failure (11)

**clitoris** in the vulva, erectile structure between the anterior ends of the labia minora (17)

**clonal selection theory** theory that explains how lymphocytes are activated by specific antigens (21)

**clone** a cell or organism that is genetically identical to another cell or organism (21)

**cochlea** in the inner ear, a bony tube that is wound into a spiral (15)

**codon** code word stored in messenger RNA during DNA replication; made of a sequence of three nucleotide bases, which specify one amino acid (20)

**collagenous fiber** a type of strong, flexible, and nonextensible fiber found in connective tissue; composed of the protein collagen (4)

**collateral ganglion** in the sympathetic division of the autonomic nervous system, ganglion that lies in a group in front of the vertebral column, near the organ it serves; also called a prevertebral ganglion (13)

**collecting tubule** tubule into which the distal convoluted tubules of many nephrons empty (12)

**columnar** epithelial cell with a columnlike shape (4)

**community** all the populations of all the species in a particular area (25)

**compact bone** extremely dense type of bone that forms the exterior surface of all bones and most of the shaft of long bones (6)

**compound** molecule containing atoms of two or more elements (2)

**conditioned reflex** simple type of learning in which the repetition of a stimulus comes to evoke automatically a particular response (14)

**cone** in the retina of the eye, neuron that functions as a photoreceptor in bright light; cone also provides color vision (15)

**conjunctiva** thin, transparent mucous membrane covering the anterior surface of the sclera and lining the eyelids (15)

**connective tissue** tissue that binds together, supports, and protects other tissues; found in all parts of the body (4)

**cornea** transparent, anterior portion of the supporting layer of the eye (15)

**corona radiata** during ovulation, the follicular cells that surround the secondary oocyte (17)

**coronary sinus** a wide venous channel that empties blood from the capillary networks of the myocardium into the right atrium of the heart (9)

**corpus callosum** myelinated fibers that connect the middle lower portions of the right and left cerebral hemispheres (14)

**corpus luteum** a ruptured ovarian follicle that has been converted by luteinizing hormone into an endocrine organ that secretes estrogen and progesterone (16, 17)

**covalent bonding** type of chemical bond in which electrons are shared more or less equally between atoms; common in organic molecules (2)

**cranial meninges** the three membranes covering the brain; continuous with the three meninges that cover the spinal cord (14)

**cranial nerve** in the peripheral nervous system, one nerve of 11 pairs of nerves that supply the head and neck or one nerve of a pair of nerves that supplies various internal organs (13)

**cranial outflow** in the parasympathetic division, preganglionic fibers that leave the brain in cranial nerves III, VII, IX, and X (13)

**craniosacral division** the parasympathetic division of the autonomic nervous system (13)

**cranium** the bony vault surrounding the brain (6)

**crista** (pl. *cristae*) fold of the inner membrane of a mitochondrion (3)

**cuboidal** epithelial cell with a cubelike shape (4)

**cyclic AMP** a special cyclic form of adenosine monophosphate; the second messenger in the second messenger mechanism (16)

**cystitis** inflammation of the bladder; usually caused by a bacterial infection (12)

**cytocidal virus** type of virus that infects a cell, resulting in the production and release of new viral particles and eventual cell death; causes such common human diseases as measles and influenza (22)

**cytokinesis** the final phase of the cell cycle, in which the cytoplasm divides (3)

**daughter cell** one of two new cells formed when a somatic cell divides (3)

**death rate** number of deaths per unit of time (25)

**decussate** to cross over; phenomenon in which ascending and descending tracts cross over as they ascend or descend the spinal cord, so that each side of the brain controls the opposite of the body (13)

**deductive reasoning** derives specific conclusions from generalizations (1)

**deep fascia** sheets of irregular dense connective tissue; forms the membranous capsules or sheaths that surround many internal organs, the brain, spinal cord, blood vessels, and muscles (4)

**dehydrogenation** biological reduction involving the removal of hydrogen atoms (2)

**dendrite** nerve fiber that usually conducts an electrical impulse toward a neuron (13)

**dense fibrous connective tissue** a type of connective tissue that is formed primarily from collagenous fibers that are arranged in compact parallel bundles running in the same basic direction; found in tendons and ligaments (4)

**dermis** the skin's inner layer of connective tissue (5)

**descending tract** in the white matter of the CNS, bundle of axons that carries motor impulses down the spinal cord (13)

**desmosome** a particularly strong junctional structure between cells; particularly common in epithelial tissues that are subject to considerable mechanical stress (3)

**diaphysis** the central shaft of long bones (6)

**diastole** in the cardiac cycle, relaxation of the heart (9)

**diastolic blood pressure** pressure maintained in the arteries during relaxation of the heart (9)

**differential permeability** in a cell, the plasma membrane's ability to determine which molecules will pass through the membrane and at what rate (3)

**digestion** process whereby the digestive tube breaks down large food particles into substances that can be absorbed by blood vessels and lymphatic vessels (11)

**digestive system** long muscular tube beginning at the lips and ending at the anus, plus glandular structures that empty their secretions into the digestive tube (4)

**digestive tract** long muscular tube consisting of the mouth, pharynx, esophagus, stomach, small intestine, and large intestine; also called the digestive tube (11)

**digestive tube** long muscular tube consisting of the mouth, pharynx, esophagus, stomach, small intestine, and large intestine; also called the digestive tract (11)

**digestive vacuole** formed by the fusion of an endocytotic vesicle and a lysosome; digests materials that diffuse through the vacuole membrane and are used as energy sources or as raw materials for biosynthesis (3)

**diploid** cell that contains two of each chromosome type; characteristic of human somatic cells (17)

**distal convoluted tubule** the final part of the renal tubule and the final portion of the nephron unit (12)

**dominant** in heredity, an allele that masks the effect of a recessive allele (10)

**dorsal cavity** cavity made up of the cranial cavity, which contains the brain, and the spinal cavity, which contains the spinal cord and roots of nerves extending from these organs (4)

**dorsal horn** one of two dorsal extensions of the butterfly-shaped region of gray matter in the spinal cord; also called a posterior horn (13)

**dorsal root** branch of the spinal nerve that enters the spinal cord at the dorsal aspect of the body (13)

**dorsal root ganglion** swelling on the dorsal root that contains all the cell bodies of all the afferent fibers of the spinal nerve (13)

**dura mater** outermost of the three meninges protecting the spinal cord (13)

**dural sinus** on the surface of the brain, cavity formed between the periosteum and dura mater (14)

**eardrum** membrane separating the outer ear from the middle ear; also called the tympanic membrane (15)

**ecological community** the population of plants, animals, and micoorganisms living and interacting in a given locality (25)

**ecosphere** the sum of the Earth's ecosystems or biomes (25)

**ecosystem** a community plus its inanimate environment (25)

**ectoderm** during the period of germ layer formation, the tall, columnar cells formed closest to the trophoblast layer of the blastocyst (18)

**ectopic pregnancy** pregnancy resulting when a fertilized egg implants in a tissue other than the endometrium of the uterus (18)

**efferent neuron** neuron that carries impulses from the CNS to muscles and glands; also called a motor neuron (13)

**ejaculation** the second of two processes involved in the deposition of the male's sperm into the female's vagina; consists of the emission and expulsion of semen (17)

**elastic cartilage** a type of cartilage in which some collagenous fibers are embedded in a dense network of elastic fibers (4)

**elastic fiber** a type of resilient and stretchable fiber found in connective tissue; composed of the protein elastin (4)

**electrocardiogram** graphic tracing of the electrical activity of the heart; also called EKG (9)

**electrolyte** a molecule that ionizes in water (2)

**electron** negatively charged particle that orbits the nucleus of an atom (2)

**electron acceptor** in oxidation-reduction reactions, the substance that gains electrons or is reduced (2)

**electron donor** in oxidation-reduction reactions, the substance that loses electrons or is oxidized (2)

**electron transport system** metabolic pathway in which reduced forms of NAD and FAD are oxidized; also called the respiratory chain (2)

**element** one of the 103 different types of neutral atoms, each of which has unique chemical and physical properties (2)

**embolism** thrombus that blocks blood flow and causes tissue death (8)

**embryonic disc** during the period of germ layer formation, the structure formed by the endoderm and ectoderm of the blastocyst (18)

**embryonic period** second of three arbitrary stages in embryonic development; extends from the period of germ layer formation through the eighth week (18)

**endocardium** innermost layer of the heart wall; also forms the valves and is continuous with the endothelial lining of blood vessels (9)

**endocrine gland** an organ that secretes hormones into the bloodstream; includes the pituitary, thyroid, parathyroids, adrenals, islet cells of the pancreas, gonads, pineal, and thymus (4, 16)

**endocrine system** all the body's endocrine glands (4)

**endocytosis** type of membrane transport mechanism in which material outside the cell is trapped in endocytotic vesicles, which detach from the plasma membrane and enter the cytoplasm; opposite of exocytosis (3)

**endocytotic vesicle** vesicle that is involved in endocytosis (3)

**endoderm** during the period of germ layer formation, the inner layer of small, cuboidal cells formed next to the cavity of the blastocyst (18)

**endogenous opioid peptide** hormone that binds to specific opiate receptors in the brain (16)

**endometrium** mucous membrane lining the body and fundus of the uterus (17)

**endomysium** the thin connective tissue sheath surrounding each muscle cell (7)

**endoneurium** fine connective tissue that binds individual nerve fibers together (13)

**endoplasmic reticulum (ER)** in a cell, a series of large, flattened, membranous, interconnecting sacs that form an extensive network of membrane-bound channels, the outer surface of which borders on the cytoplasm; two distinct types are rough ER and smooth ER (3)

**end-product** the final product of metabolism (2)

**energy-transferring molecule** a molecule that stores energy obtained from the breakdown of nutrient molecules and then releases the energy when it is needed to drive energy-requiring processes of cells (2)

**entropy** a measure of the disorder or randomness in a system; measured in kilocalories of energy (2)

**envelope** lipoprotein membrane external to the capsid of some viruses (21)

**enzyme** substance that speeds up a specific chemical reaction within a cell; also called a biological catalyst (2)

**epicardium** the outermost layer of the heart wall; made up of the serous pericardium (9)

**epidermis** the skin's outer layer of epithelial tissue (5)

**epiglottis** a cartilaginous flap projecting above the rest of the larynx on its ventral side (10)

**epimysium** tough, smooth sheath of connective tissue covering a single muscle; allows the muscle to slide freely over nearby muscles and other structures (7)

**epinephrine** hormone secreted by the chromaffin cells of the adrenal medulla in response to short-term stress; also synthesized by some neurons for use as a neurotransmitter (16)

**epiphyseal plate** region of cartilage that forms a temporary union between the primary and secondary ossification centers in growing bones; the region where bone elongation occurs as the child grows (6)

**epiphysis** (pl. *epiphyses*) the expanded end of a long bone (6)

**epithelial tissue** sheets of cells that cover the outer surfaces of the body and line the interior surfaces of body cavities and hollow body organs (4)

**erection** the first of two processes involved in the deposition of the male's sperm into the female's vagina; occurs when stimulation of the parasympathetic nerves causes blood to engorge the sinuses of the three corpora cavernosa (17)

**erythrocyte** red blood cell; also called red blood corpuscle (8)

**erythropoietin** hormone secreted by the kidneys when they lack sufficient oxygen; stimulates the production of red blood cells (16)

**esophagus** straight tube that conducts food between the pharynx and stomach (11)

**estrogen** in the female, a number of chemically related hormones necessary for the development and maintenance of the ovaries, accessory reproductive organs, and female secondary sex characteristics; produced by the ovaries and by the placenta during pregnancy (17)

**eukaryote** organism characterized by a complex cellular structure; includes all organisms except bacteria (21)

**eukaryotic cell** cell structure containing a distinct nucleus (1)

**Eustachian tube** channel connecting the middle ear to the nasopharynx; maintains air in the middle ear at atmospheric pressure (15)

**evolution** long-term adaptation to a changing environment that occurs over several generations; characteristic of living organisms (1)

**excitatory synapse** synapse in which a change in the permeability of the postsynaptic membrane initiates depolarization, triggering an action potential (13)

**exocrine gland** gland that is connected to the epithelial surface by one or more ducts, which convey secretions to the surface (4)

**exocytosis** process in which cytoplasmic vesicles fuse with the plasma membrane, releasing substances from the cell; opposite of endocytosis (3)

**external auditory canal** in the outer ear, the canal that extends into the temporal bone of the skull (15)

**extracellular fluid** fluid outside a cell (2)

**extraembryonic coelom** during the period of germ layer formation, cavity that develops between the cells that line the primary yolk sac and the trophoblast (18)

**exudation** the passage of increased amounts of plasma into the tissue fluid; causes swelling and pain symptoms of inflammation (21)

**facilitated diffusion** type of membrane transport in which substances diffuse through the cell's plasma membrane only if they are combined with specific transport proteins in the membrane; driven by a concentration gradient (3)

**Fallopian tube** one of two ducts through which ova are transported from the ovary to the uterus and where fertilization normally occurs; also known as the oviduct (17)

**fascicle** cablelike bundle of nerve fibers (13)

**fasciculus** (pl. *fasciculi*) a small bundle of muscle cells (7)

**fat cell** cell capable of storing large amount of fat in its cytoplasm; may be found in loose connective tissue (4)

**fertility** the ability of sperm to fertilize an egg (17)

**fetal period** third of three arbitrary stages in embryonic development; extends from the ninth week of development until birth (18)

**fetus** the developing embryo during the fetal period (18)

**fibrillation** a rapid, disorganized twitching of the myocardium (9)

**fibroblast** cell that secretes the fibrous and amorphous materials that constitute the intercellular substance of connective tissue (4)

**fibrous cartilage** cartilage that consists of bundles of collagenous fibers interspersed with sparse rows of very small chondrocytes; often found where tendons join another cartilaginous structure (4)

**First Law of Thermodynamics** chemical law that states that energy is neither created nor destroyed when it is converted from one form into another (2)

**fissure of Rolando** in the cerebrum, fissure dividing the frontal lobe from the parietal lobe; also called the central fissure (14)

**fissure of Sylvius** in the cerebrum, fissure separating the frontal lobe and part of the parietal lobe from the temporal lobe; also called the lateral fissure (14)

**flagellum** (pl. *flagella*) elongated, motile filamentous projection of the cell surface that propels a whole cell (3)

**fluid mosaic model** most widely accepted explanation of plasma membrane structure; envisions the structure as being made up of a double layer of lipids, in which proteins are embedded or able to diffuse laterally (3)

**flutter** condition in which the atria or ventricles contract very rapidly but in a regular rhythm (9)

**follicle stimulating hormone (FSH)** hormone that stimulates the development and maturation of germ cells; produced by the anterior lobe of the pituitary gland (16)

**follicular phase** first of two ovarian phases in the menstrual cycle, during which an ovarian follicle develops; lasts for the first 14 days of the menstrual cycle and ends at ovulation (17)

**fontanelle** unossified membrane that connects the bones forming the skull vault of an infant between birth and the middle of the second year of life (6)

**food chain** sequence of living organisms interrelated in their feeding habits (25)

**foramen magnum** opening at the base of the skull; allows the brain stem to pass into the vertebral column as the spinal cord (6)

**fovea centralis** small depression in the central region of the retina; specialized for greatest visual acuity (15)

**fraternal twin** type of twin that results when two eggs are simultaneously fertilized by two different sperm cells (18)

**gallbladder** pear-shaped sac lying inferior to the right lobe of the liver; concentrates and stores bile secreted by the liver and releases bile as needed for digestion (11)

**gallstone** stone formed within the gallbladder by the precipitation of high concentrations of cholesterol (11)

**ganglion** (pl. *ganglia*) cluster of nerve cell bodies lying outside the brain or spinal cord (13)

**gap junction** in the plasma membrane of a cell, pore or channel through which certain substances can pass, allowing communication between cells (3)

**gene** within a chromosome, a functional unit of hereditary material (17)

**genotype** the individual's genetic constitution (19)

**germ cell** a sperm or egg cell (2)

**germinal epithelium** in the walls of the seminiferous tubules, epithelial cells, lying inside and in contact with the tubule's basement membrane, that give rise to sperm cells; in an ovary, gives rise to oogonia (4, 17)

**germinal matrix** deepest part of the down-growth forming a hair follicle (5)

**gerontology** the study of aging (23)

**gill arch** during the embryonic period, ridge of tissue that later gives rise to the jaws, larynx, portions of the ear, and nearby structures; also called the pharyngeal arch (18)

**glandular epithelium** epithelial tissue that forms exocrine or endocrine glands (4)

**glans** the end of the penis, formed by the expansion of the corpus cavernosum urethrae; contains many sensory nerve endings (17)

**glaucoma** severe eye disorder resulting when blockage of the canal of Schlemm causes pressure to build inside the eye, damaging the retinal neurons; most common in people over 60 (15)

**glial cell** in the central nervous system, a cell that supports, protects, and nourishes the neurons (13)

**glomerular filtration** the production of tissue fluid at the glomeruli; the first step in the formation of urine by the kidneys (12)

**glomerulus** (pl. *glomeruli*) tuft of capillaries within a nephron in a kidney (12)

**glucagon** hormone secreted by the alpha cells of the islets of Langerhans; enhances the conversion of liver glycogen into glucose (16)

**glucocorticoids** class of steroid hormones produced by the adrenal cortex; important in physiological responses to long-term stress (16)

**gluconeogenesis** the manufacture of glucose from non-carbohydrate sources (16)

**glycolipid** lipid containing carbohydrate components (2)

**glycolysis** first of two consecutive metabolic pathways involved in oxidizing glucose (2)

**glycoprotein** protein with carbohydrate components; involved in many cellular processes, including cellular recognition phenomena (2)

**goblet cell** modified epithelial cell that secretes mucus that moistens and protects the epithelium (4)

**Golgi complex** in a cell, stack of large, disc-shaped vesicles surrounded by smaller, spherical vesicles that functions to gather and secrete proteins and lysosomes formed in the endoplasmic reticulum (3)

**gonadocorticoids** class of steroid hormones that are similar to the hormones synthesized by the gonads; produced in small amounts by the adrenal cortex (16)

**gonadotropic hormone** class of hormones that stimulate the gonads; also called a gonadotropin (17)

**gonadotropin** class of hormones that stimulate the gonads; also called gonadotropic hormone (17)

**graft rejection** the recognition and destruction of tissue or organ transplants (21)

**gray ramus communicans** in the sympathetic division of the autonomic nervous system, short nerve trunk by which preganglionic fibers that synapse with the sympathetic ganglia may rejoin the spinal nerve (13)

**growth hormone (GH)** growth-promoting hormone produced in the anterior lobe of the pituitary gland (16)

**hair follicle** down-growth of epidermis into underlying dermis or subcutaneous tissue from which hair grows (5)

**haploid** cell that contains one of each chromosome type; characteristic of germ cells (17)

**Haversian canal** in compact bone, the central channel in an Haversian system (6)

**Haversian system** in compact bone, cylindrical units of bone tissue that run parallel to each other and to the long axis of the bone (6)

**heart attack** condition in which an area of the myocardium dies when its blood supply is blocked; also called a myocardial infarction (9)

**heart block** disturbance in the conduction of electrical impulses through the AV node and the fibers that pass from it (9)

**heart failure** a consequence of third degree heart block, in which the ventricles may not beat fast enough to meet the physiological needs of the body (9)

**heart murmur** abnormal, sloshing heart sounds that result from malfunctioning heart valves (9)

**heart rate** the number of heart beats per minute (9)

**helper T-cells** T-cells that stimulate the differentiation of B-lymphocytes into antibody-secreting plasma cells (21)

**hemoglobin** protein found in red blood cells; carries oxygen from lungs to body tissues (8)

**hepatic portal system** the entire system of veins that carries blood from abdominal organs to the liver and then to the inferior vena cava (9)

**hepatitis** inflammation of the liver; most commonly caused by a viral infection or by a reaction to certain drugs (11)

**herbivore** animal that eats plants; a primary consumer occupying the second trophic level of a food chain (25)

**heterotroph** organism that must obtain some of the organic substances it needs from its diet (2, 25)

**heterozygous** a pair of genes with two different alleles (19)

**hinge joint** joint that allows movement in one plane (9)

**homeostasis** the relatively constant, but dynamic, internal environment necessary for life (1)

**homologous chromosome** one of a pair of chromosomes that have similar morphology and carry genes that determine similar traits (17)

**homozygous** a pair of genes with two identical alleles (19)

**hormone** substance that regulates and coordinates various physiological processes in different parts of the body; secreted by an endocrine gland (2, 16)

**human demography** the study of human population ecology; focuses primarily on the factors that determine the size of human populations, such as birth rates, death rates, and age structures (25)

**humoral immunity** component of immune response in which B-lymphocytes produce and secrete antibodies against specific harmful agents into the bloodstream (21)

**hyaline cartilage** cartilage composed of a network of delicate collagenous fibers embedded in a firm gel-like material; the most abundant type of cartilage (4)

**hybridoma** hybrid cell that results from fusing an activated B-cell with a myeloma cell; used to produce large amounts of a single antibody (21)

**hydrogen bond** the weak attraction between a hydrogen atom, which is covalently bonded to an oxygen or a nitrogen atom, and another oxygen or nitrogen atom in the same or a different molecule (2)

**hydrogenation** biological oxidation involving the gain of hydrogen atoms (2)

**hydrolysis** a chemical reaction in which water is used to break down a compound (2)

**hydrophobic bond** the weak attraction between nonpolar molecules or nonpolar regions of the same molecule; caused by the tendency of nonpolar entities to associate in an aqueous medium in a way that minimizes their contact with water (2)

**hymen** in the vulva, mucous membrane that partly covers the opening of the vagina prior to initial sexual intercourse (17)

**hyperosmotic solution** an aqueous solution with a solute concentration greater than that in the cytoplasm of a given cell (3)

**hypertension** high blood pressure, defined as a sustained blood pressure of 140/90 in a person at rest (9)

**hypo-osmotic solution** an aqueous solution with a solute concentration less than that in the cytoplasm of a given cell (3)

**hypophysis** in the brain, endocrine gland that secretes hormones that affect other endocrine glands as well as nonendocrine tissues; also called the pituitary gland (16)

**hypothalamus** a region of the forebrain that is critical in maintaining homeostasis; lies inferior to the thalamus (14)

**identical twin** type of twin that results when a single zygote splits into equal halves during early development (18)

**immune response** process by which the body recognizes microbes or other foreign substances and initiates events that lead to their destruction or inactivation (21)

**immune surveillance** process in which killer T-cells constantly examine cell surfaces for the appearance of new surface antigens and destroy any such cells they detect; may be involved in the detection and destruction of tumor cells (21)

**immune system** cellular elements in the blood and lymph fluid that are important in protecting the body from disease (4)

**immunization** active immunity produced when dead or weakened pathogenic microbes are injected or given orally; also called vaccination (21)

**implantation** process by which an embryo attaches to and invades the endometrium of the uterus (18)

**impotence** condition characterized by the failure to attain an erection or to sustain an erection until ejaculation occurs (17)

**incomplete dominance** in heredity, a situation in which neither allele masks the effect of the other (19)

**incus** inside the middle ear, the middle of three tiny bones that conveys vibrations of the eardrum to the oval window (15)

**inferior mesenteric ganglion** a type of collateral or prevertebral ganglion (13)

**inferior vena cava** major vein that drains blood from the lower trunk and extremities into the right atrium of the heart (9)

**inflammatory response** the response to microbes that penetrate the body's external barriers; classic symptoms are redness, heat, swelling, pain, and impairment of function (21)

**inguinal canal** on either side of the lower abdominal wall, one of two tunnels through which a testis descends into the scrotum; usually occurs by the end of the eighth month of fetal development (17)

**inhibitory synapse** synapse in which a change in the permeability of the postsynaptic membrane inhibits depolarization, blocking an action potential (13)

**insulin** hormone secreted by the beta cells of the islets of Langerhans; decreases blood sugar levels (16)

**integument** covers the external surface of the body; the skin (5)

**integumentary system** the skin and its so-called appendages, which include hair, nails, and cutaneous and sebaceous glands (4)

**interneuron** a neuron that transmits impulses from one part of the CNS to another; also called an association neuron (13)

**interphase** the first phase of the cell cycle, in which chromatin is replicated, other macromolecules are synthesized, and the cell increases in mass (3)

**interstitial fluid** aqueous, ion-containing fluid surrounding cells of tissues; also called tissue fluid (4)

**interstitial growth** type of cartilage growth in which chondrocytes proliferate within the cartilage from pre-existing ones; growth from within the cartilage (4)

**intervertebral disc** fibrous disc of cartilage between the bodies of each vertebra; absorbs vertical shock and cushions vertebrae during movement (6)

**intervertebral foramen** (pl. *foramina*) canal between adjacent vertebrae through which a spinal nerve runs (13)

**intracartilaginous ossification** process of prenatal bone development in which cartilage is transformed into bone; forms most bones of the body (6)

**intracellular fluid** fluid within a cell (2)

**intramembranous ossification** the transformation of pre-existing fibrous membrane into bone; forms the vault of the skull and some facial bones during prenatal development (6)

**ion** a charged atom or group of atoms formed when molecules are ionized (2)

**ionic balance** concentration of different ions in body fluid; important parameter of body fluids that must be maintained at constant levels to avoid cell damage or death (2)

**ionic bond** electrostatic charge resulting from ionic bonding (2)

**ionic bonding** type of chemical bond in which electrons are actually transferred from one atom to another, creating a positive and negative charge that attracts the two atoms to one another; creates an ionic bond (2)

**ionize** the process by which molecules held together by ionic bonds dissociate when dissolved in water, releasing ions (2)

**iris** the colored part of the eye that acts as a diaphragm for the lens; located in the anterior portion of vascular layer, behind the cornea and in front of the lens (15)

**irregular dense connective tissue** a type of dense fibrous connective tissue made of interwoven collagenous fibers that is able to resist moderately strong stretching in multiple directions; forms the deep fascia (4)

**irritability** the ability of all living organisms to detect and react to events that occur within the organism and its environment; tends to maximize survival (1)

**islets of Langerhans** clusters of endocrine cells scattered throughout the exocrine tissue of the pancreas (11)

**isometric contraction** muscle tension that is not accompanied by muscle shortening; occurs when an individual attempts to life an object beyond his or her capacity (7)

**iso-osmotic solution** an aqueous solution with a solute concentration equal to that in the cytoplasm of a given cell (3)

**isotonic contraction** muscle contraction accompanied by muscle shortening; occurs whenever work is done (7)

**jaundice** a yellow coloration of the skin and whites of the eye; caused by impairment of the liver's ability to eliminate bilirubin from the body (11)

**joint** the connection of two bones; also called an articulation (6)

**keratin** fibrous protein that forms a layer of tough scales over the epidermis (5)

**keratinizing epithelium** the stratified squamous cells forming the skin's epidermis; synthesizes keratin (5)

**kidney** in the urinary system, one of two organs that filter the blood and produce urine (12)

**killer T-cells** T-cells that, upon combination with an appropriate antigen, release lymphokines and interferon; primarily responsible for the destruction of fungi, parasites, cells infected with intracellular microbes, and organ transplants (21)

**Krebs cycle** the second of two metabolic pathways involved in oxidizing glucose; named for biochemist Hans Krebs (2)

**labium majus** (pl. *labia majora*) in the vulva, one of two outer skin folds that are covered with hair on their outside surface and are smooth on their inside surface (17)

**labium minus** (pl. *labia minora*) in the vulva, one of a set of smaller skin folds that are internal to the labia majora (17)

**lactation** the process by which milk is supplied to nourish a baby (18)

**lacuna** (pl. *lacunae*) in the bones, small chamber of space entombing each osteocyte (6)

**lag phase** in the basic growth pattern of all populations , the initial period of slow growth (25)

**lamella** (pl. *lamellae*) in the Haversian systems of compact bone, one of many concentric rings of bone tissue formed around the Haversian canal (6)

**large intestine** section of the digestive tube extending from the small intestine to the anus (11)

**larynx** cartilaginous structure connecting the pharynx and the trachea (10)

**lateral fissure** in the cerebrum, fissure separating the frontal lobe and part of the parietal lobe from the temporal lobe; also called the fissure of Sylvius (14)

**lateral ganglion** type of ganglion that forms a chain of 22 ganglia that lie on either side of the vertebral column from the cervical to the sacral region; also called a sympathetic or vertebral ganglion (13)

**leukemia** cancer that originates from blood cell-forming tissue in the bone marrow and lymphatic system; represents about 4 percent of all cancers (22)

**leukocyte** white blood cell; protects the body against disease and against injury by foreign substances (4, 8)

**life cycle** sequential series of events that occur during the life of an organism (1)

**ligament** dense fibrous connective tissue that holds either bones or bones and cartilage together (4)

**linguil tonsil** nodules of lymphatic tissue beneath the mucous membrane covering the root of the tongue (11)

**lipid** diverse class of chemical substances characterized by their high solubility in nonpolar organic solvents and their low solubility in water (2)

**lipoprotein** protein containing lipid components; some lipoproteins are carriers of cholesterol (2)

**liver** the body's largest gland, located in the upper part of the abdominal cavity just beneath the diaphragm; performs many important metabolic functions (11)

**locus** a specific region on a chromosome; the site of a gene (19)

**loop of Henle** the part of the renal tubule between the proximal convoluted tubule and the distal convoluted tubule (12)

**luteal phase** second of two ovarian phases in the menstrual cycle, during which the corpus luteum functions as an endocrine gland and then begins to degenerate; lasts for the 14 days following ovulation (17)

**luteinizing hormone (LH)** hormone that regulates reproductive processes in both females and males; produced by the anterior lobe of the pituitary gland (16)

**lymph** tissue fluid made up of water and dissolved substances; drains into the lymphatic capillaries and larger lymphatic vessels (8, 9)

**lymphocyte** class of white blood cells involved in identifying foreign substances or abnormal body cell types (21)

**lymphoma** solid, malignant tumors that arise in the lymphatic tissues, usually in the lymph nodes; represents about 5 percent of all cancers (22)

**lysosome** in a cell, membrane-bound sac or vesicle that contains enzymes capable of degrading proteins, nucleic acids, lipids, and polysaccharides (3)

**macromolecule** within a cell, a large organic molecule; the four types of macromolecules are carbohydrates, proteins, nucleic acids, and lipids (1)

**macrophage** large phagocytic cell that is specialized for engulfing microbes and cellular debris; formed from a monocyte (4, 21)

**malignant tumor** tumor that grows in an unregulated fashion, infiltrating and destroying normal tissues and organs; also called a cancerous tumor or cancer (22)

**malleus** inside the middle ear, the first of three tiny bones that conveys vibrations of the eardrum to the oval window (15)

**mandible** the lower jaw (6)

**marsupials** class of mammals that give birth to poorly developed young who complete their development in an external pouch (24)

**mast cell** basophil that leaves the circulation and resides in the connective tissue; releases the inflammatory substance histamine in response to tissue injury (21)

**median life span** the age at which 50 percent of the individuals in a population are still alive (23)

**medulla** 3-cm long section of the hindbrain that functions as a conduction pathway between the other parts of the brain and the spinal cord and as the site of reflex centers controlling vital functions; also called the medulla oblongata (14)

**medulla oblongata** 3-cm long section of the hindbrain that functions as a conduction pathway between the other parts of the brain and the spinal cord and as the site of reflex centers controlling vital functions; also called the medulla (14)

**medullary body** in the cerebrum, the white matter lying beneath the cerebral cortex (14)

**meiosis** cell division that gives rise to germ cells (17)

**melanocyte** cell in the deeper layers of the epidermis that produces the pigment melanin, responsible for skin color (5)

**melanocyte stimulating hormone (MSH)** function is unclear but may be related to the skin darkening that results from exposure to ultraviolet light; produced by the intermediate lobe of the pituitary gland (16)

**memory consolidation** the transfer of information from short-term to long-term memory (14)

**menarche** the beginning of menstrual periods; occurs between the ages of 11 and 14 (17)

**Mendelian genetics** principles of genetics formulated by Gregor Mendel in the nineteenth century (19)

**menopause** the cessation of menstrual cycles; occurs between the ages of 40 and 55 (17)

**menstrual cycle** the cyclic changes in the ovary and the coordinated changes in the state of the uterus that occur in the reproductively mature female (17)

**menstruation** menstrual bleeding that occurs during the menstrual cycle (17)

**mesoderm** during the period of germ layer formation, the third primary germ layer, formed when ectodermal cells migrate toward and into the primitive streak and then spread sideways between the ectodermal and endodermal layers (18)

**messenger RNA (mRNA)** RNA molecule involved in transcription during DNA replication (20)

**metabolic intermediate compound** compound formed in metabolic pathways as a result of the multistep conversion of one substance to another (2)

**metabolic pathway** lengthy sequence of chemical reactions that occurs as a result of metabolism (1)

**metabolism** all the chemical processes by which cellular material is produced, maintained, and broken down, and by which energy is made available for cell activities (1)

**metastasis** the process by which malignant tumors become disseminated in the body (22)

**microfilament** in a human cell, a bundle of thin, rod-like filaments often found beneath the plasma membrane; important in cell movement (3)

**microtubule** in a cell, cylindrical structure made up of an array of small filaments that determines cell shape by providing internal support; forms spindle fibers during cell division, forms centrioles in a nondividing cell, and provides internal support for cilia and flagella (3)

**microvillus** (pl. *microvilli*) fingerlike projection formed by the folding of the epithelial tissue lining the small intestine (4)

**midbrain** superior portion of the brain stem lying above the pons and below the thalamus (14)

**mineralocorticoids** class of steroid hormones produced by the adrenal cortex; important in the maintenance of mineral homeostasis (16)

**mitochondrion** (pl. *mitochondria*) cell structure that generates most of the cell's ATP by the aerobic oxidation of nutrients; usually rod-shaped (3)

**mitosis** the second phase of the cell cycle, in which the replicated chromatin strands are separated and enclosed in two daughter nuclei (3)

**mitral valve** heart valve that separates the left atrium from the left ventricle; also called the bicuspid valve (9)

**molecule** a stable combination of atoms (2)

**monoclonal antibody** single antibody type produced by a hybridoma (21)

**monocyte** phagocytic leukocyte that, once in the tissues, increases in size to form a macrophage (21)

**monosomy** chromosomal abnormality in which an individual has only one chromosome of one type and a pair of all other types (19)

**monotremes** class of mammals that lay eggs rather than bear live young (24)

**mons pubis** in the vulva, rounded bulge in front of the pubic symphysis; becomes covered with hair at puberty (17)

**morula** before implantation of the fertilized egg, the solid clump of cells formed by cleavages (18)

**motor cortex** area of the frontal lobe just anterior to the central fissure in the cerebral cortex; also called the primary motor area (14)

**motor neuron** neuron that carries impulses from the CNS to muscles and glands; also called an efferent neuron (7, 13)

**motor unit** a single neuron and all the muscle cells it supplies (7)

**mucosa** mucous membrane lining the digestive tube (11)

**mucous membrane** mucus-producing membrane lining all the body's internal passageways that open to the outside (4)

**multicellular** organism formed of many cells (1)

**multifactor inheritance** characteristic produced by the interaction of a number of genes; also called polygenic inheritance (19)

**muscle** tissue composed of specialized, contractile cells that move or change the shape of some body part (4)

**muscle tone** the state of partial contraction found in healthy muscles at rest; also called tonus (7)

**muscularis externa** two layers of smooth muscle tissue that generate peristaltic movements that mix and transport the contents of the digestive tube (11)

**muscularis mucosae** the outermost layer of the mucosa; consists of two layers of smooth muscle cells (11)

**muscular system** the skeletal muscles that attach to movable parts of the skeletal system (4)

**mutagen** an agent that alters the hereditary material; causes mutations (1, 20)

**mutation** alteration in nucleotide sequence of DNA (20)

**myelin** thick lipid material that sheaths myelinated neurons (13)

**myocardial infarction** condition in which an area of the myocardium dies when its blood supply is blocked; also called a heart attack (9)

**myocardium** the middle layer of the heart wall; contains cardiac muscle (9)

**myosin** in a muscle, rod-shaped protein molecule that makes up the thick filaments within a sarcomere (7)

**nail** plate of hard keratin covering the dorsal surface of the end of the fingers or toes (5)

**nail bed** epidermal cells on which a nail rests (5)

**negative feedback loop** in the regulation of hormone secretion, feedback loop in which an increase in the output of a target organ causes a reduction in the synthesis and release of a stimulatory hormone (16)

**neoplasm** any abnormal growth of tissue that lacks functional coordination with the rest of the body; also called a tumor (22)

**nephron** microscopic structural and functional unit forming the bulk of kidney tissue; serves as a miniature filtering device (12)

**nerve cell** cell that transmits electrical impulses; also called a neuron (13)

**nervous system** the brain, spinal cord, peripheral nerves, and special sense organs (4)

**nervous tissue** tissue made up of neurons (4)

**neural fold** slightly thickened edge of the neural plate (18)

**neural plate** during the embryonic period, shallow plate formed when the notochord induces the overlying ectoderm to begin differentiating into embryonic nervous tissue (18)

**neural tube** during the embryonic period, hollow tube formed when the neural folds grow together and fuse; eventually becomes the brain and the spinal cord (18)

**neuritis** inflammation of a nerve (13)

**neuroeffector junction** the juncture between efferent neurons and an effector organ (13)

**neuroepithelium** epithelium composed of nerve cells that are specialized for the reception of stimuli; present in the eye, ear, nose, and tongue (4)

**neuromuscular junction** the synapse between a neuron and a muscle cell (7)

**neuron** cell that transmits electrical impulses; also called a nerve cell (4, 13)

**neurosecretory cell** neuron specialized for the production of hormones; the cell bodies lie in the hypothalamus and the axons run down the neural stalk to the neural lobe of the pituitary gland (16)

**neurotransmission** the transmission of electrical impulses between neurons at synapses (13)

**neutron** uncharged particle in nucleus of an atom (2)

**neutrophil** finely granular phagocytic leukocyte of the blood (21)

**node of Ranvier** short, unmyelinated gap between the sheath cells that cover myelinated neurons (13)

**nonpolar molecule** molecule without positively or negatively charged poles (2)

**nonspecific defense mechanisms** the body's external and internal defense mechanisms that are not specific to any one type of microbe (21)

**norepinephrine** hormone secreted by the chromaffin cells of the adrenal medulla in response to short-term stress; also synthesized by some neurons for use as a neurotransmitter (16)

**notochord** during the embryonic period, rod formed when cells within the mesodermal layer start to aggregate along the midline of the embryonic disc; later contributes to formation of the vetebrae (18)

**nuclear envelope** two concentric membranes surrounding a cell nucleus (3)

**nuclear pore** opening in the nuclear envelope through which various substances enter and leave the nucleus (3)

**nucleolus** in a nondividing cell, a dense region of the nucleus involved in the manufacture of ribosomes (3)

**nucleus** in an atom, a dense, central mass made up of protons and neutrons (2)

**olfactory bulb** one of two masses of gray matter in which olfactory nerve fibers synapse with neurons of the olfactory tract; located underneath the frontal lobes of the cerebrum (15)

**olfactory nerve** bundle of nerve fibers formed by axons of the olfactory receptors (15)

**olfactory organ** one of two specialized regions of mucous membrane lining roof of two nasal cavities; contains olfactory receptors responsible for sense of smell (15)

**olfactory tract** one of two fiber tracts that convey impulses from the olfactory bulbs to the primary olfactory areas of the cerebral cortex (15)

**oligodendrocyte** type of glial cell that forms a cellular sheath around the unmyelinated nerve fibers in the central nervous system (13)

**omnivore** animal that occupies more than one trophic level in a food chain; animal that is both a primary and secondary consumer (25)

**oncogene** single gene that encodes a protein that is responsible for transforming normal cells into cancerous cells (22)

**oncogenic DNA virus** oncogenic virus in which the nucleic acid is DNA (22)

**oncogenic RNA virus** oncogenic virus in which the nucleic acid is RNA (22)

**oncogenic virus** type of virus that transforms host cells into tumorous ones (22)

**oogonium** (pl. *oogonia*) female diploid germ cell formed during the first three months of embryological development; gives rise to ovum (17)

**open system** a system that regularly exchanges energy and materials with its surroundings (1)

**optic chiasma** juncture of the two optic nerves at which some fibers from each optic nerve pass over to the other side of the brain (15)

**optic disc** at the back of the eye, the place where the optic nerve leaves the eye and blood vessels enter the eye; the "blind spot" of the retina (15)

**optic nerve** nerve formed by the axons of all the ganglion cells in the retina; pierces the external layers of the eye as it leaves the back of the eye (15)

**optic tract** one of two fiber tracts leaving the optic chiasma (15)

**orbit** in the skull, bony socket surrounding the eye (15)

**organ** one or more tissues that are coordinated as a structural and functional unit (4)

**organelle** within a cell, subcellular structure that is specialized for one or more metabolic functions (3)

**organic macromolecules** large organic compounds found in living organisms; the four classes are proteins, nucleic acids, polysaccharides, and lipids (2)

**organic molecule** molecule with a backbone of carbon atoms (1)

**organic waste product** organic compound found in living organisms that cannot be used further by cells and is eliminated as a waste product (2)

**organ of Corti** in the cochlea, sensory structure containing the hair cells that line the cochlear duct (15)

**orgasm** the climax of sexual excitement in males and females; simultaneous with male ejaculation (17)

**osmoreceptor** specialized neuron in the hypothalamus that is sensitive to osmotic changes in the blood; regulates the sensation of thirst (14)

**osmosis** the passive diffusion of water across a membrane in response to a concentration gradient (3)

**osteoblast** special bone-building cell that synthesizes and secretes the organic portion of the intercellular substance (6)

**osteoclast** bone cell that mediates the resorption of intercellular substance (6)

**osteocyte** a bone cell; responsible for maintaining intercellular substance of bone (4)

**oval window** the superior of two membrane-covered openings in the bone forming the common wall between the middle and inner ear; footplate of the stapes is fitted against the oval window (15)

**ovary** in the female reproductive system, one of two glands that produce ova and that synthesize the hormones estrogen and progesterone (17)

**oviduct** one of two ducts through which ova are transported from the ovary to the uterus and where fertilization normally occurs; also known as the Fallopian tube (17)

**ovum** (pl. *ova*) a female's egg cell (17)

**oxidation** in oxidation-reduction reactions, the loss of electrons from a substance termed an electron donor; opposite of reduction (2)

**oxytocin** hormone that stimulates contraction of smooth muscle cells during labor and stimulates the release of milk during nursing (16)

**pacemaker** group of specialized cardiac muscle cells, located in the region where the superior vena cava joins the right atrium, that initiates the impulse for contraction of the heart; also called the sinoatrial (SA) node (9)

**palate** the roof of the mouth; separates the nasal cavity from the mouth cavity (6)

**pancreas** glandular organ in the abdominal cavity that has both exocrine and endocrine functions (11, 16)

**papilla** (pl. *papillae*) small, fingerlike projection formed by the surface of the papillary layer of the dermis (5)

**papillary duct** in the kidney, larger vessel formed by the merging of many smaller collecting tubules (12)

**papillary layer** the thin, outer layer of the dermis (5)

**papillary muscle** in the heart, conical projection of muscle that connects reinforcing fibrous cords in the bicuspid or tricuspid valve to the muscular walls of the ventricle and keeps the bicuspid or tricuspid valve from turning inside out (9)

**parasympathetic division** in the autonomic nervous system, the parasympathetic neurons; sometimes called the craniosacral division (13)

**parathyroid gland** one of two endocrine glands embedded in the posterior side of each lateral lobe of the thyroid gland; secretes parathyroid hormone (PTH) (16)

**parathyroid hormone (PTH)** secretion of the parathyroid glands; the principle agent in regulating blood calcium homeostasis (16)

**parenchyma** specialized cells that carry out the specific function or functions of an organ (4)

**parieto-occipital fissure** in the cerebrum, fissure dividing the parietal lobe from the occipital lobe (14)

**passive diffusion** in a cell, the simplest type of membrane transport, based on the process of diffusion; driven by a concentration gradient (3)

**passive immunity** immunity that results when antibodies are transferred from one individual to another (21)

**pectoral girdle** part of appendicular skeleton that provides framework for attachment of bones of upper extremities to axial skeleton; also called shoulder girdle (6)

**pedicle** one of two short, thick projections that extend dorsally from each side of the vertebra body (6)

**pelvic cavity** lowermost portion of the abdominal cavity; contains the end of the large intestine, the urinary bladder, and some reproductive organs (4)

**pelvic girdle** the pelvis; a sturdy, bony ring that supports the trunk and attaches the lower limb bones to the body (6)

**penis** the male copulatory organ, by means of which sperm are deposited in the female's vagina (17)

**pericardial cavity** central compartment in the thoracic cavity; contains the heart (4)

**perichondrium** the connective tissue envelope surrounding cartilage; composed of an outer fibrous layer and an inner vascularized layer capable of producing additional chondrocytes (4)

**perimysium** the connective tissue surrounding each fasciculus (7)

**perineurium** relatively dense connective tissue sheath surrounding each fascicle (13)

**period of germ layer formation** first of three arbitrary stages in embryonic development; begins at fertilization and encompasses about the next two and one-half weeks (18)

**periosteum** membrane covering all bone surfaces except where bones articulate with each other (6)

**peripheral nerve** a nerve derived from one of the 43 pairs of nerves forming the peripheral nervous system (PNS) (13)

**peripheral nervous system (PNS)** 43 pairs of nerves that branch off either side of the spinal cord (13)

**peripheral resistance** the degree to which the flow of blood is impeded by muscular blood vessels, primarily the arterioles (9)

**peristalsis** the sequential contractions by which smooth muscle moves material down the digestive tract (4)

**peritoneal cavity** the area between the parietal and visceral layers of peritoneum (11)

**peritoneum** formed by the visceral and parietal peritoneum (11)

**pH scale** quantitative scale used to measure acidity or basicity of an aqueous solution (2)

**phagocytosis** process by which a cell ingests large particles (21)

**phantom pain** pain that seems to emanate from a limb or part of a limb that has been amputated (15)

**pharyngeal arch** during the embryonic period, ridge of tissue that later gives rise to the jaw, larynx, portions of the ear, and nearby structures; also called the gill arch (18)

**pharynx** short, tubelike chamber, also called the throat, that lies posterior to the nasal cavity, mouth, and superior part of the larynx; functions as a common passageway for the respiratory and digestive systems (10, 11)

**phenotype** the appearance of an individual in regard to a genetic trait (19)

**phlebitis** inflammation of veins leading to the formation of thrombi that adhere to the walls of veins (9)

**pia mater** innermost of the three meninges protecting the spinal cord (13)

**pineal gland** small, neuroendocrine structure connected to the third ventricle of the brain; secretes melatonin, an antigonadal hormone, and may be important in regulating biological clocks (16)

**pinna** in the outer ear, the flattened flap of tissue that projects from the side of the head; also called the auricle (15)

**pituitary gland** in the brain, an endocrine gland that secretes hormones that affect other endocrine glands as well as affecting other nonendocrine tissues; also called the hypophysis (16)

**pivot joint** joint that permits rotation as one bone turns about another (6)

**placental barrier** the tissue layers of the chorion and the endothelial cells that line the embryonic capillaries within the villi (18)

**placental mammals** class of mammals that give birth to relatively well-developed young that have developed inside the female and received nourishment through a placenta (24)

**plasma** fluid portion of blood containing water, ions, dissolved proteins, and other organic compounds; makes up 55 percent of total blood volume (8)

**plasma cells** B-lymphocytes that enlarge and secrete antibodies in response to the presence of foreign substances; also called B-cells (4, 21)

**plasma membrane** membrane surrounding a cell (3)

**plasmids** small pieces of circular DNA that contain genes that are not essential for normal survival and growth; used in recombinant DNA techniques (20)

**pleural cavity** one of two lateral compartments in the thoracic cavity, each of which contains a lung; the area between the visceral and parietal layers of the pleura (4, 10)

**plicae circulares** circularly or spirally disposed folds of the mucous membrane lining the small intestine (11)

**polar body** during meiosis I and II of female germ cells, the smaller of two daughter cells; its degeneration disposes of unneeded chromosomes with minimal loss of cytoplasm (17)

**polar molecule** a molecule in which one end possesses a slight positive charge and the other end possesses a slight negative charge (2)

**polygenic inheritance** characteristic produced by the interaction of a number of genes; also called multifactor inheritance (19)

**polymer** a molecule composed of many similar or identical subunits (2)

**polysaccharide** a polymer of simple sugars; important energy-storing molecule of living organisms (2)

**pons** bulbous area of the brain above the medulla oblongata (14)

**population** a group of organisms of a single species that do interbreed (25)

**positive feedback loop** in the regulation of hormone secretion, feedback loop in which an increase in the output of a target organ increases the synthesis and release of a stimulatory hormone (16)

**posterior horn** one of two dorsal extensions of the butterfly-shaped region of gray matter in the spinal cord; also called a dorsal horn (13)

**postganglionic fiber** the axon of a postganglionic neuron (13)

**postganglionic neuron** neuron in the autonomic nervous system that carries impulses from a preganglionic neuron to a visceral effector organ; second of two efferent neurons required to connect the CNS with visceral effector organs (13)

**postsynaptic neuron** the neuron that receives an impulse from another neuron at a synapse (13)

**potency** the ability to attain an erection and to sustain it long enough to ejaculate during intercourse (17)

**precarcinogen** chemical that is carcinogenic only when converted to an active form, often by the inadvertent action of liver enzymes normally involved in the detoxification of harmful substances (22)

**predator** animal that eats other animals; a secondary consumer occupying the third trophic level of a food chain (25)

**preganglionic fiber** the axon of a preganglionic neuron (13)

**preganglionic neuron** neuron in the CNS whose axon extends to a ganglion of the autonomic system; first of two efferent neurons required to connect the CNS with visceral effector organs (13)

**pressoreceptor** specialized nerve cell, located in the walls of the aortic arch and of the carotid sinuses, that registers changes in blood pressure; also called baroreceptor (9)

**presynaptic neuron** the neuron that transmits an impulse to another neuron at a synapse (13)

**prevertebral ganglion** in the sympathetic division of the autonomic nervous system, ganglion that lies in a group in front of the vertebral column, near the organ it serves; also called a collateral ganglion (13)

**primary bronchus** (pl. *bronchi*) one of two structures formed by the branching of the trachea (10)

**primary follicle** in an ovary, a primary oocyte and its follicular cells (17)

**primary motor area** area of the frontal lobe just anterior to the central fissure in the cerebral cortex; also called the motor cortex (14)

**primary oocyte** in the female reproductive system, the type of enlarged cell formed from oogonia during the third month of embryological development; undergoes meiosis to form ovum (17)

**primary sensory area** in the cerebral cortex, a band of tissue in the parietal lobe just posterior to the central fissure; also called the sensory cortex (14)

**primary yolk sac** during the period of germ layer formation, cavity formed when endodermal cells migrate from the embryonic disc and line the entire cavity of the blastocyst (18)

**primitive streak** during the period of germ layer formation, groove that develops along the long axis of the embryonic disc (18)

**product** the molecules that result from a chemical reaction; in a chemical equation, the molecules appearing on the right side of the equation (2)

**progesterone** in the female, hormone that brings about functional changes in the uterus; produced by the ovaries and by the placenta during pregnancy (17)

**prokaryote** organism with a simple cell structure; includes all bacteria (21)

**prokaryotic** relatively simple cell structure lacking a distinct nucleus; characteristic of bacteria (1)

**prolactin (PRL)** hormone that acts with other hormones to bring about full development of the mammary glands and the secretion of milk in females; produced in the anterior lobe of the pituitary gland (16)

**pronuclei** the haploid nuclei of the sperm and the egg after the sperm has penetrated the egg (18)

**prophase** the first visible stage of mitosis, in which the chromatin strands coil into chromosomes (3)

**prostacyclins** group of hormones produced by the endothelial cells that form the walls of blood vessels; prevents the disruption of blood flow by inhibiting the aggregation of blood platelets and by dilating arterioles (16)

**prostaglandins** group of hormones involved in the control of smooth muscle contraction and in a wide range of other physiological processes (16)

**prostate** in the male, a gland that produces a thin, whitish fluid containing high concentrations of enzymes that liquefy semen; released in semen upon ejaculation (17)

**proton** positively charged particle within the nucleus of an atom (2)

**proximal convoluted tubule** part of the renal tubule between Bowman's capsule and loop of Henle (12)

**pseudostratified** epithelial tissue that looks like stratified epithelium, even though all cells in the single cell layer are attached to the basement membrane (4)

**pulmonary circulation** a circulatory loop in which deoxygenated blood is pumped from the right ventricle through right and left branches of the pulmonary artery to each lung and then oxygenated blood is pumped back to the left atrium by way of the pulmonary veins (9)

**pulmonary semilunar valve** in the heart, a one-way valve at the base of the pulmonary artery (9)

**pupil** the opening in the iris; regulates the amount of light entering the eye (15)

**reactant** in a chemical reaction, the original molecules that exist before the reaction occurs; in a chemical equation, the molecules appearing on the left side of the equation (2)

**receptor site** structure in the cell's plasma membrane, usually a glycoprotein, to which viruses, hormones, drugs, and metabolic control compounds attach in order to influence the cell; site at which a virus can attach to the surface of a host cell (3, 21)

**recessive allele** in heredity, allele that is masked by a dominant allele (19)

**recognition molecule** glycoprotein or glycolipid molecule that is a component of the plasma membrane and allows a cell to recognize another cell as being of a similar or different type (3)

**rectum** section of the large intestine that empties into the anal canal (11)

**red marrow** the material that fills the cavities of some bones; produces red blood cells, white blood cells, and platelets (6)

**reduction** in oxidation-reduction reactions, the gain of electrons by a substance termed an electron acceptor; opposite of oxidation (2)

**referred pain** pain that is felt at a site that is distant from the site that is being stimulated (15)

**reflex** neural response initiated independently of the brain (13)

**reflex activity** an automatic stimulus-response reaction that does not involve the brain (13)

**reflex arc** in a reflex activity, an anatomical pathway that provides a more or less direct neural connection between receptor and effector organs (13)

**refractory period** the very short period following contraction, during which a muscle will not respond to a stimulus (7)

**renal cortex** outermost zone of the kidney; has a granular appearance in cross-section because it contains the glomeruli and most of the proximal and distal tubules (12)

**renal medulla** innermost zone of the kidney; contains 6 to 18 renal pyramids containing the loops of Henle, collecting tubules, and papillary ducts (12)

**renal tubule** hollow tubular structure within a nephron in a kidney (12)

**reproductive system** the reproductive organs and glands (4)

**respiratory chain** metabolic pathway in which reduced forms of NAD and FAD are oxidized; also called the electron transport system (2)

**respiratory control center** regulates the rate and depth of breathing; one of the reflex centers in the medulla oblongata (14)

**respiratory membrane** the entire membranous structure that separates air in the alveoli from blood in the capillaries (10)

**respiratory system** the lungs and the hollow passages through which air reaches the lungs (4)

**resting tidal volume** the amount of air that enters and leaves the lungs during normal breathing under resting conditions (10)

**reticular fiber** delicate type of fiber containing some collagen and polysaccharide; forms an extensive, intermeshing network of connective tissue (4)

**reticular formation** a complex network of nuclei and nerve fibers extending from the uppermost part of the spinal cord through the brain stem to the lower part of the thalamus; functions to arouse the brain to incoming sensory information (14)

**reticular layer** inner layer of the dermis; formed of a loose connective tissue that contains slightly denser network of collagenous fibers than does the papillary layer (5)

**retina** innermost layer of the eye; made up of an outer pigment layer and an inner neural layer (15)

**rhodopsin** the visual pigment in rods and cones (15)

**rib** in the skeleton of the thorax, one of twelve pairs of flexible arching bones encircling the chest (6)

**ribosome** in a cell, small spherical structure that is manufactured in the nucleolus and transported to the cytoplasm, where it is the site of the synthesis of proteins in the cell (3)

**rod** in the retina of the eye, neuron that functions as a photoreceptor for dim light; provides monochromatic vision (15)

**round window** the inferior of two membrane-covered openings in the bone forming the common wall between the middle and inner ear (15)

**sacral outflow** in the parasympathetic division, the pre-ganglionic fibers that leave the spinal cord in the second, third, and fourth sacral spinal nerves (13)

**salivary gland** one of three pairs of large glands that secrete saliva into the oral cavity (11)

**saprophyte** organism that obtains organic nutrients from dead or decaying organic matter; includes most fungi (21)

**sarcolemma** plasma membrane surrounding the sarco-plasm of a muscle cell (7)

**sarcoma** cancer derived from connective tissue; repre-sents about 2 percent of all cancers (22)

**sarcomere** functional unit of muscle striations that is repeated down the length of the myofibril; contracts the muscle cell (7)

**sarcoplasm** the cytoplasm of a muscle cell (7)

**sarcoplasmic reticulum** in a muscle cell, an extensive system of smooth endoplasmic reticulum surrounding each myofibril (7)

**scavenger cell** specialized cell that removes foreign substances, bacteria, and cellular debris; makes extensive use of endocytosis (3)

**Schwann cell** type of glial cell that forms a cellular sheath around the unmyelinated nerve fibers in peripheral nerves (13)

**sclera** the outer surface or supporting layer of the eye; membrane of dense connective tissue forming the "white" of the eye (15)

**scrotum** external pouch containing the testes (17)

**sebaceous gland** sebum-producing gland whose duct opens into the hair follicle (5)

**sebum** lipid secretion produced by the sebaceous glands (5)

**secondary oocyte** in the female reproductive system, the type of oocyte formed from a primary oocyte after the completion of meiosis I; cell type released at ovulation (17)

**Second Law of Thermodynamics** chemical law that states that the universe tends spontaneously toward a state of increasing entropy (2)

**second messenger mechanism** two-step process whereby some hormones alter the synthesis or ac-tivity of enzymes in cells; the hormone (the first mes-senger) attaches to a specific receptor on the target cell's plasma membrane, activating an enzyme that stim-ulates the formation of cyclic AMP (the second messen-ger), which alters the activities of other enzymes in the target cell (16)

**secretory granule** in a Golgi complex, vesicle that migrates to the plasma membrane and fuses with it, releasing vesicle contents at the cell surface (3)

**semen** fluid containing sperm cells and fluids secreted by the seminal vesicles and the prostate gland (17)

**seminal vesicle** in the male reproductive system, organ composed of convoluted tubes that secrete a viscous, yellowish fluid in which sperm cells are suspended (17)

**seminiferous tubule** within each testis, a highly coiled structure that produces sperm (17)

**sensory cortex** in the cerebral cortex, a band of tissue in the parietal lobe just posterior to the central fissure; also called the primary sensory area (14)

**sensory neuron** neuron that carries impulses from other parts of the body to the central nervous system; also called an afferent neuron (13)

**sensory receptor** nerve ending that is sensitive to a specific type of environmental sensation; five basic types include thermoreceptors, pain receptors, mechanorecep-tors, photoreceptors, and chemoreceptors (15)

**serosa** the serous membrane that forms the external layer of digestive organs that lie in the abdominal cavity below the diaphragm; also called the visceral peritoneum (11)

**serous membrane** membrane that covers various organs and lines body cavities that do not open to the outside; secretes a thin, slippery fluid (4)

**Sertoli cell** in the walls of the seminiferous tubules, a very large cell in whose plasma membrane developing germ cells are embedded; forms the blood-testis barrier (17)

**sex chromosomes** within a somatic cell, the pair of chromosomes that determines sex (19)

**sexual reproduction** reproduction that involves the fusion of germ cells produced by two individuals (1)

**shock** condition resulting from sudden drop in blood pressure, which causes inadequate flow of blood through-out the body and the disruption of vital functions (9)

**shoulder girdle** part of the appendicular skeleton that provides framework for the attachment of the bones of the upper extremities to the axial skeleton; also called the pectoral girdle (6)

**simple carbohydrate** one of the small organic com-pounds that provide energy and raw materials for the manufacture of cellular constituents (2)

**simple epithelium** epithelial tissue composed of a sin-gle cell layer (4)

**sinoatrial (SA) node** group of specialized cardiac mus-cle cells, located in the region where the superior vena cava joins the right atrium, that initiates the impulse for contraction of the heart; also called the pacemaker (9)

**sister chromatid** during prophase, the two identical parts of each chromosome (3)

**skeletal muscle** muscle that is attached to bones, facial skin, and cartilage by tendons and other connective tissue; produces rapid, powerful contractions for short times (4)

**skeletal system** the bones, cartilage, and other dense fibrous connective tissues (4)

**skin** covers the external surface of the body; also called integument (5)

**sliding filament theory** proposes that muscle contrac-tion results when thin filaments slide past the thick filaments toward the center of the sarcomere (7)

**small intestine** convoluted tube extending from the stomach to the large intestine (11)

**smooth muscle** muscle found in the walls of the abdominal organs and blood vessels; produces slow, wavelike contractions not normally under conscious control (4)

**sodium-potassium pump** an active transport system that regulates intracellular concentrations of sodium and potassium ions (3)

**solute** the dissolved substance in a solution (2)

**solution** a homogeneous mixture resulting when a solute, or substance, is dissolved in a liquid (2)

**solvent** a liquid substance capable of dissolving one or more substances; the liquid portion of a solution (2)

**somatic afferent (sensory) neuron** within the somatic nervous system, neuron that carries sensory information from the skin, skeletal muscles, and joints to the CNS (13)

**somatic cell** cell other than a germ cell (2)

**somatic efferent (motor) neuron** within the somatic nervous system, neuron that carries impulses from the CNS to the skeletal muscle to effect voluntary movement (13)

**somatic nervous system** division of the peripheral nervous system that controls voluntary functions (13)

**somatic pain** pain experienced when pain receptors in skin, muscles, tendons, and joints are stimulated (15)

**somatostatin** hormone secreted by the delta cells of the islets of Langerhans; retards the movement of nutrients from the gastrointestinal tract into the bloodstream (16)

**somites** during the embryonic period, paired segments of mesoderm that develop on either side of the notochord; contributes to the formation of the dermis, the vertebrae and ribs, and skeletal muscle (18)

**species** group of organisms that resemble each other and are capable of interbreeding in nature to produce viable and fertile offspring (25)

**specific defense mechanism** one of the body's defense mechanisms against specific harmful agents (21)

**spermatogenesis** the production of sperm (17)

**spermatozoan** (pl. *spermatozoa*) a male's sperm cell (17)

**spinal nerve** one nerve of 31 pairs of peripheral nerves that connect the body below the neck with the spinal cord (13)

**spindle apparatus** formed during prophase to pull the duplicated chromosomes apart (3)

**spindle fiber** involved in chromosome movement during cell division; made up of microtubules (3)

**spontaneous reaction** a chemical reaction that releases useful energy (2)

**squamous cell** epithelial cell with a flat, thin shape (4)

**stapes** inside the middle ear, the third of three tiny bones that conveys vibrations of the eardrum to the oval window (15)

**Starling's law of the heart** observation that, within limits, the more cardiac muscle is stretched, the more forceful is its contraction (9)

**steady state** a balanced flow of energy and materials in an open system (1)

**sternum** in the skeleton of the thorax, the breast bone; forms the front of the rib cage at the midline (6)

**stomach** expanded portion of the digestive tube that serves to store and to partially digest food (11)

**stratified epithelium** epithelial tissue composed of more than one cell layer (4)

**stroke** condition resulting when a thrombus or rupture interrupts blood flow in an artery supplying the brain; characterized by damage to brain tissue and possible paralysis (9)

**stroke volume** the amount of blood pumped by each ventricle during each contraction of the heart (9)

**stroma** connective tissue that supports the parenchymal cells of an organ (4)

**subarachnoid space** the space between the pia mater and the arachnoid; filled with cerebrospinal fluid (13)

**subcutaneous tissue** extensive fibrous membrane system of ordinary loose and adipose connective tissue that attaches the skin to underlying structures; also called the superficial fascia (4)

**submucosa** loose connective tissue that connects the overlying mucous membrane of the digestive tube to the muscularis externa (11)

**superficial fascia** extensive fibrous membrane system of ordinary loose and adipose connective tissue that attaches the skin to underlying structures; also called the subcutaneous tissue (4)

**superior mesenteric ganglion** a type of collateral or prevertebral ganglion (13)

**superior vena cava** major vein that drains blood from the head, arms, and upper parts of the trunk into the right atrium of the heart (9)

**suppressor T-cells** T-cells that inhibit or suppress the response of plasma cells to the presence of an antigen; prevents the overproduction of antibodies (21)

**suture** immovable joint connecting the bones of the skull vault; formed of thin layers of connective tissue (6)

**sympathetic chain** one of two chains of 22 sympathetic ganglia that lie on either side of the vertebral column from the cervical to the sacral region (13)

**sympathetic division** in the autonomic nervous system, the sympathetic neurons (13)

**sympathetic ganglion** type of ganglion that forms a chain of 22 ganglia that lie on either side of the vertebral column from the cervical to the sacral region; also called a lateral or vertebral ganglion (13)

**synapse** the gap between neurons or between neurons and muscle cells over which a nerve impulse passes from one cell to another (7)

**synapsis** the pairing of homologous chromosomes during the prophase I stage of meiosis I (17)

**synaptic cleft** the gap separating the plasma membranes of two neurons or the plasma membranes of a neuron and a muscle cell (7)

**synaptic knob** swelling at the end of an axon collateral; also called a terminal bulb (13)

**synovial fluid** slippery lubricating fluid secreted by the synovial membrane in a joint (6)

**synovial membrane** the inner lining of a joint capsule that is composed of connective tissue; secretes synovial fluid (6)

**system** a group of organs responsible for carrying out one general function (4)

**systemic circulation** the circulatory loop that delivers oxygenated blood from the lungs to all other tissues of the body and then returns deoxygenated blood to the heart (9)

**systole** in the cardiac cycle, contraction of the heart (9)

**systolic blood pressure** pressure generated during ventricular systole (9)

**tendon** dense fibrous connective tissue that joins muscles to bones (4)

**teratogen** an agent that causes birth defects by disrupting normal prenatal development (1, 18)

**terminal bulb** swelling at the end of an axon collateral; also called a synaptic knob (13)

**testis** (pl. *testes*) one of two male reproductive glands; produces sperm and synthesizes and secretes androgens (17)

**testosterone** male sex hormone secreted by the interstitial cells in the seminiferous tubules of the testes; required for the development, maintenance, and normal functioning of male reproductive organs (17)

**tetanic contraction** phenomenon in which a muscle remains contracted when it receives stimuli at intervals just longer than the refractory period; also called tetanus (7)

**tetanus** phenomenon in which a muscle remains contracted when it receives stimuli at intervals just longer than the refractory period; also called tetanic contraction (7)

**tetrad** the four-stranded structure created during synapsis (17)

**thalamus** a part of the forebrain that receives and relays sensory information to the cerebral cortex and is responsible for a crude type of sensory awareness (14)

**thoracolumbar division** the sympathetic division of the autonomic nervous system (13)

**thromboxane** hormone produced by blood platelets in response to chemical stimuli that are known to trigger blood clotting (16)

**thrombus** (pl. *thrombi*) a blood clot (8)

**thymosins** group of related hormones produced by the thymus gland; acts to promote the functional maturation of T-lymphocytes (16)

**thymus gland** a lymphatic organ with endocrine functions; located beneath the sternum, anterior to the trachea (16)

**thyroid gland** endocrine gland made up of two lobes of tissue that lie on either side of the upper trachea; secretes thyroid hormones (16)

**thyroid stimulating hormone (TSH)** hormone that stimulates the synthesis and secretion of thyroid hormones from the thyroid gland; produced in the anterior lobe of the pituitary gland (16)

**thyrotropin releasing factor (TRF)** hormone that stimulates the secretion of thyroid stimulating hormone (TSH); secreted from the hypothalamus (16)

**thyroxin (T$_4$)** one of two main thyroid hormones; affects growth, development, and metabolism (16)

**tight junction** way that adjoining plasma membranes interact to prevent water and other substances from leaking between cells; characteristic of epithelial tissues (3)

**tissue** organized group of cells with a specific structure and function (4)

**tissue fluid** aqueous, ion-containing fluid surrounding cells of tissues; also called interstitial fluid (4)

**T-lymphocytes** lymphocytes that are required for cell-mediated immunity (21)

**tonus** the state of partial contraction found in healthy muscles at rest; also called muscle tone (7)

**trabecula** (pl. *trabeculae*) in cancellous bone, the interconnecting struts of bone that form a latticework structure (6)

**trace element** element that is essential for life but found in only minute quantities within living organisms (2)

**trachea** the windpipe; located in the midline of the neck and extending into the thorax (10)

**transcription** during DNA replication, the first stage of protein synthesis, in which the sequence of nucleotides in a polynucleotide strand of DNA is transferred to messenger RNA (20)

**transfer RNA (tRNA)** during translation, class of RNA molecules that transports amino acids to the ribosome for assembly into proteins (20)

**transformation** with regard to cancer, the process by which a normal cell becomes a cancer cell (22)

**transitional** epithelial cell whose shape may vary when in a stretched or relaxed state (4)

**translation** in DNA replication, the actual manufacture of proteins during the second stage of protein synthesis; occurs on ribosomes in the cytoplasm (20)

**translocation** chromosomal abnormality resulting when a section of a chromosome is transferred to another region on the same chromosome or to an entirely different chromosome during meiosis (19)

**transverse tubule** in a muscle cell, tubular structure formed by the infolding of the sarcolemma (7)

**tricuspid valve** heart valve that separates the right atrium from the right ventricle (9)

**triiodothyronine (T$_3$)** one of two main thyroid hormones; affects growth, development, and metabolism (16)

**trisomy** chromosomal abnormality in which an individual has three chromosomes of one type and a pair of all other types (19)

**trophic levels** feeding levels within a food chain (25)

**trophoblast** the outer layer of cells forming the blastocyst; involved in the formation of the placenta (18)

**tropomyosin** in the thin filament of the sarcomeres, a protein molecule that runs in the groove between the two F-actin strands (7)

**troponin** in the thin filament of the sarcomeres, a protein that occurs at regular intervals along the strand in association with both the actin and tropomyosin molecules (7)

**tubular secretion** the transport of substances from the tissue fluid surrounding the renal tubules to the lumen of the tubules (12)

**tumor** any abnormal growth of tissue that lacks functional coordination with the rest of the body; also called a neoplasm (22)

**tumor promoter** substance that is not a carcinogen itself but promotes the activity of carcinogens (22)

**tympanic membrane** membrane separating the outer ear from the middle ear; also called the eardrum (15)

**ulcer** erosion of the mucosal surface of the stomach or duodenum (11)

**unicellular organism** organism composed of a single cell (1)

**ureter** tube that carries urine from the pelvis of each kidney to the urinary bladder (12)

**urethra** in females, a short tube used to excrete urine; in males, a common passageway for both urine and semen (12)

**urinary bladder** thick muscular sac lying within the pelvic cavity; stores urine prior to elimination (12)

**urinary system** the kidneys, bladder, and tubes involved in the evacuation of urine from the body (4)

**uterus** in the female reproductive system, hollow, muscular organ in which the fertilized egg is nourished and protected during embryonic development (17)

**uvula** small flap of tissue hanging down at the posterior margin of the soft palate (11)

**vaccination** active immunity produced when dead or weakened pathogenic microbes are injected or given orally; also called immunization (21)

**vacuole** in a cell, a membrane-bound sac used to store such substances as waste products, secretion products, and energy reserve compounds (3)

**vagus nerve** in the cranial outflow, cranial nerve X, which sends parasympathetic fibers to nearly all thoracic and abdominal organs (13)

**valence shell** in an atom, atomic orbital or group of orbitals that contains electrons with similar energy levels (2)

**vascular layer** the eye's middle layer, which contains blood vessels and pigmentation; consists of the choroid, ciliary body, and iris (15)

**vas deferens** in the male reproductive system, the section of the duct system between the epididymis and the ejaculatory duct (17)

**vasomotor control center** affects blood pressure by regulating the diameter of arterioles; one of the reflex centers in the medulla oblongata (14)

**vasopressin** hormone that stimulates water resorption in the kidneys; synthesized by neurosecretory cells in the hypothalamus and released by the neural lobe of the pituitary gland; also called antidiuretic hormone (ADH) (16)

**vein** blood vessel that carries blood from the venules to the heart (8)

**ventral cavity** body cavity made up of the thoracic, or chest, cavity, and the abdominal cavity (4)

**ventral horn** one of two ventral extensions of the butterfly-shaped region of gray matter in the spinal cord; also called an anterior horn (13)

**ventral root** branch of the spinal nerve that enters the spinal cord at the ventral aspect of the body (13)

**ventricle** in the heart, one of two thick-walled chambers that receive blood from the atria (9)

**venule** blood vessel that collects blood from the capillaries and drains into a vein (8)

**vertebra** (pl. *vertebrae*) one of the segments making up the vertebral column (6)

**vertebral column** the backbone; extends from the base of the skull down the dorsal midline of the body (6)

**vertebral foramen** the large opening formed by the body, pedicles, and laminae of the vertebra; contains the spinal cord (6)

**vertebral ganglion** type of ganglion that forms a chain of 22 ganglia that lie on either side of the vertebral column from the cervical to the sacral region; also called a sympathetic or lateral ganglion (13)

**villus** (pl. *villi*) tongue-or-finger-shaped projection of the mucous membrane lining the small intestine (11)

**viscera** the abdominal organs (4)

**visceral afferent (sensory) neuron** within the autonomic nervous system, neuron that conveys sensory information from visceral organs to the CNS (13)

**visceral efferent (motor) neuron** within the autonomic nervous system, neuron that carries impulses from the CNS to smooth muscle, cardiac muscle, and glands (13)

**visceral pain** pain experienced when pain receptors in the visceral organs are stimulated (15)

**vulva** the external genitalia of the female (17)

**white ramus communicans** in the sympathetic division of autonomic nervous system, the short nerve trunk by which efferent fibers leave spinal nerve and synapse with postganglionic fibers in one of two types of ganglia (13)

**X chromosome** one of two sex chromosomes; a pair of X chromosomes indicates a female (19)

**Y chromosome** a sex chromosome that is smaller than an X chromosome; males have one X chromosome and one Y chromosome (19)

**yellow marrow** red marrow that has been invaded by fat cells; functions as a fat reserve (6)

**zona pellucida** during oogenesis, thick envelope that surrounds the developing primary oocyte; secreted by both the oocyte and the follicular cells (17)

**zygote** the single diploid cell produced when the haploid nuclei of the sperm and the egg fuse (18)

# ACKNOWLEDGMENTS

## Photo Credits

Unless otherwise acknowledged, all photos are the property of Scott, Foresman and Company.
Abbreviations: (l) left, (c) center, (r) right, (t) top, (b) bottom

Cover: Kathleen Cunningham

**Chapter 1**
**3:** Dr. Florence Haseltine/ National Institutes of Health
**4:** Dr. R. C. Brinster, Lab. of Reproductive Physiology, School of Veterinary Medicine, University of Pennsylvania

**Chapter 2**
**21:** Dan McCoy/Rainbow
**33(l):** Johnny Johnson/DRK Photo
**33(c):** ANIMALS ANIMALS/Leonard Rue, Jr.
**33(r):** Mark N. Boulton/Photo Researchers
**35:** Jack Dermid/Photo Researchers

**Chapter 3**
**47(t):** Dr. Robert Slocum/Williams College, Williamstown, MA
**47(b):** Dr. Don Fawcett/Photo Researchers
**49(l):** Courtesy Dr. Robert Slocum, Williams College, Williamstown, MA.
**50:** Dr. John Cummings, Dept. of Anatomy, NY State College of Veterinary Medicine, Cornell University
**51:** Dr. Robert Slocum/Williams College, Williamstown, MA
**52(t):** Dr. Robert Slocum/Williams College, Williamstown, MA
**54:** Dr. William DeWitt/Williams College, Williamstown, MA
**56(l,r):** UPI/Bettmann Newsphotos
**60:** H. E. Buhse, Jr., and R. C. Holsen, Department of Biological Sciences, University of Illinois, Chicago

**Chapter 4**
**69:** From *TISSUES AND ORGANS: A text-atlas of scanning electron microscopy,* by Richard G. Kessel and Randy H. Kardon, W. H. Freeman & Co., copyright © 1979.
**71(t):** Dr. William DeWitt/Williams College, Williamstown, MA
**71(b):** Dr. William DeWitt/Williams College, Williamstown, MA
**72:** Dr. William DeWitt/Williams College, Williamstown, MA
**73:** Dr. William DeWitt/Williams College, Williamstown, MA
**74:** Courtesy of the Arthritis Foundation
**75:** Library of Congress

**Chapter 5**
**84:** Photo Researchers
**85(t):** U.S. Dept. of Justice/Courtesy FBI
**85(b):** Dr. William DeWitt/Williams College, Williamstown, MA
**88:** The Bettmann Archive
**91:** Dr. Burke/Photo Courtesy Lutheran General Hospital, Park Ridge, IL
**92:** Courtesy Mayo Clinic

**Chapter 6**
**96(t):** Lennart Nilsson/*BEHOLD MAN,* Little Brown & Co.
**99(c):** David Baylink.
**99(l):** David Baylink.
**101:** Photo Researchers
**112:** Courtesy Dr. Robert Meli, Naples Diagnostic Imaging Center, Ltd., Naples, FL
**113(tr):** Courtesy Dr. Louis J. Beuton, Williamstown Medical Associates
**113(bl):** Lester V. Bergman & Associates
**116:** Courtesy Dr. Robert J. Meli, Naples Diagnostic Imaging Center, Ltd., Naples, FL

**Chapter 7**
**122:** Courtesy Dr. Hugh Huxley, Medical Research Council, Cambridge, England
**124:** Wide World
**126:** Courtesy Dr. Junzo Desaki
**128:** Lennart Nilsson/*BEHOLD MAN,* Little Brown & Co.
**133:** Courtesy Dr. William DeWitt, Williams College, Williamstown, MA

**Chapter 8**
**142:** Courtesy Dr. William DeWitt, Williams College, Williamstown, MA
**144:** Courtesy Dr. Leland Clark, Children's Hospital Medical Center
**145:** Martin M. Rotker/Taurus Photos, Inc.
**147(b):** Martin M. Rotker/Taurus Photos, Inc.
**148:** Courtesy Dr. William DeWitt, Williams College, Williamstown, MA

**Chapter 9**
**152(b):** Courtesy Dr. Wallace A. McAlpine, M.D./Springer-Verlag F.R.C.S., F.R.C.S.E., from *HEART AND CORONARY ARTERIES,* Berlin, Heidelberg, New York: 1975.
**162:** Courtesy Dr. Steven Mason, Cardiology Associates Temple Medical Building, New Haven, CT
**166:** From *TISSUES AND ORGANS: A text-atlas of scanning electron microscopy,* by Richard G. Kessel and Randy H. Kardon, W. H. Freeman & Co., copyright © 1979.
**167:** Gordon T. Hewlett

**Chapter 10**
**182:** Courtesy American Lung Association
**184(b.):** Gordon T. Hewlett
**185:** Gordon T. Hewlett
**191:** Courtesy Artificial Organs Division, Deerfield, IL, Travenol Laboratories, Inc.

**Chapter 11**
**206:** From *TISSUES AND ORGANS: A text-atlas of scanning electron microscopy,* by Richard G. Kessel and Randy H. Kardon. W. H. Freeman and Company, copyright © 1979.
**213:** Sheldon, H. (1984) *Boyd's Introduction to the Study of Disease,* Lea & Febiger.
**216:** Sheldon, H. (1984). *Boyd's Introduction to the Study of Disease,* Lea & Febiger.
**218:** Dr. F. Vidal-Vanaclocha & Dr. E. Barbera-Guillem, Dept. of Histology & Cell Biology, Univ. of Basque County
**225:** Susan Rosenberg/Photo Researchers

**Chapter 12**
**234:** Lennart Nilsson/*BEHOLD MAN,* Little, Brown & Co.
**243:** Courtesy Dr. William DeWitt, Williams College, Williamstown, MA
**245:** Dan McCoy/Rainbow

**Chapter 13**
**252(tl):** Manfred Kage/Peter Arnold, Inc.

**252(tr):** Manfred Kage/Peter Arnold, Inc.

**255:** Courtesy National Multiple Sclerosis Society

**259(l):** Charles Marden Fitch/Taurus Photos, Inc.

**259(r):** The Bettmann Archive

**Chapter 14**

**286:** Lester V. Bergman & Associates

**294:** Lab of Neuroscience, NIA, Courtesy Science

**295:** Courtesy Dr. Peter J. Whitehouse, John Hopkins University

**297:** Richard Wood/Taurus Photos, Inc.

**298:** Lester V. Bergman & Associates

**Chapter 15**

**306:** The Bettmann Archive

**307:** Lew Merrim/Monkmeyer Press Photo Service

**312:** Gift of Mrs. Nickolas Muray/Museum of Modern Art

**313:** Courtesy Emanuel Rose, MD, Manchester, Royal Eye Hospital

**Chapter 17**

**362:** Courtesy Dr. David M. Phillips, The Population Council

**367:** Courtesy Dr. I. Samberg, Dept. of Obstetrics and Gynecology, Rothchild University Hospital

**376:** Centers for Disease Control, Atlanta

**Chapter 18**

**384(t):** Hank Morgan © 1987

**384(b):** Hank Morgan © 1987

**388:** Photo Researchers

**394:** John Watney/Photo Researchers

**396:** Courtesy James W. Hanson

**398:** Dr. Jason C. Birnholz, Rush-Presbyterian St. Luke Medical Center, Chicago, IL

**Chapter 19**

**411:** Courtesy Dr. Henry L. Nadler, Genetics Dept., Children's Memorial Hospital, Chicago, IL

**414(tl):** The Bettmann Archive

**414(tr):** The Bettmann Archive

**414(bl):** The Bettmann Archive

**414(br):** The Bettmann Archive

**416:** Courtesy Nancy Mueller

**418(l):** The Granger Collection, New York

**418(r):** The Granger Collection, New York

**422:** Courtesy Dr. James L. German, III

**426(all):** Courtesy E. J. Egozcue

**Chapter 20**

**450(l):** Dan McCoy/Rainbow

**450(tr):** Dan McCoy/Rainbow

**450(br):** Dan McCoy/Rainbow

**Chapter 21**

**453:** National Library of Medicine, Bethesda, MD

**455:** Courtesy of S. Lamberg. Reprinted from J. W. Rippon, *MEDICAL MYCOLOGY: THE PATHOGENIC FUNGI AND THE PATHOGENIC ACTINOMYCETES.* W. B. Saunders Co., 1974.

**456:** Courtesy of Ronald Hare and reprinted with permission from *CHEMISTRY* (now *SciQuest*), Vol. 51, no. 7, 1978, by the American Chemical Society.

**468(all):** Lennart Nilsson/*BEHOLD MAN*, Little, Brown & Co.

**470:** Courtesy Dr. John Cummings, Dept. of Anatomy, NY College of Veterinary Medicine, Cornell University

**473(all):** Courtesy Dr. A. Liepins, Memorial University, St. Johns, Newfoundland

**478:** Lennart Nilsson/Boehringer Ingelheim Zentrale GmbH

**488:** From *TISSUES AND ORGANS: A text-atlas of scanning electron microscopy,* by Richard G. Kessel and Randy H. Kardon. W. H. Freeman and Company, copyright © 1979.

**Chapter 22**

**491:** Courtesy Dr. M. B. Shimkin/University of California at San Diego School of Medicine

**493:** Biology Media/Photo Researchers

**504(l):** Lester V. Bergman & Associates

**504(r):** Lester V. Bergman & Associates

**508:** Courtesy M. Popovic, P. S. Sarin, and M. Robert Gurroff, Laboratory of Tumor Cell Biology, National Cancer Inst., Bethesda, MD

**Chapter 23**

**516(all):** Nathan Benn/Woodfin Camp & Associates

**518:** Courtesy Dr. Jerome Gross, Developmental Biology Laboratory, Mass. Gen. Hospital, Boston, MA

**523:** Photo Researchers

**Chapter 24**

**530:** Courtesy Darwin Museum/Down House and The Royal College of Surgeons of England

**536:** Courtesy Elso S. Barghorn and J. William Schopf, Dept. of Biology and Botanical Museum, Harvard University

**538:** A. W. Ambler/Photo Researchers

**Chapter 25**

**555:** Photo Researchers

**557:** George Laycock/Photo Researchers

**558:** Dave Bellak/Jeroboam, Inc.

**559:** U. S. Forest Service

**560:** John Roche/Photo Researchers

**562:** Dudley Foster/Woods Hole Oceanographic Institution

**567:** Field Museum of Natural History, Chicago

**Color photo essay THE HUMAN HEART**

**1:** Lennart Nilsson, *Behold Man,* Little Brown & Co.

**4:** Lennart Nilsson, *Behold Man,* Little Brown & Co.

**Color photo essay THE HUMAN EYE:**

**1:** Lennart Nilsson, *Behold Man,* Little, Brown & Co.

**2:** Argentum/Photo Researchers

**3(t):** Lennart Nilsson, *Behold Man,* Little, Brown & Co.

**3(bl):** Ralph C. Eagle MD/Photo Researchers

**3(br):** Alexander Tsiaras/Photo Researchers

**4:** © Mickey Pfleger 1987

**Color photo essay HUMAN DEVELOPMENT:**

**1:** Photo Researchers

**2–7:** C. Bevilacqua/CEDRI

**9(t):** Hal Stoelzle/© Mickey Pfleger 1987

**9(b):** Howard Dratch/The Image Works

## Illustration Credits

**Box on p. 6:** From "Synthetic Growth Hormone Raises Hopes of Many—And Ethical Concern Over Use," by Alan L. Otten. THE WALL STREET JOURNAL, April 9, 1987. Copyright © 1987 Dow Jones & Company, Inc. All Rights Reserved. Reprinted by permission.

**Fig. 6A, p. 99:** From "Medical Essay." Used by permission from the August, 1984, issue of the *Mayo Clinic Health Letter.* © 1984 Mayo Clinic.

**Table 9A, p. 164:** From the AMERICAN COLLEGE OF SPORTS MEDICINE *Reference Guide: Health/Fitness Instructor Workshop/Certification,* 1988. Reprinted by permission of the American College of Sports Medicine.

**Box on p. 228:** From "Nutrition and Your Health: Dietary Guidelines for Americans," 2nd ed. Printed by the U.S. Dept. of Agriculture/U.S. Dept. of Health and Human Services, 1985.

**Fig. A–3, p. 259:** From the *Chicago Tribune,* February 23, 1986. Copyright © 1986 Chicago Tribune Company. All rights reserved. Used with permission. **Fig. 14–10, p. 288:** From *The Cerebral Cortex of Man* by Wilder Penfield and Theodore Rasmussen. Copyright 1950 by Macmillan Publishing Company, renewed 1978 by Theodore Rasmussen. Reprinted by permission of Macmillan Publishing Company.

**Fig. 17–16, p. 371:** From *Textbook of Endocrine Physiology,* by C. R. Martin, fig. 21. Copyright © 1976 the Williams & Wilkins Co., Baltimore. Reprinted by permission.

**Fig. 22–8, p. 506:** G.W. Beebe, AMERICAN SCIENTIST, Vol. 70, 1982, p. 40.

**Table 22–4, p. 510:** From "Cancer Facts and Figures—1988" by The American Cancer Society. Copyright © 1988, American Cancer Society, Inc. All rights reserved. Reprinted by permission.

**Fig. 23–1, p. 514:** From "Biochemical Studies of Aging," in *Chemical & Engineering News,* August 11, 1986, p. 28. Copyright © by the American Chemical Society. Used by permission.

**Fig. 23–2, p. 514:** From *The Biology of Senescence,* 3rd ed., by A. Comfort. Copyright © 1979 by Elsevier North Holland, Inc. Reprinted by permission of Elsevier Publishing Company, Inc.

**Fig. 23A, p. 522:** "Living Will Declaration." Reprinted with permission from Concern for Dying, 250 West 57th St., New York, NY 10017.

**Fig. 24B–1, p. 535:** From *Evolution: Process and Product,* 3rd ed., p. 57. By E. O. Dodson and P. Dodson. Copyright © 1985 by PWS Publishers. Reprinted by permission of Wadsworth, Inc.

**Fig. 25–1, p. 548:** Modified from *Biology: The Unity and Diversity of Life,* 4th ed., by Starr and Taggart. Copyright © 1987 by Wadsworth, Inc. Reprinted by permission of the publisher.

**Fig. 25–2, p. 549:** From Annabelle Desmond, "How Many People Have Ever Lived on Earth?" Population Bulletin, Vol. 18, No. 1, Washington, DC: Population Reference Bureau, February 1962.

**Figs. 25–5, 25–9, 25–10, pp. 552, 556, 561:** From *Biosphere: The Realm of Life,* 2nd ed. Copyright © 1988 by Scott, Foresman and Company. All Rights Reserved. Used by permission.

**Fig. 25–15, p. 565:** From *Biology: The Science of Life,* 2nd ed., by Robert A. Wallace et al. Copyright © 1986 by Scott, Foresman and Company. All Rights Reserved. Used by permission.

**Fig. 18–2, p. 385:** From *Essential Human Anatomy* by J. E. Crouch, fig. 18–19, p. 479. Copyright © 1982 by Lea & Febiger, Publishers. Reprinted by permission.

**Fig. 18–6, p. 391:** From *Introduction to Embryonic Development* by. S. B. Oppenheimer, figs. 6–21, 6–22. Copyright © 1980 by Allyn & Bacon., Inc. Reprinted with permission.

**Fig. 18–10, p. 396:** From M. A. South and J.L. Sever. In *Teratology,* 1985, Vol. 31, p. 300. Reprinted by permission of Alan R. Liss, Inc.

**Fig. 18–11, p. 397:** From *Human Genetics,* 4th ed., by A. M. Winchester and T. R. Mertens, p. 41. Copyright © 1983 by Chas. Merrill. Used by permission.

**Box on pp. 444–445:** From "Price of Progress: Efforts to Predict Genetic Ills Pose Medical Dilemmas," by Alan L. Otten. From THE WALL STREET JOURNAL, September 14, 1987. © 1987 Dow Jones & Company, Inc. 1987. All Rights Reserved.

**Fig. 21–2, p. 454:** From *Cancer: The Misguided Cell,* by D. M. Prescott and A. S. Flexer, p. 4. Copyright © 1982 by David M. Prescott and Abraham S. Flexer. Reprinted by permission of Sinauer Associates, Inc.

**Fig. 21–21, p. 483:** From *Blood*: THE RIVER OF LIFE by Jake Page. Copyright © 1981 Torstar Books. Reprinted by permission.

**Table 22–2, p. 499:** From *"The Causes of Cancer,"* by A. E. Reif. In *Am. Scientist,* Vol. 69, p. 440, table 2. Copyright © Sigma XI, The Scientific Research Society. Used by permission.

**Fig. 22–6, p. 500:** From CANCER: SCIENCE AND SOCIETY by John Cairns. Copyright © 1978 W.H. Freeman and Company. Reprinted by permission.

**Fig. 22–7, p. 502:** From AMERICAN SCIENTIST, Vol. 69, 1981. Reprinted by permission of Arnold E. Reif.

**Fig. 22A–2, p. 505:** From "Bring Back the Parasol" by Claudia Wallis, TIME, May 30, 1983. Copyright © 1983 Time Inc. Reprinted by permission.

**Michael Goodman:** Figures 6a, 6B, 6–1, 6–2, 6–3, 6–5, 6–6, 6–10, 6–11, 6–15, 6–16.

**Teri J. McDermott:** Figures 11–16, 12–1, 12–3, 12–4, 16–4, 16–8, 16–9, 16–10, 16–14.

**Sandra McMahon:** Figures 11–4, 11–5, 11–6, 11–7, 13–14, 13–16, 14A–2, 14–1, 14–2, 14–4, 14–5, 14–6, 14–7, 14–9, 14–10.

**Precision Graphics/Karen Shannon:** Figures 1–5, 1–6, 2A–1, 2C, 2–1, 2–2, 2–3, 2–4, 2–5, 2–6, 2–8, 2–10, 2–18, 2–24, 2–26, 2–27, 2–28, 2–29, 3–1, 3–4, 3–5, 3–6, 3–7, 3–8, 3–9, 3–11, 3–12, 3–13, 3–14, 3–15, 3–17, 3–18, 4–2, 4–3, 4–4, 4–9, 4–10, 5B, 5–1, 5–4, 5–5, 6B, 6–1, 6–2, 6–3, 6–5, 6–6, 6–7, 6–8, 6–9, 6–10, 6–11, 6–15, 6–16, 7–1, 7–2, 7–3, 7–4, 7–5, 7–6, 7–7, 7–8, 7–9, 7–10, 7–12, 8–1, 8–2, 8–3, 8–6, 8–7, 9B, 9–1, 9–5A, 9–5B, 9–6, 9–7, 9–8, 9–10, 9–11, 9–13, 9–16, 10–1, 10–2, 10–3A, 10–4, 10–5, 10–8, 10–9, 10–10, 11–1, 11–3, 11–8, 11–12, 12A, 12–2, 12–5, 12–6, 12–7, 13–1A, 13–1B, 13–2, 13–3, 13–4, 13–7, 13–9, 13–10, 13–11, 13–12, 13–13, 13–15, 13–16, 14B–1, 15–4, 15–5, 15–6, 15–7, 15–9, 16A, 16–1, 16–2, 16–3, 16–5, 16–6, 16–7, 16–11, 16–12, 16–13, 16–15, 17–2, 17–3, 17–4, 17–5, 17–6, 17–8, 17–9, 17–10, 17–12, 17–14, 17–15, 17–17, 18–6, 18–7, 18–10, 18–11, 18–13, 18–14, 18–15, 19–1, 19–4, 19–5, 19–6, 19–8, 19–9, 19–11, 19–12, 19–14, 19–15, 20–2, 20–4, 20–5, 20–7, 20–8, 20–9, 20–10, 20–11, 21–2, 21–9, 21–10, 21–11, 21–13, 21–16, 21–17, 21–19, 21–20, 22–2, 22–3, 22–4, 22–5, 22–8, 22–10, 22–11, 23–1, 23–2, 23–3, 23–4, 23–5, 23–6, 23–7, 24A–2, 24–1, 24–2, 24–9.

**Thomas D. Simms:** Figure 2A–1.

**Alice Thiede:** Figures 21–A, 21B–1, 21B–2, 25–11, 25–17, 25–19.

**Troy Thomas:** Figures 13A–3, 22A–3.

**Kevin Somerville:** Figures 1–3, 4–9, 4–10, 6–7, 6–8, 7–8, 7–9, 7–10, 9–5, 9–16, 10–1, 10A, 11–1, 13–1, 13–13, 16–1, 16–13, 16A, 17–5, 17–6, 17–11, 17–12, 17–14, 18–7, 18–11, 18–13, 18–14, 21–6, 21–7, 21–8, 21–21, 21–22, 21–24, 24B–1, 24–7, 24–8, 24–10, photo essay THE HUMAN HEART: four 4/c figures of hearts.

**Sarah Forbes Woodward:** Figures 3–1, 3–4, 3–5, 3–6, 3–7, 3–8, 3–9, 4–1, 5–1, 5–4, 5–5, 5–6, 7–1, 7–2, 7–6, 7–12, 8–2, 8–6, 9–1, 9–2, 9–4, 9–6, 9–7, 9–9, 10–2, 10–3, 11–2, 11–9, 11–11, 11–14, 11–17, 13–6, 13–8, 14–3, 15–2, 15–3, 15–10, 15–11, 15–12, 15–13, 15–14, 15–15, 15–16, 17–7, 17–16, 18–1, 18–2, 18–3, 21–5, 21–12, 21–23, photo essay THE HUMAN EYE: two 4/c figures of eye and vision pathways.

# INDEX

Cochlea, *317,* 319–20, *321*
Codon, of RNA, 432–35
   termination, 435
Collagen, 517–18
   aging theories, 521
   cross-linked, 517–18, 521
Collagenous fiber, 70
Collarbone, *See* Clavicle
Collateral ganglia, 272–73, *274*
Collecting tubule, *233, 236*
Colon, 213, *214*
   spastic, 216
Color blindness, 311, 313, 417
Colostomy, 216
Colostrum, 402, 403
Columnar tissue, 67, *68*
Coma, 283
Commissural fibers, *286*
Common carotid artery, *154,* 157
Common cold, 178
Common iliac artery, *154,* 157
   vein, 155, 157
Community, 547, 548
Compact bone, 97–99, *100*
Complement system, 468–69
Compound, 18
Concentration gradient, 58
Conditional reflex, 293
   Pavlov's dogs, 293
Condom, 404
Conductive hearing loss, 320
Cones, 304, *305,* 307–8
Congenital agammaglobulinemia, 476
Congenital thymus deficiencies, 476
Congestive heart failure, 241
Conjunctiva, 304
Conjunctivitis, 311
Connective tissue, 66, 69–75
   and the aging process, 517–18
   diseases, 74–75
   types of, 70–73
Constipation, 216
Continuous ambulatory peritoneal dialysis
   (CAPD), 246
Contraceptives. *See* Birth control
Contraction, cardiac, 158–62
Contraction, muscle, 121–23, *125*
   and ATP, 123, 125
Contraction, uterine, 399, *400*
Copper, as trace element, 227
Cornea, *304*
Corneal transplant, 311
Corona radiata, 371
Coronary arteries, *156*
Coronary bypass surgery, 169–70
Coronary sinus, *156*
Coronary thrombosis, 167
Corpora cavernosa, 359
   urethrae, 359–60
Corpora quadrigemina, 282
Corpus callosum, *281,* 285
Corpus luteum, 330, 371–75
Cortex
   of adrenal gland, 335, *336,* 337–39
   renal, 234–35, *236*
Corticotropin releasing factor (CRF), 338
Cortisol, 338
Covalent bonding, 18–19
Cowper's gland, 359

Crack, 258
Cramp, 134
Cranial meninges, 278–79, *280*
Cranial nerves, 249, 268, *269, 270*
Cranial outflow, 272
Craniosacral division, 272
Cranium, 104
   bones, 103, *106*
Cretaceous Period, mammalian evolution,
   537
Cretinism, 334
Crick, Francis, 429, 430
Crista,
   of mitochondria, *51*
   of semicircular canal, 321–22
Cro-Magnons, 544
Crossing over, of chromosomes, 354, *356*
Cryptorchidism, 358
Crypts of Lieberkühn, *208,* 209, 211, *214*
C-type virus, 507
Cuboidal tissue, 67, *68*
Curare, 257
Cushing's syndrome, 339
Cutaneous gland, 89, 91
Cyanide, 42
Cyanosis, 92, 143
Cystic fibrosis, 413, 445
Cystitis, 244
Cyst, ovarian, 365
Cytocidal virus, 506
Cytokinesis, *61, 62,* 63
Cytoplasm, 50–51
Cytosine, 26, 429–39

D and C (dilation and curettage), 368
D and E (dilation and evacuation), 406
Dandruff, 84
D-antigen, 481–82
Darwin, Charles, 527, 530–31
Daughter cell, 60
DDT, 554
Deamination, *116*
Death, 523–24
Death rate, 548
Decarboxylation, 41
Deciduous teeth, *201*
Decompression
   chamber, 192–93
   sickness, 192–93
Deductive reasoning, 12
Deep sleep, 283
Defecation, 215–16
Deforestation, 558–59, 562–63
Dehydration, 141
Dehydrogenation, 37
Delivery, premature, 398
Dementia, presenile and senile, 294
Demography, human, 547
Dental pulp, *200*
Dentin, 199, *200*
Dendrite, 251, *253*
Deprenyl, 291
Dermis, *83,* 84–86
DES (Diethylstilbestrol), as carcinogen,
   501
Desensitization, 477
Desmosome, 49, *50*
Devonian period, 536
Diabetes insipidus, 242, 331

Diabetes mellitus, 238, 341, 342
   juvenile-onset, 342
   maturity-onset, 342
Diaphragm, 110
   vaginal, 404
Diaphysis, 96
Diarrhea, 216
Diastole, 160
Diastolic blood pressure, 171–72
Dicoumarol, 149
Diet
   dietary guidelines, 228–29
   eating disorders, 224–25
   energy foods, 222–23
   essential nutrients, 221–22
   minerals, 226–27
   vitamins, 226–27
Diethylstilbestrol. *See* DES
Differentiation, of cancer cells, 494
Diffusion,
   active transport, 58
   through capillaries, 140–41
   facilitated, 58
   passive, 55–56, *57*
Digestion
   products of, 211
   in small intestine, 210–11
   in stomach, 205–7
Digestive system, 65, *66,* 79, 195–216
   embryonic development of, 393
Digestive tract, 195–98
   organs of, 198–216
   structure of, 195–98
Digestive vacuole, 59–60
Digitalis, 162
*Diplococcus pneumoniae,* 409
   heredity studies, 409, *410*
Diploid cell, 351
Disaccharide, 24
Disease
   disease-producing microorganisms,
      453–65
   immune system, 469–76
   lymphatic system, 484–88
   nonspecific defense mechanisms,
      465–69
Dislocation of joint, 116
Distal convoluted tubule, *233*
Diverticulitis, 216
DNA (deoxyribonucleic acid), *21, 26,*
   29–30, *429*
   chemical structure, 429–30
   genetic disease, 439–41
   genetic engineering, 441–43, 447–49
   genetic mutation, 435–39
   heredity studies, 409, *410*
   protein synthesis, 430–35
   replication, 430, *432*
DNA polymerase, 430
Dominance, of allele, 411, 414
Dopamine, 292
Dorsal cavity, 77–78, *79*
Dorsal horn, 264–65
Dorsal root
   ganglion, *266*
   of spinal nerve, 265–66
Dowager's hump, 99
Down's syndrome, 295, 370, 422–23
Drosophila, 14

Marrow
  red, 100–1
  yellow, 100–1
Marsupials, 537
Mast cell, 70, 466
Mastectomy, 403
Materialistic theory, of life's origin, 528, 529
  simulation experiments, 529, 532–33
Maternal chromosome, 356
Matrix, 95
Maximum life span, 514
  aging theories about, 520
M-band, *120*, 121
Mechanoreceptors, 301, 302–3, 317–22
Median life span, 513
Medulla
  of adrenal gland, 335–37, 339
  renal, 234–35, *236*
Medulla oblongata, *277*, 280–81
Medullary body, *286*
Megakaryocyte, 147, 149
Meiosis
  compared with mitosis, 351, 353–54
  in human reproduction, 351, 353–54, *355*, 356
  meiosis I vs. II, 354, *355*, 356
  oogenesis, *369*, 370
Melanin, 85
  and albinism, 412–13
Melanocyte, 84
Melanocyte stimulating hormone (MSH), 327, 331
Melatonin, 327, 343
Membrane
  basement, 67, *68*
  mucous, 76
  serous, 76–77
Membrane granulosa, 370–71
Membrane, plasma, *46, 47*, 48–50
  function of, 55, 60
  permeability of, 48–49, 55
Membranous labyrinth, 318–19
Memory, 292–93
  consolidation, 292
  long-term, 292
  short-term, 292
Menarche, 375
Mendel, Gregor, 410–11
Mendelian genetics, 410–11
Meninges, 263–64, 278–79, *280*
Meningitis, bacterial, 264
Menopause, 375
  and osteoporosis, 98
Mesentery, *198*
Menstrual cycle, 371–72, *373*
  problems related to, 375, 377
  stages during, *373*, 374–75
Menstruation, 372, *373*
Mesoderm, 390–92
Mesovarium, 365
Mesozoic era, 528
Messenger RNA (mRNA), 430–35
Metabolism, 8, 32–33
Metacarpal bones, *104*, 112
Metaphase
  meiosis, 354, *355*
  mitosis, 62, 63, *352*

Metastasis, 492, 494–95
Metatarsal bones, *104, 105*, 114
Methane, 19
Methionine, 222, 224
Mevacor, 168
Micelles, 210
Microfilaments, *46*, 54
Microscope
  electron, 46
  light, 45
Microspheres, 533
Microtubules, *46*, 54
Microvilli, 69, 207, *208*, 209
Micturition, 244
Midbrain, *277, 281*, 282
Middle Ages, population growth during, 548, *549*
Middle ear, 317–18
Migraine headache, 306–7
Milk letdown, 403
Miller, Stanley, 529, 532–33
Milstein, Cesar, 483
Mineralocorticoids, 327, 337–38
Minerals, 226–27
Minoxidil, 88
Miscarriage, 398
Mitochondrion, *46, 51*
Mitosis, 61, 351, *352, 353*
  and the cell cycle, 61
  and oogenesis, *369*, 370
  stages of, 61, *62, 63, 352*
Mitral valve, 152, *153*, 154–55
Mittelschmerz, 371
Molars, *201*
Molecular genetics
  disease, 439–41
  expression of genetic material, 429–35
  gene mutation, 435–39
  genetic engineering, 441–49
Molecule, 18
Molybdenum, 227
Monet, Claude, 312
Monoclonal antibody, 483–84
Monocyte, *146*, 467–68
Monod, Jacques, 430
Mononucleosis, infectious, 147
Monosaccharides, 24
Monosomy, 422
Monotremes, 537
Mons pubis, *368*
Morula, *385*
Mosquito, *Anopheles*, 458, *459*
Motion sickness, 322
Motor areas, of cerebral cortex, 287–88
Motor cortex, 287–88
Motorneuron, 125, *126*, 260
Motor speech area, *287, 288*
Mouth, 198–202
MPP, and Parkinson's disease, 290–91
MPPP, and Parkinson's disease, 290
MPTP, and Parkinson's disease, 290–91
Mucosa, 195–96, *197*
Mucous membrane, 76
Mueller, Paul, 554
Multiple sclerosis (MS), 255
Murmur, heart, 163
Muscle, 66, 75–76
  cardiac, 76, *158*

  contraction, 121–23, 125
  origin and insertion, *129*, 130
  skeletal, 75, 119–34
  smooth, 76, 135–37
Muscle tone, 132
  and aging, *515*
Muscular dystrophy, 125
Muscularis externa, *197*
Muscularis mucosae, 196, *197*
Mutagen, 14–15, 436–37
Mutation, genetic, 30, 435–39
  causes of, 436–37
  diseases, 439–41
  types of, 435
Myasthenia gravis, 128, 260
Myelin, 252, *253*
Myeloid tissue, 73
Myocardial infarction, 167–68
Myocardium, *151*, 152
Myofibril, *120*
Myometrium, *366*, 367
Myopia, See Nearsightedness
Myosin, 121–22, *123*
Myxedema, 334

NAD (nicotinamide adenine dinucleotide), 41–42
Nail, *89*
Narcosis, nitrogen, 192
Nasal cavity, 177–78
Nasal mucosa, 177–78
Nasopharynx, 179, 202–3
  cancer of, 507
Natural childbirth, 398–99
Natural selection, 530–31
Neanderthal, 543–44
Nearsightedness, *310*
*Neisseria gonorrheae*, 378
Neo-Darwinism, 531
Neohemocytes, 144
Nephron, 231, 232, *233*
Nerve
  disorders, 261–62
  regeneration of, 262
  structure of, 261
Nerve fibers
  myelinated, 251, *253*, 255
  unmyelinated, 251, *253*, 254
Nervous system, 80
  autonomic, 268–74
  brain, 276–96
  drug action on, 274–75
  function of, 250–51
  and pain, 302–3
  and sense organs, 303–22
  structure of, 249–50
Nervous tissue, 66, 76
Neural folds, *392*
Neuralgia, 261
Neural plate, *392*
Neural stalk, 328, *329*
Neural tube, 277, *392*
Neuritic plaques, 294–95
Neuritis, 261
Neuroeffector junction, 260
Neuroepithelium, 68
Neurofibrillary tangles, 294–95
Neuromuscular junction, *126*

Neuromuscular transmission, 126–28
Neuron, 76, 249, 251–61
  action potential in, 252–55
  structure of, 251–52
  types of, 260–61
Neurosecretory cell, 328, *329*
Neurotransmission, 256–57
  diseases affecting, 260
Neurotransmitters, 257
Neutron, 17
Neutrophils, *146,* 466–67, *468*
Niacin, 226
Nipple, 401–2, *403*
Nitrogen narcosis, 192
Nitrogenous waste products, 238
Nitrosamine, 500
Nodes of Ranvier, 252, *253*
Nondisjunction, of chromosomes, 421–25
Nonpolar molecule, 19
Nonrapid eye movement sleep (NREM sleep), 283
Norepinephrine, 273, 327, 336–37
Nostril, 177
Notochord, 391–92
Novocaine, 259
Nucleic acid, 29–30
Nucleoli, *46,* 50
Nucleotides, 24, *26*
Nucleus, atom, 17
Nucleus basalis, 295
Nucleus, cell, *46,* 50–55
Nurse cell, 362
Nursing, 403
Nutrition
  energy foods, 222–23
  essential nutrients, 221–22

Obesity, 224
Occipital lobe, of brain, *285*
Oculomotor nerve (III), *269,* 270
Odontoblast, 200
Olfactory
  bulbs, 315, *316*
  cells, 315, *316*
  hairs, 315, *316*
  knob, 315, *316*
  nerve (I), *269,* 270
  receptors, 315–17
  tracts, 315, *316*
Olfactory nerve (I), *269,* 270
Oligocene epoch, 528, 538
Oligodendrocyte, 251
Omnivores, 552
Oncogene, 508–9
Oncogenic DNA virus, 506–7
  transformation by, 508
  types of, 506–7
Oncogenic RNA virus. *See* Retrovirus
Oocyte
  primary, 370, *371*
  secondary, 370, *371*
Oogenesis, 369–71
Oogonia, 370
Ontogeny, 534–35
Oparin, A. I., 528
Open system, 9
Opiates, 192, 298–99
Opsin, 308
Opthalmoscope, 313

Optic chiasma, 308, *309*
Optic disc, *304*
Optic nerve (II), *269,* 270, 304
Optic tracts, 308–9
Oral contraceptives, 404–5
Orbit, atomic, 17, 18
Orbits, of skull, 303
Order (classification), 537
Ordovician period, 536
Organelle, 46
Organism, classification of, 536–37
Organ of Corti, 319, *320, 321*
Organs, 65, *66,* 76–77
Orgasm, 363
Orgel, Leslie, *521*
Origin of life theories, 528–29
  simulation experiments, 529, 532–33
Origin, of muscle, *129,* 130
Oropharynx, 202, *203*
Osmoreceptors, 285
Osmosis, 56–58
Osteoarthritis, *116*
Osteoclast, 96
Osteocyte, 73, 95, *100*
Osteogenic cells, 97
Osteomyelitis, 115
Osteoporosis, 96, 98–99
Otitis media, 318
Otoliths, *322*
Otosclerosis, 320
Outer ear, *317*
Oval window, *317,* 318, *320*
Ovaries, 341, 356, 364, 365, *366*
  and cysts in, 365
  sex differentiation, 415–16
Oviduct, 364, *366*
Ovulation, 372, *373,* 375
  and lactation, 403
Ovum, 356–57
Oxidation, 37, 40–42
Oxygen
  metabolic role, 42–43
  transport in blood, 187–88
Oxyhemoglobin, 187
Oxytocin, 327, 331, 332, 401, 403

Pacemaker, artificial, 162
Pacemaker. *See* Sinoatrial (SV) node
Pacinian corpuscle, 302
Pain, 302
  phantom, 302
  receptors, 301, 302
  refined, 302
  somatic, 302
  visceral, 302, 303
Palate, 107, 198–99
  hard, 107, 198
  soft, 198–99
Palatine tonsil, 179, 488
Paleozoic era, 527, 528
Pancreas, *196,* 209, *217,* 221
  hormones, 325, *326,* 327, 339–41
Pancreatic juice, 209, 210
Pancreozymin, 210
Pantothenic acid, 226
Papanicolaou, George, 367
Papillae, 84, 199
Papillary duct, *233*
Papillary layer, of dermis, 84–85

Papillary muscle, *152,* 153
Papilloma virus, 506–7
Papovavirus, 506–7
Pap smear, 367–68
Paraplegia, 268
Parasympathetic division, 271–74
Parathion, 555
Parathyroid gland, 325, *326,* 327, 334, *335*
  hormones of, 327, 334–35
Parathyroid hormone (PTH), 327, 334–35
Parenchyma, 77
Parietal cell, 205
Parietal lobe, of brain, *285*
Parietal peritoneum, *198*
Parieto-occipital fissure, 286
Parkinson's disease, 260, 290–92
Parotid gland, *196,* 202
Parturition. *See* Birth
Passive immunity, 472
Patella, *104,* 113
Paternal chromosome, 356
Pauling, Linus, 441
Pavlov's dogs, 293
PCBs, 554
Pectoral girdle, *104,* 110–11
Pedicle, 108, *109*
Pelvic cavity, 78, *79*
Pelvic girdle, *104,* 112–13
Pelvic inflammatory disease, 378
Pelvis, *104,* 112–13
Pemberton, John, 258
Penicillin, *456,* 457
Penis, *357,* 359–60
  erection of, 363
Pennsylvanian period, 536
Pepsin, 205, 206, 210
Pepsinogen, 205
Peptide bond, 28
Pericardial cavity, 78, *79*
Pericarditis, 151
Pericardium, 151
  fibrous, *151*
  serous, *151*
Perichondrium, 72
Perilymph, 319
Perimysium, *119*
Perineurium, *261*
Periodontal disease, 200
Periodontal membrane, *200*
Periosteum, 73, 97, *100*
Peripheral nervous system (PNS), 249, *250,* 262–63
  nerves in, 249, *250,* 262
Peripheral resistance, 172
Peristalsis, 76, 212–13
Peritoneal cavity, *198*
Peritoneum, 198
  parietal, *198*
  visceral, *198*
Peritonitis, 198
Peritubular capillaries, 233, *234*
Pernicious anemia, 145, 205, 262
Peripheral nerves, 249, *250*
  autonomic, 262
  regeneration of, 262
  somatic, 262
Pesticides
  organophosphate, 257
  as carcinogen, 501